WILDLIFE DEMOGRAPHY: ANALYSIS OF SEX, AGE, AND COUNT DATA

Wildlife Demography: Analysis of Sex, Age, and Count Data

John R. Skalski and Kristen E. Ryding
School of Aquatic and Fishery Sciences
University of Washington
Seattle, Washington

Joshua J. Millspaugh
Department of Fisheries and Wildlife Sciences
University of Missouri
Columbia, Missouri

ELSEVIER
ACADEMIC
PRESS

AMSTERDAM • BOSTON • HEIDELBERG • LONDON
NEW YORK • OXFORD • PARIS • SAN DIEGO
SAN FRANCISCO • SINGAPORE • SYDNEY • TOKYO

Sr. Acquisitions Editor	Nancy Maragioglio
Project Manager	Brandy Lilly
Associate Editor	Kelly Sonnack
Marketing Managers	Linda Beattie
	Claire Lawlor
Cover Design	Eric Decicco
Cover Art	Sarah Conradt
Composition	SNP Best-set Typesetter Ltd., Hong Kong
Cover Printer	Phoenix Color
Interior Printer	The Maple-Vail Book Manufacturing Group

Elsevier Academic Press
30 Corporate Drive, Suite 400, Burlington, MA 01803, USA
525 B Street, Suite 1900, San Diego, California 92101-4495, USA
84 Theobald's Road, London WC1X 8RR, UK

This book is printed on acid-free paper.

Library of Congress Cataloging-in-Publication Data
Application Submitted

British Library Cataloguing in Publication Data
A catalogue record for this book is available from the British Library

ISBN-13: 978-0-12-088773-6
ISBN-10: 0-12-088773-8

For all information on all Elsevier Academic Press Publications visit our Web site at
www.books.elsevier.com

Transferred to Digital Printing 2009

Contents

Foreword

For a field wildlife ecologist beginning a study, the fundamental question often boils down to this: do I observe, count, and categorize my "critters" usually on a large spatial scale; or capture, mark, and recapture them usually on a small spatial scale? To read much of the current literature, especially in the more quantitative journals, one would think the answer was to use mark and recapture almost exclusively — which is patently ridiculous. Clearly the correct answer is the classic — it depends! This book has as its goal redressing the balance by emphasizing the importance of approaches other than mark-based wildlife population assessment methods. More power to them is my opinion!

The availability of accurate and precise estimates of the demographic parameters of wildlife populations (recruitment rates, survival rates, movement rates, population sizes, and population rate of change) is crucial to their successful conservation and management. Some populations need management because of their importance as harvested species, whereas other populations need conservation due to their rarity induced by human causes such as human overpopulation, pollution, and habitat fragmentation. One doesn't need to be a "rocket scientist" to realize that the need for reliable estimates of demographic parameters is growing more crucial by the day, whereas many wildlife and conservation agencies have been and are suffering outrageous funding cuts.

In retrospect, it is very clear that a major milestone in the field was reached in 1973 when Seber published the first edition of his book on the estimation of animal abundance and related parameters. For the first time a successful attempt had been made to synthesize the methods in a coherent pedagogy and show the strength and weaknesses of the statistical models used. Seber's book enabled statisticians and quantitative biologists to make rapid and important advances in the 30 plus years since 1973. However, these advances have been uneven. For example, for short-term studies on closed animal populations, major research and synthesis has occurred for count methods where detection is primarily related to distance (distance sampling). There has also been an explosion of methods based on marked animals. Closed population capture-recapture and removal methods are now on a sound footing. Open capture-recapture models and the robust design have also seen major advances. Sophisticated software like DISTANCE and MARK allow model selection and parameter estimation based on maximum likelihood methods.

One area that has not received as much attention has been methods for analyzing various types of counts (based on observation or removal), especially where age and sex categorizations are involved. This book has made an excellent attempt to fill that gap. Many of the methods discussed in this book were historically derived by intuitive approaches. However, for sound statistical inferences, the demographic methods must be based on solid design principles and on properly specified sampling distributions and likelihood functions. Thus, in some cases, new theory had to be developed before the synthesis could take place and the book completed. In my opinion, this is one of the key strengths of the book.

Are there other reasons why is this book important? To return to my initial theme, I believe that statisticians and quantitative biologists have tended to focus on the development of methods using marked animals. A fundamental reason for this is that marking establishes known cohorts of animals and therefore makes modeling so much simpler for the statistician. However, these methods are often expensive and suitable only for small spatial scales, whereas many wildlife agencies have emphasized the importance of other approaches that are less expensive in order to collect information on larger spatial scales. As an example, let us focus for a moment on survival rate estimation of a fish population in, say, a large lake. There are a plethora of methods used based on different types of marking schemes. For example, an open capture-recapture model (Cormack-Jolly-Seber) might be used or a tag-return model (Brownie) might be used if the population of fish is exploited. Alternately, telemetry tagging models may be used. All these approaches are very expensive and may be difficult to implement on a very large lake (they also have very strong model assumptions). Therefore, an alternative "catch curve" approach is often used by fisheries agencies because it is much less expensive. This involves taking a random sample of the age distribution (or part of the age distribution) of the population and using a catch-curve method. A critical concern, though, in the use of catch-curve methods is the validity of the fundamental concept of a stable age distribution that requires very strong assumptions of constant temporal recruitment and survival. It may also be difficult to obtain a random sample, even of a subset of the ages.

A foreword would not be complete without a few comments on where I think the field is heading. So what will be some important research thrusts? One of them is the integration of multiple approaches, which is a major focus of this volume. Integration will often give estimates that have better precision and are more robust because they allow weakening of assumptions. One of many examples in the book is the simultaneous estimation of survival rates from both age-based "catch-curve" and radiotelemetry data.

The estimation of survival rates from catch curves illustrates another very important principle, which is the integration of structural population dynamics models and statistical models to get an estimate. This means that the training of quantitative scientists working on wildlife population demography needs to include statistics, biomathematics, ecology, and the mastery of sophisticated computer software and computer programming. No longer can we afford the lack of integration and petty infighting between university departments that reduces the effectiveness of true interdisciplinary teaching and research.

Now that we have most of our models on a sound theoretical footing with well-defined likelihood functions and access to enormous computing power, it is possible to build very complex and intensive computer algorithms to compute estimates and their sampling distributions. Several major themes are emerging: one is the use of bootstrap and other resampling methods; a second is the now common usage of Monte Carlo simulation methods to study the properties of estimators; and a third is the use of Bayesian methods computed using Markov chain, Monte Carlo methods.

Dr. Kenneth H. Pollock
Professor of Zoology, Biomathematics and Statistics
Department of Zoology
North Carolina State University
Raleigh, North Carolina

Preface

In recent decades the quantitative field of wildlife demographics has increasingly focused on animal marking studies to estimate population parameters (e.g., Seber 1973). There is good reason for this trend; mark-recapture methods can provide precise and accurate estimates of survival, abundance, density, and recruitment when sufficient numbers of animals are marked, high capture probabilities can be expected, and the underlying statistical models are valid. The inherent costs and logistical difficulties of marking studies, however, often restrict the scope of these investigations to intensive localized studies. Theoretical considerations can also limit the flexibility and realism of marking models, particularly in the area of abundance estimation of both closed and open populations.

Field biologists have therefore continued to embrace extensive cost-effective methods that provide demographic information on a broader geographic scale. These methods generally rely on visual counts, sex ratios, and age-structure data commonly collected by wildlife agencies for many years. The underlying statistical and sampling models for these demographic studies, however, have generally not received the same quantitative scrutiny as statistical models used in mark-recapture theory. Moreover, there has been little attention to coupling intensive marking models with extensive count, sex ratio, and age-structure data. This union could be useful because of the statistical precision offered from intensive marking studies and geographic advantages of extensive studies. The need for cost-effective, wide-scale, and scientifically defensible demographic methods is motivating this change.

Nationwide, the Endangered Species Act and Habitat Conservation Plans have placed increasing demands on resource agencies to quantify status and trends of wildlife populations. At the same time, the human population in the United States has shifted from a predominantly rural to an urban society. Associated with this shift, the general public is viewing wildlife more and more as a nonconsumptive rather than as a consumptive resource. Consequently, the general public is requiring more justification for wildlife harvest regulations and has elevated wildlife management into the public arena. In Washington State, for example, voters impatient with state biologists have substituted legislative referenda for game management by scientific principles. Harvest apportionment between sport and Native American hunters is also requiring more detailed assessment of wildlife resources and the consequences of harvest management practices.

It is within this environment our book has been written. Our goal was to assemble the many quantitative approaches wildlife biologists commonly use to analyze and interpret count observations, and sex- and age-structure data. In assembling these methods, we often needed to reformulate them in a modern statistical framework, permitting not only parameter estimates, but also variance estimation and sample size calculations. To do so, we needed to use a level of mathematical rigor that many biologists will find challenging. We felt that model development and specification was important for our goals and to allow other investigators to extend these techniques into new situations and applications. The quantitative approach produced both a dilemma and a challenge.

The book is written foremost to be a practical guide to the analysis of sex, age, and count data. To satisfy the need for both quantitative rigor and expository value, estimation equations have been highlighted to distinguish them from model development. In addition, the book includes numerous annotated examples of the statistical methods that are clearly demarcated in the text. Biologists may wish to study a method by beginning with the examples. The more quantitatively trained readers may wish to begin with the model formulation. All readers should find the discussions of utility for each method helpful.

The book concludes with a chapter on case studies. The intent is to bring together the demographic tools presented in earlier chapters for the purposes of demographic assessment and management. The examples were chosen to illustrate different applications and joint use of multiple techniques for problem solving. The examples were not meant to be exhaustive of the demographic scenarios or solution approaches available to biologists. Instead, we hope this Preface and the chapters that follow will kindle an interest in exploring new and ever better approaches to wildlife demography for the benefit of wildlife resources everywhere.

We wish to thank Gary Brundige, Andy Cooper, Todd Farrand, Sherry Gao, Bob Gitzen, Mike Hubbard, Brian Kernohan, Mike Larson, Chad Rittenhouse, Gary Roloff, Mark Rumble, Mark Ryan, John Schulz, Frank Thompson, Brian Washburn, and Pete Zager for their timely reviews and constructive comments on the book. In addition, we thank Clait Braun for reviewing and editing the entire manuscript. His comments greatly improved the clarity, accuracy, and consistency of the book.

We would like to especially thank Rich Townsend and Jim Lady for their computer and analytical support. Peter Dillingham wrote the *Mathematica* code for Appendix D. Remington Moll carefully checked all examples contained within the book, and we are very appreciative of his efforts. Manuscript preparation was provided by Cindy Helfrich—without her help and devotion, this book could not have been produced.

Dr. Millspaugh would personally like to thank Rami for her support and patience while writing the book. He would also like to thank the many people at the University of Missouri including John Gardner, Gene Garrett, Jack Jones, Tom Payne, and Mark Ryan who supported the project and provided him with time to write and visit Seattle. He would like to thank his graduate and undergraduate students who continue to provide valuable input on presenting quantitative methods to wildlife biologists. Rather than providing an exhaustive list of influential collaborators and colleagues, he would like to acknowledge personnel at Huntington Wildlife Forest, SUNY-ESF, SUNY Cobleskill, Custer State Park, South Dakota Department of Game, Fish, and Parks, South Dakota State University, Boise Cascade Corporation, University of Washington, Raedeke and Associates, Inc., University of Missouri, Missouri Department of Conservation, National Park

Service, and the U.S. Forest Service who sparked and encouraged his interest in quantitative applications in wildlife management.

In addition, Dr. Ryding would personally like to thank the other two authors, John and Josh. She greatly appreciates family members for their support; Dee and John Fournier, Bill, Angela and Melissa Ryding, and Jan Beal. She would also like to thank friends who have been there along the way, some of whom are mentioned above, Cindy H., Andy C., Rich T., Peter D., and also Kevin Brink, Anne Avery, Sarah Hinkley, Martin Liermann, Owen Hamel, Vaughan Marable, Anna Bates, Erin Wolford, John Walker, Michelle Adams, and Ken Wieman—and finally, the graduate students of the Interdisciplinary Graduate Program for Quantitative Ecology and Resource Management at the University of Washington and the staff of Columbia Basin Research.

Finally, Dr. Skalski would personally like to thank Lauri for her encouragement, support, and patience while the book was being written. He would also like to thank the students and faculty of the School of Aquatic and Fisheries Sciences, the Interdisciplinary Graduate Program for Quantitative Ecology and Resource Management, and the Wildlife Science Program at the University of Washington for their help and support. Thanks are also extended to the Idaho Department of Fish and Wildlife and the Washington Department of Fish and Wildlife for providing examples and inspiration for this book. The School of Aquatic and Fishery Sciences provided funds to help defray some of the developmental costs of this book. Program USER, illustrated in this book, was developed with funds from the Bonneville Power Administration, Project 198910700, of the Columbia River Fish and Wildlife Program.

John R. Skalski
Kristen E. Ryding
Joshua J. Millspaugh

Introduction

1

Chapter Outline
1.1 Historical Perspectives and Current Needs
1.2 Scope of Book

1.1 Historical Perspectives and Current Needs

Early efforts to assess demographics of wild animal populations often relied on sex- and age-structure data collected from animal sightings or hunter bag checks. By use of sex- and age-structure data, biologists estimated population sex ratios (Severinghaus and Maguire 1955), productivity (Dale 1952, Hanson 1963), age-specific survival (Hayne and Eberhardt 1952), harvest mortality (Allen 1942, Petrides 1954, Selleck and Hart 1957), population abundance (Kelker 1943, DeLury 1945), and rates of population change (Kelker 1947, Cole 1954). These techniques received substantial attention early in wildlife management (Hanson 1963), perhaps because data needs were minimal, data were relatively easy to collect, and the techniques were applicable over large geographic areas. Moreover, the resulting analytical approaches used were a clever blend of life-history knowledge and available survey data a biologist could readily use and understand.

A drawback to the early heuristic procedures was lack of mathematical rigor in the derivation of the estimators. Most of these estimators lacked explicit variance expressions and statements of assumptions. Early efforts to catalogue and review these techniques offered little guidance. In the Hanson (1963) monograph, variance estimation was not addressed, presumably because of the complexity of such analyses and the space required to present them (Hanson 1963:7). Although early researchers (e.g., Davis 1960) recognized the need for variance estimates to assess the precision of demographic analyses, these estimates were widely unavailable until later (e.g., Paulik and Robson 1969). Even today, clear statements of assumptions and variance expressions are lacking for many of these techniques. Without variance expressions and a clear statement of assumptions, there is no way to evaluate accuracy or precision of the demographic parameters. Thus, biologists have no way to identify how much confidence to place in the demographic estimates. In a recent review, Skalski and Millspaugh (2002) provided variance expressions and evaluated precision of the sex-age-kill model of population reconstruction (Creed et al. 1984). They found the required level of precision in field data collection necessary to provide useful estimates of population abundance was unattainable with existing levels of effort. Numerous other techniques also need to be placed in a modern statistical framework and evaluated to examine their usefulness.

The repeated discovery, use, and modification of these sex- and age-structure procedures has resulted in a duplication of efforts and inhibited further advancements. Hanson (1963) attributed repeated rediscovery of sex- and age-structure-based demographic assessments to several factors such as omission of important formulas or derivations of those formulas in early publications, unfamiliar notation, and obscurity of the original publications. His monograph (Hanson 1963) nicely summarized the available techniques of the time to assess demographics from sex- and age-structure data and provided a few important extensions. But his monograph fell short of providing suggestions for study planning and sample size requirements, issues later addressed by Paulik and Robson (1969) for change-in-ratio techniques. A need exists today for additional guidance when planning field activities. Researchers should be aware that, for a given demographic parameter and data type, there is often more than one estimator available. Choosing the most appropriate estimator will depend on which assumptions are valid for a particular data set and study objectives.

Sex- and age-structure techniques continued to be rediscovered, extended, and used extensively throughout the late 1960s (e.g., Paulik and Robson 1969) and 1970s (e.g., Lang and Wood 1976), and further attempts have been made to summarize the available techniques (Udevitz and Pollock 1992). Roseberry and Woolf (1991) compared several alternative approaches for estimating the abundance of white-tailed deer (*Odocoileus virginianus*) by using harvest data and reconstruction techniques. Notable among their contributions was a list of important assumptions and a discussion of data requirements. Although discussed in terms of white-tailed deer, many of their suggestions apply equally well to other species. However, no variance expressions were offered, and their evaluation was strictly empirically based. Instead, they referred readers to Seber (1973), who provided more thorough mathematical descriptions, along with variance estimators for some techniques. The work by Seber (1973) is still considered the standard reference for many techniques that use sex- and age-structure data to estimate abundance. Although the utility of Seber (1973) remains, many procedures for demographic assessment were omitted, and these techniques should be advanced with the same rigor.

Despite the obvious weaknesses, demographic assessments based on sex, age, and count data are still common today. For many wildlife agencies, sex and age data collected during field surveys or through hunter check stations represent the only available demographic data to manage wildlife populations. For example, Skalski and Millspaugh (2002) noted that at least 20 state agencies use the sex-age-kill model to estimate white-tailed deer abundance. Although these techniques offer a convenient means of data collection over extensive regions, they are not free from rigorous assumptions. Because sex- and age-structure data are often opportunistically collected over large regions (e.g., at check stations), it is tempting to overlook the assumptions and sampling requirements of these data for demographic assessment. As a result, some researchers have been critical of sex- and age-structure data for demographic assessment. Johnson (1994:438) writes, "methods based on age-structure data have received much attention in the past, probably more than they merit considering their deficiencies." Mounting pressure for effective conservation strategies requires biologists to weigh the utility of different analytical and sampling options in light of accuracy, precision, and economics. Because demographic assessments based on sex- and age-structure data offer a convenient, geographically extensive, and cost-effective means to acquire data, they will continue to be used by many agencies. Consequently, many of the unresolved issues in using these techniques must be addressed.

1.2   Scope of Book

This book unifies, evaluates, updates, and illustrates methods of estimating wildlife demographic parameters from sex ratios, age structures, and count data commonly collected by wildlife biologists. The demographic parameters included are commonly used in modeling population dynamics of wildlife species: productivity, survival, harvest rates, abundance, and rates of population change. Our work focuses on estimation techniques that use sex, age, and count data because (1) these data are relatively easy to collect for game and nongame species compared with more expensive and labor-intensive techniques that require animal marking or radio tagging; (2) these data are commonly collected and used by wildlife management agencies; (3) these survey data have not received the same statistical rigor that has been focused on other field data (e.g., mark-recapture data); (4) these techniques are scattered throughout the literature with no cohesive synthesis and evaluation; and (5) once formal statistical models for the analysis of sex and age data have been developed, this information can be coupled with mark-recapture models for more advanced demographic analyses.

This book provides a variety of statistical techniques that are useful in managing wildlife resources. More efficient use of the data currently collected should aid in better evaluation of population status and trends. Further, an understanding of the nature and sources of variability should help direct data collection efforts. Greater understanding of model assumptions and the magnitude of sampling error should also improve the interpretation of information on population status and trends.

Most of the estimation techniques discussed in this book are widely scattered throughout the literature. A source is needed for the derivation, likelihood functions, variance expressions, and sample size calculations of these demographic methods. Although techniques to estimate wildlife population demographics have been the focus of several books, these books review methods that require animal marking (Seber 1982, Skalski and Robson 1992, Thompson et al. 1998, Williams et al. 2001) or radio tagging (White and Garrott 1990, Millspaugh and Marzluff 2001), or line-transect procedures (Buckland et al. 1993). This book is different because it focuses on techniques that use sex ratios, age structure, or count indices commonly collected by field biologists.

Part of the value in this work lies in presenting the estimators developed over many years, and in various outlets (e.g., journal articles, books, dissertations, unpublished agency reports), in one place, in a quantitative manner. By doing so, relationships among techniques are made and gaps in knowledge are identified and filled. An area that should be explored further is the use of auxiliary information to relax the assumptions of some of these methods. For example, age-structure analysis is often based on the assumption of a stable and stationary population, which may not be the case (e.g., Unsworth et al. 1999). Joint likelihoods using mark-recapture and age-structure data could be derived to eliminate the assumptions of stable and stationary population for some estimators, and to separately estimate natural and harvest survival probabilities.

The presentation of techniques in this book was structured so that the quantitative goals of the estimator are understood and referenced with previous developments and uses. Because some of the work has been duplicated in the literature, we are careful to report who developed each technique and how it relates to others. Next, we present the statistical models and define model parameters and variables. The statistical models are developed from first principles by modeling both sampling and life-history processes. This approach allows for the development of estimators specific to different types of data, sampling methods, and life histories. Further, by expressing estimators in mathematical terms, connections can be made among methods that were originally derived heuristically or as special cases of another model. The development of likelihood models for the age- and/or sex-structure of a population also allows these data to be combined with data from mark-recapture surveys to construct joint likelihoods and improve parameter estimation. Future developments in quantitative wildlife ecology will likely consist of combining disparate models from mark-recapture and demographic studies into a single joint assessment for improved interpretation (see Gove et al. 2002). Developing a quantitative framework for sex ratios, age structures, and count data is a first step in this process.

A discussion of model assumptions is provided for each estimator. The need for explicitly stating the assumptions can be seen by reading current wildlife journals. There are many examples in the wildlife literature in which assumptions are not stated or are readily misunderstood. For example, estimates of sex ratios that use sighting data often do not address the assumption of equal sighting probabilities between males and females (White and Lubow 2002), and regression estimators often ignore that dependent variables may be estimated quantities and, hence, have nonconstant variance (Afton and Anderson 2001). At other times, assumptions about population processes are not addressed. In a series of articles on the use of statistics in wildlife ecology in a recent issue of the *Wildlife Society Bulletin*, only passing mention was made for the need to assess biological and sampling assumptions in parameter estimation (Otis 2001).

Explicit variance expressions are derived for each of the estimators to assess uncertainty of point estimates and to aid in sample size calculations for planning studies. By reframing many of the historical approaches into maximum-likelihood or regression techniques, variance calculations are provided that did not previously exist. If estimation is based on several demographic parameters or sampling stages, the relative contributions of each element to the overall uncertainty are assessed. Analytic variance expressions for estimators are valuable for several reasons. First, they allow precision to be compared among different estimators. Second, when estimators are based on several demographic parameters, or sampling in stages, the relative contribution of each parameter or stage to the overall uncertainty can be assessed. Third, variance expressions are necessary in setting precision goals and calculating sample size requirements for planned studies. Finally, the variance estimates are required to express the uncertainty associated with parameter estimation. An understanding of the nature and sources of variability in estimation can be used to better direct the limited sampling effort available in the management of wildlife populations.

Estimation techniques within a chapter are presented sequentially as much as possible, with each model building upon a previous method by adding or modifying model assumptions. Each chapter provides annotated examples so the reader can see how to use the statistical techniques. Where feasible, sections include sample size calculations and a discussion of utility. Each chapter concludes with a summary discussion of the methods and a decision tree to help biologists select an appropriate technique.

The methods presented in this book are applicable to big game, small game, and nongame species. However, most of the numerical examples use harvested species because they have more data available. Furthermore, the wildlife literature and population dynamics models have historically focused on species of commercial or sport value. The models and estimators presented in this book have been developed to fit general life-history traits, rather than focusing on specific species. Thus, these procedures can be applied to nongame species, provided the necessary data are available.

We discuss quantitative techniques to estimate: sex ratios, productivity, survival, harvest and harvest mortality, rates of population change, and abundance. A common thread is that all techniques use sex ratios, age structures, or index counts to estimate demographic parameters. The following briefly describes the chapter topics.

Chapter 2: Primer on Population Dynamics. Population dynamics is concerned with changes in animal abundance and the factors that influence those changes (Gotelli 2001). Because immigration and emigration are commonly ignored, birth and death rates form the primary basis of a population assessment. Components of population assessment generally include examination of population status and vitality. Population status generally refers to the current state of the population and considers factors such as abundance, age and sex structure, and health (i.e., nutritional and physiological). In contrast, population vitality refers to the demographic health of the population and the ability of the population to be self-sustaining. Vitality is commonly expressed as the rate of change in population size from one year to the next.

In Chapter 2, we consider the fundamental principles of population growth, including a description of discrete and continuous time models, exponential and logistic population growth, and how to model these populations using age- and stage-structured matrices. We set the theoretical foundation for understanding density-independent and density-dependent growth for small (e.g., Allee effect parameterization) and large animal populations. In addition to the basic, female only, Leslie matrix model (Leslie 1945, Caswell 1989), we consider several special cases, including a two-sex model, harvest models, and multiple population models. Stage-based models are also considered for populations that are better represented by life-history stage rather than age (Lefkovitch 1965). We also discuss sensitivity and elasticity analyses to examine which demographic parameters are most influential in affecting the growth rate of a population. Because we use harvest data for many applications in this book, we conclude the chapter with an overview of harvest theory, including annual surplus, sustained yield, and a discussion of additive and compensatory mortality. The intent of Chapter 2 is to provide a context for the assessment of techniques presented later in this book. For more details on basic population processes, readers are referred to Caughley (1977), Caswell (1989), Berryman (1999), Donovan and Welden (2001), Gotelli (2001), and Vandermeer and Goldberg (2003).

Chapter 3: Estimating Population Sex Ratios. Sex ratio is an important parameter, because it often reflects other demographic features of a population (e.g., productivity and survival) and provides an important clue to current and future population status. Sex ratios are governed by differential survival between males and females, particularly in harvested species. Information on other demographic processes can be acquired by the analysis of sex ratios at different periods during the annual cycle. For example, estimates of productivity and survival calculated by change-in-ratio methods rely on unbiased sex ratio estimates. Further, harvest management of certain species, such as elk (*Cervus* *elaphus*) (Raedeke et al. 2002), is based on achieving a desired adult sex ratio to maximize production.

Sex ratios are defined in this book as the number of females per male, as this is the convention used in applications such as sex-age-kill models (Skalski and Millspaugh 2002) of population reconstruction. Estimates of sex ratio are derived from data obtained from direct counts, survival probabilities, and age structure in the harvest. Direct methods include sampling with and without replacement, cluster sampling, and accounting for differential detection probabilities between males and females. Sample size plots are presented to illustrate the expected precision of a sex ratio estimate based on a simple random sample of the population. Methods are presented for projecting the sex ratio of a population based on differential natural and harvest survival between males and females. The survival-based estimators also allow an assessment of the effects of different harvest management schemes on the sex ratio of the population. In Chapter 10, we project the long-term consequences of sex-specific hunting regulations on the sex ratio of a population, with the goal of understanding how much male harvest mortality is allowable given desired sex ratios.

Chapter 4: Estimating Productivity. Productivity has been defined as the number of juveniles per adult (Peterson 1955), the percentage of juveniles within a population (Robinette and Olsen 1944), the number of juveniles per territorial adult male (Wight et al. 1965), and, most commonly, the number of juveniles per breeding female or the number of juvenile females per breeding female. The definitions relevant to population dynamics models and, hence to this book, are the number of juveniles per breeding female. Fecundity, the number of juvenile females produced per adult female, is generally used in estimating and predicting population growth rates (Cole 1954, Henny et al. 1970, Caswell 1989). Estimates of fecundity can be derived from productivity estimates that use the sex ratio of juveniles.

Direct methods of estimating productivity include embryo counts, nest surveys for clutch sizes, litter counts, and direct enumeration of the number of juveniles sighted per female. All of these methods tend to be early postbreeding estimates of productivity. As juveniles and adult females become less associated later in the annual cycle, indirect methods are often used to estimate productivity. Two basic methods of indirect estimation are presented (Hanson 1963). If animals in the fall season are identifiable by sex, but not age (e.g., ring-necked pheasants [*Phasianus colchicus*]), then productivity estimates can be estimated from prebreeding and fall sex ratio estimates. Alternatively, a prebreeding sex ratio and a fall juvenile-to-adult count can be used to estimate productivity. Both of these indirect methods estimate productivity as the number of juveniles in the fall per adult female in the spring. The assumptions for each technique are clearly stated, and where possible, the estimators are modified to improve the robustness of the method. Variance expressions are derived for each of the indirect techniques, and precision curves are presented as guides to sample size calculations.

Chapter 5: Estimating Survival. Survival is defined as the probability that an animal alive at time, t, will be alive at $t + 1$. In most population dynamics models, survival is defined as the probability of surviving to the next age or stage class. The survival chapter focuses on techniques to estimate age-class survival from age-structure data. The chapter begins with age-specific radiotelemetry data to illustrate the nature of right- and left-censored data and the estimation of survivorship curves. Three alternative life-table approaches to age-specific survival estimation are presented: horizontal, vertical, and a third type of life table called depositional. For each approach, likelihood models are presented, along with maximum likelihood estimators and variance estimators. Catch-curve analyses are shown to be special cases of vertical life-table analyses when survival is assumed constant. The truncated methods of Hayne and Eberhardt (1952) and Burgoyne (1981) are, in turn, shown to be special cases of catch-curve analyses. The chapter concludes with methods for estimating survival by using regression procedures, as well as estimating juvenile survival by using change-in-ratio methods and demographic information on productivity, sex ratios, and abundance data.

Chapter 6: Estimating Harvest and Harvest Mortality. For many species, harvest data collected from hunter surveys is the only source of available information for demographic assessment. This chapter focuses on estimating total harvest from hunter survey data and the estimation of harvest mortality rates. The data used to estimate harvest mortality are kill estimates, sex and age ratios, and pre- and postharvest abundance estimates.

Regulations for reporting hunter harvest vary among states, but most require reporting the kill of larger species such as ungulates and carnivores. The reporting of small-game harvest relies heavily on voluntary information gathered from a representative sample of licensed hunters. There are several basic problems associated with hunter surveys. Some hunters may refuse to respond to surveys, or successful and unsuccessful hunters may respond with different probabilities. If successful hunters respond to surveys at a different rate than do unsuccessful hunters, harvest totals may be biased. Alternatively, a portion of the hunters may respond to a survey with only partial information. For example, a kill may be reported without the location of the harvest or sex of the animal. Methods are presented to cope with nonresponse bias. Likelihood methods are presented to estimate harvest by area when questionnaires are incompletely addressed.

Methods for estimating harvest mortality are also presented in this chapter. One method (Selleck and Hart 1957) is a change-in-ratio method using the sex ratio information from a population pre- and postharvest, as well as the sex ratio of the removals. Another method uses the differential vulnerability of the sexes and changing sex ratios over time to estimate harvest survival (Paloheimo and Fraser 1981). Finally, a method originally developed for tagging studies but applicable to general populations (Gulland 1955, Chapman 1961) can be used to partition mortality between natural and harvest sources of mortality.

Chapter 7: Estimating the Rate of Population Change. The finite rate of population change, λ, can be estimated from indices, abundance estimates, or other demographic data. The differences between the realized rate of population change and the potential rate of change are discussed. Among the simplest methods of estimating λ is to regress abundance estimates of a population at time t against abundance at time $t + 1$. The slope of the regression line is an estimate of the population growth rate, λ. Because the dependent variable, N_t, is measured with error and the presence of autocorrelation between successive estimates (Eberhardt and Simmons 1992), the resulting estimate of λ will be positively biased. Regression of abundance against time t is recommended instead. A time series approach to estimating λ is illustrated, taking into account autocorrelation in sequential abundance estimates.

Other methods of estimating population growth rates, including the Lotka equation for discrete age classes (Cole 1954, Eberhardt 1995) and population projection matrices (Leslie 1945), are discussed. These models use estimates of fecundity and survival rates to calculate finite rates of increase, bringing together topics discussed in previous chapters. The difficulty with these model approaches is obtaining a simple closed-form estimator for λ. The Lotka equation is illustrated in obtaining simple expressions for λ, thus leading to closed-form variance estimators. A computational method for estimating the variance of the finite rate of increase from a Leslie matrix model is illustrated and the computer software code provided.

Chapter 8: Analysis of Population Indices. A population index is an indirect measure that is anticipated to be proportional to the actual abundance of the population. Indices often rely on counts of sign left by animals (e.g., tracks, nests, pellet counts) or counts of a subset of the population (e.g., road kills, hunter harvest, nesting hens). In most situations, an index cannot be converted to absolute abundance or density. Yet improved insights into the nature of the index and the population(s) being monitored are possible if a mathematical structure or model for the index can be proposed. An unfortunate limitation of indices is in the role of harvest management. Animal abundance and harvest numbers are on an absolute scale, whereas indices are only measures of relative abundance. However, indices have the advantages of being easy to collect, applicable for difficult-to-handle species, relatively inexpensive, and suitable for monitoring population trends $\left(\text{i.e.,}\ \dfrac{N_{t+1}}{N_t}\right)$ when constant proportionality can be established.

In this chapter, we explore different approaches to using indices to assess population status and trends. We begin the chapter with an overview of basic sampling methods and the relationship between indices and abundance. We also describe many common indices used in wildlife studies, such as pellet counts (Neff 1968), auditory counts (McClure 1939, Gullion 1996), visual counts, catch-per-unit effort (DeLury 1947, Ricker 1958), and mark-recapture estimates (Skalski and Robson 1992). Next, we describe and offer examples of Latin-square and randomized block designs for index studies. We present methods to combine and calibrate indices using ratio estimators, regression estimators, and double-sampling techniques. Common tools used in the analysis of indices include analysis of variance (ANOVA) and regression analysis, which assume that data are normally distributed. A broader class of analyses, based on generalized linear models (GLMs), is presented for count data that may be Poisson-distributed or frequencies of occurrence that are binomially distributed.

Chapter 9: Estimating Population Abundance. An abundance estimate often forms the basis of ascertaining the status of a population. Two general approaches to abundance estimation are available, including direct estimation techniques (e.g., line-transect, mark-recapture, catch-effort) and indirect methods (e.g., population reconstruction). Much attention has been given to direct methods that estimate animal abundance by using tagging studies (Seber 1982) and visual sightings (Buckland et al. 1993). These abundance procedures have the advantage of providing a direct assessment of the status (i.e., N_i) rather than just trends in abundance $\left(\text{i.e.,}\ \dfrac{N_{i+1}}{N_i}\right)$; they allow the investigator to test model assumptions and focus on the overall parameter of interest. However, they are not applicable to many species, such as those threatened and endangered, those difficult to capture, or those in wide-ranging populations. In contrast, indirect techniques relying on sex- and age-structure data or other demographic information to estimate abundance have received much less attention. One important exception is Seber (1973), who reviewed many of the available catch-effort and change-in-ratio methods for closed and open populations.

This chapter begins with a description of visual surveys, including strip transects, bounded counts, and sightability models. A discussion of line transects, including fixed-distance and right-angle distance methods, is presented. Next, our emphasis is on abundance estimators that use harvest data commonly collected by wildlife agencies. We begin with the index-removal method (Petrides 1949, Eberhardt 1982) and then describe different change-in-ratio (Paulik and Robson 1969) and catch-effort methods (Leslie and Davis 1939, DeLury 1947, 1951, Zippin 1956, 1958). Life-history models, such as the sex-age-kill model, are presented along with variance expressions and a maximum-likelihood estimator. We conclude the chapter with the presentation of age-structured population reconstruction models, including cohort (Pope 1972) and virtual population analyses (Fry 1949, Gulland 1965), common in the fisheries literature, and end the chapter with age-at-harvest models based on maximum-likelihood methods (e.g., Gove et al. 2002).

Chapter 10: Integration of Analytical Techniques. The purpose of the last chapter is to present case studies and examples that highlight the integration of field data with techniques explored in this book for demographic assessment and management. The examples illustrate various applications of the demographic methods and the joint use of multiple techniques for problem solving. The examples are not exhaustive; rather they are intended to demonstrate a diversity of possible scenarios and how study objective(s) might be met by using multiple techniques and approaches to study design. Each example uses at least two techniques described in previous chapters.

Examples include the management of elk populations for a desired sex ratio; how to incorporate field data into a Leslie matrix model for an eastern cottontail rabbit (*Sylvilagus floridanus*) population; the integration of age-at-harvest and radiotelemetry data to estimate elk survival probabilities; combining of herd composition counts and radiotelemetry information to estimate mountain sheep (*Ovis canadensis*) abundance; a combination of change-in-ratio, catch-per-unit-effort, and mark-recapture techniques for a demographic assessment of ring-necked pheasants; and a method for partitioning harvest and natural mortality from harvest information.

Primer on Wildlife Population Dynamics

2

Chapter Outline

2.1 Introduction

Population dynamics is concerned with changes in abundance, as well as the factors that influence those changes (Gotelli 2001). Components of population assessment include an evaluation of status and vitality. Population status refers to the current state of the population and considers factors such as abundance, age and sex structure, and health (i.e., nutritional and physiological condition). In contrast, population vitality, which is commonly expressed as the relative change in population size from one year to the next, refers to the demographic health of the population and the ability of the population to be self-sustaining.

A basic understanding of the theories governing animal population dynamics is warranted before application of the sex- and age-ratio techniques discussed in this book. Because our primary interest is to assess population status and vitality from sex- and age-ratio data, we begin our discussion with basic principles of population growth. As the chapter proceeds, we build in complexity by relaxing assumptions about how populations grow. Eventually, we consider population growth of age- and stage-structured populations and how knowledge of age- or stage-specific rates can help guide management activities. Because we use harvest data for many applications, we conclude the chapter with an overview of population harvest theory, including a discussion of the concepts of the annual surplus model, sustained yield harvesting, and additive and compensatory mortality. This chapter is not meant to be exhaustive; rather, it is intended to provide context for using sex, age, and count data to evaluate the status and trends of animal populations discussed later in this book. Johnson (1994), Caughley (1977), Caswell (2001), Donovan and Welden (2001), and Gotelli (2001) offer more comprehensive discussions of these topics.

At the most fundamental level, the number of animals one time step in the future (N_{t+1}) is affected by the current population size (N_t), the number of additions to the population (i.e., number of births, B, and number of immigrants, I), and the number of reductions in the population (i.e., number of deaths, D, and number of emigrants, E). That is,

$$N_{t+1} = N_t + (B+I) - (D+E). \tag{2.1}$$

To simplify our assessment of local populations, we often assume a population closed to movement (i.e., no immigration or emigration). Thus, we drop I and E from Eq. (2.1).

2.2   Continuous Time Models

If we assume continuous population growth (i.e., births and deaths occurring continuously), the rate of change in population size (dN_t) during a given time period (dt) is (Gotelli 2001):

$$\frac{dN_t}{dt} = B - D. \tag{2.2}$$

Recognizing that total number of births is influenced by population abundance, N_t, $B = bN_t$, where b is the instantaneous birth rate or the number of births per individual per unit time (Gotelli 2001). Similarly, $D = dN_t$, where d is the instantaneous death rate. Substituting and rearranging terms in Eq. (2.2) gives (Gotelli 2001):

$$\frac{dN_t}{dt} = (b-d)N_t$$
$$= rN_t \tag{2.3}$$

where r is the intrinsic rate of increase equal to $b - d$ (Hastings 1997, Gotelli 2001). In this way, r represents the per capita rate of change or growth rate of the population. The value of r is important in our understanding of population vitality and the potential for the population abundance to be stationary ($r = 0$), increasing ($r > 0$), or declining ($r < 0$). As formulated, Eq. (2.3) assumes population closure, b and d are constant, growth is continuous with no time lags, and there is no genetic, age- or stage-structure (i.e., demographic parameters do not differ among males and females or by age or size), or environmental limitations to preclude full growth potential.

To project population size, we rearrange Eq. (2.3) to $\frac{dN_t}{N_t} = r dt$ before integrating both sides of the equation as follows (Hastings 1997:11):

$$\int \frac{1}{N_t} dN_t = \int r dt,$$
$$\ln N_t + c = rt + c', \tag{2.4}$$

or

$$N_t = e^{rt} c''. \tag{2.5}$$

Setting $t = 0$ in Eq. (2.5) yields

$$c'' = N_0; \qquad (2.6)$$

hence,

$$N_t = N_0 e^{rt}, \qquad (2.7)$$

where N_0 is the initial population size, t is time, and $e = 2.71828$ (base of the natural logarithm). To demonstrate exponential population growth under the continuous time model, consider data presented by Hailey (2000), who reported an $r = 0.137$ for Greek tortoises (*Testudo graeca*). Using $r = 0.137$ and assuming an initial population size of 25 (i.e., $N_0 = 25$), we can calculate N_t (Fig. 2.1).

Note the exponential growth form of population growth when $r = 0.137$. For comparison and to illustrate population growth under different levels of r, we also plotted N_t versus t when $r = 0$, and $r = -0.137$ (Fig. 2.1). When $r = 0$, the population growth rate is 0, thus, $N_t = 25$ for the entire period. In contrast, when $r = -0.137$, the population is declining exponentially; the more negative the value of r, the more precipitous the decline. Many populations — from yeast grown in laboratory cultures (Pearl 1927) to large mammals such as reindeer (*Rangifer tarandus*) (Scheffer 1951) — have demonstrated exponential growth for a period of time.

Caughley and Birch (1971) discussed aspects of calculating and interpreting the rate of increase for a population. They used the term r_m to describe the intrinsic rate of increase. They described r_m as the maximum rate at which a population with a stable age distribution (SAD) can increase when no resource is limiting (Caughley and Birch 1971:659). Throughout this book, we denote their r_m by the symbol r_{MAX}. The observed or realized rate of increase, denoted as r_{REAL}, is the rate of population change under prevailing demographic and environmental conditions. For example, as will be seen with

logistic growth, the realized rate of increase $r_{REAL} = r_{MAX}\left(1 - \dfrac{N_t}{K}\right)$, where N_t is the current

population abundance and K is carrying capacity (i.e., the maximum number of animals the area can support, which is related to the quantity and quality of the area). Caughley

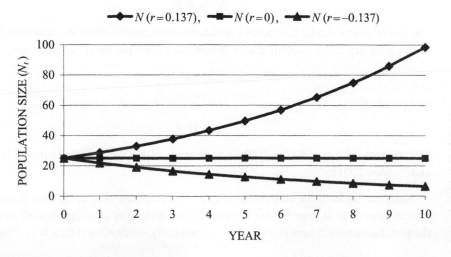

$\blacklozenge\!-\!N\,(r=0.137),\ \blacksquare\!-\!N\,(r=0),\ \blacktriangle\!-\!N\,(r=-0.137)$

Figure 2.1. Population projection of exponential population growth under the continuous time model for Greek tortoises, assuming $r = 0.137$, 0, or -0.137 (Hailey 2000) and $N_0 = 25$ over a 10-year period.

and Birch (1971) also defined r_s as the rate of increase under a SAD. Rate r_s, therefore, is calculated from prevailing survival and fecundity rates although a SAD is assumed. Throughout this book, we denote their r_s by the symbol r_{SAD}. For large mammals, which are generally not sensitive to extrinsic factors in the environment, r_{SAD} is often similar to r_{REAL}, and Caughley and Birch (1971) saw little utility to calculating r_{SAD} (although see Eberhardt and Simmons 1992). In contrast, they considered r_{MAX} important for understanding harvest, because it directly measures how quickly a population can rebuild after a reduction. This discussion is important because different data types and different analyses will provide values for either r_{MAX}, r_{SAD}, or r_{REAL}, and can dramatically influence the interpretation of population vitality and subsequent management actions.

2.3 Discrete Time Models

The previous section considered growth in a continuous fashion; however, for many populations, birth and death rates are not continuous. Rather, there are discrete birth pulses that occur one or more times a year when new animals are added after the breeding season (i.e., spring birth pulse in many wild animals). The death process may also have periods of higher or lower mortality throughout the year. Thus, population growth is better modeled by considering these pulses in a discrete time framework. Instead of using the instantaneous rate of change r, λ is used to denote the finite rate of change for populations with discrete birth pulses. The value λ expresses the proportional change in abundance from time t (i.e., N_t) to time t_{i+1} (i.e., N_{t+i}) as

$$\lambda = \frac{N_{t+1}}{N_t}, \tag{2.8}$$

or equivalently,

$$N_{t+1} = N_t \lambda. \tag{2.9}$$

In this way, λ represents the multiplicative factor by which abundance changes from one year to the next. For example, $\lambda = 1.3$ indicates the population is increasing by 30% a year. A value of $\lambda = 1$ indicates the population abundance is stationary, $\lambda > 1$ implies a growing population, and $\lambda < 1$ is a declining population. To project population size over multiple years, we use

$$N_t = \lambda^t N_0. \tag{2.10}$$

The discrete-time model resembles a continuous-time model when the time steps between birth pulses are small (Gotelli 2001). When r is calculated on an annual basis, then

$$\lambda = e^r, \tag{2.11}$$

or equivalently,

$$r = \ln \lambda. \tag{2.12}$$

The discrete-time model (Eq. 2.10) is sometimes called the geometric model (Donovan and Welden 2001).

Note that for both the continuous (Eq. 2.7) and discrete (Eq. 2.10) time models, population growth is unimpeded for the entire time growth is considered. Left unchecked, the population would continue to grow exponentially, when $r > 0$ (i.e., $\lambda > 1$). Thus, exponential population growth is considered density-independent because abundance does not affect the rate of population growth. Furthermore, these models are completely deterministic. That is, there is no variability in the growth rate of the population due to environmental stochasticity or the random fate of individuals (i.e., demographic stochasticity) (Pielou 1969, May 1974, Moller and Legendre 2001, Lande et al. 2003). For extrinsically controlled species, environmental stochasticity or the fluctuations in environmental conditions can have a dramatic effect on population trends. For example, we might expect productivity to increase when conditions are favorable because of higher juvenile survival. It might be better, therefore, to consider the expected population size under average conditions as

$$E(N_t) = N_0 e^{\bar{r}t}, \tag{2.13}$$

where $\bar{r}$ is the average value of r. The incorporation of fluctuations in r results in variability in the population-growth curve over time and the uncertainty in future abundance levels (Gotelli 2001:15). May (1974) expressed the variance of N_t at time t due to variations in r over time as

$$\sigma_{N_t}^2 = N_0^2 e^{2\bar{r}t} \left(e^{\sigma_r^2 t} - 1 \right). \tag{2.14}$$

To account for demographic stochasticity due to the random processes of births and deaths, the variability in N_t is given by Pielou (1969) as

$$\sigma_{N_t}^2 = 2N_0 bt \tag{2.15}$$

when b and d are identical or as

$$\sigma_{N_t}^2 = \frac{N_0(b+d)e^{rt}(e^{rt}-1)}{r} \tag{2.16}$$

when b and d differ. Demographic stochasticity is particularly important for small populations because random birth and death events might drive a population to extinction regardless of the value of r (Gabriel and Buerger 1992, Lande 1993, Wiegand et al. 2001). In contrast, environmental stochasticity generally affects large- and small-sized populations similarly (Lande et al. 2003).

2.4   Logistic Population Growth

Because populations cannot growth exponentially forever, as a result of resource limitations and other conditions of the environment, we must modify the basic exponential growth model. Leopold (1933), in his book *Game Management*, described numerous factors — such as predation, hunting, starvation, and overcrowding — that can cause exponential growth to become suppressed, leading to logistic population growth. Andrewartha and Birch (1954) suggested that animal populations were limited by resources such as food, inaccessibility to those resources, and effects of environmental factors (e.g., weather, competitors) (Krebs 1994). Chitty (1960) suggested that through dispersal, species could self-regulate their population size (Krebs 1978) and reach an equilibrium with their environment.

Many density-independent and density-dependent factors limit growth rates of animal populations. Density-independent factors affect population growth rates irrespective of population size. For example, density-independent factors, such as weather, ultimately dampen growth rates in some bird (e.g., Newton et al. 1998) and mammal (Georgiadis et al. 2003) populations. In Yellowstone National Park, precipitation, particularly in the spring, was an important density-independent factor regulating elk (*Cervus elaphus*) populations (Coughenour and Singer 1996). In contrast, density-dependent factors limit population growth rates as a function of population density. For example, food limitation is an important density-dependent factor in many populations, which prevents exponential growth. Often, food limitations and the competition for food in winter result in density-dependent effects on the population growth rates of ungulates (Merrill and Boyce 1991, Coughenour and Singer 1996). Early laboratory studies (Gause 1934) also showed that food was important in limiting populations.

The specific demographic parameter influenced by increasing population abundance varies among species. For example, productivity and juvenile survival were reduced as density increased in song sparrow populations (*Melospiza melodia*) (Arcese and Smith 1988). Several density-dependent mechanisms, including declines in pregnancy rates (Teer et al. 1965, Gambell 1975), increases in age at first reproduction (Fowler 1981*a*), and declines in juvenile survival (Grubb 1974, Fowler and Barmore 1979), have been reported for large terrestrial and aquatic mammals. In many species, population regulation most often affects productivity and juvenile survival rather than adult animal survival (Eberhardt 1977, Fowler 1981*a*, 1987, Coughenour and Singer 1996). Thus, several demographic parameters might be influenced as population abundance increases.

Recognition of density-dependent processes is important when trying to correctly estimate maximum sustained yield. Often, simulation models used to evaluate harvest strategies consider density-dependent reduction in reproductive rates but consider mortality independent of population density (Roelle and Bartholow 1977). This is significant because Fowler (1981*a*) reported that age-specific, density-dependent mortality and reproduction change the shape of the yield curve, which directly affects maximum sustained yield.

Regardless of the source, limitations in the environment cause a density-dependent response in population growth, whereby b decreases as N increases, and d increases as N increases (Fig. 2.2). That is, assuming a density-dependent response in a population's growth, b and d depend on the number of animals in the environment and the capability of that environment to support animals (carrying capacity, K) (Fig. 2.2).

If we assume a linear decrease in b in relation to N, we can say that

$$b = b_0 - aN, \tag{2.17}$$

where a is a constant that measures the strength of the relationship between b and N (Gotelli 2001). The higher a is, the greater the effect on b because of increasing N. Similarly, because d will increase as N increases, we can say that (Gotelli 2001)

$$d = d_0 + cN. \tag{2.18}$$

With b and d now affected by N, we can substitute Eqs. (2.17) and (2.18) into Eq. (2.3) and obtain (Gotelli 2001)

$$\frac{dN_t}{dt} = [(b_0 - aN_t) - (d_0 + cN_t)]N_t. \tag{2.19}$$

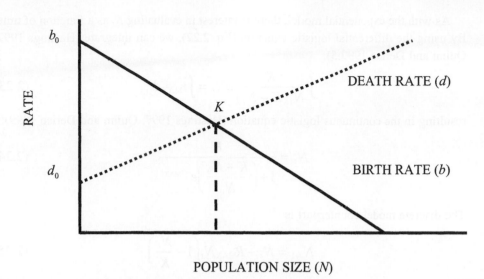

Figure 2.2. The relationship among instantaneous birth and death rates (b and d, respectively) in relation to population size (N) assuming density-dependent population growth (from Krebs 1994, Gotelli 2001). Equilibrium between birth (b) and death (d) rates is attained at carrying capacity (K).

Through rearranging these terms, this relationship reduces to

$$\frac{dN_t}{dt} = [(b_0 - d_0) - (a - c)N_t]N_t. \tag{2.20}$$

Substituting r_{MAX} for $(b_0 - d_0)$, we obtain

$$\frac{dN_t}{dt} = r_{\text{MAX}}\left[1 - \frac{(a-c)}{(b_0 - d_0)}N_t\right]N_t. \tag{2.21}$$

If $K = \dfrac{1}{\dfrac{(a-c)}{(b_0 - d_0)}}$, then rate of population change $\left(\dfrac{dN_t}{dt}\right)$ equals

$$\frac{dN_t}{dt} = r_{\text{MAX}}N_t\left(1 - \frac{N_t}{K}\right), \tag{2.22}$$

which is the familiar logistic population growth equation originally presented by Verhulst (1838) to describe growth of the human population. Equation (2.22) implies the population growth rate is a function of (1) the rate of population growth, (2) population size, and (3) how much of K is currently used (Krebs 1994). Note the rate of population change $\left(\text{i.e., } \dfrac{dN_t}{dt}\right)$ in the logistic model is similar to the exponential model, with the exception of the factor $\left(1 - \dfrac{N_t}{K}\right)$. Krebs (1994), Gotelli (2001), and others considered the factor $\left(1 - \dfrac{N_t}{K}\right)$ to be the unused portion of the carrying capacity K still available to the population. If one assumes that an area of interest can support 1,000 animals (i.e., $K = 1{,}000$) and there are already 175 animals there (i.e., $N_t = 175$), then the population is growing at $\left(1 - \dfrac{175}{1{,}000}\right)$ or 82.7% of its exponential potential. That is, because of resource limitations (e.g., food availability), the population is unable to attain exponential growth. If $N_t > K$, the growth rate will be negative and the population will decline to an equilibrium point at K.

As with the exponential model, there is interest in evaluating N_t as a function of time. By using the differential logistic equation (Eq. 2.22), we can integrate (Hastings 1997, Quinn and Deriso 1999:5).

$$\int \frac{K}{N_t(K-N_t)}dN_t = \int r_{\text{MAX}}dt, \qquad (2.23)$$

resulting in the continuous logistic equation (Hastings 1997, Quinn and Deriso 1999):

$$N_t = \frac{K}{1+\left(\dfrac{K-N_0}{N_0}\right)e^{-r_{\text{MAX}}t}}. \qquad (2.24)$$

The discrete model counterpart is

$$N_{t+1} = N_t + R_{\text{MAX}}N_t\left(1-\frac{N_t}{K}\right) \qquad (2.25)$$

or

$$N_{t+1} = N_t + (\lambda_{\text{MAX}}-1)N_t\left(1-\frac{N_t}{K}\right). \qquad (2.26)$$

To demonstrate continuous-time logistic population growth, consider the previously published data of Hailey (2000) for Greek tortoises, assuming $r_{\text{MAX}} = 0.137$ (i.e., $\lambda_{\text{MAX}} = 1.1468$ or $R_{\text{MAX}} = 0.1468$), $K = 250$, and $N_0 = 25$, the trajectory of population abundance over time results in an S-shaped curve, leveling at $K = 250$ (Fig. 2.3). This logistic growth pattern is in contrast to the exponential model (Eq. 2.7), which would project 23,597 turtles after 50 years. As demonstrated (Fig. 2.3), population size under the logistic population growth model increases relatively quickly in the early stages and closely approximates the exponential model before resources become limited. At an abundance of $\frac{K}{2}$, the growth curve (Fig. 2.3) has an inflection point at which the curve

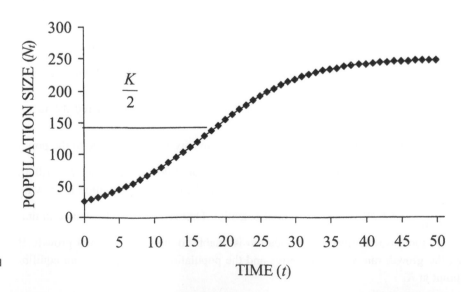

Figure 2.3. Population projection of continuous population growth of Greek tortoises (Hailey 2000) assuming logistic population growth (Eq. 2.24) with parameters $r_{\text{MAX}} = 0.137$, $K = 250$, and $N_0 = 25$ over 50 years.

goes from concave upward to concave downward. The rate of population change $\left(\text{i.e., } \dfrac{dN_t}{dt}\right)$ is greatest at a population size of $\dfrac{K}{2}$. Thereafter, the growth curve dampens, and the number of individuals added in each time step declines. Finally, population size approaches equilibrium at carrying capacity, K. Once at K, population abundance grows no further. In contrast to the exponential growth model, the per capita growth rate changes with abundance for the logistic growth model. The per capita growth rate $\left(\text{i.e., } \dfrac{N_{t+1} - N_t}{N_t}\right)$ under the logistic population growth model equals

$$\frac{N_{t+1} - N_t}{N_t} = (\lambda_{\text{MAX}} - 1)\left(1 - \frac{N_t}{K}\right), \tag{2.27}$$

of the form of a straight-line regression,

$$\frac{N_{t+1} - N_t}{N_t} = \alpha - \beta N_t.$$

That is, the per capita growth rate declines linearly under the logistic population growth model in contrast to the constant per capita growth rate assumed with the exponential model (Fig. 2.4).

An equally noteworthy feature of the logistic model is how the change in abundance (i.e., $N_{t+1} - N_t$) varies over time as population size increases. Assuming logistic growth with the same demographic parameters as in Figure 2.3 (i.e., $r_{\text{MAX}} = 0.137$, $K = 250$, and $N_0 = 25$), the change in abundance (i.e., $N_{t+1} - N_t$) versus N_t can be plotted (Fig. 2.5). Note the maximum change in abundance from one year to the next (i.e., $N_{t+1} - N_t$) occurs at half the carrying capacity $\left(\dfrac{K}{2}\right)$. If harvest began at this point in time, the maximum

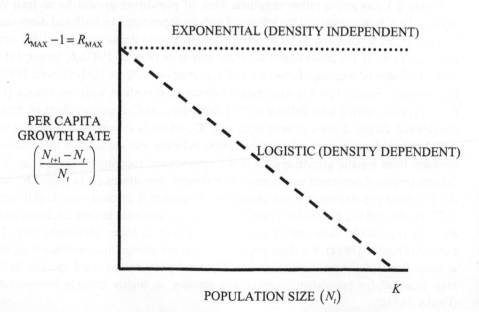

Figure 2.4. Per capita growth rates for populations that are increasing according to the exponential growth model (i.e., density-independent growth) and the logistic growth model (i.e., density-dependent growth).

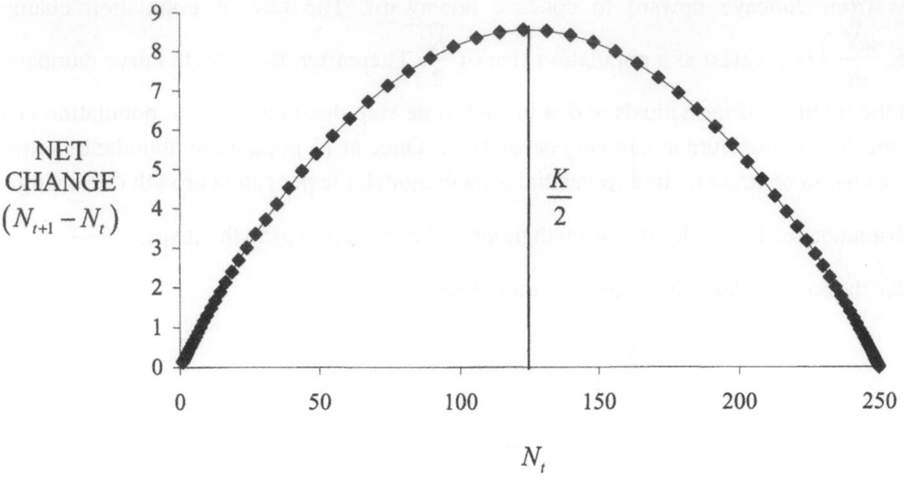

Figure 2.5. The net change in abundance, $(N_{t+1} - N_t)$ versus N_t, for an $r_{MAX} = 0.137$ reported by Hailey (2000) for Greek tortoises, and assuming $K = 250$. The corresponding plot of N_t versus t is presented in Fig. 2.3.

annual change of $(\lambda_{MAX} - 1)N_t\left(1 - \dfrac{N_t}{K}\right)\Big|_{N_t = \frac{K}{2}} = \dfrac{(\lambda_{MAX} - 1)K}{4}$ would also be the maximum

sustained yield. For example, assuming $r_{MAX} = 0.137$ (i.e., $\lambda_{MAX} = 1.1468$), which was obtained from a population in the early stages of growth, and $K = 250$, the maximum sustained yield would be

$$\frac{(\lambda_{MAX} - 1)K}{4} = \frac{(1.1468 - 1)250}{4} = 9.175$$

animals, indicating that about 9 animals could be harvested per year, provided conditions (e.g., carrying capacity) do not change. To achieve this level of harvest, the population would need to be maintained at $\left(\dfrac{K}{2}\right)$ or 125 animals.

Figure 2.3 provides a rather simplistic view of population growth for at least two aspects. First, it oversimplifies the effects of density dependence on birth and death rates. In some cases, a reduction in birth rates or an increase in death rates might not occur until conditions in the environment are extremely poor (Fowler 1981a,b; Strong 1986). Thus, nonlinear relationships between b and d, in relation to N, are likely (Fowler 1981a). For example, Fowler (1981a) described nonlinear relationships with net change (i.e., $N_{t+1} - N_t$) that resulted from delayed density dependence (i.e., negative effects of density dependence do not depress growth until near K), which is common in large mammal (i.e., k-selected species) population dynamics. Whereas the net change in abundance modeled from logistic growth results in the symmetrical recruitment curve (Fig. 2.5), delayed density dependence would result in a skewed recruitment curve (Fig. 2.6, curve B). For these populations, the net change in abundance is greatest near K (Eberhardt 1977, Fowler 1981a), but declines precipitously as K is attained. In contrast, many insect and fish populations show strong density dependence at lower abundance (Fig. 2.6, curve A) (Fowler 1981a). For these populations, the net change in abundance is greatest at lower densities. These populations roughly correspond to r-selected species, having high potential for population increase and residing in highly variable environments (Fowler 1981a).

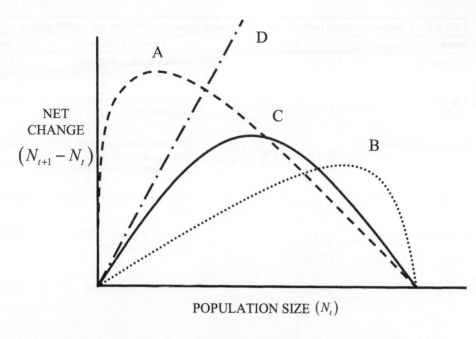

NET
CHANGE

$$(N_{t+1} - N_t)$$

POPULATION SIZE (N_t)

Figure 2.6. Net change in abundance (i.e., $N_{t+1} - N_t$) versus population size (N_t) for four types of population growth patterns (Fowler 1981a). Curve A represents net change for insects and similar populations that exhibit the greatest net population growth when N_t is small. Curve B represents net change for species such as large mammals that exhibit delayed density dependence; for these species, the net change is greatest near carrying capacity. Curve C represents the net change in abundance for the standard logistic model of population growth. Curve D represents net change in abundance for the exponential growth model.

Second, time lags might influence how populations respond to conditions in the environment that would result in a population surpassing K (McCullough 1979). May (1976) considered two factors regarding the effect of time lags on population growth. The first factor related to duration of the time lag. Second was the population's response time, which is related to its growth rate. The higher the population's growth rate, the greater the opportunity to overshoot K.

We have shown the basic logistic growth model, but there are other ways to incorporate density-dependent population growth (Table 2.1). Each variation demonstrates a different way to incorporate density-dependent growth. Consequently, each takes a different form and assumes different relationships among population growth, carrying capacity, and abundance.

To demonstrate the general form of these different density-dependent relationships reported in the ecological literature, we go back to the data of Hailey (2000). First, we begin by plotting the exponential and logistic curves for this population based on Eqs. (2.9) and (2.25) (Fig. 2.7). Although divergence in population size is not apparent until about 7 years, population size in the logistic growth curve is always less than the exponential.

Next, we consider the generalized logistic growth model (Table 2.1, Eq. 2.28), which is similar to the standard logistic, with the exception of the z parameter. The parameter modifies the density-dependent effect on population growth. When $z = 1$, the generalized logistic equation simplifies to the standard logistic model (Eq. 2.25) (Fig. 2.8). For values of $z > 1$, the density-dependent effects on population growth are dampened, and population growth more closely resembles exponential growth (Fig. 2.8). Consequently, the population reaches K more quickly (Fig. 2.8). For large values of $z \geq 10$, population abundance fluctuates about the carrying capacity. For values of $z < 1$, the density-dependent effects operate earlier, pushing the population growth rate below that of standard logistic growth. Pella and Tomlinson (1969), Gilpin et al. (1976), Gilpin and Ayala (1973), Eberhardt (1982, 1987), and Jeffries et al. (2003) have used the generalized logistic equation to model the growth of wildlife populations.

Table 2.1. Common functional forms of population growth reported in fisheries and wildlife studies.

Name	Form	General References	Equation
Exponential	$N_{t+1} = N_t \lambda_{\mathrm{MAX}}$	Malthus (1798), Cox (1923), Hastings (1997), Gotelli (2001)	(2.9)
Logistic	$N_{t+1} = N_t + N_t (\lambda_{\mathrm{MAX}} - 1)\left(1 - \dfrac{N_t}{K}\right)$	Pearl (1925), Hutchinson (1978), Kingsland (1985)	(2.25)
Generalized logistic	$N_{t+1} = N_t + N_t (\lambda_{\mathrm{MAX}} - 1)\left(1 - \left(\dfrac{N_t}{K}\right)^z\right)$	Pella and Tomlinson (1969), Gilpin et al. (1976), Gilpin and Ayala (1973), Eberhardt (1987)	(2.28)
Ricker	$N_{t+1} = N_t \lambda_{\mathrm{MAX}}^{\left(1 - \frac{N_t}{K'}\right)}$	Ricker (1954), Taper and Gogan (2002)	(2.29)
Beverton and Holt	$N_{t+1} = \dfrac{N_t \lambda_{\mathrm{MAX}}}{\left(1 + \dfrac{N_t}{K'}\right)}$	Beverton and Holt (1957), Tang and Chen (2002)	(2.30)
Hessell	$N_{t+1} = \dfrac{N_t \lambda_{\mathrm{MAX}}}{\left(1 + \dfrac{N_t}{K'}\right)^z}$	Quinn and Deriso (1999)	(2.31)
Shephard (Maynard Smith)	$N_{t+1} = \dfrac{N_t \lambda_{\mathrm{MAX}}}{1 + \left(\dfrac{N_t}{K'}\right)^z}$	Quinn and Deriso (1999)	(2.32)

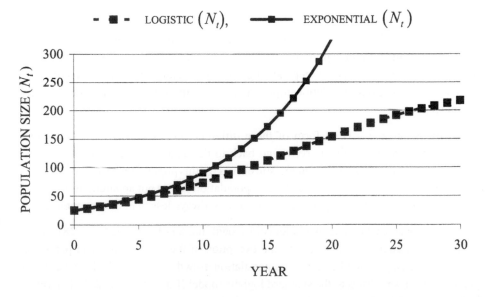

Figure 2.7. The logistic and exponential models of population growth, using $r_{\mathrm{MAX}} = 0.137$ ($\lambda_{\mathrm{MAX}} = 1.1468$), $K = 250$, and $N_0 = 25$.

The Ricker, Beverton–Holt, Hessell, and Shepherd models, similar to the logistic model, moderate exponential growth (Figs. 2.9, 2.10). However, these alternative models use different approaches to moderate population growth, and their K' parameter no longer signifies carrying capacity. The Hessell and Shepherd models degenerate to the Beverton–Holt model when z values equal one. The Beverton–Holt and Ricker models are commonly used in fisheries literature to describe the stock recruitment function (Hilborn and Walters 1992).

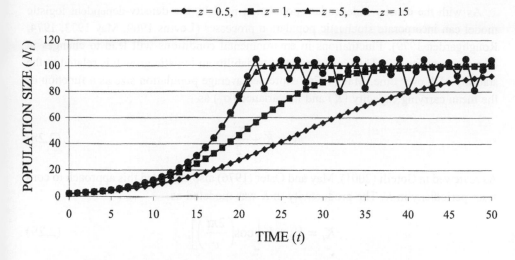

Figure 2.8. Population growth trend using the generalized logistic growth model (Eq. 2.28) for values of $z = 0.5, 1, 5, 15$; $N_0 = 2$; $K' = 100$; and $\lambda = 1.2$.

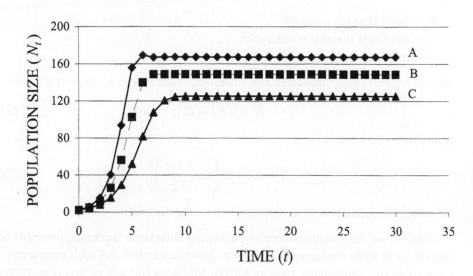

Figure 2.9. Plot of N_t versus N for the Ricker curve (Eq. 2.29). Parameters used for Curve A: $\lambda = 1.8$, $K' = 1000$; $N_0 = 2$; Curve B: $\lambda = 1.4$; $K' = 1000$; $N_0 = 2$; and Curve C: $\lambda = 1.0$; $K' = 1000$; $N_0 = 2$.

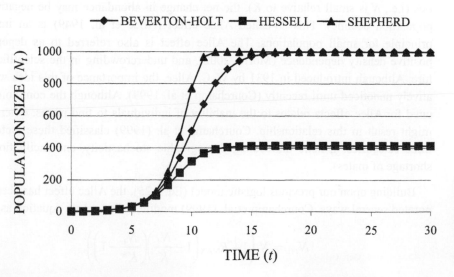

Figure 2.10. Plot of the Beverton-Holt (Eq. 2.30), Hessell (Eq. 2.31), and Shepard (Eq. 2.32) models, when $\lambda = 1.0$; $K = 1000$; $N_0 = 2$, and $z = 2$.

As with the density-independent population models, the density-dependent logistic model can incorporate stochastic population processes (Levins 1969, May 1973, 1974, Roughgarden 1979). Fluctuations in environmental conditions will lead to changes in carrying capacity (K). Thus, one source of variability in logistic growth is related to the stability of K over time. May (1974) expressed average population size as a function of the mean carrying capacity ($\overline{K}$) and its variance (σ_K^2) as

$$\overline{N} \approx \overline{K} - \frac{\sigma_K^2}{2}. \tag{2.28}$$

As reviewed in Gotelli (2001), May and Oster (1976) also extended this approach to consider periodicity in K. The productivity in K was modeled as

$$K_t = K_0 + K_1\left[\cos\left(\frac{2\pi t}{c}\right)\right], \tag{2.29}$$

where

K_1 = mean carrying capacity;
K_0 = amplitude (height) of the cycle;
c = length of the cycle.

May and Oster (1976) considered two situations. When $r \cdot c$ is small ($\ll 1.0$), then

$$\overline{N} \approx \sqrt{K_0^2 - K_1^2}; \tag{2.30}$$

and when $r \cdot c \gg 1.0$,

$$\overline{N} \approx K_0 + K_1\left[\cos\left(\frac{2\pi t}{c}\right)\right], \tag{2.31}$$

indicating the population approximately follows the fluctuations in K.

Thus far, we have assumed that as population abundance increases, there are negative effects to some demographic rates (e.g., juvenile survival and adult pregnancy). For some small size populations, such as African wild dogs (*Lycaon pictus*) (Courchamp et al. 2000), there might be an inverse density-dependent relationship where at low densities (i.e., N is small relative to K), the net change in abundance may be negative (i.e., population decline). This so-called Allee effect (Allee et al. 1949) is an important principle for small populations. The Allee effect is also referred to as depensation, positive density dependence (Morris 2002), and undercrowding in the scientific literature. Although introduced in 1931 by W. C. Allee, the importance of this idea went relatively unnoticed until recently (Courchamp et al. 1999). Although the common reason cited for Allee effects relates to the inability of individuals to find mates, other factors might result in this relationship. Courchamp et al. (1999) classified these factors into three categories: genetic inbreeding, demographic stochasticity, and facilitation (e.g., shortage of mates).

Building upon our previous logistic model (Eq. 2.25), the Allee effect has been incorporated several ways. Courchamp et al. (1999) modified the logistic equation as

$$N_{t+1} = N_t\left(1 + R_{\text{MAX}}\left(1 - \frac{N_t}{K}\right)\left(\frac{N_t}{K^*} - 1\right)\right), \tag{2.32}$$

where K^* represents the Allee threshold. Amarasakare (1998) also considered a threshold response due to Allee effects and modified the logistic equation as

$$N_{t+1} = N_t\left(1 + R_{MAX}\left(1 - \frac{N_t}{K}\right)\left(\frac{N_t}{K} - \frac{A}{K}\right)\right), \tag{2.33}$$

where A $(0 < A < K)$ represents population size when rate of change switches from positive to negative. Stephens and Sutherland (1999) used a scalar to adjust the severity of the Allee effect as

$$N_{t+1} = N_t\left(1 + R_{MAX}\left(1 - \frac{N_t}{K}\right) - \left(\frac{\alpha\theta}{\theta + N_t}\right)\right), \tag{2.34}$$

where θ scales the Allee effect and α adjusts the severity of the Allee effect. Thus, all of these variations build on the generic logistic model (Eq. 2.25).

The Allee effect is important to population growth because it produces an unstable equilibrium at low densities, thus increasing the probability of extinction (Courchamp et al. 1999). It also can reduce recruitment at all population densities, except the very highest (Amarasekare 1998, Courchamp et al. 1999, Stephens and Sunderland 1999) (Figs. 2.11, 2.12). Thus, harvesting from populations experiencing Allee effects may be precarious, because a reduction in density might not result in higher yields as assumed in the standard logistic population model. Instead, the population might respond with unexpectedly low recruitment (Fig. 2.12). Also, population growth models that do not consider Allee effects often underestimate the risk of extinction for small populations, particularly for those populations exhibiting spatial structure (i.e., metapopulations) (Elmhagen and Angerbjorn 2001, Hanski and Ovaskainen 2003). Fowler and Baker (1991) suggested the Allee effect would be most evident when populations are at 10% of carrying capacity (i.e., $0.1K$). In addition to abundance in relation to K, factors such as the breeding system of the animal should be considered, because monogamous species would be affected more severely than polygamous breeders at lower densities.

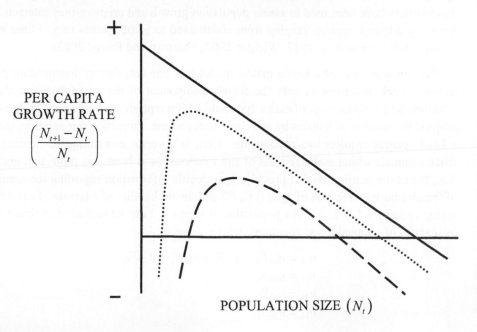

PER CAPITA
GROWTH RATE
$\left(\dfrac{N_{t+1} - N_t}{N_t}\right)$

POPULATION SIZE (N_t)

Figure 2.11. The per capita rate of population growth $\left(\dfrac{N_{t+1} - N_t}{N_t}\right)$ in relation to population size (N_t). The standard logistic model (unbroken line) (Eq. 2.25) demonstrates a linear decline in the per capita growth rate as population size increases. The lower, dashed line indicates a severe Allee effect (after Liermann and Hilborn 1997, Stephens and Sunderland 1999), whereas the dotted line indicates a more mild, yet important, effect of abundance. Reprinted with permission from Elsevier.

Figure 2.12. Net change (i.e., $N_{t+1} - N_t$) versus population size for the standard logistic growth curve (unbroken line) and populations experiencing the Allee effect (dotted and dashed lines). The lower dashed line indicates a greater response to the Allee effect. Due to the Allee effect, net change is delayed until a certain number of animals are present in the population. Abundance is lower when an Allee effect exists than when logistic growth is operating; consequently, a population does not attain its biotic potential (Stephens and Sunderland 1999). Reprinted with permission from Elsevier.

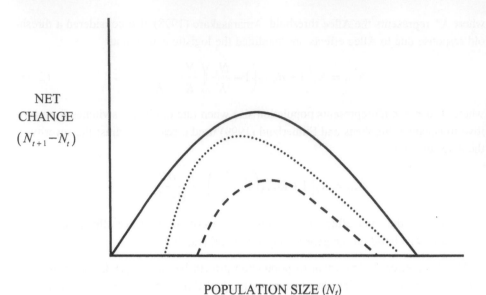

NET CHANGE $(N_{t+1} - N_t)$

POPULATION SIZE (N_t)

2.5   Age Structure Models

There is often interest in examining which demographic features of a population most influence the realized rate of increase or how changes in demographic parameters would ultimately influence population growth. Wildlife managers might have some control over survival rates, for example, by limiting the number of animals harvested. Thus, it would be useful to know whether management actions, such as reductions in harvest, would appreciably affect population growth. Because survival and fecundity change with age of the animal (e.g., reproduction in long-lived vertebrates is delayed for 1 or more years), use of a projection matrix, such as a Leslie matrix, is helpful in modeling the effects of age-specific factors on population growth (Bernadelli 1941, Leslie 1945, 1948). Projection matrices have been used to assess population growth and predator-prey interactions for many different species, ranging from endangered to game species (e.g., Flipse and Veling 1984, Crouse et al. 1987, Wielgus 2002, Sherman and Runge 2003).

The simplest form of a Leslie matrix model is a one-sex, density-independent projection model. It represents only the female component of the population and, thus, assumes there is no shortage of males that could inhibit reproductive potential. The model projects the number of females by age class through time. Projection through time using a Leslie matrix requires two components. First, is a vector containing the number of female animals within each age class of age x denoted as $\underline{n}$. Next, is a projection matrix (i.e., the Leslie matrix, **M**) that provides age-specific information regarding the number of females born to a female of age i (i.e., F_i) and the probability of a female of age i surviving to age $i + 1$ (i.e., S_i). The population at time $t = 0$ can be projected to time $t = 1$ by a series of coupled linear equations, where

$$n_{01} = n_{00}F_0 + n_{10}F_1 + n_{20}F_2 + n_{30}F_3,$$
$$n_{11} = n_{00}S_0,$$
$$n_{21} = n_{10}S_1,$$
$$n_{31} = n_{20}S_2,$$

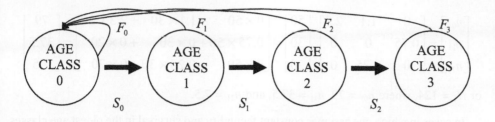

Figure 2.13. Schematic of a one-sex, density-independent, age-structure, matrix model with four age classes and age-dependent survival (S) and fecundity (F).

and where

n_{ij} = number of females in age class i ($i = 0, 1 \ldots, 3$) at time j ($j = 0, 1$).

Following the approach of Caswell (2001), a four-age-class matrix population model is represented by a flow model (Fig. 2.13).

Age classes 1, 2, and 3 (Fig. 2.13) contribute back to age class 0 via reproduction (F_i). Individuals then move with some probability (S_i) to the next age class each year. The system of equations can be expressed as a matrix equation where

$$\begin{bmatrix} n_{01} \\ n_{11} \\ n_{21} \\ n_{31} \end{bmatrix} = \begin{bmatrix} F_0 & F_1 & F_2 & F_3 \\ S_0 & 0 & 0 & 0 \\ 0 & S_1 & 0 & 0 \\ 0 & 0 & S_2 & 0 \end{bmatrix} \cdot \begin{bmatrix} n_{00} \\ n_{10} \\ n_{20} \\ n_{30} \end{bmatrix}$$

or

$$\underset{\sim}{n_1} = \mathbf{M} \cdot \underset{\sim}{n_0}.$$

The vector $\underset{\sim}{n_t}$ denotes abundance by age class at time t. The Leslie matrix $\mathbf{M}$ contains the age-specific fecundities (F_i) in the top row. The elements immediately below the diagonal, in the subdiagonal, are the age-specific survivals (S_i). In this one-sex model, fecundity parameters (i.e., F_i) are the age-specific fecundities expressed in terms of the net number of female offspring produced per female to the anniversary date of the projections. The projection matrix projects the age distribution that exists at time t to time $t + 1$.

To project population size, one multiplies the initial vector of population abundance by the projection matrix to obtain $\underset{\sim}{n_{t+1}}$. For example, consider a three-age-class model with initial conditions of $N_0 = 100$, where $n_{00} = 50$, $n_{10} = 30$, and $n_{20} = 20$. In matrix notation, this initial age-specific abundance would be

$$\underset{\sim}{n_0} = \begin{bmatrix} 50 \\ 30 \\ 20 \end{bmatrix}.$$

Next, consider a Leslie matrix $\mathbf{M}$ for this population, taking the form

$$\mathbf{M} = \begin{bmatrix} 0 & 1.1 & 2.3 \\ 0.75 & 0 & 0 \\ 0 & 0.25 & 0 \end{bmatrix}.$$

To estimate $\underset{\sim}{n_1}$, we multiply $\mathbf{M} \cdot \underset{\sim}{n_0}$ and obtain the resultant vector

$$\begin{bmatrix} n_{01} \\ n_{11} \\ n_{21} \end{bmatrix} = \begin{bmatrix} 0 & 1.1 & 2.3 \\ 0.75 & 0 & 0 \\ 0 & 0.25 & 0 \end{bmatrix} \cdot \begin{bmatrix} 50 \\ 30 \\ 20 \end{bmatrix} = \begin{bmatrix} 0 \times 50 & +1.1 \times 30 & +2.3 \times 20 \\ 0.75 \times 50 + 0 \times 30 & +0 \times 20 \\ 0 \times 50 & +0.25 \times 30 + 0 \times 20 \end{bmatrix} = \begin{bmatrix} 79 \\ 37.5 \\ 7.5 \end{bmatrix}$$

or $N_1 = 124$, where $n_{01} = 79$, $n_{11} = 37.5$, and $n_{21} = 7.5$.

In cases in which one assumes constant fecundity and survival in the oldest age classes (i.e., when all animals over a certain age are assumed to have the same fecundity and survival rate), it is possible to collapse the older age classes to simplify the model (Caswell 2001). Consider a population matrix that has constant survival in the last two age classes and fecundity in the last three age classes, in which case, the Leslie model can be expressed as

$$\begin{bmatrix} n_{0,t+1} \\ n_{1,t+1} \\ n_{2,t+1} \\ n_{3,t+1} \\ n_{4,t+1} \end{bmatrix} = \begin{bmatrix} F_0 & F_1 & F & F & F \\ S_0 & 0 & 0 & 0 & 0 \\ 0 & S_1 & 0 & 0 & 0 \\ 0 & 0 & S & 0 & 0 \\ 0 & 0 & 0 & S & 0 \end{bmatrix} \cdot \begin{bmatrix} n_{0t} \\ n_{1t} \\ n_{2t} \\ n_{3t} \\ n_{4t} \end{bmatrix}.$$

Here, the probability of surviving from age class 4 to age class 5 is zero. This same population can also be modeled by collapsing the older age classes with constant survival and fecundity as

$$\begin{bmatrix} n_{0,t+1} \\ n_{1,t+1} \\ n_{2+,t+1} \end{bmatrix} = \begin{bmatrix} F_0 & F_1 & F \\ S_0 & 0 & 0 \\ 0 & S_1 & S \end{bmatrix} \cdot \begin{bmatrix} n_{0t} \\ n_{1t} \\ n_{2+,t} \end{bmatrix}.$$

In this latter case, the assumptions are now slightly different. The survival probability is S between age classes 2 and 3, classes 3 and 4, infinitum. The distinction between the two matrix formulations is quite small, if S is not large.

A population with constant age-specific fecundity and survival rates through time will asymptotically establish a SAD (Mertz 1970, Caughley 1977) that does not depend on its initial age structure. Thus, when a population reaches a SAD, the relative number of animals in each age class will remain constant. To find the proportion of the population in each age class i, first consider the total population abundance as

$$N_t = N_0 S_0 + N_0 \lambda^{-1} S_{01} + N_0 \lambda^{-2} S_{02} + \cdots$$

or

$$N_t = \sum_{x=0}^{w} N_0 \lambda^{-x} S_{0x},$$

where the number of animals in age class i is expressed as

$$N_t = N_0 \lambda^{-i} S_{0i},$$

and where S_{0x} = probability an animal survives age classes 0 through x.

Thus, the proportion of the population in each age class i is

$$\frac{N_i}{N_t} = \frac{N_0 \lambda^{-i} S_{0i}}{\sum\limits_{x=0}^{w} N_0 \lambda^{-x} S_{0x}},$$

which simplifies to

$$C_i = \frac{S_{0i} \lambda^{-1}}{\sum\limits_{x=0}^{w} S_{0x} \lambda^{-x}}. \tag{2.35}$$

Equation (2.35) is equivalent to that of Mertz (1970), Johnson (1994), and Gotelli (2001) but is expressed in terms of λ instead of r. To calculate the number of animals in age class i relative to the newborns, begin with the number of newborns where

$$N_0 = F_0 N_0 + F_1 S_0 N_0 \lambda^{-1} + \cdots,$$

or equivalently,

$$N_0 = \sum_{x=0}^{w} F_x N_0 S_{0x} \lambda^{-x}. \tag{2.36}$$

Now, the number of animals in age class i relative to the newborns is

$$\frac{N_i}{N_0} = \frac{N_0 S_{0i} \lambda^{-i}}{\sum\limits_{x=0}^{w} F_x N_0 S_{0x} \lambda^{-x}},$$

or, more simply,

$$\frac{N_i}{N_0} = \frac{S_{0i} \lambda^{-i}}{\sum\limits_{x=0}^{w} F_x S_{0x} \lambda^{-x}}. \tag{2.37}$$

Euler's equation (also known as Lotka's equation) (Lotka 1925),

$$1 = \sum_{x=0}^{w} F_x S_{0x} \lambda^{-x} \tag{2.38}$$

can be used to estimate λ when annual age-specific survival and fecundity rates are available. The value of λ is the largest positive eigenvalue of the projection matrix **M**. The corresponding SAD for that population is calculated from the eigenvector associated with eigenvalue λ. A stable and stationary population is a special situation in which a population has a SAD and in which abundance is constant across time (i.e., in which $\lambda = 1$). In the case of a stable and stationary population, the abundance counts by age class will be the same in both relative and absolute terms over time.

The basic matrix model formulated above does not consider density-dependent responses. That is, no resource limitations are considered. Once the population has achieved a SAD, abundance will change by the factor λ each year. It is relatively straightforward to incorporate density dependence in the Leslie matrix by using the equations described above. For example, Miller et al. (1995) used a density-dependent Leslie matrix to assess gray wolf (*Canis lupus*) recovery in Michigan. The model used standard age-specific survival and fecundity values but also incorporated carrying capacity, which depressed population abundance.

Miller et al. (1995) began with the basic logistic model (Eq. 2.25) and used the analogous matrix model equation:

$$\underset{\sim}{n}_{t+1} = \underset{\sim}{n}_t + D_{(N)t}(\mathbf{M} - \mathbf{I})\underset{\sim}{n}_t, \tag{2.39}$$

where $\mathbf{I}$ is the identity matrix, $(\mathbf{M} - \mathbf{I})$, analogous to $(\lambda - 1)$. $D_{(N)t}$ is the density-dependent function (Jensen 1995), defined as

$$D_{(N)t} = \left(1 - \frac{N_{T(t)}}{K}\right), \tag{2.40}$$

where $N_{T(t)}$ is the total number of animals over all age classes at time t. They also included a random variable in this matrix model to account for environmental stochasticity.

The Leslie matrix model, as discussed thus far, only considers females. An extension of the Leslie matrix is the two-sex model that allows tracking of the abundance of both males and females in the population (Caswell 2001). In this case, let the vector

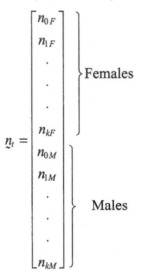

denote the numbers of females (n_{iF}) and males (n_{iM}) of age i ($i = 1, \ldots, k$) in the population at time t. The projection matrix for the two-sex model is written as

$$\left[\begin{array}{cccc|cccc} F_{0F} & F_{1F} & \cdots & F_{kF} & 0 & \cdots & & 0 \\ S_{0F} & & & & & & & \\ & S_{1F} & & & \vdots & \ddots & & \vdots \\ & & & S_{k-1F}\,0 & 0 & \cdots & & 0 \\ \hline F_{0M} & F_{1M} & \cdots & F_{kM} & 0 & \cdots & & 0 \\ 0 & & \cdots & 0 & S_{0M} & & & \\ & & & & & S_{1M} & & \\ \vdots & & & \vdots & & & \ddots & \\ 0 & & \cdots & 0 & & & & S_{k-1,M}\,0 \end{array}\right].$$

The projection matrix for the two-sex model is of size $2(k + 1) \times 2(k + 1)$ when $(k + 1)$ unique age classes are involved. This matrix can be subdivided into four $(k + 1) \times (k + 1)$ submatrices. The upper-left submatrix is the projection matrix for the standard female-only model. The upper-right submatrix is composed of all zeros. The lower-left submatrix has the fecundities of juvenile males per female in the first row. The lower-right submatrix has the male age-specific survival probabilities in the subdiagonal. Consider, for example, a two-age-class population model with males and females. The initial condition is

$$
n_{00} = \begin{bmatrix} n_{0F} \\ n_{1F} \\ n_{0M} \\ n_{1M} \end{bmatrix} = \begin{bmatrix} 62 \\ 20 \\ 83 \\ 25 \end{bmatrix},
$$

and the age-sex-specific survival rates are assumed to be $S_{0F} = 0.30$ and $S_{0M} = 0.25$. The litters are anticipated to be of the sex ratio (male to female) of $0.53 : 0.47$ with age-specific fecundity values of 6 and 7 for age classes 0 and 1, respectively. Thus, the age-specific fecundity levels would be: $f_{1F} = 2.82 \ (= 6 \times 0.47)$, $f_{1M} = 3.18$, $f_{2F} = 3.29$, and $f_{2M} = 3.71$; the matrix equation becomes

$$
\begin{bmatrix} n_{0F} \\ n_{1F} \\ n_{0M} \\ n_{1M} \end{bmatrix} = \left[\begin{array}{cc|cc} 2.82 & 3.20 & 0 & 0 \\ 0.30 & 0 & 0 & 0 \\ \hline 3.18 & 3.71 & 0 & 0 \\ 0 & 0 & 0.25 & 0 \end{array}\right] \cdot \begin{bmatrix} 62 \\ 20 \\ 83 \\ 25 \end{bmatrix}.
$$

Thus, the projected population size next year would be

$$
\begin{bmatrix} n_{0F} \\ n_{1F} \\ n_{0M} \\ n_{1M} \end{bmatrix} = \begin{bmatrix} 241 \\ 19 \\ 270 \\ 21 \end{bmatrix}.
$$

For harvested populations, one can also include a harvest matrix to permit population projections in the presence of exploitation (Caswell 2001). The form of the modified Leslie matrix equation can be written as

$$
\underset{\sim}{n}_{t+1} = \mathbf{MH}\underset{\sim}{n}_{t}, \tag{2.41}
$$

where $\mathbf{H} =$ a diagonal matrix of dimension $(k + 1) \times (k + 1)$. The harvest matrix for a single-sex model is

$$
\mathbf{H} = \begin{bmatrix} h_{0F} & & & \\ & h_{1F} & & \\ & & \ddots & \\ & & & h_{kF} \end{bmatrix},
$$

where h_{iF} = the probability of surviving the harvest for age class i females $(i = 0, \ldots, k)$. For a two-sex model, the harvest matrix of dimension $2(k + 1) \times 2(k + 1)$ is of the form:

$$\mathbf{H} = \begin{bmatrix} h_{0F} & & & & & & & \\ & h_{1F} & & & & & & \\ & & \ddots & & & & 0 & \\ & & & h_{kF} & & & & \\ & & & & h_{0M} & & & \\ & & & & & h_{1M} & & \\ & & 0 & & & & \ddots & \\ & & & & & & & h_{kM} \end{bmatrix}.$$

Harvest yield is computed as $y_t = \underset{\sim}{1}'(\mathbf{I} - \mathbf{H})\mathbf{M}\underset{\sim}{n}_t$, where

$$\underset{\sim}{1} = \begin{bmatrix} 1 \\ 1 \\ 1 \\ \cdot \\ \cdot \\ \cdot \\ 1 \end{bmatrix}_{2(k+1)}$$

for a two-sex model. A sustained yield y exists if there is a matrix $\mathbf{H}$, such that $y_t > y$ for all years $(t = 0, 1, \ldots)$. Note the harvest matrix $\mathbf{H}$ may vary with time. Maximum sustained yield is found by finding a matrix $\mathbf{H}$ such that the expression $|\mathbf{H^*M} - \lambda\mathbf{I}| = 0$ has 1.0 as the dominant eigenvalue. With this harvest matrix, $\mathbf{H^*}$, the population is stationary with a SAD. The maximum sustained yield is then equal to $1'(\mathbf{I} - \mathbf{H^*})\mathbf{M}n^*$.

It is also possible to consider multiple populations within the Leslie matrix framework (Caswell 2001). The simplest multiple population model is a two-population model with no interchange between populations (i.e., no immigration or emigration). Let the vector equal

$$\underset{\sim}{n}_t = \begin{bmatrix} x_{0t} \\ x_{1t} \\ \cdot \\ \cdot \\ \cdot \\ x_{kt} \\ y_{0t} \\ y_{1t} \\ \cdot \\ \cdot \\ \cdot \\ y_{kt} \end{bmatrix} \begin{array}{l} \left.\rule{0pt}{6em}\right\} \text{Population 1 females} \\ \left.\rule{0pt}{6em}\right\} \text{Population 2 females} \end{array}$$

where

x_{it} = number of females of age class i ($i = 0, \ldots, k$) at time t in the first population;
y_{it} = number of males of age class i ($i = 0, \ldots, k$) at time t in the second population.

In the case of two noncommunicating populations, the projection matrix can be written as

$$\mathbf{M} = \begin{bmatrix} \mathbf{M}_1 & 0 \\ 0 & \mathbf{M}_2 \end{bmatrix}_{2(k+1) \times 2(k+1)}.$$

Matrices $\mathbf{M}_1$ and $\mathbf{M}_2$ are the standard female-only population matrices of size $(k + 1) \times (k + 1)$ for populations 1 and 2, respectively.

In the case in which movement occurs between the two populations, the projection matrix can be written as

$$\mathbf{M} = \begin{bmatrix} \mathbf{M}_1 - \mathbf{A}_{12} & \mathbf{A}_{21} \\ \mathbf{A}_{12} & \mathbf{M}_2 - \mathbf{A}_{21} \end{bmatrix}_{2(k+1) \times 2(k+1)}.$$

Matrix $\mathbf{A}_{21}$ is the movement matrix of females from population 2 to population 1 of dimension $(k + 1) \times (k + 1)$ with probabilities of movement $m_{i(21)}$ in the subdiagonal and zeros elsewhere, such that

$$\mathbf{A}_{21} = \begin{bmatrix} 0 & & & & \\ m_{0(21)} & 0 & & 0 & \\ & m_{1(21)} & 0 & & \\ & & m_{2(21)} & 0 & \\ 0 & & & \vdots & 0 \\ & & & & m_{k-1(21)} & 0 \end{bmatrix}.$$

The value $m_{i(21)}$ is the probability a female of age class i in population 2 moves to population 1. Matrix $\mathbf{A}_{12}$ is defined analogously. Keyfitz and Murphy (1967) discuss the properties of the resultant migration model.

Sensitivity analysis (or perturbation analysis) (Caswell 2001) offers a useful tool for assessing how past and future changes in demographic parameters would influence the status and vitality of a population. Sensitivity analysis can be used to evaluate how changes in fecundities or survival affect λ when all other elements in the Leslie matrix remain constant. These analyses allow researchers to examine how plans to manage a population might influence the vitality of the population (Caswell 2001). For example, if we reduced harvest mortality by 20%, how would λ change? Alternatively, given changes in λ, how much of these changes can be attributable to specific demographic parameters such as survival or fecundity (Caswell 2001)? Field sampling designs can also be refined by identifying which demographic parameters need to be most precisely estimated. Several sources, including Wisdom and Mills (1997), Mills et al. (1999), de Kroon et al. (2000), Caswell (2000, 2001), and Wisdom et al. (2000), provide a description of these procedures.

Conducting a sensitivity analysis involves three steps (Donovan and Welden 2001): (1) simulate the population until a SAD has been attained; (2) calculate the stable age structure, which is given by vector $\underset{\sim}{w}$ (i.e., this vector of proportions is called a "right" eigenvector) (Donovan and Welden 2001); and (3) calculate the reproductive value (Caswell 2001) given by vector $\underset{\sim}{v}$ (i.e., this vector is also called a "left" eigenvector) (Donovan and Welden 2001). The right eigenvectors $\underset{\sim}{w}_i$ of **M** are defined by the equation:

$$\mathbf{M}\underset{\sim}{w}_i = \lambda_i \underset{\sim}{w}_i.$$

Alternatively, the left eigenvectors $\underset{\sim}{v}_i$ of **M** are defined by the equation:

$$\underset{\sim}{v}_i^* \mathbf{M} = \lambda \underset{\sim}{v}_i^*,$$

where $\underset{\sim}{v}_i^*$ = the complex conjugate transpose of v_i. The eigenvalues are the same for the left and right eigenvectors. However, the left and right eigenvectors satisfy the relationship $\underset{\sim}{v}_i' \underset{\sim}{w}_j = 0$ for $i \neq j$. The reproductive value considers individuals in terms of their future contribution to the population (Caswell 2001, Donovan and Welden 2001).

To assess the sensitivity (s_{ij}) of λ to a change in the demographic parameter (a_{ij}) of a Leslie matrix, Caswell (2001:209) uses the partial derivative of λ with respect to a_{ij}:

$$s_{ij} = \frac{\partial \lambda}{\partial a_{ij}} = \frac{v_i w_j}{\underset{\sim}{v}' \underset{\sim}{w}}, \tag{2.42}$$

where v_i and w_j relate to the ith and jth elements of the reproductive value $\underset{\sim}{v}$ and stable stage distribution $\underset{\sim}{w}$ vectors. Hence, the relative sensitivity of λ to changes in element a_{ij} is proportional to the product of the ith element of the left eigenvector and the jth element of the right eigenvector.

Interpreting sensitivity results can be complicated because matrix elements are measured on different scales (Donovan and Welden 2001). That is, survival is expressed as a probability, whereas fecundity is expressed in terms of the number of female offspring per female. To aid in interpretation of **M**, elasticity analysis is often used. Elasticity (e_{ij}), which is the proportional change in λ for a proportional change in a matrix element (a_{ij}), is calculated (Caswell 2001:226) as

$$e_{ij} = \frac{\left(\dfrac{a_{ij}}{\partial a_{ij}}\right)}{\left(\dfrac{\lambda}{\partial \lambda}\right)}$$

$$= \frac{a_{ij} s_{ij}}{\lambda}. \tag{2.43}$$

Equation (2.43) expresses the proportional contribution of each a_{ij} to the value of λ.

Mills et al. (1999) describe how to analyze situations when the demographic parameters of interest are not explicitly incorporated in the Leslie models. For example, in stage-based models, the annual survival rate is not explicitly incorporated in the Leslie matrix. Instead, the matrix elements consider the joint probability of surviving and transitioning to the next stage. In these situations, researchers use the expression (Caswell 2001:232):

$$\frac{x}{\lambda} \frac{\partial \lambda}{\partial x} = \frac{x}{\lambda} \sum_{i,j} \frac{\partial \lambda}{\partial a_{ij}} \frac{\partial a_{ij}}{\partial x} \tag{2.44}$$

to calculate the elasticity associated with a particular demographic rate, such as annual survival. Mills et al. (1999) describe the use and calculation of elasticities in wildlife ecology studies, including a useful flowchart that demonstrates approaches to using elasticity measures in applied ecology situations. In general, the choice between using sensitivity or elasticity of λ to evaluate contributions of individual parameters depends on the scale. If interest lies in additive perturbations, one should use sensitivities; if perturbations are proportional, elasticities should be used (Caswell 2001:244).

2.6 Stage Structure Models

For some species, the size or stage of development better represents differences in demographics than does age. For example, differences in amphibian survival or fertility might be affected more by their size than age, because size better characterizes these processes. Lefkovitch (1965) provides a modification to the Leslie matrix to model stages rather than age. Instead of ages, transition in the Lefkovitch matrix (i.e., the projection matrix) considers transition to the next stage. Because it is possible to remain within a stage, which is not possible in an age-structured model, that probability is also included. This is done through use of a modified Leslie matrix (Lefkovitch 1965), where

$$\begin{bmatrix} n_{01} \\ n_{11} \\ n_{21} \\ n_{31} \end{bmatrix} = \begin{bmatrix} F_0 & F_1 & F_2 & F_3 \\ S_{01} & S_{11} & 0 & 0 \\ 0 & S_{12} & S_{22} & 0 \\ 0 & 0 & S_{23} & S_{33} \end{bmatrix} \cdot \begin{bmatrix} n_{00} \\ n_{10} \\ n_{20} \\ n_{30} \end{bmatrix}.$$

In the stage-based model, demographic parameters are grouped by biologically relevant stages. The first row represents the stage-specific fecundities. In the Lefkovitch matrix, values in the subdiagonal (e.g., S_{01}) represent the probability of an individual surviving *and* transitioning into the next stage class at the time of the next annual survey. Note, however, the values in the diagonal (e.g., S_{11}) now represent the probability of an individual surviving to the next annual survey but remaining within that stage class. Although an animal might age one year, it may not gain enough mass or size to move into the next stage. In the Lefkovitch matrix, it is assumed that all individuals within a particular stage share the same survival and fecundity values. As before, the initial vector (i.e., $\underline{n}_t$) is multiplied by the projection matrix to estimate abundance in subsequent years.

Donovan and Welden (2001) described a hypothetical stage-based model for loggerhead sea turtles (*Caretta caretta*). Their model, built around published data and the Lefkovitch matrix, took the following form:

$$\begin{pmatrix} 0 & 0 & 0 & 4.665 & 61.896 \\ 0.675 & 0.703 & 0 & 0 & 0 \\ 0 & 0.047 & 0.657 & 0 & 0 \\ 0 & 0 & 0.019 & 0.682 & 0 \\ 0 & 0 & 0 & 0.061 & 0.8091 \end{pmatrix}.$$

A flow diagram (Fig. 2.14) can help describe the processes of their stage-structured model. In evaluating the Lefkovitch matrix from Donovan and Welden (2001) and Figure 2.14, note that each adult female would contribute 61.896 female offspring each year and

Figure 2.14. Schematic of a one-sex, density-independent, Lefkovitch matrix model with five life stages [Hatchling (hatchs.), Small juveniles (S. juvs.), Large juveniles (L. juvs.), Subadult, and Adult] (from Donovan and Welden 2001).

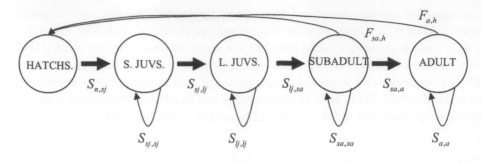

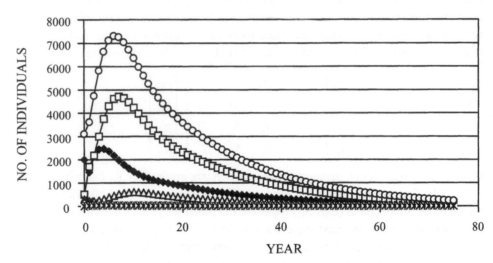

Figure 2.15. Projection of a hypothetical sea turtle population provided by Donovan and Welden (2001). In this example, the population maintains $\lambda \approx 0.95$ after it reaches a stable age distribution.

the subadult females would contribute 4.665 female offspring each year; individuals in other stages do not reproduce. The value of 0.675 in the subdiagonal above indicates there is a 67.5% chance that a hatchling will survive and transition to the small juvenile stage by the time of the next annual survey. Numbers in the direct diagonal have a slightly different meaning. For example, 0.703 implies there is a 70.3% chance that a small juvenile will survive and remain in that stage class at the time of the next annual survey. Assuming $N_0 = 3101$ divided into the stages

$$
n_0 = \begin{pmatrix} 2000 \\ 500 \\ 300 \\ 300 \\ 1 \end{pmatrix},
$$

a projection of the population abundance by stage class through time results (Fig. 2.15). For the sea turtle population, the finite rate of annual change will be $\lambda \approx 0.95$ when a stable stage structure is attained. In Chapter 7, we will discuss techniques to estimate rates of change.

2.7   Harvest Management Theory

Animals are harvested for many reasons, from controlling pests to reduce their damage to other resources, to providing food and recreational hunting opportunities (Caughley and Sinclair 1994). Caughley (1977) considered sustainable harvest and population control two of the main problems encountered by wildlife managers. Early harvest management attempted to prevent overexploitation (Bolen and Robinson 1995). Modern harvest management attempts to balance consumptive interests with nonconsumptive interests (e.g., wildlife viewing), which has become an increasingly difficult challenge as our society becomes more urbanized. Hunting licenses continue to be an important source of revenue for wildlife management agencies. Additional issues, such as subsistence harvest, continue to challenge harvest management (Byers and Dickson 2001).

From a biological perspective, wildlife harvest can have tremendous effects on the sex- and age-structures of populations, particularly if certain classes of animals are favored by hunters. In the case of many heavily exploited ungulate populations, skewed sex ratios can influence demographic parameters such as productivity. For example, Squibb et al. (1991) reported higher pregnancy rates of elk in an area in northern Utah with a sex ratio of 41 bulls per 100 cows preseason versus another with 15 bulls per 100 cows. The effects of differential harvest on productivity will be even greater for monogamous species. For this reason, harvest management often focuses on maintenance of appropriate sex ratios to ensure maximum reproduction (Raedeke et al. 2002).

Several interrelated factors must be considered when developing a harvest management strategy. Issues such as population status and vitality, recreational objectives, effects of harvest on species behavior, and demographic processes must be considered (Strickland et al. 1996). Social factors, such as tradition (e.g., season opening dates) and subsistence harvest, also have an important role in harvest strategies. For many species, game management units are established that provide the spatial extent of area in which management efforts are directed. For example, the state of Missouri has 59 deer management units (Fig. 2.16). These units were based on grouping areas with similar habitat conditions and identifiable road boundaries (roads help hunters identify boundaries more readily). Two of these units have special status because of their proximity to Kansas City and St. Louis. Thus, in Missouri, deer management units consider both biological and political factors. In contrast, the federal government has jurisdiction over the harvest of migratory birds, although states and provinces can also regulate hunting of migratory birds. For example, federal regulations allow for harvest of mourning doves (*Zenaida macroura*), but many states, such as Iowa, New York, and Minnesota, do not permit it.

Setting harvest goals is another important consideration. The establishment of population abundance objectives helps game managers institute harvest regulations to meet those objectives and evaluate the success of the management program. Often harvest objectives include issues such as stabilization of population abundance, control of animal damage to other resources, and providing recreational opportunities. For example, the state of Missouri's deer management objective is to "maintain good deer numbers without causing large amount of public conflict." For wild turkeys (*Meleagris gallopavo*), the goal

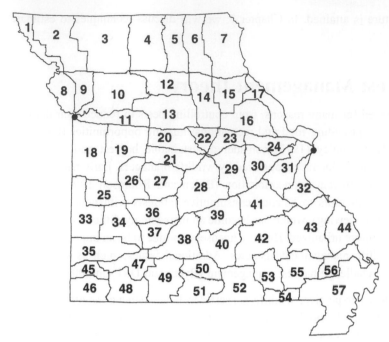

Figure 2.16. White-tailed deer (*Odocoileus virginianus*) management units in Missouri (map courtesy of the State of Missouri). Units were assigned based on habitat conditions and roads to assist hunters in identifying unit boundaries. Units 58 and 59 (•) are not visible on the map; they are urban management units around St. Louis and Kansas City, respectively.

is to "provide quality turkey hunting for Missouri sportsmen." In this case, quality is defined as "good opportunity to harvest adult males during the spring season."

The basis for harvest regulations in most states includes tradition (e.g., in Wisconsin, the white-tailed deer season is always the Saturday before through the Sunday after Thanksgiving Day), local population size, public attitudes, and the belief that harvest does not negatively influence populations (i.e., there is surplus stock). Some populations require additional information due to public attitudes and specific harvest goals for the management unit. Thus, harvest regulations, including season setting and the number of animals that can be legally harvested, are affected by biological and sociological factors as well as political reality.

In this book, we will often use harvest-related data as the basis for demographic assessments. Thus, it is important to understand the advantages and disadvantages associated with such data. The most obvious advantage is that hunters aid in data collection efforts. Harvest data can be collected over extensive areas for less cost than can other field survey methods, because hunters supply the information compared with biologists conducting costly surveys. For this reason, harvest data still provide the basis for many harvest management decisions. Rupp et al. (2000) found that 80% of the states responding to their survey used hunter data to estimate white-tailed deer harvest. In addition, at least 40% of the states used hunter harvest data to assess white-tailed deer population trends. These data are commonly collected from check stations, mail questionnaires, telephone surveys, and report cards (Cada 1985, Steinert et al. 1994, Rupp et al. 2000). Annually, states spent at least $3.5 million to collect deer harvest information.

Despite these obvious advantages, there are several important disadvantages that must be considered when using hunter harvest data in any population assessment. Hunter behavior, hunting regulations, weather factors, as well as population abundance can influence which and how many animals are harvested (e.g., Parker et al. 2002). Thus, vulnerability may not be constant across years or between age classes or genders. Often, in the case of ungulate harvest, large antlered males are favored, whereas young-of-the-year are generally not (Raedeke et al. 2002). Conversely, female deer are often harvested proportionally by age class (Roseberry and Woolf 1998). Thus, doe harvest might provide useful age-structure data, while buck harvest may not. Even with mandatory reporting systems, low return rates of report cards, nonresponse, and biased responses all reduce the utility of these data (Coe et al. 1980, Pendleton 1992, Kohlmann et al. 1999). For example, Strickland et al. (1996) report three biases in using mail surveys: (1) respondents often overestimate success, (2) unsuccessful hunters are less likely to respond than are successful hunters, and (3) nonhunting sportsmen are less likely to respond.

2.7.1   Annual Surplus Model

The annual surplus model of wildlife harvest assumes that hunters remove a proportion of animals that would have succumbed to some other form of mortality had they not been harvested. Often annual surplus relates to r-selected species (i.e., small game animals having high potential for population increase with high annual mortality rates). These populations have a great capacity to rebound when conditions become favorable because of their high reproductive potential. However, these species sustain a large amount of natural mortality annually and are largely influenced by density-independent factors such as weather. For example, many upland game bird populations have a 0.50 annual natural mortality rate (e.g., Johnsgard 1973). For many small mammals, such as squirrels (*Sciurus* spp.), annual mortality rates of 0.40 and greater have been reported (e.g., Mosby 1969). Annual mortality rates of eastern cottontail rabbits (*Sylvilagus floridanus*) might be as high as 0.80 (Trent and Rongstad 1974).

To understand this harvest philosophy, it is important to understand the basic demographic life cycle of r-selected animals, commonly harvested under the annual surplus model. We begin with a general annual cycle of an r-selected species, such as that presented in Quick (1963). There are four important time periods in the annual life cycle (Fig. 2.17). First, spring increases due to natality result in the highest population size of the year. Second, incremental reductions in abundance occur because of natural mortality, with heavy losses during the spring and early summer. Small reductions in abundance in July and August occur because of predation, disease, and other such factors. Last, there are deaths in winter, often due to weather factors and carrying capacity. The assumption underlying the annual surplus model is that population size at the beginning and end of the annual life cycle would be the same whether animals are harvested or not.

Central to this harvest theory is the idea of compensatory mortality, whereby hunting mortality is compensated for by a reduction in natural mortality. Hunter harvest simply replaces natural mortality, and the annual mortality rate remains constant. The theory surrounding compensatory and additive mortality is hotly debated among ecologists because it is an important component of management of harvested populations. The root of the concept came from Errington (1946, 1956), who described predation mortality in muskrats (*Ondatra zibethicus*). Others (e.g., Allen 1954) applied the general principles outlined by Errington (1945, 1946) to consider the effects of harvest on animal populations. The predominant thought put forth by Errington (1945) was that predation removed the "doomed surplus" of individuals that would have succumbed to some other form of

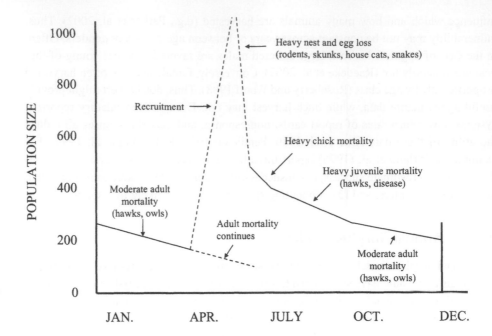

Figure 2.17. Changes in population of valley quail during one year (from Quick 1963).

mortality when winter carrying capacity was exceeded. Later, Anderson and Burnham (1976) offered two hypotheses regarding harvest influences on total mortality. Since that time, many others have described compensatory mortality as it relates to harvest of birds and mammals to aid in evaluation of natural and human harvest of wild populations (e.g., Burnham and Anderson 1984, Clark 1987, Boyce et al. 1999).

If hunting mortality is completely compensatory, as hunting mortality increases, the natural mortality rate would decline, resulting in a constant annual mortality rate (Anderson and Burnham 1976). In contrast, if the annual mortality rate increases as hunting mortality is added, hunting mortality is considered an additive form of mortality (Anderson and Burnham 1976). Following the approach of Anderson and Burnham (1976), the annual survival probability (S_T) equals

$$S_T = S_N(1 - bP_H), \tag{2.45}$$

where

    S_T = annual survival rate in the absence of hunting;
    b = slope of the linear relationship between the annual survival rate and annual kill rate P_H;
  P_H = probability of mortality due to harvest.

This relationship between S_T and P_H (Anderson and Burnham 1976) is illustrated in Figure 2.18. Until P_H reaches a level of J (Fig. 2.18), harvest is a compensatory form of mortality because the annual survival probability does not decline. Beyond that point, harvest is an additive form of mortality, because S_T declines as P_H increases.

Expressed in terms of S_N, the natural survival probability (Fig. 2.19), as P_H increases, there is a 1 : 1 decline (i.e., $b = -1$) in S_N as P_H increases up to point J (i.e., $b = -1$). Thus, up to point J (Fig. 2.19), harvest mortality is compensated by decreases in the

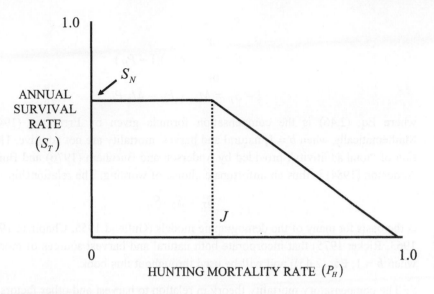

Figure 2.18. Annual total survival probability (S_T) as a function of the hunting mortality rate (P_H) based on concepts in Peek (1986:287).

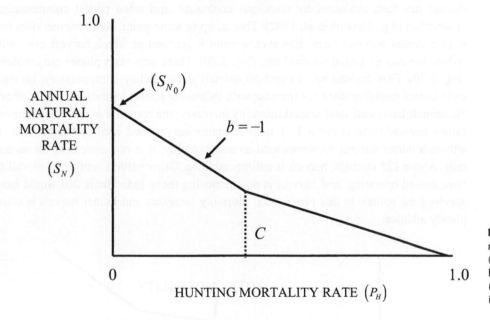

Figure 2.19. Annual natural mortality rate (S_T) as a function of the hunting mortality rate (P_H) based on concepts in Peek (1986:287).

natural survival probability. However, beyond that point, as P_H increases, S_T decreases more rapidly, indicating that harvest mortality is additive.

When $b = 0$ in Eq. (2.45), there is complete compensation for mortality due to hunting (i.e., $S_T = S_N$) (Anderson and Burnham 1976, Burnham and Anderson 1984). When $b = 1$, Anderson and Burnham (1976) and Burnham and Anderson (1984) consider harvest mortality to be completely additive. However, if we rearrange Eq. (2.45) in terms of the complement of $P_H = 1 - S_H$, then

$$S_T = S_N (1 - bP_H) = S_N (1 - b(1 - S_H)),$$

and when $b = 1$,

$$S_T = S_N S_H$$
or
$$S_T = (1 - M_N)(1 - P_H)$$
or
$$1 - S_T = M_N + P_H - M_N P_H, \tag{2.46}$$

where Eq. (2.46) is the compensation formula given by Errington (1945, 1946). Mathematically, when $b = 1$, natural and harvest mortality are not additive. The description of "total additivity" provided by Anderson and Burnham (1976) and Burnham and Anderson (1984) is thus an unfortunate choice of wording. The relationship

$$S_T = S_N \cdot S_H \tag{2.47}$$

is the basis for many of the demographic models (Gulland 1955, Chapman 1961, Paulik 1963, Ricker 1975) that incorporate both natural and harvest sources of mortality (i.e., when $b = 1$; Eq. (2.45)) and will be used throughout this book.

The compensatory mortality theory in relation to harvest and other factors (e.g., predation) has been evaluated by numerous ecologists, and often partial compensation is observed (e.g., Herkert et al. 1992). That is, up to some point, some harvest does not reduce annual survival rates. However, a point is reached at which harvest rates will reduce the overall annual survival rate (Fig. 2.20). Three important phases are involved (Fig. 2.20). First, harvest up to about 50 animals is completely compensatory, because total annual mortality does not increase with increasing hunter harvest. However, above 50 animals harvested, total annual mortality increases; the response is partially compensatory because there is not a 1 : 1 ratio in hunter harvest and total mortality. That is, although hunter harvest increases total annual mortality, it is not increasing incrementally. Above 125 animals, harvest is entirely additive. Other natural sources of mortality have ceased operating, and harvest is now removing those individuals that would have survived the winter. In this phase, total mortality increases and hunter harvest is completely additive.

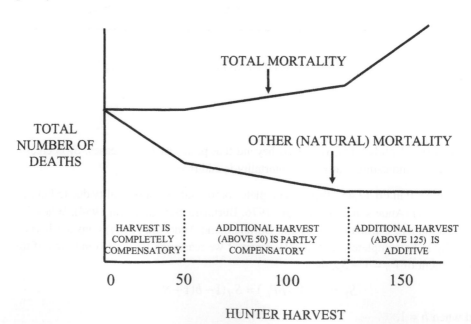

Figure 2.20. The relationship of hunter harvest, natural and total mortality under the compensatory mortality hypothesis of Bailey (1984). Reprinted with permission from John Wiley & Sons, Inc.

Some research suggests that harvest mortality less than 50% of the total annual mortality is completely compensatory (Mosby and Overton 1950, Hickey 1955, Roseberry 1979), whereas other researchers suggest hunting mortality is entirely additive (e.g., Bergerud 1985). Still others suggest higher harvest rates might be permissible. Peterle and Fouch (1969) and Nixon et al. (1974) suggested that fox squirrels (*Sciurus niger*) could withstand a 60% annual harvest rate. For northern bobwhite (*Colinus virginianus*), Vance and Ellis (1972) considered annual harvest rates of 70% to have little impact on quail population size. Immigration appears to play an important role in minimizing harvest effects in localized regions. Small et al. (1991) found that size of a ruffed grouse (*Bonasa umbellus*) population was not affected by high harvest rates. They observed, however, that habitat fragmentation may inhibit grouse movements to areas of low density.

It is apparent the ability to estimate harvest mortality and partition total mortality into component sources is important in understanding the dynamics of exploited populations. The ability to partition mortality sources depends on the assumptions of additive or compensatory harvest mortality. Radio telemetry can provide cause-specific mortality rates, but usually at a high cost. Nevertheless, the functional relationship between natural and harvest mortality usually cannot be ascertained from a single observational study. At a minimum, repeated observations under alternative harvest rates are necessary. However, to date, virtually all analytical models assume natural and harvest sources operate according to Eqs. (2.46) and (2.47).

2.7.2   Sustained Yield Model

In contrast to the annual surplus model typically used for r-selected species, some form of sustained yield modeling is most commonly used for k-selected species. Permissible harvest levels are often based on the logistic growth curve. The logistic curve is used because for k-selected species (i.e., large game animals with low annual mortality rates), the rate of population increase slows as the population approaches its carrying capacity (McCullough 1984). Based on the logistic growth curve, the greatest net change in abundance (i.e., $N_{t+1} - N_t$) occurs at half the carrying capacity $\left(\text{i.e., } \dfrac{K}{2} \right)$. For abundance beyond $\dfrac{K}{2}$, the annual net change in abundance declines, until at K, the change is zero. Thus, a population with abundance $\dfrac{K}{2}$ produces the maximum sustained yield. If harvest began at this point in time, the maximum annual change in abundance of $\dfrac{(\lambda - 1)K}{4}$ would be the maximum sustained yield. By using empirical models from the George Reserve deer herd in Michigan, McCullough (1979, 1984) found the maximum sustainable yield to be 56% of K. In reality, population levels are not maintained or managed at $\dfrac{K}{2}$, because K is difficult to calculate and subject to fluctuations caused by climate and habitat changes. Thus, the actual available harvest is at a level below the maximum sustained yield.

When basing harvest on the sustained yield curve, one could sustain the same harvest numbers at two different population sizes (Fig. 2.21). Often, these two related points

Figure 2.21. Net change (i.e., $N_{t+1} - N_t$) versus N_t for an r_{MAX} = 0.137 (i.e., λ_{MAX} = 1.1468) for Greek tortoises, reported by Hailey (2000), assuming K = 250 and N_0 = 25. Given the symmetrical nature of the curve, there are two different abundance levels that can result in the same sustained yield.

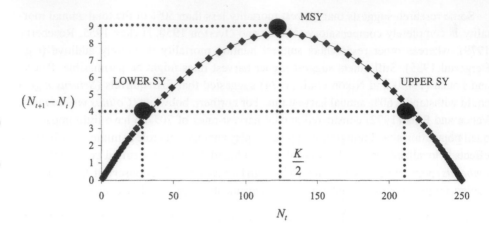

along the curve are called sustained yield pairs (Fig. 2.21) (Caughley and Sinclair 1994). The maximum sustained yield (Fig. 2.21) occurs at roughly 9 individuals per year, if the population is maintained at $\frac{K}{2}$ or 125 animals. Alternatively, harvest can be sustained at 4 individuals when population abundance levels are maintained at either 38 or 215 individuals. McCullough (1984) discusses population stability in relation to harvest levels. If harvest exceeds maximum sustained yield, the population will go extinct, unless other mechanisms (e.g., refugia, immigration) maintain the population. If harvest is consistently less than maximum sustained yield and the population is above $\frac{K}{2}$ (Fig. 2.21), the population is more stable and less prone to extinction. Should harvest numbers be maintained at those levels, population abundance would increase and eventually come into equilibrium at the point on the upper portion of the sustained yield curve where annual production and population abundance match.

For example, assume a take of 4 animals (Fig. 2.21) when abundance (N) is 180 individuals. Because the level of harvest is below the sustained yield, population abundance would increase because more animals are added to the population through reproduction than are removed through harvest and natural mortality. If harvest continued at 4 animals, population abundance would eventually increase to about 210 animals. At this point, equilibrium would be met, and a sustained harvest of 4 animals would be achieved from a population size of 210. However, when $N < \frac{K}{2}$ and harvest is less than the sustained yield, the population is unstable. If harvest is fixed below the sustained yield curve, then the population will likely find a new equilibrium on the right-hand side of the curve. However, if harvest is above the sustained yield curve, then the population will likely go extinct. Thus, it is risky to harvest near the maximum sustained yield and to manage harvests along the left-hand side of the sustained yield curve. In principle, it is best to manage population abundance above $\frac{K}{2}$ at the upper point of the sustained yield pair.

Other benefits, such as more game for hunters and nonconsumptive users to view, would be evident if harvest was managed at the upper sustained yield pair.

There are some operational problems with this straightforward evaluation. First, we have assumed that harvests are at a fixed level, which is rarely the case. Second, K will shift depending on habitat and climate conditions, thus altering the shape of the sustained yield curve. Third, maximum sustained yield of a population is inherently affected by age structure, as well as the sex and age composition of the harvest. For example, the growth of a population will be influenced more if the prime-age individuals are harvested rather than young-of-the-year (DeMaster 1981). Fourth, delayed density dependence (Fowler 1981a,b) in ungulates and inverse density dependence in small populations (Courchamp et al. 1999) would influence the shape of the sustained yield curve (Fig. 2.6). Fifth, stochastic variability in recruitment and survival may shift population trajectories and growth potential. Last, a manager does not know where on the curve the population resides unless both N and K are known. Lancia et al. (1988) describe proportional harvesting, opposed to harvesting a fixed number of individuals, because it is inherently more stable. However, McCullough et al. (1990) suggest the Lancia et al. (1988) strategy requires accurate estimates of N_t.

Several other sustainable harvest strategies have been proposed. McCullough et al. (1990) reported use of a linked-sex harvest strategy that optimizes harvests by altering female harvest rates as a result of observed differences in male harvests. The harvest model does not require survival or productivity information. In reviewing the linked-sex harvest strategy, Lubow et al. (1996) did not support its use and considered the annual survival rate to be more important when developing harvest management strategies. McCullough (1996) later described how spatial controls can help avoid problems of overexploitation. Hatter (1998) outlined a Bayesian approach to moose (*Alces alces*) population assessment and argued the importance of considering risks associated with different harvest regimes.

For marine mammal populations, harvest is based on the optimum sustainable population (OSP), which was defined as the abundance between carrying capacity and the abundance that provides the maximum net productivity level (MNPL) (Eberhardt 1977, DeMaster 1981, Gerrodette and DeMaster 1990). Annual net production is calculated as the difference in population sizes from N_{t+1} and N_t (Jeffries et al. 2003), which is maximized at $\left(\text{i.e., } \dfrac{K}{2}\right)$ (Fig. 2.21). Expressed in terms of population abundance over time (Fig. 2.22), the OSP range occurs above 0.5 carrying capacity $\left(\text{i.e., } \dfrac{K}{2}\right)$ and carrying capacity (i.e., K). Assuming delayed density dependence, as is common in marine mammals (Fowler 1981b), annual production would likely be greatest above 0.5 carrying capacity $\dfrac{K}{2}$ (Fig. 2.22). Taylor and DeMaster (1993) reported that MNPL is between about 50% and 85% of carrying capacity (Fig. 2.23).

Eventually, issues with the OSP approach led to the idea of potential biological removals (PBRs) (Wade 1998). Wade (1998:5) defines PBR as "the maximum number of animals, not including natural mortalities, that may be removed from a marine mammal stock while allowing that stock to reach or maintain its optimum sustainable population." Thus, PBR has the goal of allowing populations to reach or maintain the OSP. The PBRs are calculated as

$$\text{PBR} = N_{\text{MIN}} \frac{1}{2} R_{\text{MAX}} F_R, \tag{2.48}$$

Figure 2.22.
Relationship between annual net change in abundance (i.e., $N_{t+1} - N_t$) and population size (N_t) for a hypothetical marine mammal population. The optimum sustainable population (OSP) size occurs within a range above the maximum sustained yield (MSY) level (Gerrodette and DeMaster 1990) and carrying capacity. This curve roughly corresponds to "Curve B" (Fig. 2.6). Due to delayed density dependence, the net change is greatest near K. The MNPL level is above MSY based on the arguments of Eberhardt (1977) and Fowler (1981a,b).

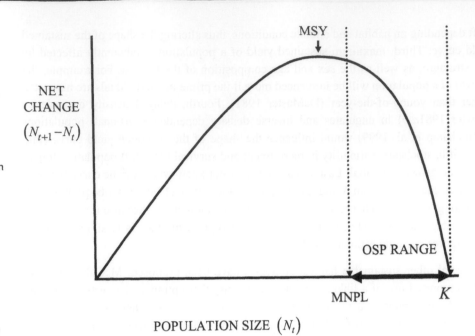

Figure 2.23.
Relationship among population abundance (i.e., N_t) versus time for a hypothetical marine mammal population. The optimum sustainable population (OSP) size occurs within a range above $\frac{K}{2}$ (Gerrodette and DeMaster 1990) and carrying capacity (i.e., K). Taylor and DeMaster (1993) reported that MNPL is between about 50% and 85% of carrying capacity. OSP will be closest to K when the extent of delayed density dependence is strongest.

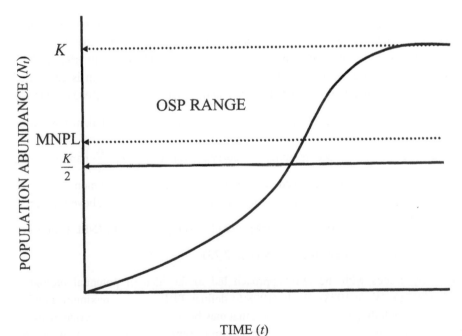

where

N_{MIN} = minimum population estimate;

R_{MAX} = maximum theoretical or estimated net productivity rate (i.e., $1 - \lambda_{MAX} = R_{MAX}$);

F_R = a recovery factor between 0.1 and 1 (Wade 1998).

Wade (1998:6) described F_R as a factor that accelerates growth of reduced populations and a "safety" factor that accounts for uncertainties. The minimum population estimate is based on the best available information, which considers variability around that estimate, thus assuring the stock is at least that size or greater (Wade 1998).

2.8 Summary

We have discussed general patterns of population growth. The first, exponential growth, assumes that population size over time is not restricted by resources. Under the exponential growth model, population growth is rapid and unchecked (Table 2.1, Fig. 2.1), thus, the net change ($N_{t+1} - N_t$) is linear and increasing over time (Fig. 2.6), whereas the per capita growth rate $\left(\dfrac{N_{t+1} - N_t}{N_t} \right)$ is constant over time (Fig. 2.4). In contrast, the logistic growth model assumes that as the number of animals in the population increases, birth rates decline and death rates increase, until the population growth reaches and maintains itself at carrying capacity (Fig. 2.3). Under the logistic population model, initial growth of the population is nearly exponential, reaches a maximum growth rate at $\dfrac{K}{2}$, and then begins to decline as the population reaches and stabilizes at K (Fig. 2.5).

The per capita growth rate declines linearly over time (Fig. 2.4) because of the dampening effect of carrying capacity. Other variations on the logistic model are commonly used in fisheries, and even simple forms of these models, such as the Beverton–Holt, have been found to be adequate in describing stock recruitment relationships (Hilborn and Walters 1992). Table 2.2 provides a summary of demographic characteristics associated with exponential and logistic growth models.

Several other factors, including demographic stochasticity, environmental stochasticity, time lags, delayed density dependence (Fig. 2.6), inverse density dependence (i.e., Allee effect) (Figs. 2.11, 2.12), and removals (e.g., hunter harvest), will influence population size and number of animals recruited into the population. The effect these factors will have on population growth will depend on population abundance and resilience,

Table 2.2. Characteristics of discrete-time exponential and logistic growth models.

Population Feature	Exponential	Logistic	Equations
Population size (N_{t+1})	$N_t \lambda_{MAX}$	$N_t + (\lambda_{MAX} - 1)N_t\left(1 - \dfrac{N_t}{K}\right)$	(2.9), (2.27)
Rate of change $\left(\dfrac{dN}{dt}\right)$	$r_{MAX}N_t$	$r_{MAX}N_t\left(1 - \dfrac{N_t}{K}\right)$	(2.3), (2.22)
Net annual change ($N_{t+1} - N_t$)	$N_t(\lambda_{MAX} - 1)$	$N_t(\lambda_{MAX} - 1)\left(1 - \dfrac{N_t}{K}\right)$	
Per capita growth rate $\left(\dfrac{N_{t+1} - N_t}{N_t}\right)$	$\lambda_{MAX} - 1 = R_{MAX}$	$(\lambda_{MAX} - 1)\left(1 - \dfrac{N_t}{K}\right)$	(2.27)

which is related to its capacity to increase (i.e., maximum rate of increase), quality of the habitat (i.e., carrying capacity), and effects of harvest on demographic rates (e.g., reductions in productivity because of altered sex ratios, age compositions).

Matrix models are useful tools to assess age- and stage-structured population growth. The simplest form of a Leslie matrix model is a one-sex, density-independent projection model that considers only the female component of the population. Extensions to this model make it possible to consider density dependence, male population projections, harvest, and movement between populations. Sensitivity and elasticity analyses are useful in helping identify the parameters that most influence population growth.

Wildlife harvest can have important effects on sex- and age-structures of animal populations. Because harvest data are relatively easily to collect compared with field intensive techniques (e.g., mark-recapture, radio-tagging), these data are commonly used in demographic assessments. We discussed two long-held theories concerning the harvest of wildlife populations. The first, the annual surplus model, which relates to r-selected species (Fig. 2.17), assumes that hunter harvest simply removes a proportion of the population that would have succumbed to some other form of mortality (i.e., that hunting is a compensatory form of mortality) (Figs. 2.18, 2.19). Thus, harvest does not decrease the annual survival rate. In practice, partial compensation is likely. That is, low harvest rates may not decrease the annual survival rate of a population. However, at some higher harvest rate, the annual survival rate will decrease (Fig. 2.20). At that point, harvest contributes to the overall annual mortality rate.

In contrast to the annual surplus model, the sustained yield model is based on the logistic growth curve (Fig. 2.21). The maximum sustained yield occurs at $\frac{K}{2}$. If harvest began at this point in time, then the maximum annual change of $\frac{(\lambda_{\text{MAX}} - 1)K}{4}$ would also be the maximum sustained yield. When managing on the sustained yield curve, it is better to manage on the right-hand side of the curve because overharvest is less likely to drive the population to extinction.

For marine mammal populations, harvest is based on the OSP, defined as the abundance between carrying capacity (K) and the abundance that provides the MPNL (Eberhardt 1977). Today MSY is treated more as a threshold not to be exceeded than a target harvest rate to be attained (Lovejoy 1996, Strickland et al. 1996, Punt and Smith 2001).

Estimating Population Sex Ratios

3

Chapter Outlines

3.1 Introduction

Sex ratio generally refers to the ratio of males to females in a population. The ratio is commonly expressed as the number of males per 100 females (Bolen and Robinson 1995). However, for some applications (e.g., sex-age-kill model of population reconstruction) (Skalski and Millspaugh 2002), the ratio of females to males is used. The sex ratio depends on the life-history stage of interest. Thus, this definition has been further refined to consider the sex ratio at fertilization (primary), birth or hatching (secondary), juveniles (tertiary), and adults (quaternary) (Bolen and Robinson 1995). Coupled with species-specific information on life-history strategy, such as mating systems, the sex ratio at each life-history stage is important in understanding the past, present, and future of a population (Bolen and Robinson 1995). For example, sex ratio data may be used to convert some field data (e.g., crowing counts) to population indices (Trautman 1982). Comparison of prehunting to posthunting sex ratios allows one to examine the effect of hunting regulations on sex ratios. Examination of sex ratios can also provide insight on the effects of hunting regulations on gender-specific survival rates (Bender and Miller 1999, Hatter 1999). Data on sex ratios often provide an index on the well-being of a population. For example, a ratio of 25 adult bull elk (*Cervus elaphus*) per 100 adult cow elk is considered "healthy" (Raedeke et al. 2002). Regulations exist to manage the adult sex ratios for such species.

Several factors ultimately affect sex ratios. These include, but are not limited to, temperature and other climatic factors (Mysterud et al. 2000, Godfrey and Mrosovsky 2001), population density (Kruuk et al. 1999), competitive ability of the sexes (Oddie 2000), nutritional status of the breeding female (Raedeke et al. 2002), predation (Johnson and Sargeant 1977), and hunter harvest (Raedeke et al. 2002). Often, management actions such as harvest have profound influences on sex ratios. In many large mammal populations, antlered males are either the preferred or the only allowable game. In sexually dimorphic, small game species such as ring-necked pheasant (*Phasianus colchicus*) and mallard ducks (*Anas platyrhynchos*), hunter harvest also focuses on the male segment of the population. The extent of the effect of selective harvest on the sex ratio will depend on both the degree of differential harvest and the life expectancy of the species. The longer the life expectancy and the greater the selection bias, the more skewed the sex ratio of the population. Consequently, many heavily exploited large mammal populations are predominantly female in composition. Such changes in sex ratio may influence population productivity. The magnitude of the effect is related to how disparate the sex ratio is and, most importantly, the type of breeding system for a particular species.

It has long been recognized that changes in sex ratio more dramatically influence dynamics of monogamous species than polygamous breeders (Leopold 1933). In monogamous breeders such as the northern bobwhite (*Colinus virginianus*), productivity is maximized at an exact 1 : 1 sex ratio; any disparity decreases productivity. For polygamous breeders, there is a more complex demographic relationship between productivity and sex ratio, i.e., productivity will not be maximized at a 1 : 1 ratio because not all males are necessary for successful reproduction. Part of the male population is surplus reproductive stock that may compete with females for resources. Thus, maximum productivity would be favored by fewer males, and a larger breeding stock of females. However, too skewed a sex ratio will ultimately reduce productivity in polygamous species.

3.1.1 Statistical Notation

Throughout this chapter, the following demographic parameters will be used in expressing sex ratios:

N_M = abundance of adult males;
N_F = abundance of adult females;
N_{MJ} = abundance of juvenile males;
N_{FJ} = abundance of juvenile females;
$N_{MP} = N_M + N_{MJ}$ = abundance of males, both juveniles and adult males in the population;

$N_{FP} = N_F + N_{FJ}$ = abundance of females, both juvenile and adult females in the population;

$N_A = N_M + N_F$ = total abundance of adults;

$N = N_{MP} + N_{FP}$.

Additional notation will be introduced as necessary for special cases in subsequent sections of the chapter.

3.1.2   Alternative Definitions of Sex Ratio

Throughout this chapter, we express the sex ratio as the number of females to the number of males in the population because several later applications require this value (e.g., sex-age-kill model of reconstruction). In a most basic sense, the sex ratio of the population is expressed mathematically as

$$R_P = \frac{N_{FP}}{N_{MP}} = \frac{N_F + N_{FJ}}{N_M + N_{MJ}}. \tag{3.1}$$

In many applications, such as the sex-age-kill model (Eberhardt 1960, Creed et al. 1984, Hansen 1998, Bender and Spencer 1999, Skalski and Millspaugh 2002), only the sex ratio of adults is considered. Using the above notation, the adult sex ratio ($R_{F/M}$) is expressed as

$$R_{F/M} = \frac{N_F}{N_M}, \tag{3.2}$$

where $R_{F/M}$ = number of adult females per adult males in the population. Alternatively, the juvenile sex ratio (R_J) is written as

$$R_J = \frac{N_{FJ}}{N_{MJ}} \tag{3.3}$$

where R_J = number of juvenile females per juvenile males in the population. Both ratio estimators (3.2) and (3.3) are age specific; i.e., both the numerator and the denominator consider the same age class(es). Choice of parameter should depend on the objectives of the population monitoring.

For investigators interested in estimating the sex ratio of males to females, i.e.,

$$R_{M/F} = \frac{N_M}{N_F}, \tag{3.4}$$

we recommend using the formula as presented but changing the notation in the equations for the two genders. While it is true that

$$R_{M/F} = \frac{1}{R_{F/M}},$$

and it follows that

$$\hat{R}_{M/F} = \frac{1}{\hat{R}_{F/M}}, \tag{3.5}$$

variance calculations are best performed using the formula as presented with the notation reversed. Alternatively, the delta method can be used to approximate the variance, where

$$\text{Var}(\hat{R}_{M/F}) \doteq \frac{\text{Var}(\hat{R}_{F/M})}{(R_{F/M})^4}.$$ (3.6)

3.1.3 Direct Versus Indirect Methods of Estimating Sex Ratios

This chapter begins with methods that directly measure or estimate sex ratios based on survey counts of animals. These methods assume that sample observations of the sexes are representative of the actual sex ratio of the population. A variety of survey sampling techniques (for reviews, see Cochran 1977, Thompson 1992) can be used to canvass a population to estimate sex ratios assuming the males and females have an equal probability of detection. Simple random sampling techniques with and without replacement will be described in addition to the analysis of single and replicated surveys. Stratified and cluster sampling techniques to estimate sex ratios will also be examined along with issues concerning unequal detection probabilities.

For populations difficult to survey and in which detection probabilities are known to vary between males and females, this chapter also presents indirect methods of sex ratio estimation. These methods estimate the sex ratio within populations based on gender-specific survival and harvest probabilities (Severinghaus and Maguire 1955, Beddington 1974, Lang and Wood 1976). Estimates of survival and/or harvest probabilities can be derived from either age-structure data or marking studies. Alternatively, indirect methods for projecting sex ratios can also be used to assess the effects of different harvest and survival scenarios on population structure. The importance of indirect sex ratio methods resides more in population management than as an alternative to direct measurement techniques for estimating sex ratios.

3.2 Direct Sampling Techniques

3.2.1 Single Sample Survey without Replacement

We begin our description of survey sampling techniques to estimate sex ratios with the most basic approach, a single sample of n observations from a population in which all animals have equal detection probabilities. The basic premise is a random sample of individuals. If individual animals can be counted only once during the course of a survey sample, then sampling can be considered without replacement.

Sampling without replacement can occur in numerous ways. For example, waterfowl wings are often submitted to assess the ratio of males and females in the harvested group (Hopper and Funk 1970). Similarly, data collected at hunter check stations or during certain research activities (e.g., small mammal snap trapping) to estimate sex ratios may be considered sampling without replacement. In these cases, sampling is without replacement because the selected individuals are not available for future sampling.

There are many other instances when sampling will be conducted without replacement because of species-specific life-history characteristics or decisions made by biologists. If biologists checked the sex of cubs in a bear den, there would be no need to revisit the den, and consequently, sampling would be without replacement. In some instances, identification of unique individuals can be facilitated through marking sampled individuals (e.g., animals are marked as they are sampled). It also may be possible to differentiate individuals by using natural markings (e.g., killer whales, *Orcinus orca*) (Lyrholm et al. 1987). Alternatively, if an investigator traverses a region with widely spaced and nonoverlapping travel routes, individual animals will likely be sighted at most once during the course of the survey, and sampling may be considered without replacement.

Model Development

When sampling without replacement, all animals are assumed to have an equal probability of selection, but the probabilities of selection change as animals are withdrawn from the sampling frame. Thus, sampling is modeled using a hypergeometric likelihood. For a single sample without replacement, the likelihood model can be written as

$$L(N_F, N_M | n, f, m) = \frac{\binom{N_M}{m}\binom{N_F}{f}}{\binom{N_A}{n}},$$

where

f = number of females collected in a sample of size n;

m = number of males collected in a sample of size n;

or more parsimoniously as

$$L(N_M, R_{F/M} | n, f, m) = \frac{\binom{N_M}{m}\binom{N_M \cdot R_{F/M}}{f}}{\binom{N_M(R_{F/M}+1)}{n}}. \tag{3.7}$$

The sex ratio of the population is estimated by the maximum likelihood estimator (MLE) as

$$\hat{R}_{F/M} = \frac{f}{m}, \tag{3.8}$$

which is unbiased to the first term of a Taylor series expansion.

The variance for the sex ratio estimator, $\hat{R}_{F/M}$, Eq. (3.8) is derived using the delta method and is approximated by the expression

$$\text{Var}(\hat{R}_{F/M}) \doteq \frac{np(1-p)(N_A-n)}{n^2(1-p)^2(N_A-1)}\left(1+\frac{np}{n(1-p)}\right)^2,$$

where $p = \dfrac{N_F}{N_A} = \dfrac{R_{F/M}}{1+R_{F/M}}$ and $1-p = \dfrac{N_M}{N_A} = \dfrac{1}{1+R_{F/M}}$.

Simplifying the equation yields the variance expression

$$\text{Var}(\hat{R}_{F/M}) \doteq \frac{R_{F/M}(1+R_{F/M})^2}{n} \cdot \frac{(N_A-n)}{(N_A-1)}, \tag{3.9}$$

with estimated variance of

$$\widehat{\text{Var}}\left(\hat{R}_{F/M}\right) = \frac{\hat{R}_{F/M}\left(1+\hat{R}_{F/M}\right)^{2}}{n} \cdot \frac{\left(N_{A}-n\right)}{\left(N_{A}-1\right)} \qquad (3.10)$$

and an estimated standard error of $\widehat{\text{SE}}(\hat{R}_{F/M}) = \sqrt{\widehat{\text{Var}}(\hat{R}_{F/M})}$. Note that knowledge of the total number of animals in the population is required to calculate the variance. The right-hand term in Eq. (3.10) is the finite population correction (fpc). The fpc is an adjustment for the amount of information contained in a sample of size n from a population of size N_{A}. As more observations are drawn in relation to N_{A}, the amount of information about the population contained in the n samples increases. Hence, variance estimates decrease as sample sizes increase. Ignoring the fpc will overestimate the variance of $\hat{R}_{F/M}$. As sample size n approaches N_{A}, the fpc goes to zero as does as the variance of $\hat{R}_{F/M}$. Use of the fpc requires that N_{A} be known.

Model Assumptions

The assumptions associated with likelihood (Eq. 3.7), estimator (Eq. 3.8), and variance (Eq. 3.9) include the following:

1. All individuals have equal and independent probability of detection/collection.
2. The sample is representative of the population of interest.
3. Sampling is without replacement.
4. The interval of sampling is short relative to the survival process, or both sexes have equal probability of survival during the duration of the sample survey.

The first assumption implies there is no sex-specific behavior that makes one sex easier or harder to detect or collect than the other. The second assumption refers to the representativeness of the sampling frame. If the sex ratio is estimated from harvest data, the harvested animals must be representative of the entire population for the estimate to be valid. Further, if only a sample of the harvested animals is examined rather than all harvested individuals, the sample must also be representative of the harvest. If these assumptions do not hold, the estimates of sex ratio will be biased. In addition, if the sampling process is prolonged and the mortality rates are sex specific, the estimated sex ratio will not be representative of either the beginning or the end of the sampling event.

In most instances, population abundance (i.e., N_{A}) from which the sample of size n was drawn is unknown. The common approach under such circumstances is to ignore the fpc term:

$$\left(\frac{N_{A}-n}{N_{A}-1}\right)$$

in variance formula (Eq. 3.10). Should the fpc be ignored, sampling variance will be overestimated, resulting in wider confidence interval estimates. Should the fpc < 0.10, the standard error of the estimate will be changed little whether the fpc is included or excluded from the variance calculations. Ignoring the fpc in Eq. (3.10) eliminates the precision advantages of sampling without replacement and reduces its perceived precision to no better than sampling with replacement.

Example 3.1: Estimating Sex Ratio from Northern Bobwhite Hunter Bag Check

A hunter bag check yielded a sample of 72 male and 86 female northern bobwhite. From the sample, the sex ratio is estimated by Eq. (3.8) to be

$$\hat{R}_{F/M} = \frac{f}{m}$$

$$= \frac{86}{72} = 1.1944 \text{ females/male.}$$

Ignoring the fpc in Eq. (3.10), the variance of $\hat{R}_{F/M}$ is estimated to be

$$\widehat{\text{Var}}(\hat{R}_{F/M}) = \frac{\hat{R}_{F/M}(1 + \hat{R}_{F/M})^2}{n}$$

$$= \frac{1.1944(1 + 1.1944)^2}{158} = 0.0364$$

or $\widehat{\text{SE}}(\hat{R}_{F/M}) = 0.1908$.

Discussion of Utility

For the sampling design to be valid, males and females must be discernable, which will depend on the species and the time of year when sexual dimorphism is most evident. Seasonal changes in animal behavior associated with breeding may also violate the assumption of equal detection probabilities. Therefore, researchers should carefully consider the logistics and timing of sex ratio surveys. This is particularly true if harvest data are used to estimate sex ratios. Often, selective harvest by hunters (e.g., preference for antlered males) or hunting regulations may violate the equal detection assumption.

3.2.2   Single Sample Survey with Replacement

Numerous examples of this approach to sampling for birds, mammals, and herptiles exist in the wildlife literature. Roadside counts (Trautman 1982), spotlight surveys (Melchiors et al. 1985), aerial surveys (Bordage et al. 1998), or counts of animals along a walked transect (Sun et al. 2001) are common single sample surveys with replacement. Sampling with replacement is applicable in situations in which visual counts are made and there is a possibility of reobserving or recounting the same individuals more than once. This situation can occur when the route traversed by the observer crisscrosses or overlaps or the route is traveled multiple times. The intent of recanvassing survey routes is to increase the opportunity of observing new individuals and to increase sample size.

Model Development

Assuming a random sample of individuals within a population, the sex ratio is easily calculated. A maximum likelihood estimate of the sex ratio can be derived by modeling a sample taken with replacement using a binomial distribution. The likelihood can be written as

$$L(N_M, N_F | n, m, f) = \binom{n}{m, f} \left(\frac{N_M}{N_M + N_F} \right)^m \left(\frac{N_F}{N_M + N_F} \right)^f,$$

or equivalently,

$$L(R_{F/M} | n, m, f) = \binom{n}{m, f} \left(\frac{1}{1 + R_{F/M}} \right)^m \left(\frac{R_{F/M}}{1 + R_{F/M}} \right)^f \qquad (3.11)$$

where $R_{F/M} = N_F/N_M$ and

  n = number of animals examined for gender;
  m = number of males in the sample size of n;
  f = number of females in the sample size of n.

The MLE for the sex ratio is the same as reported earlier for sampling without replacement (Eq. 3.8) where,

$$\hat{R}_{F/M} = \frac{f}{m}, \qquad (3.12)$$

which is the ratio of the number of females to the number of males observed in the sample. The variance of $\hat{R}_{F/M}$ can be approximated using the delta method as

$$\mathrm{Var}(\hat{R}_{F/M}) \doteq \frac{R_{F/M}(1 + R_{F/M})^2}{n}, \qquad (3.13)$$

with estimated variance,

$$\widehat{\mathrm{Var}}(\hat{R}_{F/M}) = \frac{\hat{R}_{F/M}(1 + \hat{R}_{F/M})^2}{n} = \frac{nf}{m^3}. \qquad (3.14)$$

Variance formula (Eq. 3.13) can be used for sample size calculations in the planning of field studies. The standard error for the ratio estimator is expressed as

$$\widehat{\mathrm{SE}}(\hat{R}_{F/M}) = \sqrt{\widehat{\mathrm{Var}}(\hat{R}_{F/M})}.$$

An asymptotic $(1 - \alpha)100\%$ confidence interval estimate for $\hat{R}_{F/M}$ can be constructed using the variance estimator (Eq. 3.14) and the expression as follows:

$$\mathrm{CI}\left(\hat{R}_{F/M} - Z_{1-\frac{\alpha}{2}} \sqrt{\widehat{\mathrm{Var}}(\hat{R}_{F/M})} \leq R_{F/M} \leq \hat{R}_{F/M} + Z_{1-\frac{\alpha}{2}} \sqrt{\widehat{\mathrm{Var}}(\hat{R}_{F/M})} \right) = 1 - \alpha, \quad (3.15)$$

where $Z_{1-\frac{\alpha}{2}}$ = standard normal deviate corresponding to $P\left(|Z| > Z_{1-\frac{\alpha}{2}} \right) = \alpha$ (e.g., $Z = 1.96$ for $\alpha = 0.05$). A nominal $(1 - \alpha)100\%$ confidence interval is often better achieved using a ln-transformation assuming $\ln \hat{R}_{F/M} = \ln f - \ln m$ is normally distributed. In this case, a $(1 - \alpha)100\%$ confidence interval would be calculated using the formula

$$\mathrm{CI}\left(\hat{R}_{F/M} e^{-Z_{1-\frac{\alpha}{2}} \sqrt{\frac{\widehat{\mathrm{Var}}(\hat{R}_{F/M})}{\hat{R}_{F/M}^2}}} \leq R_{F/M} \leq \hat{R}_{F/M} e^{+Z_{1-\frac{\alpha}{2}} \sqrt{\frac{\widehat{\mathrm{Var}}(\hat{R}_{F/M})}{\hat{R}_{F/M}^2}}} \right) = 1 - \alpha. \qquad (3.16)$$

A numerically more rigorous approach using profile likelihood confidence interval methods and likelihood model (3.11) will provide an interval estimate with still better nominal coverage when sample sizes are small (Appendix C, Program USER).

Model Assumptions

This basic method relies on several key assumptions:

1. All individuals have equal and independent probability of detection.
2. The sample is representative of the population of interest.
3. Sampling is with replacement.
4. The interval of sampling is short relative to the survival process or both sexes have equal probability of survival during the duration of the sample survey.

When these assumptions are met, the sex ratio estimator is unbiased to the first term of a Taylor series expansion. The assumptions that all animals have equal and independent probabilities of detection and that sampling is with replacement allow use of a binomial likelihood. The binomial model is robust if the probabilities of detection are independent but unequal among individuals as long as the distribution of the probabilities is the same for both sexes. In this case, variance formula (3.13) will overestimate the true variance (Feller 1968:230–231). If sampling is conducted without replacement, the binomial variance will overestimate the true variance.

The assumption violation with the greatest impact on estimators (3.8) and (3.12) is when detection probabilities are not equally distributed among males and females. In this case, the ratio estimators (Eqs. 3.8 and 3.12) are biased. Letting p_F be the probability of detecting females and p_M be the probability of detecting males, the expected value of $\hat{R}_{F/M}$ is approximately

$$E\left(\hat{R}_{F/M}\right) \doteq R_{F/M} \cdot \frac{p_F}{p_M}.$$

Simply collecting count data will typically be inadequate in identifying unequal detection probabilities or subsequently providing a bias correction to the sex ratio estimator.

Example 3.2: Estimating Sex Ratio from Ring-Necked Pheasant Drive Survey

A drive survey was repeated three times over the course of 1 day (i.e., early morning, mid-morning, and evening) to count rooster and hen ring-necked pheasants. The repeated surveys during the day were used to account for any diel difference in pheasant behavior and to increase the chances of detecting birds. The results of the daily survey yielded sightings of 369 females and 217 males for an estimated sex ratio (Eq. 3.12) of

$$\hat{R}_{F/M} = \frac{f}{m}$$

$$= \frac{369}{217} = 1.7005 \text{ females/male},$$

with an associated variance estimate (Eq. 3.14) of

$$\widehat{\text{Var}}(\hat{R}_{F/M}) = \frac{nf}{m^3}$$

$$= \frac{586(369)}{217^3} = 0.02116,$$

or equivalently,

$$\widehat{\text{SE}}(\hat{R}_{F/M}) = 0.1455.$$

Confidence interval calculation using Eq. (3.15), based on the assumption $\hat{R}_{F/M}$ is normally distributed,

$$\hat{R}_{F/M} \pm Z_{1-\frac{\alpha}{2}} \sqrt{\widehat{\text{Var}}(\hat{R}_{F/M})}$$

$$1.7005 \pm 1.645\sqrt{0.02116},$$

provides CI(1.4612 < $R_{F/M}$ < 1.9398) = 0.90. Alternatively, assuming $\ln\hat{R}_{F/M}$ is normally distributed, a resulting 90% confidence interval (Eq. 3.16) is calculated as

$$\hat{R}_{F/M} e^{\pm Z_{1-\frac{\alpha}{2}} \sqrt{\frac{\widehat{\text{Var}}(\hat{R}_{F/M})}{\hat{R}_{F/M}^2}}}$$

$$1.7005 e^{\pm 1.645 \sqrt{\frac{0.02116}{(1.7005)^2}}},$$

or CI(1.4773 ≤ $R_{F/M}$ ≤ 1.9574) = 0.90. The profile likelihood confidence interval estimate calculated numerically from Eq. (3.11) is CI(1.4783 ≤ $R_{F/M}$ ≤ 1.9592) = 0.90. Note both Eq. (3.16) and the profile likelihood method shift the confidence interval estimate to the right relative to Eq. (3.15).

Precision of the Sex Ratio Estimator

The precision of the sex ratio estimator will be defined as

$$P\left(\left|\ln\left(\frac{\hat{R}_{F/M}}{R_{F/M}}\right)\right| < \varepsilon\right) \geq 1 - \alpha. \tag{3.17}$$

In this form, precision of the sex ratio estimates is invariant to whether $R_{F/M}$ or $R_{M/F}$ is estimated. Skalski and Robson (1992) used the same definition of precision when estimating the relative abundance of one population (N_1) to that of another (N_2) $\left(\text{i.e., } \frac{N_2}{N_1}\right)$. The precise definition can be re-expressed as

$$P\left(e^{-\varepsilon} < \frac{\hat{R}_{F/M}}{R_{F/M}} < e^{\varepsilon}\right) \geq 1 - \alpha,$$

or similarly,

$$P\left(e^{-\varepsilon} - 1 < \frac{\hat{R}_{F/M} - R_{F/M}}{R_{F/M}} < e^{\varepsilon} - 1\right) \geq 1 - \alpha. \tag{3.18}$$

Equation (3.18) can be interpreted as the desire for the relative error in estimation $\left(\text{i.e., } \dfrac{\hat{R}_{F/M} - R_{F/M}}{R_{F/M}}\right)$ to be between $e^{-\varepsilon} - 1$ and $e^{\varepsilon} - 1$, $(1 - \alpha)100\%$ of the time. For example, letting $1 - \alpha = 0.90$ and $\varepsilon = 0.10$, then Eq. (3.18) becomes

$$P\left(-0.0952 < \frac{\hat{R}_{F/M} - R_{F/M}}{R_{F/M}} < 0.1052\right) = 0.90.$$

That is, the desire is to avoid underestimating $R_{F/M}$ by more than 9.52% or overestimating it by more than 10.52%, 90% of the time. Figure 3.1 plots ε as a function of sample size n for a range of values of $R_{F/M}$ when $1 - \alpha = 0.90$ or 0.95. Assuming $\ln \hat{R}_{F/M}$ is normally distributed and using precision expression (3.17), the required sample size is

$$n = \frac{(1 + R_{F/M})^2}{R_{F/M}} \cdot \frac{Z_{1-\frac{\alpha}{2}}^2}{\varepsilon^2}.$$

Sample sizes are smallest at a value of $R = 1$ and increases as the sex ratio (i.e., $R_{F/M}$ or $R_{M/F}$) becomes more disparate (see Fig. 3.1).

Discussion of Utility

This design is most appropriate for species in which males and females are readily distinguishable. For example, sex can be readily discerned for certain game birds, including pheasants and many species of ducks. On the other hand, for species such as mourning dove (*Zenaida macroura*), males and females are indistinguishable during observational surveys. For many other species, the ability to distinguish between males and females differs by season. For many ungulates, males are most easily distinguishable from females during breeding seasons because of antlers. The season in which the survey is conducted is also important because seasonal differences in behavior may violate the equal detection assumption. For example, male ungulates may be more visible than females as the males advertise their presence in an attempt to attract mates (Geist 1982). Similarly females may be less visible during calving season as they become more secretive (Geist 1982). Similar factors need to be considered for each species when designing surveys.

3.2.3   Multiple Samples with Replacement

In the previous survey approach, sex ratio was estimated using data from one sampling event from the population of interest. Multiple sampling events often occur within an area to account for animal movements and intermixing of individuals. This extension allows for multiple surveys of the same area or population. Multiple surveys have the capability to observe more individuals and increase sample size to enhance precision. In contrast to the previous sections (i.e., Sections 3.2.1 and 3.2.2), there are at least two distinct sampling events that are combined to estimate the population sex ratio. For example, roadside counts of pheasants may include multiple samples along the same survey route (Trautman 1982). Another example is sex ratio surveys conducted during bird migrations, in which surveys conducted over time may sample different groups of a single population. The sampling modification could also be applicable to gregarious animals

a.

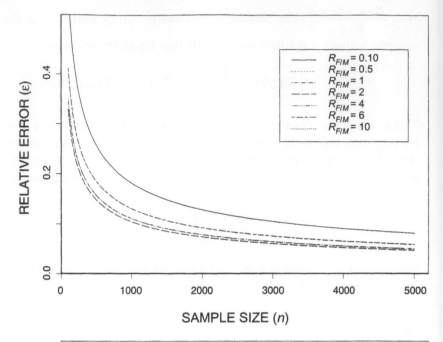

b.

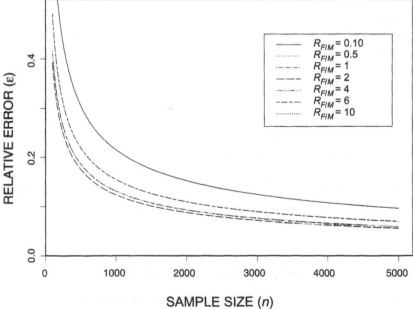

Figure 3.1. Precision curves for estimating a sex ratio using the binomial variance formula (3.13) and precision expression (3.18) as a function of the true sex ratio R_{FIM} and sample size n at (a) $1 - \alpha = 0.90$ and (b) $1 - \alpha = 0.95$.

when, for example, the sex ratio is estimated per group (e.g., herds of elk) and there is interest in a population level estimate of the sex ratio.

Model Development

The general form of the likelihood equation (from Section 3.2.2) can be used if more than one survey is conducted of the same population. If each survey is indexed by i, the likelihood for k different surveys of the same population can be written as

$$L(N, R_{F/M}, p_i | m_i, f_i) = \prod_{i=1}^{k} \binom{N}{m_i, f_i} \left(\frac{p_i}{1 + R_{F/M}}\right)^{m_i} \left(\frac{R_{F/M} \, p_i}{1 + R_{F/M}}\right)^{f_i} (1 - p_i)^{N - (m_i + f_i)},$$

(3.19)

where

f_i = number of females observed in the ith survey ($i = 1, \ldots k$);
m_i = number of females observed in the ith survey ($i = 1, \ldots k$);
p_i = probability of detecting an animal during the ith survey ($i = 1, \ldots k$);
n_i = total number of animals observed (i.e., $f_i + m_i$) in the ith survey ($i = 1, \ldots k$).

Similar to the likelihood for a single sample survey (Eq. 3.11), conditioning on the sample size of each survey, n_i, removes the nuisance parameters p_i from the likelihood. The conditional likelihood of $\underset{\sim}{m}$ and $\underset{\sim}{f}$, given $\underset{\sim}{n}$, can then be written as

$$L(N, R_{F/M} | \underset{\sim}{m}, \underset{\sim}{f}) = \prod_{i=1}^{k} \frac{L(N, R_{F/M}, p_i, | m_i, f_i)}{L(N, p_i | n_i)},$$

or

$$L(R_{F/M} | \underset{\sim}{m}, \underset{\sim}{f}) = \prod_{i=1}^{k} \binom{n_i}{m_i, f_i} \left(\frac{1}{1 + R_{F/M}}\right)^{m_i} \left(\frac{R_{F/M}}{1 + R_{F/M}}\right)^{f_i}.$$

The MLE is easily solved by noting the following relationship:

$$L(R_{F/M} | \underset{\sim}{m}, \underset{\sim}{f}) \propto \left(\frac{1}{1 + R_{F/M}}\right)^{\sum_{i=1}^{n} m_i} \left(\frac{R_{F/M}}{1 + R_{F/M}}\right)^{\sum_{i=1}^{n} f_i}.$$

(3.20)

The MLE of the sex ratio, $R_{F/M}$, is

$$\hat{R}_{F/M} = \frac{\sum_{i=1}^{k} f}{\sum_{i=1}^{k} m} = \frac{\bar{f}}{\bar{m}}.$$

(3.21)

Thus, the sex ratio can be expressed in terms of the total number of males and females observed, or as a ratio of the average number of females observed to the average number of males observed. However, the estimator *is not* the average of the individual ratio estimators. The estimated variance of $\hat{R}_{F/M}$ is

$$\widehat{\text{Var}}(\hat{R}_{F/M}) = \frac{\sum_{i=1}^{k} (f_i - \hat{R}_{F/M} m_i)^2}{k \bar{m}^2 (k - 1)},$$

(3.22)

where $\bar{m}$ = average number of males observed across all samples. Note that both the number of surveys (k) and the number of animals observed in each survey, as indicated by $\bar{m}$, influence the variance estimate. The estimated variance of $\hat{R}_{F/M}$ (Eq. 3.22) can be used in conjunction with Eq. (3.16) for confidence interval construction.

Model Assumptions

The assumptions of the single sample survey (Section 3.2.2) also apply to the repeated surveys. In addition, there are two more assumptions:

1. The sample surveys are independent of each other.
2. The sex ratio $R_{F/M}$ is invariant over the course of the replicate surveys.

If the surveys are not independent, the effective sample size is actually smaller than perceived $\left(\text{i.e., } \sum_{i=1}^{k} n_i\right)$ and the sampling variance is larger than calculated by Eq. (3.22).

Proper temporal and spatial stratification should help avoid any violation of independence. The probability of detecting animals can differ from one survey to the next. However, the likelihood model (3.20) assumes all replicate surveys estimate the same sex ratio. Should the sex ratio shift over time or differ among subpopulations, the estimator (3.21) is inappropriate and the approach in Section 3.2.4 is required.

Discussion of Utility

Pooling replicate survey data is appropriate when individual surveys are independent and identically distributed. This situation most likely occurs when the same population is resurveyed over a short time frame. The repeated surveys increase sample size and detect new individuals through movement and intermixing of individuals within the population. For cases in which different segments of the population are sampled during different survey events, this approach may not be applicable. Only if the different subpopulations have the same sex ratio in expectation (i.e., $R_{F/M}$) is Eq. (3.21) appropriate. Chi-square contingency tests of homogeneity can be used to compare the proportion of males and females across different subpopulations (Zar 1996:485–503). If sex ratios are homogeneous, Eq. (3.21) can be used to estimate the overall sex ratio of a population. Nevertheless, caution is needed when basing pooling decisions on chi-square tests of homogeneity. Significance of the test depends on sampling sizes and should not be a substitute for biological judgment.

3.2.4   Stratified Random Sampling

For cases in which a single random sample of an entire population is impractical or impossible, a stratified random sample may be drawn. The various strata could be spatially distinct segments of a population or temporally distinct views of a population as it migrates through an area. Within each stratum, the animals will be assumed to be randomly sampled without replacement, regardless of the situation.

Let the following terms be defined:

L = number of population strata;
n_i = number of animals sampled for gender in the ith stratum ($i = 1, \ldots L$);
m_i = number of males observed from a sample of size n_i in the ith stratum ($i = 1, \ldots L$);
$f_i = n_i - m_i$ = number of females observed from a sample of size n_i in the ith stratum ($i = 1, \ldots L$);

N_i = animal abundance in the ith stratum ($i = 1, \ldots L$);

P_{F_i} = probability of a female in the ith stratum ($i = 1, \ldots L$);

$P_{M_i} = 1 - P_{F_i}$ = probability of a male in the ith stratum ($i = 1, \ldots L$).

The overall sex ratio in the population can then be expressed as

$$R_{F/M} = \frac{\sum_{i=1}^{L} N_i P_{F_i}}{\sum_{i=1}^{L} N_i P_{M_i}} = \frac{\sum_{i=1}^{L} N_i \left(\frac{R_{F/M,i}}{1 + R_{F/M,i}} \right)}{\sum_{i=1}^{L} N_i \left(\frac{1}{1 + R_{F/M,i}} \right)} \tag{3.23}$$

where

$$R_{F/M,i} = \frac{P_{F_i}}{P_{M_i}} = \text{sex ratio in the } i\text{th stratum } (i = 1, \ldots, L).$$

In practice, animal abundance in the ith stratum (N_i) will be unknown and will need to be estimated ($\hat{N}_i$). Expression (3.23) can be estimated by the formula:

$$\hat{R}_{F/M} = \frac{\sum_{i=1}^{L} \hat{N}_i \left(\frac{f_i}{n_i} \right)}{\sum_{i=1}^{L} \hat{N}_i \left(\frac{m_i}{n_i} \right)}. \tag{3.24}$$

Alternatively, if a reliable index (I_i) to abundance (N_i) exists where

$$I_i = CN_i \ \forall i$$

for some unknown value of C, then Eq. (3.24) can be rewritten as

$$\hat{R}_{F/M} = \frac{\sum_{i=1}^{L} I_i \left(\frac{f_i}{n_i} \right)}{\sum_{i=1}^{L} I_i \left(\frac{m_i}{n_i} \right)}. \tag{3.25}$$

Inspection of Eqs. (3.24) and (3.25) reveals an indicator of animal abundance is necessary along with the sex ratio data whenever stratified sampling is conducted. Stratified sampling is useful when the original population is too large to sample as a single unified sampling frame or when the sex ratio is known to vary across the sampling frame.

When stratified sampling is performed, there are three situations when estimates of animal abundance are not necessary in obtaining valid estimates of the population sex ratio.

1. The sex ratio is constant (i.e., $P_{F_i} = P_F \ \forall i$) across all strata in which case

$$R_{F/M} = \frac{\sum_{i=1}^{L} N_i P_F}{\sum_{i=1}^{L} N_i P_M} = \frac{P_F}{P_M} \text{ and estimated by } \hat{R}_{F/M} = \frac{\sum_{i=1}^{L} f_i}{\sum_{i=1}^{L} m_i}.$$

2. Abundance is constant (i.e., $N_i = N_i \ \forall i$) across all strata in which case

$$R_{F/M} = \frac{\sum_{i=1}^{L} NP_{F_i}}{\sum_{i=1}^{L} NP_{M_i}} = \frac{\sum_{i=1}^{L} P_{F_i}}{\sum_{i=1}^{L} P_{M_i}} = \frac{\overline{P}_F}{\overline{P}_M} \text{ and estimated by } \hat{R}_{F/M} = \frac{\sum_{i=1}^{L} \left(\frac{f_i}{n_i}\right)}{\sum_{i=1}^{L} \left(\frac{m_i}{n_i}\right)}.$$

3. A self-weighted sample is drawn so that sample size is proportional to strata size (i.e., $cn_i = N_i$) in which case

$$\hat{R}_{F/M} = \frac{\sum_{i=1}^{L} cn_i \left(\frac{f_i}{n_i}\right)}{\sum_{i=1}^{L} cn_i \left(\frac{m_i}{n_i}\right)} = \frac{\sum_{i=1}^{L} f_i}{\sum_{i=1}^{L} m_i}. \tag{3.26}$$

In this third situation, sample sizes n_i ($i = 1, \ldots, L$) are assumed to be reliable indices of animal abundance in the strata. Basically, Eq. (3.26) assumes the number of animals sighted/collected in the ith stratum is a measure of "catch-per-unit-effort." Special considerations are necessary if a self-weighted sample is to be drawn. The detection probability will need to be constant across the entire population, and to do so, sampling effort will need to be proportional to strata size (Skalski and Robson 1992:35–36) for sample size to be proportional to abundance (i.e., $n_i \propto N_i \ \forall i$).

In the general case in which strata weights are estimated by independent survey values of $\hat{N}_i$, the variance can be approximated by the expression

$$\text{Var}\left[\frac{\sum_{i=1}^{L} \hat{N}_i \left(\frac{f_i}{n_i}\right)}{\sum_{i=1}^{L} \hat{N}_i \left(\frac{m_i}{n_i}\right)}\right] \doteq \sum_{i=1}^{L} \frac{p_i(1-p_i)}{n_i}\left(\frac{N_i - n_i}{N_i - 1}\right)\left[\frac{N_i \sum_{i=1}^{L} N_i}{\left[\sum_{i=1}^{L} N_i(1-p_i)\right]^2}\right]^2$$

$$+ \sum_{i=1}^{L} \text{Var}(\hat{N}_i|N_i)\left[\frac{p_i \sum_{i=1}^{L} N_i - \sum_{i=1}^{L} N_i p_i}{\left[\sum_{i=1}^{L} N_i(1-p_i)\right]^2}\right]^2 \tag{3.27}$$

where $p_i = \dfrac{N_{F_i}}{N_{M_i} + N_{F_i}}$ for the ith stratum. Should indices of abundance be used (i.e., I_i) in estimating the population-wide sex ratio (Eq. 3.25), the variance can be expressed by substituting I_i for N_i in Eq. (3.27).

For the case in which a self-weighted sample is drawn and estimator (3.26) is used in estimating the sex ratio, the variance can be expressed by the approximation

$$\text{Var}\left(\frac{\sum_{i=1}^{L} f_i}{\sum_{i=1}^{L} m_i}\right) \doteq \left(\frac{\sum_{i=1}^{L} n_i}{\sum_{i=1}^{L} (n_i - f_i)^2}\right)^2 \sum_{i=1}^{L} \frac{(N_i - n_i)}{(N_i - 1)} n_i p_i (1 - p_i). \tag{3.28}$$

3.2.5   Unequal Detection Probabilities among Sexes

The previous techniques for estimating sex ratios assumed that detection probabilities for males and females were equal. Behavioral differences between males and females will often result in differential sighting or collection probabilities. For example, males may be more visible during the breeding season in an attempt to attract mates, whereas females may be more secretive during the birthing season. Males and females may have different detection probabilities if they occur differentially in aggregates or herds of different sizes (Samuel et al. 1992). For example, if males tend to be solitary and females occur in clusters, males will be visually more difficult to detect. Even differences in resource selection between the sexes may ultimately influence detection probabilities. All of these situations might contribute to differential detection rates between the sexes. Other less obvious factors such as differences in appearance or coloration can also contribute to unequal detection rates if one sex is more visible than the other during surveys. For example, the cryptic coloration of female waterfowl versus the bright plumage of males might influence detection. These factors can, and will, change throughout the annual cycle, making timing of surveys important.

To account for differences in detection between males and females, sex-specific sighting probabilities must be incorporated into the likelihood model. As previously discussed, an unbiased estimate of the sex ratio based solely on survey counts is not possible, if differences in sighting probabilities exist.

Model Development

Incorporating sex-specific detection probabilities into a likelihood model is illustrated in the case of a single survey. Let the likelihood for a single survey be written as follows,

$$L(N_F, N_M, R_{F/M}, p_m, p_f | m, f) = \binom{N_F + N_M}{m, f} \left(\frac{1}{1 + R_{F/M}} \cdot p_m \right)^m \left(\frac{R_{F/M}}{1 + R_{F/M}} \cdot p_f \right)^f$$
$$\cdot \left(1 - \left(\frac{1}{1 + R_{F/M}} \cdot p_m + \frac{R_{F/M}}{1 + R_{F/M}} \cdot p_f \right) \right)^{N-(f+m)}$$

(3.29)

where p_m = probability of sighting a male during the survey,

p_f = probability of sighting a female during the survey.

The MLE for $R_{F/M}$ in this case is then

$$\hat{R}_{F/M} = \frac{f}{m} \cdot \frac{p_m}{p_f}.$$

(3.30)

Unfortunately, in the absence of auxiliary information on the detection rates, $\hat{R}_{F/M}$ is not estimable. Closer examination of Eq. (3.30) indicates the MLE is a ratio of abundance estimates where

$$\hat{R}_{F/M} = \frac{\left(\dfrac{f}{p_f} \right)}{\left(\dfrac{m}{p_m} \right)} = \frac{\hat{N}_F}{\hat{N}_M}.$$

(3.31)

Thus, in populations with different detection probabilities for males and females, an unbiased estimate of the sex ratio using survey counts alone is not possible. Equation (3.30) illustrates the relative bias in the ratio estimator $\dfrac{f}{m}$ is $\left(\dfrac{p_f}{p_m}-1\right)\times100\%$. Inspection of Eq. (3.30) indicates the sex ratio estimator $\dfrac{f}{m}$ is not robust to any violation of the assumption $p_m = p_f$.

Distance sampling methods (Buckland et al. 1993) can be used in conjunction with sex ratio counts to estimate sex-specific detection probabilities and associated abundance estimates. Sighting distance data should be routinely collected along with sex ratio counts to confirm that detection probabilities are equal for both sexes. Chi-square contingency table tests of homogeneity (Zar 1984:61–64) or Kolmogorov-Smirnov tests of equal distribution (Conover 1980:368–373) can be used to test for equality of detection-distance distributions. However, only formalized distance sampling methods (Buckland et al. 1993) can test for equality and, if rejected, provide reliable estimates of detection probabilities and sex-specific densities or abundance estimates.

3.2.6   Cluster Sampling Methods

For animal species that occur in aggregates or herds, survey data on sex ratios are likely to be obtained in the form of group counts. The survey data would report group size n_i (i.e., herd or flock) as well as numbers of males (m_i) and females (f_i) for all individuals within aggregates observed. Bowden et al. (1984) considered the special case of simple random sampling without replacement of the aggregates within a population (labeled SRSG). In this situation, the individual groups of animals are the sampling elements, which are assumed to have equal probabilities of detection. In this special case of ratio estimation, the estimate of the sex ratio would be

$$\hat{R}_{F/M} = \frac{\displaystyle\sum_{i=1}^{g} f_i}{\displaystyle\sum_{i=1}^{g} m_i} \tag{3.32}$$

where g = number of groups or aggregates of animals surveyed. The variance of (3.32) can be approximated (Cochran 1977:153–155) by

$$\mathrm{Var}\left(\hat{R}_{F/M}\right) = \frac{\left(1-\dfrac{g}{G}\right)}{g\overline{N}_M^2}\left[\frac{\displaystyle\sum_{i=1}^{G}\left(f_i - R_{F/M}m_i\right)^2}{(G-1)}\right] \tag{3.33}$$

where

G = total number of aggregates in the population;

$$\overline{N}_M = \frac{\displaystyle\sum_{i=1}^{G} m_i}{G} = \text{mean number of males per aggregate.}$$

In turn, the variance (3.33) can be estimated by

$$\widehat{\text{Var}}(\hat{R}_{F/M}) = \frac{\left(1 - \frac{g}{G}\right)}{g\bar{m}^2} \left[\frac{\sum_{i=1}^{g} f_i^2 + \hat{R}_{F/M}^2 \sum_{i=1}^{g} m_i^2 - 2\hat{R}_{F/M} \sum_{i=1}^{g} f_i m_i}{(g-1)} \right]. \tag{3.34}$$

The Bowden et al. (1984) method assumes all aggregates have an equal probability of detection regardless of size. For animals that associate in groups, however, the probability of detection is likely a function of the size of the flock or herd in which it resides. These unequal detection probabilities have no effect on the accuracy of the sex ratio estimator (3.32) as long as the sex composition is independent of aggregate size. However, if both the sex composition and probability of detection are related to aggregate size, estimator (3.32) will be biased.

Steinhorst and Samuel (1989) and Samuel et al. (1992) consider the case of size-biased sampling. They developed a Horvitz–Thompson (Horvitz and Thompson 1952, Cochran 1977:259–261) type estimator for the case of two-stage sampling. Their estimator considered sampling of landscape units and aggregates of animals within a landscape unit. Here, we consider the simpler case of a single landscape unit and the sampling of aggregates with unequal probabilities of detection. The sex ratio estimator in this case can be written as

$$\hat{R}_{F/M} = \frac{\sum_{i=1}^{g} \frac{f_i}{\hat{\Pi}_i}}{\sum_{i=1}^{g} \frac{m_i}{\hat{\Pi}_i}} = \frac{\hat{N}_F}{\hat{N}_M}. \tag{3.35}$$

where $\hat{\Pi}_i$ = estimated probability of detecting the ith aggregate of animals ($i = 1, \ldots g$).

The detection probabilities $\hat{\Pi}_i$ could, in turn, be derived from a sightability model (Section 9.2.4) that estimates the probability of detecting an animal group characterized by environmental covariates represented by a vector x_i. Sightability models have been developed for aerial surveys of elk in Idaho (Samuel et al. 1987); Yellowstone National Park, Wyoming (Eberhardt et al. 1998); and California (Bleich et al. 2001).

The variance of Eq. (3.35) is quite involved, but an approach analogous to Samuel et al. (1992) can be used to derive an approximation. The general form of the variance comes from the delta method approximation where

$$\widehat{\text{Var}}(\hat{R}_{F/M}) = \hat{R}_{F/M}^2 \left[\frac{\widehat{\text{Var}}(\hat{N}_F)}{\hat{N}_F^2} + \frac{\widehat{\text{Var}}(\hat{N}_M)}{\hat{N}_M^2} - \frac{2\widehat{\text{Cov}}(\hat{N}_F, \hat{N}_M)}{\hat{N}_F \hat{N}_M} \right]. \tag{3.36}$$

The variance of $\hat{N}_F$ can be found in stages where

$$\text{Var}(\hat{N}_F) = E_{f_i}\left[\text{Var}_{\hat{\Pi}_i}\left(\sum_{i=1}^{g} \frac{f_i}{\hat{\Pi}_i} \middle| f_i \right) \right] + \text{Var}_{f_i}\left[E_{\hat{\Pi}_i}\left(\sum_{i=1}^{g} \frac{f_i}{\hat{\Pi}_i} \middle| f_i \right) \right]$$

$$\doteq E_{f_i}\left[\sum_{i=1}^{g} \frac{f_i^2}{\hat{\Pi}_i^4} \text{Var}(\hat{\Pi}_i | \Pi_i) \right] + \text{Var}_{f_i}\left[\sum_{i=1}^{g} \frac{f_i}{\Pi_i} \right]$$

$$= \frac{g}{G} \sum_{i=1}^{G} \frac{f_i^2}{\hat{\Pi}_i^4} \text{Var}(\hat{\Pi}_i | \Pi_i) + \sum_{i=1}^{G} \frac{(1 - \Pi_i)}{\Pi_i} f_i^2 + 2\sum_{i=1}^{G} \sum_{\substack{j=1 \\ j>i}}^{G} \frac{(\Pi_{ij} - \Pi_i \Pi_j)}{\Pi_i \Pi_j} f_i f_j,$$

and estimated by the expression

$$\widehat{\operatorname{Var}}(\hat{N}_F) = \sum_{i=1}^{g} \frac{f_i^2}{\hat{\Pi}_i^4} \widehat{\operatorname{Var}}(\hat{\Pi}_i | \Pi_i) + \sum_{i=1}^{g} \left[\frac{(1 - \hat{\Pi}_i)}{\hat{\Pi}_i^2} f_i^2 \right] + 2 \sum_{i=1}^{g} \sum_{\substack{j=1 \\ j>i}}^{g} \left[\frac{(\hat{\Pi}_{ij} - \hat{\Pi}_i \hat{\Pi}_j)}{\hat{\Pi}_i \hat{\Pi}_j \hat{\Pi}_{ij}} f_i f_j \right].$$

(3.37)

A similar expression to that of Eq. (3.37) can be used for the variance of $\hat{N}_M$. In Eq. (3.37), the estimated variance of $\hat{\Pi}_i$ [i.e., $\widehat{\operatorname{Var}}(\hat{\Pi}_i | \Pi_i)$] would come from the logistic regression model for sighting probabilities (Section 9.2.4). The parameter Π_{ij} is the joint probability both groups i and j would be drawn in a sample size of g. Samuel et al. (1992) ignored the third term as a matter of computational convenience (Cochran 1977:262). Rao (1965) shows that $\Pi_{ij} - \Pi_i \Pi_j > 0$ for all $i \neq j$, in which case, ignoring the third term will produce a positively biased variance estimate.

The covariance term in Eq. (3.36) is again found in stages, where

$$\operatorname{Cov}(\hat{N}_F, \hat{N}_M) = E_{f,m} \left[\operatorname{Cov}_{\hat{\Pi}_i} \left(\sum_{i=1}^{g} \frac{f_i}{\hat{\Pi}_i}, \sum_{i=1}^{g} \frac{m_i}{\hat{\Pi}_i} \middle| f_i, m_i \right) \right]$$

$$+ \operatorname{Cov}_{f,m} \left[E_{\hat{\Pi}_i} \left(\sum_{i=1}^{g} \frac{f_i}{\hat{\Pi}_i} \middle| f_i \right), E_{\hat{\Pi}_i} \left(\sum_{i=1}^{g} \frac{m_i}{\hat{\Pi}_i} \middle| m_i \right) \right]$$

$$\doteq E_{f,m} \left[\sum_{i=1}^{g} \sum_{j>i}^{g} \frac{\operatorname{Cov}(\hat{\Pi}_i, \hat{\Pi}_j)}{\hat{\Pi}_i^2 \hat{\Pi}_j^2} f_i m_i \right] + \operatorname{Cov}_{f,m} \left[\sum_{i=1}^{g} \frac{f_i}{\hat{\Pi}_i}, \sum_{i=1}^{g} \frac{m_i}{\hat{\Pi}_i} \right],$$

leading to

$$\widehat{\operatorname{Cov}}(\hat{N}_F, \hat{N}_M) = \sum_{i=1}^{g} \sum_{j>i}^{g} \frac{\operatorname{Cov}(\hat{\Pi}_i \hat{\Pi}_j)}{\hat{\Pi}_i^2 \hat{\Pi}_j^2} f_i m_i + \sum_{i=1}^{g} \frac{(1 - \hat{\Pi}_i)}{\hat{\Pi}_i^2} f_i m_i. \qquad (3.38)$$

Example 3.3: Estimating Sex Ratios from an Elk Aerial Survey, Adjusting for Differential Sightability

Six groups of elk were visually observed during an aerial survey, with group size (n_i), number of males (m_i), number of females (f_i), and estimates of the probability of sighting calculated from a pre-existing sightability model (Table 3.1). The larger herds of animals were composed primarily of cow elk, whereas smaller groups were typically composed of males. The estimated sex ratio for the population, calculated by Eq. (3.35) (data in Table 3.1), is

$$\hat{R}_{F/M} = \frac{\sum_{i=1}^{g} \frac{f_i}{\hat{\Pi}_i}}{\sum_{i=1}^{g} \frac{m_i}{\hat{\Pi}_i}} = \frac{\left(\frac{32}{0.173}\right) + \left(\frac{4}{0.063}\right) + \left(\frac{14}{0.101}\right) + \left(\frac{0}{0.007}\right) + \left(\frac{7}{0.087}\right) + \left(\frac{2}{0.021}\right)}{\left(\frac{5}{0.173}\right) + \left(\frac{0}{0.063}\right) + \left(\frac{2}{0.101}\right) + \left(\frac{1}{0.007}\right) + \left(\frac{2}{0.087}\right) + \left(\frac{0}{0.021}\right)}$$

$$\hat{R}_{F/M} = \frac{562.775}{194.949} = 2.887 \text{ females/male}.$$

Note, if raw counts had been used without adjustment for probability of detection (i.e., Eq. 3.8), the sex ratio would have been incorrectly estimated as 5.9 females/male.

Table 3.1. Herd size (n_i) for each group of elk sighted along with male (m_i) and female (f_i) counts and estimated probability of detection ($\hat{\Pi}_i$).

Group	Herd size (n_i)	Males (m_i)	Females (f_i)	Estimated Detection Probability ($\hat{\Pi}_i$)
1	37	5	32	0.173
2	4	0	4	0.063
3	16	2	14	0.101
4	10	1	0	0.007
5	9	2	7	0.087
6	2	0	2	0.021

Discussion of Utility

In many cases in which sex ratios are estimated from visual counts of live animals, observations will be in the form of clusters or aggregates of animals. Ignoring the form of the count data runs the risk of unreliable point estimates and/or variance estimates. Ignoring the aggregate nature of the data and treating individuals as independent Bernoulli events will underestimate the true variance, resulting in a perceived precision unwarranted by the sampling scheme. Horvitz–Thompson type estimators are an invaluable tool when sampling with unequal probabilities of detection. Sightability models need to be developed in conjunction with or as a separate element of an integrated visual survey plan. Some form of radio telemetry is almost inevitably part of any sightability study to identify the circumstances under which animals were and were not detected. Samuel et al. (1987), Steinhorst and Samuel (1989), Eberhardt et al. (1998), Anderson et al. (1998), Cogan and Diefenbach (1998), and Bleich et al. (2001) review methods to construct wildlife sightability models.

3.3   Indirect Methods of Estimating Sex Ratios

Because of hunting regulations and hunter preference, the adult sex ratio observed in the harvest is not necessarily representative of the population (Burgoyne 1981, Raedeke et al. 2002). The resulting biased estimate of the adult sex ratio should not be used in subsequent population reconstruction models or demographic analysis. Despite the potential for biased sex ratio data from hunter harvest data, age ratios in the harvest data may still be representative of the age structure in the population. For example, hunting regulations may preclude the harvest of hen ring-neck pheasants in places such as South Dakota, but a check of the hunter bag may still provide representative information about the age structure of male pheasants. That is, hunters are likely to harvest male pheasants irrespective of age, providing a potentially useful index to male age structure in the population. Using relationships between age structure and survival, it is possible to derive an unbiased estimate of the population sex ratio using sex-ratio-biased harvest data.

The use of age composition data to calculate sex ratios was first developed by Severinghaus and Maguire (1955) for white-tailed deer (*Odocoileus virginianus*) populations in New York State. They used the ratio of the proportion of yearling males to yearling females to estimate the adult sex ratio of the population (Severinghaus and Maguire 1955:242). Their primary assumption was that during open hunting seasons (in which both antlered and antlerless harvests were allowed), the proportion of yearlings in the harvest was representative of the population as a whole (Severinghaus and Maguire 1955).

Lang and Wood (1976) applied the Severinghaus and Maguire (1955) approach to estimate the sex ratio of adult white-tailed deer in Pennsylvania. Their estimate of the proportion of female and male yearlings was derived from harvest checks. They also used a male-to-female fetus ratio to adjust for uneven sex ratios at birth. One important assumption of their method was that survival of male and female fawns was equal (Lang and Wood 1976), an assumption that may not hold true in some ungulate populations (Smith and Anderson 1998).

3.3.1   Sampling without Replacement

The Severinghaus and Maguire (1955) method of estimating a population sex ratio is based on two independent samples, one for each of the male and female subpopulations. It is assumed that whatever method of data collection is used, representative samples of the age structures of the male and female populations are collected. Sampling rates do not need to be the same for both sexes. This approach is ideal in situations in which hunter bag checks are used to collect age composition data and in which hunting regulations may favor the harvest of one sex over that of the other. Behavioral differences between sexes may also contribute to differential harvest rates. Nevertheless, sampling is assumed unbiased with regard to animal age.

For each animal examined, age is classified as either yearling (i.e., 1.5-year-olds for deer) or older (i.e., 2.5+ years for deer). Age determination is therefore relatively simple; yearlings only need to be differentiated from older individuals. Age determination by tooth eruption and wear is most accurate for usually the first two or three age classes of ungulates (Jacobson and Renier 1989). Differentiating yearlings from older animals therefore optimally uses check station aging methods to its advantage.

Model Development

Define

> N_M = abundance of males in the population, yearlings and older;
> Y_M = abundance of yearling males in the population;
> N_F = abundance of females in the population, yearlings and older;
> Y_F = abundance of yearling females in the population;
> m = number of males examined for age;
> y_m = number of yearling males identified from a sample of m males;
> f = number of females examined for age;
> y_f = number of yearling females identified from a sample of f females.

A sample survey of harvested animals is performed. Of m males examined for age, y_m are yearling males. Similarly, of f females examined for age, y_f are yearling females. Given carcasses are sampled without replacement, the hypergeometric distribution can be used to model the observed number of yearlings in each of the male and female samples. The joint likelihood for the number of male and female yearling is

$$L(Y_M, N_M, Y_F, N_F | m, f, y_m, y_f) = \frac{\binom{Y_M}{y_m}\binom{N_M - Y_M}{m - y_m}}{\binom{N_M}{m}} \cdot \frac{\binom{Y_F}{y_f}\binom{N_F - Y_F}{f - y_f}}{\binom{N_F}{f}}.$$

Next, it will be assumed that the production of male and female yearlings is the same and that they were subject to harvest for the first time as yearlings, in which case $Y_M = Y_F = Y$. Female abundance will also be reparameterized in terms of the sex ratio where $N_F = R_{F/M}N_M$. The joint likelihood can then be rewritten as

$$L(Y, N_M, R_{F/M} | m, f, y_m, y_f) = \frac{\binom{Y}{y_m}\binom{N_M - Y}{m - y_m}}{\binom{N_M}{m}} \cdot \frac{\binom{Y}{y_f}\binom{R_{F/M}N_M - Y}{f - y_f}}{\binom{R_{F/M}N_M}{f}}. \quad (3.39)$$

Under this model,

$$E(y_m) = m\left(\frac{Y}{N_M}\right)$$

and

$$E(y_f) = f\left(\frac{Y}{R_{F/M}N_M}\right).$$

These expected values lead to the method-of-moment estimator and MLE of

$$\hat{R}_{F/M} = \frac{\left(\frac{y_m}{m}\right)}{\left(\frac{y_f}{f}\right)} = \frac{p_{YM}}{p_{YF}}. \quad (3.40)$$

The variance of $\hat{R}_{F/M}$ is derived as

$$\text{Var}(\hat{R}_{F/M}) = \text{Var}\left(\frac{y_m f}{y_f m}\right)$$

$$= \left(\frac{f}{m}\right)^2 \text{Var}\left(\frac{y_m}{y_f}\right)$$

$$= \left(\frac{f}{m}\right)^2 \left[E(y_m)^2 \text{Var}\left(\frac{1}{y_f}\right) + E\left(\frac{1}{y_f}\right)^2 \text{Var}(y_m) + \text{Var}(y_m) \cdot \text{Var}\left(\frac{1}{y_f}\right)\right]$$

$$\text{Var}(\hat{R}_{F/M}) \doteq \hat{R}_{F/M}^2 \left[CV(\hat{p}_{YM})^2 + CV(\hat{p}_{YF})^2 + CV(\hat{p}_{YM})^2 \cdot CV(\hat{p}_{YF})^2\right]. \quad (3.41)$$

The variances for $\hat{p}_{YF}$ and $\hat{p}_{YM}$ are

$$\text{Var}(\hat{p}_{YF}) = \frac{p_{YF}(1 - p_{YF})}{f}\left(\frac{N_F - f}{N_F - 1}\right)$$

and

$$\text{Var}(\hat{p}_{YM}) = \frac{p_{YM}(1-p_{YM})}{m}\left(\frac{N_M-m}{N_M-1}\right)$$

with estimated variances,

$$\widehat{\text{Var}}(\hat{p}_{YF}) = \frac{\hat{p}_{YF}(1-\hat{p}_{YF})}{f-1}\left(\frac{N_F-f}{N_F}\right) \tag{3.42}$$

and

$$\widehat{\text{Var}}(\hat{p}_{YM}) = \frac{\hat{p}_{YM}(1-\hat{p}_{YM})}{m-1}\left(\frac{N_M-m}{N_M}\right), \tag{3.43}$$

as given by Cochran (1977:51–52). The estimated variance is calculated, following Goodman (1960), as

$$\widehat{\text{Var}}(\hat{R}_{F/M}) = \hat{R}_{F/M}^2\left[\widehat{\text{CV}}(\hat{p}_{YM})^2 + \widehat{\text{CV}}(\hat{p}_{YF})^2 - \widehat{\text{CV}}(\hat{p}_{YM})^2 \cdot \widehat{\text{CV}}(\hat{p}_{YF})^2\right], \tag{3.44}$$

where

$$\widehat{\text{CV}}(\hat{p}_{YM})^2 = \frac{(1-\hat{p}_{YM})}{(m-1)\hat{p}_{YM}}\left(\frac{N_M-m}{N_M}\right) \tag{3.45}$$

and where $\widehat{\text{CV}}(\hat{p}_{YF})$ is defined analogously.

Model Assumptions

1. All males have an equal probability of collection/harvest.
2. All females have an equal probability of collection/harvest.
3. Collection/harvest probabilities can be different for males and females.
4. Yearling production is equal for males and females at the time of harvest.
5. Sampling is without replacement.
6. Age determination is accurate for differentiating yearling (i.e., 1.5 years) and older animals (2.5+ years).

The assumption of equal harvest probabilities for yearling and older animals is necessary to obtain a representative sample of the age structure for the two sexes. Hunting might skew the age distribution given preference for older, larger animals. However, opportunistic behavior of most hunters for many species will tend to dominate hunter harvests (Coe et al. 1980). Differential behavior of animals subjected to hunting for the first time may also bias age composition (Coe et al. 1980). The other key assumption is equal recruitment of male and female yearlings into the population. This assumption implies equal sex ratios at time of birth and equal survival probabilities of the young until recruitment as yearlings into the population subject to hunting.

This model can also be used for small game species. Instead of yearlings and older animals, data would be collected on young-of-the-year and 1+ year olds. All the other assumptions of the model would apply. Schierbaum and Foley (1957) found young-of-the-year waterfowl to be more vulnerable to harvest than older birds. They suggested the inexperienced animals were more prone to hunter harvest than were animals that had previously been hunted.

Example 3.4: Estimating Sex Ratio by using the Severinghaus and Maguire (1955) Method from Deer Check Station Data

Consider the case in which $m = 325$ male deer carcasses were examined at a check station, of which $y_m = 110$ were yearlings. Similarly, of $f = 215$ female deer carcasses examined, $y_f = 35$ were yearlings. From these data, the sex ratio is estimated by Eq. (3.40) to be

$$\hat{R}_{F/M} = \frac{p_{YM}}{p_{YF}}.$$

From the sample, the proportion of yearling males is estimated to be

$$\hat{p}_{YM} = 110/325 = 0.3385,$$

and the proportion of yearling females is estimated to be

$$\hat{p}_{YF} = 35/215 = 0.1628.$$

The adult female-to-male sex ratio is then estimated to be

$$\hat{R}_{R/M} = \frac{0.3385}{0.1628} = 2.08 \, \text{females/male}.$$

Assuming the carcasses were from a large population such that $N_M \gg m$ and $N_F \gg f$, the terms $\left(\frac{N_M - m}{N_M}\right)$ and $\left(\frac{N_F - f}{N_F}\right)$ would be near one and can thus be ignored. The CV^2 for $\hat{p}_{YM}$ is calculated (Eq. 3.45) as

$$CV(\hat{p}_{YM})^2 = \frac{(1 - 0.3385)}{(325 - 1)(0.3385)} = 0.006032,$$

with the CV^2 for $\hat{p}_{YF}$ calculated analogously, where

$$CV(\hat{p}_{YF})^2 = \frac{(1 - 0.1628)}{(215 - 1)(0.1628)} = 0.024030.$$

Then by using Eq. (3.44), the variance is estimated as

$$\widehat{Var}(\hat{R}_{F/M}) = \hat{R}^2_{F/M}\left[\widehat{CV}(\hat{p}_{YM})^2 + \widehat{CV}(\hat{p}_{YF})^2 - \widehat{CV}(\hat{p}_{YM})^2 \cdot \widehat{CV}(\hat{p}_{YF})^2\right]$$

$$= (2.08)^2[0.006032 + 0.024030 - (0.006032)(0.024030)] = 0.1294$$

with $SE(\hat{R}_{FM}) = 0.3598$ and $CV(\hat{R}_{FM}) = 0.1730$. An approximate 95% confidence interval would be calculated (Eq. 3.15) as $2.08 \pm 1.96 \, (0.3598)$, or from 1.37 to 2.79 females per male. A better assumption is to assume $\ln \hat{R}_{FM}$ is normally distributed (Eq. 3.16), in which case a 95% confidence interval is between 1.48 and 2.92 females/male.

Discussion of Utility

The Severinghaus and Maguire (1955) method is most useful when harvest data are readily available, yet the harvest of adult males and females is skewed because of hunter preference, differences in vulnerability, or hunting regulations. This approach is best suited in areas where hunter bag checks are mandatory to reduce the likelihood of biased reporting. The assumption that the ratio of the proportions of yearling males to yearling females is the adult sex ratio depends strongly on equal survival for all young. This approach is well suited for species that mature quickly so that differential juvenile survival does not have time to alter the juvenile sex ratios. If through hunter selection, adult males are harvested in greater numbers than yearling males, then p_m will be estimated artificially low, and the resulting bias of $\hat{R}_{F/M}$ will be negative (underestimated). Alternatively, if adult females are harvested in greater numbers than yearling females, the adult sex ratio will be overestimated.

3.3.2   Sampling with Replacement

The hypergeometric distribution is used for finite sampling (Eq. 3.39), in which the number of males and females in the population, N_F and N_M, are known and sampling is conducted without replacement. However, for most species, sample sizes (m and f) will be small compared with the population, and the fpc trivial. For example, both age class (i.e., juvenile or adult) and sex are identifiable for many waterfowl species using wing samples. However, the number of wings sampled may be small in comparison to the number harvested or the population size. In these situations, the sampling process can be modeled assuming binomial distributions. The binomial likelihood is used to model the probabilities of selecting yearlings and adults from the subpopulations of males and females. The likelihood for the number of observed yearlings from the population of males is

$$L(Y_M, N_M | m, y_m) = \binom{m}{y_m} \left(\frac{Y_M}{N_M}\right)^{y_m} \left(1 - \frac{Y_M}{N_M}\right)^{m-y_m},$$

and the likelihood for the number of juvenile females selected or observed is

$$L(Y_F, N_F | f, y_f) = \binom{f}{y_f} \left(\frac{Y_F}{N_F}\right)^{y_f} \left(1 - \frac{Y_F}{N_F}\right)^{f-y_f}.$$

Assuming males and females have independent probabilities of selection, letting $Y_F = Y_M = Y$ and noting $N_F = R_{F/M} N_M$, the joint likelihood can be rewritten as

$$L = \binom{m}{y_m} \left(\frac{Y}{N_M}\right)^{y_m} \left(1 - \frac{Y}{N_M}\right)^{m-y_m}$$
$$\cdot \binom{f}{y_f} \left(\frac{Y}{R_{F/M} N_M}\right)^{y_f} \left(1 - \frac{Y}{R_{F/M} N_M}\right)^{f-y_f}$$

By reparameterizing $\dfrac{Y}{N_M} = p_{YM}$, the joint likelihood can be reexpressed as

$$L = \binom{m}{y_m} p_{YM}{}^{y_m} (1 - p_{YM})^{m-y_m} \cdot \binom{f}{y_f} \left(\frac{p_{YM}}{R_{F/M}}\right)^{y_f} \left(1 - \frac{p_{YM}}{R_{F/M}}\right)^{f-y_f}. \quad (3.46)$$

The MLEs are

$$\hat{p}_{YM} = \frac{y_m}{m},\tag{3.47}$$

and

$$\hat{R}_{F/M} = \frac{\left(\dfrac{y_m}{m}\right)}{\left(\dfrac{y_f}{f}\right)} = \frac{\hat{p}_{YM}}{\hat{p}_{YF}}.\tag{3.48}$$

Thus, the adult sex ratio is again estimated by the ratio of the proportion of yearling males to yearling females in the population (Eq. 3.40).

The variance of $\hat{R}_{F/M}$ can be approximated using the exact variance for a product and the delta method (Seber 1982:11). The estimated variance for $\hat{R}_{F/M}$ can be expressed by Eq. (3.44), and where

$$\widehat{CV}(\hat{p}_{YM})^2 = \frac{\text{Var}(\hat{p}_{YM})}{(\hat{p}_{YM})^2} = \frac{(1 - \hat{p}_{YM})}{m\hat{p}_{YM}}.\tag{3.49}$$

Similarly,

$$\widehat{CV}(\hat{p}_{YF})^2 = \frac{(1 - \hat{p}_{YF})}{f\hat{p}_{YF}},\tag{3.50}$$

where $\hat{p}_{YF} = \dfrac{y_f}{f}$.

Model Assumptions

The assumptions of this method are identical to those of Section 3.3.1 with the exception that sampling is with replacement. Use of Eqs. (3.49) and (3.50) to estimate the variance of $\hat{R}_{F/M}$ when sampling is actually conducted without replacement will overestimate the true magnitude of sampling error. In situations in which the sampling fractions $\left(\text{i.e., } \dfrac{m}{N_m}, \dfrac{f}{N_f}\right)$ are less than 0.10, there will be little difference in estimated standard errors whether the fpc is used or not.

3.3.3   Juvenile Sex Ratio $\neq$ 1

There may be situations in which the juvenile sex ratio $\dfrac{Y_F}{Y_M} = R_J$ is not equal to one at the time of initial harvest. If the value of R_J is known or estimable, the joint likelihood model (3.46) can be rewritten as follows

$$L(Y_M, N_M, R_{F/M}, R_J | m, f, y_m, y_f) = \binom{m}{y_m}\left(\frac{Y_M}{N_M}\right)^{y_m}\left(1 - \frac{Y_M}{N_M}\right)^{m - y_m}$$
$$\binom{f}{y_f}\left(\frac{Y_M R_J}{R_{F/M} N_M}\right)^{y_f}\left(1 - \frac{Y_M R_J}{R_{F/M} N_M}\right)^{f - y_f}$$

Again, reparameterizing $p = \dfrac{Y_M}{N_M}$, the likelihood can be expressed as

$$L = \binom{m}{y_m} p^{y_m} (1-p)^{m-y_m} \cdot \binom{f}{y_f} \left(\frac{pR_J}{R_{F/M}}\right)^{y_f} \left(1 - \frac{pR_J}{R_{F/M}}\right)^{f-y_f}.$$

The resulting maximum likelihood estimate of the adult sex ratio is then

$$\tilde{R}_{(F/M)} = \frac{\left(\dfrac{y_m}{m}\right)}{\left(\dfrac{y_f}{f} \hat{R}_J\right)} = \frac{\hat{p}_{YM}}{\hat{p}_{YF} \hat{R}_J}. \tag{3.51}$$

Thus, if the yearling sex ratio is known, and other than one, estimator (3.51) can be adjusted accordingly. The variance of $\tilde{R}_{F/M}$ can also be derived using the exact variance formula (Seber 1982) and the delta method by noting that $\tilde{R}_{F/M} = \hat{R}_{F/M} \cdot \dfrac{1}{\hat{R}_J}$. The variance of $\tilde{R}_{F/M}$ is written as

$$\mathrm{Var}(\tilde{R}_{F/M}) = R_{F/M}^2 \left[\mathrm{CV}(\hat{R}_{F/M})^2 + \mathrm{CV}\left(\frac{1}{\hat{R}_J}\right)^2 + \mathrm{CV}(\hat{R}_{F/M})^2 \cdot \mathrm{CV}\left(\frac{1}{\hat{R}_J}\right)^2 \right]$$

and estimated by Goodman (1960) as

$$\widehat{\mathrm{Var}}(\hat{R}_{F/M}) = \tilde{R}_{F/M}^2 \left[\widehat{\mathrm{CV}}(\hat{R}_{F/M})^2 + \widehat{\mathrm{CV}}(\hat{R}_J)^2 - \widehat{\mathrm{CV}}(\hat{R}_{F/M})^2 \cdot \widehat{\mathrm{CV}}(\hat{R}_J)^2 \right]. \tag{3.52}$$

The variance estimator assumes the yearling sex ratio is estimated independently of the harvest data and $\hat{p}_{YM}$, $\hat{p}_{YF}$. Inspection of Eq. (3.51) indicates the sex ratio estimator (Eq. 3.48) is not robust to any violation of the assumption that $R_J = 1$.

3.4 Sex Ratio Projections Based on Survival and Harvest Rates

Beddington (1974) was among the first to project sex ratios based on survival probabilities. Ratios of males to females in the population were expressed as a function of the juvenile sex ratio, and the natural and harvest survival for males and females. Beddington (1974) assumed a stable and stationary population, and nonconstant survival across age classes. Projecting sex ratios is particularly useful when assessing the effects of alternative management activities on population structure. Furthermore, these techniques allow projections for the entire population or for just the juvenile or adult segments of the population.

All sex ratio projections presented in this section are derived under the assumptions of a stable and stationary population. Namely, the population has constant annual recruitment and age-specific survival, and harvest probabilities that are constant over time. It is therefore important to make a distinction between these model projections and direct estimation techniques. The projected sex ratios will only be achieved if and when a population reaches a stable and stationary condition. Hence, the projections are an asymptotic condition at best and do not necessarily represent the actual sex ratio of the population at any specific time.

The most general expression for the population sex ratio is the quotient

$$E(R_{F/M}) = \frac{N_{F0} + N_{F1} + N_{F2} + N_{F3} + \cdots + N_{Fl}}{N_{M0} + N_{M1} + N_{M2} + N_{F3} + \cdots + N_{Fl}}$$

where

N_{Fi} = number of females in the population of age class i ($i = 0, \ldots, l$);
N_{Mi} = number of males in the population of age class i ($i = 0, \ldots, l$).

Evoking the assumption of constant recruitment and constant age-specific survival, the sex ratio of a population can be expressed as follows:

$$E(R_{F/M}) = \frac{N_0 + N_0 S_{F1} + N_0 S_{F1} S_{F2} + N_0 S_{F1} S_{F2} S_{F3} + \cdots}{N_0 + N_0 S_{M1} + N_0 S_{M1} S_{M2} + N_0 S_{M1} S_{M2} S_{M3} + \cdots} \tag{3.53}$$

where

N_0 = number of male and number of female recruits into the population each year
   (i.e., $N_{F0} = N_{M0} = N_0$);
S_{Fi} = total annual survival of females from age $i - 1$ to i;
S_{Mi} = total annual survival of males from age $i - 1$ to i.

A further refinement to the model is to partition total annual survival (S_T) into components associated with natural (S_N) and harvest sources of mortality (S_H), where

$$S_T = S_N S_H,$$

assuming stochastic independence between mortality factors. This partitioning of total annual survival (S_T) into its natural and harvest components can be done for each age class and sex in Eq. (3.53). Subsequent sections examine special cases of Eq. (3.53) in projecting the sex ratios of wild populations.

3.4.1   Constant Annual Survival for All Age Classes

Using Eq. (3.53) and making the additional assumption that survival is constant across all age classes within a sex but different between sexes, then

$$E(R_{F/M}) = \frac{N_0 + N_0 S_F + N_0 S_F^2 + N_0 S_F^3 + \cdots + N_0 S_F^l}{N_0 + N_0 S_M + N_0 S_M^2 + N_0 S_M^3 + \cdots + N_0 S_M^l}, \tag{3.54}$$

where S_i = annual survival rate for the ith gender (i = female, male).
Equation (3.54) can be rewritten as

$$E(R_{F/M}) = \frac{N_0(1 + S_F + S_F^2 + S_F^3 + \cdots + S_F^l)}{N_0(1 + S_M + S_M^2 + S_M^3 + \cdots + S_M^l)}. \tag{3.55}$$

Equation (3.55) simplifies to the expression

$$E(R_{F/M}) = \frac{\sum_{i=0}^{l} S_F^{\,i}}{\sum_{i=0}^{l} S_M^{\,i}}. \tag{3.56}$$

The sum of an infinite sequence,

$$\sum_{i=0}^{\infty} S^i = \frac{1}{1-S}, \quad \text{for } 0 < S < 1. \tag{3.57}$$

Combining Eqs. (3.56) and (3.57), the sex ratios can be approximated by

$$E(R_{F/M}) \doteq \frac{\left(\dfrac{1}{1-S_F}\right)}{\left(\dfrac{1}{1-S_M}\right)} = \frac{1-S_M}{1-S_F}, \tag{3.58}$$

for l large and/or S_F, S_M small. Expression (3.58) can be estimated by

$$\hat{R}_{F/M} \doteq \frac{1-\hat{S}_M}{1-\hat{S}_F}. \tag{3.59}$$

Because mortality is the complement of survival, i.e., $M_i = 1 - S_i$, Eq. (3.59) can also be expressed as the ratio of annual sex specific mortality probabilities, where

$$\hat{R}_{F/M} \doteq \frac{\hat{M}_M}{\hat{M}_F} \tag{3.60}$$

and M_M = annual mortality rate of males;
   M_F = annual mortality rate of females.

The variance of $\hat{R}_{F/M}$, (Eq. 3.59) can be calculated using the delta method. The estimated variance of $\hat{R}_{F/M}$ can be expressed as

$$\widehat{\text{Var}}(\hat{R}_{F/M}) \doteq (\hat{R}_{F/M})^2 \cdot \left(\frac{\widehat{\text{Var}}(\hat{S}_M)}{(1-\hat{S}_M)^2} + \frac{\widehat{\text{Var}}(\hat{S}_F)}{(1-\hat{S}_F)^2} - \frac{2\widehat{\text{Cov}}(\hat{S}_M, \hat{S}_F)}{(1-\hat{S}_M)\cdot(1-\hat{S}_F)} \right). \tag{3.61}$$

In the common case in which male ($\hat{S}_M$) and female ($\hat{S}_F$) survival probabilities are estimated independently, variance estimator (3.61) reduces to

$$\widehat{\text{Var}}(\hat{R}_{F/M}) = (\hat{R}_{F/M})^2 \left[\frac{\widehat{\text{Var}}(\hat{S}_M)}{(1-\hat{S}_M)^2} + \frac{\widehat{\text{Var}}(\hat{S}_F)}{(1-\hat{S}_F)^2} \right], \tag{3.62}$$

or more simply expressed as

$$\widehat{\text{Var}}(\hat{R}_{F/M}) = (\hat{R}_{F/M})^2 \left[\widehat{\text{CV}}(\hat{M}_M)^2 + \widehat{\text{CV}}(\hat{M}_F)^2 \right] \tag{3.63}$$

where $\widehat{\text{CV}}(\hat{M}_M)$ is the coefficient of variation (CV) for the estimate of the probability of death for males with an analogous CV for females. If female and male survival probabilities are estimated independently, Eq. (3.62) follows from Eq. (3.61) because $\text{Cov}(\hat{S}_F, \hat{S}_M) = 0$.

For an exploited species, the population sex ratio, $R_{F/M}$ can be estimated by the expression

$$\hat{R}_{F/M} = \frac{1 - \hat{S}_{N-M} \cdot \hat{S}_{H-M}}{1 - \hat{S}_{N-F} \cdot \hat{S}_{H-F}} \tag{3.64}$$

where $\hat{S}_{i-j}$ is the estimated survival of the jth sex ($j = $ M, F) for the ith mortality source ($i = $ natural [N], hunting [H]). Equation (3.64) can be used to project the long-term consequences of sex-specific hunting regulations on the sex ratio of a population.

The variance of $\hat{R}_{F/M}$ (Eq. 3.64) can be derived by the delta method and estimated by the quantity

$$\widehat{\text{Var}}(\hat{R}_{F/M}) = \left(\frac{\hat{S}_{H-M}\hat{S}_{N-M}}{(1 - \hat{S}_{N-F} \cdot \hat{S}_{H-F})} \right)^2 \left(\widehat{\text{CV}}(\hat{S}_{N-M})^2 + \widehat{\text{CV}}(\hat{S}_{H-M})^2 \right)$$

$$+ \left(\frac{\hat{R}_{F/M}\hat{S}_{H-F}\hat{S}_{N-F}}{(1 - \hat{S}_{N-F} \cdot \hat{S}_{H-F})} \right)^2 \left(\widehat{\text{CV}}(\hat{S}_{N-F})^2 + \widehat{\text{CV}}(\hat{S}_{H-F})^2 \right). \tag{3.65}$$

Again, all survival probabilities are assumed to be estimated independently; hence, all covariance terms are equal to zero.

Model Assumptions

The use of Eqs. (3.59) and (3.64) have a number of restrictive assumptions:

1. The number of male and female juveniles produced annually (N_0) is constant.
2. The sex-specific annual survival probabilities (i.e., S_F and S_M) from all sources are constant across years.
3. The sex ratio of juveniles (animals entering the population) is equal to 1, i.e., $R_J = 1$.
4. Total annual survival is a product of natural and harvest survival probabilities (i.e., $S_T = S_N S_H$).
5. The population has reached a stable and stationary state.

Robustness of the sex ratio estimators (Eqs. 3.59, 3.64) to violations of the stable and stationary assumptions will depend on the extent of the violation. If both survival and juvenile recruitment experience small variation, the stable and stationary assumptions are reasonably approximated. However, the stable and stationary assumption may not hold true. For three mule deer (*Odocoileus hemionus*) populations, Unsworth et al. (1999:323) reported a process standard deviation of 0.034 and 0.217 for adult female and fawn survival, respectively. In addition to process variation, changes in harvest regulations may also act to violate the stable and stationary assumption (White 1999). Managers using the models in the following sections must carefully consider whether their data meet the stable and stationary assumption, and recognize that biased estimates may result if these assumptions are violated.

Converting the expected sex ratio from a ratio of finite sums (Eq. 3.56) to a ratio of infinite sums (Eq. 3.58) has mathematical and biological consequences. The use of an infinite sum that converges has obvious mathematical convenience. There are also biological considerations. The number of observed age classes (l) in a sample of n animals is a function of sample size and magnitude of the annual survival probabilities (S). It would be inherently wrong to assume that if we only observe animals up to age class l, survival from l to $l + 1$ must be zero. Animals with lower annual survival will have fewer observed age classes, and populations with high survival will have a correspondingly greater number of age classes. However, the infinite sum (Eq. 3.57) allows animals of ever increasing age classes with exponentially smaller likelihoods of occurrence, but not zero.

In some species, senescence can play an important role in demographics. Survival of ungulates dramatically declines after dental wear reaches the gum line (Raedeke et al. 2002). Eberhardt et al. (1982) found wild burro (*Equus asinus*) population dynamics were governed by the age of senescence. If only l age classes exist, the finite sum (Eq. 3.56) can be simplified, as the following set of equations show. The infinite sum can be written as

$$\frac{1}{1-S} = 1 + S + S^2 + \cdots + S^l + S^{l+1} + S^{l+2} + \cdots + S^{\infty}$$

or reexpressed as

$$\frac{1}{1-S} = 1 + S + S^2 + \cdots + S^l + S^{l+1}(1 + S + S^2 + S^{\infty})$$

$$= 1 + S + S^2 + \cdots + S^l + \frac{S^{l+1}}{1-S}$$

which leads to the relationship

$$\frac{1 - S^{l+1}}{1 - S} = 1 + S + S^2 + \cdots + S^l. \tag{3.66}$$

The above expression (Eq. 3.66) could be applied to the sex ratio of a population (Eq. 3.55) as

$$E(R_{F/M}) = \frac{(1 + S_F + S_F^2 + S_F^3 + \cdots + S_F^l)}{(1 + S_M + S_M^2 + S_M^3 + \cdots + S_M^l)} = \frac{\dfrac{1 - S_F^{l+1}}{1 - S_F}}{\dfrac{1 - S_M^{l+1}}{1 - S_M}},$$

yielding a modified estimator of the population sex ratio of

$$\hat{R}_{F/M} = \frac{1 - \hat{S}_M}{1 - \hat{S}_F}\left(\frac{1 - \hat{S}_F^{l+1}}{1 - \hat{S}_M^{l+1}}\right), \tag{3.67}$$

which is an unbiased estimator of the population sex ratio to the first term of a Taylor series expansion. Hence, the term $\left(\dfrac{1 - \hat{S}_F^{l+1}}{1 - \hat{S}_M^{l+1}}\right)$ could be considered as a bias correction for age-truncated samples. The assumptions of constant animal survival probabilities (i.e., assumptions 3 and 4) can be relaxed as seen in subsequent sections (i.e., Section 3.4.2). Furthermore, the age of senescence, or truncation, for males and females need not be the same.

Example 3.5: Projecting Sex Ratios from Elk Radiotelemetry Survival Data

Using radio telemetry, annual survival of bull elk was estimated to be $\hat{S}_M = 0.9275$ ($\widehat{SE} = 0.0572$); for cow elk, $\hat{S}_F = 0.9716$ ($\widehat{SE} = 0.0369$). The projected sex ratio of the population is then estimated by Eq. (3.59) to be

$$\hat{R}_{F/M} \doteq \frac{1 - \hat{S}_M}{1 - \hat{S}_F}$$

$$= \frac{1 - 0.9275}{1 - 0.9716} = \frac{0.0725}{0.0284} = 2.25 \text{ females/male.}$$

Using Eq. (3.62), the variance of $\hat{R}_{F/M}$ is calculated to be

$$\widehat{\mathrm{Var}}(\hat{R}_{F/M}) = (\hat{R}_{F/M})^2 \left[\frac{\widehat{\mathrm{Var}}(\hat{S}_M)}{(1 - \hat{S}_M)^2} + \frac{\widehat{\mathrm{Var}}(\hat{S}_F)}{(1 - \hat{S}_F)^2} \right]$$

$$= (2.5528)^2 \left[\frac{(0.0572)^2}{(1 - 0.9275)^2} + \frac{(0.0369)^2}{(1 - 0.9716)^2} \right] = 15.0581,$$

or equivalently,

$$\widehat{\mathrm{SE}}(\hat{R}_{F/M}) = 3.8805,$$

assuming a covariance of zero between $\hat{S}_M$ and $\hat{S}_F$. Note despite quite precise survival estimates (i.e., $\widehat{\mathrm{CV}}(\hat{S}_M) = 0.0617, \widehat{\mathrm{CV}}(\hat{S}_F) = 0.0380$), the precision of the sex ratio estimator is quite poor (i.e., $\widehat{\mathrm{CV}}(\hat{R}_{F/M}) = 1.5201$). What is more important when estimating the sex ratio, however, is the precision of the mortality probabilities (i.e., $\widehat{\mathrm{CV}}(\hat{M}_M) = 0.7890, \widehat{\mathrm{CV}}(\hat{M}_F) = 1.2993$).

Example 3.6: Effect of Harvest on the Sex Ratio of *r*- versus
***k*-selected Species**

A species life history (*r*- vs. *k*-selected) will affect how harvest mortality rates change adult sex ratios. Two species with different life histories, harvested at the same rate, will have different sex ratios. Consider the case in which male and female elk (*k*-selected) have a natural survival rate of 0.95, a male elk hunting survival rate of 0.75, and a female hunting survival rate of 1.0 (i.e., no hunting of females); then, $R_{F/M}$ is projected using Eq. (3.64) to be

$$\hat{R}_{F/M} = \frac{1 - \hat{S}_{N-M} \cdot \hat{S}_{H-M}}{1 - \hat{S}_{N-F} \cdot \hat{S}_{H-F}}$$

$$= \frac{1 - 0.95 \cdot 0.75}{1 - 0.95 \cdot 1.0} = \frac{0.2875}{0.05} = 5.75 \text{ females/male.}$$

Now, assume that in a pheasant (*r*-selected) population, males and females have a natural survival rate of 0.30, and hunting pressures are the same as those of the elk population: a male hunting survival rate of 0.75 and a female hunting survival rate of 1.0. Now, $R_{F/M}$ is projected to be

$$\hat{R}_{F/M} = \frac{1 - 0.30 \cdot 0.75}{1 - 0.30 \cdot 1.0}$$

$$= \frac{0.775}{0.7}$$

$$= 1.11 \, \text{females/male}.$$

Note that similar harvest rates can have different effects on adult sex ratios because of differences in natural mortality rates. Thus, r-selected species, such as pheasant, will be less impacted by harvest of males than will k-selected species, such as elk. If maintenance of sex ratios is a management concern, hunting regulations can be more liberal in harvesting r-selected species without jeopardizing sex ratio objectives of a population.

Precision of Sex Ratio Estimates

Using objective function Eq. (3.17), or its equivalent, Eq. (3.18), the precision of estimator (3.59) can be expressed in terms of desired levels of $(1 - \alpha)$ and ε. Using variance formula (3.63), the anticipated level of ε can be calculated as a function of the CV in mortality estimates (i.e., $CV(\hat{M})$) for each sex. Assuming a common CV,

$$\varepsilon = \sqrt{2} \cdot Z_{1-\frac{\alpha}{2}} \cdot CV(\hat{M}), \tag{3.68}$$

or conversely, the required precision for $\hat{M}$ is

$$CV(\hat{M}) = \frac{\varepsilon}{\sqrt{2} \cdot Z_{1-\frac{\alpha}{2}}}. \tag{3.69}$$

Values of ε change linearly as a function of $CV(\hat{M})$ (Fig. 3.2).

For example, letting $1 - \alpha = 0.90$ and $\varepsilon = 0.1054$ such that

$$P\left(e^{-0.1054} - 1 < \frac{\hat{R}_{F/M} - R_{F/M}}{R_{F/M}} < e^{0.1054} - 1 \right) = 0.90,$$

or equivalently,

$$P\left(-0.1000 < \frac{\hat{R}_{F/M} - R_{F/M}}{R_{F/M}} < 0.1111 \right) = 0.90,$$

the required coefficients of variation for male and female mortality studies must be

$$CV(\hat{M}) = \frac{0.1054}{\sqrt{2} \cdot 1.645} = 0.0453.$$

Thus, for the sex ratio estimate to be within approximately 10% to 11% of the true value, 90% of the time, mortality needs to be estimated with a CV of 4.53%.

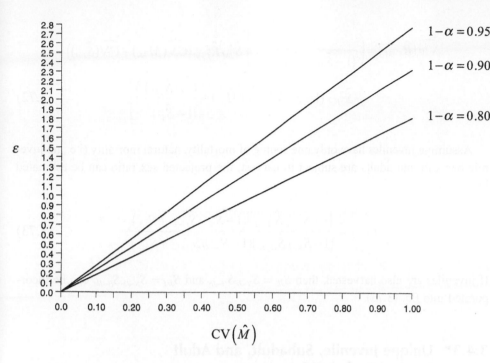

Figure 3.2. Precision curves for estimating a sex ratio based on variance formula (3.63) and precision expression (3.18) as a function of the CV($\hat{M}$) for various values of $1 - \alpha$.

3.4.2   Unique Juvenile Survival Probabilities

Juvenile survival probabilities are often not equal to adult survival probabilities, i.e., $S_J \neq S_A$, which would violate assumption 2 (Section 3.4.1, Model Assumptions). The expression for the expected value of $R_{F/M}$ can be modified to account for differential survival of the juvenile age class, where

$$E(R_{F/M}) = \frac{N_0 + N_0 S_{JF} + N_0 S_{JF} S_F + N_0 S_{JF} S_F^2 + N_0 S_{JF} S_F^3 + \cdots + N_0 S_{JF} S_F^l \cdots}{N_0 + N_0 S_{JM} + N_0 S_{JM} S_M + N_0 S_{JM} S_M^2 + N_0 S_{JM} S_M^3 + \cdots + N_0 S_{JM} S_M^l \cdots},$$

(3.70)

and where

S_{Ji} = juvenile survival rate for the ith sex (i = female (F), male (M));

S_i = annual adult survival rate for the ith sex (i = female (F), male (M)); and

N_0 = number of males and females recruited into the population annually.

Using the same approach, the expected value of the population sex ratio is

$$E(R_{F/M}) \doteq \frac{(1 - S_M)(1 - S_F + S_{JF})}{(1 - S_F)(1 - S_M + S_{JM})}.$$

In turn, the sex ratio can be estimated by the quotient

$$\hat{R}_{F/M} = \frac{\left(1 - \hat{S}_M\right)\left(1 - \hat{S}_F + \hat{S}_{JF}\right)}{\left(1 - \hat{S}_F\right)\left(1 - \hat{S}_M + \hat{S}_{JM}\right)},$$

(3.71)

where survival probabilities are estimated from field studies.

The variance of $R_{F/M}$ can be derived using the delta method. For the estimator in Eq. (3.71), observing that $\text{Var}(S_F) = \text{Var}(M_F)$ and $\text{Var}(S_M) = \text{Var}(M_M)$, the variance estimator is

$$\widehat{\mathrm{Var}}(\hat{R}_{F/M}) = \left(\frac{1}{(1-\hat{S}_M+\hat{S}_{JM})}\right)^2 \cdot \left[\hat{S}_{JM}^2 \hat{R}_{F/M}^2 \left(\widehat{\mathrm{CV}}(\hat{M}_M)+\widehat{\mathrm{CV}}(\hat{S}_{JM})\right)\right.$$

$$\left. +\hat{S}_{JF}^2 \cdot (1-\hat{S}_M)^2 \left(\widehat{\mathrm{CV}}(\hat{M}_F)+\frac{\widehat{\mathrm{CV}}(\hat{S}_{JM})}{(1-\hat{S}_F)^2}\right)\right]. \tag{3.72}$$

Assuming juveniles have only one source of mortality, natural mortality (i.e., no juvenile harvest), but adults are subject to harvest, the projected sex ratio can be estimated by

$$\hat{R}_{F/M} = \frac{(1-\hat{S}_{N-M}\hat{S}_{H-M})(1-\hat{S}_{N-F}\hat{S}_{H-F}+\hat{S}_{FJ})}{(1-\hat{S}_{N-F}\hat{S}_{H-F})(1-\hat{S}_{N-M}\hat{S}_{H-M}+\hat{S}_{MJ})}. \tag{3.73}$$

If juveniles are also harvested, then $S_{JF} = S_{N-JF}S_{H-JF}$ and $S_{JM} = S_{N-JM}S_{H-JM}$ can be incorporated into Eq. (3.73).

3.4.3 Unique Juvenile, Subadult, and Adult Survival Probabilities

Projection of the sex ratio of a population can be extended to the case in which juveniles (S_J), subadults (S_S), and adults (S) each have unique survival probabilities. In this situation

$$E(R_{F/M}) = \frac{N_0 + N_0 S_{JF} + N_0 S_{JF} S_{SF} + N_0 S_{JF} S_{SF} S_F + N_0 S_{JF} S_{SF} S_F^2 + \cdots}{N_0 + N_0 S_{JM} + N_0 S_{JM} S_{SM} + N_0 S_{JM} S_{SM} S_M + N_0 S_{JM} S_{SM} S_M^2 + \cdots}$$

$$= \frac{1 + S_{JF} + S_{JF} S_{SF}(1 + S_F + S_F^2 + S_F^3 + \cdots)}{1 + S_{JM} + S_{JM} S_{SM}(1 + S_M + S_M^2 + S_M^3 + \cdots)}. \tag{3.74}$$

As the number of age classes goes to infinity, $E(R_{F/M})$ has the limit

$$E(R_{F/M}) \doteq \frac{(1-S_M)(1-S_F+S_{JF}-S_{JF}S_F+S_{JF}S_{SF})}{(1-S_F)(1-S_M+S_{JM}-S_{JM}S_M+S_{JM}S_{SM})}. \tag{3.75}$$

Similarly, the general equation (3.53) can be used to examine other special cases of age-specific survival patterns.

3.4.4 Juvenile Sex Ratio ≠ 1 : 1

If the sex ratio of juveniles is not 1 : 1 but the number of males and females produced each year is constant, the expected value of the population sex ratio, $R_{F/M}$, can be written as follows:

$$E(R_{F/M}) = \frac{N_{FJ} + N_{FJ}S_F + N_{FJ}S_F^2 + N_{FJ}S_F^3 + \cdots + N_{FJ}S_F^l}{N_{MJ} + N_{MJ}S_M + N_{MJ}S_M^2 + N_{MJ}S_M^3 + \cdots + N_{MJ}S_M^l}, \tag{3.76}$$

or more simply,

$$E(R_{F/M}) = \frac{N_{JF}\left(1 + S_F + S_F^2 + S_F^3 + \cdots + S_F^l\right)}{N_{JM}\left(1 + S_M + S_M^2 + S_M^3 + \cdots + S_M^l\right)}.$$

Writing the juvenile sex ratio $\dfrac{N_{JF}}{N_{JM}}$ as R_J and simplifying the expression, the estimator of $R_{F/M}$ written in terms of total survival is

$$\tilde{R}_{F/M} = \hat{R}_J \cdot \frac{1 - \hat{S}_M}{1 - \hat{S}_F} = \hat{R}_J \cdot \hat{R}_{F/M}. \tag{3.77}$$

The variance of Eq. (3.74) can be estimated by the formula

$$\widehat{\mathrm{Var}}(\hat{R}_{F/M}) = \hat{R}_{F/M} \cdot \widehat{\mathrm{Var}}(\hat{R}_J) + \hat{R}_J \cdot \widehat{\mathrm{Var}}(\hat{R}_{F/M}) - \widehat{\mathrm{Var}}(\hat{R}_{F/M}) \cdot \widehat{\mathrm{Var}}(\hat{R}_J). \tag{3.78}$$

3.5   Summary

Sex ratios are among the most basic of demographic parameters and provide an indication of both the relative survival of females and males and the future breeding potential of a population. The observed sex ratio is a consequence of natural selection on the sexes and any anthropogenic effects of harvest. This allows wildlife managers to regulate animal harvests to try to maintain desired sex ratios. Conversely, by monitoring sex ratios, wildlife managers can assess how harvest regulations may be influencing the relative mortality rates of the male and female segments of a population.

In this chapter, two general approaches to estimating sex ratios were addressed (Fig. 3.3). The most general approach used direct field observations to estimate the population sex ratio. Finite sampling methods were used to estimate these ratios under a variety of scenarios. Choosing among these methods should begin with the structure and nature of the population under investigation. Species that form aggregates or herds need to be surveyed differently than populations of solitary individuals. Cluster sampling techniques were therefore presented. Unequal probabilities of detection are possible when making visual counts of animals. This consideration is important when probabilities of detecting animal aggregations of different sizes are expected to be a function of group size. The advantage of direct estimation methods for calculating sex ratios is that no assumption is required about the dynamics of the population being studied. Instead, adherence to the principles of probabilistic sampling is all that is needed to ensure reliable estimation of population sex ratios. Although easily said, the validity of the demographic study relies totally on the design and conduct of the sample survey.

Indirect methods of estimating population sex ratios (see Fig. 3.3), on the other hand, are analytically simple but rely heavily on the assumed dynamics of the population. Using sex-specific survival rates, the population sex ratio can be estimated assuming a stable and stationary population. These indirect methods may be viewed more realistically as projections rather than as estimation techniques. Given sex-specific values of survival, indirect methods project the long-term sex ratio of a population under stable and stationary conditions. These projections may nevertheless have value. First, they may be used as preliminary estimates for demographic modeling until empirical field

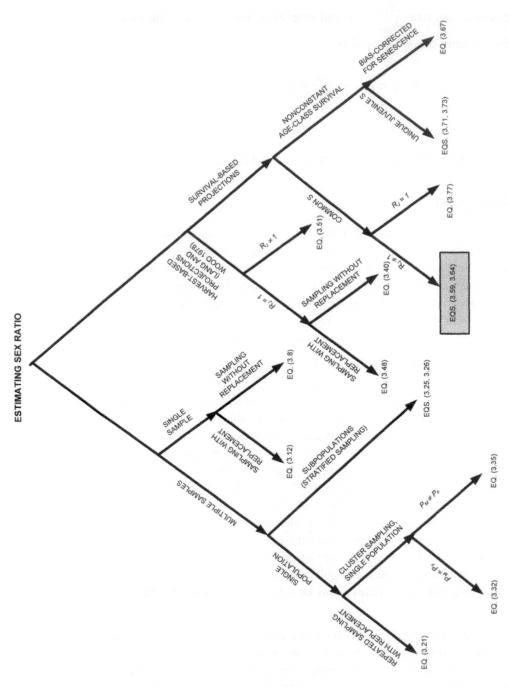

Figure 3.3.   Decision tree for the methods of estimating sex ratios. Equations with shading have corresponding sample size calculations available.

sampling results are available. Second, the projections provide a long-term perspective of the population should current management practices persist. And finally, a comparison of projections with empirically based estimates of the sex ratio may provide an indication of a population not in equilibrium with itself and foretell demographic changes yet to come.

Information on age and sex ratios will be used in subsequent chapters to estimate population abundance and productivity. Change-in-ratio methods will provide a useful and convenient means of converting changes in demographic ratios to estimates of demographic parameters. Population reconstruction methods will also use sex and age ratios to estimate juvenile or female adult abundance from estimates of adult male abundance.

Estimating Productivity

4

4.1 Introduction

The number of individuals recruited into a population is essential in understanding the dynamics of wildlife populations. The specific form of the recruitment estimate varies depending on the question of interest, the model being used, the population under study, and the available data. The broadest definition of productivity is an increase in biomass due to reproduction and growth (Bolen and Robinson 1995). In *Game Management*, Leopold (1933:452) defined productivity as "the rate at which mature breeding stock produces other mature stock, or mature removable crop." Pimlott (1959:393) defined productivity as the "percentage of the population that can be removed yearly without diminishing the population" and "the rate at which breeding stock produces a removable crop or additional breeding stock." Pimlott (1959) further defined gross productivity as the rate if all embryos at the end of the breeding season survived to a harvestable age. In contrast, net productivity refers to the rate at which new harvestable animals enter the population. These definitions reflect the focus on consumptive uses of wildlife (i.e., harvest) early in wildlife management. More recently, Bolen and Robinson (1995:525) defined productivity as "the number of surviving offspring produced during a specific period of time, usually expressed per year." They defined net productivity as "the annual gain or loss between births and deaths in a population." However, Krausman (2001:410) used Leopold's (1933) definition. Despite these varied definitions, all productivity measures consider reproduction and survival to some point in time.

Productivity can be expressed either in terms of absolute numbers, a ratio, or a proportion. Estimating the absolute number of births, young at a particular point in time, or new harvestable animals has some disadvantages. First, it is hard to collect reliable estimates of absolute numbers, particularly for juveniles. Second, and more importantly, population dynamics models often require productivity to be expressed as a ratio or a proportion, rather than as abundance. Absolute numbers can also be hard to place into perspective without additional data, e.g., 100 new juveniles this year versus 80 last year. However, expressing the same information as a ratio, 2 juveniles per adult this year versus 1.6 juveniles per adult last year, is more useful in many applications because the latter explicitly considers adult population size. Consequently, throughout this chapter, productivity will be expressed as the ratio of juveniles produced per adult. This ratio can be expressed either in terms of juveniles per adult, juveniles per adult female, or juvenile females per adult female. Although varied, these ratios consider the sex and age of the animals under investigation. The best way to express productivity will depend on the intended use of the estimate. For example, in Leslie matrices, the measure of productivity used is the net number of juvenile females produced per adult female that survive to the anniversary of the model (i.e., fecundity) (Caswell 1989).

As with other demographic parameters, productivity is sometimes age or stage specific. Eberhardt (1985) provided age-specific reproductive rates for a few species, and we provide age-specific productivity relationships for moose (*Alces alces*) (Ericsson et al. 2001), sea otters (*Enhydra lutris*) (Bodkin et al. 1993), and raccoons (*Procyon lotor*) (Fritzell et al. 1985) (Fig. 4.1). Eberhardt (1985) described three important reproductive

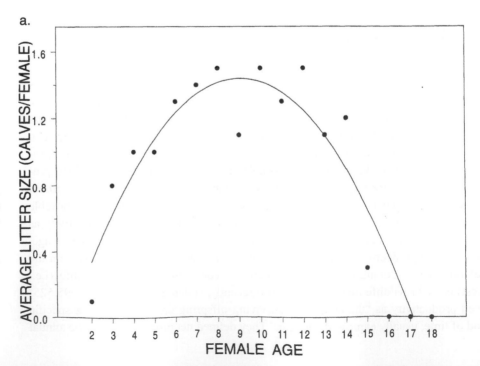

Figure 4.1. Plots of (a) litter size versus age of moose in Sweden (Ericsson et al. 2001) using a simple second-degree polynomial of the form $P = \alpha + \beta_1\ age + \beta_2\ age^2$. (b) pregnancy versus age of sea otter in Alaska (Bodkin et al. 1993) using a cubic polynomial expression, $P = \alpha + \beta_1\ age + \beta_2\ age^2 + \beta_3\ age^3$; and (c) litter size versus age of raccoon in Missouri (Fritzell et al. 1985), using a simple third-degree polynomial equation of the form $P = \alpha + \beta_1\ age + \beta_2\ age^2 + \beta_3\ age^3$ that provided the best fit of the relationship between age and average litter size.

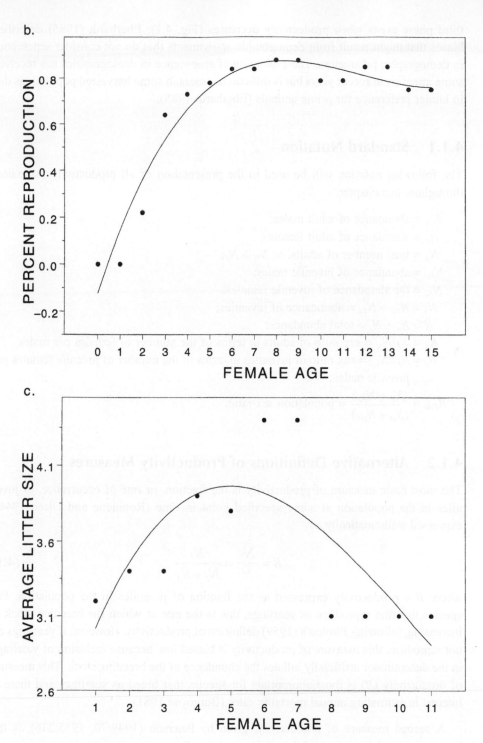

Figure 4.1. *Continued.*

phases for long-lived vertebrates (examples in Fig. 4.1). The first phase of productivity considers early reproduction. Typically, reproductive rates are lower in juvenile animals. For example, all relationships show lower productivity during the first few years of life than during prime years (Fig. 4.1). The second phase refers to reproduction during the prime years of adulthood. During this time, productivity is generally maximized, such is the case for moose, sea otter, and raccoon (Fig. 4.1). For many species in older ages, a third phase exists when productivity decreases (Fig. 4.1). Eberhardt (1985) described biases that might result from demographic assessments that do not consider senescence in demographic parameters. The evaluation of senescence in demographics has received some attention in recent years but is difficult to assess in some harvested populations due to hunter preference for prime animals (Eberhardt 1985).

4.1.1   Standard Notation

The following notation will be used in the presentation of all productivity estimators throughout this chapter.

N_M = abundance of adult males;
N_F = abundance of adult females;
N_A = total number of adults, or $N_M + N_F$;
N_{MJ} = abundance of juvenile males;
N_{FJ} = the abundance of juvenile females;
$N_J = N_{MJ} + N_{FJ}$ = abundance of juveniles;
$N = N_A + N_J$ = total abundance;
$R_A = N_F/N_M$ = sex ratio of adults in terms of the number of females per males;
$R_J = N_{FJ}/N_{MJ}$ = sex ratio of juveniles in terms of the number of juvenile females per juvenile males;
$R_{F/M} = \dfrac{(N_F + N_{FJ})}{(N_M + N_{MJ})}$ = population sex ratio.

4.1.2   Alternative Definitions of Productivity Measures

The most basic measure of productivity is the fraction, or rate of occurrence, of juveniles in the population at some specified point in time (Robinette and Olsen 1944), expressed mathematically as

$$B = \frac{N_J}{N} = \frac{N_J}{N_J + N_A} \tag{4.1}$$

where B = productivity expressed as the fraction of juveniles in the population. For species that first reproduce as yearlings, this is the rate at which the breeding stock is increasing, following Pimlott's (1959) definition of productivity. However, if yearlings do not reproduce, this measure of productivity is biased low because inclusion of yearlings in the denominator artificially inflates the abundance of the breeding stock. This measure of productivity (B) is most appropriate for species that breed as yearlings and there is interest in estimating annual mortality rates (Burgoyne 1981).

A second measure of productivity, given by Peterson (1949:70, 1955:216), is the "average annual increment of the adult population," or

$$U = \frac{N_J}{N_A} \tag{4.2}$$

which is the ratio of juveniles to adults in the population (U). This measure is most appropriate when the desired metric is to relate juvenile production in terms of the total breeding population.

These two productivity measures are functions of the total adult abundance. However, expressing productivity in terms of the number of adult females rather than total adult numbers is more widely used in population dynamics models (Caswell 1989). We will define productivity (P) as the ratio of juveniles to adult females, which is expressed mathematically as

$$P = \frac{N_J}{N_F} = \frac{N_{FJ} + N_{MJ}}{N_F}. \tag{4.3}$$

Productivity can also be expressed in terms of the number of female juveniles per adult female, which may be a more accurate representation of the increase in breeding stock (e.g., Leslie matrix) (Caswell 1989). This ratio

$$F = \frac{N_{FJ}}{N_F} \tag{4.4}$$

will be termed fecundity (F) in this book.

Productivity (P) can be related to fecundity (F) by considering the juvenile sex ratio, where

$$F = P \cdot \left(\frac{R_J}{1 + R_J} \right), \tag{4.5}$$

and where $R_J = \frac{N_{FJ}}{N_{MJ}}$.

For this chapter, productivity (P) will be defined as the total number of juvenile animals, male plus female, produced per adult (breeding) female (Eq. 4.3). This definition is consistent with various applications, such as the sex-age-kill model of population reconstruction, commonly used for large mammals (Skalski and Millspaugh 2002). The estimates of productivity implicitly include reference to the time when recruitment into the population is measured. For harvest-related models, productivity is often measured at the time new animals enter the harvestable population. In Leslie matrices, fecundity (Eq. 4.5) is used to model the population dynamics of the female component of the population to the anniversary date of the model.

4.1.3   Direct Versus Indirect Methods of Estimating Productivity

Estimates of productivity may be calculated using either direct or indirect methods of data collection. We consider direct methods as those techniques in which productivity is directly observed, such as repeated observations on the number of juveniles per litter, or the number of juveniles and number of females in the population. For example, Follis et al. (1972) and Harder and Kirkpatrick (1994) discuss the calculation of productivity from laparotomy examinations of females. Indirect methods estimate productivity from data other than simple age-structure information. An indirect method to estimate productivity may include reconstructing productivity based on changes in sex ratios (Dale 1952, Hanson 1963). The sex ratios themselves do not provide an estimate of the number of juveniles per female, but an estimate can be calculated through an intermediate step.

Data collection for direct methods is best viewed in the context of finite sampling theory (Cochran 1977, Thompson 2002). Data may be collected on the number of eggs per nest in an area (assuming one nest per female) or on the number of young per female. Independent observations of these data would result in an estimate of the average number of juveniles per female. Alternatively, the observed number of juveniles and adult females in an aggregation or group of animals can be used to estimate the ratio of juveniles per females.

Generally, direct methods are used early in the breeding season when there is still close affinity and direct association between parents and offspring, thus facilitating accurate counts. As observational periods occur later in the year, there is less association between juveniles and adult females, and direct estimation techniques become less feasible until ultimately they are no longer possible. In contrast, indirect estimation techniques are often more applicable later in the breeding season. In our description of productivity estimators, we will first consider direct estimation techniques, followed by indirect estimation procedures later in the chapter.

4.2 Direct Sampling Techniques

4.2.1 Single Survey

Sample survey techniques (Cochran 1977, Thompson 2002) can be used to estimate the average clutch size or litter size per female, assuming each adult female successfully breeds. Let y_i equal the number of juveniles observed with the ith female. Productivity can then be estimated by the average

$$\hat{P} = \bar{y} = \frac{\sum_{i=1}^{n} y_i}{n}, \tag{4.6}$$

where y_i = number of eggs or young in the ith clutch or litter ($i = 1, \ldots, n$);
   n = number of clutches or litters sampled.

The variance of $\hat{P}$ will depend on how the observations were sampled. If estimates were based on observational surveys, in which a clutch or litter may be observed more than once, sampling is with replacement and the variance is estimated by the quantity

$$\widehat{\text{Var}}(\hat{P}) = \frac{s^2}{n}, \tag{4.7}$$

where $s^2 = \dfrac{\sum_{i=1}^{n}(y_i - \bar{y})^2}{n-1}$, which is the empirical variance of the y_i's ($i = 1, \ldots, n$). If sample observations are taken without replacement, the variance of $\hat{P}$ is estimated by

$$\widehat{\text{Var}}(\hat{P}) = \left(1 - \frac{n}{N_F}\right)\frac{s^2}{n}. \tag{4.8}$$

The term $\left(1 - \dfrac{n}{N_F}\right)$ is the finite population correction (fpc). Use of variance estimators (4.7) and (4.8) assumes that the data are based on a simple random sample of brood production in the population. Thus, the brood of each female has an equal probability of observation. More complex survey sampling designs can also be used to directly estimate productivity based on stratification and multistage sampling (Cochran 1977, Thompson 2002).

An excellent example of directly estimating productivity based on juvenile counts in a finite population using direct methods was presented by Jumber et al. (1957). They estimated the number of mourning dove (*Zenaida macroura*) juveniles produced per nest using a multistaged, stratified sampling design. The study site was divided into 14 strata, and each stratum into blocks consisting of about 40 trees. Sampling occurred in randomly chosen blocks every 2 weeks. Estimates of productivity, the mean number of juveniles per nest, and variances were based on multistage stratified sampling of a finite population (Cochran 1977).

4.2.2 Repeated Surveys

Productivity can be estimated by repeated surveys of the ratio of juveniles to adult females in the population. Repeated surveys may be used to improve the precision of the productivity estimates. The estimator of productivity is the ratio estimator,

$$\hat{P} = \frac{\sum_{i=1}^{n} y_i}{\sum_{i=1}^{n} f_i} = \frac{\bar{y}}{\bar{f}}, \tag{4.9}$$

where y_i = number of juveniles in the ith sample ($i = 1, \ldots, n$);
$\quad\; f_i$ = number of females in the ith sample ($i = 1, \ldots, n$).

The terms $\bar{y}$ and $\bar{f}$ are the average juvenile and female counts across all surveys, respectively. The variance of the ratio estimator (Eq. 4.9) can be calculated, assuming sampling was conducted with replacement by the expression

$$\widehat{\mathrm{Var}}(\hat{P}) = \frac{\sum_{i=1}^{n}(y_i - \hat{P}f_i)^2}{n\bar{f}^2(n-1)}. \tag{4.10}$$

If sampling was conducted without replacement, the fpc term is added to the variance estimator

$$\widehat{\mathrm{Var}}(\hat{P}) = \frac{(1-C)\sum_{i=1}^{n}(y_i - \hat{P}f_i)^2}{n\bar{f}^2(n-1)}, \tag{4.11}$$

where $1 - C$ is the fraction of the total population not surveyed (i.e., fpc). If individuals are indistinguishable and may be counted more than once, sampling is usually assumed to be with replacement.

Example 4.1: Estimating Productivity from Repeated Surveys of Snowshoe Hares (*Lepus americanus*), Lake Alexander Area, Minnesota

The estimator in Eq. (4.9) can be used for repeated surveys in a single breeding season, or to estimate an average productivity rate across several years when productivity can be assumed stationary over time. An example comes from a study on snowshoe hare

(Green and Evans 1940). Data were collected on the total number of embryos from a sample of pregnant hares (Table 4.1) each year for $n = 7$ years. The results of the annual surveys of embryo counts are presented (Table 4.1). By use of Eq. (4.9), the average productivity rate across all years is estimated to be

$$\hat{P} = \frac{\sum\limits_{i=1}^{7} y_i}{\sum\limits_{i=1}^{7} f_i} = \frac{406}{140} = 2.90 \text{ embryos per female.}$$

This is not the same as taking the average of yearly pregnancy rates. For this example, the variance is calculated using Eq. (4.10), because sampling was conducted with replacement, where

$$\widehat{\text{Var}}(\hat{P}) = \frac{\sum\limits_{i=1}^{n} (y_i - \hat{P}f_i)^2}{n\bar{f}^2(n-1)} = \frac{295.8}{7 \cdot 20^2 \cdot (7-1)} = 0.018 \text{ or } \widehat{\text{SE}}(\hat{P}) = 0.133.$$

An asymptotic 90% confidence interval (CI) is calculated as $2.90 \pm 1.943 \, (0.133)$, yielding CI $(2.64 \leq P \leq 3.16) = 0.90$ using a t-statistic with 6 degrees of freedom.

Discussion of Utility

The direct estimator of productivity (Eq. 4.6) assumes there is one-to-one correspondence between observed clutches or litters and the adult females that produced the progeny. Hence, productivity is estimated simply by the mean size of the clutch or litter and its variance (Eqs. 4.7, 4.8) characterized by the empirical variance among litter groups. However, estimator (4.6) may be positively biased if unsuccessful females are not included in the calculations. By focusing on observed clutches or litters, unsuccessful breeding attempts may go unnoticed. This is the case with the snowshoe hare example, in which only pregnant females were included in the sample.

After young-of-the-year and their parents have aggregated into groups or herds, estimator (4.9) and its associated variance estimator (4.10) are more applicable. The repeated observations in Eq. (4.9) could be different segments of the population or repeated obser-

Table 4.1. Number of embryos in a sample of pregnant snowshoe hares from Minnesota during 1932–1938 (Green and Evans 1940).

Year (i)	Number of pregnant hares (f_i)	Total number of embryos (y_i)	Average number of embryos per pregnant hare
1932	23	73	3.17
1933	12	43	3.58
1934	34	95	2.79
1935	35	89	2.54
1936	20	61	3.05
1937	11	29	2.64
1938	5	16	3.20
Total	$\sum\limits_{i=1}^{7} f_i = 140$	$\sum\limits_{i=1}^{7} y_i = 406$	
Mean	$\bar{f} = 20$	$\bar{y} = 58$	

vations of the same group. If both successful and unsuccessful adult females are present, Eq. (4.9) should provide an accurate estimate of actual productivity. If only successful clutches are surveyed, a proper estimate of productivity must take into account the probability of successful breeding.

4.2.3 Productivity Adjusted for Breeding Success

Cowardin and Johnson (1979) defined breeding success for female waterfowl as the number of fledged females per hen in the breeding population (i.e., fecundity). Fecundity, F, can be calculated from an estimate of the probability a breeding pair successfully rears a brood to fledging stage and the estimated average size of a brood. The estimator of F can then be calculated as

$$\hat{F} = \frac{\hat{H} \cdot \bar{y}}{2}, \tag{4.12}$$

where $\hat{F}$ = number of juvenile females fledged per adult female;
$\hat{H}$ = estimated probability an adult female successfully produces a fledged brood;
$\bar{y}$ = estimated average size of the fledged brood for a successful nesting pair.

The above estimator, Eq. (4.12) assumes a $1:1$ sex ratio for the fledglings. The mean $\bar{y}$ (Eq. 4.6) is the average fledged brood size per successful nesting female under the assumption of 1 female per nesting pair. Alternatively, an estimator of productivity, P, the number of total juveniles, males and females, produced per adult female can be written as

$$\hat{P} = \hat{H} \cdot \bar{y}. \tag{4.13}$$

Following the approach of Ryding (2002), the variance of $\hat{F}$ can be expressed as

$$\widehat{\text{Var}}(\hat{F}) = \text{Var}\left(\frac{\hat{H}\bar{y}}{2}\right) = \frac{1}{4}\text{Var}(\hat{H}\bar{y}) \tag{4.14}$$

with exact variance estimator

$$\widehat{\text{Var}}(\hat{F}) = \hat{F}^2\left[\widehat{\text{CV}}(\hat{H})^2 + \widehat{\text{CV}}(\bar{y})^2 - \widehat{\text{CV}}(\hat{H})^2 \cdot \widehat{\text{CV}}(\bar{y})^2\right] \tag{4.15}$$

assuming $\hat{H}$ and $\bar{y}$ are estimated independently. The estimated variance of $\hat{P}$ is similar to that of $\hat{F}$ and is

$$\widehat{\text{Var}}(\hat{P}) = \hat{P}^2\left[\widehat{\text{CV}}(\hat{H})^2 + \widehat{\text{CV}}(\bar{y})^2 - \widehat{\text{CV}}(\hat{H})^2 \cdot \widehat{\text{CV}}(\bar{y})^2\right], \tag{4.16}$$

where CV denotes the coefficient of variation $\left[\text{i.e., } \widehat{\text{CV}}(\hat{\theta}) = \frac{\widehat{SE}(\hat{\theta})}{\hat{\theta}}\right]$.

Both productivity (P) and fecundity (F) implicitly refer to some time in the annual cycle. In Leslie matrix models, the required fecundity is a measure of the net number of juvenile females produced per adult female to the anniversary date of the model. The time during the annual cycle the population is being modeled, i.e., the anniversary date, is an arbitrary choice but critical to the proper calculation of fecundity. If field observations on clutch or litter size are made before the anniversary date, the previous estimates of productivity (Eq. 4.13) or fecundity (Eq. 4.12) need to be adjusted by the juvenile survival probability from date of observation to the anniversary date. In this case, the fecundity estimator can be rewritten as

$$\hat{F} = \frac{\hat{H} \cdot \bar{y}}{2} \cdot \hat{S}_J \tag{4.17}$$

or productivity as

$$\hat{P} = \hat{H} \cdot \bar{y} \cdot \hat{S}_J \tag{4.18}$$

where $\hat{S}_J$ = probability a juvenile survives from time of the clutch or litter size observations to anniversary date of the demographic model.

Again, Eqs. (4.17) and (4.18) assume a 1:1 sex ratio for juveniles to the anniversary date being modeled.

The variance of $\hat{F}$ (Eq. 4.17), can be estimated by

$$\widehat{Var}(\hat{F}) = \hat{F}^2 \left[\widehat{CV}(\hat{H})^2 + \widehat{CV}(\hat{S}_J)^2 + \widehat{CV}(\bar{y})^2 \right] \tag{4.19}$$

for independence of $\hat{H}$, $\hat{S}_J$, and $\bar{y}$ (Ryding 2002). Alternatively, an exact variance estimator for $\hat{F}$ (Eq. 4.17) when $\hat{H}$, $\hat{S}_J$, and $\bar{y}$ are estimated independently is

$$\widehat{Var}(\hat{F}) = \hat{F}^2 \Big[\widehat{CV}(\hat{H})^2 + \widehat{CV}(\hat{S}_J)^2 + \widehat{CV}(\bar{y})^2 - \widehat{CV}(\hat{H})^2 \cdot \widehat{CV}(\hat{S}_J)^2$$
$$- \widehat{CV}(\hat{H})^2 \cdot \widehat{CV}(\bar{y})^2 - \widehat{CV}(\hat{S}_J)^2 \cdot \widehat{CV}(\bar{y})^2 + \widehat{CV}(\hat{H})^2 \cdot \widehat{CV}(\hat{S}_J)^2 \cdot \widehat{CV}(\bar{y})^2 \Big]. \tag{4.20}$$

The approximate and exact variance estimators for $\hat{P}$ (Eq. 4.18) are, respectively,

$$\widehat{Var}(\hat{P}) \doteq \hat{P}^2 \left[\widehat{CV}(\hat{H})^2 + \widehat{CV}(\hat{S}_J)^2 + \widehat{CV}(\bar{y})^2 \right] \tag{4.21}$$

and

$$\widehat{Var}(\hat{P}) = P^2 \Big[\widehat{CV}(\hat{H})^2 + \widehat{CV}(\hat{S}_J)^2 + \widehat{CV}(\bar{y})^2 - \widehat{CV}(\hat{H})^2 \cdot \widehat{CV}(\hat{S}_J)^2$$
$$- \widehat{CV}(\hat{H})^2 \cdot \widehat{CV}(\bar{y})^2 - \widehat{CV}(\hat{S}_J)^2 \cdot \widehat{CV}(\bar{y})^2 + \widehat{CV}(\hat{H})^2 \cdot \widehat{CV}(\hat{S}_J)^2 \cdot \widehat{CV}(\bar{y})^2 \Big]. \tag{4.22}$$

4.2.4 Productivity Adjustment for Renesting

Essential to the calculation of Eqs. (4.17) and (4.18) is estimating the value of H, the probability that a female will successfully breed. Nests observed during the breeding season may not be successful, and heavy nesting losses may delay the final hatching of a successful brood by breeding pairs (Errington and Hamerstrom 1937). Using surveys on the success rate of nesting throughout a breeding season, estimates of the probability of nesting success given repeated nesting attempts might be calculated. Using the notation of Cowardin and Johnson (1979), we define the following:

H = probability a hen will successfully produce a brood to hatch during one of her attempts;

P = probability the nest hatches during an attempt;

r_i = probability hen attempts an ith nest given the previous $(i - 1)$ nesting attempts have failed.

Thus, the probability of successful nesting given i attempts is

$$H = r_1 P + r_1(1 - P)r_2 P + r_1(1 - P)r_2(1 - P)r_3 P + \cdots$$

$$= P \cdot \sum_{i=1}^{\infty} \left[(1 - P)^{i-1} \prod_{j=1}^{i} r_j \right] \tag{4.23}$$

assuming P is constant across time.

Cowardin and Johnson (1979) also assumed that

1. $r_i > r_{i+1}$ for all i,
2. $\lim_{i \to \infty} r_i = 0$,
3. r_i are inversely proportional to P, where
   $r_1 = \alpha$ for $0 \le \alpha \le 1$ and
   $$r_i = \frac{\alpha(1 - P)}{(i - 1)} \quad i > 1,$$

leading to the simplification

$$H = P \sum_{i=1}^{\infty} \frac{(1 - P)^{i-1} \alpha^i (1 - P)^{i-1}}{(i - 1)!}$$

$$= \alpha P \cdot \sum_{i=1}^{\infty} \frac{\left[\alpha(1 - P)^2 \right]^{i-1}}{(i - 1)!},$$

and then noting $\sum_{i=1}^{\infty} \dfrac{\left[\alpha(1 - P)^2 \right]^{i-1}}{(i - 1)!} = e^{\alpha(1-P)^2}$, such that

$$H = \alpha P e^{\alpha(1-P)^2} \tag{4.24}$$

However, Eq. (4.24) is based on numerous simplifications and assumptions that might not be supported by empirical data.

The probability of successful nesting in each period can alternatively be estimated by using a multinomial likelihood model (Ryding 2002), in which the cell probabilities are the terms of Eq. (4.23). The model is written as

$$L(r_i P_i | \underset{\sim}{x}, f) = \binom{f}{\underset{\sim}{x}} (r_1 P_1)^{x_1} (r_1(1 - P_1)r_2 P_2)^{x_2} (r_1(1 - P_1)r_2(1 - P_2)r_3 P_3)^{x_3}$$

$$\cdots \left(r_n P_n \prod_{i=1}^{n-1} r_i(1 - P_i) \right)^{x_n} \left(1 - \sum_{i=1}^{n} r_i P_i \left(\prod_{j=1}^{i-1} r_j(1 - P_j) \right) \right)^{f - \sum_{i=1}^{n} x_i}, \tag{4.25}$$

where x_i = the number of females establishing an ith nest;

  f = the total number of females in a sample; and

each of the cell probabilities is the probability of success for the ith nesting attempt. Reparameterizing the cell probabilities as h_i, the probability of nesting success in the ith attempt can be rewritten as

$$L(h_i \mid \underset{\sim}{x}, f) = \binom{f}{\underset{\sim}{x}} (h_1)^{x_1} (h_2)^{x_2} (h_3)^{x_3} \ldots (h_n)^{x_n} \left(1 - \sum_{i=1}^{n} h_i\right)^{f - \sum_{i=1}^{n} x_i}. \tag{4.26}$$

The maximum likelihood estimator (MLE) for each of the cell probabilities h_i is

$$\hat{h}_i = \frac{x_i}{f}. \tag{4.27}$$

The estimated variance for the conditional probability of success (h_i) is

$$\widehat{\mathrm{Var}}(\hat{h}_i) = \frac{\hat{h}_i(1 - \hat{h}_i)}{f}, \tag{4.28}$$

with estimated covariance between cell probabilities

$$\widehat{\mathrm{Cov}}(\hat{h}_i, \hat{h}_j) = \frac{-\hat{h}_i \hat{h}_j}{f}. \tag{4.29}$$

The overall probability of success is estimated by

$$\hat{H} = \frac{\sum_{i=1}^{n} x_i}{f},$$

with associated variance estimate

$$\widehat{\mathrm{Var}}(\hat{H}) = \frac{\hat{H}(1 - \hat{H})}{f}.$$

The total net productivity can be estimated from a weighted mean of the average brood sizes from each nesting period, or

$$\hat{P} = \hat{h}_1 \bar{l}_1 + \hat{h}_2 \bar{l}_2 + \hat{h}_3 \bar{l}_3 + \cdots \hat{h}_n \bar{l}_n$$
$$= \sum_{i=1}^{n} \bar{l}_i \hat{h}_i, \tag{4.30}$$

where

$\bar{l}_i$ = average brood size of the ith nesting attempt ($i = 1, \ldots, n$) calculated by

$$\bar{l}_i = \frac{1}{x_i} \sum_{j=1}^{x_i} l_{ij},$$

and where

l_{ij} = brood size for the ith nesting attempt ($i = 1, \ldots, n$) for the jth adult female ($j = 1, \ldots, x_i$).

Note that $\hat{P}$ in Eq. (4.30) can be substituted for $\hat{H} \cdot \bar{y}$ in Eqs. (4.17) and (4.18).

The variance of $\hat{P}$ can be derived using the delta method and estimated by the quantity

$$\widehat{\mathrm{Var}}(\hat{P}) = \sum_{i=1}^{n} \hat{h}_i^2 \, \widehat{\mathrm{Var}}(\bar{l}_i) + \sum_{i=1}^{n} \bar{l}_i^2 \, \widehat{\mathrm{Var}}(\hat{h}_i) + 2 \sum_{i=1}^{n} \sum_{i \neq j}^{n} \bar{l}_i \bar{l}_j \, \widehat{\mathrm{Cov}}(\hat{h}_i, \hat{h}_j). \qquad (4.31)$$

Example 4.2: Estimating Productivity Adjusting for Nesting Attempts, Wild Turkeys (*Meleagris gallopavo silvestris*), Wisconsin

In a study of the reproductive ecology of eastern wild turkeys (Paisley et al. 1998), 27 subadult females ($f = 27$) were radio-marked and followed during spring 1992 to estimate breeding success. Data were presented on the number of attempted, successful, and failed nests for two nesting periods. Hens hatching one or more poults were considered successful. The number of first attempts was 18, with one successful nest ($x_1 = 1$). Five hens attempted a second nest, and of these one was successful ($x_2 = 1$). It is assumed that nesting attempts observed in the second period were by hens that also attempted nesting in the first period. From these data, the joint probability of a subadult female establishing an ith nest and being successful, h_i, can be estimated by Eq. (4.27):

$$\hat{h}_i = \frac{x_i}{f},$$

$$\hat{h}_1 = \frac{1}{27} = 0.037, \text{ and}$$

$$\hat{h}_2 = \frac{1}{27} = 0.037.$$

The variance and covariances of $\hat{h}_1$ and $\hat{h}_2$ are calculated as

$$\widehat{\mathrm{Var}}(\hat{h}_i) = \frac{\hat{h}_i(1 - \hat{h}_i)}{f},$$

$$\widehat{\mathrm{Var}}(\hat{h}_1) = \widehat{\mathrm{Var}}(\hat{h}_2) = \frac{0.037(1 - 0.037)}{27} = 0.00132,$$

and

$$\widehat{\mathrm{Cov}}(\hat{h}_1, \hat{h}_2) = \frac{-\hat{h}_1 \hat{h}_2}{f}$$

$$\widehat{\mathrm{Cov}}(\hat{h}_1, \hat{h}_2) = \frac{-(0.037)(0.037)}{27} = -0.0000508.$$

Paisley et al. (1998) provided an average clutch size for subadult hens of $\widehat{\mathrm{CS}} = 10.3$ ($\mathrm{SE} = 0.479$) with an average hatching rate of $\widehat{\mathrm{HR}} = 0.871$ ($\widehat{\mathrm{SE}} = 0.038$). Hence, only an average brood size for all attempts is calculated. From the information provided, the average brood size for each successful nest is estimated as

$$\bar{l} = \widehat{\mathrm{CS}} \cdot \widehat{\mathrm{HR}}$$
$$= 10.3(0.871) = 8.971 \text{ poults per successful nest}$$

No covariances between the average clutch size ($\widehat{\text{CS}}$) and average hatching rates ($\widehat{\text{HR}}$) were available. Hence, the variance of $\bar{l}$ can be estimated by the exact product formula (Goodman 1960), where

$$\widehat{\text{Var}}(\bar{l}) = \widehat{\text{CS}}^2 \cdot \widehat{\text{Var}}(\widehat{\text{HR}}) + \widehat{\text{HR}}^2 \cdot \widehat{\text{Var}}(\widehat{\text{CS}}) - \widehat{\text{Var}}(\widehat{\text{HR}}) \cdot \widehat{\text{Var}}(\widehat{\text{CS}})$$

$$= 10.3^2(0.038)^2 + 0.871^2(0.479)^2 - (0.038)^2(0.479)^2 = 0.3269.$$

Total net productivity for subadult females in the 1992 breeding season is estimated by Eq. (4.30) as

$$\hat{P} = \hat{h}_1\bar{l}_1 + \hat{h}_2\bar{l}_2$$

$$= 0.037(8.971) + 0.037(8.971) = 0.6639 \text{ poults/subadult female.}$$

The estimated variance of $\hat{P}$ is calculated by Eq. (4.31). The variance of $\hat{P}$ is estimated by

$$\widehat{\text{Var}}(\hat{P}) = \hat{h}_1^2\widehat{\text{Var}}(\bar{l}_1) + \hat{h}_2^2\widehat{\text{Var}}(\bar{l}_2) + \bar{l}_1^2\widehat{\text{Var}}(\hat{h}_1) + \bar{l}_2^2\widehat{\text{Var}}(\hat{h}_2)$$

$$+ 2\bar{l}_1\bar{l}_2\widehat{\text{Cov}}(\hat{h}_1\hat{h}_2)$$

$$= (0.037)^2(0.3269) + (0.037)^2(0.3269)$$

$$+ (8.917)^2(0.00132) + (8.917)^2(0.00132)$$

$$+ 2(8.917)(8.917)(-0.0000508)$$

$$\widehat{\text{Var}}(\hat{P}) = 0.2027,$$

with a standard error of $\widehat{\text{SE}}(\hat{P}) = 0.4503$.

The probability estimates presented here differ from those presented by Paisley et al. (1998) for two reasons. First, h_i is a conditional probability of a successful nest given an ith attempt. Second, the estimated probabilities are not independent. That is, both attempt and success probabilities resulted from following the same subadult hens for the entire breeding season. Thus, calculating renesting attempt probabilities as separate binomial probabilities would be inappropriate.

4.2.5   Estimating Nesting Success: Mayfield (1975) Method

Ideally, nesting success (H) would be estimated from direct observations of nests from initiation to fledgling. If each nest has an equal and identical probability of success, the number of successful fledglings (x) for f females is binomially distributed with parameters f and H. Nesting success can be estimated by the fraction

$$\hat{H} = \frac{x}{f} = \frac{\sum_{i=1}^{n} x_i}{f}, \tag{4.32}$$

with an estimated variance of

$$\widehat{\text{Var}}(\hat{H}) = \frac{\hat{H}(1-\hat{H})}{f}. \tag{4.33}$$

If the nests have independent but unequal probabilities of success with mean μ_H, Eq. (4.32) remains valid. However, variance estimator (4.33) will overestimate the true variance (Feller 1968:230–232).

Unfortunately, nests are typically not found until incubation is underway and some nests have already failed. In practice, nest success studies should consider left-censored data (Kleinbaum 1996:7). Ignoring early nest failures will result in $\hat{H}$ (Eq. 4.32) having a positive bias.

Mayfield (1961, 1975) suggested estimating overall nest success based on the following assumptions:

1. Survival rate is constant for all nests.
2. Daily survival rate is constant throughout incubation.
3. Nest-days are conditionally independent observations.

Therefore, let

d_i = number of days a nest is observed from first sighting to failure or success;
n = number of nests observed;
$I_i = \begin{cases} 1 \text{ if nest fails during observation} \\ 0 \text{ if nest is successful.} \end{cases}$

The daily survival rate (S_d) for a nest is estimated by the quantity

$$\hat{S}_d = 1 - \frac{\sum\limits_{i=1}^{n} I_i}{\sum\limits_{i=1}^{n} d_i}. \tag{4.34}$$

Hensler and Nichols (1981) demonstrated that Eq. (4.34) is the MLE of S_d with the associated variance estimator of

$$\widehat{\text{Var}}(\hat{S}_d) = \frac{\hat{S}_d(1 - \hat{S}_d)}{\sum\limits_{i=1}^{n} d_i}. \tag{4.35}$$

The overall probability of success for a nest is then estimated by

$$\hat{H} = \hat{S}_d^{\,\hat{L}} \tag{4.36}$$

where L = true incubation time. In practice, L may be estimated by some observed average

$$\hat{L} = \bar{l} = \frac{\sum\limits_{i=1}^{n'} l_i}{n'}$$

where

l_i = total number of days to incubation for a nest observed from onset of laying;
n' = total number of nests with complete incubation durations.

In this case, nest success would be estimated by the quantity

$$\hat{H} = \hat{S}_d^{\bar{l}}. \tag{4.37}$$

By use of the delta method (Seber 1982), an approximate variance estimator can be expressed as

$$\widehat{\text{Var}}(\hat{H}) = \widehat{\text{Var}}(\hat{S}_d)(\bar{l}\hat{S}_d^{\bar{l}-1})^2 + \widehat{\text{Var}}(\bar{l})(\hat{S}_d^{\bar{l}} \cdot \ln \hat{S}_d)^2, \tag{4.38}$$

where

$$\widehat{\text{Var}}(\hat{S}_d) = \frac{\hat{S}_d(1-\hat{S}_d)}{\displaystyle\sum_{i=1}^{n} d_i},$$

$$\widehat{\text{Var}}(\bar{l}) = \frac{s_{l_i}^2}{n'}.$$

Example 4.3: Estimating Nesting Success Using the Mayfield (1961) Method for Kirtland's Warbler (*Dendroica kirtlandii*), Northern Michigan

Mayfield (1961) reported a study of Kirtland's warbler in which 35 nests were lost during 878 nest-days of incubation. The daily survival rate is estimated from Eq. (4.34) to be

$$\hat{S}_d = 1 - \frac{\displaystyle\sum_{i=1}^{n} I_i}{\displaystyle\sum_{i=1}^{n} d_i}$$

$$= 1 - \frac{35}{878} = 0.9601367.$$

The variance of $\hat{S}_d$ is estimated by Eq. (4.35), where

$$\widehat{\text{Var}}(\hat{S}_d) = \frac{\hat{S}_d(1-\hat{S}_d)}{\displaystyle\sum_{i=1}^{n} d_i}$$

$$= \frac{0.9601(1-0.9601)}{878} = 0.00004359,$$

or a standard error of $\widehat{\text{SE}}(\hat{S}_d) = 0.0066$. Assuming a 14-day incubation period, overall probability of nest success is then estimated to be

$$\hat{H} = \hat{S}_d^{\bar{l}}$$

$$= 0.9601367^{14} = 0.5658.$$

Discussion of Utility

The Mayfield (1961, 1975) method provides a ready means to estimate nest success from incomplete observations. However, the method assumes a constant daily rate of survival, which may not be true. Daily nest visits can be labor-intensive and increase the risk of nest predation (Bart 1977). In calculating nest success, a successful nest is defined as any nest with one or more successful hatchings. Partial predation of a nest is not considered in the calculation of $\hat{H}$ but will be reflected in mean clutch size ($\bar{y}$). Mayfield (1975) suggested using similar calculations when calculating the survival rate of hatched chicks to fledglings. Overall survival from egg laying to fledgling can be estimated by the product of the stage-specific survival probabilities.

Bart and Robson (1982) suggested an alternative MLE for the Mayfield method. They also assumed a constant daily survival rate but permitted nest visits on an irregular schedule. Bart and Robson (1982) discussed the problems of visitor impact on nest success and the potential bias associated with "unrecorded visits." By that, they referred to observer behavior in which if a nest is suspected of failure, it is immediately inspected; otherwise, observers must stay away from a nest to avoid disturbing it. In these cases, no visit is recorded and possible failures go unnoticed.

Manly and Schmutz (2001) compared the statistical behavior of the Mayfield method with several alternative estimators. When deaths are recorded within time intervals rather than daily, they recommend a generalization of the Klett and Johnson (1982) method that must be iteratively solved. Although their iterative approach performed better in simulation studies, the maximum likelihood method provides a theoretical framework for incorporating covariates and allowing for modeling heterogeneity in survival probabilities (Manly and Schmutz 2001). Bart and Robson (1982) provided explicit likelihood equations and examples of more complex models of daily survival probabilities.

4.3   Estimating Productivity from Sex and Age Ratios

4.3.1   Stokes (1954)–Hanson (1963) Method

A dual-ratio method for estimating productivity (P) was first introduced by Stokes (1954) and later discussed by Hanson (1963). It is suitable when juvenile and adult animals are distinguishable but assignment of individuals to sex is not readily possible in the juvenile age class. For example, some upland game birds, including quail and grouse, would meet these criteria.

This dual-ratio method is based on two survey periods. The first survey estimates the sex ratio of adults from counts of males (m_1) and females (f_1) in the breeding population. After recruitment, when juveniles become subadults, a second survey is conducted. This second survey estimates the ratio of subadults (y_2) to adults (a_2) in the population. Productivity is then estimated as the net number of juveniles produced at the time of the second survey per adult female.

Model Development

The likelihood for the first survey, conditional on the total number of adult males and females observed (i.e., $x_1 = m_1 + f_1$) can be written as

$$L(N_M, N_F | x_1, m_1, f_1) = \binom{x_1}{m_1, f_1} \left(\frac{N_M}{N_M + N_F} \right)^{m_1} \left(\frac{N_F}{N_M + N_F} \right)^{f_1}. \qquad (4.39)$$

The second likelihood models the age structure of the data from the second survey. The conditional likelihood of a_2 and y_2, given a sample size of x_2, can be written as

$$L(S, N_F, N_M, N_{FJ}, N_{MJ} | x_2, a_2, y_2) = \binom{x_2}{a_2, y_2} \left(\frac{S(N_F + N_M)}{S(N_F + N_M) + (N_{FJ} + N_{MJ})} \right)^{a_2}$$

$$\cdot \left(\frac{(N_{FJ} + N_{MJ})}{S(N_F + N_M) + (N_{FJ} + N_{MJ})} \right)^{y_2}. \tag{4.40}$$

The number of juveniles produced can be expressed as a function of adult female abundance, where $N_J = N_F P$. Substituting this expression into the likelihood for the second survey, and letting adult survival, $S = 1$, Eq. (4.40) becomes

$$L(N_F, N_M, P | x_2, a_2, y_2) = \binom{x_2}{a_2, y_2} \left(\frac{N_F + N_M}{(N_F + N_M) + N_F P} \right)^{a_2} \left(\frac{N_F P}{(N_F + N_M) + N_F P} \right)^{y_2}. \tag{4.41}$$

The joint likelihood for the first (Eq. 4.39) and second (Eq. 4.41) surveys can then be written following the approach of Ryding (2002) as

$$L = \binom{x_1}{m_1, f_1} \left(\frac{N_M}{N_M + N_F} \right)^{m_1} \left(\frac{N_F}{N_M + N_F} \right)^{f_1}$$

$$\cdot \binom{x_2}{a_2, y_2} \left(\frac{N_M + N_F}{N_M + N_F + N_F P} \right)^{a_2} \left(\frac{N_F \cdot P}{N_M + N_F + N_F P} \right)^{y_2}.$$

This joint likelihood, in turn, can be rewritten in terms of the adult sex ratio ($R_{F/M}$), where

$$L(R_{F/M}, P | x_1, x_2, m_1, f_1, a_2, y_2) = \binom{x_1}{m_1, f_1} \left(\frac{1}{1 + R_{F/M}} \right)^{m_1} \left(\frac{R_{F/M}}{1 + R_{F/M}} \right)^{f_1}$$

$$\cdot \binom{x_2}{a_2, y_2} \left(\frac{1 + R_{F/M}}{1 + R_{F/M} + R_{F/M} P} \right)^{a_2} \tag{4.42}$$

$$\cdot \left(\frac{R_{F/M} P}{1 + R_{F/M} + R_{F/M} P} \right)^{y_2}.$$

The maximum likelihood estimates for the model parameters are

$$\hat{R}_{F/M} = \frac{f_1}{m_1}$$

$$\hat{P} = \frac{\hat{R}_2 + \hat{R}_2 \hat{R}_1}{\hat{R}_1}, \tag{4.43}$$

where $\hat{R}_1 = \frac{f_1}{m_1}$, and $\hat{R}_2 = \frac{y_2}{a_2}$. Productivity can also be expressed as

$$\hat{P} = \frac{y_2 x_1}{a_2 f_1}. \tag{4.44}$$

The variance of $\hat{P}$ can be derived using the delta method (Ryding 2002). The variance of $\hat{P}$ (Eq. 4.44) expressed in terms of $R_{F/M}$ and P is

$$\text{Var}(\hat{P}) \doteq \frac{P}{R_{F/M}} \left[\frac{P}{x_1} + \frac{(1 + R_{F/M} + PR_{F/M})^2}{x_2(R_{F/M} + 1)} \right]. \tag{4.45}$$

The estimated variance for $\hat{P}$, in terms of the observed ratio estimators $\hat{R}_1$ and $\hat{R}_2$, is

$$\widehat{\text{Var}}(\hat{P}) \doteq \frac{\hat{R}_2(1 + \hat{R}_1)^2}{\hat{R}_1^2} \left[\frac{\hat{R}_2}{x_1\hat{R}_1} + \frac{(1 + \hat{R}_2)^2}{x_2} \right], \tag{4.46}$$

or, more simply,

$$\widehat{\text{Var}}(\hat{P}) \doteq \frac{y_2 x_1^2}{a_2^2 f_1^2} \left[\frac{y_2 m_1}{x_1 f_1} + \frac{x_2}{a_2} \right]. \tag{4.47}$$

The ratios $\hat{R}_1$ and $\hat{R}_2$ alone are not sufficient to calculate the variance; information on sample sizes x_1 and x_2 is also required.

Assuming an equal sex ratio among juveniles, fecundity, expressed as the number of juvenile females per adult female, can be estimated as

$$\hat{F} = \frac{\hat{P}}{2}. \tag{4.48}$$

The variance of the fecundity estimator is then

$$\text{Var}(\hat{F}) = \text{Var}\left(\frac{\hat{P}}{2}\right) = \frac{\text{Var}(\hat{P})}{4}. \tag{4.49}$$

Model Assumptions

There are five assumptions specific to the Stokes (1954)–Hanson (1963) method of estimating productivity.

1. All animals, males and females, have independent and equal probabilities of detection in the first survey.
2. All animals, adults and subadults, have independent and equal probabilities of detection during the second survey.
3. Detection probabilities can be different between survey periods.
4. Survival probability of adults is one (i.e., $S = 1$) between the two surveys.
5. Juveniles recruit into the population with a sex ratio of $R_J = 1$.

The assumption of equal and independent probability of seeing individuals in the first and second surveys may not hold because of differences in behavior and spatial use of resources by adult males and females and juveniles. If the animals aggregate in coveys or herds, individuals may have neither equal nor independent detection probabilities. If the assumption of independence among animals is violated, the effective sample sizes are less than x_1 and x_2, and variance formula (Eq. 4.45) will underestimate the true variance. Detection probabilities may also be related to group size. If size-biased sampling occurs and animals aggregate differentially by category, sex, or age, ratio estimates may be biased, leading to biased estimates of productivity.

This method assumes that adults and subadults are observed or harvested with the same probability during the second survey. If game-bag surveys are used for the second survey, greater vulnerability of juveniles to harvest may bias estimates of productivity. If this occurs, productivity will be overestimated because the number of juveniles will be artificially inflated. An alternative would be to conduct the second survey before harvest, using a method that better ensures equal detection of all individuals in the population.

The estimator of productivity (Eq. 4.44) is also sensitive to any violation of the assumption that adult survival probability is 1 (i.e., $S = 1$) between the time of the first and second survey periods. If $S < 1$, productivity will be overestimated by $(1 - S)100\%$. Consequently, the duration between the two survey periods can influence the validity of the method. If adult survival is not 1, the likelihood model (Eq. 4.42) needs to be reformulated (Section 4.3.2) and an independent estimate of S incorporated in the joint likelihood model.

Example 4.4: Estimating Productivity Using the Stokes (1954)–Hanson (1963) Method for Gambel Quail (*Callipepla gambelii*)

The following example is from a hypothetical population of Gambel quail provided by Hanson (1963:14–15). A survey of birds conducted in early August yielded counts of $m_1 = 325$ adult males and $f_1 = 250$ adult females ($x_1 = 575$). Although some juveniles were observed, none were counted in the first survey. The second survey of harvested quail, conducted in early October after juveniles matured, yielded $a_2 = 210$ adults and $y_2 = 835$ subadults ($x_2 = 1045$). The estimated productivity calculated from Eq. (4.43) is

$$\hat{R}_1 = \frac{250}{325} = 0.7692, \text{ and}$$

$$\hat{R}_2 = \frac{835}{210} = 3.9762,$$

leading to

$$\hat{P} = \frac{\hat{R}_2 + \hat{R}_2 \hat{R}_1}{\hat{R}_1}$$

$$= \frac{3.9762 + 3.9762 \cdot (0.7692)}{0.7692} = 9.1452 \text{ young/adult female.}$$

The estimated variance of $\hat{P}$ using Eq. (4.46) is

$$\widehat{\text{Var}}(\hat{P}) \doteq \frac{\hat{R}_2 (1 + \hat{R}_1)^2}{\hat{R}_1^2} \left[\frac{\hat{R}_2}{x_1 \hat{R}_1} + \frac{(1 + \hat{R}_2)^2}{x_2} \right]$$

$$= \frac{3.9761 \cdot (1 + 0.7692)^2}{0.7692^2} \cdot \left[\frac{3.9761}{(575) \cdot 0.7692} + \frac{(1 + 3.9761)^2}{(1045)} \right] = 0.6875,$$

or equivalently, $\widehat{\text{SE}}(\hat{P}) = 0.8292$. An asymptotic 90% CI estimate of P is $9.1452 \pm 1.645 \cdot (0.8292)$, or CI $(7.7812 \leq P \leq 10.5092) = 0.90$. The form of Eq. (4.44) suggests the estimator may be more appropriately assumed to be log-normally distributed. Thus, the 90% CI would be calculated as

$$\hat{P}e^{\pm z_{1-\frac{\alpha}{2}}\sqrt{\frac{\widehat{Var}(\hat{P})}{\hat{P}^2}}}$$

$$9.1452e^{\pm 1.645\sqrt{\frac{0.6875}{(9.1452)^2}}},$$

or CI $(7.8784 \le P \le 10.6164) = 0.90$. The use of Program USER to analyze these data and calculate a profile likelihood CI for P is illustrated in Appendix C.

Relative Precision of the Productivity Estimator

Precision curves are presented to illustrate how sample sizes during the two surveys affect the precision of the productivity estimate. Precision of the productivity estimate will be defined by the expression

$$P\left(\left|\frac{\hat{P}-P}{P}\right| < \varepsilon\right) = 1 - \alpha$$

where the relative error in estimation $\left|\dfrac{\hat{P}-P}{P}\right|$ is desired to be less than ε, $(1-\alpha)100\%$ of the time. Assuming P is normally distributed,

$$\varepsilon = Z_{1-\frac{\alpha}{2}}\widehat{CV}(\hat{P}).$$

For $\alpha = 0.10$ and productivity of $P = 2$, 5, and 10, the relative error (ε) in estimation is plotted against per-period sample size ($x_1 = x_2 = x$) (Fig. 4.2), based on variance formula (4.45). The precision of $\hat{P}$ is relatively invariant to values of $R_{F/M}$ over the range depicted.

Discussion of Utility

Estimation of productivity using the Stokes (1954)–Hanson (1963) method is applicable to populations in which juveniles mature rapidly before the fall hunting season. That is, between the first and second surveys, it is assumed that juveniles mature and become equally vulnerable to sampling as adults. In the first survey, it is necessary that only adults be surveyed. If subadults are present and indistinguishable from adults, this would result in an underestimate of productivity. Investigators need to consider behavioral differences among classes of animals that could bias survey results. A final consideration in using the method is the assumption that adults have a survival probability of $S = 1$ between survey events. The method is sensitive to this model violation and will produce positively biased estimates of productivity if untrue. Investigators should use this method only for those species with high adult survival or otherwise consider the methods in Section 4.3.2.

4.3.2   Generalized Stokes (1954)–Hanson (1963) Method

In the previous dual-ratio method for estimating productivity (Stokes 1954, Hanson 1963), it was assumed adult survival was 1 (i.e., $S = 1$) between the survey periods.

a. P = 2

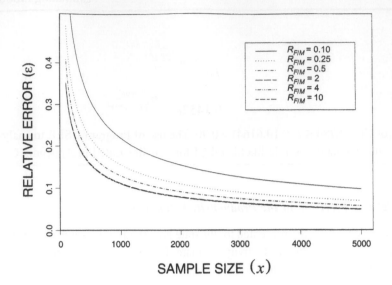

b. P = 5

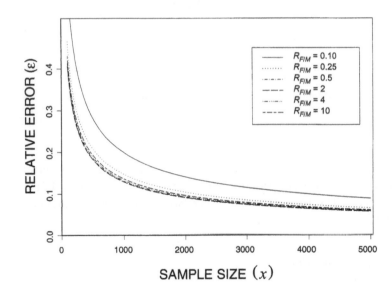

c. P = 10

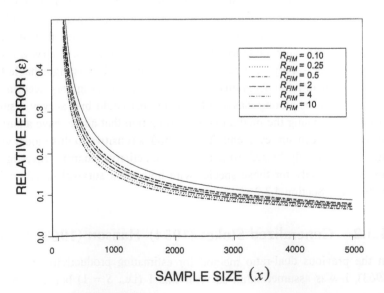

Figure 4.2. Plots of sample size ($x_1 = x_2 = x$) versus sampling precision expressed as relative error (ε) in estimating productivity (P) when $\alpha = 0.10$ by using the Stokes (1954)–Hanson (1963) method for various values of R_{FIM} and (a) $P = 2$, (b) $P = 5$, and (c) $P = 10$.

If $S < 1$, the number of adult females in the population is no longer constant over time. Hence, the productivity estimate now depends on time. The original Stokes (1954)–Hanson (1963) estimator expressed productivity as the ratio of juveniles at the time of the second survey (N_{J2}) to that of the constant adult female abundance (N_F), where

$$P = \frac{N_{J2}}{N_F}.$$

Now that survival may be less than one in the adult population, productivity can be expressed in terms of the adult female abundance at the time of the first (N_{F1}) or the second (N_{F2}) survey. Two productivity parameters will therefore be defined:

$$P_1 = \frac{N_{J2}}{N_{F1}}$$

and

$$P_2 = \frac{N_{J2}}{N_{F2}}.$$

The choice of which parameter to estimate depends on how $\hat{P}$ will be used in subsequent demographic analyses.

As with the original Stokes (1954)–Hanson (1963) procedure, two survey samples will be performed. During the first survey at the beginning of the breeding season, the number of adult males (m_1) and adult females (f_1) observed are enumerated. A postbreeding survey records the number of adults (a_2) and juveniles (y_2) observed. However, in addition to these survey counts, an adult survival study provides an unbiased estimate of $\hat{S}$ between survey periods and a corresponding variance estimate, $\widehat{\text{Var}}(\hat{S})$. The subsequent method proposed by Ryding (2002) relaxes the assumption $S = 1$, thereby permitting valid estimates of both P_1 and P_2. This modification allows for cases in which time between surveys is sufficiently long that $S < 1$. For example, consider an unusually harsh spring event that adversely affects adult survival in upland game birds before the second survey. In this case, $S < 1$, and the estimate of productivity will be overestimated by $(1 - S)100\%$ if Eq. (4.43) is used and not adjusted for the adult mortality.

Model Development

In the modification of the Stokes (1954)–Hanson (1963) method by Ryding (2002), the number of adults at the time of the second survey is $S(N_F + N_M)$, and thus, the joint likelihood can be written as follows:

$$L = \binom{x_1}{m_1, f_1} \left(\frac{N_M}{N_M + N_F}\right)^{m_1} \left(\frac{N_F}{N_M + N_F}\right)^{f_1}$$
$$\cdot \binom{x_2}{a_2, y_2} \left(\frac{S(N_M + N_F)}{S(N_M + N_F) + N_F P}\right)^{a_2} \left(\frac{N_F \cdot P}{S(N_M + N_F) + N_F P}\right)^{y_2},$$

or, in terms of the sex ratio of adults,

$$L(R_{F/M},S,P|x_1,x_2,m_1,f_1,a_2,y_2) = \binom{x_1}{m_1,f_1}\left(\frac{1}{1+R_{F/M}}\right)^{m_1}\left(\frac{R_{F/M}}{1+R_{F/M}}\right)^{f_1}$$
$$\cdot\binom{x_2}{a_2,y_2}\left(\frac{S(1+R_{F/M})}{S(1+R_{F/M})+R_{F/M}P}\right)^{a_2} \quad (4.50)$$
$$\cdot\left(\frac{R_{F/M}P}{S(1+R_{F/M})+R_{F/M}P}\right)^{y_2}\cdot$$

Letting $\hat{R}_1 = \dfrac{f_1}{m_1}$, and $\hat{R}_2 = \dfrac{y_2}{a_2}$, the MLE can be expressed as

$$\hat{P}_1 = \hat{S}\cdot\frac{\hat{R}_2 + \hat{R}_2\hat{R}_1}{\hat{R}_1}. \quad (4.51)$$

The estimate $\hat{P}_1$ expresses productivity in terms of the number of juveniles measured at time of the second survey (N_{J2}) and number of adult females measured at time of the first survey (N_{F1}). If $S < 1$, the original Stokes (1954)–Hanson (1963) estimator (Eq. 4.43),

$$\hat{P}_2 = \frac{\hat{R}_2 + \hat{R}_2\hat{R}_1}{\hat{R}_1},$$

expresses productivity in terms of the number of juveniles at the end of the second survey (N_{J2}) divided by the number of adult females at the time of the second survey (N_{F2}). Hence, for situations in which the adult population is subject to mortality, a decision must be made with regard to how fecundity will be expressed. Both estimators, $\hat{P}_1$ and $\hat{P}_2$, are valid and have useful roles in demographic analyses. The proper choice depends on the application and inference sought.

Adult survival (S) can be obtained from reported values from previous studies or estimated by using independent auxiliary information from radiotelemetry or mark-recapture studies. In the latter case, the auxiliary study will have an associated likelihood. Assuming the survival study is independent of the productivity surveys, this third likelihood can be multiplied by the two other likelihoods to form a joint likelihood, written as

$$L = \binom{x_1}{m_1,f_1}\left(\frac{1}{1+R_{F/M}}\right)^{m_1}\left(\frac{R_{F/M}}{1+R_{F/M}}\right)^{f_1}$$
$$\cdot\binom{x_2}{a_2,y_2}\left(\frac{S(1+R_{F/M})}{S(1+R_{F/M})+R_{F/M}P}\right)^{a_2}\left(\frac{R_{F/M}P}{S(1+R_{F/M})+R_{F/M}P}\right)^{y_2} \quad (4.52)$$
$$\cdot L(S|\sim),$$

where $L(S|\sim)$ is the likelihood of the auxiliary survival study.

The exact variance of $\hat{P}_1$ when $\hat{S}$ is independently estimated is as

$$\text{Var}(\hat{P}_1) = S\cdot\text{Var}(\hat{P}_2) + P_2\cdot\text{Var}(\hat{S}) + \text{Var}(\hat{P}_2)\cdot\text{Var}(\hat{S}), \quad (4.53)$$

where $\text{Var}(\hat{P}_2)$ is taken from Eq. (4.45). This variance can be estimated by the expression

$$\widehat{\text{Var}}(\hat{P}_1) = \hat{S}\cdot\widehat{\text{Var}}(\hat{P}_2) + \hat{P}_2\cdot\widehat{\text{Var}}(\hat{S}) - \widehat{\text{Var}}(\hat{P}_2)\cdot\widehat{\text{Var}}(\hat{S}) \quad (4.54)$$

using Eq. (4.46).

Model Assumptions

The assumptions for the dual-ratio method when adult survival $S \neq 1$ are similar to those of the dual-ratio method first described by Stokes (1954). The five assumptions of the modified Stokes (1954)–Hanson (1963) estimator proposed by Ryding (2002) include:

1. All animals, males and females, have independent and equal probabilities of detection in the first survey.
2. All animals, adults and subadults, have independent and equal detection probabilities during the second survey.
3. Detection probabilities can be different between survey periods.
4. An unbiased estimate of adult survival between the first and second survey events is available.
5. Juveniles are recruited into the population with a sex ratio of $R_J = 1$.

Behavioral differences between males and females, and subadults and adults, could result in unequal probabilities of detection. Differential harvest vulnerability during the second survey between adults and subadults could bias the estimates of productivity. The assumption of independent detection probabilities will be violated if animals occur in groups (e.g., quail coveys). Violation of the independence assumption would result in effectively smaller sample sizes with the variance formula underestimating the magnitude of the error. If the demographic composition of the groups varies and probability of detection is related to group size, both the estimator of productivity and its variance will be biased.

Example 4.5: Estimating Productivity, Adjusting for Adult Mortality, when using the Stokes (1954)–Hanson (1963) Method for Gambel Quail

The data from the hypothetical population of Gambel quail discussed earlier (Hanson 1963) will be analyzed again. This time, we further assume that adult survival was unbiasedly estimated to be $\hat{S} = 0.80$ and $Var(\hat{S}) = 0.03$. Using the same count data analyzed in Example 4.4, $m_1 = 325$, $f_1 = 250$, $a_2 = 210$, and $y_2 = 835$, sex and age ratios are estimated to be

$$\hat{R}_1 = \frac{250}{325} = 0.7692 \text{ and,}$$

$$\hat{R}_2 = \frac{835}{210} = 3.9762.$$

Productivity $\left(P_1 = \frac{N_{J2}}{N_{F1}} \right)$ is then estimated using Eq. (4.51) as

$$\hat{P}_1 = \hat{S} \cdot \frac{\hat{R}_2 + \hat{R}_2 \hat{R}_1}{\hat{R}_1}$$

$$= 0.80 \left(\frac{3.9762 + 3.9762 \cdot (0.7692)}{0.7692} \right)$$

$$= 0.80(9.1452) = 7.3162$$

young at the time of the second survey per female at time of breeding. The variance of $\hat{P}_1$ is estimated using Eq. (4.54) as

$$\widehat{\text{Var}}(\hat{P}_1) = \hat{S} \cdot \widehat{\text{Var}}(\hat{P}_2) + \hat{P}_2 \cdot \widehat{\text{Var}}(\hat{S}) - \widehat{\text{Var}}(\hat{P}_2) \cdot \widehat{\text{Var}}(\hat{S})$$

$$= 0.80(0.6875) + 9.1452(0.03) - 0.6875(0.03) = 0.8037$$

or $\widehat{\text{SE}}(\hat{P}_1) = 0.8965$. The value of $\hat{P}_2 = 9.1452(\widehat{\text{SE}}(\hat{P}_2) = 0.8292)$ measures productivity as the ratio of juvenile abundance to adult female abundance when both are measured at the time of the second survey.

Discussion of Utility

Incorporation of an adult survival estimate ($\hat{S}$) will result in less precision in the estimate of productivity than the original Stokes (1954)–Hanson (1963) method (Fig. 4.2). However, accuracy rather than precision should be the first concern. Sample size calculations should incorporate the additional uncertainty associated with estimating adult survival ($\hat{S}$) as expressed in variance formula (4.53).

Biologists must carefully consider the intended use of the productivity measure to identify which estimator, $\hat{P}_1$ or $\hat{P}_2$, is most appropriate. For example, $\hat{P}_1$ would be appropriate when using Leslie matrix-type models. In contrast, the estimator $\hat{P}_2$ would be appropriate to use in some life-history-based models (Section 9.7).

4.3.3 Dale (1952)–Stokes (1954) Method

This change-in-ratio (CIR) method for estimating productivity was first introduced by Dale (1952) and Stokes (1954) and later expanded on by Hanson (1963) and Paulik and Robson (1969). Dale (1952) measured productivity as the number of young per female. In contrast, Stokes (1954), Hanson (1963), and Paulik and Robson (1969) expressed productivity as the number of young per adult in the population. This method of estimating productivity is primarily used on species that are sexually dimorphic and where adults are indistinguishable from juveniles. For example, during a fall aerial survey of waterfowl, an observer may not be able to distinguish between juveniles and adults, yet males and females may be discerned.

The estimator presented is based on a likelihood model for two sighting surveys assuming binomial sampling distributions. The first survey, conducted before the breeding season, provides a ratio of adult females (f_1) to adult males (m_1). The second survey, conducted after the breeding season, is assumed to include the adult population from the first survey plus those juveniles recruited into the population. During the second survey, f_2 females and m_2 males are observed to provide a new estimate of the population sex ratio. An observed shift in the sex ratio following recruitment permits estimation of productivity.

Model Development

We define:

N_M = abundance of adult males during first survey;
N_F = abundance of adult females during first survey;
N_1 = total abundance of animals during first survey = $N_M + N_F$.

Let us assume a sample survey is conducted during the first period during which m_1 males and f_1 females were observed. By conditioning on the total sample size, $x_1 = f_1 + m_1$, the number of males and females in the sample is modeled by the likelihood

$$L(N_M, N_F | x_1, m_1, f_1) = \binom{x_1}{m_1, f_1} \left(\frac{N_M}{N_M + N_F} \right)^{m_1} \left(\frac{N_F}{N_M + N_F} \right)^{f_1}.$$

The above likelihood can be reparameterized using the adult sex ratio $R_{F/M} = \dfrac{N_F}{N_M}$, which yields

$$L(R_{F/M} | x_1, m_1, f_1) = \binom{x_1}{m_1, f_1} \left(\frac{1}{1 + R_{F/M}} \right)^{m_1} \left(\frac{R_{F/M}}{1 + R_{F/M}} \right)^{f_1}. \qquad (4.55)$$

By the time of the second survey, recruitment of juveniles that are indistinguishable from adults has altered the population sex ratio. For the second survey, let:

$x_2 = f_2 + m_2$;
N_{MJ} = abundance of juvenile males at time of the second survey;
N_{FJ} = abundance of juvenile females at time of the second survey;
N_2 = total abundance of animals during the second survey
$\quad = S(N_M + N_F) + N_{MJ} + N_{FJ}$;
S = adult survival between the first and second surveys.

Conditioning on the total sample size for the second survey, $x_2 = f_2 + m_2$, the number of males (m_2) and females (f_2), observed can be modeled by the likelihood,

$$L(S, N_M, N_{MJ}, N_{FJ} | x_2, m_2, f_2) = \binom{x_2}{m_2, f_2} \left(\frac{SN_M + N_{MJ}}{S(N_M + N_F) + N_{MJ} + N_{FJ}} \right)^{m_2}$$
$$\cdot \left(\frac{SN_F + N_{FJ}}{S(N_M + N_F) + N_{MJ} + N_{FJ}} \right)^{f_2}. \qquad (4.56)$$

Next, defining the number of juveniles in terms of the number of adult females (N_F) and productivity (P),

$$N_{MJ} = \frac{1}{2} N_F \cdot P, \text{ and}$$

$$N_{FJ} = \frac{1}{2} N_F \cdot P$$

where P = productivity, the number of juveniles (male and female) at the time of the second survey produced per adult female measured at the time of the first survey,

$$= \frac{(N_{MJ} + N_{FJ})}{N_F}.$$

The likelihood (Eq. 4.56) can then be rewritten as

$$L(S, R_{F/M}, P | x_2, m_2, f_2) = \binom{x_2}{m_2, f_2} \left(\frac{S + \frac{1}{2} R_{F/M} \cdot P}{S(1 + R_{F/M}) + R_{F/M} \cdot P} \right)^{m_2}$$

$$\cdot \left(\frac{S \cdot R_{F/M} + \frac{1}{2} R_{F/M} \cdot P}{S(1 + R_{F/M}) + R_{F/M} \cdot P} \right)^{f_2}.$$

(4.57)

The joint likelihood for the CIR survey is a product of Eqs. (4.55) and (4.57), such that

$$L(R_{F/M}, P, S | x_1, x_2, m_1, m_2, f_1, f_2) = L(R_{F/M}, x_1 | m_1, f_1)$$
$$\cdot L(R_{F/M}, S, P, x_2 | m_2, f_2).$$

(4.58)

The joint likelihood (Eq. 4.58) has two minimum sufficient statistics (m_1, m_2) and three parameters ($R_{F/M}$, S, F), making parameter estimation impossible. One approach to reducing the parameterization is to assume $S = 1$ for adults between the first and second surveys. Thus, the joint likelihood can be written as

$$L = \binom{x_1}{m_1, f_1} \left(\frac{1}{1 + R_{F/M}} \right)^{m_1} \left(\frac{R_{F/M}}{1 + R_{F/M}} \right)^{f_1}$$

$$\cdot \binom{x_2}{m_2, f_2} \left(\frac{1 + \frac{1}{2} R_{F/M} \cdot P}{(1 + R_{F/M}) + R_{F/M} \cdot P} \right)^{m_2} \left(\frac{R_{F/M} + \frac{1}{2} R_{F/M} \cdot P}{(1 + R_{F/M}) + R_{F/M} \cdot P} \right)^{f_2}.$$

(4.59)

The resulting MLEs are

$$\hat{R}_{F/M} = \frac{f_1}{m_1}, \text{ and}$$

$$\hat{P} = \frac{2(\hat{R}_1 - \hat{R}_2)}{\hat{R}_2 \hat{R}_1 - \hat{R}_1},$$

(4.60)

where

$$\hat{R}_1 = \frac{f_1}{m_1},$$

and

$$\hat{R}_2 = \frac{f_2}{m_2}.$$

The estimator can also be expressed in terms of the counts of males and females from the first and second survey, where

$$\hat{P} = \frac{2(f_1 m_2 - f_2 m_1)}{f_1 (f_2 - m_2)}.$$

(4.61)

Using the delta method and the binomial sampling models of Eq. (4.59), the variance for the estimator $\hat{P}$ (Ryding 2002) is

$$\mathrm{Var}(\hat{P}) \doteq \frac{(4+2P)}{R_{F/M}(R_{F/M}-1)^2}$$

$$\cdot \left[\frac{(1+R_{F/M})^2\left(1+\dfrac{P}{2}\right)}{x_1} + \frac{(1+R_{F/M}+R_{F/M}P)^2\left(1+\dfrac{R_{F/M}P}{2}\right)}{x_2} \right]. \tag{4.62}$$

The variance of the estimator is presented in terms of model parameters $R_{F/M}$ and P and can be used for sample size calculations when planning surveys. A more general expression for the variance of $\hat{P}$, written in terms of the estimates $\hat{R}_1$ and $\hat{R}_2$, is

$$\widehat{\mathrm{Var}}\,(\hat{P}) = \frac{4\hat{R}_2^2}{\hat{R}_1^2(\hat{R}_2-1)^2} \cdot \left[\widehat{\mathrm{CV}}(\hat{R}_1)^2 + \widehat{\mathrm{CV}}(\hat{R}_2)^2\left(\frac{1-\hat{R}_1}{\hat{R}_2-1}\right)^2 \right]. \tag{4.63}$$

Equation (4.63) is applicable for both sampling with and without replacement.

Fecundity, the number of juvenile females produced per adult female in the population (Bolen and Robinson 1995), is productivity divided by two, assuming an equal sex ratio among the juveniles. The MLE for fecundity ($\hat{F}$) is

$$\hat{F}_1 = \frac{\hat{P}}{2} = \frac{\hat{R}_1-\hat{R}_2}{\hat{R}_2\hat{R}_1-\hat{R}_1}. \tag{4.64}$$

The variance for the fecundity estimator $\hat{F}$ can be expressed as

$$\mathrm{Var}(\hat{F}) = \mathrm{Var}\left(\frac{\hat{P}}{2}\right) = \frac{\mathrm{Var}(\hat{P})}{4}. \tag{4.65}$$

Model Assumptions

For the productivity estimator ($\hat{P}$) to be valid, a number of assumptions must be fulfilled with regard to sampling and population processes:

1. All individuals have equal and independent probability of detection in the first period.
2. All individuals have equal and independent probability of detection in the second period.
3. Detection probabilities can be different between the first and second periods.
4. There is a $1:1$ sex ratio for juveniles at the time of recruitment into the adult population.
5. Adult survival probability is one (i.e., $S=1$) between the first and second periods.

The assumption of equal and independent observation probability for all individuals in the first and second surveys may not always be met. Differences in individual behavior during the breeding season could result in unequal observation probabilities. For example, females may be more secretive than breeding males (e.g., leks) (Crawford and Bolen 1975). Vegetative cover within a survey area and differential preference for habitat types could influence observation probabilities (Samuel et al. 1987). Although juveniles may look like adults during the second survey, behavioral differences could lead to unequal observation probabilities between age classes (e.g., Fisher et al. 2000). If juveniles are less likely to be observed, the second sample may not adequately reflect the sex ratio during the period of the second survey. In this case, productivity will be underestimated.

In the derivation of Eq. (4.59), a binomial likelihood was used to model a sampling process with replacement. If individuals can be counted at most once, then sampling is conducted without replacement and can be modeled by using a hypergeometric likelihood. The MLE Eqs. (4.60) and (4.61) for productivity derived from a hypergeometric model will be the same as from a binomial model, but the variance will be smaller than Eq. (4.62). Variance estimator (4.63) is applicable for sampling both with and without replacement.

The assumption that adult survival probability is one (i.e., $S = 1$) between the first and second survey is essential for valid use of estimator (4.60); otherwise, productivity will be overestimated by $(1 - S)100\%$. The amount of time that elapses between the first and second surveys may affect how well this assumption holds. The longer the time interval between the two surveys, the more likely that adult survival will be <1. However, the productivity estimator could be modified by incorporating an independent survival estimate (Section 4.3.4).

For most species, the initial sex ratio of juveniles is considered near $1:1$ (Fisher 1930, Hardy 1997). But differential survival from birth to time of the second survey for male and female juveniles could alter the sex ratio, resulting in a biased estimate of productivity. Again, the likelihood model can be modified to account for unequal juvenile sex ratios (Section 4.3.4).

Example 4.6: Estimating the Productivity of Mallards (*Anas platyrhynchos*) using the Dale (1952)–Stokes (1954) Method

A hypothetical survey of mallards conducted before the breeding season yielded counts of $m_1 = 23$ males and $f_2 = 57$ females. During the second survey of the same population, when juveniles are indistinguishable from adults, $m_2 = 37$ males and $f_2 = 68$ females were observed. The estimate of productivity using Eq. (4.60) is

$$\hat{P} = \frac{2(\hat{R}_1 - \hat{R}_2)}{\hat{R}_2\hat{R}_1 - \hat{R}_1}$$

$$= \frac{2\left(\dfrac{57}{23} - \dfrac{68}{37}\right)}{\left(\dfrac{68}{37} \cdot \dfrac{57}{23} - \dfrac{57}{23}\right)} = 0.6169 \text{ young per adult female.}$$

Note $\hat{P}$ estimates the relative number of juveniles produced at the time of the second survey to the adult female abundance assumed constant over the first and second periods. The variance is estimated from Eq. (4.63) as

$$\widehat{\mathrm{Var}}(\hat{P}) = \frac{4\hat{R}_2^2}{\hat{R}_1^2(\hat{R}_2 - 1)^2} \cdot \left[\widehat{\mathrm{CV}}(\hat{R}_1)^2 + \widehat{\mathrm{CV}}(\hat{R}_2)^2\left(\frac{1-\hat{R}_1}{\hat{R}_2 - 1}\right)^2\right]$$

$$= \frac{4\left(\dfrac{68}{37}\right)^2}{\left(\dfrac{57}{23}\right)^2\left(\dfrac{68}{37} - 1\right)^2}\left[\frac{\left(1+\dfrac{57}{23}\right)^2}{\left(\dfrac{57}{23}\right)(57+23)} + \frac{\left(1+\dfrac{68}{37}\right)^2}{\left(\dfrac{68}{37}\right)(68+37)}\left(\frac{1-\dfrac{57}{23}}{\dfrac{68}{37} - 1}\right)^2\right]$$

$$= 0.59835,$$

or equivalently, $\widehat{\mathrm{SE}}(\hat{P}) = 0.7735$. An asymptotic 95% CI is calculated as $0.6169 \pm 1.96(0.7735)$, yielding $\mathrm{CI}(0 \le \hat{P} \le 1.5161) = 0.95$.

Example 4.7: Estimating Productivity Using the Dale (1952)–Stokes (1954) Method for Ring-Necked Pheasant (*Phasianus colchicus*)

Hanson (1963:12) presents a hypothetical study in which only the estimated sex ratios are given. In this example involving ring-necked pheasants, an August field survey indicated a sex ratio of 4 mature females per mature male ($\hat{R}_1 = 4.0$). A second survey was conducted before the hunting season, during the last week of October when juveniles were mature in appearance. The sex ratio during the second survey was 1.6 females per male ($\hat{R}_2 = 1.6$). Using Eq. (4.60), productivity is estimated to be

$$\hat{P} = \frac{2(\hat{R}_1 - \hat{R}_2)}{\hat{R}_2 \hat{R}_1 - \hat{R}_1}$$

$$\hat{P}_1 = \frac{2(4.0 - 1.6)}{(1.6)(4.0) - 4.0} = \frac{4.8}{2.4} = 2.0 \text{ young per female.}$$

Note the ratios alone are insufficient to calculate the variance; the total numbers observed (i.e., x_1 and x_2) are also required (Eq. 4.63).

Precision of the Productivity Estimator

Precision curves are presented (Fig. 4.3) to illustrate how sample sizes during the first and second surveys affect the precision of the productivity estimates. Precision of the productivity estimates will be defined by the expression

$$P\left(\left|\frac{\hat{P} - P}{P}\right| < \varepsilon\right) = 1 - \alpha$$

where the relative error in estimation $\left|\dfrac{\hat{P} - P}{P}\right|$ is desired to be less than ε, $(1 - \alpha) \cdot 100\%$ of the time. Assuming $\hat{P}$ is normally distributed,

$$\varepsilon = Z_{1-\frac{\alpha}{2}} \widehat{\mathrm{CV}}(\hat{P})$$

where CV is expressed as a decimal. Thus, you can expect to be within $\varepsilon \cdot 100\%$ of the true value of P, $(1 - \alpha)100\%$ of the time. For instance, you can expect to be within $\approx \pm 1.645 \, \mathrm{CV}(\hat{P})$ 100% of the true value of P, 90% of the time.

The expected percentage error in estimation (ε) about P was plotted against sample sizes ($x_1 = x_2 = x$) for various values of adult sex ratio during the first survey period (i.e., $R_{F/M}$) and for productivity values of $P = 2, 5, 10$ juveniles per female at $\alpha = 0.10$ (Fig. 4.3). For convenience, equal sample sizes were assumed for the first and second surveys, although this assumption is not necessary. In this case, sample sizes on the x-axis (Fig. 4.3) are for each survey. The plots illustrate that relatively large sample sizes are needed for the Dale (1952)–Stokes (1954) method to provide precise estimates of P.

Also noteworthy is that as $R_{F/M}$ moves further from the value 1, precision increases (i.e., ε decreases). In other words, the larger the difference between $R_{F/M}$ and a value of $R_J = 1$

a. $P = 2$

b. $P = 5$

c. $P = 10$

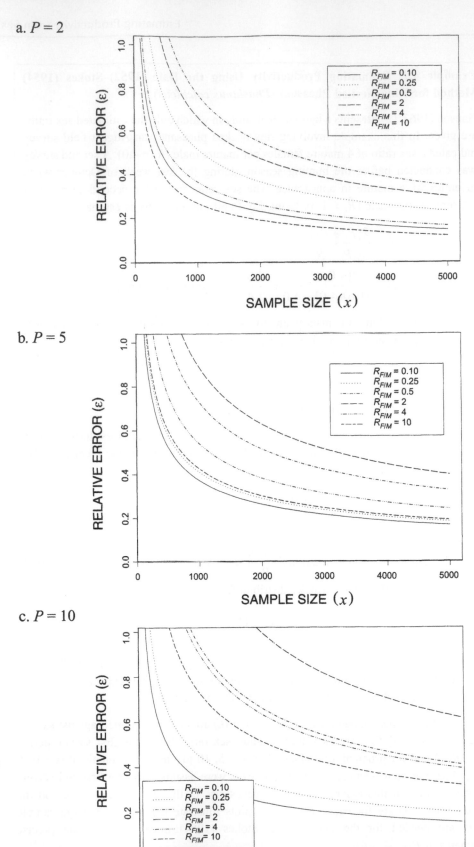

Figure 4.3. Sample size ($x_1 = x_2 = x$) versus sampling precision (i.e., ε) for estimating productivity (P) using the Dale (1952)–Stokes (1954) change-in-ratio method when $\alpha = 0.10$ for various values of R_{FIM} and (a) $P = 2$, (b) $P = 5$, and (c) $P = 10$.

for juveniles entering the population, the more precise the technique. Should $R_{F/M} = 1$, this CIR method for estimating productivity becomes ineffective and inappropriate.

Discussion of Utility

Estimation of productivity using the Dale (1952), Stokes (1954), Hanson (1963), and Paulik and Robson (1969) methods pertains to those populations in which juveniles mature rapidly (subadults <1 year), thus becoming indistinguishable from adults during the fall hunting season. Also, males and females must be discernible in the field, and the sex ratio of the entering juveniles must be 1:1. For the method to provide valid estimates, all animals — males and females, juveniles and adults — must have equal detection probabilities. This method of estimating productivity should be applicable for many gallinaceous birds, waterfowl, and small game mammals. However, for other species such as geese that do not reach adulthood until 2 years of age and that are not sexually dimorphic, this method is inappropriate.

4.3.4   Generalized Dale (1952)–Stokes (1954) Method

In the previous section on CIR methods for estimating productivity (Dale 1952, Stokes 1954, Hanson 1963, Paulik and Robson 1969), it was assumed that $S_A = 1$ and $R_J = 1$. If either assumption is wrong, estimating productivity using Eq. (4.60) will be biased. The productivity estimator (4.60) is not robust to violations of either assumption.

Although researchers generally assume a 1:1 sex ratio within the recruited juvenile population, this assumption may be violated. At least two factors, both related to maternal condition, influence the sex ratio at birth. First, population density has been shown to affect the sex ratio at birth (Kruuk et al. 1999, Mysterud et al. 2000). For example, in red deer (*Cervus elaphus*) populations, more male offspring were produced during periods of abnormally high animal density (Kruuk et al. 1999). Second, environmental conditions can accentuate differential sex ratios. Years following larger than average snowstorms were found to produce greater numbers of male red deer (Kruuk et al. 1999).

Disparate survival between precocial young during the first few months of life has also been reported in numerous bird species. For example, Korschgen et al. (1996) noted that female canvasback (*Aythya valisineria*) ducklings had lower survival rates than their male counterparts in northwest Minnesota. Similarly, Kear (1965) noted greater survival of male mallard ducklings than female ducklings. Juvenile greater sage-grouse (*Centrocercus urophasianus*) males have been noted to suffer higher mortality than females (Swenson 1986). A diversity of hypotheses have been used to explain the differential sex ratio of juveniles, including predation, physiological differences, and food availability, but no conclusive evidence has been presented.

In the original Dale (1952)–Stokes (1954) method, adult survival is assumed to be one (i.e., $S = 1$) between the first and second surveys. Consequently, the abundance of adult males and females is constant, and changes in the sex ratio of the population are entirely the result of juvenile recruitment. An equal survival probability among adult males and females would not alter the adult sex ratio, but the number of females in the population at the time of the second survey (N_{F2}) would be reduced. The result is an over-estimation of productivity. Because the abundance of adult females is changing over time, there is no single measure of productivity. Productivity can be expressed as the ratio of juvenile abundance at the time of the second survey (N_{J2}) to adult female abundance (N_{F1}) at the time of the first survey, i.e.,

$$P_1 = \frac{N_{J2}}{N_{F1}}.$$

Productivity can also be expressed as the ratio of juvenile abundance at the time of the second survey (N_{J2}) to adult female abundance (N_{F2}) at the time of the second survey, i.e.,

$$P_2 = \frac{N_{J2}}{N_{F2}}.$$

As with the original Dale (1952)–Stokes (1954) method, two sample surveys are performed. During the first survey, before recruitment, the number of adult males (m_1) and adult females (f_1) observed are recorded. A postbreeding survey estimates the sex ratio based on the numbers of males (m_2) and females (f_2) observed. In addition to the original two surveys, a third survey to estimate the juvenile sex ratio is incorporated into the statistical model. This modification relaxes the assumption that $R_J = 1$ and allows the juvenile sex ratio to be empirically estimated. The juvenile sex ratio can be estimated from a third independent survey of female (f_3) and male (m_3) juveniles, or parameter estimates obtained from the literature. To relax the assumption that adult $S = 1$, it is assumed an unbiased estimate of adult survival and associated variance $\left(\widehat{\text{Var}}(\hat{S})\right)$ are available.

Model Development

The estimator presented accounts for differential juvenile sex ratios and adult mortality by modifying the likelihood model for the original Dale (1952)–Stokes (1954) method. The likelihood from the first survey of breeding adults (Eq. 4.55) remains unchanged. The likelihood for the second survey is modified to relax the assumptions. Let the ratio of juvenile males to juvenile females be expressed as $R_J = \frac{N_{FJ}}{N_{MJ}}$. The fractions of juvenile females and males in terms of the sex ratio R_J would then be $\frac{R_J}{1+R_J}$ and $\frac{1}{1+R_J}$, respectively. The number of juvenile males and females in the population can then be expressed in terms of productivity (P) and the number of adult females (N_F) as

$$N_{FJ} = P \cdot N_F \cdot \left(\frac{R_J}{1+R_J}\right), \text{ and}$$

$$N_{MJ} = P \cdot N_F \cdot \left(\frac{1}{1+R_J}\right).$$

Substituting these expressions into Eq. (4.56) leads to the generalized joint likelihood:

$$L = \binom{x_1}{m_1, f_1}\left(\frac{1}{1+R_{F/M}}\right)^{m_1}\left(\frac{R_{F/M}}{1+R_{F/M}}\right)^{f_1} \cdot \binom{x_2}{m_2, f_2}$$

$$\cdot \left(\frac{S + \dfrac{R_{F/M}P}{1+R_J}}{S(1+R_{F/M}) + \dfrac{R_{F/M}P}{1+R_J} + \dfrac{R_{F/M}PR_J}{1+R_J}}\right)^{m_2}\left(\frac{SR_{F/M} + \dfrac{R_{F/M}PR_J}{1+R_J}}{S(1+R_{F/M}) + \dfrac{R_{F/M}P}{1+R_J} + \dfrac{R_{F/M}PR_J}{1+R_J}}\right)^{f_2}.$$

$$(4.66)$$

The MLE for $P_1 = \dfrac{N_{J2}}{N_{F1}}$ is

$$\hat{P}_1 = \hat{S} \cdot \frac{(\hat{R}_1 - \hat{R}_2)(1+\hat{R}_J)}{\hat{R}_1(\hat{R}_2 - \hat{R}_J)} = \hat{S} \cdot \hat{P}_2, \qquad (4.67)$$

where

$$\hat{R}_1 = \frac{f_1}{m_1},$$

$$\hat{R}_2 = \frac{f_2}{m_2},$$

$$\hat{R}_J = \frac{f_3}{m_3} = \text{estimate of the juvenile sex ratio.}$$

The estimator

$$\hat{P}_2 = \frac{(\hat{R}_1 - \hat{R}_2)(1+\hat{R}_J)}{\hat{R}_1(\hat{R}_2 - \hat{R}_J)} \qquad (4.68)$$

measures productivity as the ratio of juvenile abundance at the time of the second survey (N_{J2}) to adult female abundance at the time of the second survey (N_{F2}) $\left(\text{i.e., } P_2 = \dfrac{N_{J2}}{N_{F2}}\right)$.
The variance of $\hat{P}_1$ can be approximated by the delta method, where

$$\widehat{\text{Var}}(\hat{P}_1) = \hat{P}_1^2\left[\widehat{\text{CV}}(\hat{S})^2 + \widehat{\text{CV}}(\hat{R}_1)^2\left(\frac{\hat{R}_2}{\hat{R}_1 - \hat{R}_2}\right)^2 + \widehat{\text{CV}}(\hat{R}_2)^2\left(\frac{\hat{R}_2(\hat{R}_1 + \hat{R}_J - 2\hat{R}_2)}{(\hat{R}_2 - \hat{R}_J)(\hat{R}_1 - \hat{R}_2)}\right)^2\right.$$

$$\left. + \widehat{\text{CV}}(\hat{R}_J)^2\left(\frac{\hat{R}_J(\hat{R}_2 + 1)}{(\hat{R}_2 - \hat{R}_J)(1+\hat{R}_J)}\right)^2\right].$$

$$(4.69)$$

Similarly, the variance of $\hat{P}_2$ can be expressed as

$$\widehat{\text{Var}}(\hat{P}_2) = \hat{P}_2^2\left[\widehat{\text{CV}}(\hat{R}_1)^2\left(\frac{\hat{R}_2}{\hat{R}_1 - \hat{R}_2}\right)^2 + \widehat{\text{CV}}(\hat{R}_2)^2\left(\frac{\hat{R}_2(\hat{R}_1 + \hat{R}_J - 2\hat{R}_2)}{(\hat{R}_2 - \hat{R}_J)(\hat{R}_1 - \hat{R}_2)}\right)^2\right.$$

$$\left. + \widehat{\text{CV}}(\hat{R}_J)^2\left(\frac{\hat{R}_J(\hat{R}_2 + 1)}{(\hat{R}_2 - \hat{R}_J)(1+\hat{R}_J)}\right)^2\right].$$

$$(4.70)$$

Assumptions

Six assumptions must hold for the productivity estimator ($\hat{P}_1$) to be valid:

1. All individuals have equal and independent probability of detection in the first period.
2. All individuals have equal and independent probability of detection in the second period.
3. Detection probabilities may differ between periods.
4. The survey samples are independent between periods.
5. An unbiased estimate of the juvenile sex ratio ($\hat{R}_J$) and associated variance $\left(\text{i.e., } \widehat{\text{Var}}\left(\hat{R}_J\right)\right)$ are available.
6. An unbiased estimate of adult survival ($\hat{S}$) and associated variance $\left(\text{i.e., } \widehat{\text{Var}}\left(\hat{S}\right)\right)$ are available.

Example 4.8: Estimating Productivity for a Waterfowl Population, Adjusting for the Juvenile Sex Ratio

Based on a hypothetical example of a duck population (Hanson 1963), the first survey of adults, conducted during early August, yielded counts of $m_1 = 610$ males and $f_1 = 405$ females. In a second survey conducted in late September after juveniles had matured and were indistinguishable from adults, $m_2 = 1951$ males and $f_2 = 1622$ females were observed. A previous study of hunters' bags consistently indicated that 53% of the juveniles were male and 47% were female, in which case

$$\hat{R}_J = \frac{0.47}{0.53} = 0.8868.$$

However, no sample size or variance estimate is provided. Assuming adult survival is one, then the general estimators for $\hat{P}_1 = \hat{P}_2$ are Eqs. (4.67) and (4.68), where

$$\hat{P} = \frac{\left(\hat{R}_1 - \hat{R}_2\right)\left(1 + \hat{R}_J\right)}{\hat{R}_1\left(\hat{R}_2 - \hat{R}_J\right)}$$

$$= \frac{\left(\dfrac{405}{610} - \dfrac{1622}{1951}\right)\left(1 + \dfrac{0.47}{0.53}\right)}{\dfrac{405}{610}\left(\dfrac{1622}{1951} - \dfrac{0.47}{0.53}\right)} = 8.585 \text{ juveniles per adult female.}$$

Assuming sampling with replacement, the variance of $\hat{R}_1$ is estimated by

$$\widehat{\text{Var}}\left(\hat{R}_1\right) = \frac{\hat{R}_1\left(1 + \hat{R}_1\right)^2}{x_2}$$

$$= \frac{\dfrac{405}{610}\left(1 + \dfrac{405}{610}\right)^2}{1015} = 0.001811.$$

Similarly, $\widehat{\text{Var}}\left(\hat{R}_2\right)$ is estimated to be 0.000780. In the case in which $S = 1$, then Var(S) = 0 and the appropriate variance estimator is Eq. (4.70). No sample sizes were given for the estimate of the juvenile sex ratio. For purposes of illustration only, it is assumed R_J is estimated with a CV = 0.05. The variance (4.70) is calculated as

$$\widehat{\text{Var}}(\hat{P}_2) = \hat{P}_2^2 \left[\widehat{\text{CV}}(\hat{R}_1)^2 \left(\frac{\hat{R}_2}{\hat{R}_1 - \hat{R}_2} \right)^2 + \widehat{\text{CV}}(\hat{R}_2)^2 \left(\frac{\hat{R}_2(\hat{R}_1 + \hat{R}_J - 2\hat{R}_2)}{(\hat{R}_2 - \hat{R}_J)(\hat{R}_1 - \hat{R}_2)} \right)^2 \right.$$

$$\left. + \widehat{\text{CV}}(\hat{R}_J)^2 \left(\frac{\hat{R}_J(\hat{R}_2 + 1)}{(\hat{R}_2 - \hat{R}_J)(1 + \hat{R}_J)} \right)^2 \right]$$

$$= (8.585)^2 \left[\frac{0.001811}{(0.66393)^2} \left(\frac{0.83137}{0.66393 - 0.83137} \right)^2 \right.$$

$$+ \frac{0.000780}{(0.83137)^2} \left(\frac{0.83137(0.66393 + 0.8868 - 2(0.83137))}{(0.83137 - 0.8868)(0.66393 - 0.83137)} \right)^2$$

$$\left. + (0.05)^2 \left(\frac{0.8868(0.83137 + 1)}{(0.83137 - 0.8868)(1 + 0.8868)} \right)^2 \right]$$

$$\widehat{\text{Var}}(\hat{P}_2) = 60.2686,$$

or a standard error of $\widehat{\text{SE}}(\hat{P}) = 7.7633$. If R_J was known without error, the standard error of $\hat{P}$ would decline to a value of $\widehat{\text{SE}}(\hat{P}) = 3.9797$. Thus, even small uncertainty (i.e., CV = 0.05) in the value of R_J can have a large effect on the precision of the productivity estimate.

4.4   Summary

This chapter began with productivity estimators based on the analysis of juvenile and adult count data (Sections 4.2.1, 4.2.2). These direct counts are appropriate for species in which juveniles are readily differentiated from adults at time of the survey. These direct counts also rely on the ability to observe relatively large numbers of individuals with equal probabilities of detection (Fig. 4.4).

Productivity estimates can also be derived from litter or brood counts (Sections 4.2.1, 4.2.2). However, these direct observations may need to be adjusted for nesting success (Sections 4.2.3, 4.2.5) and renesting attempts (Section 4.2.4). As in the case of all direct counting methods, investigators need to consider when the survey is conducted and when productivity is to be inferred during the annual cycle. If there is a time difference between the occurrence of the count surveys and the date of inference, additional adjustments for juvenile survival during the interim may be necessary.

Change-in-ratio methods (Section 4.3) provide a convenient means to estimate productivity (Fig. 4.4). However, precision of both the Stokes (1954)–Hanson (1963) and the Dale (1952)–Stokes (1954) methods relies on appreciable shifts in ratio estimators between survey events. The Dale (1952)–Stokes (1954) method is inappropriate for species in which adults naturally have a sex ratio near 1:1. Sample size charts (Figs. 4.2, 4.3) should be considered, or sample size calculations performed before any serious attempts are made to use the ratio methods. In the case of open populations in which adult survival changes over the course of the study, attention to the nature of the productivity parameter being estimated is important. Auxiliary studies to estimate juvenile sex ratios or adult survival may be necessary to insure valid estimates of productivity (Sections 4.3.2, 4.3.4).

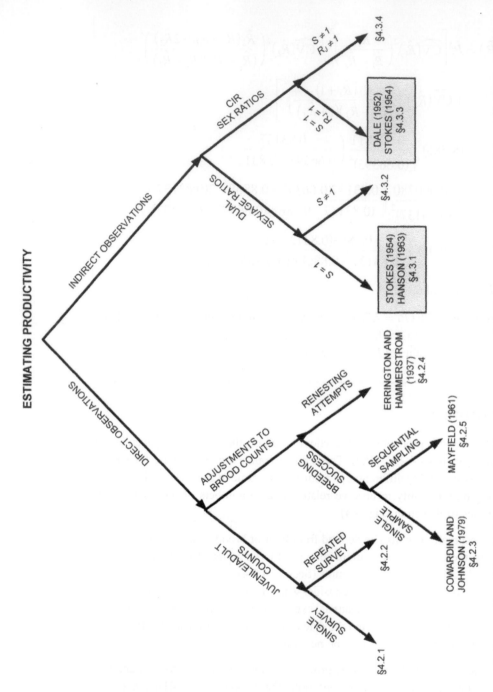

Figure 4.4. The relationship between productivity estimators presented in Chapter 4, with sample size calculations indicated by shading.

Throughout this chapter, there has been the implicit assumption that the age or sex of the observed animals is known without error. However, during data collection and observations, there may be misclassification or nonidentifiability of animals to age or sex category. These problems can greatly complicate data analysis and, in some instances, render a survey useless. Should misclassification be a random process, point estimates will not be biased, although variance calculations will underestimate the true uncertainty of the sampling process. Systematic misclassification of animals will result in both biased point estimates and variance calculations.

The inability to properly classify individual animals to sex or age category can also impact study performance. In the case in which the probabilities of nonclassification are the same for all categories, point estimates remain unbiased. However, effective sample size is reduced, and sampling precision is degraded. A much more troublesome case exists if the probabilities of nonclassification differ between the sex or age categories. In this situation, sex or age ratios will be biasedly estimated. Furthermore, the information needed to provide bias correction, the probabilities of nonclassification, will not be available by using just the data collected from standard observational surveys. Instead, field trials with known sex or age animals would need to be conducted to rigorously estimate the category-specific probabilities of nonclassification.

Estimating Survival

5

Chapter Outline

5.1 Introduction

Productivity is viewed as the mechanism by which individuals are added to a population, mortality is its complement, the mechanism by which animals are removed from a population. In the absence of immigration and emigration, births and deaths need to be equal for population abundance to remain constant. Mortality expresses the likelihood or probability that an animal will die within some defined time interval. Survival, on the other hand, is the complement to mortality and expresses the probability an animal will live through that same time interval.

Survival is a continuous process measured from one time event to any other critical time event of interest. Commonly, survival is measured from one year to the next (i.e., annual survival). However, the most informative frame of reference is survival from birth to death for all individuals within a cohort. This trend in survival over time is known as a survivorship curve or survivorship function (Ferguson 2002, Stolen and Barlow 2003). Theoretically, all information regarding the nature of the survival process is captured by the survival curve. Unfortunately, in wildlife studies, the ability to follow the fate of an entire cohort from birth to its demise is rare. Instead, snapshots of the survival process from time t to time $t + 1$ or from age x to age $x + 1$ are the only data available. With enough of these brief glimpses, the survival function can be reconstructed.

A wide range of methods and data structures will be used for estimating survival probabilities in this chapter, and emphasis is placed on demographic methods that use age- and sex-structure data. The chapter begins with age-specific radiotelemetry data to illustrate the basic nature of right- and left-censored data and estimation of survivorship curves. From there, life-table analysis is used to reconstruct survivorship curves from age-structure information. Three types of life tables are presented: cohort (i.e., horizontal), time specific (i.e., vertical), and a third type we call "depositional," along with associated error variances. Catch curves (Chapman and Robson 1960, Robson and Chapman 1961) will be shown to be special cases of vertical life tables. A variety of other techniques commonly used by wildlife biologists (e.g., Hayne and Eberhardt 1952, Burgoyne 1981) for estimating annual survival rates from age-structure data will also be formally derived and shown to be special cases of the catch-curve analysis. Last, regression techniques and change-in-ratio (CIR) procedures to estimate survival will be considered.

5.2 Basic Concepts and Notation

5.2.1 Basic Concepts

Because survivorship is continuous and operates at the level of the individual, it is most appropriately characterized by the individual lifetimes of the animals. Let X denote the random variable for the lifetime of an animal; x, the observed value. In survival analyses, this variable is also called age-at-death, age-at-failure, or failure time. The cumulative distribution function (cdf) for this random variable can be denoted as

$$F_X(x) = P(X \le x). \tag{5.1}$$

The cdf expresses the probability that death occurred on or before age x and is known as the lifetime distribution or the failure-time distribution (Elandt-Johnson and Johnson 1980:50). The complement to $F_X(x)$ is the survival function denoted as

$$S_X(x) = 1 - F_X(x) = P(X > x). \tag{5.2}$$

The survival function $S_X(x)$ is the probability an animal is alive at age x.

Age-specific survival is a conditional probability of surviving to age $x + 1$ given that the animal is alive at age x. This conditional survival probability can be expressed as a quotient of the survival function, where

$$P(X > x+1|x) = \frac{S_X(x+1)}{S_X(x)} = S_x. \tag{5.3}$$

These age-specific survivals (S_x) can, in turn, be used to reconstruct the survivorship function over discrete time steps. For example, the probability of surviving from age $x - 1$ to age $x + 2$ can be expressed as

$$P(X > x+2|x-1) = \frac{S_X(x+2)}{S_X(x-1)}$$
$$= \frac{S_X(x)}{S_X(x-1)} \cdot \frac{S_X(x+1)}{S_X(x)} \cdot \frac{S_X(x+2)}{S_X(x+1)}$$
$$S_{x-1,x+2} = S_{x-1} \cdot S_x \cdot S_{x+1}. \tag{5.4}$$

Thus, the probability of surviving from age $x - 1$ to age $x + 2$ is a product of age-specific survival probabilities S_{x-1}, S_x, and S_{x+1}. Equation (5.4) illustrates a general principle when expressing survival and mortality processes. It is typically easier to work with conditional survival probabilities than with conditional mortality probabilities. For example, the conditional mortality probability from age $x - 1$ to age $x + 2$ can be expressed as

$$M_{x-1,x+2} = 1 - S_{x-1,x+2} = 1 - S_{x-1} \cdot S_x \cdot S_{x+1}, \tag{5.5}$$

or

$$M_{x-1,x+2} = 1 - (1 - M_{x-1})(1 - M_x)(1 - M_{x+1})$$
$$= M_{x-1} + M_x + M_{x+1} - M_{x-1}M_x - M_{x-1}M_{x+1} - M_xM_{x+1} + M_{x-1}M_xM_{x+1}. \tag{5.6}$$

Comparison of Eqs. (5.5) and (5.6) illustrates why the mathematics of characterizing mortality focuses on the analysis of survival.

Conditional or age-specific survival can also be interpreted as the proportion of the cohort alive at age x that is expected to be still alive at age $x + 1$, where

$$S_x = E\left[\frac{N_{x+1}}{N_x}\right], \tag{5.7}$$

and where

N_x = abundance of cohort at age x,
N_{x+1} = abundance of cohort at age $x + 1$.

Equation (5.7) illustrates the potential connection between abundance and survival estimation. The abundance of a cohort in consecutive years provides a basis to estimate age-specific survival rates. What Eq. (5.7) ignores and a survivorship curve considers is the intervening pattern of mortality events between ages x and $x + 1$. The seasonal pattern can reveal much about the nature of survival factors affecting the fates of animals within a year.

Instantaneous mortality rates convert the discrete time conditional survival probabilities to continuous time responses. The mortality rate from time t to $t + 1$, the complement to the survival rate, can be expressed as

$$M_t = 1 - \frac{N_{t+1}}{N_t} = 1 - S_t.$$

If mortality is considered a Poisson process in time, then

$$\frac{N_{t+1}}{N_t} = S_t = e^{-\mu t} \tag{5.8}$$

where μ is defined as the instantaneous mortality rate. The Poisson process assumes that the expected number of deaths is constant over time and that death events are independent from one instant to another. Solving for μ in Eq. (5.8) yields

$$\mu = \frac{\ln S_t}{-t}$$

or, equivalently,

$$\mu = \frac{\ln N_{t+1} - \ln N_t}{-t}. \tag{5.9}$$

The instantaneous mortality rate is an explicit function of the units of time for t and can be expressed in terms of days, months, or years. The survival probability S_t ignores the pattern of survival over time interval t, whereas the instantaneous mortality rate assumes the survival process is constant over time t.

Example 5.1: Instantaneous Mortality Rate

Consider an annual survival rate of $S = 0.76$. The instantaneous mortality rate is computed as

$$\mu = \frac{\ln 0.76}{-1} = 0.27444$$

on an annual basis. On a monthly basis, the instantaneous mortality rate is computed as

$$\mu = \frac{\ln 0.76}{-12} = 0.02287,$$

and, on a daily basis,

$$\mu = \frac{\ln 0.76}{-365} = 0.00075.$$

5.2.2   Notation

The following parameters, observed random variables, and their associated definitions will be used throughout this chapter.

x = age of an individual in either days, weeks, or months, but most typically in years;
l_x = number of individuals alive at time x;
d_x = number of individuals that died in the interval x, $x + 1$;
t_i = death time for the ith individual;
c_i = time of censoring for the ith individual;
w = oldest age class in the sample;
S_x = probability of surviving from age x to $x + 1$;
$M_x = 1 - S_x$ = probability of dying between age x and $x + 1$;
N = animal abundance.

Other variables and terms will be defined as necessary for specific methods.

5.3   Survival Curve Analysis

The purpose of survival curve analysis is to characterize the amplitude and shape of the survival function $S_X(x)$ over the lifetime of a cohort of animals. Both parametric and nonparametric survival curve analyses are possible. Either approach requires information on age of the animals at time of entry to the study and their subsequent age at death. Parametric approaches assume a specific family of curves that can be used to model the survivorship data. The nonparametric methods of Kaplan-Meier (or product-limit estimator) (Kaplan and Meier 1958) and Nelson (1972)–Aalen (1978) assume no functional form of the survivorship curve, instead allowing the raw observations to dictate its shape. A nonparametric approach may be preferred in wildlife investigations in which seasonal factors can dramatically affect survival rates over time. Extrinsic factors of weather and climate will affect survival rates both within and between years. Seasonal harvest patterns of game species also make the assumption of a smooth survivorship curve unrealistic in some instances. Alternatively, a well-fitted parametric survivorship curve can help characterize the general form of the mortality process over time and can be used to efficiently extract information on conditional and unconditional survival probabilities, life expectancy, and relative risks.

5.3.1   Kaplan-Meier (1958) or Product-Limit Estimator

Pollock et al. (1989) were among the first to apply this epidemiological method to the analysis of wildlife data. They used the Kaplan-Meier (1958) method to characterize the survivorship curve of northern bobwhite (*Colinus virginianus*). This nonparametric approach, also called the product-limit estimator (Parmar and Marchin 1995:26–30, Klein and Moeshberger 1997:83–124), monitors a cohort through time, and at each death, the time and number of animals remaining in the study are recorded. The entire animal cohort can begin the study at the same time, or the cohort can be comprised of individuals entering the study at different times (i.e., staggered entry) (Pollock et al. 1989).

The Kaplan-Meier analysis is a discrete time approach to survival estimation. At time zero (t_0), survivorship $S_X(0) = 1.0$ (i.e., 100%). Each declining step in the survivorship curve represents the time of death for an individual. The survival process continues until the time when the last animal (L) in the cohort dies (i.e., $S_x(L) = 0$).

The Kaplan-Meier method can be approached from two different perspectives: the timeline can be discretized into convenient regular time intervals, or the timeline can be subdivided according to the actual death events. The use of regular subdivisions of time is an approximation of the actual survivorship curve. In any interval from time t to $t + 1$, the conditional survival probability can be estimated as

$$\hat{P}(X > t+1 | t) = \hat{S}_t = \frac{l_t - d_t}{l_t}, \tag{5.10}$$

where

l_t = number of animals alive and at risk of death at time t to $t + 1$;
d_t = number of animals that died in the time interval t to $t + 1$.

These binomial proportions have the estimated variance of

$$\widehat{\text{Var}}\left(\hat{S}_t\right) = \frac{\hat{S}_t(1 - \hat{S}_t)}{l_t}. \tag{5.11}$$

The product-limit estimate acquired its name because estimating the survival probability from time 0 to t is simply a matter of multiplication, where

$$\begin{aligned} \hat{S}_X(t) &= \hat{S}_{0,t} \\ &= \hat{S}_0 \cdot \hat{S}_1 \cdot \hat{S}_2 \cdots \hat{S}_{t-1}. \end{aligned} \tag{5.12}$$

Specifically, the estimator of $S_x(t)$ can be written as

$$\hat{S}_X(t) = \hat{S}_{0,t} = \prod_{i=0}^{t-1}\left(\frac{l_i - d_i}{l_i}\right) = \prod_{i=0}^{t-1}\hat{S}_i. \tag{5.13}$$

The asymptotic variance of $\hat{S}_X(t)$ provided by Greenwood (1926) can be represented as

$$\widehat{\text{Var}}\left(S_X(t)\right) = S_X(t)^2 \sum_{i=0}^{t-1}\frac{d_i}{l_i(l_i - d_i)} \tag{5.14}$$

for some specified time t.

The estimators (5.10) and (5.13) are applicable regardless of whether censoring has occurred or not. However, the simple form of the estimators and associated variance errors make them particularly attractive in the presence of staggered entry (i.e., left-censored) and right-censored data. The biggest limitation of the approach is that precision of the survival estimates $\hat{S}_t$ depends entirely on the number of animals at risk within a sampling period (i.e., l_t).

Assumptions

There are at least six assumptions associated with the Kaplan-Meier analysis:

1. The time of death for each individual is known and correctly recorded.
2. The cohort is representative of the population at time of inference.

3. Fates of individual animals are independent.
4. If staggered-entry data are used, animal ages at time of entry into the study are known.
5. If right-censored data are used, animals are known to be alive at the time of censoring and that time is known.
6. If making inferences to other cohorts, the probabilities of survival depend on age but not time or location.

The most common use of the Kaplan-Meier method in wildlife biology has been the analysis of radiotelemetry data. Nonparametric survivorship curves using radiotelemetry data have been constructed for Texas tortoise (*Gopherus berlandieri*) (Hellgren et al. 2000), ring-necked pheasants (*Phasianus colchicus*) (Warner et al. 2000), little penguins (*Eudyptula minor*) (Renner and Davis 2001), wild turkeys (*Meleagris gallopavo*) (Thogmartin and Johnson 1999), and juvenile guanaco (*Lama guanicoe*) (Sarno et al. 1999).

Example 5.2: Product-Limit Estimator for Grizzly Bears (*Ursus arctos*), Northern Idaho

Idaho Department of Fish and Game (W. L. Wakkinen, Idaho Department of Fish and Game, personal communication) conducted a staggered-entry study of grizzly bear survival in the panhandle of Northern Idaho. Twenty-eight bears were radio-marked between the ages of 1.5 and 28.5 years during 1983–1993. Bears were tracked from 1 to 7 years. At the end of the tracking session, fates of individual bears were recorded as alive at time of radio failure (i.e., right-censored), dead because of natural causes, or dead because of human-induced mortality (e.g., poaching) (Fig. 5.1).

Radio tracking provided only a portion of the roughly 30-year lifespan of the bears, and the data are both left-truncated and right-censored. Only two bears entered the study at its onset, i.e., age 1.5 years. All other bears had survived 1 or more years before inclusion in the study. Ignoring the staggered-entry nature of the data would bias the subsequent survival analysis. For instance, a bear entering the study at age 10 years actually represents many individuals not reaching that age. Failing to monitor individuals from the onset (i.e., age 1.5 years) of the study is known as left-truncated or staggered entry. In addition, because of radio-collar failure (e.g., battery power loss), there was right censoring. Thus, bears exited the study before their time of death could be recorded.

The Kaplan-Meier method easily copes with both left truncation and right censoring. The survival between ages x and $x + 1$ is based solely on those animals known to be alive and at risk at age x. For example, consider the estimated survival probability from age 10.5 to 11.5 years. A total of $l_{10.5} = 9$ animals was available for observation throughout the age class, of which $d_{10.5} = 1$ animal died, for a survival estimate of $S_{10.5} = \dfrac{9-1}{9} = 0.8889$. The estimated variance associated with this survival estimate is

$$\widehat{\mathrm{Var}}\left(\hat{S}_{10.5}\right) = \frac{\left(\dfrac{8}{9}\right)\left(\dfrac{1}{9}\right)}{9} = 0.0110 \text{ or } \widehat{\mathrm{SE}}\left(\hat{S}_{10.5}\right) = 0.1048.$$

Note that sample sizes are small for any particular age class (i.e., $0 \leq l \leq 9$) and the estimates of age-specific survival probabilities vary considerably (Table 5.1). The product-limit estimate is impeded by the small sample sizes and, in particular, age classes with sample sizes of $l_x = 0$. Survival probabilities for the missing age classes had to be presumed to construct the survivorship curve. The curve was constructed beginning with age 2.5 because the two 1.5-year-olds in the study died in their first year. The survivorship curve (Fig. 5.2) is constructed by progressively multiplying consecutive age-specific survival estimates together. For example, the estimated survival from age 2.5 to 5.5 years is calculated (Eq. 5.13) as

$$\hat{S}_{2.5,5.5} = \left(\frac{7-0}{7}\right) \cdot \left(\frac{7-2}{7}\right) \cdot \left(\frac{7-1}{7}\right) = 1(0.7142)(0.8571) = 0.6122.$$

By use of the Greenwood (1926) variance estimator (Eq. 5.14),

$$\widehat{\mathrm{Var}}\left(\hat{S}_X(t)\right) = \hat{S}_X(t)^2 \sum_{i=0}^{t-1} \frac{d_i}{l_i(l-d_i)}$$

$$\widehat{\mathrm{Var}}\left(\hat{S}_{2.5,5.5}\right) = (0.6122)^2 \left[\frac{0}{7(7-0)} + \frac{2}{7(7-2)} + \frac{1}{7(7-1)}\right] = 0.0303$$

with a standard error of $\widehat{\mathrm{SE}}(\hat{S}_{2.5,5.5}) = 0.1742$.

5.3.2 Nelson (1972)–Aalen (1978) Estimator

The Kaplan-Meier (1958) method for estimating the survivorship curve can behave poorly with small sample sizes. The product-limit estimator could not be used starting with age class 1.5 in the grizzly bear example because of too many initial mortalities. In the case of the grizzly bear age class 1.5, the two observed animals died in the initial interval. The product-limit estimate of the next time period would then be

$$S_{1.5,3.5} = S_{2.5} \cdot S_{1.5}$$
$$= S_{2.5} \cdot \left(\frac{2-2}{2}\right) = 0.$$

Table 5.1. Age-specific survival data used in the Kaplan-Meier analysis of northern Idaho grizzly bears, 1983–1993.

Age	1.5	2.5	3.5	4.5	5.5	6.5	7.5	8.5	9.5	10.5	11.5	12.5	13.5	14.5
At risk, l_i	2	7	7	7	7	6	6	9	8	9	5	2	3	2
Deaths, d_i	2	0	2	1	1	0	1	0	0	1	0	0	0	0
S_i	0	1	0.7142	0.8571	0.8571	1	0.8333	1	1	0.8889	1	1	1	1
ΠS_i		1	0.7142	0.6121	0.5247	0.5247	0.4372	0.4372	0.4372	0.3886	0.3886	0.3886	0.3886	0.3886

Age	15.5	16.5	17.5	18.5	19.5	20.5	21.5	22.5	23.5	24.5	25.5	26.5	27.5	28.5
At risk, l_i	3	1	N/A	1	1	2	2	2	N/A	N/A	2	2	1	1
Deaths, d_i	1	0	N/A	0	0	0	0	1	N/A	N/A	0	1	0	0
S_i	0.6667	1	N/A 1[a]	1	1	1	1	0.5	N/A 0.75[b]	N/A 0.75[b]	1	0.5	1	1
ΠS_i	0.2591	0.2591	0.2591	0.2591	0.2591	0.2591	0.2591	0.1295	0.0972	0.0729	0.0729	0.0364	0.0364	0.0364

Age-specific survival estimates (S_i) and cumulative survival from age 2.5 (i.e., ΠS_i) calculated by age class.
[a]Assumed $S_{17.5} = 1$.
[b]Assumed $S_{23.5} = S_{24.5} = 0.75$.
(Data from W. L. Wakkinen, Idaho Department of Fish and Game, personal communication.)

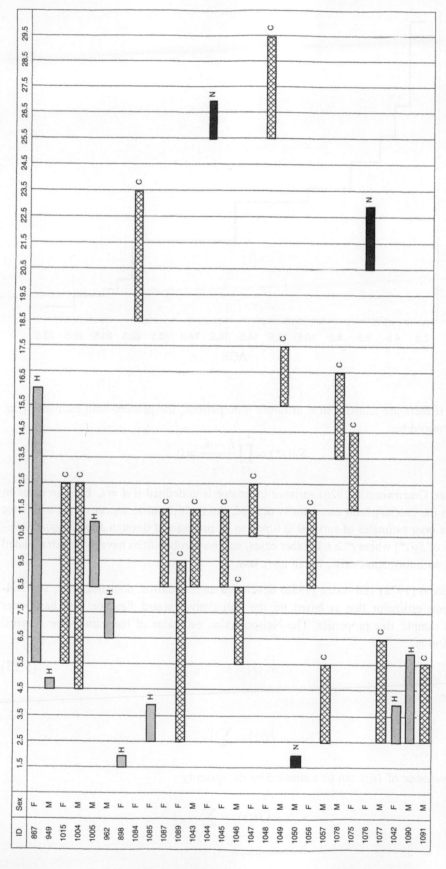

Figure 5.1. Individual tracking records for radio-marked grizzly bears in northern Idaho (1983–1993). C denotes bear was alive when censored by radio failure; H, human-induced mortality; and N, natural mortality. (Data from W. L. Wakkinen, Idaho Department of Fish and Game, personal communication.)

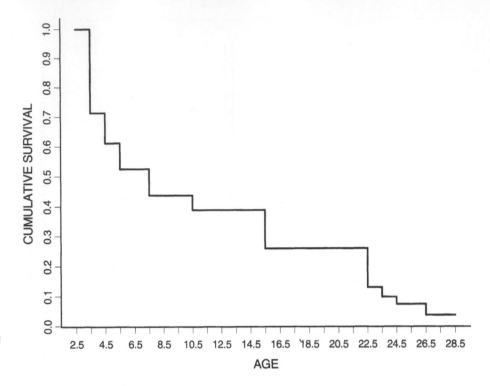

Figure 5.2. Kaplan-Meier survivorship curve for northern Idaho grizzly bears age 2.5 years and older, 1983–1993. (Data from W. L. Wakkinen, Idaho Department of Fish and Game, personal communication.)

Even if there are animals alive in future time periods, the product-limit estimator (Eq. 5.13) would be

$$S_X(t^*) = \prod_{1.5}^{t^*-1} \frac{l_i - d_i}{d_i} = 0.$$

Further, Greenwood's (1926) variance estimator is undefined if $d_i = l_i$. Even in cases in which the product-limit estimator is defined, the multiplicative aspect of the estimator causes poor estimates of survival at time t to be propagated through all subsequent estimates of $S_X(t^*)$ where $t^* \geq t$. In other cases, survival probabilities have to be extrapolated when no information exists for an age class.

Nelson (1972) and Aalen (1978) developed an alternative, nonparametric, survival-function estimator that is based on the cumulative hazard function and has better small sample size properties. The Nelson-Aalen estimator of the cumulative survival function is

$$\hat{S}_X(t) = e^{-\hat{H}(t)}, \tag{5.15}$$

where

$$\hat{H}(t) = \sum_{i=0}^{t} \frac{d_i}{l_i}.$$

The variance of $\hat{H}(t)$ can be estimated by the quantity

$$\widehat{\text{Var}}(\hat{H}(t)) = \sum_{i=0}^{t} \frac{d_i}{l_i^2}.$$

In turn, by use of the delta method, the variance for estimator (5.15) can be approximated by

$$\widehat{\mathrm{Var}}\left(\hat{S}_{X(t)}\right) = \widehat{\mathrm{Var}}\left(\hat{H}(t)\right)\cdot\hat{S}_X(t)^2. \tag{5.16}$$

In the specific case of staggered-entry survival analysis common in wildlife studies (e.g., DelGiudice et al. 2002), additional modifications of the Nelson-Aalen estimator have been suggested by Tsai et al. (1987), Uzunogullari and Wang (1992), and Pan and Chappell (1998).

Example 5.3: Nelson-Aalen Estimator for Grizzly Bears, Northern Idaho

By using the radiotelemetry survival data (Table 5.1) for grizzly bears in northern Idaho, 1983–1993, the Nelson-Aalen estimator of the cumulative survival function is calculated (Table 5.2) with use of Eq. (5.15). In this reanalysis of the grizzly bear data, the survival function was calculated for age classes 1.5 and older. The two bear mortalities between ages 1.5 and 2.5 do not hinder the Nelson-Aalen method (Eq. 5.15) as they did the Kaplan-Meier method (Eq. 5.13). For the first four age classes, the calculations are as follows:

$$\hat{S}_X(t) = e^{-\left(\sum_{i=0}^{t}\frac{d_i}{l_i}\right)}$$

$$\hat{S}_X(1.5) = 1$$

$$\hat{S}_X(2.5) = e^{-\left(\frac{2}{2}\right)} = 0.3679$$

$$\hat{S}_X(3.5) = e^{-\left(\frac{2}{2}+\frac{0}{7}\right)} = 0.3679$$

$$\hat{S}_X(4.5) = e^{-\left(\frac{2}{2}+\frac{0}{7}+\frac{2}{7}\right)} = 0.2765.$$

5.3.3 Nonparametric Test for Comparing Survival Curves

Survival curves estimated by the Kaplan-Meier method from two or more samples can be compared by using nonparametric methods. The need to compare survivorship curves can arise when survival studies are conducted in different locations (Warner et al. 2000),

---

Table 5.2. Age-specific survival data used in the Nelson (1972)–Aalen (1978) estimates of the cumulative survival function for the northern Idaho grizzly bears, 1983–1993.

Age	1.5	2.5	3.5	4.5	5.5	6.5	7.5	8.5	9.5	10.5	11.5	12.5	13.5	14.5
At risk, l_i	2	7	7	7	7	6	6	9	8	9	5	2	3	2
Deaths, d_i	2	0	2	1	1	0	1	0	0	1	0	0	0	0
$H_x(t)$	1	1	1.2857	1.4286	1.5714	1.5714	1.7381	1.7381	1.7381	1.8492	1.8492	1.8492	1.8492	1.8492
$S_x(t)$	0.3679	0.3679	0.2765	0.2397	0.2077	0.2077	0.1759	0.1759	0.1759	0.1574	0.1574	0.1574	0.1574	0.1574

Age	15.5	16.5	17.5	18.5	19.5	20.5	21.5	22.5	23.5	24.5	25.5	26.5	27.5	28.5
At risk, l_i	3	1	N/A	1	1	2	2	2	N/A	N/A	2	2	1	1
Deaths, d_i	1	0	N/A	0	0	0	0	1	N/A	N/A	0	1	0	0
$H_x(t)$	2.1825	2.1825	2.1825	2.1825	2.1825	2.1825	2.1825	2.6825	2.6825	2.6825	2.6825	3.1825	3.1825	3.1825
$S_x(t)$	0.1128	0.1128	0.1128	0.1128	0.1128	0.1128	0.1128	0.0684	0.0684	0.0684	0.0684	0.0415	0.0415	0.0415

(Data from W. L. Wakkinen, Idaho Department of Fish and Game, personal communication.)

in different years (Pollock et al. 1989), or for different age classes (Thogmartin and Johnson 1999). Lee (1992:104–130) discussed five nonparametric methods for comparing survival curves. Two of the methods are generalizations of the two-sample Wilcoxon test (Zar 1984:153–156) and the others are non-Wilcoxon-based tests (Lee 1992:116). The log-rank test is perhaps the most versatile because it handles either censored or noncensored data and multiple samples (Kleinbaum 1996:62–64) and can be modified for a staggered-entry design (Pollock et al. 1989).

The log-rank test is similar to a contingency table analysis conducted within each time interval. For all time intervals, death times t, are ranked in order from the start of the study to its conclusion, and at each time, the number of animals that died within the interval and the number still alive at the end of the interval for each of the i groups are recorded. The expected number of animals dying in the ith group during the tth interval $E(d_{it})$ is also recorded and is calculated by the following:

$$E(d_{it}) = \left(\sum_{i=1}^{n} d_{it} \right) \left(\frac{r_{it}}{\sum_{i=1}^{n} r_{it}} \right),$$

where

d_{it} = number of animals dying in the ith group ($i = 1, 2, 3, \ldots, n$) during the tth interval ($t = 1, 2, 3, \ldots, T$);

r_{it} = number of animals still alive (or at risk) in the ith group during the tth interval.

The log-rank test is based on the difference between the observed and expected number of deaths in each group at each interval. The test statistic is written as

$$\chi_1^2 = \frac{(O_i - E_i)^2}{\text{Var}(O_i - E_i)},$$

where

$O_i = \sum\limits_{t=0}^{T} d_{it}$ = all deaths in the ith group;

$E_i = \sum\limits_{t=0}^{T} E(d_{it})$ = total expected number of deaths in the ith group;

χ_1^2 = chi-squared random variable with 1 degree of freedom.

The group selected for calculating the statistic, i.e., $i = 1$ or $i = 2$, may be chosen arbitrarily, with no effect on the result. The variance of $(O_i - E_i)$ is calculated by

$$\text{Var}(O_i - E_i) = \sum_{t=0}^{T} \frac{d_t(r_t - d_t)r_{1t}r_{2t}}{r_t^2(d_t - 1)}.$$

This result can be generalized for $n > 2$ groups (Kleinbaum 1996:62–64), although the algebra becomes more complex. In addition, there may be a desire to emphasize differences in certain characteristics of the survival function. This may be accomplished through modified test statistics that are based on weighted differences between observed and expected deaths. That is, $O_i - E_i = \sum_{t=0}^{T}(d_{it} - E(d_{it}))$ is replaced by

$$\frac{\sum_{i=1}^{n} w_i(d_{it} - E(d_{it}))}{\sum_{i=1}^{n} w_i}$$

for some weight w_i (Klein and Moeshberger 1997:191–202). For example, the Peto-Wilcoxon test uses weights equal to an estimate of the survival function, emphasizing differences at the beginning of the survival curve that may be important in particular studies (Klein and Moeshberger 1997:199).

5.3.4   Parametric Survival Curve Analysis

A parametric survival curve analysis is less subject to the sensitivities of small sample sizes and stochastic variability observed in the Kaplan-Meier nonparametric analysis. Instead of allowing the data to solely dictate the shape of the survival curve, parametric methods assume a particular family of curves and let the data specify the member. Accuracy of the method depends on the flexibility of the family of curves to conform to the data. Precision relies on the number and distribution of the death time events to reliably estimate model parameters. Parametric survival analysis is widely used in the fields of epidemiology and quality control of manufactured goods (Elandt-Johnson and Johnson [1980] and Parmar and Marchin [1995] provide general references).

Practically any distribution that has the properties

$$F_X(0) = 1 \text{ and } F_X(\infty) = 0,$$

can be used in survival analysis. However, certain distributions have useful and desirable properties; among them are the following:

1. Uniform —

$$S_X(x) = \begin{cases} 1 & \text{if } x \leq \alpha \\ \dfrac{\beta - x}{\beta - \alpha} & \text{if } \alpha < x \leq \beta \\ 0 & \text{if } x > \beta \end{cases}$$

   with associated probability density function (pdf) of

$$f_X(x) = \begin{cases} \dfrac{1}{\beta - \alpha} & \alpha \leq x \leq \beta \\ 0 & \text{elsewhere.} \end{cases}$$

2. Exponential —

$$S_X(x) = e^{-\frac{x}{\alpha}}$$

   with associated pdf of $f_X(x) = \dfrac{1}{\alpha} e^{-\frac{x}{\alpha}}$ for $x > 0$.

3. Two-parameter Weibull —

$$S_X(x) = e^{-\left(\frac{x}{\alpha}\right)^\beta}. \tag{5.17}$$

with associated pdf of

$$f_X(x) = \left[\frac{\beta x^{\beta-1}}{\alpha^\beta}\right] e^{-\left(\frac{x}{\alpha}\right)^\beta} \tag{5.18}$$

and the cdf of

$$F_X(x) = 1 - e^{-\left(\frac{x}{\alpha}\right)^\beta}. \tag{5.19}$$

4. Three-parameter Weibull —

$$S_X(x) = e^{-\left(\frac{x-\gamma}{\alpha}\right)^\beta}$$

with associated pdf of

$$f_X(x) = e^{-\left(\frac{x-\gamma}{\alpha}\right)^\beta} \left(\frac{\beta}{\alpha^\beta}\right)(x-\gamma)^{\beta-1}.$$

5. Logistic —

$$S_X(x) = \frac{1}{1 + e^{(\beta x - \alpha)}}$$

with associated pdf of

$$f_X(x) = \frac{\beta e^{\alpha - \beta x}}{[1 + e^{\alpha - \beta x}]^2}.$$

6. Gamma —

$$S_X(x) = \frac{1}{\Gamma(\alpha)} \int_{\frac{x}{\beta}}^{\infty} t^{\alpha-1} e^{-t}\, dt$$

with associated pdf of $f_X(x) = \dfrac{1}{\beta^\alpha \Gamma(\alpha)} x^{\alpha-1} e^{-\frac{x}{\beta}}.$

7. Gompertz —

$$S_X(x) = e^{\frac{\beta}{\alpha}(1 - e^{\alpha x})}$$

with associated pdf of

$$f_X(x) = \beta e^{\frac{\beta}{\alpha}(1 - e^{\alpha x}) + \alpha x}.$$

Siler (1979) proposed a competing risk model composed of three components corresponding to the early, middle, and later life of an animal. In each phase of life, the hazard rate has a different trajectory. Emlen (1970) suggested that hazards decrease as the animal adjusts to its environment and matures, hazards remain constant in midlife because of those factors to which an animal does not adjust, and hazards increase as the animal enters senescence. Corresponding to the three life phases, Siler (1979) proposed a survival function of the form

$$S_X(x) = l_J(x) \cdot l_C(x) \cdot l_S(x)$$

where the juvenile component has the Gompertz function

$$l_J(x) = e^{-\frac{a_1}{b_1}\left(1 - e^{-b_1 x}\right)},$$

and the adult component with constant hazard has the exponential component

$$l_C(x) = e^{-a_2 x},$$

and the component describing senescence by a Gompertz model has the form

$$l_S(x) = e^{\frac{a_3}{b_3}\left(1 - e^{b_3 x}\right)}.$$

The resultant five-parameter survival function can then be written as

$$S_X(x) = e^{-\frac{a_1}{b_1}\left(1 - e^{-b_1 x}\right) - a_2 x + \frac{a_3}{b_3}\left(1 - e^{b_3 x}\right)}.$$

Siler (1979) applied his competing risk model to a variety of life-table data, including the Dall sheep (*Ovis dalli*) data of Deevey (1947). Eberhardt (1985) further described the application of the Siler (1979) model to the analysis of wildlife data, including grizzly bears in Yellowstone National Park. Barlow (1991) applied the Siler (1979) model to analyze marine mammal survivorship data. Stolen and Barlow (2003) used the Siler (1979) model to fit a survivorship curve to data from bottlenose dolphins (*Tursiops truncatus*). In the analysis of marine mammal data, the model was modified where

$$S_X(x) = e^{-\frac{a_1}{b_1}\left(1 - e^{-b_1 \frac{x}{\Omega}}\right) - a_2 \left(\frac{x}{\Omega}\right) + \frac{a_3}{b_3}\left(1 - e^{b_3 \frac{x}{\Omega}}\right)},$$

and where Ω is the lifespan of the animal, assumed known, such that $\frac{x}{\Omega}$ is the fraction of longevity lived by the animal. Because a closed form of the pdf for the Siler (1979) model does not exist, Barlow (1991) suggested the following approximation:

$$f_X(x) = \frac{S_X(x)}{\displaystyle\sum_{i=0}^{w} S_X(i)}$$

where $w = 1.5\,\Omega$, such that $S_X(1.5\,\Omega)$ is most assuredly zero.

Cumulative, $S_X(x)$, and age-specific survival, S_X, are two ways to quantify survival. The hazard function, $h(x)$, is another way to characterize the survival process over the lifetime of an individual, and is expressed mathematically as

$$h(x) = \frac{f_X(x)}{S_X(x)},$$

where $S_X(x)$ and $f_X(x)$ are given (above) for several of the more common distributions. The hazard function describes the rate at which a death event will occur in the next instant, given survival to time x. Note that mathematically the function $h(x)$ is inversely proportional to cumulative survival. The hazard rate can be viewed as the opposite of either cumulative or age-specific survival in which higher survival probabilities are associated with lower hazard rates. In addition, the hazard function concerns the event death, whereas the cumulative survival and age-specific survival are probabilities of that event *not* occurring. The only restriction on the hazard rate is that it be greater than zero. Often, distributions are chosen to model survival by the shape and nature of the hazard function, which can constantly increase or decrease over the entire lifetime, or increase during a portion of the life span and decrease at other points. In addition, knowing the shape of the survival curve, and the corresponding hazard and age-specific survival curves, can help guide the choice of parametric distribution for modeling the survival process.

We present five different parametric cumulative survivorship curves (Fig. 5.3), each with its associated hazard and age-specific survival curve. Choosing from among the different parametric distributions for modeling survival data is based on the shape of these survival and hazard functions. Consequently, the distribution used to model survival data and the shape of survival curves can be discussed simultaneously.

The uniform distribution has limited applicability; at best, modeling survival over only a short duration or the midlife of long-lived individuals. A constantly decreasing survival function, $S_X(x)$ (Fig. 5.3a), characterizes the resulting survival curve of a uniform distribution. The associated hazard function, $h(x)$, for a uniform distribution shows an increasing hazard as the individual ages. Age-specific survival, S_x, is defined by the proportion of the individuals of a cohort surviving to the next age classes. To maintain a constant cumulative survival, the proportion dying in each age class becomes larger each year, resulting in decreasing age-specific survival probability over time for a uniform distribution.

The exponential distribution is based on the assumption that both the hazard rate and age-specific survival probabilities are constant over time (Fig. 5.3b). Exponential distributions are often termed "memoryless," meaning that no matter how long an individual has lived, the probability of dying within any interval remains constant. Both the hazard rate and age-specific survival curves illustrate this property.

The Weibull distribution (Pinder et al. 1978) is a generalization of the exponential model that allows the hazard rate to change over time (Fig. 5.3c). The survival curve (Fig. 5.3c) was drawn from a Weibull distribution with parameters $\alpha = 5$ and $\beta = 3$. The hazard function increases over the lifetime, with a corresponding decrease in age-specific survival. Thus, an individual has a high survival probability early in life, which then decreases over the lifetime of the individual. Although flexible, Weibull distributions are not well suited for modeling survival of species that experience high juvenile mortality.

The Gompertz (1825) model has a long history and is based on the assumption that accumulative hazards escalate over time. Similar to the first part of the Weibull survival curve, the Gompertz curve (Fig. 5.3d) has high early survival (low hazard), but it also has a more rapid decrease in survival later in life that does not level off. This curve is

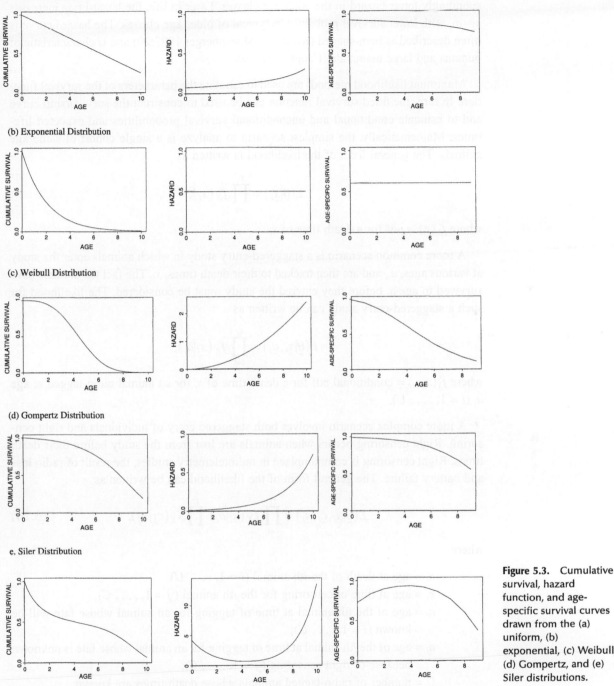

Figure 5.3. Cumulative survival, hazard function, and age-specific survival curves drawn from the (a) uniform, (b) exponential, (c) Weibull, (d) Gompertz, and (e) Siler distributions.

not well suited to most wildlife species and is more appropriate for modeling the failure time of machine parts, including radiotelemetry transmitters.

The set of survival curves derived from the Siler (1979) model perhaps most accurately characterizes the survival patterns of many wildlife species (Fig. 5.3e). The curve is characterized by early low survival and associated higher hazard, which is typical of juvenile age classes, followed by a period of increased age-specific survival with corre-spondingly lower hazard in the adult age classes. Later in life, the hazard rate increases again, with lower survival probabilities typical of older age classes. The hazard curve is often described as bath-shaped (Klein and Moeshberger 1997:29) and is characteristic of humans and large mammals (Caughley 1966).

Maximum likelihood methods are used to estimate the parameters of the survival function. In turn, the fitted survival function can be used to construct the survivorship curve and to estimate conditional and unconditional survival probabilities and expected lifetimes. Mathematically, the simplest scenario to analyze is a single cohort of same age animals. The general form of the likelihood is written as

$$L(\theta|\underaccent{\tilde}{x}_i) = \prod_{i=1}^{U} f_X(x_i),$$

where $f_X(x_i)$ = pdf for a death time of x_i.

A more common scenario is a staggered-entry study in which animals enter the study at various ages, a_i, and are then tracked to their death times, x_i. The fact the animals have survived to age a_i before they entered the study must be considered. The likelihood for such a staggered-entry study can be written as

$$L(\theta|\underaccent{\tilde}{x}_i,\underaccent{\tilde}{a}_i) = \prod_{i=1}^{U} f_X(x_i|a_i),$$

where $f_X(x_i|a_i)$ = conditional pdf for a death time of x_i for an animal radio-tagged at age a_i $(i = 1, \ldots, U)$.

A more complex scenario involves both staggered entry of individuals and right censoring. Right censoring occurs when animals are lost from the study before their death times. Right censoring is commonplace in radiotelemetry studies, the result of radio loss and battery failure. The general form of the likelihood can be written as

$$L(\theta|\underaccent{\tilde}{x}_i,\underaccent{\tilde}{c}_i,\underaccent{\tilde}{a}_i) = \prod_{i=1}^{U} f_X(x_i|a_i) \cdot \prod_{i=1}^{C} S_X(c_j|a_j), \tag{5.20}$$

where

> x_i = age at death of the ith animal $(i = 1, \ldots, U)$;
> c_j = age at time of censoring for the jth animal $(j = 1, \ldots, C)$;
> a_i = age of the ith animal at time of tagging for an animal whose fate will be known $(i = 1, \ldots, U)$;
> a_j = age of the jth animal at time of tagging for an animal whose fate is unknown because of right censoring $(j = 1, \ldots, C)$;
> U = number of radio-tagged animals whose death times are known;
> C = number of radio-tagged animals whose death times are unknown because of censoring;
> $S_X(c_j|a_j)$ = conditional survival function for a right-censored animal to time c_j, given it was tagged at age $a_j (j = 1, \ldots, C)$.

By conditioning on the ages of the animals at time of tagging, $f_X(x_i|a_i)$, staggered entry is taken into account. The conditional survival function $S_X(c_j|a_j)$, in turn, considers those animals that were known to be alive at age c_j when they were last observed (i.e., right-censored).

Assumptions

Most of the assumptions of the parametric survival analyses (Eq. 5.20) are similar to those of the Kaplan-Meier analysis. A major difference is the additional assumption that the form of the survivorship function is known. The assumptions are as follows:

1. The time of death for each individual is known and correctly recorded.
2. The cohort is representative of the population at the time of inference.
3. Fates of the individual animals are independent.
4. If left-censored or staggered-entry data are used, animal ages at time of entry into the study are known.
5. If right-censored data are used, the animals are known to be alive at the time censoring occurs and that time is known.
6. When making inferences to other cohorts, the survival probabilities depend on age but not time or location.
7. The form of the survival distribution is known and correctly chosen.

It is important in selecting a survival function that the family of curves be robust to adequately characterize the observed data.

Example 5.4: Fitting a Weibull Survivorship Curve for Grizzly Bears, Northern Idaho

The previous analysis of 28 radio-collared grizzly bears from the panhandle of northern Idaho will be repeated by using a two-parameter Weibull analysis (Eqs. 5.17–5.19). The grizzly bear data (Table 5.3) were characterized by being both left- and right-censored. Right censoring occurred because of radio failure before the bear died. Left censoring occurred when the bears were tagged after time 0, in this case, beyond age 1.5 years. The presence of both types of censoring requires the likelihood model be constructed by using both the conditional pdf and conditional survival functions. Hence, the likelihood for a particular animal will be

$$L(x_i, c_j | a_i, a_j) = \begin{cases} f_X(x_i | a_i) & \text{if not right-censored} \\ S_X(c_j | a_j) & \text{if right-censored.} \end{cases} \quad (5.21)$$

The conditional pdf for a left-censored animal can be written as

$$\begin{aligned} f_X(x_i | a_i) &= P(X = x_i | X > a_i) \\ &= \frac{P(X = x_i)}{P(X > a_i)} \\ &= \frac{f_X(x_i)}{S_X(a_i)}. \end{aligned} \quad (5.22)$$

Assuming a two-parameter Weibull distribution, Eq. (5.22) can be parameterized by using Eqs. (5.17) and (5.18) as

$$f_X(x_i | a_i) = \frac{\left[\dfrac{\beta x_i^{\beta-1}}{\alpha^\beta} e^{-\left(\frac{x_i}{\alpha}\right)^\beta} \right]}{e^{-\left(\frac{a_i}{\alpha}\right)^\beta}}. \quad (5.23)$$

The conditional survival function for a right-censored animal can be written as

$$
\begin{aligned}
S_X(c_j|a_j) &= P(X > c_j | X > a_j) \\
&= \frac{P(X > c_j)}{P(X > a_j)} \\
&= \frac{S_X(c_j)}{S_X(a_j)},
\end{aligned}
\tag{5.24}
$$

and, again assuming a two-parameter Weibull distribution (Eq. 5.17),

$$
\begin{aligned}
S_X(c_j|a_j) &= \frac{e^{-\left(\frac{c_j}{\alpha}\right)^\beta}}{e^{-\left(\frac{a_j}{\alpha}\right)^\beta}} \\
&= e^{\left(\frac{a_j}{\alpha}\right)^\beta - \left(\frac{c_j}{\alpha}\right)^\beta}.
\end{aligned}
\tag{5.25}
$$

Substituting in Eqs. (5.23) and (5.25) into the general form of the likelihood (5.20) produces the likelihood model:

$$
L(\beta,\alpha|x_i,c_j,a_i,a_j) = \prod_{i=1}^{U}\left[\frac{\beta x_i^{\beta-1}}{\alpha^\beta} e^{\left(\frac{a_i}{\alpha}\right)^\beta - \left(\frac{x_i}{\alpha}\right)^\beta}\right] \cdot \prod_{j=1}^{C}\left[e^{\left(\frac{a_j}{\alpha}\right)^\beta - \left(\frac{x_j}{\alpha}\right)^\beta}\right].
\tag{5.26}
$$

Survival data for all bears 1.5 years and older were used in the analysis. All death times were assumed to occur midyear, and all tagging and censoring occurred at the beginning of each age class (Table 5.3). The maximum likelihood estimates of the Weibull parameters were $\hat{\alpha} = 4.2647$ ($\widehat{SE} = 3.8471$) and $\hat{\beta} = 0.5725$ ($\widehat{SE} = 0.2357$). The resulting survivorship function is

$$
S_X(x_i) = e^{-\left(\frac{x_i}{4.2647}\right)^{0.5725}}.
$$

The survivorship curve shows a steep decline over ages 1 to 10 years followed by relatively stable survival probabilities over older age classes (Fig. 5.4). The parametric survivorship curve agrees with the nonparametric Nelson-Aalen curve (Fig. 5.4). Thus, a 1.5-year-old bear has a 0.381 probability of living to age 5.5, a 0.216 probability to age 10.5, a 0.139 probability to age 15.5, and a 0.095 probability to age 20.5 (Fig. 5.4).

For a two-parameter Weibull distribution, the expected lifetime for a Weibull random variable can be written as

$$
E(x_i) = \alpha\Gamma\left(\frac{\beta+1}{\beta}\right)
\tag{5.27}
$$

where Γ is the gamma function. The expected lifetime for a 1.5-year-old grizzly bear in northern Idaho is estimated by using Eq. (5.27) to be an additional 6.84 years.

Table 5.3. Individual records for radio-collared grizzly bears in northern Idaho, 1983–1993, indicating age at tagging and their subsequent fate regarding natural mortality (*N*), human-related mortality (*H*), or censored (*C*) owing to radio failure.

ID	Sex	Age at tagging	Age at death or censoring	Fate
867	F	5.5	16	H
898	F	1.5	2	H
949	M	4.5	5	H
962	M	6.5	8	H
1004	M	4.5	12.5	C
1005	M	8.5	11	H
1015	F	5.5	12.5	C
1042	F	2.5	4	H
1043	M	8.5	11.5	C
1044	F	25.5	27	N
1045	F	8.5	11.5	C
1046	M	5.5	8.5	C
1047	F	10.5	12.5	C
1048	F	25.5	29.5	C
1049	M	15.5	17.5	C
1050	M	1.5	2	N
1056	F	8.5	11.5	C
1057	M	2.5	5.5	C
1075	F	11.5	14.5	C
1076	F	20.5	23	N
1077	M	2.5	6.5	C
1078	M	13.5	16.5	C
1084	F	18.5	23.5	C
1085	F	2.5	4	H
1087	F	8.5	11.5	C
1089	F	2.5	9.5	C
1090	M	2.5	6	H
1091	M	2.5	5.5	C

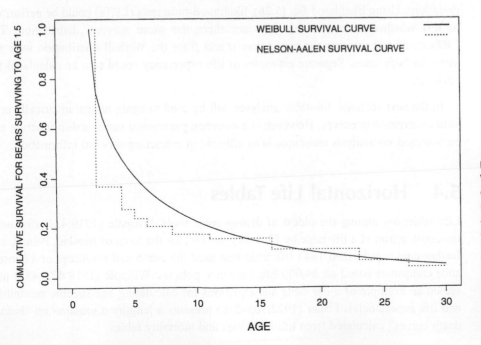

Figure 5.4. Fitted Weibull survivorship curve and the nonparametric Nelson-Aalen curve to the 1.5 age and older grizzly bear data (Table 5.3) from northern Idaho, 1983–1993. (Data from W. L. Wakkinen, Idaho Department of Fish and Game, personal communication.)

Discussion of Utility

Survivorship curves, whether parametric or nonparametric, require relatively high degrees of data sophistication. In the northern Idaho grizzly bear example, age at tagging was obtained by extracting a tooth and counting cementum annuli. Without age-at-tagging information, the survivorship curve could not have been constructed. Age classification is not necessary, however, if a single known age cohort (e.g., yearlings) was tagged *en masse* and monitored through their lifetime. A higher degree of invasiveness is required of tagging studies if the animals enter the study in a staggered manner. In big-game species, the staggered entry design depicted in the grizzly bear example will be the norm. Tagging *en masse* and tracking a known age cohort to its demise will likely be limited in most cases to small-game or nongame species in which numbers are large and lifetimes short.

The ability to perform the grizzly bear survivorship analysis was made possible by the use of radio-collars with motion sensors. The motion sensors emitted a special radio signal if the radio-collar did not move for more than 24 hours. Motion sensors and constant vigilance allowed investigators to distinguish mortalities from radio failure and to assign cause of death. A key element of right-censored studies is the requirement to differentiate death from censor events. In the grizzly bear example, known-age animals and radio tracking worked together to provide a successful study.

The advantage of parametric analyses over the Kaplan-Meier product-limit and Nelson-Aalen estimators is that the former uses more of the survivorship information collected. Regardless of what age an animal enters the study, information is provided on the same parameters of the parametric distribution as all other animals. Hence, a parametric survivorship curve can be constructed even when certain age classes may be missing from the study. This is possible because it is assumed that all animals share the same parametric survivorship process and parameters throughout their lives.

The parametric survival analysis also lends itself to additional analyses and interpretations. Using likelihood Eq. (5.26), likelihood-ratio tests (LRTs) could be performed to test whether male and female bears share the same survival distribution. The LRTs could test whether the parameters α and β for the Weibull distribution were the same for both sexes. Separate estimates of life expectancy could also be calculated for each sex.

In the next sections, life-table analyses will be used to again nonparametrically construct survivorship curves. However, if a common parametric survivorship function can be assumed, no analysis technique is as efficient in extracting survival information.

5.4   Horizontal Life Tables

Life tables are among the oldest of demographic tools. Whipple (1919:431) discussed the construction of a life table by E. Halley in 1692 for the town of Breslau, Poland, and further cites a published 1843 life table that used the combined resources of 17 insurance companies based on 84,000 life insurance policies. Whipple (1919:422–435) presented an example of a life table and equations for calculating age-specific mortalities and life expectancies. Fisher (1922:105–216) presents a lengthy discussion on "human death curves" calculated from life-table data and mortality tables.

Table 5.4.   Data columns for horizontal (i.e., age-specific) and vertical (i.e., time-specific) life tables.

Age interval	No. of individuals alive at age x	No. of individuals that died between ages x and x + 1 (d_x)	Survival probability between ages x and x + 1 (S_x)
0–1	l_0	$d_0 = l_0 - l_1$	S_0
1–2	l_1	$d_1 = l_0 - l_1 - l_2$	S_1
2–3	l_2	$d_2 = l_0 - l_1 - l_2 - l_3$	S_2
3–4	l_3	$d_3 = l_0 - \ldots - l_4$	S_3
⋮	⋮	⋮	⋮
w – 1 to w	l_{w-1}	$d_{w-1} = l_0 - \sum_{x=1}^{w} l_x$	S_{w-1}
w	l_w	$d_w = l_w$	0

Table 5.5.   The expected values for the number of deaths (d_x) in each time interval expressed in terms of the number of animals alive at time 0 (i.e., l_0) and age-specific survival probabilities for a horizontal life table.

Age class	Age interval	$E(d_x \mid l_0, S_x)$
0	0–1	$E(d_0) = l_0 (1 - S_0)$
1	1–2	$E(d_1) = l_0 S_0 (1 - S_1)$
2	2–3	$E(d_2) = l_0 S_0 S_1 (1 - S_2)$
3	3–4	$E(d_3) = l_0 S_0 S_1 S_2 (1 - S_3)$
⋮	⋮	
w – 1	w – 1 to w	$E(d_{w-1}) = l_0 S_0 S_1 S_2 \ldots S_{w-2} (1 - S_{w-1})$
w	w to w + 1	$E(d_w) = l_0 S_0 S_1 S_2 \ldots S_{w-1}$

Simply stated, a life table summarizes age- or stage-specific survival probabilities of a population. Also known as a dynamic, age-specific, cohort, or d_x-series table, a horizontal life table is constructed from data on the number of age-specific deaths (i.e., d_x) within a cohort from birth through its demise (Table 5.4). Data for the horizontal life table originate from recording the number of individuals in a cohort at the beginning of a study and the number of deaths occurring within specified intervals. Construction of the horizontal life table requires an investigator to monitor a single cohort of animals through time and accurately record the number of deaths within each time interval. The relationship between l_x and d_x is $l_{x+1} = l_x - d_x$. Thus, the number of animals alive at time $x + 1$ is equal to the number of animals alive at time x minus mortalities. The sum of all deaths among the individuals in a cohort is equal to the cohort's size at time 0, or

$l_0 = \sum_{x=0}^{w} d_x$. The expected value for the number of deaths within an interval from time x to $x + 1$, d_x, may be written as a function of l_0 and age-specific survival probabilities, S_x ($x = 0, 1, \ldots, w - 1$) (Table 5.5).

5.4.1   Standard Life-Table Analysis

Let w denote the oldest age class such that $S_w = 0$. Treating the number of deaths (d_x) within an interval as a random variable, a multinomial distribution can be constructed from the expected values (Table 5.5). The probability distribution for the number of deaths in each age class can be written as

$$f(d_0, d_1, \ldots, d_w | l_0, \underset{\sim}{S}_x) = \binom{l_0}{\underset{\sim}{d}_x}(1-S_0)^{d_0}(S_0(1-S_1))^{d_1}(S_0 S_1(1-S_2))^{d_2} \ldots \left(\prod_{i=0}^{w-1} S_i\right)^{d_w}$$

$$(5.28)$$

or

$$f(d_0, d_1, \ldots, d_w | l_0, \underset{\sim}{S}_x) = \binom{l_0}{\underset{\sim}{d}_x}\left[\prod_{x=0}^{w-1}\left(\prod_{i=0}^{x-1} S_i(1-S_x)\right)^{d_x}\right]\left(\prod_{i=0}^{w-1} S_x\right)^{d_w}. \quad (5.29)$$

The maximum likelihood estimator (MLE) for the probability of survival in the interval x to $x+1$ is

$$\hat{S}_x = \frac{l_0 - \sum\limits_{i=0}^{x} d_i}{l_0 - \sum\limits_{i=0}^{x-1} d_i} = \frac{l_{x+1}}{l_x}. \quad (5.30)$$

Variance estimators for $\hat{S}_x$ can be derived by using the information matrix (Casella and Berger 1990:336). However, the analytic variance expression for the estimator of survival $\hat{S}_x$ is more easily approximated by using the delta method. Thus, the variance is derived as

$$\text{Var}(\hat{S}_x) = \sum_{i=0}^{x} \text{Var}(d_i)\left(\frac{\partial \hat{S}_x}{\partial d_i}\right)^2_{|E(d_i)} + 2\left(\sum_{i=0}^{x-1}\sum_{\substack{j=0 \\ i<j}}^{x-1} \hat{S}_{0i}\text{Cov}(d_i, d_j)\left(\frac{\partial \hat{S}_x}{\partial d_i}\right)_{|E(d_i)}\left(\frac{\partial \hat{S}_x}{\partial d_j}\right)_{|E(d_j)}\right).$$

$$(5.31)$$

Substituting values for the first derivates, variances, and covariances, Ryding (2002) derived the following variance estimator:

$$\widehat{\text{Var}}(\hat{S}_x) = \frac{(1-\hat{S}_x)^2}{l_0 \hat{S}_{0x}^2}\left[\frac{\hat{S}_{0x}(1-\hat{S}_{0x}(1-\hat{S}_x))}{(1-\hat{S}_x)} + \sum_{i=0}^{x-1}\hat{S}_{0i}(1-\hat{S}_i)(1-\hat{S}_{0i}(1-\hat{S}_i))\right.$$
$$\left. -2\sum_{i=0}^{x-1}\sum_{\substack{j=0 \\ i<j}}^{x-1}\hat{S}_{0i}(1-\hat{S}_i)\hat{S}_{0j}(1-\hat{S}_j) - 2\hat{S}_{0x}\sum_{j=0}^{x-1}\hat{S}_{0j}(1-\hat{S}_j)\right].$$

$$(5.32)$$

The variance estimator for $\hat{S}_0$ is a special case of Eq. (5.32) where

$$\widehat{\text{Var}}(\hat{S}_0) = \frac{\hat{S}_0(1-\hat{S}_0)}{l_0}. \quad (5.33)$$

Chiang (1960a) modeled the number of deaths in each age class as a binomial distribution conditioned on l_x. The joint probability for the number of deaths in each age class (d_x) based on the conditional binomial distributions can be written as

$$f(d_0, d_1, \ldots, d_w | \underset{\sim}{l}_x, \underset{\sim}{S}_x) = \prod_{x=1}^{w}\binom{l_x}{d_x}S_x^{l_x - d_x}(1-S_x)^{d_x}. \quad (5.34)$$

A major difference between Eqs. (5.28) and (5.34) is that the values l_x are parameters of the binomial distribution in Eq. (5.34) but treated as random variables (function of the numbers of deaths, d_x) in Eq. (5.28). Although MLEs of survival are the same for both likelihoods, the associated variance estimates will differ. The variance expression Eq. (5.32) differs from that derived by Chiang (1960a, 1960b) and subsequently presented in Seber (1982:408–409). Chiang (1960a, 1960b) derived the variance of $\hat{S}_x$ from the expression $\hat{S}_x = \dfrac{l_{x+1}}{l_x}$, using the binomial distribution for l_{x+1} conditioned on l_x. However, l_x is a random variable and contributes to the overall variance of $\hat{S}_x$. The variance expressions in Chiang (1960a) and Seber (1982:429) underestimate the true variance of $\hat{S}_x$ because the variability in l_x is not considered.

The cumulative probability of survival from time zero (t_0) to age x is estimated by the expression,

$$\hat{S}_{0x} = \prod_{i=0}^{x-1} \hat{S}_i, \tag{5.35}$$

or can be expressed in terms of l_0 and d_x, where

$$\hat{S}_{0x} = \frac{l_x}{l_0} = \frac{l_0 - \displaystyle\sum_{i=0}^{x-1} d_i}{l_0}. \tag{5.36}$$

The variance of $\hat{S}_{0x}$ is expressed exactly by using the variance of a sum as

$$\mathrm{Var}\left(\hat{S}_{0x}\right) = \frac{1}{l_0}\left[\sum_{i=0}^{x-1} S_{0i}(1 - S_i)[1 - S_{0i}(1 - S_i)] - 2\sum_{i=0}^{x-1}\sum_{\substack{j=0 \\ i<j}}^{x-1} S_{0i}(1 - S_i)S_{0j}(1 - S_j) \right],$$

with estimated variance of

$$\widehat{\mathrm{Var}}\left(\hat{S}_{0x}\right) = \frac{1}{l_0}\left[\sum_{i=0}^{x-1} \hat{S}_{0i}(1 - \hat{S}_i)[1 - \hat{S}_{0i}(1 - \hat{S}_i)] - 2\sum_{i=0}^{x-1}\sum_{\substack{j=0 \\ i<j}}^{x-1} \hat{S}_{0i}(1 - \hat{S}_i)\hat{S}_{0j}(1 - \hat{S}_j) \right]. \tag{5.37}$$

The estimator for the probability of surviving between any two ages, x to $x + j$, for example, is derived in the same way as surviving from age 0 to age x. By use of age-specific survival probabilities, $\hat{S}_{x,x+j}$ can be written as

$$\hat{S}_{x,x+j} = \hat{S}_x \cdot \hat{S}_{x+1} \cdot \hat{S}_{x+2} \cdot \ldots \cdot \hat{S}_{x+j-1} = \prod_{i=x}^{x+j-1} \hat{S}_i, \tag{5.38}$$

or, in terms of the data,

$$\hat{S}_{x,x+j} = \frac{l_0 - \displaystyle\sum_{i=0}^{x+j-1} d_i}{l_0 - \displaystyle\sum_{i=0}^{x-1} d_i}. \tag{5.39}$$

The variance of $\hat{S}_{x,x+j}$ can be derived by using the delta method and noting that the calculation only includes the variances and covariances associated with ages x to $x+j$. Thus, the variance estimator of $\hat{S}_{x,x+j}$ is

$$\widehat{\mathrm{Var}}\left(\hat{S}_{x,x+j}\right) = \frac{\left(1-\hat{S}_x\right)}{\left(l_0\hat{S}_{0x}\right)^2}\left[\sum_{i=x}^{x+j} l_0\left(1-\hat{S}_x\right)\hat{S}_{0i}\left(1-\hat{S}_i\right)\left(1-\hat{S}_{0i}\left(1-\hat{S}_i\right)\right)\right.$$
$$\left. +2\sum_{\substack{i=x \\ i<j}}^{x+j-1}\sum_{j=x}^{x+j-1} l_0\left(1-\hat{S}_x\right)S_{0i}\left(1-\hat{S}_i\right)\hat{S}_{0j}\left(1-\hat{S}_j\right)\right]. \tag{5.40}$$

Often, the l_x and, subsequently, d_x, columns in the life table are scaled so the number of animals in the 0 age class, l_0, is 1000 or 10,000. The resulting scaled values l_x and d_x, denoted d'_x and l'_x, can be expressed mathematically as

$$l'_x = l_x \cdot \frac{1000}{l_0} \tag{5.41}$$

and

$$d'_x = l'_x - l'_{x+1} = \frac{1000}{l_0}\left(l_x - l_{x+1}\right). \tag{5.42}$$

The data columns in the life-table analysis that Caughley (1977:85–96) discusses are based on l'_x and d'_x. Many authors present cohort life tables with the 0 age class standardized to a reference number, often 1000 individuals (Anderson 1985:35, Krebs 1994:169). However, without knowledge of the original cohort size, l_0, the variances are inestimable. Variances estimated from the arbitrarily scaled numbers are invalid.

Assumptions

The horizontal life table has the fewest assumptions (i.e., four) of all life-table methods presented in this chapter because the data are direct records of the death times of individual animals. The assumptions of the horizontal life table are as follows:

1. The number of individuals alive and dead at each time interval is correctly enumerated.
2. The cohort is representative of the population at the time of inference.
3. Fates (death times) of animals are independent between individuals.
4. When making inferences to other cohorts, the probabilities of survival depend on age but not time or locale.

No assumptions concerning the long-term structure or dynamics of the population are necessary when using a horizontal life table. Survival probabilities will be biased if the numbers of animals alive and dead at each interval are miscounted. The above assumptions (2) and (4) imply that if only a sample of a cohort is followed, the sample must be representative for the survival estimates to be valid.

Estimating survival using a horizontal life table requires identifying members of a cohort, monitoring those individuals through time, and recording the number alive at each interval, which is often difficult for wild animals. If tagging data are used, the capture probabilities must be one. Radio telemetry is the only practical tagging method for horizontal life-table analysis for which this assumption may be true. However, radio telemetry can be cost-prohibitive, and radio failure can result in right-censored data. It is also implicitly assumed the cohort is closed to immigration and emigration, and any harvest loss is accurately recorded.

Manly and Seber (1973) present a method for constructing a life table by using a multiple mark-recapture method estimating the number of animals alive that entered the population during the recapture periods and survived to a subsequent period. Cohort and time-specific survival probabilities can then be calculated from the abundance estimates, although the main objective of the method is population reconstruction. Except for the method of Manly and Seber (1973), in which recapture probabilities are estimated, tagging data are not appropriate for life-table analysis because of confounding between survival and recapture probabilities. Both Seber (Seber 1982:475) and Eberhardt (1969) warn against the use of bird band returns in life-table analyses.

5.4.2   Constant Survival Across Age Classes

If a constant survival probability is assumed among all age groups, the likelihood for the horizontal life table, Eq. (5.28) simplifies as follows:

$$L(S|l_0, \underset{\sim}{d}_x) \propto S^{\sum_{x=0}^{w} x d_x} (1-S)^{\sum_{x=0}^{w-1} d_x}. \tag{5.43}$$

The MLE of S is easily solved and is

$$\hat{S} = \frac{\sum_{x=0}^{w} x d_x}{\sum_{x=0}^{w-1} (x+1) d_x + w d_w}, \tag{5.44}$$

which is often referred to as the average, or pooled, survival rate across all age classes (Seber 1982:413, 545). Because Eq. (5.43) has only one parameter, S, the variance is easily derived by using the asymptotic variance formula

$$\widehat{Var}(\hat{S}) = \frac{-1}{E\left[\dfrac{\partial^2 \ln L(S)}{\partial S^2}\right]_{S=\hat{S}}},$$

where

$$Var(\hat{S}) = \frac{S(1-S)}{l_0\left[1 + S^w(w-1) + (1-S)\sum_{x=0}^{w-1} x S^x\right]}, \tag{5.45}$$

with an estimated variance of

$$\widehat{Var}(\hat{S}) = \frac{\hat{S}(1-\hat{S})}{\sum_{x=0}^{w-1} (x+1) d_x + w d_w}. \tag{5.46}$$

Sample Size Calculations

The simplicity of likelihood Eq. (5.43) and the subsequent variance (Eq. 5.45) for $\hat{S}$ make this model ideal for sample size calculations. The precision of the survival estimate $\hat{S}$ will be defined as

$$P(|\hat{S} - S| < \varepsilon) \geq 1 - \alpha,$$

where the absolute error in estimation (i.e., $|\hat{S} - S|$) is less than ε, $(1 - \alpha) \cdot 100\%$ of the time. Assuming the MLE $\hat{S}$ is normally distributed,

$$\varepsilon = Z_{1-\frac{\alpha}{2}} \cdot \text{SE}(\hat{S}).$$

For $\alpha = 0.10$ and $S = 0.3$, 0.5, 0.7, or 0.9, the absolute error in estimation (ε) is plotted as a function of initial cohort size (l_0) (Fig. 5.5).

The maximum number of age classes (w) has little effect on the sample size of l_0 (Fig. 5.5). However, the required sample size for the number of animals in a cohort to follow (l_0) decreases as annual survival (S) increases. The suggested sample sizes (Fig. 5.5) may also be considered the minimum sizes needed for a horizontal life table. Precise estimation of unique age-specific survival probabilities in a life table can be expected to require sample sizes in excess of those calculated in Fig. 5.5. Precision will be much less (i.e., greater standard errors) when unique age-specific survivals are calculated than when a common survival is estimated across age classes.

Example 5.5: Horizontal Life-Table Analysis for Gray Squirrels (*Sciurus carolinensis*), Crumpacker Woods, Virginia

This example of a horizontal life-table analysis (Table 5.6) uses data from a study on the dynamics of a gray squirrel population in an isolated woodlot closed to hunting (Mosby 1969). A cohort of squirrels was captured as immatures (age 0), marked, and recaptured several times a year from 1952–1960. Because of the high trapping effort and isolation of the population, the assumptions of a detection probability near one and known fates may be reasonable. It must be further assumed that the capture-recapture process did not alter observed survival probabilities.

Age-specific survival probabilities are estimated by Eq. (5.30). The survival probabilities for age classes 0 and 1 are calculated, respectively, as follows:

$$\hat{S}_0 = \frac{42 - 22}{42} = \frac{20}{42} = 0.4762$$

and

$$\hat{S}_1 = \frac{42 - (22 + 10)}{42 - 22} = \frac{10}{20} = 0.5000.$$

The variance for S_0 is estimated by using Eq. (5.33) as

$$\widehat{\text{Var}}(\hat{S}_0) = \frac{\hat{S}_0(1 - \hat{S}_0)}{l_0} = \frac{0.4762(1 - 0.4762)}{42} = 0.00594$$

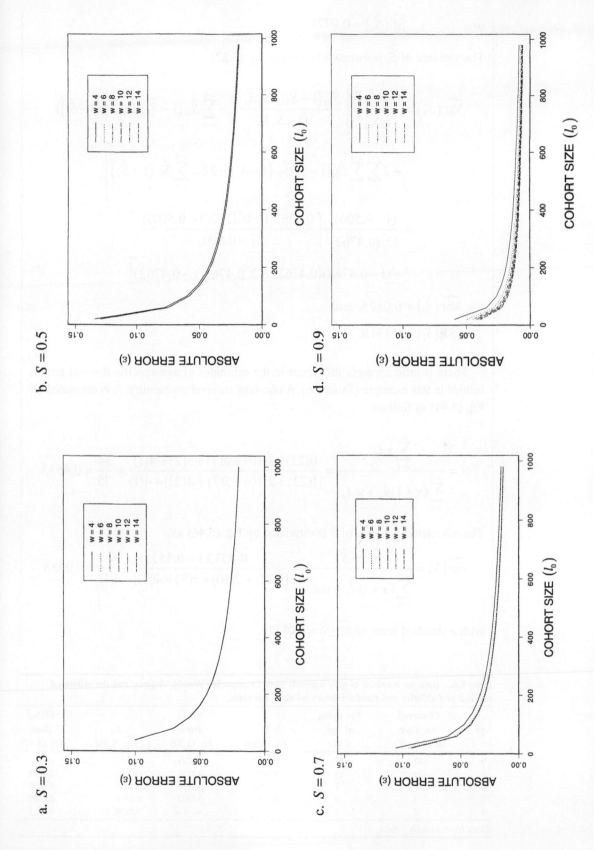

Figure 5.5. Cohort size (l_0) versus sampling precision expressed in terms of absolute error (ε) in estimating a common survival probabilty S from horizontal life-table data for $S = 0.3, 0.5, 0.7,$ and 0.9 at $1 - \alpha = 0.90.$

a. $S = 0.3$

b. $S = 0.5$

c. $S = 0.7$

d. $S = 0.9$

or $\qquad \widehat{SE}(\hat{S}_0) = 0.0771.$

The variance of $\hat{S}_1$ is estimated by using Eq. (5.32)

$$\widehat{Var}(\hat{S}_x) = \frac{(1-\hat{S}_x)^2}{l_0\hat{S}_{0x}^2}\left[\frac{\hat{S}_{0x}(1-\hat{S}_{0x}(1-\hat{S}_x))}{(1-\hat{S}_x)} + \sum_{i=0}^{x-1}\hat{S}_{0i}(1-\hat{S}_i)(1-\hat{S}_{0i}(1-\hat{S}_i))\right.$$

$$\left. -2\sum_{\substack{i=0 \\ i<j}}^{x-1}\sum_{j=0}^{x-1}\hat{S}_{0i}(1-\hat{S}_i)\hat{S}_{0j}(1-\hat{S}_j) - 2\hat{S}_{0x}\sum_{j=0}^{x-1}\hat{S}_{0j}(1-\hat{S}_j)\right]$$

$$= \frac{(1-0.500)^2}{42\cdot(0.4762)^2}\left[\frac{0.4762(1-0.4762(1-0.500))}{(1-0.500)}\right.$$

$$\left. +(1-0.4762)(0.4762) - 2\cdot0.4762(1-0.4762)\right]$$

$\widehat{Var}(\hat{S}_1) = 0.0125,$ and

$\widehat{SE}(\hat{S}_1) = 0.1118.$

There is little apparent difference in the estimates of age-specific survival probabilities in this example (Table 5.6). A constant survival probability, $\hat{S}$, is estimated by Eq. (5.44) as follows:

$$\hat{S} = \frac{\sum_{x=0}^{w}xd_x}{\sum_{x=0}^{w-1}(x+1)d_x + wd_w} = \frac{0(22)+1(10)+2(7)+3(2)+4(1)}{[1(22)+2(10)+3(7)+4(2)]+4(1)} = \frac{34}{75} = 0.4533$$

The associated variance of $\hat{S}$ is estimated by Eq. (5.46) as

$$\widehat{Var}(\hat{S}) = \frac{\hat{S}(1-\hat{S})}{\sum_{x=0}^{w-1}(x+1)d_x + wd_w} = \frac{0.4533(1-0.4533)}{[1(22)+2(10)+3(7)+4(2)]+4(1)} = 0.0033$$

with a standard error of $\widehat{SE}(\hat{S}) = 0.0575.$

Table 5.6. Data for a cohort of gray squirrels from Crumpacker Woods, Virginia, and the estimated survival probabilities and standard errors for each age class.

Age (x)	Observed no. alive at age x (l_x)	No. dying at age x (d_x)	S_x Eq. (5.30)	$\widehat{SE}(\hat{S}_x)$ from Eq. (5.32)	$\hat{S}_{0x}$ Eq. (5.36)	$\widehat{SE}(\hat{S}_{0x})$ from Eq. (5.37)
0	42	22	0.4762	0.0771	1	—
1	20	10	0.5000	0.1118	0.4762	0.0771
2	10	7	0.3000	0.1449	0.2381	0.0657
3	3	2	0.3333	0.2722	0.0714	0.0397
4	1	1	0	—	0.0238	0.0235

(Data from Mosby 1969.)

The assumption of constant age-specific survival can be tested by using a LRT of the form

$$\chi_3^2 = -2\left(\ln L\left(\hat{S}|\underline{d}_x, l_0\right) - \ln L\left(\hat{\underline{S}}_x, |\underline{d}_x, l_0\right)\right).$$

Using the log-likelihood values for models (5.28) and (5.43), the LRT yields

$$= -2(-51.6589 - (-50.9457)) = 1.4264,$$

which gives a P-value of $P\left(\chi_3^2 \geq 1.4264\right) = 0.6994$. Hence, age-specific survival probabilities are not significantly different, and a constant age class survival of 0.4533 may be assumed for this cohort.

Note that the standard error of the common survival probability $\hat{S}$ is much smaller than the standard errors of the age-specific survival probabilities (Table 5.6). This is often the case when estimating several parameters from limited data. A likelihood equation with fewer parameters will yield more precise parameter estimates, but at the potential cost of decreased accuracy. Hence, there is a trade-off between accuracy and precision that must be balanced by the realism and appropriateness of the model under investigation.

5.5   Vertical Life Tables

The vertical life table, also called time-specific, current, static, or a l_x-series table, is based on an instantaneous sample of the age structure of a population. The analysis presented here treats the number of individuals alive by age class as the random variables of the sampling distribution. Chiang (1960b) in deriving the estimators for a vertical life table considered age-specific mortality $(1 - S_x)$ as the random variable when, in fact, it is the parameter being estimated. Caughley (1977:85–100) did not discuss the statistical properties of the vertical life table, or the sampling distribution of the age structure data. In this section, the estimates of age-specific survival are derived by modeling the age-structure data used to build a vertical life table. The nature of the data are discussed, and variances are derived for each estimator along with a list of assumptions.

The age-structure data for a vertical life table most often come from a sample of the age frequencies in a population. If the sample is representative, the age structure of the sample should reflect that of the population. Because the population age structure may change over time, the observed sample frequencies are specific to that sampling time; hence, the term "time-specific" is used. Seber (1982:400) considered the vertical life table less accurate than the horizontal life table but noted the data were easier to obtain.

The two most common sources of age-structure data for the construction of a vertical life table come from harvest data and catastrophic events that cause mortality. Both data sources form a l_x series. Note that just because the sample of age frequencies comes from dead animals, it is not necessarily a d_x series. Use of the l_x-series data assumes the sample is representative of the age structure of the live population. However, unequal vulnerability and hunter selection may result in a nonrepresentative sample and biased survival estimates. In big-game harvests, the bias is often an under-representation of younger age classes as the result of hunter preference or regulations (Caughley 1966). For waterfowl hunting, the youngest age classes may be over-represented because of their greater vulnerability to harvest (Schierbaum and Foley 1957).

5.5.1   Standard Life-Table Analysis

The columns in the vertical life table are the same as those of the horizontal life table (Table 5.4). However, the expected values of the number of individuals alive at time x differ from those in a horizontal life table. We define the following:

N_j = number of individuals recruited into the jth cohort at age 0;
S_{xj} = probability of surviving from age x to $x + 1$ for the jth cohort;
p_j = probability of observing an individual from the jth cohort;

where the expected values of the l_x series are presented in Table 5.7. If a random sample of the population age structure has been taken, then p_j should be equal for all age classes (i.e., $p_j = p$, $\forall j$). Two additional assumptions would simplify the vertical life table. The first is a stationary population, where annual recruitment is constant over time, i.e., $N_1 = N_2 = N_j = N_0$. The second assumption is that survival is age-specific but time-independent, i.e., $S_{i1} = S_{i2} = \ldots S_i$. The expected values for the number of individuals observed in each age class can thus be expressed in terms of N_0, S_x, and p (Table 5.8).

The number of individuals observed in each age class can be modeled by using the Poisson distribution, or

$$P(L_x = l_x) = \frac{\varphi_x^{l_x} e^{-\varphi_x}}{l_x!} \tag{5.47}$$

Table 5.7.   The expected values for l_x counts used in the vertical life table for each age group expressed in terms of N_j, S_{xj} and p_j.

Age class	Age interval	l_x	No. in population at age x	$E(l_x)$
0	0–1	l_0	N_0	$N_0 p_0$
1	1–2	l_1	$N_0 S_{01}$	$N_1 S_{01} p_1$
2	2–3	l_2	$N_2 S_{02} S_{12}$	$N_2 S_{02} S_{12} p_2$
3	3–4	l_3	$N_3 S_{03} S_{13} S_{23}$	$N_3 S_{03} S_{13} S_{23} p_3$
$\vdots$	$\vdots$	$\vdots$	$\vdots$	$\vdots$
w	w to $w + 1$	l_w	$N_w S_{0,w} S_{1,w} \ldots S_{w-1,w}$	$N_w S_{0,w} S_{1,w} \ldots S_{w-1,w} p_w$

Table 5.8.   The expected values of l_x for a vertical life table assuming a stable and stationary population.

Age class	Age interval	No. in sample of age x (l_x)	No. in population of age x	$E(l_x)$
0	0–1	l_0	N_0	$N_0 p_0$
1	1–2	l_1	$N_0 S_0$	$N_0 S_0 p$
2	2–3	l_2		$N_0 S_0 S_1 p$
3	3–4	l_3	$N_0 S_0 S_1 S_2$	$N_0 S_0 S_1 S_2 p$
$\vdots$	$\vdots$	$\vdots$		$\vdots$
w	w to $w + 1$	l_w	$N_0 S_0 S_1 \ldots S_{w-1}$	$N_0 S_0 S_1 \ldots S_{w-1} p$

where $\varphi_x = E(l_x) = N_0 \left(\prod\limits_{i=0}^{x-1} S_i \right) p$. Because the sum of Poisson random variables is also Poisson-distributed, the total number of individuals in the sample has the distribution:

$$\varphi. = E(l.) = pN_0 \sum_{x=0}^{w} \left(\prod_{i-0}^{x-1} S_i \right)$$

where

$$l. = \sum_{x=0}^{w} l_x,$$

$$P(L. = l.) = \frac{\varphi_.^l e^{-\varphi.}}{l.!}.$$

By use of the joint likelihood for all age classes and conditioning on the total sample size, $l.$, the likelihood is

$$L(\varphi_x | l., l_x) = \frac{\prod\limits_{x=0}^{w} \dfrac{\varphi_x^{l_x} e^{-\varphi_x}}{l_x!}}{\dfrac{\varphi_.^l e^{-\varphi.}}{l.!}},$$

or, equivalently,

$$L(\varphi_x | l., l_x) = \binom{l.}{l_x} \prod_{x=0}^{w} \left(\frac{\varphi_x}{\varphi.} \right)^{l_x}. \tag{5.48}$$

By rewriting the likelihood in terms of the survival parameters, Eq. (5.48) yields the following multinomial model for the sampling distribution of the l_x series:

$$L(S_x | l., l_x) = \binom{l.}{l_x} \left(\frac{1}{\Psi} \right)^{l_0} \left(\frac{S_0}{\Psi} \right)^{l_1} \left(\frac{S_0 S_1}{\Psi} \right)^{l_2} \cdots \left(\frac{S_0 S_1 \dots S_{w-1}}{\Psi} \right)^{l_w} \tag{5.49}$$

where

$$\Psi = 1 + S_0 + S_0 S_1 + S_0 S_1 S_2 + \cdots + S_0 S_1 S_2 \dots S_{w-1} \text{ or}$$

$$\Psi = 1 + S_0 + S_0 S_1 + S_0 S_1 S_2 + \cdots + \prod_{x=0}^{w-1} S_x.$$

Likelihood Eq. (5.49) is equivalent to that presented by Seber (1982:429). The maximum likelihood estimate of the age-specific survival probabilities is

$$\hat{S}_x = \frac{l_{x+1}}{l_x}. \tag{5.50}$$

Neither the likelihood model, Eq. (5.49), nor the survival estimator, Eq. (5.50), is expressed in terms of the number of deaths. In vertical life tables generated from l_x series, deaths are not observed or recorded. Survival estimates from a vertical life table (i.e., Eq. 5.50) are the same as those of the horizontal life table (i.e., Eq. 5.30). However, neither the sampling distributions nor the associated variances for the estimators are equivalent for the vertical and horizontal life tables.

The variance of $\hat{S}_x$ is approximated by using the delta method as follows:

$$\text{Var}(\hat{S}_x) \doteq S_x^2 \left[\frac{\text{Var}(l_{x+1})}{E(l_{x+1})^2} + \frac{\text{Var}(l_x)}{E(l_x)^2} - \frac{2\text{Cov}(l_x, l_{x+1})}{E(l_{x+1})E(l_x)} \right], \tag{5.51}$$

where

$$\text{Var}(l_x) = l \cdot \frac{\prod_{i=0}^{x-1} S_i}{\Psi} \left(1 - \frac{\sum_{i=0}^{x-1} S_i}{\Psi} \right),$$

$$\text{Var}(l_{x+1}) = l \cdot \frac{\prod_{i=0}^{x} S_i}{\Psi} \left(1 - \frac{\sum_{i=0}^{x} S_i}{\Psi} \right),$$

$$\text{Cov}(l_x, l_{x+1}) = -l \cdot \frac{\prod_{i=0}^{x-1} S_i}{\Psi} \cdot \frac{\prod_{i=0}^{x} S_i}{\Psi}.$$

Substituting the expressions for the variances and covariance of l_x and l_{x+1} into Eq. (5.51), the variance of $\hat{S}_x$ is approximated by

$$\text{Var}(\hat{S}_x) \doteq \left(\frac{\Psi S_x (1 + S_x)}{l \prod_{i=0}^{x-1} S_i} \right). \tag{5.52}$$

The estimated variance of $\hat{S}_x$ can then be expressed by

$$\widehat{\text{Var}}(\hat{S}_x) = \left(\frac{\hat{\Psi} \hat{S}_x (1 + \hat{S}_x)}{l \prod_{i=0}^{x-1} \hat{S}_i} \right) \tag{5.53}$$

where $\hat{\Psi} = 1 + \hat{S}_0 + \hat{S}_0 \hat{S}_1 + \hat{S}_0 \hat{S}_1 \hat{S}_2 + \cdots + \prod_{x=0}^{w-1} \hat{S}_x$. Equations (5.52) and (5.53) are equivalent to the variance expressions provided by Seber (1982:429).

The cumulative probability of survival from age 0 to age x, $\hat{S}_{0x}$, is estimated by

$$\hat{S}_{0x} = \hat{S}_0 \cdot \hat{S}_1 \cdot \hat{S}_2 \cdot \ldots \cdot \hat{S}_{x-1}$$

or, equivalently,

$$\hat{S}_{0x} = \frac{l_1}{l_0} \cdot \frac{l_2}{l_1} \cdot \frac{l_3}{l_2} \cdots \frac{l_x}{l_{x-1}} = \frac{l_x}{l_0}, \tag{5.54}$$

which is the same estimator as (5.36) for a horizontal life table. The variance estimator for $\hat{S}_{0x}$ is approximated as

$$\widehat{\text{Var}}\left(\hat{S}_{0x}\right) = \frac{\hat{\Psi}\prod\limits_{i=0}^{x-1}\hat{S}_i\left(1+\prod\limits_{i=0}^{x-1}\hat{S}_i\right)}{l_.} \tag{5.55}$$

(Ryding 2002). The estimator for S_{0x} from both the horizontal and vertical life tables is the product of all survival probabilities from 0 to age x. However, the variances are different. The estimator for $\hat{S}_{0x}$, derived from a vertical life table, is the ratio of two random variables, whereas for the horizontal life table, the denominator (i.e., l_0) is a parameter of the sampling distribution.

Assumptions

There are six assumptions associated with vertical life tables:

1. Age distribution is stable with age-specific survivals constant over time.
2. Population is stationary with constant abundance (N_0) over time.
3. All individuals in the population have the same probability of selection.
4. Fates of all animals are independent.
5. Ages of all animals in the sample are measured without error.
6. The probabilities of survival depend on age but not time or locale if inferences to other populations are being made.

The variances calculated above for $\hat{S}_x$ and $\hat{S}_{0x}$ incorporate the multinomial sampling error of the data. However, if measurement error exists because of inaccuracies in the age classification of animals, additional variation not accounted for in the variance formula will exist. If there are systematic aging errors, the vertical life-table estimates of survival will be biased. An important source of bias may come from violations of the assumptions of a stable and stationary population. Typically, populations have neither a stable age structure nor stationary abundance (e.g., Unsworth et al. 1999). Caughley (1966) lists five methods for testing the assumption of a stationary population but states that none of the tests is effective. In fact, it is impossible to assess these assumptions with a single vertical age-structured sample (Caughley 1966, Johnson 1996). In the analysis of age-structure data of white-tailed deer (*Odocoileus virginianus*), McCullough (1979:22) stated that the assumptions of a stable and stationary population are rarely met in practice. Instead, McCullough (1979:221) suggested using a vertical life table to point out year classes with higher productivity. However, life-table analysis is unnecessary to detect dominant year classes. Instead, a simple age-frequency histogram should suffice. However, from a single sample, it may be difficult to tell whether a year class has a higher recruitment or simply a higher detection rate. Multiple years of age-frequency data are necessary to assess whether the stronger year class persists over time.

Seber (1982:400–401) suggested that even though a population may not be stable or stationary overall, it may have a stable age distribution for a segment of the population during some portion of the year. For seasonal breeders, a stationary age distribution may be present only for some older age classes. Sampling periods should be chosen in the annual cycle when the assumption of stationarity may be most realistic.

Unfortunately, the observed age frequency of a population age structure is not always monotonically decreasing, meaning that some of the later age classes may have more individuals than do younger age classes. As a result, some survival estimates will be greater than one (Eq. 5.50). Several factors working together or separately may result in nonmonotonically decreasing age-structure data. First, the population may be neither stable nor stationary. Second, sampling techniques used to obtain the sample age distribution may be biased for some age classes. Third, there may be either a dominant or weak recruitment year because of environmental variability. Finally, stochasticity in the sampling and survival processes can result in some older age classes having a higher frequency of individuals than do younger age classes by chance, even when all assumptions are met.

By using a simulation study, Polacheck (1985) examined the distribution of survival estimates from life-table analysis using sample sizes of 250 and 1000 individuals. Results of the study illustrated the high variability in survival estimates and the potential for the survival estimates to be greater than one. Polacheck (1985) focused entirely on sample size in the discussion of precision. However, the variance of survival estimates from a vertical life table will depend on both sample size and number of parameters estimated from the data. Precision in survival estimates will decrease as more parameters are estimated from the same data.

Reasons for survival estimates greater than one were discussed earlier. Polacheck (1985) mentions only a nonstable age structure and sampling variability as reasons for nonmonotonically decreasing age frequencies. Smoothing the age-frequency data or the resulting survival estimates is mentioned as a possible solution. However, a better alternative may be to set that survival estimate equal to one or to assume that survival is constant across several age classes. These assumptions can be easily tested by using LRTs. Pooling data across age classes is discussed in the next section.

Occasionally, the vertical life table is changed to resemble a horizontal life table by summing the number of individuals in successive age classes, to produce an age-class distribution that is monotonically decreasing through time. The expected values (Table 5.9) of the cumulative counts can be calculated by assuming a Poisson distribution for each age class. However, Eq. (5.50) cannot be used to provide unbiased estimates of survival probabilities from the cumulative counts. For example, the estimator of survival from age 4 to age 5, S_4, using the cumulative counts, has the expected value:

Table 5.9. The expected values of a vertical life table when age frequencies are accumulated to form a monotonically decreasing series.

Age class	l_x	$E(l_x)$	Cumulative	E (Cumulative)
0	l_0	$Np = \varphi_0$	$l_5 + l_4 + l_3 + l_2 + l_1 + l_0$	$\varphi_5 + \varphi_4 + \varphi_3 + \varphi_2 + \varphi_1 + \varphi_0$
1	l_1	$NpS_0 = \varphi_1$	$l_5 + l_4 + l_3 + l_2 + l_1$	$\varphi_5 + \varphi_4 + \varphi_3 + \varphi_2 + \varphi_1$
2	l_2	$NpS_0S_1 = \varphi_2$	$l_5 + l_4 + l_3 + l_2$	$\varphi_5 + \varphi_4 + \varphi_3 + \varphi_2$
3	l_3	$NpS_0S_1S_2 = \varphi_3$	$l_5 + l_4 + l_3$	$\varphi_5 + \varphi_4 + \varphi_3$
4	l_4	$NpS_0S_1S_2S_3 = \varphi_4$	$l_5 + l_4$	$\varphi_5 + \varphi_4$
5	l_5	$NpS_0S_1S_2S_3S_4 = \varphi_5$	l_5	φ_5

$$E(\hat{S}_4) = E\left(\frac{l_5}{l_4}\right) \doteq E\left(\frac{\varphi_5}{\varphi_4 + \varphi_5}\right)$$

$$= \frac{NpS_0S_1S_2S_3S_4}{NpS_0S_1S_2S_3 + NpS_0S_1S_2S_3S_4}$$

$$= \frac{S_4}{1 + S_4}.$$

The estimator $\hat{S}_4$ is biased, as will be the other estimates (Table 5.9). Hence, using cumulative totals (Table 5.9) from a vertical life table will not yield the survival estimates desired or intended.

Example 5.6: Vertical Life-Table Analysis for White-Tailed Deer, Michigan

A sample of age-structure data (Table 5.10) from harvested male white-tailed deer will be used to illustrate the analyses of a vertical life table. The data from the Upper Peninsula of Michigan were collected at a check station in 1959 (Eberhardt 1969). Only animals age 1.5 years and older are used in the calculations because young-of-the-year were considered more vulnerable to hunting, invalidating the assumption of equal probability of selection among all age classes. Therefore, the 0 age class is 1.5-year-olds (yearlings), and recruitment into this age class is assumed constant.

Age-specific survival probabilities for the first two age classes are estimated by using Eq. (5.50) as follows:

$$\hat{S}_x = \frac{l_{x+1}}{l_x}$$

$$\hat{S}_0 = \frac{274}{425} = 0.6447$$

$$\hat{S}_1 = \frac{149}{274} = 0.5438.$$

The estimated variances for $\hat{S}_0$ and $\hat{S}_1$ are calculated by using Eq. (5.53), where $\hat{\Psi} = 2.2046$, the sum of the values in the $\hat{S}_{0x}$ column, and $l. = 937$, where

$$\widehat{\text{Var}}(\hat{S}_0) = \frac{\hat{\Psi}\hat{S}_X(1+\hat{S}_X)}{l.\prod_{i=0}^{x-1}\hat{S}_x} = \left(\frac{2.2046 \cdot (0.6447) \cdot (1 + 0.6447)}{937}\right) = 0.00249,$$

or $\widehat{\text{SE}}(\hat{S}_0) = 0.0499$ and

$$\widehat{\text{Var}}(\hat{S}_1) = \left(\frac{2.2046 \cdot (0.5438) \cdot (1 + 0.5438)}{937 \cdot (0.6447)}\right) = 0.003063$$

or $\widehat{\text{SE}}(\hat{S}_1) = 0.0553$. Repeating the estimation process for all age classes, a nonparametric survivorship curve (Fig. 5.6) for the white-tailed deer population can be constructed.

Table 5.10.　Results for the vertical life table analysis of age-structure data for harvested male white-tailed deer from a check station in Michigan during the 1959 hunting season.

Age	Coded age (x)	l_x	$\hat{S}_x$ Eq. (5.50)	$\widehat{SE}(\hat{S}_x)$ from Eq. (5.51)	$\hat{S}_{0x}$ Eq. (5.52)	$\widehat{SE}(\hat{S}_{0x})$ from Eq. (5.53)
1.5	0	425	0.6447	0.0500	1.0000	0
2.5	1	274	0.5437	0.0553	0.6447	0.0499
3.5	2	149	0.3557	0.0569	0.3505	0.0334
4.5	3	53	0.3207	0.0894	0.1247	0.0182
5.5	4	17	0.4706	0.2018	0.0400	0.0099
6.5	5	8	0.7500	0.4051	0.0188	0.0067
7.5	6	6	0.5000	0.3536	0.0141	0.0058
8.5	7	3	0.3333	0.3849	0.0071	0.0041
9.5	8	1	1.000	1.4145	0.0024	0.0024
10.5	9	1	0	0	0.0024	0.0024

(Data from Eberhardt 1969:473.)

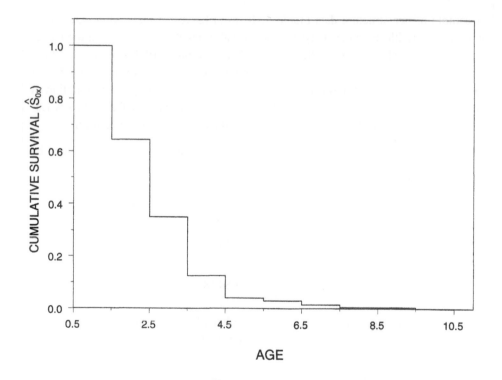

Figure 5.6.
Nonparametric survivorship curve resulting from vertical life-table analysis of male white-tailed deer in Michigan (Eberhardt 1969).

Example 5.7: Weibull Analysis of Vertical Life-Table Data for Grizzly Bears, Yellowstone National Park, Wyoming

A parametric survival curve can be constructed from age-specific data used in vertical life-table analysis. Craighead et al. (1974) presented an age-specific life table for grizzly bears from the Yellowstone National Park, Wyoming, for the period 1959–1967. The numbers alive in each age class were derived from data collected over the entire period of the study, and were considered more representative of the true age distribution than actual census figures (Craighead et al. 1974:12–13). To illustrate parametric methods, a Weibull distribution was fit to the vertical life-table data of male grizzly bears.

The likelihood for the Weibull distribution is

$$L(\alpha, \beta | x_i) = \prod_{i=0}^{n} \left(\frac{\beta x_i^{\beta-1}}{\alpha^\beta} \right) e^{-\left(\frac{x_i}{\alpha}\right)^\beta}$$

where x_i = age at death of the ith individual. The x_i's were calculated for the 19 deaths by age interval. Because precise time of death is unknown, the midpoint of the interval was chosen as the x_i for all individuals (Table 5.11). From the life-table data, estimates and standard errors for the parameters of the Weibull distribution are

$$\hat{\alpha} = 4.7532, \text{ with } \widehat{SE}(\hat{\alpha}) = 0.1801$$

and

$$\hat{\beta} = 1.0069, \text{ with } \widehat{SE}(\hat{\beta}) = 1.1403.$$

When $\beta = 1$, the Weibull distribution is equivalent to the exponential distribution with parameter α. A LRT can be used to test the null hypothesis $H_o : \beta = 1$. The LRT statistic is

$$\chi_1^2 = -2(-48.55216 - (-48.55117)) = 0.00198,$$

which gives a P-value of $P(\chi_1^2 > 0.00198) = 0.9645$. Hence, the grizzly bear data can be simply modeled by using an exponential survival curve (Fig. 5.7) with $\hat{\alpha} = 4.7368$ ($\widehat{SE} = 1.0867$). The expected lifetime for an exponential random variable is $E(x_i) = \alpha$. Thus, the survival analysis estimates a Yellowstone grizzly bear of age 0.5 year to have a life expectancy of approximately 4.7 years.

Table 5.11. Age structure of grizzly bears in Yellowstone National Park, Wyoming, between 1959 and 1967.

Age class	No. (l_x)	Deaths in interval	Midpoint (x_i)
0.5	19.5		
1.5	14.5	5.0	1.0
2.5	9.9	5.0	2.0
3.5	8.5	1.0	3.0
4.5	7.0	2.0	4.0
5.5	3.6	3.0	5.0
6.5	3.4		
7.5	3.2		
8.5	3.1		
9.5	3.0	1.0	9.5
10.5	2.9		
11.5	2.8		
12.5	2.7		
13.5	2.4		
14.5	2.1		
15.5	1.8		
16.5	1.2	1.0	16.5
17.5	1.0		
18.5	0.8		
19.5	0.6		
20.5	0.5		
21.5	0.4		
22.5	0.3		
23.5	0.2	1.0	23.0
24.5	0.1		
25.5	0.1		

(Data from Craighead et al. 1974.)

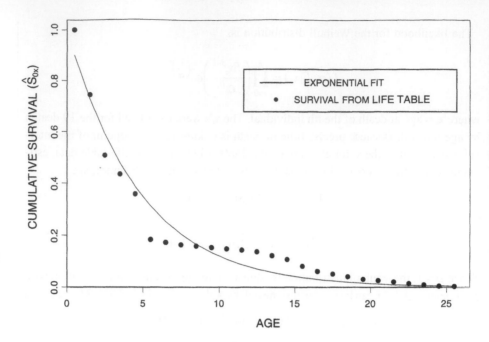

Figure 5.7. Survival curve for grizzly bears in Yellowstone National Park, Wyoming (Craighead et al. 1974), derived from the age structure of live animals.

5.5.2 Estimating Survival with Truncated Age Classes

Reliable age-frequency data are only available for a few younger age classes for some wildlife species. The multinomial likelihood model (5.49) can be modified so that age-specific survival for a subset of age classes is estimated. Derivation of the truncated model is the same as for the full model. The frequency of the youngest age class in the sample will be denoted as l_0, the next oldest age class frequency as l_1, and so on. The last reliably classified age class will have a count of l_n.

The likelihood model for truncated age classes is analogous to Eq. (5.49). The likelihood for the younger age classes only is

$$L(S_x | l., l_x) = \binom{l.}{l_0, l_1, \ldots, l_n} \left(\frac{1}{\Psi}\right)^{l_0} \left(\frac{S_0}{\Psi}\right)^{l_1} \left(\frac{S_0 S_1}{\Psi}\right)^{l_2} \cdots \left(\frac{S_0 S_1 \ldots S_{n-1}}{\Psi}\right)^{l_n} \quad (5.56)$$

The age-specific survival estimates are given by Eq. (5.50). The associated variances are the same as those for the full l_x-series life table (Eq. 5.52), with estimated variances in Eq. (5.53).

5.5.3 Vertical Life Table with Constrained Survival

The models presented thus far have assumed age-specific survival. However, it may be more reasonable to assume constant survival across adult age classes. To do so, one merely adds the assumption of a common survival probability for older ages to the likelihood model for the l_x series (Eq. 5.49). The method is also useful when the number of animals in successive age classes is not monotonically decreasing as a result of sampling error.

As an example, assume that survival is age-specific for the first two age classes (i.e., S_0, S_1) and constant (S_c) for age classes 3 to $w - 2$, and that survival is age-specific for the last two age classes (i.e., S_{w-2}, S_{w-1}). The likelihood for the l_x series can be written as

$$L(S_x|l_\cdot, l_x) = \binom{l_\cdot}{l_x}\left(\frac{1}{\Psi}\right)^{l_0}\left(\frac{S_0}{\Psi}\right)^{l_1}\left(\frac{S_0 S_1}{\Psi}\right)^{l_2}\left(\frac{S_0 S_1 S_c}{\Psi}\right)^{l_3}\left(\frac{S_0 S_1 S_c^2}{\Psi}\right)^{l_4}\cdots$$

$$\left(\frac{S_0 S_1 \ldots S_c^{w-4}}{\Psi}\right)^{l_{w-2}}\left(\frac{S_0 S_1 \ldots S_c^{w-4} S_{w-2}}{\Psi}\right)^{l_{w-1}}\left(\frac{S_0 S_1 \ldots S_c^{w-4} S_{w-2} S_{w-1}}{\Psi}\right)^{l_w}$$

$$(5.57)$$

where $\Psi = 1 + S_0 + S_0 S_1 + \cdots + S_0 S_1 \ldots S_c^{w-4} S_{w-2} S_{w-1}$.

The MLEs for S_0, S_1, S_{w-2}, and S_{w-1} are the same as in Eq. (5.50), i.e., $\hat{S}_x = \frac{l_{x+1}}{l_x}$.
However, the point estimate and associated variance estimate for S_c need to be obtained numerically.

5.6   Depositional Life Tables

Caughley (1977:92) and Seber (1982:407) mention that a vertical life table can be constructed from the frequency of ages at death originating from carcasses accumulated over time as a result of natural causes. Examples of such data include horns collected from Dall sheep (Deevey 1947) and sea otter (*Enhydra lutris*) carcasses found in spring (Udevitz and Ballachey 1998). The number of animals found dead of age class x will be denoted by c_x. The biological and sampling processes that generate this c_x series of counts is different from both vertical and horizontal life tables.

The collected carcasses or other remains may have accumulated over several years (k) in a c_x series. Hence, the animals in any individual age class will be represented by k different cohorts. Thus, the observed counts are a function of mortality processes, recruitment, and carcass decay. To isolate the survival process, it is necessary to assume the population was stable and stationary with constant numbers of new recruits N_0 each year. There is the additional assumption of a constant decay or loss rate of carcasses for all age classes, so that k is equal for all age classes. Under these assumptions, the expected values of the c_x series (Table 5.12) are similar to the d_x values in a horizontal life table (Table 5.5).

Table 5.12.   The expected number of animals in each age class of a depositional life table, assuming a stable and stationary population.

Age class	Observed no. of animals of age $x(c_x)$	Expected no. in any cohort that died at age x	$E(c_x)$
0	c_0	$N_0(1 - S_0)$	$kN_0 p(1 - S_0)$
1	c_1	$N_0 S_0(1 - S_1)$	$kN_0 p S_0(1 - S_1)$
2	c_2	$N_0 S_0 S_1(1 - S_2)$	$kN_0 p S_0 S_1(1 - S_2)$
3	c_3	$N_0 S_0 S_1 S_2(1 - S_3)$	$kN_0 p S_0 S_1 S_2(1 - S_3)$
$\vdots$	$\vdots$	$\vdots$	$\vdots$
$w - 1$	c_{w-1}	$N_0 S_0 S_1 \ldots S_{w-2}(1 - S_{w-1})$ or $N_0\left(\prod_{x=0}^{w-2} S_x\right)(1 - S_{w-1})$	$kN_0 p S_0 S_1 \ldots S_{w-2}(1 - S_{w-1})$ or $kN_0 p\left(\prod_{x=0}^{w-2} S_x\right)(1 - S_{w-1})$
w	c_w	$N_0 S_0 S_1 \ldots S_{w-1}(1 - S_w)$ or $N_0\left(\prod_{x=0}^{w-1} S_x\right)$	$kN_0 p S_0 S_1 \ldots S_{w-1}(1 - S_w)$ or $kN_0 p\left(\prod_{x=0}^{w-1} S_x\right)$

5.6.1   Standard Life-Table Analysis

The number of carcasses observed in each age class can be modeled as a Poisson distribution where

$$P(C_x = c_x) = \frac{\varphi_x^{c_x} e^{-\varphi_x}}{c_x!}$$

c_x = number of carcasses of animals that died at age x;

$$\varphi_x = E(c_x) = kN_0 p \left(\prod_{i=0}^{x-1} S_i \right)(1 - S_x);$$

k = number of cohorts contributing carcasses;
p = probability a carcass is recovered.

Because the sum of independent Poisson random variables is also Poisson-distributed, the total number of individuals in the sample has the distribution

$$P(C. = c.) = \frac{\varphi.^{c.} e^{-\varphi.}}{c.!},$$

where

$$c. = \sum_{x=0}^{w} c_x, \text{ and}$$

$$\varphi. = E(c.) = \sum_{i=0}^{w} E(c_x) = kN_0 p \sum_{i=0}^{w} \left(\left(\prod_{i=0}^{x-1} S_i \right)(1 - S_x) \right).$$

Note the probability of surviving past age w is assumed to be zero (i.e., $S_w = 0$). By using the joint likelihood of all age classes and conditioning on the total sample size, $c.$, the conditional likelihood is

$$L(\varphi_x | c., \underset{\sim}{c}_x) = \frac{\prod\limits_{i=0}^{w} \dfrac{\varphi_x^{c_x} e^{-\varphi_x}}{c_x!}}{\dfrac{\varphi.^{c.} e^{-\varphi.}}{c.!}},$$

or, equivalently,

$$L(\varphi_x | c., \underset{\sim}{c}_x) = \binom{c.}{\underset{\sim}{c}_x} \prod_{x=0}^{w} \left(\frac{\varphi_x}{\varphi.} \right)^{c_x}. \tag{5.58}$$

Rewriting the likelihood in terms of the survival parameters, Eq. (5.58), yields a multinomial model for the sampling distribution of the c_x series as follows:

$$L(\underset{\sim}{S}_x | c., \underset{\sim}{c}_x) = \binom{c.}{\underset{\sim}{c}_x} (1 - S_0)^{c_0} (S_0(1 - S_1))^{c_1} (S_0 S_1 (1 - S_2))^{c_2} \cdots \left(\prod_{x=0}^{w-1} S_x \right)^{c_w}. \tag{5.59}$$

Likelihood equation (5.59) is the same as that of a horizontal life table (Eq. 5.28). Thus, the parameter estimates and associated variances are equivalent to Eqs. (5.30) and (5.32).

Assumptions

The depositional likelihood, Eq. (5.59), has the following assumptions:

1. The population was stable and stationary during the period of carcass accumulation represented in the sample.
2. The age distribution of the population was stable.
3. All carcasses have the same probability of selection.
4. The sample is representative of the age structure of naturally dying animals in the population.
5. The fate of each animal is independent.
6. All ages were recorded accurately (no measurement error).
7. Carcass decay rates are the same for all age classes.
8. The number of cohorts available for sampling, k, is equal for all age classes.

This method has the greatest number of assumptions of all life-table analyses presented thus far. If data are from a population in which carcasses are only identifiable for a year or less after death, then $k = 1$ and the data can still be represented by the likelihood, Eq. (5.59). The above assumptions 1, 7, and 8 ensure that individuals in each age class have an equal probability of being included in the sample, so that survival, S_x, and detection, p, probabilities are not confounded. Accurately classifying the age of animals, assumption 6, may be difficult, thus limiting the utility of this method to unique situations and species such as Dall sheep, as in the example below.

Example 5.8: Depositional Life-Table Analysis for Dall Sheep, Alaska

Adolph Murie collected horns from 608 Dall sheep found dead in Mt. McKinley National Park (Deevey 1947). Rings on the sheep horns were used to classify age of the animals. Assuming a stable and stationary population over the years of accumulation and a constant decay rate for all cohorts, age-specific survival and associated variances can be estimated from these data by using Eqs. (5.30) and (5.32), respectively (Table 5.13).

Survival estimates for the first two age classes, 0 and 1, respectively, are as follows:

$$\hat{S}_0 = \frac{c. - \sum\limits_{i=0}^{0} c_i}{c.} = \frac{c. - c_0}{c.} = \frac{608 - 33}{608} = 0.9457,$$

$$\hat{S}_1 = \frac{c. - \sum\limits_{i=0}^{1} c_i}{c. - \sum\limits_{i=0}^{0} c_i} = \frac{c. - (c_1 + c_0)}{c. - c_0} = \frac{608 - (33 + 88)}{608 - 33} = 0.8470.$$

All other survival estimates are calculated similarly. Variance calculations for the first two survival estimates follow from Eqs. (5.33) and (5.32), respectively, where

$$\widehat{\text{Var}}(\hat{S}_0) = \frac{\hat{S}_0(1 - \hat{S}_0)}{c.} = \frac{0.9457(1 - 0.9457)}{608} = 0.000084,$$

$$\widehat{\text{SE}}(\hat{S}_0) = 0.0092$$

and

$$\widehat{Var}(\hat{S}_x) = \frac{(1-\hat{S}_x)^2}{c.\hat{S}_{0x}^2}\left[\frac{\hat{S}_{0x}(1-\hat{S}_{0x}(1-\hat{S}_x))}{(1-\hat{S}_x)} + \sum_{i=0}^{x-1}\hat{S}_{0i}(1-\hat{S}_i)(1-\hat{S}_{0i}(1-\hat{S}_i))\right.$$

$$\left. -2\sum_{\substack{i=0 \\ i<j}}^{x-1}\sum_{j=0}^{x-1}\hat{S}_{0i}(1-\hat{S}_i)\hat{S}_{0j}(1-\hat{S}_j) - 2\hat{S}_{0x}\sum_{j=0}^{x-1}\hat{S}_{0j}(1-\hat{S}_j)\right],$$

$$\widehat{Var}(\hat{S}_1) = \frac{(1-0.8470)^2}{608\cdot(0.9457)^2}\left[\frac{0.9457(1-0.9457(1-0.8470))}{(1-0.8470)}\right.$$

$$\left. + (1-0.9457)0.9457 - 2\cdot0.9457(1-0.9457)\right]$$

$$\widehat{Var}(\hat{S}_1) = 0.00023,$$

or $\widehat{SE}(\hat{S}_1) = 0.0150.$

By using similar analyses, age-specific survival estimates for age classes 0 to 13 years can be calculated for the Dall sheep population (Table 5.13). Siler (1979) assumed a 10% to 20% decay loss for the skulls of juveniles, ages 0 to 6 months, thereby adjusting the c_x counts and subsequent survival estimates.

5.6.2 Nonstationary Populations: Udevitz and Ballachey (1998)

Udevitz and Ballachey (1998) developed modifications to the depositional (c_x-series) and vertical (i.e., l_x-series) life tables that relax the assumption of a stationary population. Instead of assuming a stationary population, population abundance was assumed to change annually at a constant finite rate of change (i.e., $N_{t+1} = \lambda N_t$, $\forall t$). Their approach

Table 5.13. The Dall sheep carcass count data for a c_x-series life table with age-specific survival estimates and associated standard errors.

Age	Age class	c_x	$\hat{S}_x$ from Eq. (5.30)	$\widehat{SE}(\hat{S}_x)$ from Eq. (5.32)	$\hat{S}_{0x}$ from Eq. (5.36)	$\widehat{SE}(\hat{S}_{0x})$ from Eq. (5.37)
0–0.5	0	33	0.9460	0.0092	1	—
0.5–1	1	88	0.8470	0.0151	0.9457	0.0092
1–2	2	7	0.9856	0.0054	0.8013	0.0162
2–3	3	8	0.9833	0.0057	0.7897	0.0165
3–4	4	7	0.9852	0.0056	0.7765	0.0169
4–5	5	18	0.9613	0.0089	0.7650	0.0168
5–6	6	28	0.9374	0.0115	0.7354	0.0179
6–7	7	29	0.9308	0.0124	0.6894	0.0188
7–8	8	42	0.8923	0.0157	0.6417	0.0194
8–9	9	80	0.7701	0.0225	0.5726	0.0201
9–10	10	114	0.5746	0.0302	0.4409	0.0201
10–11	11	95	0.3831	0.0392	0.2534	0.0176
11–12	12	55	0.0678	0.0327	0.0971	0.0120
12–13	13	2	0.5000	0.2500	0.0066	0.0033
13–14	14	2	0.0000	0.0000	0.0033	0.0023

(Data from Deevey 1947.)

is similar to that (method 6) of Caughley (1977:86–95). They developed estimates of age-specific survival from depositional and vertical life tables when λ is assumed known. They also used the joint likelihood of both the depositional and vertical life tables to estimate λ when it is not known. A more detailed description of the development is presented here than provided by the original paper.

By use of the finite rate of population change, λ, the numbers of individuals of age 0 for each cohort of a l_x series can be written as a function of N_{0t}, the number of recruits at the current time t. For example, let the population growth rate between the current cohort, N_{0t}, and previous year's age 0 cohort, N_{0t-1}, be written as

$$\frac{N_{0,t}}{N_{0,t-1}} = \lambda_0 \quad \text{or} \quad N_{0,t-1} = \frac{N_{0t}}{\lambda_0}.$$

The expected number of recruits to age class 0 for each cohort of a l_x series, written as a function of N_{0t} and λ_t, is illustrated (Table 5.14). The expected values of a l_x series (Table 5.14) are written in terms of survival and recovery probabilities, p.

Under the assumption of constant age-specific survival probabilities, the expected value of l_x can be written as

$$\varphi_x = E(l_x) = \frac{N_{0,t}S_0S_1 \ldots S_{x-1}p}{\lambda_0 \ldots \lambda_{x-1}}.$$

Construction of the likelihood for this l_x series is the same as presented previously for the vertical life table (Section 5.5). The number of animals observed in each age class is modeled as a Poisson random variable and written as

$$P(L_x = l_x) = \frac{\varphi_x^{l_x} e^{-\varphi_x}}{l_x!}. \tag{5.60}$$

Because the sum of independent Poisson random variables is also Poisson-distributed, the total number of individuals in the sample has the distribution:

$$P(L. = l.) = \frac{\varphi^{l.} e^{-\varphi.}}{l.!},$$

where

Table 5.14. The expected values for the l_x counts used in the vertical life table for each age class expressed in terms of $N_{0,t}$, S_i, and p_j for a nonstationary population.

Age interval	l_x	No. of age 0 entering each cohort	$E(l_x)$
0–1	l_0	$N_{0,t}$	$N_{0,t}p$
1–2	l_1	$N_{0,t-1} = \dfrac{N_{0,t}}{\lambda_0}$	$\dfrac{N_{0,t}}{\lambda_0}S_0p$
2–3	l_2	$N_{0,t-2} = \dfrac{N_{0,t}}{\lambda_1\lambda_0}$	$\dfrac{N_{0,t}}{\lambda_1\lambda_0}S_0S_1p$
3–4	l_3	$N_{0,t-3} = \dfrac{N_{0,t}}{\lambda_2\lambda_1\lambda_0}$	$\dfrac{N_{0,t}}{\lambda_2\lambda_1\lambda_0}S_0S_1S_2p$
$\vdots$	$\vdots$	$\vdots$	$\vdots$
$w-1$ to w	l_{w-1}	$N_{0,t-(w-1)} = \dfrac{N_{0,t}}{\lambda_{w-2}\ldots\lambda_0}$	$\dfrac{N_{0,t}}{\lambda_{w-2}\ldots\lambda_0}S_0S_1S_2\ldots S_{w-2}p$

$$l. = \sum_{x=0}^{w} l_x,$$

$$\varphi. = E(l.) = N_{0,t} p \sum_{x=0}^{w} \left(\frac{\prod_{i=0}^{x-1} S_i}{\prod_{i=0}^{x-1} \lambda_i} \right).$$

The joint likelihood for all age classes, conditional on total sample size $l.$, can be written as

$$L(\varphi_x | l., l_x) = \frac{\prod_{x=0}^{w} \frac{\varphi_x^{l_x} e^{-\varphi_x}}{l_x!}}{\frac{\varphi.^{l.} e^{-\varphi.}}{l.!}},$$

or, equivalently,

$$L(\varphi_x | l., l_x) = \binom{l.}{l_x} \prod_{x=0}^{w} \left(\frac{\varphi_x}{\varphi.} \right)^{l_x}. \tag{5.61}$$

Rewriting the likelihood, Eq. (5.61), in terms of the demographic parameters yields a multinomial model for the sampling distribution of the l_x series

$$L(S_x | l., l_x) = \binom{l.}{\underset{\sim}{l_x}} \left(\frac{1}{\Psi} \right)^{l_0} \left(\frac{S_0}{\lambda_0 \Psi} \right)^{l_1} \left(\frac{S_0 S_1}{\lambda_0 \lambda_1 \Psi} \right)^{l_2} \cdots \left(\frac{S_0 S_1 \ldots S_{(w-1)}}{\lambda_0 \ldots \lambda_{w-1} \Psi} \right)^{l_w}$$

where

$$\Psi = 1 + \frac{S_0}{\lambda_0} + \frac{S_0 S_1}{\lambda_0 \lambda_1} + \frac{S_0 S_1 S_2}{\lambda_0 \lambda_1 \lambda_2} + \cdots + \frac{\prod_{x=0}^{w-1} S_x}{\prod_{x=0}^{w-1} \lambda_x}.$$

Further assuming a constant annual finite growth rate, λ, the likelihood can be rewritten as

$$L(S_x | l., l_x) = \binom{l.}{\underset{\sim}{l_x}} \left(\frac{1}{\Psi'} \right)^{l_0} \left(\frac{S_0}{\lambda \Psi'} \right)^{l_1} \left(\frac{S_0 S_1}{\lambda^2 \Psi'} \right)^{l_2} \cdots \left(\frac{S_0 S_1 \ldots S_{w-1}}{\lambda^w \Psi'} \right)^{l_w} \tag{5.62}$$

where

$$\Psi' = 1 + \frac{S_0}{\lambda} + \frac{S_0 S_1}{\lambda^2} + \frac{S_0 S_1 S_2}{\lambda^3} + \cdots + \frac{\prod_{x=0}^{w-1} S_i}{\lambda^w}.$$

In a similar manner, the likelihood for carcass counts, c_x, can also be rewritten for a population with constant age-specific survival probabilities and a constant rate of population increase, where

$$L(\underline{S}_x|c.,\underline{c}_x) = \binom{c.}{\underline{c}_x}\left(\frac{1-S_0}{\Upsilon}\right)^{c_0}\left(\frac{S_0(1-S_1)}{\lambda\Upsilon}\right)^{c_1}\left(\frac{S_0 S_1(1-S_2)}{\lambda^2\Upsilon}\right)^{c_2}\cdots\left(\frac{\prod_{x=0}^{w-1}S_x}{\lambda_w\Upsilon}\right)^{c_w} \quad (5.63)$$

and

$$\Upsilon = 1 + S_0 + \frac{S_0(1-S_1)}{\lambda} + \frac{S_0 S_1(1-S_2)}{\lambda^2} + \cdots + \frac{(1-S_{w-1})\prod_{i=0}^{w-2}S_i}{\lambda^{w-1}} + \frac{\prod_{i=0}^{w-1}S_i}{\lambda^w}.$$

The assumptions associated with likelihood (5.63) are the same as those for Eq. (5.59), except the assumption of a stationary population has been relaxed to allow a constant annual rate of growth of λ.

Udevitz and Ballachey (1998) derived estimates of S_x for both the l_x series and the c_x series, assuming λ is known. For the l_x series, Udevitz and Ballachey (1998) provided the estimator

$$\hat{S}_{x(UB)} = \frac{l_{x+1}}{l_x}\lambda. \quad (5.64)$$

Assuming λ is independently estimated, the variance of $\hat{S}_{x(UB)}$ can be derived by using the exact product formula (Goodman 1960), noting that Eq. (5.64) is equal to $\hat{S}_x \cdot \lambda$, where $\hat{S}_x$ is estimated by Eq. (5.50). The associated variance estimator (5.64) is

$$\widehat{\mathrm{Var}}\left(\hat{S}_{x(UB)}\right) = \hat{\lambda}^2\,\widehat{\mathrm{Var}}\left(\hat{S}_x\right) + \hat{S}_x^2\,\widehat{\mathrm{Var}}\left(\hat{\lambda}\right) - \widehat{\mathrm{Var}}\left(\hat{S}_x\right)\widehat{\mathrm{Var}}\left(\hat{\lambda}\right), \quad (5.65)$$

where

$$\widehat{\mathrm{Var}}\left(\hat{S}_x\right) = \left(\frac{\hat{\Psi}'\hat{S}_x(1+\hat{S}_x)}{l.\prod_{i=0}^{x-1}\hat{S}_i}\right)$$

and

$$\hat{\Psi}' = 1 + \frac{\hat{S}_0}{\hat{\lambda}} + \frac{\hat{S}_0\hat{S}_1}{\hat{\lambda}^2} + \frac{\hat{S}_0\hat{S}_1\hat{S}_2}{\hat{\lambda}^3} + \cdots + \frac{\prod_{i=0}^{w-1}\hat{S}_i}{\hat{\lambda}^w}.$$

For depositional data from a c_x series, Udevitz and Ballachey (1998) derived the survival estimator

$$\hat{S}_{x(UB)} = 1 - \frac{c_x\lambda^x}{\sum_{i=x}^{w}c_i\lambda^i} \quad (5.66)$$

with associated variance estimator

$$\widehat{Var}\left(\hat{S}_x\right) = \frac{\left(1-\hat{S}_x\right)^2}{c.\hat{S}_{0x}{}^2}\left\{\frac{\hat{S}_x\hat{S}_{0x}\left(\lambda^x\hat{\Psi}'-\hat{S}_{0x}\left(1-\hat{S}_x\right)\right)}{\left(1-\hat{S}_x\right)}\right.$$

$$+ \sum_{i=x+1}^{w}\left[\hat{S}_{0i}\left(1-\hat{S}_i\right)\left(\lambda^x\hat{\Psi}'-\hat{S}_{0i}\left(1-\hat{S}_i\right)\right)+2\hat{S}_x\hat{S}_{0x}\hat{S}_{0i}\left(1-\hat{S}_i\right)\right] \quad (5.67)$$

$$\left. - 2\sum_{i=x+1}^{w-1}\sum_{j=i+1}^{w}\hat{S}_{0i}\left(1-\hat{S}_{0i}\right)\hat{S}_{0j}\left(1-\hat{S}_{0j}\right)\right\},$$

where

$$c. = \sum_{i=0}^{w}c_i,$$

$$\hat{\Psi}' = 1 + \frac{\hat{S}_0}{\lambda} + \frac{\hat{S}_0\hat{S}_1}{\lambda^2} + \frac{\hat{S}_0\hat{S}_1\hat{S}_2}{\lambda^3} + \cdots + \frac{\prod_{i=0}^{w-1}\hat{S}_i}{\lambda^w}.$$

If $\hat{\lambda}$ is estimated independently with an associated variance, $\widehat{Var}(\hat{\lambda})$, the term

$$\left[\frac{\left(1-\hat{S}_x\right)}{\hat{\lambda}\hat{S}_{0x}}\left(\sum_{i=x+1}^{w}\hat{S}_{0i}\right)\right]^2\widehat{Var}(\hat{\lambda})$$

is added to Eq. (5.67).

If λ is unknown, neither likelihoods (5.62) nor (5.63) can estimate age-specific survivals alone. However, by using their joint likelihood

$$L(\underline{S}_x,\lambda|\underline{l}_x,\underline{c}_x,l.,c.) = L(\underline{S}_x,\lambda|\underline{l}_x,l.)\cdot L(\underline{S}_x,\lambda|\underline{c}_x,c.), \quad (5.68)$$

λ, as well as the survival probabilities, are estimable. The survival probabilities and λ can only be estimated by using iterative methods because the number of parameters in Eq. (5.68) is $w + 1$ and the number of minimally sufficient statistics is $2w$.

Assumptions

The assumptions of the joint likelihood model (5.68) are a composite of the assumptions for both vertical and depositional life tables. In addition, there is the further assumption the finite rate of annual change (λ) is constant across years.

Inclusion of the finite rate of change parameter (λ) will reduce the bias associated with violation of the assumption of a stationary population in vertical and depositional life-table analyses. However, violating the assumption of a constant λ can also introduce bias in the survival estimates of various magnitudes and direction. Udevitz and Ballachey (1998) assert that if the geometric mean of λ is near one, it would be best to assume a stationary population. However, it is impossible to test the assumption of constant λ using solely the data in Eqs. (5.62) and (5.63).

5.7 l_x-Series Data with Abbreviated or Pooled Age Classes

Estimates of a common adult survival probability across age classes have been developed by numerous wildlife investigators (Heincke 1913, Hayne and Eberhardt 1952, Hesselton et al. 1965, Severinghaus 1969, Burgoyne 1981) assuming a stable, stationary, and closed population. In all of these cases, age-class data have either been collapsed or limited to a few age classes to simplify the computations or logistics of the data collection. These early method-of-moment estimators typically lacked variance estimates and a careful description of assumptions. A formal review of some of these approaches is provided below.

5.7.1   Hayne and Eberhardt (1952)

Hayne and Eberhardt (1952) provided an estimator of an annual survival probability common across all age classes, S, based on the number of animals in each of the first three age classes of a population. The method was originally derived for the number of deer harvested in the 1.5-, 2.5-, and 3.5-year-old age classes.

The Hayne and Eberhardt (1952) estimator is

$$\hat{S} = \frac{1-A}{1+A},$$

where

$$A = \frac{x_{1.5} - x_{3.5}}{x_{1.5} + 2x_{2.5} + x_{3.5}},$$

and x_i = number of deer harvested of age class i (i = 1.5, 2.5, 3.5). However, the estimator may be more simply expressed as

$$\hat{S} = \frac{x_{2.5} + x_{3.5}}{x_{1.5} + x_{2.5}}. \tag{5.69}$$

Without specifying a sampling distribution, no variance estimator for Eq. (5.69) was given by Hayne and Eberhardt (1952).

By use of a vertical life-table approach to model the harvest numbers of 1.5-, 2.5-, and 3.5-year-old deer in a l_x series, one can derive the MLE and associated variance of annual survival. Keeping with life-table notation, let

$$x_{1.5} = l_0,$$
$$x_{2.5} = l_1,$$
$$x_{3.5} = l_2.$$

The conditional likelihood of $\underline{l}$, given $l. = l_0 + l_1 + l_2$, can be written as

$$L(S|l., l_0, l_1, l_2) = \binom{l.}{l_0, l_1, l_2} \left(\frac{1}{1+S+S^2} \right)^{l_0} \left(\frac{S}{1+S+S^2} \right)^{l_1} \left(\frac{S^2}{1+S+S^2} \right)^{l_2}. \tag{5.70}$$

This likelihood model has one parameter (S) and two minimum sufficient statistics (l_0 and l_1). The MLE for S is the positive root of the quadratic equation

$$(2l_0 + l_1)S^2 + (l_0 - l_2)S + (-l_1 - 2l_2) = 0,$$

or, more specifically,

$$\hat{S} = \frac{(l_2 - l_0) + \sqrt{(l_0 - l_2)^2 + 4(2l_0 + l_1)(l_1 + 2l_2)}}{2(2l_0 + l_1)}. \tag{5.71}$$

An asymptotic variance for Eq. (5.71) is

$$\mathrm{Var}(\hat{S}) = \frac{S(1 + S + S^2)^3}{l.(1 + 6S^2 + 13S^3 + S^4)}, \tag{5.72}$$

with an estimated variance of

$$\widehat{\mathrm{Var}}(\hat{S}) = \frac{\hat{S}(1 + \hat{S} + \hat{S}^2)^3}{l.(1 + 6\hat{S}^2 + 13\hat{S}^3 + \hat{S}^4)}. \tag{5.73}$$

Assumptions

The assumptions of the Hayne and Eberhardt (1952) method are more restrictive than those of the vertical life table:

1. There is a stable age distribution with constant annual survival over time and across age classes.
2. The population is stationary with constant abundance (N_0) across years.
3. All animals have an equal probability of selection, which also implies a constant harvest rate across all age classes if harvest data are used.
4. The age-structure data are a random sample of the population (representative sample).
5. Fates of all animals are independent.
6. All ages are recorded accurately.

The key difference in assumptions between the Hayne and Eberhardt (1952) method and a vertical life table is the added requirement for a constant survival probability across age classes. In practice, any three consecutive age classes for which survival is considered constant could be used for data analysis. With big-game species, there may be several options regarding which age classes to use. However, one practical motivation for using age classes 1.5 through 3.5 is the accurate aging of deer by using tooth eruption and wear techniques (Jacobson and Renier 1989) in this age range. For most species, accuracy of the tooth-wear technique decreases with age (Jacobson and Renier 1989). Use of tooth annuli methods would possibly eliminate some of the concern.

Example 5.9: Hayne–Eberhardt (1952) Estimator of Survival for White-Tailed Deer

The following example illustrates the difference between the estimates obtained by using the Hayne and Eberhardt (1952) method-of-moments estimator, Eq. (5.69) and MLE Eq. (5.71). Consider a check station where the following hypothetical harvest counts were recorded:

$$x_{1.5} = l_0 = 97,$$
$$x_{2.5} = l_1 = 73,$$
$$x_{3.5} = l_2 = 58.$$

The Hayne and Eberhardt (1952) estimator, Eq. (5.69), yields

$$\hat{S} = \frac{x_{2.5} + x_{3.5}}{x_{1.5} + x_{2.5}}$$

$$= \frac{73 + 58}{97 + 73} = 0.7706.$$

The MLE (Eq. 5.71) is

$$\hat{S} = \frac{(l_2 - l_0) + \sqrt{(l_0 - l_2)^2 + 4(2l_0 + l_1)(l_1 + 2l_2)}}{2(2l_0 + l_1)}$$

$$= \frac{(58 - 97) + \sqrt{(97 - 58)^2 + 4(2(97) + 73)(73 + 2(58))}}{2(2(97) + 73)} = 0.7715.$$

The estimated variance of $\hat{S}$, Eq. (5.73), is

$$\widehat{\text{Var}}(\hat{S}) = \frac{\hat{S}(1 + \hat{S} + \hat{S}^2)^3}{l.(1 + 6\hat{S}^2 + 13\hat{S}^3 + \hat{S}^4)}$$

$$= \frac{0.7715(1 + 0.7715 + 0.7715^2)^3}{228(1 + 6 \cdot 0.7715^2 + 13 \cdot 0.7715^3 + 0.7715^4)} = 0.004117$$

or $\widehat{\text{SE}}(\hat{S}) = 0.0642$. Use of Program USER to analyze this data set is illustrated in Appendix C.

5.7.2   Modified Hayne and Eberhardt (1952) for Unequal Juvenile Survival

The likelihood model for the original Hayne and Eberthardt (1952) estimator can be modified to allow differential survival between juveniles (i.e., $l_{1.5}$) and adults (i.e., $l_{2.5}, l_{3.5}$) using the same age-structure data. The likelihood with unique juvenile and adult survivals, S_0 and S, respectively, can be written as

$$L(S_0, S | l., l_0, l_1, l_2) = \binom{l.}{l_0, l_1, l_2} \left(\frac{1}{1 + S_0 + S_0 S}\right)^{l_0} \left(\frac{S_0}{1 + S_0 + S_0 S}\right)^{l_1} \left(\frac{S_0 S}{1 + S_0 + S_0 S}\right)^{l_2}$$

$$(5.74)$$

where

S_0 = annual survival probability for age class 1.5;
S = annual survival probability among older adult age classes.

The MLEs for both S_0 and S are similar to those derived from the vertical life table for all age classes and are

$$\hat{S}_0 = \frac{l_1}{l_0},$$

(5.75)

and

$$\hat{S} = \frac{l_2}{l_1},$$

(5.76)

respectively. Variances for the survival estimators can be derived by using the delta method, where

$$\widehat{\mathrm{Var}}(\hat{S}_0) = \frac{\hat{S}_0(1+\hat{S}_0)(1+\hat{S}_0+\hat{S}_0\hat{S})}{l.},$$

(5.77)

and

$$\widehat{\mathrm{Var}}(\hat{S}) = \frac{\hat{S}}{l.\hat{S}_0}\left[(1+\hat{S}_0)+\hat{S}(1+\hat{S}_0\hat{S})+2\hat{S}_0\hat{S}\right].$$

(5.78)

Example 5.10: Modified Hayne-Eberhardt (1952) Estimator of Survival for White-Tailed Deer

By using the same harvest numbers as in Example 5.9, the estimate of juvenile survival can be calculated by Eq. (5.75) as

$$\hat{S}_0 = \frac{l_1}{l_0} = \frac{73}{97} = 0.7526,$$

and annual adult survival (Eq. 5.76) as

$$\hat{S} = \frac{l_2}{l_1} = \frac{58}{73} = 0.7945.$$

The variances for $\hat{S}_0$ and $\hat{S}$ are estimated as

$$\widehat{\mathrm{Var}}(\hat{S}_0) = \frac{\hat{S}_0(1+\hat{S}_0)(1+\hat{S}_0+\hat{S}_0\hat{S})}{l.}$$

$$= \frac{0.7526(1+0.7526)(1+0.7526+0.7526\cdot0.7945)}{228} = 0.01360,$$

or $\widehat{\mathrm{SE}}(\hat{S}_0) = 0.1166$

and

$$\widehat{\text{Var}}(\hat{S}) = \frac{\hat{S}}{l \cdot \hat{S}_0} \left[(1 + \hat{S}_0) + \hat{S}(1 + \hat{S}_0 \hat{S}) + 2\hat{S}_0 \hat{S} \right]$$

$$= \frac{0.7945}{228(0.7526)} [(1 + 0.7526) + 0.7945(1 + 0.7526 \cdot (0.7945))$$

$$+ 2 \cdot (0.7526)(0.7945)]$$

$$\widehat{\text{Var}}(\hat{S}) = 0.01954$$

or $\quad \widehat{\text{SE}}(\hat{S}) = 0.1398.$

The difference between estimates of a common adult survival based on likelihood Eq. (5.70) and separate juvenile and adult survivals based on the model in Eq. (5.74) appears small. Testing for equality between the adult and juvenile survival is best done by using a LRT, written as

$$\chi_1^2 = -2 \left[\ln L(\hat{S} | l., l_0, l_1, l_2) - \ln L(\hat{S}_0, \hat{S} | l., l_0, l_1, l_2) \right].$$

Substituting the log-likelihood values based on the MLEs for the survival probabilities for likelihood Eqs. (5.70) and (5.74) yields

$$\chi_1^2 = -2(-245.4524 - (-245.4345))$$
$$= 0.0358$$

which gives a P-value of $P(\chi_1^2 \geq 0.0358) = 0.8499$. Thus, the survival probabilities between juveniles and adults are not significantly different at $P = 0.8499$. For these data, the annual survival across all age classes is estimated to be a common value of $\hat{S} = 0.7715$ ($\widehat{\text{SE}} = 0.0642$).

5.7.3 Modified Hayne and Eberhardt (1952) Pooling Older Age Classes

Another extension of the original Hayne and Eberhardt (1952) method is to consider the count of all animals 3.5 years and older in the third age class. The same assumptions of a constant harvest and natural survival probabilities are used, but the last cell in the multinomial likelihood model incorporates the sum of age classes 3.5 and older. This method is useful when age classification becomes more difficult as animals grow older. Tooth wear and eruption techniques can often accurately age only the first several age classes of ungulates (Jacobson and Renier 1989). Later age classes often need to be pooled because of subsequent inaccuracies in the age classification technique. Caughley (1977:8) noted similarly that for most bird species, only the first few age classes can be accurately assigned.

The likelihood model with age classes 3.5 and older pooled can be written as

$$L(S | l_0, l_1, l_{2+}) = \binom{l.}{l_0, l_1, l_2} \left(\frac{1}{1 + S + S^2 + \cdots + S^{w-1}} \right)^{l_0}$$

$$\cdot \left(\frac{S}{1 + S + S^2 + \cdots + S^{w-1}} \right)^{l_1} \left(\frac{S^2 + \cdots + S^{w-1}}{1 + S + S^2 + \cdots + S^{w-1}} \right)^{l_{2+}},$$

(5.79)

where l_{2+} = sum of all individuals in the third age class and older (i.e., ≥3.5). Allowing w to go to ∞, then

$$\lim_{w\to\infty}\left[\sum_{i=0}^{w-1}S^i\right]=\sum_{i=0}^{\infty}S^i=\frac{1}{1-S},\ -1<S<1,\tag{5.80}$$

and the likelihood (5.79) can be rewritten as

$$L(S|l_0,l_1,l_{2+})=\binom{l.}{l_0,l_1,l_{2+}}(1-S)^{l_0}(S(1-S))^{l_1}(S^2)^{l_{2+}}.\tag{5.81}$$

The MLE for S is

$$\hat{S}=\frac{l_1+2l_{2+}}{l_0+2(l_1+l_{2+})}.\tag{5.82}$$

The asymptotic variance of $\hat{S}$, based on likelihood Eq. (5.81), is

$$\mathrm{Var}(\hat{S})=\frac{S(1-S)}{l.(1+S)},$$

with estimated variance

$$\widehat{\mathrm{Var}}(\hat{S})=\frac{\hat{S}(1-\hat{S})}{l.(1+\hat{S})}.\tag{5.83}$$

Assumptions

Assumptions of the extended Hayne and Eberhardt (1952) method for all age classes are very similar to the original method:

1. The population is stationary.
2. The population has a stable age structure.
3. All animals have an equal probability of selection, which also implies a constant harvest rate across all age classes if harvest data are used.
4. The sample is representative of the population of interest.
5. Fates of all animals are independent.
6. Ages are recorded accurately for the first two age classes and accurately binned for animals 3.5 years and older.
7. Infinite age classes are possible ($w=\infty$).

Some of the accuracy of the estimator rests on the behavior of the infinite sum in Eq. (5.80). The smaller the survival probability, the fewer age classes needed for a reasonable approximation. As survival probabilities approach 1, more age classes are required for the series to converge. This method allows the number of age classes to be unbounded, resulting in a reasonable approximation. The use of all age classes to estimate total annual survival, rather than only the first three, increases the sample size and reduces the variance of the survival estimator (Eq. 5.83). One potential concern in using estimator Eq. (5.82) is whether survival probabilities will decrease for the oldest age classes (i.e., senescence).

Example 5.11: Hayne-Eberhardt (1952) Estimator of Survival with Pooled Age Classes Using Cow Elk (*Cervus elaphus*) Harvest Data, Northern Idaho

Age-at-harvest data reported by Gove et al. (2002) for cow elk in Idaho are used to illustrate estimation of annual survival probabilities (Table 5.15). Harvest occurred in October 1991, and reporting by hunters was mandatory but incomplete. The number of cow elk in age classes 0, 1, and 2+ were $l_0 = 39$, $l_1 = 32$, and $l_{2+} = 71$. Assuming constant survival across all age classes, S is estimated by Eq. (5.82) as

$$\hat{S} = \frac{l_1 + 2l_{2+}}{l_0 + 2(l_1 + l_{2+})}$$

$$= \frac{32 + 2 \cdot (71)}{39 + 2(32 + 71)} = 0.7102.$$

The variance is estimated by Eq. (5.83) as

$$\widehat{\mathrm{Var}}(\hat{S}) = \frac{\hat{S}(1 - \hat{S})}{l_\cdot(1 + \hat{S})}$$

$$= \frac{0.7102(1 - 0.7102)}{142 \cdot (1 + 0.7102)} = 0.000848,$$

$$\widehat{\mathrm{SE}}(\hat{S}) = 0.0291.$$

Table 5.15. Age-structure data for cow elk from Northern Idaho.

Age	Age class	Observed frequency
1	0	39
2	1	32
3	2	12
4	3	17
5	4	15
6	5	10
7	6	5
8	7	6
9	8	6

(Data from Gove et al. 2002.)

5.7.4 Modified Hayne and Eberthardt (1952) with Pooled Age Classes and Unique Juvenile Survival

Likelihood model (5.79) can be readily modified to allow unique juvenile and adult survival as follows:

$$L(S_0, S | l_\cdot, l_0, l_1, l_{2+}) = \binom{l_\cdot}{l_0, l_1, l_{2+}} \left(\frac{1}{1 + S_0 + S_0 S + \cdots + S_0 S^{w-2}} \right)^{l_0} \cdot$$

$$\left(\frac{S_0}{1 + S_0 + S_0 S + \cdots + S_0 S^{w-2}} \right)^{l_1}$$

$$\left(\frac{S_0 S + \cdots + S_0 S^{w-2}}{1 + S_0 + S_0 S + \cdots + S_0 S^{w-2}} \right)^{l_{2+}},$$

where l_{2+} = number of all individuals in the third age class and older (i.e., ≥ 3.5). Allowing $w \to \infty$, the likelihood can be reexpressed as

$$L(S_0, S | l_., l_0, l_1, l_{2+}) = \binom{l_.}{l_0, l_1, l_{2+}} \left(\frac{1-S}{1+S_0-S} \right)^{l_0} \left(\frac{S_0(1-S)}{1+S_0-S} \right)^{l_1} \left(\frac{S_0 S}{1+S_0-S} \right)^{l_{2+}}.$$
(5.84)

The respective MLEs for S_0 and S are

$$\hat{S}_0 = \frac{l_1}{l_0}$$
(5.85)

and

$$\hat{S} = \frac{l_{2+}}{l_.-l_0}.$$
(5.86)

The estimated variance of $\hat{S}_0$ can be approximated by using the delta method as

$$\widehat{\text{Var}}(\hat{S}_0) = \frac{\hat{S}_0(\hat{S}_0+1)(1+\hat{S}_0-\hat{S})}{l_.(1-\hat{S})}.$$
(5.87)

Again, the delta method can be used to derive a variance estimator for $\hat{S}$, where

$$\widehat{\text{Var}}(\hat{S}) = \frac{l_. \hat{S}_0 \hat{S}(1-\hat{S})}{(l_.-l_0)^2 (1+\hat{S}_0-\hat{S})}.$$
(5.88)

Example 5.12: Modified Hayne-Eberhardt (1952) Estimator of Survival with Pooled Age Classes and Unique Juvenile Survival Using Cow Elk Harvest Data, Northern Idaho

The cow elk data (Gove et al. 2002) (Table 5.15) can be analyzed under likelihood model (5.84), where the survival probabilities for age classes 1.5 and older animals are now estimated as

$$\hat{S}_0 = \frac{l_1}{l_0} = \frac{32}{39} = 0.8205$$

and

$$\hat{S} = \frac{l_{2+}}{l_.-l_0} = \frac{71}{142-39} = 0.6893.$$

The variances of $\hat{S}_0$ and $\hat{S}$ are estimated by Eqs. (5.87) and (5.88), respectively, as follows:

$$\widehat{\text{Var}}(\hat{S}_0) = \frac{\hat{S}_0(\hat{S}_0+1)(1+\hat{S}_0-\hat{S})}{l_.(1-\hat{S})}$$

$$= \frac{0.8205(0.8205+1)(1+0.8205-0.6893)}{142(1-0.6893)} = 0.0383$$

or $\widehat{\text{SE}}(\hat{S}_0) = 0.1957$,

and

$$\widehat{\mathrm{Var}}(\hat{S}) = \frac{l \cdot \hat{S}_0 \hat{S}(1-\hat{S})}{(l. - l_0)^2 (1 + \hat{S}_0 - \hat{S})}$$

$$= \frac{142 \cdot (0.8205)(0.6893)(1-0.6893)}{(142-39)^2 (1+0.8205-0.6983)} = 0.00208$$

or $\widehat{\mathrm{SE}}(\hat{S}) = 0.0456$.

The assumption of equal survival probabilities for both 1.5 and older elk can be tested by using a LRT. By use of the log-likelihood values from likelihood models (5.81) and (5.84), the resulting LRT is

$$\chi_1^2 = -2(-147.4824 - (-147.2947)) = 0.3754,$$

which results in a P-value of $P(\chi_1^2 \geq 0.3754) = 0.5400$. Thus, the survival probabilities between age class 1.5 and older cow elk are not statistically different. For this data set, the annual survival across all age classes can be concluded to be a common value of 0.7102 ($\widehat{\mathrm{SE}}(\hat{S}) = 0.0291$).

5.7.5   Heincke (1913) and Burgoyne (1981)

Under conditions of a stable and stationary population closed to immigration and emigration, the number of juveniles entering a population should equal the number of animals dying each year. Heincke (1913) and Burgoyne (1981) used this fundamental relationship to develop a heuristic estimator of annual survival.

Annual survival is estimated from the proportion of juveniles in a sample based on the likelihood function

$$L(S|l., l_0) = \binom{l.}{l_0} \left(\frac{1}{1+S+S^2+\cdots+S^{w-2}} \right)^{l_0} \left(\frac{S+S^2+\cdots+S^{w-2}}{1+S+S^2+\cdots+S^{w-2}} \right)^{l_{1+}}, \quad (5.89)$$

where $l_{1+} = l. - l_0$. Assuming $w \to \infty$, Eq. (5.89) simplifies to

$$L(S|l., l_0) = \binom{l.}{l_0} (1-S)^{l_0} (S)^{l_{1+}}. \quad (5.90)$$

The MLE of annual survival, S, is

$$\hat{S} = \frac{l_{1+}}{l.} = \frac{l. - l_0}{l.} \quad (5.91)$$

or its converse, mortality, estimated by

$$\hat{M} = 1 - \hat{S} = \frac{l_0}{l.}.$$

The variance of $\hat{S}$ is the variance of a binomial proportion where

$$\mathrm{Var}(\hat{S}) = \frac{S(1-S)}{l.}$$

with estimated variance

$$\widehat{\text{Var}}(\hat{S}) = \frac{\hat{S}(1-\hat{S})}{l.}. \qquad (5.92)$$

Assumptions

The assumptions of Heincke (1913) and Burgoyne (1981) estimator are essentially the same as those of the Hayne and Eberhardt (1952) method and will not be repeated.

Example 5.13: Heincke (1931)–Burgoyne (1981) Survival Estimate Using Cow Elk Harvest Data, Northern Idaho

Total annual survival (Table 5.15) is estimated by the fraction

$$\hat{S} = \frac{l_{1+}}{l.} = \frac{103}{142} = 0.7254,$$

with an estimated variance and standard error, respectively,

$$\widehat{\text{Var}}(\hat{S}) = \frac{\hat{S}(1-\hat{S})}{l.}$$
$$= \frac{0.7254(1-0.7254)}{142} = 0.00140,$$

and $\qquad\qquad \widehat{\text{SE}}(\hat{S}) = 0.0375.$

Discussion of Utility

The relationship between the variance for the Heincke (1913)–Burgoyne (1981) estimator ($\text{Var}_{HB}(\hat{S})$) given by Eq. (5.92) and the modified Hayne and Eberhardt (1952) model ($\text{Var}_{HE}(\hat{S})$), given by Eq. (5.83), is

$$\text{Var}_{HB}(\hat{S}) = (1+S)\text{Var}_{HE}(\hat{S}).$$

Although the Heincke (1913)–Burgoyne (1981) estimator is less precise than is Hayne and Eberhardt (1952), this method is useful when only juveniles can be correctly identified to age.

Sample Size Calculations

The precision of the Heincke (1913)–Burgoyne (1981) survival estimator $\hat{S}$ will be defined as

$$P(|\hat{S} - S| < \varepsilon) \geq 1 - \alpha$$

where the absolute error in estimation (i.e., $|\hat{S} - S|$) is less than ε, $(1 - \alpha)100\%$ of the time. Assuming the MLE is normally distributed, then

$$\varepsilon = Z_{1-\frac{\alpha}{2}} \cdot \text{SE}(\hat{S}).$$

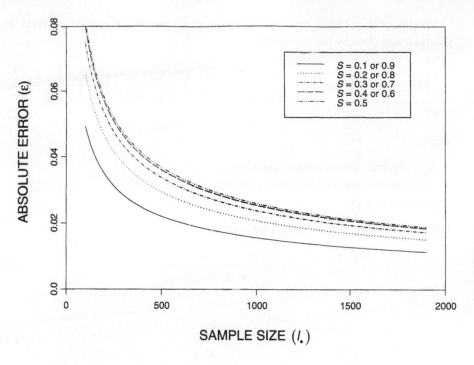

Figure 5.8. Sample size $l.$ versus sampling precision as absolute error (ε) in estimating a common survival probabilty by using the Heincke (1913)–Burgoyne (1981) estimator for $S = 0.1, 0.2, \ldots, 0.9$ at $1 - \alpha = 0.90$.

For $\alpha = 0.10$, the required sample size $l.$ for the Heincke (1913)-Burgoyne (1981) survival estimator is presented in Fig. 5.8 for a range of survival values of $S = 0.20, \ldots, 0.90$. The required sample size, in general, decreases as the anticipated value of S moves closer to either zero or one, and is maximal at $S = 0.50$ because of the binomial error structure (Eq. 5.92).

5.8 c_x-Series Data with Abbreviated or Pooled Age Classes

Methods of using only the first few age classes to estimate survival probabilities can be extended to the analysis of c_x-series data (Ryding 2002). These methods, similar to those in Section 5.7, are necessitated by the inability to accurately identify individuals in the older age classes. Instead, only data for the first few age classes will be analyzed, or data from older age classes may be pooled.

5.8.1 c_x-Series with the First Three Age Classes

The likelihood for the first three age classes (0, 1, and 2) of a c_x series can be written as

$$L(S|c., c_0, c_1, c_2) = \binom{c.}{c_0, c_1, c_2} \left(\frac{1 - S}{1 - S^3} \right)^{c_0} \left(\frac{S(1 - S)}{1 - S^3} \right)^{c_1} \left(\frac{S^2(1 - S)}{1 - S^3} \right)^{c_2}. \quad (5.93)$$

Noting that $(1 - S^3) = (1 - S)(1 + S + S^2)$, Eq. (5.93) is equivalent to likelihood Eq. (5.70). Hence, the MLE for S based on the first three age classes (0, 1, and 2) of a c_x series is the positive root of Eq. (5.71), with associated variance Eq. (5.72) and estimated variance Eq. (5.73).

Likelihood Eq. (5.93) can also be modified to permit differential survival in the youngest age class by the expression:

$$L(S_0, S | c., c_0, c_1, c_2) = \binom{c.}{c_0, c_1, c_2}\left(\frac{1-S_0}{1-S_0 S^2}\right)^{c_0}\left(\frac{S_0(1-S)}{1-S_0 S^2}\right)^{c_1}\left(\frac{S_0 S(1-S)}{1-S_0 S^2}\right)^{c_2}.$$

(5.94)

where

S_0 = survival of the youngest age class;

S = common survival probability for age classes 1 and 2;

$c. = c_0 + c_1 + c_2$.

The estimators for $\hat{S}_0$ and $\hat{S}$ are, respectively,

$$\hat{S}_0 = \frac{c_1^2}{c.(c_1 - c_2) + c_2^2}$$

(5.95)

and

$$\hat{S} = \frac{c_2}{c_1}.$$

(5.96)

The variance estimator for $\hat{S}_0$ is

$$\widehat{Var}(\hat{S}_0) = \frac{\hat{S}_0}{c.(1-\hat{S})^3}\left[(1 + \hat{S}_0\hat{S} - \hat{S}_0 - \hat{S}_0\hat{S}^2)(1 - 2\hat{S} + \hat{S}_0\hat{S}^2)^2\right.$$
$$+ \hat{S}(1 - \hat{S}_0\hat{S})(1 - 2\hat{S}_0\hat{S} + \hat{S}_0\hat{S}^2)^2$$
$$\left. - 2\hat{S}_0\hat{S}(1 - \hat{S})(1 - 2\hat{S} + \hat{S}_0\hat{S}^2)(1 - 2\hat{S}_0\hat{S} + \hat{S}_0\hat{S}^2)\right].$$

(5.97)

The variance of $\hat{S}$ is estimated by the expression:

$$\widehat{Var}(\hat{S}) = \frac{\hat{S}(1 + \hat{S})(1 - \hat{S}_0\hat{S}^2)}{c.\hat{S}_0(1 - \hat{S})}.$$

(5.98)

Assumptions

Assumptions for likelihoods (5.93) and (5.94) are the same as those used previously for the depositional life-table analysis of carcass count data (Section 5.6).

Example 5.14: Estimating Survival from Depositional Data for Impalas (*Aepyceros melampus*), Akagera National Park, Rwanda

Between July 1968 and September 1969, 458 impala skulls were recovered (Table 5.16) in Akagera National Park, Rwanda (Spinage 1972). Mortalities were not harvest-related or the result of a natural disaster, so data may be regarded as the ages-at-natural-death. All skulls collected were estimated to have been deposited within the last 2 years (Spinage 1972). By use of the data from female impala skulls, total annual survival of female adults and juveniles can be estimated.

By using only the first three ages classes from the female impala skull data, constant total survival S and the associated estimated variance are calculated analogous to Eqs. (5.71) and (5.73), respectively, as

$$\hat{S} = \frac{(c_2 - c_0) + \sqrt{(c_0 - c_2)^2 + 4(2c_0 + c_1)(c_1 + 2c_2)}}{2(2c_0 + c_1)}$$

$$= \frac{(3 - 23) + \sqrt{(23 - 3)^2 + 4(2(23) + 6)(6 + 2 \cdot 3)}}{2(2 \cdot 23 + 6)} = 0.3251$$

$$\widehat{\text{Var}}(\hat{S}) = \frac{\hat{S}(1 + \hat{S} + \hat{S}^2)^3}{l.(1 + 6\hat{S}^2 + 13\hat{S}^3 + \hat{S}^4)}$$

$$= \frac{0.3251 \cdot (1 + 0.3251 + 0.3251^2)^3}{32(1 + 6 \cdot 0.3251^2 + 13 \cdot 0.3251^3 + 0.3151^4)} = 0.01422,$$

and $\widehat{\text{SE}}(\hat{S}) = 0.1193$.

Estimates of $\hat{S}_0$ and $\hat{S}$, assuming unique survival in the first age class, are calculated by Eqs. (5.95) and (5.96) respectively, as

$$\hat{S}_0 = \frac{c_1^2}{c.(c_1 - c_2) + c_2^2}$$

$$= \frac{6^2}{32(6 - 3) + 3^2} = 0.3429$$

and

$$\hat{S} = \frac{c_2}{c_1}$$

$$= \frac{3}{6} = 0.5000.$$

The estimated variances of $\hat{S}_0$ and $\hat{S}$ are calculated by Eqs. (5.97) and (5.98), respectively, as

$$\widehat{\text{Var}}(\hat{S}_0) = \frac{\hat{S}_0}{c.(1 - \hat{S})^3} \Big[(1 + \hat{S}_0\hat{S} - \hat{S}_0 - \hat{S}_0\hat{S}^2)(1 - 2\hat{S} + \hat{S}_0\hat{S}^2)^2$$

$$+ \hat{S}(1 - \hat{S}_0\hat{S})(1 - 2\hat{S}_0\hat{S} + \hat{S}_0\hat{S}^2)^2$$

$$- 2\hat{S}_0\hat{S}(1 - \hat{S})(1 - 2\hat{S} + \hat{S}_0\hat{S}^2)(1 - 2\hat{S}_0\hat{S} + \hat{S}_0\hat{S}^2) \Big]$$

$$\widehat{\text{Var}}(\hat{S}_0) = \frac{0.3429}{32(1 - 0.5000)^3} \Big[(1 + 0.3429(0.5000) - 0.3429 - 0.3429(0.5000)^2) \cdot$$

$$(1 - 2(0.5000) + 0.3429(0.5000)^2)^2 + (0.5000)(1 - 0.3429(0.5000)) \cdot$$

$$(1 - 2(0.3429)(0.5000) + 0.3429(0.5000)^2)^2 - 2(0.3429)(0.5000) \cdot$$

$$(1 - 0.5000)(1 - 2(0.5000) + 0.3429(0.5000)^2) \cdot$$

$$(1 - 2(0.3429)(0.5000) + 0.3429(0.5000)^2) \Big]$$

$$= 0.01913,$$

or, equivalently, $\widehat{SE}(\hat{S}_0) = 0.1383$, and

$$\widehat{Var}(\hat{S}) = \frac{\hat{S}(1+\hat{S})(1-\hat{S}_0\hat{S}^2)}{c.\hat{S}_0(1-\hat{S})}$$

$$= \frac{0.5000(1+0.5000)(1-0.3429(0.5000)^2)}{32(0.3429)(1-0.5000)} = 0.12498,$$

or, equivalently, $\widehat{SE}(\hat{S}) = 0.3535$.

As with the earlier example, the assumption of equal survival probabilities for both juveniles and adults can be tested by using a LRT of the form

$$\chi_1^2 = -2\left(\ln L(\hat{S}|c.,c_0,c_1,c_2) - \ln L(\hat{S},\hat{S}_0|c.,c_0,c_1,c_2)\right).$$

Substituting the log-likelihood values based on the MLEs for the survival probabilities, the LRT is

$$\chi_1^2 = -2(-24.9466 - (-24.7408)) = 0.4116,$$

which gives a P-value of $P(\chi_1^2 \geq 0.4116) = 0.5212$. Based on the result of the LRT, there is no significant difference in survival between the three age classes.

Table 5.16.   Depositional life table
for female impala, Rwanda.

Age	Age class	c_x
0–1	0	23
1–2	1	6
2–3	2	3
3–4	3	9
4–5	4	9
5–6	5	23
6–7	6	16
7–8	7	12
8–9	8	4
9–10	9	4
10–11	10	4
11–12	11	3

(Data from Spinage 1972.)

5.8.2   Older Age Classes Pooled

If carcass age is not identifiable beyond the third age class, the count of all animals in the third age class and older may be pooled, provided survival is constant. The subsequent likelihood is

$$L(S|c.,c_0,c_1,c_{2+}) = \binom{c.}{c_0,c_1,c_{2+}}(1-S)^{c_0}(S(1-S))^{c_1}(S^2)^{c_{2+}}. \tag{5.99}$$

The MLE for the survival parameter in likelihood Eq. (5.99) is

$$\hat{S} = \frac{c_1 + 2c_{2+}}{c_0 + 2(c_1 + c_{2+})}, \tag{5.100}$$

which has an asymptotic variance of

$$\text{Var}(\hat{S}) = \frac{(1-S)S}{c.(1+S)}$$

and estimated variance of

$$\widehat{\text{Var}}(\hat{S}) = \frac{(1-\hat{S})\hat{S}}{c.(1+\hat{S})}, \tag{5.101}$$

where $c. = c_0 + c_1 + c_{2+}$.

Likelihood Eq. (5.99) can also be modified to account for differential age-class survival in the earliest age class by the expression

$$L(S|c., c_0, c_1, c_{2+}) = \binom{c.}{c_0, c_1, c_{2+}} (1-S_0)^{c_0} (S_0(1-S))^{c_1} (S_0 S(1-S))^{c_{2+}}. \tag{5.102}$$

The survival estimators for age class 0 and older animals, respectively, are

$$\hat{S}_0 = \frac{c. - c_0}{c.} \tag{5.103}$$

and

$$\hat{S} = \frac{c_{2+}}{c_1 + c_{2+}}. \tag{5.104}$$

These estimators are the same as those for horizontal life-table analysis. For $\hat{S}_0$, the variance estimator is

$$\widehat{\text{Var}}(\hat{S}_0) = \frac{\hat{S}_0(1-\hat{S}_0)}{c.}. \tag{5.105}$$

The variance of $\hat{S}$ is estimated by the expression

$$\widehat{\text{Var}}(\hat{S}) = \frac{\hat{S}(1-\hat{S})(1-\hat{S}_0\hat{S}+\hat{S}_0\hat{S}^2)}{c.\hat{S}_0}. \tag{5.106}$$

Example 5.15: Estimating Survival Using Depositional Data and Pooling Older Age Classes for Impalas in Akagera National Park, Rwanda

By pooling the impala data (Table 5.16) for all animals in age classes 2 and older, a common annual survival probability can be estimated with greater precision if the model holds true. The age-class frequencies for the 2+ model (Eq. 5.100) are $c_0 = 23$, $c_1 = 6$, and $c_{2+} = 87$. The estimate of a constant age class survival (Eq. 5.100) is

$$\hat{S} = \frac{c_1 + 2c_{2+}}{c_0 + 2(c_1 + c_{2+})}$$

$$= \frac{6 + 2 \cdot 87}{23 + 2(6 + 87)} = 0.8612,$$

with an estimated variance calculated by Eq. (5.101) as

$$\widehat{\text{Var}}(\hat{S}) = \frac{(1-\hat{S})\hat{S}}{c.(1+\hat{S})}$$

$$= \frac{(1-0.8612)0.8612}{116(1+0.8612)} = 0.000554$$

with a standard error of $\widehat{\text{SE}}(\hat{S}) = 0.0235$.

Alternatively, first age-class survival (S_0) can be estimated separately from that of older animals (S), based on likelihood Eq. (5.102). The estimate for survival of the first age class, S_0, is calculated by Eq. (5.103) as

$$\hat{S}_0 = \frac{c. - c_0}{c.}$$

$$= \frac{116 - 23}{116} = 0.8017,$$

with estimated variance (Eq. 5.105)

$$\widehat{\text{Var}}(\hat{S}_0) = \frac{\hat{S}_0(1-\hat{S}_0)}{c.}$$

$$= \frac{0.8017(1-0.8017)}{116} = 0.00137,$$

and a standard error of $\widehat{\text{SE}}(\hat{S}_0) = 0.0370$. Annual survival for older age classes is estimated by Eq. (5.104) as

$$\hat{S} = \frac{c_{2+}}{c_1 + c_{2+}}$$

$$= \frac{87}{6 + 87} = 0.9355$$

with an estimated variance calculated by Eq. (5.106)

$$\widehat{\text{Var}}(\hat{S}) = \frac{\hat{S}(1-\hat{S})(1 - \hat{S}_0\hat{S} + \hat{S}_0\hat{S}^2)}{c.\hat{S}_0}$$

$$= \frac{0.9355(1-0.9355)(1 - 0.8017\cdot 0.9355 + 0.8017(0.9355^2))}{116\cdot 0.8017}$$

$$= 0.000617,$$

with associated standard error $\widehat{\text{SE}}(\hat{S}) = 0.0025$. The assumption of equal survival probabilities for all age classes can be tested by using a LRT of the form

$$\chi_1^2 = -2\big(\ln L(\hat{S}|c., c_0, c_1, c_{2+}) - \ln L(\hat{S}, \hat{S}_0|c., c_0, c_1, c_{2+})\big).$$

Substituting the log-likelihood values based on the MLEs for the survival probabilities, the LRT is

$$\chi_1^2 = -2(-84.1640 - (-80.0155)) = 8.2970,$$

which gives a P-value of $P(\chi_1^2 \geq 8.297) = 0.0040$. Clearly, the assumption of constant survival for young and older animals is untenable when the pooled age-class model is used. The estimate of $\hat{S}_0 = 0.8017$, for juvenile survival is similar to $\hat{S}_0 = 0.805$ from the life-table analysis (Spinage 1972). Total adult survival, estimated as $\hat{S} = 0.9355$, is similar to that of the first few adult age classes only.

Choosing among the analytical techniques (i.e. vertical life table, three age-class, or pooled age-class models) should be based on selection of the most parsimonious model that adequately describes the data. However, accuracy in estimation is paramount, followed by precision.

Spinage (1972) considered the impala survival estimates obtained from a life-table analysis as crude approximations. In a life-table analysis, the population is assumed stationary. Spinage (1972) also adjusted the number of skulls in the young age classes slightly upward because of concerns over lower probabilities of detection. Although some correction for the younger age classes was likely necessary, the particular adjustment factor was ad hoc and subjective.

5.9 Catch-Curve Analyses

The catch-curve analyses of Chapman and Robson (1960) and Robson and Chapman (1961) were originally developed to estimate annual survival probabilities by using age-at-catch data from fisheries hauls. The data are essentially a l_x series, whether they came from a commercial fisheries or a sport harvest of game animals. The purpose of the analysis is to estimate a common or constant survival probability across adult age classes. The name "catch-curve" comes from a plot of the l_x counts against age classes x. The survival estimate is related to the slope of the descending arm of the catch curve (Fig. 5.9).

The survival estimate derived from a catch-curve analysis is based on the probability of observing a sample of animal ages from the population. The ages of the animals in a l_x series are the random variables for the Chapman and Robson (1960) estimator. In a life-table analysis, numbers of animals in each age class are the random variables. The result of the catch-curve analysis is a unique, minimum variance, unbiased estimator (UMVUE) of survival. Chapman and Robson (1960) also present an estimator when age-class survival is constant only for a segment of the age-structure data (i.e., when the sample is left- and right-truncated). Finally, Robson and Chapman (1961) provided an estimator of survival when the older age classes in the age-structure data are pooled. The Hayne and Eberhardt (1952) and the Heincke (1913)–Burgoyne (1981) methods are special cases of the Robson and Chapman (1961) estimator.

5.9.1 Estimating Survival Using All Age Classes (Chapman and Robson 1960)

Assuming a constant survival probability across all age classes, the number of animals in the age classes of a vertical life table can be written as a multinomial distribution, where

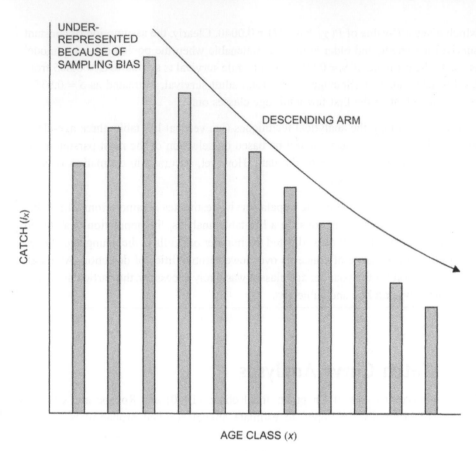

Figure 5.9. Schematic of the catch-vs.-age plot, which is the basis of the catch-curve analysis.

$$L(S|l., l_0, l_1, l_2, \ldots, l_w) = \binom{l.}{l_0, l_1, l_2, \ldots, l_x} (1-S)^{l_0} (S(1-S))^{l_1} \cdot$$
$$(S^2(1-S))^{l_2} \cdots (S^x(1-S))^{l_w}. \tag{5.107}$$

The probability of sampling an animal of age x from the above distribution follows a geometric distribution written as

$$f_X(x) = (1-S)S^x. \tag{5.108}$$

The multinomial likelihood, Eq. (5.107), was previously derived as a conditional distribution of Poisson random variables based on their sum. Chapman and Robson (1960) also showed the same derivation of the multinomial model and, hence, the geometric probability mass function (pmf) by using a Poisson model.

The geometric distribution (Eq. 5.108) can also be derived from the proportion of age x animals in the population, which is expressed by

$$P(X = x) = \frac{N_x}{N}. \tag{5.109}$$

Assuming a stationary and stable population, Eq. (5.109) can be rewritten as

$$P(X = x) = \frac{N_0 S^x}{N_0 \left(\sum_{x=0}^{\infty} S^x \right)} = (1 - S)S^x, \tag{5.110}$$

the pmf for a geometric distribution. The pmf (5.110) is not the probability of an animal surviving to age x but, rather, the probability of drawing an individual of age x from a population with age frequencies represented by the multinomial distribution (Eq. 5.107).

Chapman and Robson (1960) showed the joint probability of observing the vector of ages in a sample of l. animals is

$$P(X_0 = x_0, X_1 = x_1, \ldots X_{l.} = x_{l.}) = \prod_{i=0}^{l.} [(1 - S)S^{x_i}]$$

$$= (1 - S)^{l.} S^{\sum_{i=0}^{l.} x_i}, \tag{5.111}$$

where $\sum_{i=0}^{l.} x_i$ = sum of the ages of all individuals in the sample of size l. and is usually

denoted by T. In turn, the sum of the ages in the sample, $T = \sum_{i=0}^{l.} x_i$ has a negative

binomial distribution

$$f(T) = \binom{l. + T - 1}{T} (1 - S)^{l.} S^T. \tag{5.112}$$

The unique minimum variance unbiased estimator (UMVUE) of S, if it exists, should be some function $h(T)$ of the minimally sufficient statistic T. Setting the expected value of $h(T)$ equal to S and solving for $h(T)$ yields an unbiased estimator for survival that satisfies the conditions for an UMVUE. The expected value is written as

$$E(h(T)) = \sum_{T=0}^{\infty} h(T) \binom{l. + T - 1}{T} (1 - S)^{l.} S^T = S.$$

Solving for $h(T)$ yields the estimator

$$\hat{S} = \frac{T}{l. + T - 1}. \tag{5.113}$$

The asymptotic variance of $\hat{S}$ is given as

$$\mathrm{Var}(\hat{S}) \doteq S \left(S - \frac{T - 1}{l. + T - 2} \right), \tag{5.114}$$

or written in terms of the sample size l. and survival S,

$$\mathrm{Var}(\hat{S}) = \left(\frac{l.S}{l. + S - 1} \right) \left(\frac{l.S}{l. + S - 1} - \frac{l.S + S - 1}{l. - 2(1 - S)} \right), \tag{5.115}$$

with estimated variance

$$\widehat{\text{Var}}(\hat{S}) = \hat{S}\left(\hat{S} - \frac{T-1}{l.+T-2}\right). \tag{5.116}$$

Assumptions

The assumptions of the Chapman and Robson (1960) catch-curve estimator (5.113) include the following:

1. There is a stable age structure.
2. The population is stationary.
3. All animals have an equal probability of selection.
4. The sample is representative of the population of interest.
5. Fates of all animals are independent.
6. All ages are recorded accurately.
7. Annual survival probability is constant across all age classes.

If harvest effort is used to collect the data, the assumption of a constant harvest probability across all age classes is also needed for the age structure of the sample to be representative of the population. The assumption that all ages are recorded accurately implies there is no measurement error that could systematically bias the estimator or inflate the variance.

Eberhardt (1985) demonstrated that if the mortality rate of older animals increased because of senescence, the Chapman and Robson (1960) method will underestimate the annual survival probability. A similar bias was also suggested by de la Mare (1985). In the presence of senescence, it is better to use a truncated age range and use an alternative catch-curve estimator which is presented in Section 5.9.3.

Discussion of Utility

Chapman and Robson (1960) noted that for a large sample size, the variance of $\hat{S}$ in Eq. (5.114) can be approximated by

$$\text{Var}(\hat{S}) \doteq \frac{S(1-S)^2}{n}. \tag{5.117}$$

Hence, the variance of the Chapman and Robson (1960) estimator is a factor $(1-S)$ smaller than that of the Heincke (1913)–Burgoyne (1981) estimator. However, the Chapman and Robson (1960) estimator requires that ages of all animals be accurately classified. In contrast, the Heincke (1913)–Burgoyne (1981) estimator (Eq. 5.91) depends only on accurately enumerating the number of juveniles in a sample, which may be less resource-intensive than is assigning ages to all animals, particularly if the sample is large or age classification is difficult.

Example 5.16: Catch-Curve Analysis Using Black Bear (*Ursus americanus*) Harvest Data, Washington State

The Chapman and Robson (1960) method is used to estimate annual survival with the age-at-harvest data of male black bears (Table 5.17). For a sample of $l_. = 874$ bears with cumulative age of

$$T = 0(217) + 1(79) + \cdots + 17(5) = 3314,$$

annual survival is estimated to be

$$\hat{S} = \frac{T}{(l_. + T - 1)} = \frac{3314}{874 + 3314 - 1} = 0.7915.$$

Note T is calculated based on the coded age classes (Table 5.17) of the animals (i.e., $x_i = 0, 1, 2, \ldots$). The associated variance and standard error are estimated as

$$\widehat{\mathrm{Var}}(\hat{S}) = \hat{S}\left(\hat{S} - \left(\frac{T-1}{l_. + T - 2}\right)\right) = 0.7915\left(0.7915 - \left(\frac{3314 - 1}{874 + 3314 - 2}\right)\right)$$

$$= 0.0000414$$

$$\widehat{\mathrm{SE}}(\hat{S}) = 0.0064.$$

Table 5.17. Age-at-harvest data for male black bears, Washington State, 1989.

Age in years	Coded age class	Frequency in sample
1.5	0	217
2.5	1	79
3.5	2	122
4.5	3	101
5.5	4	53
6.5	5	85
7.5	6	32
8.5	7	26
9.5	8	32
10.5	9	53
11.5	10	16
12.5	11	5
13.5	12	16
14.5	13	11
15.5	14	11
16.5	15	5
17.5	16	5
18.5	17	5

(Data from Bender 1998.)

5.9.2 Survival Estimation with Older Age Classes Pooled (Robson and Chapman 1961)

Often, older animals are difficult to accurately classify by age, increasing the measurement error and potentially biasing the estimates of survival (Chapman and Robson 1960). To account for the difficulty in classifying ages of older animals, the frequencies of animals in age classes $k + 1$ and older are pooled into a single age class. Annual survival is estimated from the following equation:

$$\hat{S} = \frac{T}{l. - l_{k+1} + T}$$

(5.118)

where

$$l_{k+1} = \sum_{x=k+1}^{\infty} l_x = \text{number of individuals age } k + 1 \text{ and older;}$$

$$T = \left[\sum_{x=0}^{k} x l_x \right] + (k+1) l_{k+1}.$$

The variance of $\hat{S}$ is

$$\text{Var}(\hat{S}) = \frac{S(1-S)^2}{l.(1 - S^{k+1})},$$

(5.119)

with estimated variance

$$\widehat{\text{Var}}(\hat{S}) = \frac{\hat{S}(1-\hat{S})^2}{l.(1 - \hat{S}^{k+1})}.$$

(5.120)

Accuracy of the survival estimate may be increased by reducing measurement errors, although some precision may be lost by pooling. The assumptions for estimator Eq. (5.118) are essentially the same as those of Eq. (5.113), in which all animals are classified according to age.

Discussion of Utility

Both the modified Hayne and Eberhardt (1952) and the Heincke (1913)–Burgoyne (1981) estimators are special cases of the Robson and Chapman (1961) estimator, in which $k = 1$ and 0, respectively. For the Hayne and Eberhardt (1952) 2+ age class model, in which all animals age classes 2 and older are pooled, Eq. (5.118) can be rewritten as

$$\hat{S} = \frac{T}{l. - l_{k+1} + T} = \frac{(0 \cdot l_0 + 1 l_1 + 2 l_{2+})}{l. - l_{2+} + (0 \cdot l_0 + 1 l_1 + 2 l_{2+})}.$$

Noting that $l. = l_0 + l_1 + l_2$, the equation can be rewritten as

$$\hat{S} = \frac{l_1 + 2 l_{2+}}{l_0 + 2(l_1 + l_{2+})},$$

which is the same estimator of $\hat{S}$ derived from the likelihood for the modified Hayne and Eberhardt (1952) Eq. (5.82). The variance of $\hat{S}$, Eq. (5.119), for $k = 1$ is written as

$$\widehat{\text{Var}}(\hat{S}) = \frac{\hat{S}(1-\hat{S})^2}{l(1 - \hat{S}^2)} = \frac{\hat{S}(1-\hat{S})}{l.(1 + \hat{S})},$$

which is the same as variance estimator Eq. (5.83). For $k = 0$, the estimator Eq. (5.118) reduces to the Heincke (1913)–Burgoyne (1981) estimator Eq. (5.91) and the variance expression (Eq. 5.119) is equivalent to the variance estimator Eq. (5.92). Thus, the Robson and Chapman (1961) estimator, Eq. (5.118), can be considered the general case for estimating S when pooling older age classes.

Example 5.17: Catch-Curve Analysis with Pooled Older Age Classes Using Cow Elk Harvest Data, Northern Idaho

The age-structure data from harvested cow elk in northern Idaho (Table 5.15) (Gove et al. 2002) is used to demonstrate the use of the pooled Robson and Chapman (1961) estimator (Eq. 5.118). Animals in age classes 2 and older (i.e., $k = 1$) will be pooled (i.e., animals age 3 and older will be pooled).

Annual survival is estimated by using Eq. (5.118) as

$$\hat{S} = \frac{T}{l_. - l_{k+1} + T}$$

$$= \frac{174}{142 - 71 + 174} = 0.7102$$

for

$$l_. = 142,$$
$$l_{k+1} = 12 + 7 + 15 + 10 + 5 + 6 + 6 = 71,$$
$$T = 0(39) + 1(32) + 2(71) = 174.$$

The estimated variance of $\hat{S}$, Eq. (5.120), and standard error are

$$\widehat{\text{Var}}(\hat{S}) = \frac{\hat{S}(1 - \hat{S})^2}{l_.(1 - \hat{S}^{k+1})}$$

$$= \frac{0.7102(1 - 0.7102)^2}{142(1 - 0.7102^2)} = 0.000848, \text{ and}$$

$$\widehat{\text{SE}}(\hat{S}) = 0.0291.$$

As expected, the estimate and associated variance of $\hat{S}$ are the same as those outlined from the modified Hayne and Eberhardt (1952) method, Eqs. (5.82) and (5.83), respectively.

5.9.3 Estimator for Left- and Right-Truncated Data: Chapman and Robson (1960) and Robson and Chapman (1961)

At times, inclusion of both the youngest and oldest age classes may be inappropriate when estimating S. Harvest regulations or hunter preference can result in differential selection probabilities for these age classes. The youngest and oldest age classes may also have inherently different survival probabilities than do prime-age adults. Alternatively, it may be impossible to accurately classify ages of older animals. In these situations, the younger age classes, older age classes, or both are disregarded. When the younger age classes are dropped from the analysis, the data are left-truncated. If the older age classes are removed from analysis, the data are right-truncated. In a study of elk, Kimball and Wolfe (1974) used the double-truncated estimator, because assumptions of constant survival and equal detection for both younger and older age classes were not met.

The multinomial likelihood for a double-truncated age-structure sample can be written as

$$l(S|l., l_0, l_1, \ldots, l_k) = \binom{l.}{l_0, l_1, \ldots, l_k} \left(\frac{1}{\sum_{x=0}^{k} S^x} \right)^{l_0} \left(\frac{S}{\sum_{x=0}^{k} S^x} \right)^{l_1} \left(\frac{S^2}{\sum_{x=0}^{k} S^x} \right)^{l_2} \cdots \left(\frac{S^k}{\sum_{x=0}^{k} S^x} \right)^{l_k},$$

where x is a coded age, beginning at zero, the lowest age class in the analysis, and continuing to k, the oldest age class in the analysis. The probability of observing an animal of age x when sampling from this distribution is

$$f(x) = \frac{S^x}{\sum_{x=0}^{k} S^x}.$$

Chapman and Robson (1960) give the MLE of S as the solution to the equation

$$\frac{T}{l.} = \left(\frac{S}{1-S} \right) - \frac{(k+1)S^{k+1}}{(1-S^{k+1})}, \qquad (5.121)$$

where $T = \sum_{x=0}^{k} x l_x$, and the term $\frac{T}{l.}$ is the average age in the sample.

The asymptotic variance for the estimator of S is

$$\mathrm{Var}(\hat{S}) = \frac{1}{l.} \left[\frac{1}{S(1-S)^2} - \frac{(k+1)^2 S^{k-1}}{(1-S^{k+1})^2} \right]^{-1}, \qquad (5.122)$$

with estimated variance

$$\widehat{\mathrm{Var}}(\hat{S}) = \frac{1}{l.} \left[\frac{1}{\hat{S}(1-\hat{S})^2} - \frac{(k+1)^2 \hat{S}^{k-1}}{(1-\hat{S}^{k+1})^2} \right]^{-1}. \qquad (5.123)$$

Chapman and Robson (1960) mention a possible solution for Eq. (5.121) is $\hat{S} > 1$, in which case, the MLE should be taken as $\hat{S} = 1$. They do not offer an alternative variance formula for this case when the estimated variance is undefined. When the solution of Eq. (5.121) is $\hat{S} \geq 1$, a profile-likelihood confidence interval could be used to provide an interval estimate (Kalbfleisch and Sprott 1970).

Example 5.18: Catch-Curve Analysis Using Cohort Data for Black Bears, Washington State

The cohort harvest data reported by Bender (1998) for 3.5-year-old bears in 1988 (Table 5.18) will be used to illustrate the survival estimator in Eq. (5.121). The estimator of S is a solution to the equation

$$\frac{939}{460} = \left(\frac{S}{1-S}\right) - \left(\frac{(6+1)S^{6+1}}{1-S^{6+1}}\right),$$

where $l. = 114 + 101 + \cdots + 26 + 19 = 460$ and $T = 0(114) + 1(101) + \cdots + 5(26) + 6(19) = 939$. The solution to the above equation is

$$\hat{S} = 0.7772,$$

with an estimated variance (Eq. 5.123) of

$$\widehat{Var}(\hat{S}) = \frac{1}{l.}\left[\frac{1}{\hat{S}(1-\hat{S})^2} - \frac{(k+1)^2 \hat{S}^{k-1}}{(1-\hat{S}^{k+1})^2}\right]^{-1}$$

$$= \frac{1}{460}\left[\frac{1}{0.7772(1-0.7772)^2} - \frac{(6+1)^2(0.7772)^{6-1}}{(1-0.7772^{6+1})^2}\right]^{-1} = 0.000382,$$

or $\widehat{SE}(\hat{S}) = 0.0196$.

Note that with use of cohort age-structure data, the assumptions of a stable and stationary population are not necessary. Any of the Chapman and Robson (1960) estimators can be used with cohort data without loss of generality as long as survival is constant across age classes.

Assumptions

The easiest way to assess if the assumption of constant survival probability is valid is by plotting the age code against the natural log of the age-class frequencies. Under the assumption of constant survival across age classes, the data should scatter around a straight line (Fig. 5.10). If only a portion of the age classes form a line, then survival may be constant for only those age classes, and the appropriate truncated model (i.e., left, right, or truncated at both ends) should be used for estimating survival. It should be

Table 5.18. Harvest records of the 3.5-year-old, 1988 cohort of black bears in Washington state, 1988–1994.

Year	Age	Coded age	Harvest (x_i)
1988	3.5	0	114
1989	4.5	1	101
1990	5.5	2	67
1991	6.5	3	72
1992	7.5	4	61
1993	8.5	5	26
1994	9.5	6	19

(Data from Bender 1998.)

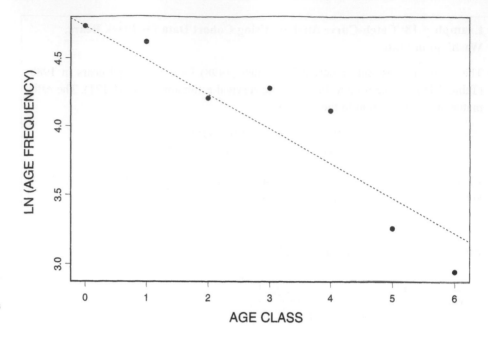

Figure 5.10. The relationship between age classes and ln (l_x) for Washington State black bears (Bender 1998) (Table 5.18) under the assumption of constant survival across all age classes.

noted that testing the significance of the regression coefficient is not equivalent to a goodness-of-fit test for constant survival. As long as age-class frequencies decrease with increasing age, the slope of the line will tend to be statistically significant.

The assumption of a constant survival probability across all age classes can be tested by using a chi-square goodness-of-fit test. The equation for the test is

$$\chi^2_{w-1} = \sum_{x=0}^{w} \frac{(l_x - E(l_x))^2}{E(l_x)},$$

where w = the oldest age class. If S was estimated by Eq. (5.113), the expected value of l_x is calculated by the expression

$$E(l_x) = l.((1 - S)S^x).$$

If the sample is truncated and survival estimated by Eq. (5.121), then $E(l_x)$ is calculated as

$$E(l_x) = l. \left(\frac{S^x}{\sum_{x=0}^{k} S^x} \right),$$

where k = the oldest age class represented in the sample. The assumption of equal survival probability is rejected if the chi-square test indicates significant lack-of-fit.

Chapman and Robson (1960) gave particular attention to the assumption of a geometric distribution (constant survival) with regard to the 0 age class. If harvest vulnerability is different for this age class, survival estimates will be biased. If the initial age class is underrepresented (lower vulnerability), then $l.$ will be artificially small and $\hat{S}$ will be overestimated. The bias will be negative if this age class is overrepresented (artificially large $l.$). Chapman and Robson (1960) derived a hypergeometric distribution based on the Heincke (1913)–Burgoyne (1981) estimator, Eq. (5.91), which provides an exact test for model validity with regard to survival in the first age class. The pmf for the hypergeometric distribution is

$$P(l_0|T,l.) = \frac{\binom{l.}{l.-l_0}\binom{T-1}{T-l.+l_0}}{\binom{l.+T-1}{T}}.$$

If $P(l_0|T,l.) < \frac{\alpha}{2}$ or $P(l_0|T,l.) > 1-\frac{\alpha}{2}$, the assumption of equal survival can be rejected for this initial age class. For a large sample, the binomial test comparing $\hat{S}$ estimated from Eq. (5.91) (Heincke 1913, Burgoyne 1981) with that of Eq. (5.113) can be used and is written as

$$Z = \frac{\left(\dfrac{l.-l_0}{l.} - \dfrac{T}{l.+T-1}\right)}{\sqrt{\dfrac{T(T-1)(l.-1)}{l.(l.+T-1)^2(l.+T-2)}}}, \tag{5.124}$$

where $Z \sim N(0,1)$. If $Z < -Z_{1-\frac{\alpha}{2}}$ or $Z > Z_{1-\frac{\alpha}{2}}$, the assumption of equal survival for the 0 age class and older animals can be rejected.

Sampling Precision

Required sample size ($l.$) for the Chapman and Robson (1960) estimator (Eq. 5.113) of survival to have a precision defined by

$$P(|\hat{S} - S| < \varepsilon) = 1 - \alpha$$

was calculated by using variance estimator (5.115). Assuming $\hat{S}$ is normally distributed,

$$\varepsilon = Z_{1-\frac{\alpha}{2}}\sqrt{\mathrm{Var}(\hat{S})}.$$

Precision curves were calculated plotting ε versus $l.$ for a range of survival probabilities $S = 0.2, 0.3, 0.5, 0.7, 0.8,$ and 0.9 (Fig. 5.11). For survivals greater than 0.5, required sample sizes decrease as animal survival increases. For survival probabilities between 0.2 and 0.5, there is little difference in absolute precision for a given sample size ($l.$). Regarding absolute sample size, gains in precision for the Chapman and Robson (1960) estimator (Eq. 5.113) are not generally appreciable beyond a sample of 500 (Fig. 5.11).

Robson and Chapman (1961) considered the effect of pooling age classes on the variance $\hat{S}$ (Eq. 5.119). For a given sample size, the relationship between the variances of the survival estimates when ages of all animals are assigned and when the older age classes are pooled is $\mathrm{Var}(\hat{S}_{\text{All aged}})(1 - S^{k+1}) = \mathrm{Var}(\hat{S}_{\text{Pooled}})$ (Robson and Chapman 1961). By inspection of the pooled variance equation, Eq. (5.119), it is clear that precision of $\hat{S}$ will

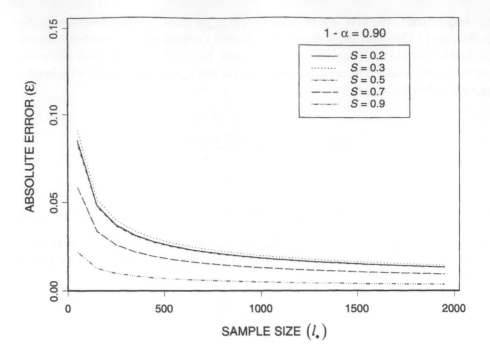

Figure 5.11. Required sample size *l.* for the Chapman and Robson (1960) estimator (Eq. 5.113) to have an absolute precision of ε, 90% of the time.

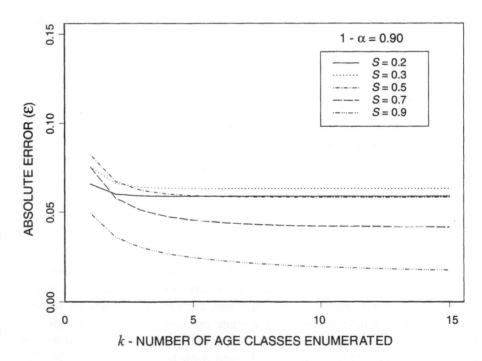

Figure 5.12. The effect of increasing the number of age classes enumerated on the precision of the Chapman and Robson (1960) estimator, (Eq. 5.118) for several values of *S*.

increase as k increases (Fig. 5.12). Robson and Chapman (1961) mention that for populations with low survival probabilities, there is little precision gained by assigning age to more than the first three classes. Inspection of Fig. 5.12 indicates there is little gain in precision by assigning age to more than five age classes for a wide range of survival values.

Robson and Chapman (1961) reported that to achieve the same absolute error in estimation, $\dfrac{100}{(1-S^{k+1})}$ animals would have to be used for the pooled estimate or for every 100 animals classified to age. Choosing how many age classes to enumerate should also depend on the accuracy and capability of the age-classification techniques used.

5.10   Regression Techniques

Regression analysis has been a predecessor and alternative to many of the maximum-likelihood methods used in survival estimation. Regression techniques can be used to estimate a common annual survival probability from l_x-series data, based on the same assumptions as the catch-curve analysis (Chapman and Robson 1960). Alternatively, repeated surveys of the cohort abundance can be regressed against time to estimate survival. The repeated survey design has few inherent assumptions and can provide estimates of survival between any two points in time or can be used to characterize survival trends over time. These regression techniques provide an appealing and robust approach that is not predicated on the multinomial sampling models of previous sections.

5.10.1   Regression on Age-Structure Data

Regression methods can be used to estimate a common annual survival probability across age classes from the l_x-series data of a catch curve. The expected number of individuals in age class x, assuming constant annual survival (S) and recruitment (N_0), can be written as

$$E(l_x|x) = N_0 p S^x \qquad (5.125)$$

where p is the probability of an animal entering the sample. Use of a logarithmic transformation on both sides of Eq. (5.125) yields

$$\ln E(l_x|x) = \ln(N_0 p) + x \ln S, \qquad (5.126)$$

which is a linear equation of the form

$$\ln(l_x) = \alpha + \beta x,$$

where $\beta = \ln S$. If model (5.125) is correct, a plot of the log-frequencies (i.e., $\ln l_x$) versus x should be a straight-line relationship. This data plot is a good check of the appropriateness of the regression model as well as the catch-curve analysis. The slope of the fitted regression model estimates $\ln S$, producing the following estimator of survival:

$$\hat{S} = e^{\hat{\beta}}. \qquad (5.127)$$

The delta method provides an approximate variance estimator,

$$\widehat{\operatorname{Var}}(\hat{S}) \doteq \widehat{\operatorname{Var}}(\hat{\beta})\left(e^{\hat{\beta}}\right)^2$$
$$\doteq \hat{S}^2\, \widehat{\operatorname{Var}}(\hat{\beta}). \qquad (5.128)$$

Data transformations will bias regression results because the linear equation (5.126) estimates $\widehat{\ln S}$ rather than $\hat{S}$. The relationship between the expected value of a transformation and the transformation of the expected value is only correct to the first term of a

Taylor series. An alternative is to estimate S directly by using iteratively reweighted non-linear regression of the form

$$SS = \sum_{x=0}^{w} \frac{(l_x - \alpha S^x)^2}{\text{Var}(l_x|x)}.$$ (5.129)

By minimizing the sum of squares (SS), the expected value of the nonlinear equation (5.125) is modeled directly. Assuming the l_x are Poisson-distributed, the variance of l_x can be approximated by $\text{Var}(l_x|x) = E(l_x|x) = \alpha S^x$ in Eq. (5.129).

Assumptions

Assumptions of the regression methods are the same as those of the Chapman and Robson (1960) estimators:

1. Probability of selection is equal for all individuals.
2. The sample is representative of the population of interest.
3. All observations are independent (uncorrelated errors).
4. All ages are recorded accurately.
5. Annual survival probabilities are constant across all age classes.
6. The population is stable and stationary if a l_x series is used.

If linear regression of the form in Eq. (5.126) is used, there is the assumption of multiplicative errors on the original scale, which become additive after data transformation. The nonlinear least-squares (NLLS) method (Eq. 5.129) has the assumption of additive errors on the original scale.

Recognizing the correct error structure is important in obtaining unbiased estimates. Without a large number of observations, there is no guarantee that estimates obtained from NLLS will be unbiased, have minimum variance, or be normally distributed (Neter et al. 1996:548). Simulation results show that estimates of $\hat{S}$ obtained from NLLS are biased when the data have multiplicative errors. When the data do have additive errors, reasonable estimates of $\hat{S}$ are obtained by using NLLS. However, errors in l_x will likely be multiplicative. Unless the data are known to have additive errors, Eq. (5.126) should be used. Although the estimate $\hat{S}$ obtained from Eq. (5.126) will still be biased, it will be less biased when the data have multiplicative errors and NLLS is used.

Chapman and Robson (1960) mention that regression estimation of survival should be used when age-class frequencies, survival among individuals, or probabilities of observing individual animals are not independent. Under those circumstances, the l_x-series counts are not multinomially distributed, and maximum likelihood methods will underestimate the sampling variance.

If sampling is not random but the assumptions of independent survival and Poisson-distributed age-class data hold, Chapman and Robson (1960) advise using the expression $\log(l_x) - \left(\frac{1}{(l_x + 1)} \right)$ as the dependent variable to reduce the bias caused by the transformation of the age-frequency data. They further suggested eliminating age classes with fewer than five individuals. By using the data transformation suggested, the regression coefficients, α and β, are estimated by minimizing the expression

$$SS = \sum_{x=0}^{w}\left[\left(\ln(l_x) - \frac{1}{(l_x + 1)}\right) - (\alpha + \beta x)\right]^2. \tag{5.130}$$

Caughley (1977:96) suggested polynomial regression as a method to smooth age-frequency data, with age-specific survivals estimated from the fitted values. Linear models would result in constant or age-independent survival, and higher-order polynomials would estimate age-dependent survival. Crowe (1975) used a linear model in estimating survival probabilities in a bobcat (*Lynx rufus*) population, and Boer (1988) used a second-degree polynomial model to estimate age-dependent mortality rates of moose (*Alces alces*). The regression used in both studies assumed a constant variance across all observations, which may not be true.

Example 5.19: Regression Estimators of Survival for Black Bears, Washington State

Black bear harvest data (Bender 1998) from 1989 in Washington State (Table 5.17) are used to illustrate the regression estimators and compare them with the Chapman and Robson (1960) catch-curve estimator. A plot of the natural log of the counts (i.e., $\ln(l_x)$) in each age class versus age appears linear, suggesting constant survival across all age classes (Fig. 5.13).

By using the linear-regression model (Eq. 5.126), the fitted equation to the age-structure data of the harvested black bears is

$$\ln(l_x) = 5.0712 - 0.2141x$$

with estimated standard errors for $\hat{\alpha}$ and $\hat{\beta}$ of $\widehat{SE}(\hat{\alpha}) = 0.1985$ and $\widehat{SE}(\hat{\beta}) = 0.0199$, respectively. Annual survival is estimated to be

$$\hat{S} = e^{\hat{\beta}} = e^{-0.2141} = 0.8073$$

with a variance and standard error of

$$\widehat{Var}(\hat{S}) = \hat{S}^2 \widehat{Var}(\hat{\beta})$$
$$= 0.8073^2(0.0199^2) = 0.000258, \text{ and}$$
$$\widehat{SE}(\hat{S}) = 0.0161,$$

with an asymptotic 90% confidence interval of (0.7808, 0.8338).

By using the Chapman and Robson (1960) modified regression estimator, Eq. (5.130), where age-frequency data are transformed to $\ln(l_x) - \left(\frac{1}{l_x + 1}\right)$, the fitted equation is

$$\left(\ln(l_x) - \left(\frac{1}{l_x + 1}\right)\right) = 5.0924 - 0.2240x$$

with estimated standard errors for $\hat{\alpha}$ and $\hat{\beta}$ of

$$\widehat{SE}(\hat{\alpha}) = 0.2093,$$
$$\widehat{SE}(\hat{\beta}) = 0.0210,$$

respectively. Survival is estimated as

$$\hat{S} = e^{\hat{\beta}} = e^{-0.2240} = 0.7993.$$

The estimated variance and standard error of $\hat{S}$ are, respectively,

$$\widehat{Var}(\hat{S}) = \hat{S}^2 \widehat{Var}(\hat{\beta}_1)$$
$$= 0.7993^2(0.0210^2) = 0.000282, \text{ and}$$
$$\widehat{SE}(\hat{S}) = 0.0168,$$

with an asymptotic 90% confidence interval of (0.7717, 0.8269).

By using the Chapman and Robson (1960) estimator, annual survival was estimated from Eq. (5.113) as $\hat{S} = 0.7915$ with an associated standard error of $\widehat{SE}(\hat{S}) = 0.0064$ and an asymptotic 90% confidence interval of (0.7811, 0.8017).

Although the point estimates of annual survival are similar, the variances are quite different. The variance for the annual survival estimate using the Chapman and Robson (1960) estimator is smaller because the variance incorporates only multinomial sampling error. The regression estimators incorporate sampling error, process error, and lack-of-fit, and will tend to be larger and more robust than the variance estimators from MLE.

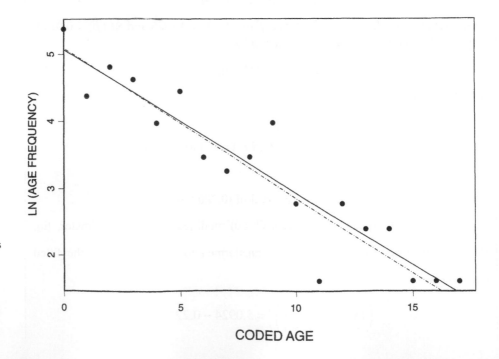

Figure 5.13. The natural log of the number of male black bears in each age class versus age (Bender 1998) (Table 5.17). The solid line is the fitted linear regression (Eq. 5.126), and the dashed line is the modified regression (Eq. 5.130).

Example 5.20: Regression Estimators of Survival for Black Bear Cohort Data, Washington State

The same regression techniques can also be applied to the analysis of cohort data. By using the data from Bender (1998) (Table 5.18), annual survival is estimated from the linear-regression model (5.126) as

$$\ln(l_x) = 4.8976 - 0.2923x$$

with estimated standard errors for $\hat{\alpha}$ and $\hat{\beta}$ of $\widehat{SE}(\hat{\alpha}) = 0.1727$ and $\widehat{SE}(\hat{\beta}) = 0.0479$, respectively. Annual survival is estimated to be

$$\hat{S} = e^{\hat{\beta}} = e^{-0.2923} = 0.7465,$$

with a variance and standard error of

$$\widehat{Var}(\hat{S}) = \hat{S}^2 \widehat{Var}(\hat{\beta})$$
$$= 0.7465^2(0.0479^2) = 0.001279, \text{ and}$$
$$\widehat{SE}(\hat{S}) = 0.0358,$$

with an asymptotic 90% confidence interval of (0.6876, 0.8054).

By use of the Chapman and Robson (1960) modified regression estimator (Eq. 5.130), where the age-frequency data are transformed to $\ln(l_x) - \left(\dfrac{1}{l_x + 1}\right)$, the fitted equation is

$$\left(\ln(l_x) - \left(\frac{1}{l_x + 1}\right)\right) = 4.8955 - 0.2987x,$$

with estimated standard errors for $\hat{\alpha}$ and $\hat{\beta}$ of

$$\widehat{SE}(\hat{\alpha}) = 0.1780, \text{ and}$$
$$\widehat{SE}(\hat{\beta}) = 0.0494,$$

respectively. The survival estimate is calculated to be

$$\hat{S} = e^{\hat{\beta}} = e^{-0.2987} = 0.7418.$$

The estimated variance and standard error of $\hat{S}$ are, respectively,

$$\widehat{Var}(\hat{S}) = \hat{S}^2 \widehat{Var}(\hat{\beta})$$
$$= 0.7418^2(0.0494^2) = 0.001343, \text{ and}$$
$$\widehat{SE}(\hat{S}) = 0.0366$$

with an asymptotic 90% confidence interval of (0.6816, 0.8020). The Chapman and Robson (1960) estimator of annual survival for a truncated sample was 0.7772 ($\widehat{SE}(\hat{S}) = 0.0196$).

5.10.2 Regression on Abundance Estimates

One direct approach to estimating survival of a cohort is to repeatedly survey cohort abundance over time. Let N_t be the abundance of that cohort at time t. Then survival from time t to time $t + 1$ can be expressed as

$$S_t = \frac{N_{t+1}}{N_t}$$

and estimated by

$$\hat{S}_t = \frac{\hat{N}_{t+1}}{\hat{N}_t}.$$

This approach is appropriate regardless of the nature of the survival process from time t to time $t + 1$. Variance of the survival estimate can be approximated by the expression

$$\widehat{\mathrm{Var}}\left(\hat{S}_t\right) = \hat{S}_t^2 \left[\frac{\widehat{\mathrm{Var}}\left(\hat{N}_{t+1}\right)}{\hat{N}_{t+1}^2} + \frac{\widehat{\mathrm{Var}}\left(\hat{N}_t\right)}{\hat{N}_t} - \frac{2\widehat{\mathrm{Cov}}\left(\hat{N}_t, \hat{N}_{t+1}\right)}{\hat{N}_t \cdot \hat{N}_{t+1}} \right]. \tag{5.131}$$

If survey events are independent, the third term in Eq. (5.131) can be omitted. The expected value of $\hat{S}_t$ to the first term of a Taylor series expansion is

$$E\left(\hat{S}_t\right) = S_t \left(1 + \frac{\widehat{\mathrm{Var}}\left(\hat{N}_t | N_t\right)}{N_t^2}\right) = S_t\left(1 + \mathrm{CV}\left(\hat{N}_t\right)^2\right).$$

Hence, $\hat{S}_t$ will have a positive bias that increases with the variance or coefficient of variation (CV) of $\hat{N}_t$. For example, a CV(N_t) of 30% will result in positive bias, inflating the survival estimate by a factor of approximately 1.09.

Alternatively, assuming the survival process is constant over a period of several survey events, a common survival parameter can be estimated through time. The expected cohort abundance at time t (i.e., N_t) will be a function of initial abundance N_0, where

$$E(N_t) = N_0 e^{-rt}$$

is of the form of an exponential decay. The relationship can be reexpressed in the form

$$E\left(\frac{N_t}{N_0}\right) = e^{-rt},$$

leading to the log-linear model

$$\ln\left(\frac{N_t}{N_0}\right) = -rt. \tag{5.132}$$

The relationship in Eq. (5.132) is a straight-line regression through the origin of the form

$$y_t = \beta t,$$

where $y_t = \ln\left(\dfrac{\hat{N}_t}{\hat{N}_0}\right)$. Alternatively, NLLS of the form

$$\sum_{i=1}^{t} \left(\frac{\hat{N}_i}{\hat{N}_0} - e^{-ri} \right)^2$$

could be used to estimate the instantaneous mortality rate (r) to avoid bias induced by data transformation. However, as with age-structure data, NLLS will be biased if the error structure is not additive.

The variance of $\frac{\hat{N}_t}{\hat{N}_0}$ will be proportional to the variance of $\hat{N}_t$ for a given $\hat{N}_0$, i.e., $\text{Var}\left(\frac{\hat{N}_t}{\hat{N}_0} \middle| \hat{N}_0 \right) \propto \widehat{\text{Var}}(\hat{N}_t)$. In turn, the $\widehat{\text{Var}}(\hat{N}_t)$ will often be proportional to N_t for most survey techniques. In the case of multiplicative errors, log-linear regression of the form in Eq. (5.132) should be used to obtain an unbiased estimate of r. However, ignoring the variance of $\ln\left(\frac{\hat{N}_t}{\hat{N}_0} \right)$ will result in estimates of $\hat{r}$ that do not have minimum variance. A weighted linear regression will alleviate this problem, with weights proportional to

$$\frac{1}{\left[\frac{\widehat{\text{Var}}(\hat{N}_t)}{\hat{N}_t^2} \right]} = CV(\hat{N}_t)^{-2}.$$

An estimate of the survival probability from time $t = 0$ to time t is then

$$\hat{S}_t = e^{-\hat{r}t}. \tag{5.133}$$

By using the delta method, the variance for $\hat{S}_t$ can be estimated by the expression

$$\widehat{\text{Var}}(\hat{S}_t) = \widehat{\text{Var}}(\hat{r})t^2 e^{-2\hat{r}t}. \tag{5.134}$$

Given the difficulty and expense of obtaining abundance estimates in wildlife sciences, this method of survival estimation is often of limited use. Further complicating this method is the assumption that all changes in abundance are solely the result of mortality.

5.11   Estimating Juvenile Survival

The array of methods to estimate adult survival is largely unsuitable for estimating survival probabilities of juveniles. First, sampling probabilities for young animals are usually different from those of adults. The behavior of juveniles makes them less detectable in many visual surveys. Depending on harvest regulations, juveniles also may be more or less vulnerable to harvest than are adults, resulting in selection bias. Second, juvenile survival rates are rarely the same as survival rates of older adult age classes. Finally, appreciable mortality has usually occurred before juveniles have recruited into the harvestable, trappable, or viewable population. These factors have prompted wildlife investigators to use a combination of age- and sex-structure data to estimate juvenile survival. The following section reviews some of the more common approaches for estimating juvenile survival, based on age- and sex-structure data.

5.11.1   Two-Sample Change-in-Ratio Methods of Hanson (1963) and Paulik and Robson (1969)

Change-in-ratio methods offer an alternative to the standard life table or catch-curve analysis commonly used for estimating adult survival from age-structure data. Hanson (1963) and Paulik and Robson (1969) suggested using a two-sample CIR method to estimate juvenile survival between survey periods. At each sampling occasion, the ratio of juveniles to adults is quantified. The change in this ratio between periods provides an estimate of juvenile survival under specific sampling conditions.

For the first survey event, the following terms are defined:

a_1 = number of adults observed in the first survey;
y_1 = number of juveniles observed in the first survey;
$x_1 = a_1 + y_1$;
p_1 = probability of detecting an animal during the first survey;
N_J = abundance of juveniles during the first survey;
N_A = abundance of adults during the first survey;
N_1 = abundance of all animals during the first survey = $N_A + N_J$.

For the second survey event, the remaining terms are defined as follows:

a_2 = number of adults observed during the second survey;
y_2 = number of juveniles observed during the second survey;
$x_2 = a_2 + y_2$;
p_2 = probability of detecting an animal during the second survey;
N_2 = abundance of animals during the second survey = $S_A N_A + S_J N_J$;
S_A = adult survival rate between the first and second surveys;
S_J = juvenile survival rate between the first and second surveys.

From the first survey, the number of juveniles and adults observed can be modeled as a multinomial distribution, where

$$L(N_1, p_1 | a_1, y_1) = \binom{N_1}{a_1, y_1} \left(\frac{p_1 N_A}{N_A + N_J} \right)^{a_1} \left(\frac{p_1 N_J}{N_A + N_J} \right)^{y_1} (1 - p_1)^{N_1 - x_1} \quad (5.135)$$

and where the total number counted (i.e., $x_1 = a_1 + y_1$) has the binomial distribution

$$L(N_1, p_1 | x_1) = \binom{N_1}{x_1} p_1^{x_1} (1 - p_1)^{N_1 - x_1}. \quad (5.136)$$

The conditional likelihood of a_1 and y_1, given x_1, can be written as

$$L(N_A, N_J | a_1, y_1, x_1) = \frac{\binom{N_1}{a_1, y_1} \left(\frac{p_1 N_A}{N_A + N_J} \right)^{a_1} \left(\frac{p_1 N_J}{N_A + N_J} \right)^{y_1} (1 - p_1)^{N_1 - x_1}}{\binom{N_1}{x_1} p_1^{x_1} (1 - p_1)^{N_1 - x_1}}$$

$$= \binom{x_1}{a_1} \left(\frac{N_A}{N_A + N_J} \right)^{a_1} \left(\frac{N_J}{N_A + N_J} \right)^{y_1}. \quad (5.137)$$

Written in terms of the juvenile-to-adult ratio, i.e., $R_{J/A} = N_J/N_A$, Eq. (5.137) can be rewritten as

$$L(R_{J/A}|a_1,y_1,x_1) = \binom{x_1}{a_1}\left(\frac{1}{1+R_{J/A}}\right)^{a_1}\left(\frac{R_{J/A}}{1+R_{J/A}}\right)^{y_1}. \tag{5.138}$$

During the second survey, differential survival probabilities of juveniles and adults shift the age ratio of the sample. For the second survey, the likelihood for the juvenile (y_2) and adult (a_2) counts, conditioned on the sum $x_2 = a_2 + y_2$, can be written as

$$L(S_A,N_A,N_J|a_2,y_2,x_2) = \binom{x_2}{a_2}\left(\frac{S_A N_A}{S_A N_A + S_J N_J}\right)^{a_2}\left(\frac{S_J N_J}{S_A N_A + S_J N_J}\right)^{y_2}. \tag{5.139}$$

Reparameterizing Eq. (5.139) in terms of the ratio of juveniles to adults and letting adult survival, $S_A = 1$, the likelihood can be written as

$$L(S_J,R_{J/A}|a_2,y_2,x_2) = \binom{x_2}{a_2}\left(\frac{1}{1+S_J R_{J/A}}\right)^{a_2}\left(\frac{S_J R_{J/A}}{1+S_J R_{J/A}}\right)^{y_2}. \tag{5.140}$$

The joint likelihood for the first and second surveys can be written as

$$\begin{aligned} L = &\binom{x_1}{a_1}\left(\frac{1}{1+R_{J/A}}\right)^{a_1}\left(\frac{R_{J/A}}{1+R_{J/A}}\right)^{y_1}. \\ &\binom{x_2}{a_2}\left(\frac{1}{1+S_J R_{J/A}}\right)^{a_2}\left(\frac{S_J R_{J/A}}{1+S_J R_{J/A}}\right)^{y_2}. \end{aligned} \tag{5.141}$$

The MLE for juvenile survival is

$$\hat{S}_J = \frac{a_1 y_2}{y_1 a_2} \tag{5.142}$$

where $\hat{R}_{J/A} = \dfrac{y_1}{a_1}$. The variance for $\hat{S}_J$ can be approximated by using the delta method, where

$$\mathrm{Var}(\hat{S}_J) \doteq \frac{S_J}{R_{J/A}}\left(\frac{S_J(1+R_{J/A})^2}{x_1} + \frac{(1+S_J R_{J/A})^2}{x_2}\right) \tag{5.143}$$

with estimated variance

$$\widehat{\mathrm{Var}}(\hat{S}_J) = \frac{\hat{S}_J}{\hat{R}_{J/A}}\left(\frac{\hat{S}_J(1+\hat{R}_{J/A})^2}{x_1} + \frac{(1+\hat{S}_J \hat{R}_{J/A})^2}{x_2}\right). \tag{5.144}$$

Assumptions

The assumptions associated with juvenile survival estimator (5.142) include the following:

1. All animals, juveniles and adults, have independent and equal probabilities of detection in the first survey.

2. All animals, juveniles and adults, have independent and equal probabilities of detection in the second survey.
3. The survival rate of adults is one (i.e., $S_A = 1$) between the two surveys.
4. All juveniles had equal and independent probabilities of survival.
5. Sampling within a period is with replacement.

If the adult survival probability is not equal to one, the juvenile survival estimator has an approximate expected value of

$$E(\hat{S}_J) = \frac{S_J}{S_A},$$

which is the estimator of relative juvenile-to-adult survival described by Hanson (1963) and Paulik and Robson (1969). Inspection of the expected value of $\hat{S}_J$ indicates if $S_A < 1$, the juvenile survival estimate will be positively biased.

The above assumptions 1 and 2 may not hold if detection probabilities for adults and juveniles are affected by behavioral differences. In deriving a CIR abundance estimator, Chapman (1955) allowed for differential detection rates between juveniles (p_j) and adults (p_a). However, Chapman (1955) assumed the ratio of the detection probabilities was the same across the two surveys, i.e., $\frac{p_{j1}}{p_{a1}} = \frac{p_{j2}}{p_{a2}} = \eta$. By using the same assumption, the survival estimator (Eq. 5.142) is unchanged. Thus, the CIR estimator for survival is robust with regard to assumptions 1 and 2, as long as the ratio of the detection probability is constant across surveys.

The CIR method for estimating juvenile survival can also be expressed in terms of the ratios of juveniles per adult female ($R_{J/F}$) rather than in terms of all adults. Enumerating numbers of adult females instead of total adults is often more convenient, considering juveniles are often closely associated with females. The sampling process remains the same, where the joint likelihood, Eq. (5.141), is rewritten in terms of $R_{J/F}$. The resulting estimators of $R_{J/F}$ and S_J are

$$\hat{R}_{J/F} = \frac{y_1}{f_1} \text{ and}$$

$$\hat{S}_J = \frac{f_1 y_2}{f_2 y_1}. \tag{5.145}$$

This is the estimator of juvenile survival given in Paulik and Robson (1969). Recalling that productivity is defined as the ratio of juveniles to adult females $\left(\text{i.e., } P = \frac{N_J}{N_F} \right)$, the juvenile survival estimate in Eq. (5.145) is the ratio of two estimates of productivity, P, measured at times t and $t + 1$, where

$$\hat{S}_J = \frac{\hat{P}_{t+1}}{\hat{P}_t}. \tag{5.146}$$

The associated variance of survival estimator (5.146) is approximated by

$$\widehat{\text{Var}}(\hat{S}_J) = \hat{S}_J^2 \left(\widehat{\text{CV}}(\hat{P}_{t+1})^2 + \widehat{\text{CV}}(\hat{P}_t)^2 \right), \tag{5.147}$$

where the CV is defined as

$$\widehat{CV}(\hat{P}) = \frac{\widehat{SE}(\hat{P})}{\hat{P}}.$$

Example 5.21: Estimating Juvenile Survival of Bighorn Sheep (*Ovis canadensis*) using CIR, Custer State Park, South Dakota

As part of a study to examine lamb production and summer mortality of bighorn sheep in Custer State Park, South Dakota (Merwin 2000), monthly counts of lambs and ewes were taken of several groups within the Park from April 1998 to January 1999. For purposes of illustration, only counts from July and September 1998 will be used. During the July survey of the East End herd, 150 lambs ($y_1 = 150$) and 273 ewes ($a_1 = 273$) were observed. The September survey recorded 19 lambs ($y_2 = 19$) and 158 ewes ($a_2 = 158$). From these data, $\hat{R}_{J/F}$ is estimated as

$$\hat{R}_{J/F} = \frac{y_1}{f_1}$$

$$= \frac{150}{273} = 0.5495,$$

and juvenile survival from July to September is estimated by Eq. (5.145) as

$$\hat{S}_J = \frac{f_1 y_2}{f_2 y_1}$$

$$= \frac{273(19)}{158(150)} = 0.2189.$$

The variance of $\hat{S}_J$ is calculated analogous to Eq. (5.144) as

$$\widehat{Var}(\hat{S}_J) = \frac{\hat{S}_J}{\hat{R}_{J/F}} \left(\frac{\hat{S}_J(1+\hat{R}_{J/F})^2}{x_1} + \frac{(1+\hat{S}_J\hat{R}_{J/F})^2}{x_2} \right)$$

$$= \frac{0.2189}{0.5495} \left(\frac{0.2189(1+0.5495)^2}{423} + \frac{(1+0.2189(0.5495))^2}{177} \right)$$

$$= 0.00332,$$

with a standard error of $\widehat{SE}(\hat{S}_J) = 0.0576$ and an asymptotic 90% confidence interval of $CI(0.1241 \leq S_J \leq 0.3137) = 0.90$.

Sampling Precision

Defining sampling precision for the juvenile survival estimate as

$$P(|\hat{S}_J - S_J| < \varepsilon) = 1 - \alpha$$

and assuming $\hat{S}_J$ is normally distributed, sample size charts are presented for the two-sample CIR method (Fig. 5.14). For simplicity, sample sizes x_1 and x_2 were assumed to

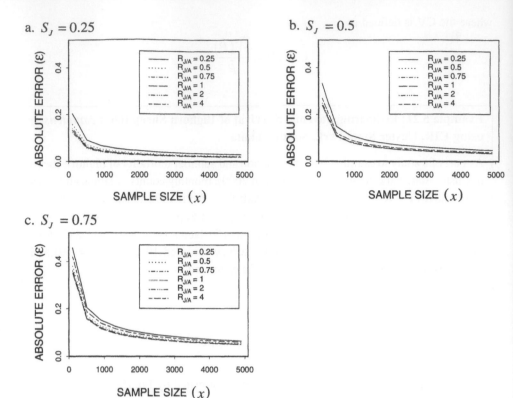

Figure 5.14. The absolute error ε in estimating juvenile survival (S_J) using the two-smaple change-in-ratio method for three different values of S_J over a range of values $0.25 \leq R_{J/A} \leq 4.0$ for $1 - \alpha = 0.90$.

be equal for the two sampling periods. In general, as the juvenile survival decreases, the change in the juvenile : adult ratio becomes more discernable between sampling periods, requiring fewer observations to precisely characterize S_J. Sample size requirements (Fig. 5.14) are reasonably robust to the initial juvenile : adult ratio over the range 0.25 to 4.0.

5.11.2 Three-Sample Change-in-Ratio Methods of Selleck and Hart (1957) and White et al. (1996)

In Section 5.11.1, juvenile survival was estimated assuming adult survival $S_A = 1.0$ between survey periods. Should this assumption not be true, the juvenile survival estimate will be positively biased by $\left(\dfrac{1 - S_A}{S_A} \right) \cdot 100\%$. To relax this assumption, additional information is needed to estimate adult survival and derive a bias-corrected juvenile survival estimate.

Selleck and Hart (1957) first presented a three-sample CIR method to estimate differential harvest survival of males and females. Their method of using carcass counts from a postwinter survey is similar to the method of estimating juvenile survival presented below. White et al. (1996) provide an estimator for calculating adult and juvenile survival from age ratios by adding this additional carcass count survey.

In the first period, a survey sample of the population is taken and the number of juveniles and adults observed is enumerated. A second sample of the population is taken later in the annual cycle, at the end of the period for which juvenile survival is estimated. However, during this time, adult survival may not be one. Consequently, a carcass count of animals that died between the first and second period is conducted, recording the number of juveniles and adults in the sample. The carcass survey requires that all carcasses are a result of natural mortality between the first and second survey, and juveniles are identifiable. Carcass count surveys are often conducted in spring, after adults have experienced winter mortality and while winter-killed animals are still identifiable (White et al. 1996).

The conditional likelihood for the first sampling period is the same as Eq. (5.138), where

$$L(R_{J/A}|a_1, y_1, x_1) = \binom{x_1}{a_1}\left(\frac{1}{1+R_{J/A}}\right)^{a_1}\left(\frac{R_{J/A}}{1+R_{J/A}}\right)^{y_1}.$$

The conditional likelihood for the second sampling period (Eq. 5.139) includes adult survival and can be written as

$$L(S_A, R_{J/A}|a_2, y_2, x_2) = \binom{x_2}{a_2}\left(\frac{S_A}{S_A+S_J R_{J/A}}\right)^{a_2}\left(\frac{S_J R_{J/A}}{S_A+S_J R_{J/A}}\right)^{y_2}. \quad (5.148)$$

The likelihood for the carcass count survey is derived in the same manner as Eq. (5.138) and Eq. (5.139), and is written as

$$L(S_A, S_J, N_A, N_J, p_3|a_3, y_3) = \binom{N_3}{a_3, y_3}\left(\frac{p_3(1-S_A)N_A}{(1-S_A)N_A+(1-S_J)N_J}\right)^{a_3}$$
$$\cdot\left(\frac{p_3(1-S_J)N_J}{(1-S_A)N_A+(1-S_J)N_J}\right)^{y_3}\cdot(1-p_3)^{N_3-x_3},$$

where

a_3 = number of adults carcasses found;
y_3 = number of juvenile carcasses found;
$x_3 = a_3 + y_3$.

The likelihood for the total number of carcasses found, x_3, can be written as

$$L(N_3, p_3|x_3) = \binom{N_3}{x_3}p_3^{x_3}(1-p_3)^{N_3-x_3},$$

where p_3 = probability of observing an animal carcass. The conditional likelihood of a_3 and y_3, given x_3, can be written as

$$L(S_A, S_J, N_A, N_J|a_3, y_3, x_3) = \binom{x_3}{a_3}\left(\frac{(1-S_A)N_A}{(1-S_A)N_A+(1-S_J)N_J}\right)^{a_3}$$
$$\cdot\left(\frac{(1-S_J)N_J}{(1-S_A)N_A+(1-S_J)N_J}\right)^{y_3}. \quad (5.149)$$

Reparameterizing Eq. (5.149) in terms of the ratio of juvenile to adults, $R_{J/A}$, the likelihood can be written as

$$L(S_A, S_J, R_{J/A} | a_3, y_3, x_3) = \binom{x_3}{a_3} \left(\frac{(1-S_A)}{(1-S_A)+(1-S_J)R_{J/A}} \right)^{a_3}$$
$$\cdot \left(\frac{(1-S_J)R_{J/A}}{(1-S_A)+(1-S_J)R_{J/A}} \right)^{y_3}.$$

The joint likelihood from all three surveys is then written as

$$L = \binom{x_1}{a_1} \left(\frac{1}{1+R_{J/A}} \right)^{a_1} \left(\frac{R_{J/A}}{1+R_{J/A}} \right)^{y_1} \cdot$$
$$\binom{x_2}{a_2} \left(\frac{S_A}{S_A + S_J R_{J/A}} \right)^{a_2} \left(\frac{S_J R_{J/A}}{S_A + S_J R_{J/A}} \right)^{y_2} \cdot \tag{5.150}$$
$$\binom{x_3}{a_3} \left(\frac{(1-S_A)}{(1-S_A)+(1-S_J)R_{J/A}} \right)^{a_3} \left(\frac{(1-S_J)R_{J/A}}{(1-S_A)+(1-S_J)R_{J/A}} \right)^{y_3}.$$

The MLEs of survival from Eq. (5.150) are

$$\hat{S}_A = \frac{\hat{R}_3 - \hat{R}_1}{\hat{R}_3 - \hat{R}_2} \tag{5.151}$$

and

$$\hat{S}_J = \frac{\hat{R}_2}{\hat{R}_1} \left(\frac{\hat{R}_3 - \hat{R}_1}{\hat{R}_3 - \hat{R}_2} \right), \tag{5.152}$$

where

$$\hat{R}_i = \frac{y_i}{a_i}$$

for $i = 1, \ldots, 3$. The estimators (5.151) and (5.152) are the same as those given by White et al. (1996:39).

White et al. (1996) derived the variances for $\hat{S}_A$ and $\hat{S}_J$ by using the delta method, where

$$\widehat{Var}(\hat{S}_A) = \frac{1}{(\hat{R}_2 - \hat{R}_3)^2} \left(\widehat{Var}(\hat{R}_1) + \left(\frac{\hat{R}_3 - \hat{R}_1}{\hat{R}_2 - \hat{R}_3} \right)^2 \widehat{Var}(\hat{R}_2) \right.$$
$$\left. + \left(\frac{\hat{R}_1 - \hat{R}_2}{\hat{R}_2 - \hat{R}_3} \right)^2 \widehat{Var}(\hat{R}_3) \right), \tag{5.153}$$

and

$$\widehat{Var}(\hat{S}_J) = \left(\frac{\hat{R}_2 \hat{R}_3}{\hat{R}_1(\hat{R}_2 - \hat{R}_3)} \right)^2 \left(\widehat{CV}(\hat{R}_1)^2 + \left(\frac{\hat{R}_3 - \hat{R}_1}{\hat{R}_2 - \hat{R}_3} \right)^2 \widehat{CV}(\hat{R}_2)^2 \right.$$
$$\left. + \left(\frac{\hat{R}_1 - \hat{R}_2}{\hat{R}_2 - \hat{R}_3} \right)^2 \widehat{CV}(\hat{R}_3)^2 \right), \tag{5.154}$$

for $\widehat{\text{Var}}(\hat{R}) = \dfrac{\hat{R}(1+\hat{R})^2}{x}$ and $\widehat{\text{CV}}(\hat{R})^2 = \dfrac{(1+\hat{R})^2}{\hat{R}x}$. These variance estimators are expressed in terms of the estimates of the ratios, rather than the parameters. Consequently, they are cumbersome for sample size calculations. The variance of $\hat{S}_A$ expressed in terms of parameters $R_{J/A}$, S_A, and S_J and sample sizes x_1, x_2, and x_3 is written as

$$\text{Var}(\hat{S}_A) = \frac{S_A(1-S_A)}{R_{J/A}(S_A-S_J)^2}\left[\frac{(1+R_{J/A})^2 S_A(1-S_A)}{x_1} + \frac{S_J(1-S_A)(S_A+R_{J/A}S_J)^2}{x_2}\right.$$
$$\left. + \frac{S_A(1-S_J)((1-S_A)+(1-S_J)R_{J/A})^2}{x_3}\right].$$
(5.155)

The variance estimator for $\hat{S}_J$ is written as

$$\text{Var}(\hat{S}_J) \doteq \frac{1}{R_{J/A}(S_A-S_J)^2}\left[\frac{(1+R_{J/A})^2 S_J^2(1-S_J)^2}{x_1} + \frac{S_J^3(1-S_A)^2(S_A+R_{J/A}S_J)^2}{x_2 S_A}\right.$$
$$\left. + \frac{S_A(1-S_J)(1-S_A)((1-S_A)+(1-S_J)R_{J/A})^2}{x_3}\right].$$
(5.156)

Assumptions

In deriving the survival estimators (5.151) and (5.152), White et al. (1996) allowed sighting probabilities for adults and juveniles to differ within a sampling period, but the ratio of sighting probabilities was assumed constant across surveys. Thus, the CIR survival estimates are robust to the assumption that all animals have equal detection probability within a period as long as the detection bias is constant. However, it may be unlikely the same detection bias transcends across both visual surveys of live animals and carcass surveys of dead animals.

Assumptions associated with the third carcass survey are similar to those of the two live animal surveys with the additions:

1. All carcasses are identifiable to age (i.e., juvenile vs. adult).
2. Death was a result of natural mortality within the period of interest.

Although the joint likelihood model (5.150) was written in terms of sampling with replacement, the estimators and associated variances are applicable for sampling with or without replacement. When sampling with replacement, the variance of $\hat{R}_i$ can be written as

$$\widehat{\text{Var}}(\hat{R}_i) = \frac{\hat{R}_i(1+\hat{R}_i)^2}{x_i}$$

and, in the case of sampling without replacement,

$$\widehat{\text{Var}}(\hat{R}_i) = \frac{\hat{R}_i(1+\hat{R}_i)^2}{x_i}\left(\frac{N_i-x_i}{N_i-1}\right).$$

For each survey period, the variance of $\hat{R}_i$ can be selected as appropriate.

Example 5.22: Estimating Juvenile and Adult Survival using a Three-Sample CIR for Mule Deer (*Odocoileus hemionus*), Southcentral Wyoming

A study to estimate the mortality of mule deer was conducted in southcentral Wyoming during winter 1988–1989 by White et al. (1996). A prewinter ground survey was conducted during December, in which $a_1 = 3465$ adults and $y_1 = 2238$ juveniles were counted. An independent postwinter survey in April yielded $a_2 = 3768$ adults and $y_2 = 1726$ juveniles. In a separate survey for winter-kill carcasses conducted in early May, $a_3 = 22$ adults and $y_3 = 58$ juveniles were found. From these data, juvenile and adult survivals, $\hat{S}_J$ and $\hat{S}_A$, can be estimated by using Eqs. (5.152) and (5.151), respectively, where

$$\hat{R}_{J/A} = \hat{R}_1 = \frac{y_1}{a_1} = \frac{2283}{3465} = 0.65887,$$

$$\hat{R}_2 = \frac{y_2}{a_2} = \frac{1726}{3768} = 0.45807,$$

$$\hat{R}_3 = \frac{y_3}{a_3} = \frac{58}{22} = 2.6364,$$

and where it follows

$$\hat{S}_J = \frac{\hat{R}_2}{\hat{R}_1}\left(\frac{\hat{R}_3 - \hat{R}_1}{\hat{R}_3 - \hat{R}_2}\right)$$

$$= \frac{0.45807}{0.65887}\left(\frac{2.6364 - 0.65887}{2.6364 - 0.45807}\right) = 0.6311$$

and

$$\hat{S}_A = \frac{\hat{R}_3 - \hat{R}_1}{\hat{R}_3 - \hat{R}_2}$$

$$= \left(\frac{2.6364 - 0.65887}{2.6364 - 0.45807}\right) = 0.9078.$$

The estimated variances of $\hat{S}_J$ and $\hat{S}_A$ are calculated from Eqs. (5.154) and (5.153):

$$\widehat{\mathrm{Var}}\left(\hat{R}_i\right) = \frac{\hat{R}_i\left(1 + \hat{R}_i\right)^2}{x_i},$$

$$\widehat{\mathrm{Var}}\left(\hat{R}_1\right) = \frac{0.65887(1 + 0.65887)^2}{3465 + 2283} = 0.000315,$$

$$\widehat{\mathrm{Var}}\left(\hat{R}_2\right) = \frac{0.45807(1 + 0.45807)^2}{3768 + 1726} = 0.0001773,$$

$$\widehat{\mathrm{Var}}\left(\hat{R}_3\right) = \frac{2.6364(1 + 2.6364)^2}{22 + 58} = 0.43578,$$

such that

$$\widehat{Var}(\hat{S}_J) = \left(\frac{\hat{R}_2 \hat{R}_3}{\hat{R}_1(\hat{R}_2 - \hat{R}_3)}\right)^2 \left(\widehat{CV}(\hat{R}_1)^2 + \left(\frac{\hat{R}_3 - \hat{R}_1}{\hat{R}_2 - \hat{R}_3}\right)^2 \widehat{CV}(\hat{R}_2)^2 \right.$$

$$\left. + \left(\frac{\hat{R}_1 - \hat{R}_2}{\hat{R}_2 - \hat{R}_3}\right)^2 \widehat{CV}(\hat{R}_3)^2 \right)$$

$$= \left(\frac{0.45807 \cdot 2.6364}{0.65887(0.45807 - 2.6364)}\right)^2 \left(\frac{0.000315}{0.65887^2} + \left(\frac{2.6364 - 0.65887}{0.45807 - 2.6364}\right)^2 \right.$$

$$\left. \cdot \left(\frac{0.0001773}{0.45807^2}\right) + \left(\frac{0.65887 - 0.45807}{0.45807 - 2.6364}\right)^2 \frac{0.43578}{2.6364^2} \right)$$

$$\widehat{Var}(\hat{S}_J) = 0.001384 \text{ or } \widehat{SE}(\hat{S}_J) = 0.0372,$$

and

$$\widehat{Var}(\hat{S}_A) = \frac{1}{(\hat{R}_2 - \hat{R}_3)^2} \left(\widehat{Var}(\hat{R}_1) + \left(\frac{\hat{R}_3 - \hat{R}_1}{\hat{R}_2 - \hat{R}_3}\right)^2 \widehat{Var}(\hat{R}_2) \right.$$

$$\left. + \left(\frac{\hat{R}_1 - \hat{R}_2}{\hat{R}_2 - \hat{R}_3}\right)^2 \widehat{Var}(\hat{R}_3)\right)$$

$$= \frac{1}{(0.45807 - 2.6364)^2} \left(0.000315 + \left(\frac{2.6364 - 0.65887}{0.45807 - 2.6364}\right)^2 \cdot 0.0001773 \right.$$

$$\left. + \left(\frac{0.65887 - 0.45807}{0.45807 - 2.6364}\right)^2 \cdot 0.43578\right)$$

$$\widehat{Var}(\hat{S}_A) = 0.000878 \text{ or } \widehat{SE}(\hat{S}_A) = 0.0296.$$

It is interesting to note, assuming adult survival is $S_A = 1$ and using only pre- and postwinter surveys, juvenile survival is estimated (Eq. 5.142) to be

$$\hat{S}_J = \frac{a_1 y_2}{y_1 a_2} = \frac{\hat{R}_2}{\hat{R}_1} = \frac{0.45807}{0.65887} = 0.6952.$$

This estimate of juvenile survival is unadjusted for any adult mortality that may have occurred. Multiplying this unadjusted estimate of juvenile survival by the estimate of adult survival, i.e., 0.6952 (0.9078) = 0.6311, yields the adjusted estimate of juvenile survival from Eq. (5.152).

5.11.3   Life-History Methods of Keith and Windberg (1978)

Keith and Windberg (1978) presented three different methods for estimating juvenile survival by using information on juvenile:adult ratios, productivity, adult survival probabilities, and abundance estimates. These methods may most appropriately be considered as life-history methods, because they use models based on life-history processes in deriving the estimates of juvenile survival. Analogous methods can be easily envisioned by using similar processes and available information for a wide variety of species. The methods of Keith and Windberg (1978) are presented so that their general approach may be adapted to other field survey scenarios and other life histories of interest to the reader.

Method 1

By using the notation of Keith and Windberg (1978), let $\hat{S}_{Ji}$ denote juvenile survival from birth to time i and let birth time be denoted as zero. Then

$$\hat{S}_{Ji} = \frac{\hat{R}_{(J/A)i} \cdot \hat{S}_{Ai}}{\hat{U}_0} \tag{5.157}$$

where

$\hat{R}_{(J/A)i}$ = estimated ratio of juveniles to adults at time i;

$\hat{S}_{Ai}$ = estimated adult survival from time 0 to i;

$\hat{U}_0$ = estimated numbers of juveniles born (time 0) per adult $\left(\text{i.e., as defined in}\right.$

Chapter 4, $U = \frac{N_{J0}}{N_{A0}}\Big)$.

To see the reasonableness of the estimator, Eq. (5.157), the expected value of $\hat{S}_{Ji}$ can be approximated as

$$E\left(\hat{S}_{Ji}\right) \doteq \frac{E\left(\hat{R}_{(J/A)i}\right) \cdot E\left(\hat{S}_{Ai}\right)}{E(U_0)}$$

$$= \frac{\left(\dfrac{N_{Ji}}{N_{Ai}}\right)\dfrac{N_{Ai}}{N_{A0}}}{\dfrac{N_{J0}}{N_{A0}}}$$

$$= \frac{N_{Ji}}{N_{J0}}$$

where

N_{Ji} = number of juveniles at time i;

N_{J0} = number of juveniles at time 0;

N_{Ai} = number of adults at time i;

N_{A0} = number of adults at time 0.

Hence, to the first term of a Taylor series expansion, the juvenile survival estimator is unbiased.

Assuming all three components of the survival estimator (5.157) are independent, the variance of $\hat{S}_{Ji}$ can be written as

$$\widehat{\text{Var}}\left(\hat{S}_{Ji}\right) = \left(\hat{S}_{Ji}\right)^2\left[\widehat{\text{CV}}\left(\hat{R}_{(J/A)i}\right)^2 + \widehat{\text{CV}}\left(\hat{S}_{Ai}\right)^2 + \widehat{\text{CV}}\left(\hat{U}_0\right)^2\right] \tag{5.158}$$

where

$$\widehat{\text{CV}}(\theta)^2 = \frac{\widehat{\text{Var}}\left(\hat{\theta}\right)}{\hat{\theta}^2}.$$

Productivity is defined (see Chapter 4) as the number of juveniles per adult female. Therefore, it may be more natural to present Eq. (5.157) in terms of juveniles per adult female or

$$\hat{S}_{Ji} = \frac{\hat{P}_i \cdot \hat{S}_{Ai}}{\hat{P}_0} \qquad (5.159)$$

where

$\hat{P}_0$ = number of juveniles per female at time 0;
$\hat{P}_i$ = number of juveniles per female at time i.

Productivity is more easily estimated by direct techniques at the beginning of the reproductive period, because of the close association between females and juveniles. As the interval between birth and time i increases, juveniles either are less closely associated with females or become indistinguishable from adults. Consequently, CIR techniques may be the only way to estimate $\hat{P}_i$. Note that under the assumption $S_{Ai} = 1$, Eq. (5.159) is equivalent to Eq. (5.146). As written, Eq. (5.159) is equivalent to the White et al. (1996) estimator, Eq. (5.152).

Method 2

Keith and Windberg (1978) present an alternative juvenile survival estimator, based on the use of population abundance estimates and expressed as

$$\hat{S}_{Ji} = \frac{\dfrac{\hat{N}_i}{\hat{N}_{A0}} \cdot \hat{p}_{Ji}}{\hat{U}_0} \qquad (5.160)$$

where

$\hat{N}_{A0}$ = estimated adult abundance at time 0 when juveniles are born;
$\hat{N}_i$ = estimated population abundance at time i, including both adults and juveniles;
$\hat{p}_{Ji}$ = estimated proportion of population at time i that are juveniles, $\left(\dfrac{N_{Ji}}{N_{Ai} + N_{Ji}} \right)$.

To understand the nature of this estimator, its expected value to the first term of a Taylor series expansion is

$$E(\hat{S}_{Ji}) \doteq \frac{\dfrac{E(\hat{N}_i)}{E(\hat{N}_{A0})} \cdot E(\hat{P}_{Ji})}{E(\hat{U}_0)}$$

$$= \frac{\left(\dfrac{N_{Ai} + N_{Ji}}{N_{A0}} \right) \cdot \left(\dfrac{N_{Ji}}{N_{Ai} + N_{Ji}} \right)}{\left(\dfrac{N_{J0}}{N_{A0}} \right)}$$

$$= \frac{N_{Ji}}{N_{J0}}.$$

An approximate variance estimator for the survival estimator (5.160), assuming independent data sources, is

$$\widehat{\text{Var}}(\hat{S}_{Ji}) = (\hat{S}_{Ji})^2 \left[\widehat{\text{CV}}(\hat{N}_i)^2 + \widehat{\text{CV}}(\hat{N}_{A0})^2 + \widehat{\text{CV}}(\hat{p}_{Ji})^2 + \widehat{\text{CV}}(\hat{U}_0)^2 \right].$$ (5.161)

Method 3

The third estimator of Keith and Windberg (1978) uses abundance estimates of the population taken at time 0 and i and the associated adult survival rate for this period. The juvenile survival estimator is expressed as

$$\hat{S}_{Ji} = \frac{\dfrac{\hat{N}_i}{\hat{N}_{A0}} - \hat{S}_{Ai}}{\hat{U}_0}$$ (5.162)

where $\hat{S}_{Ai}$ = estimated adult survival probability from time 0 to time i. The expected value of $\hat{S}_{Ji}$ to the first term of Taylor series is

$$E(\hat{S}_{Ji}) \doteq \frac{\dfrac{E(\hat{N}_i)}{E(\hat{N}_{A0})} - E(\hat{S}_{Ai})}{E(\hat{U}_0)}$$

$$= \frac{\dfrac{N_{Ai} + N_{Ji}}{N_{A0}} - \dfrac{N_{Ai}}{N_{A0}}}{\dfrac{N_{J0}}{N_{A0}}}$$

$$= \frac{N_{Ji}}{N_{J0}}.$$

Based on the delta method, an approximate variance estimator for $\hat{S}_{Ji}$ is

$$\widehat{\text{Var}}(\hat{S}_{Ji}) = \frac{\widehat{\text{Var}}(\hat{N}_i)}{(\hat{N}_{A0}\hat{U}_0)^2} + \frac{\widehat{\text{Var}}(\hat{S}_{Ai})}{\hat{U}_0^2} + \frac{\hat{N}_i^2 \cdot \widehat{\text{Var}}(\hat{N}_{A0})}{(\hat{N}_{A0}\hat{U}_0)^2} + \frac{\hat{S}_{Ji}^2 \cdot \widehat{\text{Var}}(\hat{U}_0)}{\hat{U}_0^2}.$$ (5.163)

Example 5.23: Estimating Juvenile Survival using Keith and Windberg (1978) Methods for Snowshoe Hares (*Lepus americanus*), Alberta, Canada

Keith and Windberg (1978) presented an example of their juvenile estimator based on method 1 (Eq. 5.157) by using data on snowshoe hares from Alberta, Canada. Survival of juvenile hares was based on a juvenile : adult ratio of $\hat{R}_{(J/A)i} = 2.64$, adult survival of $\hat{S}_{Ai} = 0.656$, and $\hat{U}_0 = 5.96$, yielding a value of

$$\hat{S}_{Ji} = \frac{\hat{R}_{(J/A)i} \cdot \hat{S}_{Ai}}{\hat{U}_0}$$

$$= \frac{2.64(0.656)}{5.96} = 0.291.$$

Unfortunately, no associated variance estimates were provided for these input parameters, so no variance of $\hat{S}_{Ji}$ can be calculated.

The same study on snowshoe hares yielded estimates on the population abundance at time i ($\hat{N}_i = 1163$), the juvenile fraction in the population at time i ($\hat{p}_{Ji} = 0.73$), the number of adults at time of birth ($\hat{N}_{A0} = 635$), and the number of juveniles born per adult ($\hat{U}_0 = 5.96$). The estimate of $\hat{S}_{Ji}$ from these values is calculated by using the Keith and Windberg (1978) method 2 (Eq. 5.160), as

$$\hat{S}_{Ji} = \frac{\dfrac{\hat{N}_i}{\hat{N}_{A0}} \cdot \hat{p}_{Ji}}{\hat{U}_0}$$

$$= \frac{\dfrac{1163}{635} \cdot 0.73}{5.96} = 0.224.$$

The variance of $\hat{S}_{Ji}$ cannot be calculated because no variance estimates for $\hat{N}_i$, $\hat{p}_{Ji}$, $\hat{N}_{A0}$, and $\hat{U}_0$ are available.

As a final illustration, Keith and Windberg (1978) estimated juvenile survival based on their method 3 (Eq. 5.162) by using adult abundance at time of birth ($\hat{N}_{A0} = 635$), population abundance at time i ($\hat{N}_i = 1163$), adult survival from time of birth to time i ($\hat{S}_{Ai} = 0.656$), and the number of juveniles born per adult ($\hat{U}_0 = 5.96$). Juvenile survival for the study period is then estimated as

$$\hat{S}_{Ji} = \frac{\dfrac{\hat{N}_i}{\hat{N}_{A0}} - \hat{S}_{Ai}}{\hat{U}_0}$$

$$= \frac{\dfrac{1163}{635} - 0.656}{5.96} = 0.197.$$

5.12   Discussion

The sheer number of methods (Fig. 5.15) for survival estimation conveys the relative importance of understanding the role of survival in demographic studies. Beyond the age-structure methods for survival analysis presented in this book are the vast number of additional tag-recovery and tag-recapture methods available for survival estimation (e.g., Cormack 1964, Burnham et al. 1987, Smith 1991, Lebreton et al. 1992, Hoffmann 1993). However, little effort has been focused on coupling the information from age-structure and tagging studies into a joint survival analysis. We have suggested situations in which auxiliary information on adult survival from tagging studies, for instance, may be used to obtain more reliable estimates of juvenile survival (see Sections 5.11.1, 5.11.3). We also suggested situations in which adult survival estimates from tagging studies could be used in conjunction with CIR methods (Section 4.3) to obtain more reliable estimates of productivity. More coordinated uses of tagging studies with other demographic surveys are assuredly the wave of the future. By developing likelihood models for age-structure-based survival analyses, the opportunity to perform joint tagging and age-based analyses has become a little easier.

Survival analyses generally fall into one of two categories, either cohort or time-specific methods. The vast majority of the age-structure-based methods are time

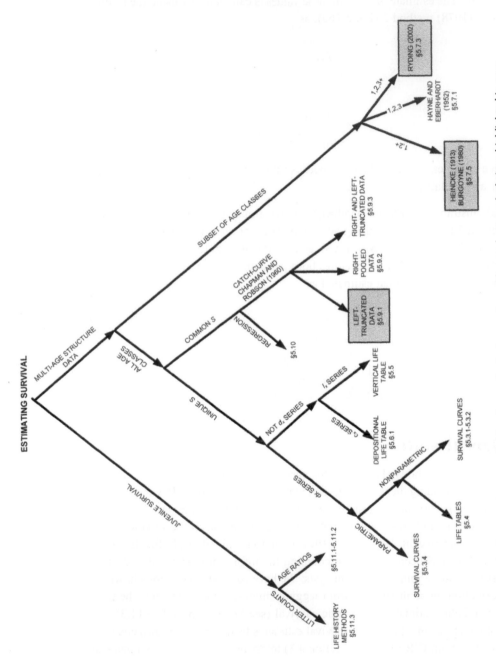

Figure 5.15. Decision tree for survival estimation methods. Approaches with sample size calculations highlighted in gray.

specific in nature. When rapid demographic assessments are necessary, these time-specific methods provide the first line of analytical attack. Time-specific samples of the age structure of a population are often easier and faster to collect than are the protracted approaches of cohort-based methods. However, there is an inherent price in terms of reliability to be paid by using time-specific methods. These time-specific methods usually require restrictive and often unrealistic assumptions of a stable and stationary population. However, populations in crisis requiring immediate analyses are rarely stable or stationary. Hence, the time-specific methods must be considered to be, at best, first-order approximations to the truth and, at worst, flawed. The λ-adjusted vertical life-table analysis of Udevitz and Ballachey (1998) is one of the few possible exceptions. Yet, survival and productivity information may be necessary to estimate the finite rate of change.

A choice between life-table analyses that yield age-specific survival estimates or catch-curve analyses and their special cases (i.e., Hayne and Eberhardt 1952, Burgoyne 1981) depends on whether a common survival across ages can be assumed. For long-lived species, estimating a common survival during midlife may be both reasonable and prudent. By simplifying model structure and using fewer parameters, survival may be more precisely estimated. In this chapter, we have shown the methods of Hayne and Eberhardt (1952) and Heincke (1913)–Burgoyne (1981) are special cases of catch-curve analyses. In general, the more age classes of data available, the more precisely the common adult survival probability can be estimated. However, logistics and costs may mandate pooling or truncating data on older animals. In which case, the simpler data requirements of the Hayne and Eberhardt (1952) or Heincke (1913)–Burgoyne (1981) methods may overshadow the statistical considerations of precision.

The initial demographic analyses using time-specific methods will often be replaced by cohort-specific approaches as time permits. The commitment to follow a cohort or cohorts through time can be considerable. The reward is survival estimates no longer predicated on assumptions of a stable and stationary population. These more realistic survival estimates should, in turn, provide more robust demographic assessments. Many time-specific methods presented in this chapter can also use cohort-specific data and benefit from a relaxation of model assumptions. A good example was the use of cohort data from black bears (Bender 1998) in conjunction with catch-curve analyses (Section 5.9.3) and regression estimators (Section 5.10.1).

Proper age classification is key to both cohort and age-specific survival analyses. The reconstructed grizzly bear survivorship curve from doubly censored radiotelemetry observations (i.e., Section 5.3.2) was predicated on knowing the animal age at the time of tagging. In the absence of age-at-tagging information, only a common annual survival probability can be estimated. Special animal care provisions may be necessary to extract teeth samples for age classification purposes. Yet, more detailed and realistic demographic models may more than compensate for hardships and risks to individual animals. For animals examined during hunter-bag surveys, age classification should be a standard practice. Over time, annual samples of age distribution can be used for both time-specific and cohort-based survival estimates. These annual estimates of age structure (Chapter 9) may also be used for population reconstruction and abundance estimation.

Estimating Harvest and Harvest Mortality

6

Chapter Outline

6.1   Introduction

Among the objectives of wildlife management is control or regulation of factors that effect productivity and survival (Eberhardt 1969). Habitat and harvest management are two common tools used to manipulate productivity, survival, and dispersal. For many game species, hunter harvest is often the most important factor affecting survival. Differences in male and female harvest will impact sex ratios, subsequently influencing productivity rates (Raedeke et al. 2002). Harvest management goals consider the maintenance of certain population sizes or densities. An assessment of the effectiveness of a harvest management action requires knowledge of removal numbers and a resulting measure of the population density or abundance.

Total annual survival includes survival of both natural causes and anthropogenic effects, specifically harvest. If the mortality sources act independently, the instantaneous morality rate (μ) can be written as the sum of the instantaneous natural mortality rate (μ_N) and the instantaneous harvest mortality rate (μ_H), where

$$\mu = \mu_N + \mu_H. \tag{6.1}$$

Overall survival from time 0 to t may be expressed as

$$
\begin{aligned}
S(t) &= e^{-(\mu_N + \mu_H)t} \\
&= e^{-\mu_N t} \cdot e^{-\mu_H t}
\end{aligned}
\tag{6.2}
$$

$$S(t) = S_N \cdot S_H \tag{6.3}$$

where

S_H = harvest survival probability;
S_N = probability of surviving natural sources of mortality.

Equations (6.1) through (6.3) imply the probability an animal is harvested is independent of an animal's likelihood of dying of natural causes. Thus, the probability of harvesting an animal is independent of that animal's ability to survive natural sources of mortality. Equations (6.2) and (6.3) by their form imply that total mortality (M_T) can be expressed as

$$
\begin{aligned}
M_T &= 1 - S_N \cdot S_H \\
&= 1 - (1 - M_N)(1 - M_H) \\
&= M_N + M_H - M_N M_H.
\end{aligned}
\tag{6.4}
$$

Equation (6.4) is a direct consequence of the assumptions of stochastic independence between mortality sources. Errington (1945, 1956) was among the first to express compensatory mortality in the form of Eq. (6.4). Expression (6.4) shows that mortality can originate from either natural causes or harvest, but not both. The term $-M_N M_H$ should be considered more correctly as stochastic compensation under the property of independent mortality sources.

Harvest and natural mortality sources may not operate independently. For example, a trophy hunter by definition seeks to harvest the fittest of animals, in which case,

$$M_T = M_N + M_H - \beta M_N M_H$$

where $\beta < 1$. The result is less compensation than proscribed if the sources of mortality were independent (i.e., $\beta = 1$). In the midwestern United States, where white-tailed deer (*Odocoileus virginianus*) use winter "yards," the process may be reversed. If hunters purposefully select yearlings during years of poor winter range when the chance of natural mortality is higher for young animals, then $\beta > 1$. By selecting the weakest individuals, more compensation would result than if the sources of mortality were operating independently (i.e., $\beta = 1$). Opportunistic hunting, which characterizes the majority of hunters, suggests random processes (Eqs. 6.3, 6.4).

6.2 Analysis of Harvest Records

State wildlife agencies use a variety of methods to collect harvest information from hunters. Rupp et al. (2000) provide a review of deer harvest survey techniques. Some agencies have mandatory reporting of big-game animals harvested. Other state agencies randomly sample hunters after the season to obtain information on numbers and locations of harvested game. In some jurisdictions, a combination of hunter check stations and postseason surveys are conducted (Rupp et al. 2000). In virtually all cases, information is collected in an attempt to estimate total harvest and allocate that harvest to game management units. The appropriate approach to analysis depends on the sampling scheme used to generate the survey data. Only the simplest of approaches are presented here. However, a valid statistical inference can only be drawn from a population probabilistically sampled. Readers are encouraged to review the survey sampling literature for details on finite sampling methods (e.g., Cochran 1977, Thompson 2002).

6.2.1   Locker and Field Checks

Hunter response is often incomplete in states with mandatory reporting requirements for big-game harvest. Simply using the reported numbers would underestimate total harvest. For this reason, agencies often conduct independent surveys to estimate reporting rates (r) (Geis and Atwood 1961, Martinson and McCann 1966, Henny 1967). Total harvest (C) is then estimated by the quantity

$$\hat{C} = \frac{c}{\hat{r}}$$

(6.5)

where

$\hat{C}$ = estimated total harvest;

c = number of harvested animals reported by hunters;

$\hat{r}$ = estimated reporting rate, i.e., the fraction of successful hunters that reported their harvest.

Estimator (6.5) assumes average hunter success is the same for those hunters who did and did not complete their end-of-season report cards.

Two common and analogous approaches have been used to estimate reporting rates. Both approaches are based on surveys of permit-tagged and harvested animals. A permit-tag (i.e., clip-tag, carcass-tag) is attached to the animal carcass by the hunter after harvest to identify it as a legally taken animal. These tagged animals can be subsequently crosschecked with annual report cards submitted by hunters to estimate reporting rates. In a locker check, agency personnel go to commercial meat lockers and record identification numbers from permit-tags attached to animal carcasses waiting to be processed (e.g., Pietz 1972). These identification numbers taken from animals at meat lockers are then crosschecked against filed report cards submitted by hunters. In field checks, agency personnel record the permit-tag numbers of animals as successful hunters leave the field. These permit-tag numbers can also be checked against subsequent report cards submitted by hunters. We define

H = total number of licensed hunters;

Q = number of permit-tagged carcasses recorded by agency personnel during field checks or locker checks;

q = number of corresponding mandatory reporting cards submitted from among the Q permit-tagged animals canvassed.

The binomial proportion

$$\hat{r} = \frac{q}{Q} \qquad (6.6)$$

can be used to estimate the reporting rate (r) with associated variance of

$$\text{Var}(\hat{r}) = \left(\frac{H-Q}{H-1}\right)\frac{r(1-r)}{Q}.$$

The variance of $\hat{r}$ can be estimated by

$$\widehat{\text{Var}}(\hat{r}) = \left(1 - \frac{Q}{H}\right)\frac{\hat{r}(1-\hat{r})}{Q-1}. \qquad (6.7)$$

The first term in Eq. (6.7) is the finite population correction (fpc), which goes to zero as the sampling function $\frac{Q}{H}$ goes to one. In most cases, the sampling fraction will be small, $\frac{Q}{H} < 0.10$. In which case, the simpler binomial variance estimate can be used,

$$\widehat{\text{Var}}(\hat{r}) = \frac{\hat{r}(1-\hat{r})}{Q}. \qquad (6.8)$$

The variance for the estimated total harvest from Eq. (6.5) can be found in stages, where

$$\text{Var}(\hat{C}) = \text{Var}_1\left[E_2\left(\frac{c}{\hat{r}}\Big|1\right)\right] + E_1\left[\text{Var}_2\left(\frac{c}{\hat{r}}\Big|1\right)\right],$$

and where stage 1 denotes the sample reporting of c of C and stage 2 denotes the estimation of the reporting rate r. In the case of one hunter and the harvest of at most one big-game animal, the variance of the total harvest is approximated by the expression

$$\text{Var}(\hat{C}) \doteq \frac{C(1-r)}{r} + \frac{C^2 \cdot \text{Var}(\hat{r})}{r^2} \qquad (6.9)$$

and estimated by

$$\widehat{\text{Var}}(\hat{C}) = \frac{c(1-\hat{r})}{\hat{r}^2} + \frac{c^2(1-\hat{r})}{\hat{r}^3 Q}. \qquad (6.10)$$

Assumptions

The key assumptions of the reporting-rate-corrected harvest estimate are as follows:

1. Hunters accurately record their harvest on the annual report cards.
2. Hunter success is independent of whether a report card is submitted or not.

Both assumptions may be readily violated. Hunters may accurately recall big-game harvests, but numbers and species of small game are often approximated by hunters at the end of each hunting season. Various psychologies may also influence hunters to overestimate or underreport their harvest numbers (Miller and Anderson 2002). Similarly, various psychologies may result in the average success differing between hunters who do or do not submit their annual reporting cards (Bellrose 1947, MacDonald and Dillman

1968). Estimator (6.5) assumes average hunter success is the same for both respondents and nonrespondents. Variance estimator Eq. (6.9) is appropriate only for the special case in which hunter performance is measured as either zero or one animal harvested.

Example 6.1: Reporting-Rate–Corrected Harvest Estimate

Consider a state with 100,000 licensed white-tailed deer hunters. At the end of the season, $c = 20,000$ hunters who submitted a report card signifying a deer harvest. A field check recorded $Q = 1000$ clip-tagged deer, of which $q = 800$ were subsequently also reported by report cards. The reporting rate would be estimated as

$$\hat{r} = \frac{q}{Q} = \frac{800}{1000} = 0.80,$$

with a corresponding total harvest estimate of

$$\hat{C} = \frac{c}{\hat{r}} = \frac{20,000}{0.80} = 25,000 \text{ deer.}$$

Using variance estimator Eq. (6.10),

$$\widehat{\mathrm{Var}}(\hat{C}) = \frac{c(1-\hat{r})}{\hat{r}^2} + \frac{c^2(1-\hat{r})}{\hat{r}^3 Q}$$

$$= \frac{20,000(1-0.80)}{(0.80)^2} + \frac{20,000^2(1-0.80)}{(0.80)^3(1000)} = 162,500$$

or $\widehat{\mathrm{SE}}(\hat{C}) = 403.11$.

Discussion of Utility

Field or locker checks are an essential element in estimating reliable harvest totals in states where only successful hunters are required to file records. Implicit with this reporting system is the assumption no report card means no harvest. However, hunter compliance with mandatory reporting requirements is both incomplete and variable between locales and over time. For the adjusted harvest estimate to be reliable, field checks or locker surveys need to be representative of the hunter population. Isolated or erratic surveys of hunter compliance run the risk of misrepresenting the intended target population and biasing estimates of total harvest.

6.2.2   Random Sample of Hunter Responses

A probabilistic sample of hunters and their bag success provides a direct means of estimating harvest numbers. Stratified random sampling will be used to illustrate estimators of common harvest statistics such as total harvest, mean bag per hunter, proportion of successful hunters, and mean bag per successful hunter. In illustrating these calculations, it is assumed either that a random sample of hunters was drawn with complete compliance, or that hunters who submitted annual report cards are representative of all hunters. The issues of nonreporting bias will be addressed in succeeding sections.

For purpose of illustration, a state will be stratified into management units. Within each management unit, a random sample of n_h of N_h total hunters will submit harvest results by questionnaire ($h = 1, \ldots, L$ strata). For the hth stratum (i.e., management unit), the total number of animals harvested, C_h, is estimated by

$$\hat{C}_h = \frac{N_h}{n_h} \sum_{i=1}^{n_h} c_{hi} \tag{6.11}$$

where

c_{hi} = number of animals harvested by the ith hunter ($i = 1, \ldots, n_h$) in the hth stratum ($h = 1, \ldots, L$);

N_h = numbers of permits in the hth stratum ($h = 1, \ldots, L$);

n_h = numbers of questionnaires submitted in the hth stratum ($h = 1, \ldots, L$).

The estimated variance of $\hat{C}_h$ is

$$\widehat{\text{Var}}(\hat{C}_h) = \frac{N_h^2 \left(1 - \dfrac{n_h}{N_h}\right) s_h^2}{n_h}, \tag{6.12}$$

where $\displaystyle s_h^2 = \frac{\sum_{i=1}^{n_h} (c_{hi} - \bar{c}_h)^2}{(n_h - 1)}$;

$$\bar{c}_h = \frac{\sum_{i=1}^{n_h} c_{hi}}{n_h}.$$

The estimate of the total harvest across all strata L is

$$\hat{C} = \sum_{h=1}^{L} \hat{C}_h = \sum_{h=1}^{L} \frac{N_h}{n_h} \sum_{i=1}^{n_h} c_{hi} \tag{6.13}$$

with associated variance estimator

$$\widehat{\text{Var}}(\hat{C}) = \sum_{h=1}^{L} \widehat{\text{Var}}(\hat{C}_h) = \sum_{h=1}^{L} \left[\frac{N_h^2 \left(1 - \dfrac{n_h}{N_h}\right) s_h^2}{n_h} \right]. \tag{6.14}$$

Estimation of mean bag per hunter in the hth stratum is estimated by

$$\hat{\bar{C}}_h = \frac{1}{n_h} \sum_{i=1}^{n_h} c_{hi},$$

with the estimated variance

$$\widehat{\text{Var}}\left(\hat{\bar{C}}_h\right) = \frac{\left(1 - \dfrac{n_h}{N_h}\right) s_h^2}{n_h}. \tag{6.15}$$

The average number of animals harvested per hunter across all strata would be estimated as a weighted average of the strata means, or

$$\hat{\bar{C}} = \frac{\sum_{h=1}^{L} N_h \hat{\bar{C}}_h}{N} \tag{6.16}$$

where $N = \sum_{h=1}^{L} N_h$, with estimated variance

$$\widehat{\mathrm{Var}}(\hat{\bar{C}}) = \sum_{h=1}^{L}\left[\left(\frac{N_h}{N}\right)^2 \frac{\left(1-\dfrac{n_h}{N_h}\right)s_h^2}{n_h}\right]. \tag{6.17}$$

Another performance measure is the proportion of hunters successful in harvesting at least one game animal of interest. Let y_h = number of successful hunters among the n_h responses in the hth stratum. The proportion successful in the hth stratum would be estimated by

$$\hat{P}_h = \frac{y_h}{n_h} \tag{6.18}$$

with estimated variance

$$\widehat{\mathrm{Var}}(\hat{P}_h) = \frac{\hat{P}_h(1-\hat{P}_h)}{(n_h-1)}\left(1-\frac{n_h}{N_h}\right). \tag{6.19}$$

The proportion of the total hunter population that was successful would be estimated as a weighted average of the $\hat{P}_h$, where across strata

$$\hat{P} = \frac{\sum_{h=1}^{L} N_h\left(\dfrac{y_h}{n_h}\right)}{N} = \frac{\sum_{h=1}^{L} N_h\hat{P}_h}{N}, \tag{6.20}$$

with associated variance estimate

$$\widehat{\mathrm{Var}}(\hat{P}) = \sum_{h=1}^{L}\left[\left(\frac{N_h}{N}\right)^2 \frac{\left(1-\dfrac{n_h}{N_h}\right)\hat{P}_h(1-\hat{P}_h)}{(n_h-1)}\right]. \tag{6.21}$$

An estimate of the number of successful hunters readily follows from the estimate of $\hat{P}_h$. The estimate of the number of successful hunters within the hth stratum is calculated as

$$\hat{Y}_h = N_h \cdot \hat{P}_h \tag{6.22}$$

with associated variance estimate

$$\widehat{\mathrm{Var}}(\hat{Y}_h) = N_h^2 \cdot \widehat{\mathrm{Var}}(\hat{P}_h) = \frac{N_h^2\hat{P}_h(1-\hat{P}_h)}{(n_h-1)}\left(1-\frac{n_h}{N_h}\right). \tag{6.23}$$

The population-wide estimate of the total number of successful hunters would be the sum of the strata estimates, where

$$\hat{Y} = \sum_{h=1}^{L} \hat{Y}_h = \sum_{h=1}^{L} N_h \hat{P}_h \tag{6.24}$$

with associated variance estimator

$$\widehat{\text{Var}}(\hat{Y}) = \sum_{h=1}^{L} \left[N_h^2 \frac{\hat{P}_h(1-\hat{P}_h)}{(n_h-1)} \left(1 - \frac{n_h}{N_h}\right) \right]. \tag{6.25}$$

An estimate of the mean bag size per successful hunter, denoted by $\hat{\bar{C}}_Y$, can be estimated by the ratio

$$\hat{\bar{C}}_Y = \frac{\hat{C}}{\hat{Y}}. \tag{6.26}$$

The variance of $\hat{\bar{C}}_Y$ can be approximated by using the delta method, where

$$\widehat{\text{Var}}\left(\hat{\bar{C}}_Y\right) \doteq \left(\hat{\bar{C}}_Y\right)^2 \left[\frac{\displaystyle\sum_{h=1}^{L}\left[\frac{N_h^2\left(1-\frac{n_h}{N_h}\right)s_h^2}{n_h} \right]}{\hat{C}^2} + \frac{\displaystyle\sum_{h=1}^{L}\left[\frac{N_h^2\hat{P}_h(1-\hat{P}_h)\left(1-\frac{n_h}{N_h}\right)}{(n_h-1)} \right]}{\hat{Y}^2} \right. $$
$$\left. - \frac{\frac{2}{N}\displaystyle\sum_{h=1}^{L}N_h^2\left(1-\frac{n_h}{N_h}\right)\frac{\hat{\bar{C}}_h(1-\hat{P}_h)}{(n_h-1)}}{\hat{Y}\hat{C}} \right]. \tag{6.27}$$

A common measure of catch-per-unit effort (CPUE) in wildlife science is the number of animals harvested per hunter-day (Lancia et al. 1996). Defining d_{hi} as the number of days hunted by the ith hunter ($i = 1, \ldots, n_h$) in the hth stratum ($h = 1, \ldots, L$), CPUE for the ith hunter in the hth stratum is

$$\text{CPUE}_{hi} = \frac{c_{hi}}{d_{hi}}.$$

For hunters in the hth stratum, the mean CPUE would be estimated as the ratio

$$\widehat{\text{CPUE}}_h = \frac{\displaystyle\sum_{i=1}^{n_h} c_{hi}}{\displaystyle\sum_{i=1}^{n_h} d_{hi}} = \frac{\bar{c}_h}{\bar{d}_h} \tag{6.28}$$

with approximate variance

$$\widehat{\text{Var}}\left(\widehat{\text{CPUE}}_h\right) = \frac{\left(1 - \dfrac{n_h}{N_h}\right)}{n_h \bar{d}_h^2} \left(\frac{\displaystyle\sum_{i=1}^{n_h} c_{hi}^2 - 2\widehat{\text{CPUE}}_h \sum_{i=1}^{n_h} c_{hi} d_{hi} + \widehat{\text{CPUE}}_h^2 \sum_{i=1}^{n_h} d_{hi}^2}{(n_h - 1)}\right), \quad (6.29)$$

and where

$$\bar{d}_h = \frac{1}{n_h} \sum_{i=1}^{n_h} d_{hi}.$$

For the entire population of hunters, the mean CPUE would be estimated as

$$\widehat{\text{CPUE}} = \frac{\displaystyle\sum_{h=1}^{L} \frac{N_h}{n_h} \sum_{i=1}^{n_h} c_{hi}}{\displaystyle\sum_{h=1}^{L} \frac{N_h}{n_h} \sum_{i=1}^{n_h} d_{hi}} = \frac{\displaystyle\sum_{h=1}^{L} \hat{C}_h}{\displaystyle\sum_{h=1}^{L} \hat{D}_h} = \frac{\hat{C}}{\hat{D}}. \quad (6.30)$$

The variance of $\widehat{\text{CPUE}}$ can be found by using the delta method as

$$\widehat{\text{Var}}\left(\widehat{\text{CPUE}}\right) = \widehat{\text{CPUE}}^2 \left[\frac{\displaystyle\sum_{h=1}^{L} \widehat{\text{Var}}(\hat{C}_h)}{\hat{C}^2} + \frac{\displaystyle\sum_{h=1}^{L} \widehat{\text{Var}}(\hat{D}_h)}{\hat{D}^2} - \frac{2\displaystyle\sum_{h=1}^{L} \text{Cov}(\hat{C}_h, \hat{D}_h)}{\hat{C}\hat{D}}\right]$$

$$= \widehat{\text{CPUE}}^2 \left[\frac{\displaystyle\sum_{h=1}^{L} N_h^2 \left(1 - \frac{n_h}{N_h}\right) \frac{s_h^2}{n_h}}{\hat{C}^2} + \frac{\displaystyle\sum_{h=1}^{L} N_h^2 \left(1 - \frac{n_h}{N_h}\right) \frac{s_{d_h}^2}{n_h}}{\hat{D}^2}\right]$$

$$- \frac{2\displaystyle\sum_{h=1}^{L} N_h^2 \left(1 - \frac{n_h}{N_h}\right) \frac{\widehat{\text{Cov}}(c_{hi}, d_{hi})}{n_h}}{\hat{C}\hat{D}}\right] \quad (6.31)$$

where

$$\widehat{\text{Cov}}(c_{hi}, d_{hi}) = \frac{\displaystyle\sum_{i=1}^{n_h} (c_{hi} - \bar{c}_h)(d_{hi} - \bar{d}_h)}{(n_h - 1)}.$$

Assumptions

It is assumed that surveyed hunters are a random sample of all licensed hunters and all surveyed hunters responded. Alternatively, if only a subset of the hunters responded to the survey, it must be assumed the responses elicited are independent of the willingness of a hunter to respond to the survey. However, should willingness to respond to the survey depend on hunter success, the resultant estimates of harvest and hunter success will be biased. Results from several studies found that inaccuracies in hunter reporting were more a factor in over- or underestimating harvest than was nonresponse bias (Sen 1971, Wright 1978, Barker 1991).

6.2.3 Resampling for Nonresponse

The problem of nonresponse bias has been known for some time and is common in surveys where respondents are self-selected (Fillion 1980). Even if the original sample is representative of the hunting population, there is the possibility that respondents are not. The magnitude of the bias in a parameter μ caused by nonresponse can be expressed mathematically (Cochran 1977:361) as

$$\text{Bias}(\mu) = P_2(\mu_1 - \mu_2),$$

where

P_2 = proportion of nonrespondents in the population of hunters;
μ_1 = mean value for respondents;
μ_2 = mean value for nonrespondents.

The magnitude of the bias will depend on the fraction of the population that is composed of nonrespondents, and the difference in mean values between hunters who do and do not respond to the survey. Unfortunately, from a single sample, it is impossible to judge the size of the difference $\mu_1 - \mu_2$. Repeated or double sampling of the nonrespondents is the usual method for arriving at an estimate of μ_2. Typically, wildlife agencies will conduct two or three repeated surveys, surveying nonrespondents from the previous stage to obtain information from the nonrespondents and correct for any associated biases.

Follow-up surveys can be used to try and obtain the required information from initial nonrespondents to questionnaires (e.g., Taylor et al. 2000). That information can be used to either dispel concern over nonresponse bias or adjust estimates for such bias. However, there is no reason to expect that individuals who do not respond to the first mail survey would respond to subsequent mail surveys. An alternative to the mail survey is phone interviews. Cada (1985) found that mail and phone surveys produced significantly different results, with telephone surveys being more accurate. Barker (1991) combined mail surveys with follow-up telephone surveys of individuals who did not respond to the initial mail survey. We now outline a method of double sampling from Cochran (1977:333, Eq. 12.25).

In the initial survey, a sample of k hunters is selected, of which k_1 responded to the survey and k_2 did not respond (Fig. 6.1); thus, $k = k_1 + k_2$. The proportion of people who responded to the survey, p_1, is expressed as

$$p_1 = \frac{k_1}{k},$$

and the proportion of nonrespondents is

$$p_2 = (1 - p_1) = \frac{k_2}{k}.$$

In the second stage, a subsample of size k_2' from among the k_2 nonrespondents is resurveyed by using a different method than the first survey. The subsample is chosen such

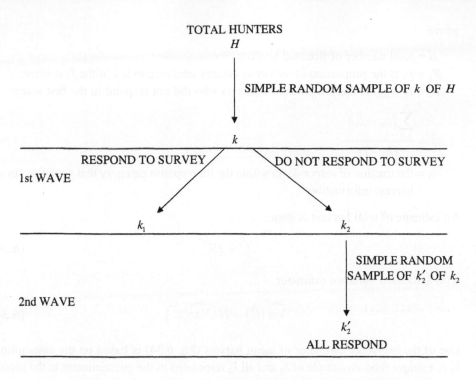

TOTAL HUNTERS
H

SIMPLE RANDOM SAMPLE OF k OF H

k

RESPOND TO SURVEY   DO NOT RESPOND TO SURVEY

1st WAVE

k_1 k_2

SIMPLE RANDOM
SAMPLE OF k_2' OF k_2

2nd WAVE

k_2'

ALL RESPOND

Figure 6.1. Schematic for double sampling with nonresponse by using two different survey methods.

that $k_2' \leq k_2$. The results of the second survey are used to estimate the mean bag for non-respondents to the first survey. The mean harvest per hunter in each stage, $\hat{\bar{c}}_1$ and $\hat{\bar{c}}_2$, is then estimated by

$$\hat{\bar{c}}_1 = \frac{c_1}{k_1} \qquad (6.32)$$

and

$$\hat{\bar{c}}_2 = \frac{c_2}{k_2'}, \qquad (6.33)$$

where

c_1 = the reported total harvest from the k_1 responses of the first-wave survey;

c_2 = the reported total harvest from the k_2' responses of the second-wave survey.

The unbiased estimate of mean harvest across all hunters, $\overline{C}$, is then

$$\hat{\overline{C}} = p_1\hat{\bar{c}}_1 + (1 - p_1)\hat{\bar{c}}_2$$

$$= \left(\frac{k_1}{k}\right)\hat{\bar{c}}_1 + \left(\frac{k_2}{k}\right)\hat{\bar{c}}_2. \qquad (6.34)$$

Cochran (1977:333, eq. 12.25) provides an unbiased variance estimator of

$$\widehat{\mathrm{Var}}\left(\hat{\overline{C}}\right) = \frac{k(H-1)}{H(k-1)}\left[\sum_{i=1}^{2}\left(\frac{1}{kv_i} - \frac{1}{H}\right)P_i s_i^2\right.$$

$$\left. + \frac{(H-k)}{k(H-1)}\sum_{i=1}^{2} s_i^2\left(\frac{P_i}{H} - \frac{1}{kv_i}\right) + \frac{(H-k)}{k(H-1)}\sum_{i=1}^{2} P_i\left(\hat{\bar{c}}_i - \hat{\overline{C}}\right)^2\right], \qquad (6.35)$$

where

H = total number of licensed hunters;

$P_1 = p_1$ is the proportion of surveyed hunters who responded in the first wave;

$P_2 = (1 - p_1)$ is the proportion of hunters who did not respond in the first wave;

$$s_i^2 = \frac{\sum_{h=1}^{k_i}\left(c_{hi} - \hat{\bar{c}}_i\right)^2}{(k_i - 1)};$$

v_i = the fraction of respondents within the ith response category that actually provide harvest information.

An estimate of total harvest is then

$$\hat{C} = H\hat{\bar{C}} \tag{6.36}$$

with associated variance estimator

$$\widehat{\mathrm{Var}}(\hat{C}) = H^2\,\widehat{\mathrm{Var}}(\hat{\bar{C}}). \tag{6.37}$$

Use of the two-stage estimator of mean harvest (Eq. 6.34) is based on the assumptions k_2' is a simple random sample of k_2 and all k_2' responded to the questionnaire in the second wave of sampling. However, nonresponse behavior can persist into the second wave of sampling and beyond.

Concern over nonrespondents centers around the possibility that hunter response to the survey may be correlated with hunter success. If hunter success is uncorrelated with the likelihood of a person responding to a survey, the data from the first wave of cooperative respondents would be adequate for unbiased estimation. However, if average success varies with the propensity of a person to respond to the survey, follow-up surveys of nonrespondents are necessary to obtain unbiased estimates of harvest numbers. The process of repeated sampling of nonrespondents might continue until assurance that progressive sampling waves will not alter the conclusion. Mean catch per hunter can be compared between successive survey waves using standard t-tests or F-tests for equality of means. Should successive mean values be equal, the repeated surveying could be halted.

Should appreciable nonresponse exist during the second survey wave, a third wave of sampling may be performed (Fig. 6.2). In this multiple-wave sampling scheme, all non-respondents in one wave are resurveyed in the next wave. Extension of Eq. (6.34) to a third wave of sampling can be written as

$$\hat{\bar{C}} = p_1\hat{\bar{c}}_1 + p_2(1 - p_1)\hat{\bar{c}}_2 + (1 - p_2)(1 - p_1)\hat{\bar{c}}_3 \tag{6.38}$$

where

$\hat{\bar{c}}_i$ = mean number of animals harvested by hunters responding in the ith survey wave ($i = 1, \ldots, 3$);

p_i = fraction of survey recipients who responded in the ith survey wave ($i = 1, \ldots, 3$).

In a similar manner, mean harvest using a fourth wave (Fig. 6.2) of nonresponse sampling can be estimated by the expression

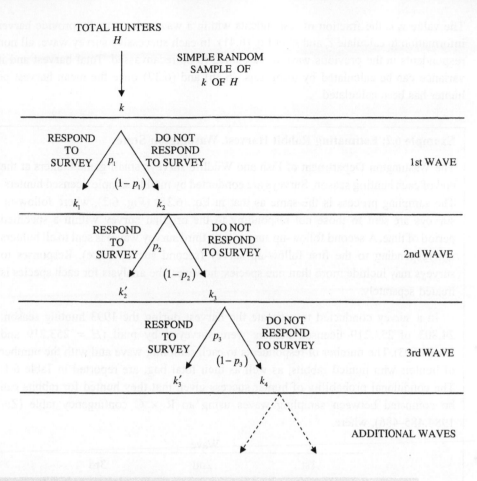

TOTAL HUNTERS
H

SIMPLE RANDOM
SAMPLE OF
k OF H

k

| RESPOND TO SURVEY | p_1 | DO NOT RESPOND TO SURVEY | 1st WAVE |

$(1-p_1)$

k_1 k_2

| RESPOND TO SURVEY | p_2 | DO NOT RESPOND TO SURVEY | 2nd WAVE |

$(1-p_2)$

k_2' k_3

| RESPOND TO SURVEY | p_3 | DO NOT RESPOND TO SURVEY | 3rd WAVE |

$(1-p_3)$

k_3' k_4

ADDITIONAL WAVES

Figure 6.2. Schematic for multiple waves of resampling for nonresponse to game harvest surveys.

$$\hat{\bar{C}} = p_1\hat{\bar{c}}_1 + p_2(1-p_1)\hat{\bar{c}}_2 + p_3(1-p_2)(1-p_1)\hat{\bar{c}}_3 + (1-p_3)(1-p_2)(1-p_1)\hat{\bar{c}}_4. \quad (6.39)$$

Equations (6.38) and (6.39) can be rewritten in terms of the proportion of respondents in the original sample who responded in a particular survey wave by noting the relationships:

$$p_1 = \frac{k_1}{k}, \ p_2 = \frac{k_2'}{k_2} = \frac{k_2'}{k - k_1'}, \dots, \ p_i = \frac{k_i'}{k - k_1' - \cdots - k_{i-1}'},$$

where k_i' = respondents in the ith wave. Hence, the average kill per hunter can be written in terms of the proportion of respondents in each wave as

$$\hat{\bar{C}} = \sum_{i=1}^{r} P_i\hat{\bar{c}}_i, \quad (6.40)$$

where P_i = proportion of hunters responding in each wave out of the total sample k where $P_1 = p_1$, $P_2 = p_2(1 - p_1)$, $P_3 = p_3(1 - p_1)(1 - p_2)$, and $P_4 = (1 - p_1)(1 - p_2)(1 - p_3)$ in the case of four survey waves. The variance for estimator (6.40) can be estimated by a generalization of Eq. (6.35) (Eq. 12.25 in Cochran 1977)

$$\widehat{\mathrm{Var}}(\hat{\bar{C}}) = \frac{k(H-1)}{H(k-1)}\left[\sum_{i=1}^{r}\left(\frac{1}{kv_i} - \frac{1}{H}\right)P_i s_i^2\right.$$
$$\left. + \frac{(H-k)}{k(H-1)}\sum_{i=1}^{r} s_i^2\left(\frac{P_i}{H} - \frac{1}{kv_i}\right) + \frac{(H-k)}{k(H-1)}\sum_{i=1}^{r} P_i\left(\hat{\bar{c}}_i - \hat{\bar{C}}\right)^2\right]. \quad (6.41)$$

The value v_i is the fraction of respondents within a wave who actually provide harvest information to calculate $\hat{\bar{c}}_i$ and s_i^2 in Eq. (6.41). In each successive survey wave, all non-respondents in the previous wave are assumed to be recanvassed. Total harvest and its variance can be calculated by using Eqs. (6.36) and (6.37) once the mean harvest per hunter has been calculated.

Example 6.2: Estimating Rabbit Harvest, Washington State

The Washington Department of Fish and Wildlife surveys small game hunters at the end of each hunting season. Surveys are conducted by mail to sample licensed hunters. The sampling process is the same as that in Eq. (6.38) (Fig. 6.2), where follow-up surveys are sent to those not responding to the original survey within a specified period of time. A second follow-up survey (i.e., third survey wave) is sent to all hunters not responding to the first follow-up survey (second survey wave). Responses to surveys may include more than one species; however, the analysis for each species is treated separately.

In a survey conducted to estimate the harvest during the 1993 hunting season, 24,803 of 253,219 licensed hunters were surveyed by mail ($H = 253,219$ and $k = 24,803$). The number of respondents to each sampling wave and with the number of hunters who hunted rabbits, as well as their total bag, are reported in Table 6.1. The conditional probability of hunter success given that they hunted for rabbits can be compared between sampling waves using an R × C contingency table (Zar 1984:485–486), where

	Wave		
	1st	2nd	3rd	
Successful	127 (0.4922)	38 (0.5588)	19 (0.6786)	184
Nonsuccessful	131 (0.5078)	30 (0.4412)	9 (0.3214)	170
	258	68	28	354

resulting in a nonsignificant P-value of $P(\chi_2^2 \geq 4.0272) = 0.1335$.

Mean bag sizes by sampling wave (Table 6.2) suggest no relationship between the wave in which a hunter responded and the reported value. An estimate of mean number of rabbits harvested by licensed hunters would typically be calculated by using Eq. (6.39). However, the mean harvest rate for nonrespondents after the third sampling wave is unknown. If we assume these nonrespondents have the same success rate as do the respondents during the third and last sampling wave, then Eq. (6.39) reduces to Eq. (6.38), where

---

Table 6.1. The 1993 Washington Department of Fish and Wildlife hunter survey results on rabbit harvest during three waves of sampling for hunter nonresponse.

Wave	No. surveyed	No. of respondents	p_i	No. that hunted rabbits	No. of successful hunters	No. harvested
1	24,803	4,929	0.1987	258	127	561
2	19,874	1,319	0.0664	68	38	280
3	18,555	599	0.0323	28	19	74

Table 6.2. The unconditional and conditional hunter success for rabbits from the 1993 game survey in Washington State, with hunter response rates by sampling wave.

Wave	Probability of responding by wave (p_i)	Average harvest by all hunters ($\hat{\bar{c}}_i$)	Probability rabbit hunter successful	Average harvest by rabbit hunters
1	0.1987	0.1138	0.4922	2.1744
2	0.0664	0.2123	0.5588	4.1176
3	0.0323	0.1235	0.6786	2.6429

$$\hat{\bar{C}} = p_1\hat{\bar{c}}_1 + p_2(1-p_1)\hat{\bar{c}}_2 + (1-p_2)(1-p_1)\hat{\bar{c}}_3.$$

The estimated mean bag per hunter is calculated as

$$\hat{\bar{C}} = (0.1987)(0.1138) + 0.0664(1-0.1987)(0.2133)$$
$$+ (1-0.0664)(1-0.1987)(0.1235)$$
$$= 0.1263 \text{ rabbits/licensed hunter}.$$

The variance of $\hat{\bar{C}}$ is estimated using Eq. (6.41). The within-response-wave variances were not provided in the state records examined. However, for purposes of illustrating the variance calculation, the within-wave variances were approximated by a negative binomial mean-to-variance relationship, yielding values of $\hat{s}_1^2 = 0.5727$, $\hat{s}_2^2 = 3.1973$, and $\hat{s}_3^2 = 3.8235$. The variance of $\hat{\bar{C}}$ is calculated as

$$\widehat{\text{Var}}(\hat{\bar{C}}) = \frac{24803(253129-1)}{253129(24803-1)}.$$

$$\left\{ \left(\left(\frac{1}{24803(1)} - \frac{1}{253129} \right)(0.1987)(0.5727) \right. \right.$$

$$+ \left(\frac{1}{24803(1)} - \frac{1}{253129} \right)(0.0532)(3.1973)$$

$$+ \left(\frac{1}{24803(0.0323)} - \frac{1}{253129} \right)(0.7481)(3.8235) \right)$$

$$+ \frac{(253129-24803)}{24803(253129-1)} \left((0.5757) \left(\frac{0.1987}{253129} - \frac{1}{24803(1)} \right) \right.$$

$$+ (3.1973) \left(\frac{0.0532}{253129} - \frac{1}{24803(1)} \right)$$

$$+ (3.8235) \left(\frac{0.7481}{253129} - \frac{1}{24803(0.0323)} \right) \right)$$

$$+ \frac{(253129-24803)}{24803(253129-1)} \left((0.1987)(0.1138-0.1263)^2 \right.$$

$$+ (0.0532)(0.2123-0.1263)^2$$

$$+ \left. \left. (0.7481)(0.1235-0.1263)^2 \right) \right\}$$

$$\widehat{\text{Var}}(\hat{\bar{C}}) = 0.003569,$$

with an associated standard error of $\widehat{\text{SE}}(\hat{\bar{C}}) = 0.0597$ rabbits per hunter.

An estimate of total rabbits harvested by licensed hunters is calculated by using Eq. (6.36) as

$$\hat{C} = H \cdot \hat{\bar{C}}$$

$$= 253{,}219(0.1263) = 31{,}981.6 \text{ rabbits.}$$

The variance of the total harvest, $\hat{C}$, is calculated from Eq. (6.37) as

$$\text{Var}(\hat{C}) = H^2 \widehat{\text{Var}}(\hat{\bar{C}})$$

$$= 253129^2(0.003569) = 228{,}681{,}114$$

with an associated standard error of $\widehat{\text{SE}}(\hat{C}) = 15{,}122$ rabbits.

Discussion of Utility

The above example illustrates one way to assess differences in success probabilities between individuals who respond and those who do not respond to the first mailing. In this example, average hunter bag increased after the first survey wave. Hence, estimates of total harvest based on the first survey alone would have underestimated harvest. The low response in successive survey waves suggest that the multiple-mail survey method may not adequately address the problem of nonresponse bias. Unfortunately, this is a problem that cannot be solved by data analysis alone.

Alternatively, nonresponse bias in hunter surveys has been estimated by regressing the cumulative number of returned surveys against the cumulative reported harvest (Chapman et al. 1959, Sen 1971, Fillion 1975). Total kill is estimated by extrapolating the regression line to include all hunters. Often, the origin is added as a fourth data point so that a quadratic curve can be fitted to the points. Unfortunately, the typical number of respondents in the third survey does not extend the curve much farther than do the cumulative responses from the first and second surveys. Thus, there is little difference between the second and third data points. This method of estimating harvest totals is subject to the assumptions of ordinary least-squares regression, which may or may not hold (e.g., the assumption of uncorrelated responses).

6.3 Estimating Harvest by Area

Statewide surveys of game harvest attempt to estimate total harvest numbers and harvest by game management units within state boundaries. Allocation of harvest to specific regions of the state is important because game management typically occurs at a more local level. Harvest numbers are also required at the local level because hunting regulations often operate at spatial scales smaller than a state. Fine-scaled information must be extracted from harvest data collected at the statewide level to monitor hunter success and the impact of hunting regulations on local populations.

The ability to obtain estimates of harvest numbers at the local level depends on the nature of the hunter survey, specificity of the survey questions, and willingness of hunters to fully and honestly answer survey questions. It is imperative the questionnaire be constructed to yield the resolution needed by game managers. It is equally important that biometricians identify beforehand how the collected information will be specifically used in harvest estimation and variance calculations. Failure to incorporate the biological or statistical requirements into a questionnaire from the onset will likely doom the usefulness of the survey. We present alternative approaches to survey analysis for estimating local harvest numbers. As will be seen, the nature of the data reported by hunters will have a major effect on statistical analyses, assumptions, and data extraction.

6.3.1   Common Reporting and Success Probabilities (White 1993)

White (1993) derived a multinomial model to estimate harvest by reporting region or locale using hunter survey responses on the annual game reporting cards. The objective was to estimate the number of game harvested in a specific area. The analysis can then be repeated for successive areas, one at a time. The technique uses information from both complete and incomplete surveys to estimate harvest in an area. Even when hunters respond to surveys, some questions regarding area of harvest, sex, or age of the animal may be left unanswered intentionally or unintentionally. This multinomial approach allows available information to be used to estimate harvest numbers.

This model derived by White (1993) assumes common harvest probabilities and reporting rates for all areas. The model uses the following terms:

H = number of observed (licensed) hunters;
k = number of hunters surveyed;
b = number of hunters who were active among the hunters surveyed;
p_h = probability that a licensed hunter actually hunted;
p_a = probability a hunter hunts in area A of interest;
V = probability a hunter is successful;
r = probability a hunter reports the area hunted;
h_a = number of hunters who harvested an animal in area A and reported the area in the survey;
h_o = number of hunters who harvested an animal in another area and reported the other area on the survey;
h_u = number of hunters who successfully harvested an animal but did not specify the area on the survey.

The White (1993) harvest reporting model has a dichotomous branching process (Fig. 6.3). The joint likelihood for the number of active hunters surveyed (b), h_a, h_o, and h_u is the model

$$L(p_a,V,r|b,h_a,h_o,h_u) \cdot L(p_h|k,b) = \binom{b}{h_a,h_o,h_u}(p_aVr)^{h_a}((1-p_a)Vr)^{h_o} \cdot (V(1-r))^{h_u}$$

$$\cdot (1-V)^{b-h_a-h_o-h_u} \cdot \binom{k}{b}(p_h)^b(1-p_h)^{k-b}.$$

$$(6.42)$$

The maximum likelihood estimator (MLE) for the probability that a licensed holder actually hunted, p_h, is

$$\hat{p}_h = \frac{b}{k},$$
$$(6.43)$$

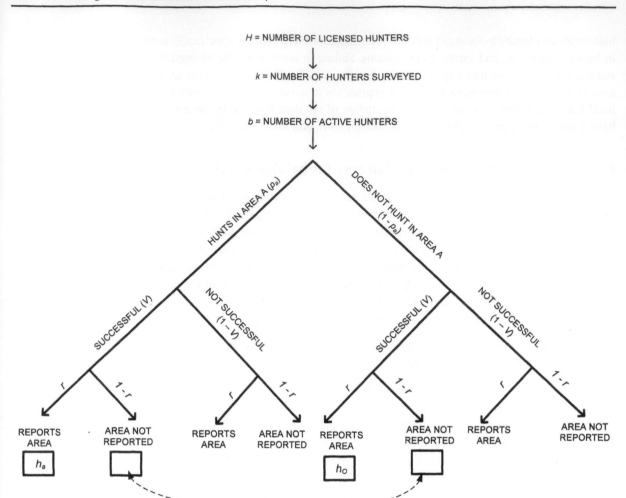

Figure 6.3. Dichotomous branching process associated with the White (1993) multinomial harvest model for harvest apportionment to game management areas, where p_a is the probability a hunter hunts in area A, V is the probability of hunter success, and r is the probability the hunter reports the harvest area.

with an associated estimated variance of

$$\widehat{\text{Var}}(\hat{p}_h) = \left(1 - \frac{k}{H}\right)\frac{\hat{p}_h(1 - \hat{p}_h)}{k}. \tag{6.44}$$

The MLE for p_a is

$$\hat{p}_a = \frac{h_a}{h_a + h_o} \tag{6.45}$$

with associated variance estimator

$$\widehat{\text{Var}}(\hat{p}_a) = \left(1 - \frac{k}{H}\right)\frac{\hat{p}_a(1 - \hat{p}_a)}{b\hat{V}\hat{r}}. \tag{6.46}$$

The hunter success probability is estimated by

$$\hat{V} = \frac{h_a + h_o + h_u}{b} \tag{6.47}$$

with associated variance estimator

$$\widehat{\mathrm{Var}}\,(\hat{V}) = \frac{\hat{V}(1-\hat{V})}{b}\left(1-\frac{k}{H}\right). \tag{6.48}$$

The probability a hunter reported the area hunted is estimated by

$$\hat{r} = \frac{h_a + h_o}{h_a + h_o + h_u} \tag{6.49}$$

with associated variance estimator

$$\widehat{\mathrm{Var}}\,(\hat{r}) = \left(1-\frac{k}{H}\right)\frac{\hat{r}(1-\hat{r})}{b\hat{V}}\Big[(1-\hat{r})(1-\hat{p}_a)\big(1-(1-\hat{p}_a)\hat{V}\hat{r}\big) \\ + \hat{p}_a(1-\hat{r})(1-\hat{p}_a\hat{V}\hat{r}) + \hat{r} + 2\hat{V}\hat{r}^2(1-\hat{r})^2(1-\hat{p}(1-\hat{p}))\Big]. \tag{6.50}$$

The estimate of total game harvest in area A is

$$\hat{C}_a = H\hat{p}_h\hat{p}_a\hat{V}, \tag{6.51}$$

or, equivalently,

$$\hat{C}_a = \hat{C}\cdot\hat{p}_a,$$

where $\hat{C}$ is an estimate of total harvest. Under the assumption of independence between $\hat{p}_h$ and the term $\hat{p}_a\hat{V}$, the variance of $\hat{C}_a$ is obtained by the exact variance formula for a product (Goodman 1960),

$$\widehat{\mathrm{Var}}\,(\hat{C}_a) = H^2\Big(\big(\hat{p}_a\hat{V}\big)^2\,\widehat{\mathrm{Var}}\,(\hat{p}_h) + \hat{p}_h^2\,\widehat{\mathrm{Var}}\,(\hat{p}_a\hat{V}) - \widehat{\mathrm{Var}}\,(\hat{p}_a\hat{V})\widehat{\mathrm{Var}}\,(\hat{p}_h)\Big), \tag{6.52}$$

where the variance of $\hat{p}_h$ is estimated by Eq. (6.43). The variance of $\hat{p}_a\hat{V}$ is estimated by the delta method as

$$\widehat{\mathrm{Var}}\,(\hat{p}_a\hat{V}) = \left(1-\frac{k}{H}\right)\frac{\hat{p}_a\hat{V}}{b\hat{r}}\Big[(1-\hat{p}_a\hat{V}\hat{r})(\hat{p}_a(1-\hat{r})+1)^2 \\ + \hat{p}_a(1-\hat{p}_a)(1-\hat{r})\big(1-(1-\hat{p}_a)\hat{V}\hat{r}\big) + \hat{p}_a\hat{r}(1-\hat{r})\big(1-\hat{V}(1-\hat{r})\big) \\ + 2\hat{p}_a\hat{V}\hat{r}\big((1-\hat{p}_a)^2 - \hat{p}_a^2\hat{r}\big)\Big]. \tag{6.53}$$

Assumptions

There are at least five assumptions associated with the estimator of area A harvest total (Eq. 6.51):

1. Harvest is limited to one big-game animal per hunter.
2. Hunters act independently.
3. The probability of hunter success is the same across areas.
4. The probability a hunter reports the area hunted is the same across areas and the same for successful and unsuccessful hunters.
5. All hunters surveyed submit their annual harvest report card, or nonresponse is independent of hunter success and area hunted.

The White (1993) model is based on the situation where only successful hunters report the area of harvest. Without areal reporting by both successful and unsuccessful hunters, White (1993) was limited to a likelihood model that necessitated assumptions 4 and 5. These assumptions will be relaxed in the next section when information from both successful and unsuccessful hunters becomes available.

Example 6.3: Using Responses from Only Successful Hunters to Estimate Deer Harvest by Management Unit

An example provided by White (1993) used hypothetical responses to a hunter survey for a deer harvest. The number of licensed hunters is $H = 25{,}000$ and $k = 2500$ are randomly sampled, of which $b = 2000$ actively hunted. The number of hunters who reported hunting in the area of interest was 200 with a harvest of 34 ($h_a = 34$). Among the other respondents, there were 1600 reported hunting in other areas, with a harvest of 346 ($h_o = 346$), and 200 did not report the area of hunting but did report a harvest of 33 individuals ($h_u = 33$). From these data, the probabilities of success, hunting in the area of interest, and reporting can be estimated by using likelihood Eq. (6.42). Subsequently, Eq. (6.51) can be used to estimate the total harvest in the area of interest, $\hat{C}_a$.

The probability that a license holder actually hunted is estimated by Eq. (6.43) as

$$\hat{p}_h = \frac{b}{k}$$

$$= \frac{2000}{2500} = 0.80,$$

with estimated variance (Eq. 6.44)

$$\widehat{\mathrm{Var}}(\hat{p}_h) = \left(1 - \frac{k}{H}\right)\frac{\hat{p}_h(1 - \hat{p}_h)}{k}$$

$$= \left(1 - \frac{2500}{25000}\right)\frac{0.80\,(1 - 0.80)}{2500} = 0.0000576.$$

The reporting probability is estimated (Eq. 6.49) as

$$\hat{r} = \frac{h_a + h_o}{h_a + h_o + h_u}$$

$$= \frac{34 + 346}{34 + 346 + 33} = 0.9201.$$

The estimate and associated variance for the probability of success, $\hat{V}$, are calculated by Eqs. (6.47) and (6.48), respectively, as

$$\hat{V} = \frac{h_a + h_o + h_u}{b}$$

$$= \frac{34 + 346 + 33}{2000} = 0.2065,$$

and

$$\widehat{\mathrm{Var}}(\hat{V}) = \frac{\hat{V}(1-\hat{V})}{b}\left(1-\frac{k}{H}\right)$$

$$= \frac{0.2065\,(1-0.2065)}{2000}\left(1-\frac{2500}{25000}\right) = 0.000074,$$

with a standard error of $\widehat{\mathrm{SE}}(\hat{V}) = 0.0086$. The probability of hunting in the area of interest, p_a, is estimated by Eq. (6.45) as follows:

$$\hat{p}_a = \frac{h_a}{h_a + h_o}$$

$$= \frac{34}{34+346}$$

$$= 0.0895.$$

The associated variance of $\hat{p}_a$ is estimated using $\hat{r}$ and $\hat{V}$ by Eq. (6.46) as

$$\widehat{\mathrm{Var}}(\hat{p}_a) = \left(1-\frac{k}{H}\right)\frac{\hat{p}_a(1-\hat{p}_a)}{b\hat{V}\hat{r}}$$

$$= \left(1-\frac{2500}{25000}\right)\frac{0.0895\,(1-0.0895)}{2000(0.2065)(0.9201)} = 0.000193,$$

with an associated standard error of $\widehat{\mathrm{SE}}(\hat{p}_a) = 0.0139$.

Total harvest from area A, $\hat{C}_a$, is calculated by Eq. (6.51) as

$$\hat{C}_a = H\hat{p}_h\hat{p}_a\hat{V}$$

$$= 25000(0.80)(0.0895)(0.2065) = 369.64\ \text{deer}.$$

The variance of $\hat{C}_a$ is calculated by using Eq. (6.52) as

$$\widehat{\mathrm{Var}}(\hat{C}_a) = H^2\left((\hat{p}_a\hat{V})^2\,\widehat{\mathrm{Var}}(\hat{p}_h) + \hat{p}_h^2\,\widehat{\mathrm{Var}}(\hat{p}_a\hat{V}) - \widehat{\mathrm{Var}}(\hat{p}_a\hat{V})\widehat{\mathrm{Var}}(\hat{p}_h)\right)$$

$$= (25000)^2\left((0.0895\cdot0.2065)^2(0.0000576) + (0.80)^2(0.00001055)\right.$$

$$\left. - (0.00001055)(0.0000576.)\right)$$

$$\widehat{\mathrm{Var}}(\hat{C}_a) = 4231.9169,$$

where the variance of $(\hat{p}_a\hat{V})$ is calculated by Eq. (6.53) as

$$\widehat{\mathrm{Var}}(\hat{p}_a\hat{V}) = \left(1-\frac{2500}{25000}\right)\frac{0.0895\cdot0.2065}{2000(0.9201)}$$

$$\cdot\left[(1-0.0895\cdot0.2065\cdot0.9201)(0.0895(1-0.2065)+1)^2\right.$$

$$+ 0.0895(1-0.0895)(1-0.9201)(1-(1-0.0895)0.2065\cdot0.9201)$$

$$+ 0.0895\cdot0.9201(1-0.9201)(1-0.2065(1-0.9201))$$

$$\left. + 2\cdot0.0895\cdot0.9201\cdot0.2065\left((1-0.0895)^2 - 0.0895^2\cdot0.9201\right)\right]$$

$$\widehat{\mathrm{Var}}(\hat{p}_a\hat{V}) = 0.00001055.$$

The standard error of $\hat{C}_a$ is $\widehat{\mathrm{SE}}(\hat{C}_a) = 65.05$.

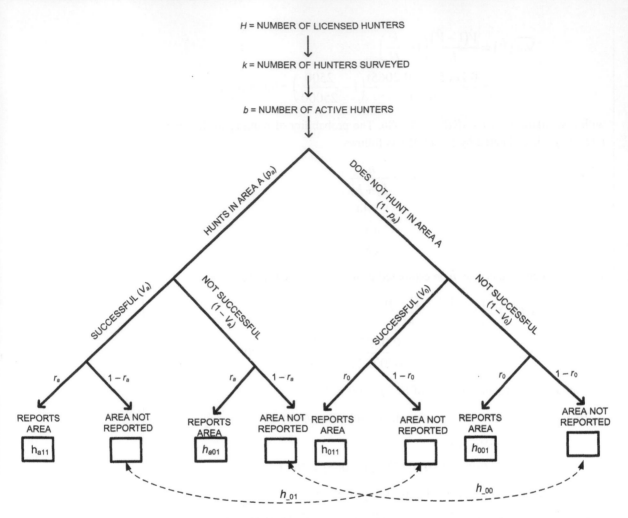

Figure 6.4. Dichotomous branching process for harvest apportionment to game management areas with unique success rates (i.e., V_a, V_0) and unique reporting rates (i.e., r_a, r_0) by area.

6.3.2   Unique Reporting and Success Probabilities

When the hunter survey questionnaire is designed so that both successful and unsuccessful hunters report the area in which they hunted, a more flexible and realistic analysis model is possible than was available in the scenario addressed by White (1993). Area-specific reporting and success probabilities can be estimated by using a similar branching modeling (Fig. 6.4).

The expanded model using information from both successful and unsuccessful hunters has the terms:

k = number of hunters surveyed;
b = number of active hunters;
p_a = probability a hunter hunts in area A of interest;
p_h = probability a licensed hunter actually hunts;
V_a = probability of hunter success in area A;
r_a = probability a hunter who hunts in area A reports the area;
V_o = probability of hunter success in other areas outside area A;

r_o = probability a hunter who hunts outside area A reports the area;

h_{ijk} = number of hunters in area i ($i = a$ for area A, zero otherwise) that had success status j ($j = 1$ for successful, zero otherwise) and reporting status k ($k = 1$ for reported area hunted, zero otherwise);

$h_{\_10} = h_{a10} + h_{o10}$ = number of hunters who were successful but failed to indicate the area hunted;

$h_{\_00} = h_{a00} + h_{o10}$ = number of hunters who were unsuccessful and failed to indicate the area hunted.

The dichotomous branching process (Fig. 6.4) is useful to display the harvest and areal reporting by both successful and unsuccessful hunters. The multinomial model for the number of surveyed hunters (i.e., h_{ijk}) can be written as

$$
L(p_a, \underset{\sim}{V}, \underset{\sim}{r} | b, \underset{\sim}{h}_{ijk}) \cdot L(p_h | k, b) = \binom{b}{h_{ijk}} (p_a V_a r_a)^{h_{a11}} (p_a (1 - V_a) r_a)^{h_{a01}} \cdot
$$

$$
((1 - p_a) V_o r_o)^{h_{011}} ((1 - p_a)(1 - V_o) r_o)^{h_{001}} \cdot
$$

$$
(p_a V_a (1 - r_a) + (1 - p_a) V_o (1 - r_o))^{h_{\_01}} \cdot \qquad (6.54)
$$

$$
(p_a (1 - V_a)(1 - r_a) + (1 - p_a)(1 - V_o)(1 - r_o))^{h_{\_00}} \cdot
$$

$$
\binom{k}{b} (p_h)^b (1 - p_h)^{k-b}.
$$

The probability that a license holder actually hunted, p_h, is estimated by Eq. (6.43), with associated variance estimated by Eq. (6.44).

The MLE for hunter success in area A (i.e., V_a) is

$$
\hat{V}_a = \frac{h_{a11}}{h_{a01} + h_{a11}} \qquad (6.55)
$$

with associated variance estimator

$$
\widehat{\mathrm{Var}}(\hat{V}_a) = \frac{\hat{V}_a (1 - \hat{V}_a)}{b \hat{p}_a \hat{r}_a} \left(1 - \frac{k}{H}\right). \qquad (6.56)
$$

The estimator of hunter success in other areas is

$$
\hat{V}_o = \frac{h_{011}}{h_{001} + h_{011}} \qquad (6.57)
$$

with associated variance estimator

$$
\widehat{\mathrm{Var}}(\hat{V}_o) = \frac{\hat{V}_o (1 - \hat{V}_o)}{b(1 - \hat{p}_a) \hat{r}_o} \left(1 - \frac{k}{H}\right). \qquad (6.58)
$$

The probability hunters in area A reported the area hunted is estimated by

$$
\hat{r}_a = \frac{h_{a11} + h_{a01}}{b \hat{p}_a}. \qquad (6.59)
$$

Outside of area A, the probability a hunter reported the area in which hunting occurred is estimated by

$$\hat{r}_o = \frac{h_{011} + h_{001}}{b(1 - \hat{p}_a)}. \tag{6.60}$$

The proportion of hunters who hunted in area A is estimated by

$$\hat{p}_a = \frac{(h_{010} + h_{011} + h_{a11}) - b\hat{V}_o}{b(\hat{V}_a - \hat{V}_o)}. \tag{6.61}$$

The variance of $\hat{p}_a$ is approximated using the delta method as

$$
\begin{aligned}
\widehat{\mathrm{Var}}(\hat{p}_a) =& \\
&\frac{H-k}{Hb(\hat{V}_a - \hat{V}_o)^2}\Bigg[\frac{\hat{p}_a\hat{V}_a}{\hat{r}_a}\Big[(1 - \hat{p}_a\hat{V}_a\hat{r}_a)(\hat{V}_a + \hat{r}_a - 1)^2 + \hat{V}_a(1 - \hat{V}_a)(1 - \hat{p}_a(1 - \hat{V}_a\hat{r}_a))\Big] \\
&+ \frac{(1 - \hat{p}_a)\hat{V}_o}{\hat{r}_o}\Big[(1 - (1 - \hat{p}_a)\hat{V}_o\hat{r}_o)(\hat{V}_o + \hat{r}_o - 1)^2 + \hat{V}_o(1 - \hat{V}_o)(1 - (1 - \hat{p}_a)(1 - \hat{V}_o)\hat{r}_o)\Big] \\
&+ \Big[(\hat{p}_a\hat{V}_a(1 - \hat{r}_a) + (1 - \hat{p}_a)\hat{V}_o(1 - \hat{r}_o))(1 - \hat{p}_a\hat{V}_a(1 - \hat{r}_a) - (1 - \hat{p}_a)\hat{V}_o(1 - \hat{r}_o))\Big] \\
&+ 2\Big\{(1 - \hat{p}_a)\hat{V}_o(\hat{V}_o + \hat{r}_o - 1)[(1 - \hat{p}_a)\hat{V}_o(2 - \hat{V}_o - \hat{r}_o) + \hat{p}_a\hat{V}_o] \\
&+ \hat{p}_a\hat{V}_a(\hat{V}_a + \hat{r}_a - 1)[\hat{V}_o(1 - \hat{p}_a)(2 - \hat{V}_o - \hat{r}_o) + \hat{p}_a\hat{V}_a(2 - \hat{V}_a - \hat{r}_a)] \\
&+ (1 - \hat{p}_a)\hat{V}_o(1 - \hat{V}_o)[\hat{p}_a\hat{V}_a(2 - \hat{V}_a - \hat{r}_a) + (1 - \hat{p}_a)\hat{V}_o(1 - \hat{r}_o)] \\
&+ \hat{p}_a\hat{V}_a(1 - \hat{V}_a)(\hat{p}_a\hat{V}_a(1 - \hat{r}_a) + (1 - \hat{p}_a)\hat{V}_o(1 - \hat{r}_o))\Big\}\Bigg].
\end{aligned}
\tag{6.62}
$$

The estimate of total harvest in area A is calculated by

$$\hat{C}_a = H\hat{p}_h\hat{p}_a\hat{V}_a. \tag{6.63}$$

Under the assumption of independence between $\hat{p}_h$ and the term $\hat{p}_a\hat{V}_a$, the variance of $\hat{C}_a$ is estimated by the exact product formula (Goodman 1960)

$$\widehat{\mathrm{Var}}(\hat{C}_a) = H^2\big((\hat{p}_a\hat{V}_a)^2\,\widehat{\mathrm{Var}}(\hat{p}_h) + \hat{p}_h^2\,\widehat{\mathrm{Var}}(\hat{p}_a\hat{V}_a) - \widehat{\mathrm{Var}}(\hat{p}_a\hat{V}_a)\widehat{\mathrm{Var}}(\hat{p}_h)\big), \tag{6.64}$$

where the variance of $\hat{p}_h$ is estimated by Eq. (6.44). The variance of the term $\hat{p}_a\hat{V}_a$ is approximated by the delta method as

$$\widehat{\mathrm{Var}}(\hat{p}_a\hat{V}_a) = (\hat{p}_a\hat{V}_a)^2\left[\frac{\widehat{\mathrm{Var}}(\hat{V}_a)}{\hat{V}_a^2} + \frac{\widehat{\mathrm{Var}}(\hat{p}_a)}{\hat{p}_a^2} + \frac{2\widehat{\mathrm{Cov}}(\hat{p}_a\hat{V}_a)}{\hat{p}_a\hat{V}_a}\right], \tag{6.65}$$

where the variances and covariances can be obtained numerically. The covariance is approximated by the delta method as

$$
\begin{aligned}
\widehat{\mathrm{Cov}}(\hat{p}_a\hat{V}_a) =& \\
&\left(\frac{(H-k)\hat{V}_a^2}{Hb(\hat{V}_a - \hat{V}_o)^2}\right) \\
&\{\hat{p}_a(\hat{r}_a - \hat{V}_o(1 - \hat{V}_a))(\hat{V}_o(1 - \hat{V}_a\hat{p}_a) + \hat{p}_a\hat{V}_a(1 - \hat{r}_a)) + \hat{V}_o^2\hat{p}_a(1 - \hat{p}_a)(1 - \hat{V}_a)(1 - \hat{V}_o) \\
&+ ((1 - \hat{p}_a)\hat{V}_o(\hat{V}_o + \hat{r}_o - 1)[\hat{V}_o(\hat{p}_a(1 - \hat{V}_a) + (1 - \hat{p}_a)(2 - \hat{V}_o - \hat{r}_o)) + \hat{p}_a\hat{V}_a(1 - \hat{r}_a)]) \\
&+ \hat{V}_o[\hat{p}_a\hat{V}_a(1 - \hat{r}_a) + (1 - \hat{p}_a)\hat{V}_o(1 - \hat{r}_o)](\hat{p}_a(1 - \hat{V}_a) + (1 - \hat{p}_a)(1 - \hat{V}_o))\}.
\end{aligned}
\tag{6.66}
$$

Example 6.4: Using Responses from Both Successful and Unsuccessful Hunters to Estimate Deer Harvest by Management Unit

Consider a situation similar to Example 6.3, in which $k = 2500$ deer hunters are surveyed out of $H = 25{,}000$ licensed hunters. Of the 2500 respondents, there were $b = 2000$ active hunters. However, this time, more detailed survey information is obtained for estimating success and reporting probabilities. Two hundred hunters reported using area A, with 120 successful ($h_{a11} = 120$) and 80 unsuccessful ($h_{a01} = 80$). In all other areas, there were $h_{011} = 192$ successful and $h_{001} = 288$ unsuccessful hunters who reported the area of hunting. The number of hunters who did not report the area of hunting was 1320, with $h_{\_10} = 568$ successful, and $h_{\_00} = 752$ unsuccessful hunters. The probability of success and reporting in area A and in other areas can be estimated from these data.

The probabilities of success in area A and in other areas are estimated by Eqs. (6.55) and (6.57), respectively, as

$$\hat{V}_a = \frac{h_{a11}}{h_{a01} + h_{a11}}$$

$$= \frac{120}{80 + 120} = 0.60$$

and

$$\hat{V}_o = \frac{h_{011}}{h_{001} + h_{011}}$$

$$= \frac{192}{192 + 288} = 0.4000.$$

The proportion of hunters who hunted in area A is calculated by using the probabilities of success (Eq. 6.61) as

$$\hat{p}_a = \frac{(h_{010} + h_{011} + h_{a11}) - b\hat{V}_o}{b(\hat{V}_a - \hat{V}_o)}$$

$$= \frac{(120 + 192 + 568) - 2000(0.4000)}{2000(0.6000 - 0.4000)} = 0.2000.$$

The probability of a hunter in area A reported the area of hunting is estimated by Eq. (6.59) as

$$\hat{r}_a = \frac{h_{a11} + h_{a01}}{b\hat{p}_a}$$

$$= \frac{120 + 80}{2000(0.2000)} = 0.5000.$$

Outside of area A, the probability a hunter reported the area in which hunting occurred is estimated by Eq. (6.60) as

$$\hat{r}_o = \frac{h_{011} + h_{001}}{b(1 - \hat{p}_a)}$$

$$= \frac{192 + 288}{2000(1 - 0.2000)}$$

$$= 0.3000.$$

The variance of the success probability $\hat{V}_a$ is estimated by Eq. (6.56) using $\hat{r}_a$ and $\hat{p}_a$ as follows:

$$\widehat{Var}(\hat{V}_a) = \frac{\hat{V}_a(1 - \hat{V}_a)}{b\hat{p}_a\hat{r}_a}\left(1 - \frac{k}{H}\right)$$

$$= \frac{0.6000(1 - 0.6000)}{2000(0.2000)(0.5000)}\left(1 - \frac{2500}{25000}\right) = 0.00108$$

with standard error $\widehat{SE}(\hat{V}_a) = 0.0329$. The variance of $\hat{V}_o$ is estimated by Eq. (6.58) as

$$\widehat{Var}(\hat{V}_o) = \frac{\hat{V}_o(1 - \hat{V}_o)}{b(1 - \hat{p}_a)\hat{r}_o}\left(1 - \frac{k}{H}\right)$$

$$= \frac{0.4000(1 - 0.4000)}{2000(1 - 0.2000)(0.3000)}\left(1 - \frac{2500}{25000}\right) = 0.00045,$$

with a standard error of $\widehat{SE}(\hat{V}_o) = 0.0212$. Total harvest from the area of interest is estimated from Eq. (6.63) as

$$\hat{C}_a = H\hat{p}_h\hat{p}_a\hat{V}_a$$

$$= 25000(0.80)(0.2000)(0.6000) = 2,400 \text{ deer.}$$

The variance of $\hat{C}_a$ is estimated by Eq. (6.64) as

$$\widehat{Var}(\hat{C}_a) = H^2\left((\hat{p}_a\hat{V}_a)^2 \widehat{Var}(\hat{p}_h) + \hat{p}_h^2 \widehat{Var}(\hat{p}_a\hat{V}_a) - \widehat{Var}(\hat{p}_a\hat{V}_a)\widehat{Var}(\hat{p}_h)\right)$$

$$= (25000)^2 \cdot \left[\left((0.2 \cdot 0.6)^2(0.0000576) + (0.8)^2(0.001930)\right.\right.$$

$$\left.\left. - (0.001930)(0.0000576)\right)\right]$$

$$\widehat{Var}(\hat{C}_a) = 772,448.92,$$

where $\widehat{Var}(\hat{p}_h)$ is estimated by Eq. (6.44) and $\widehat{Var}(\hat{V}_a\hat{p}_a)$ is calculated by Eq. (6.65). By using $\widehat{Var}(\hat{V}_a) = 0.00108$, as estimated above, $\widehat{Var}(\hat{p}_a) = 0.006$, estimated by Eq. (6.62), and $\widehat{Cov}(\hat{p}_a\hat{V}_a) = -0.0005395$, as calculated by Eq. (6.66), the $\widehat{Var}(\hat{V}_a\hat{p}_a)$ is estimated as

$$\widehat{Var}(\hat{p}_a\hat{V}_a) = (\hat{p}_a\hat{V}_a)^2\left[\frac{\widehat{Var}(\hat{V}_a)}{\hat{V}_a^2} + \frac{\widehat{Var}(\hat{p}_a)}{\hat{p}_a^2} + \frac{2\widehat{Cov}(\hat{p}_a\hat{V}_a)}{\hat{p}_a\hat{V}_a}\right]$$

$$= (0.2(0.6))^2\left[\frac{0.00108}{(0.6)^2} + \frac{0.0056}{(0.2)^2} + \frac{2(-0.0005395)}{(0.2)(0.6)}\right] = 0.001930.$$

The standard error of the harvest is $\widehat{\text{SE}}(\hat{C}_a) = 878.83$ deer. Although calculated by analytic expressions, estimators, variances, and covariances for a likelihood as complex as Eq. (6.54) are more easily obtained numerically. See Appendix C on the use of Program USER to compute maximum likelihood estimates for this example.

Discussion of Utility

Comparison of likelihood models (Eqs. 6.42, 6.54) and the difference in specifying which harvest parameters can be estimated indicate the importance of questionnaire design. Unless simplified assumptions regarding common harvest rates across areas can be justified, *a priori*, questionnaires need to be designed from the onset to provide detailed regional information. This information must include the hunting activity of both successful and unsuccessful hunters. The questionnaire design is not complete until the required parameters needed by game managers have been shown to be estimable from the resulting survey data. There is often a disconnect between questionnaire designers, game managers, and biometricians. This disconnect unnecessarily alienates the public with useless intrusion and frustrates many managers with the inadequacy of the information collected.

The survey models (Section 6.3) consider only the case of one hunter–one harvested game animal. The design of the questionnaire and subsequent analysis become more involved if the harvest can involve multiple harvested animals across multiple game management areas, as in the case of small-game species. The specificity of the information requested in such cases (i.e., harvest numbers and effort by area) may tax the patience of the average survey respondent. The more demanding the questionnaire, the larger the expected percentage of nonrespondents and the larger the expected percentage of missing information. The art of questionnaire design, therefore, lies in asking what *must* be known and omitting what is of secondary curiosity. Unfortunately, what one investigator finds irrelevant is essential information to another. Agency goals should guide the design of hunter questionnaires, with review by all parties, including managers, biologists, and biometricians.

6.4   Direct Estimation of Harvest Mortality

Harvest survival, S_H, is the probability an animal will survive the process of harvest; harvest mortality, $M_H = 1 - S_H$, is its complement. The most direct estimator of S_H is the proportion of the population at the start of the hunting season that is still alive after harvest, and can be calculated by the expression

$$\hat{S}_H = \frac{\hat{N}_{\text{Post-harvest}}}{\hat{N}_{\text{Pre-harvest}}}, \tag{6.67}$$

where

$\hat{N}_{\text{Pre-harvest}}$ = estimated total number of individuals alive before harvest;
$\hat{N}_{\text{Post-harvest}}$ = estimated total number of individuals alive after harvest.

If the two population surveys were conducted independently, the variance of $\hat{S}_H$ is approximated by the delta method as

$$\widehat{\mathrm{Var}}\left(\hat{M}_H\right)=\widehat{\mathrm{Var}}\left(\hat{S}_H\right)=\hat{S}_H^2\left(\widehat{\mathrm{CV}}(\hat{N}_{\text{Post-harvest}})^2+\widehat{\mathrm{CV}}(\hat{N}_{\text{Pre-harvest}})^2\right). \quad (6.68)$$

This estimator of S_H assumes all mortalities during the period are a result of harvest (i.e., natural survival, $S_N=1$). Survival probabilities for any subgroup of the population, males, females, or juveniles can also be estimated by using Eq. (6.67). For populations with short harvest seasons and high mortality rates, Eq. (6.67) may be appropriate. However, the onus of estimating survival has been turned into a problem of estimating abundance, and not once, but twice. Thus, direct estimation using Eq. (6.67) is impractical and rarely used.

If harvest numbers are either known (C) or estimated ($\hat{C}$), the harvest survival rate can be estimated by

$$\hat{S}_H=1-\frac{\hat{C}}{\hat{N}_{\text{Pre-harvest}}}, \quad (6.69)$$

based on a preharvest abundance survey, with associated variance estimator

$$\widehat{\mathrm{Var}}\left(\hat{M}_H\right)=\widehat{\mathrm{Var}}\left(\hat{S}_H\right)=\left(1-\hat{S}_H\right)^2\left[\widehat{\mathrm{CV}}(\hat{C})^2+\widehat{\mathrm{CV}}(\hat{N}_{\text{Pre-harvest}})^2\right]. \quad (6.70)$$

If C is known without error, the variance (6.70) reduces to

$$\widehat{\mathrm{Var}}\left(\hat{M}_H\right)=\widehat{\mathrm{Var}}\left(\hat{S}_H\right)=\left(1-\hat{S}_H\right)^2\cdot\widehat{\mathrm{CV}}(\hat{N}_{\text{Pre-harvest}})^2. \quad (6.71)$$

If there is only a postharvest abundance survey, harvest survival can be estimated by

$$\hat{S}_H=\frac{\hat{N}_{\text{Post-harvest}}}{\hat{N}_{\text{Post-harvest}}+\hat{C}} \quad (6.72)$$

with associated variance estimator

$$\widehat{\mathrm{Var}}\left(\hat{M}_H\right)=\widehat{\mathrm{Var}}\left(\hat{S}_H\right)=\hat{S}_H^2\left(1-\hat{S}_H\right)^2\left[\widehat{\mathrm{CV}}(\hat{C})^2+\widehat{\mathrm{CV}}(\hat{N}_{\text{Post-harvest}})^2\right]. \quad (6.73)$$

It is important to note that a single postharvest survey results in a smaller variance estimate (Eq. 6.73) (i.e., $(1-S_H^2)100\%$ smaller) than a preharvest survey Eq. (6.70), given that all other considerations are equal.

The direct estimators of harvest (Eqs. 6.67, 6.69, 6.72) have translated the problem of survival estimation into that of abundance estimation. Given the difficulties in estimating abundance and their typically large sampling errors, investigators have conceived of less laborious methods of estimating harvest mortality probabilities.

6.5　Estimating Harvest Mortality from Sex Ratios

Estimators (Chapter 3) have been presented for the expected sex ratio of a population based on male and female survival probabilities, under the assumptions of a stable and stationary population. In terms of both natural (N) and harvest (H) survival probabilities for males (M) and females (F), an estimator of the population sex ratio, $\hat{R}_{F/M}$, can be written as (Eq. 3.64)

$$\hat{R}_{F/M} \doteq \frac{1 - \hat{S}_{N-M}\hat{S}_{H-M}}{1 - \hat{S}_{N-F}\hat{S}_{H-F}}. \tag{6.74}$$

Equation (6.74) is based on the assumption of common natural and harvest survival probabilities for all age classes. Special cases of Eq. (6.74) were developed earlier (Chapter 3). Rearranging the terms in Eq. (6.74), harvest survival for males and females can be estimated, respectively, as

$$\hat{S}_{H-M} = \frac{1 - \hat{R}_{F/M}(1 - \hat{S}_F)}{\hat{S}_{N-M}} \tag{6.75}$$

and

$$\hat{S}_{H-F} = \frac{\hat{R}_{F/M} + \hat{S}_M - 1}{\hat{R}_{F/M}\hat{S}_{N-F}}, \tag{6.76}$$

where

$\hat{S}_F$ = total survival for females, $\hat{S}_F = \hat{S}_{N-F} \cdot \hat{S}_{H-F}$;
$\hat{S}_M$ = total survival for males, $\hat{S}_M = \hat{S}_{N-M} \cdot \hat{S}_{H-M}$.

Both estimators (6.75) and (6.76) require independent estimates of the sex ratio, $\hat{R}_{F/M}$, the annual survival of the other sex, and the natural survival probability for the sex of interest.

Alternatively, two assumptions can be made:

1. Natural survival probabilities of both sexes are equal ($S_{N-F} = S_{N-M} = S_N$).
2. Only males are harvested (i.e., $S_{H-F} = 1$).

This is the most common scenario in which this sex ratio technique is used. Under this set of assumptions, the probability of surviving harvest for males can be estimated as

$$\hat{S}_{H-M} = \frac{1 - \hat{R}_{F/M}(1 - \hat{S}_N)}{\hat{S}_N} \tag{6.77}$$

where $\hat{S}_N$ = estimated natural survival probability for the population. Assuming that $R_{F/M}$ and S_N are estimated independently, the variance of $\hat{S}_{H-M}$ can be estimated by using the delta method, where

$$\widehat{Var}(\hat{S}_{H-M}) = \left(\frac{1}{\hat{S}_N}\right)^2 \left[\widehat{Var}(\hat{R}_{F/M})(1 - \hat{S}_N)^2 + \frac{\widehat{Var}(\hat{S}_N)(\hat{R}_{F/M} - 1)^2}{\hat{S}_N^2}\right]. \tag{6.78}$$

A variation on Eq. (6.77) is use of the Heincke (1913)–Burgoyne (1981) estimator of annual mortality for females, where

$$(1 - \hat{S}_{N-F}) = \hat{p}_{YF}$$

and $\hat{p}_{YF}$ = proportion of the female population composed of yearlings (based on Eq. (5.91)). Therefore, harvest mortality can be estimated by the expression

$$\hat{S}_{H-M} = \frac{1 - \hat{R}_{F/M}\hat{p}_{YF}}{(1 - \hat{p}_{YF})}, \qquad (6.79)$$

with associated variance estimator

$$\widehat{Var}\left(\hat{S}_{H-M}\right) = \widehat{Var}\left(\hat{R}_{F/M}\right)\left(\frac{\hat{p}_{YF}}{1 - \hat{p}_{YF}}\right)^2 + \widehat{Var}\left(\hat{p}_{YF}\right)\left(\frac{\left(1 - \hat{R}_{F/M}\right)^2}{\left(1 - \hat{p}_{YF}\right)^4}\right). \qquad (6.80)$$

Assumptions

Estimators (6.77) and (6.79) and their associated variances are based on four assumptions:

1. The population is stable and stationary.
2. All age classes have the same annual survival probabilities within a sex.
3. Only males experience a harvest mortality of $(1 - \hat{S}_{H-M})$ (i.e., $\hat{S}_{H-F} = 1$ for females).
4. The probabilities of surviving natural and harvest mortality sources are independent.

Whether natural survival is estimated from age-structure data or auxiliary tagging studies, the underlying principles of the sex ratio estimator (Eq. 6.74) is a stable and stationary population.

Example 6.5: Estimating Female Black Bear (*Ursus americanus*) Harvest Mortality, Western Massachusetts

Between August 1980 and April 1986, 15 males and 33 female black bears age 1.0 year and older were radio-marked in western Massachusetts and relocated each week (Elowe et al. 1991). Annual survival of 0.59 for males ($\hat{S}_M = 0.59$) and natural survival of 0.97 ($\hat{S}_{N-F} = 0.97$) were reported for female black bears during the study. The investigators also reported a population sex ratio of $\hat{R}_{F/M} = 2.33$, based on applying survival probabilities to a cohort of juveniles with an even sex ratio (Elowe et al. 1991). From the estimated survival probabilities and the independent estimate of the sex ratio, harvest survival of female black bears can be estimated from Eq. (6.76).

Female harvest survival (Eq. 6.76) is estimated as

$$\hat{S}_{H-F} = \frac{\hat{R}_{F/M} + \hat{S}_M - 1}{\hat{R}_{F/M}\hat{S}_{N-F}}$$

$$= \frac{(2.33) + 0.59 - 1}{(2.33)0.97} = 0.85.$$

No variance of the sex ratio is available, or estimable, because the way in which $\hat{R}_{F/M}$ was obtained. Hence, the variance of $\hat{S}_{H-F}$ cannot be calculated.

Elowe et al. (1991) reported a female harvest survival of 0.94, higher than the 0.85 obtained by Eq. (6.76). Possible reasons for the difference include a population that was not stable and stationary, varying harvest rates over time, or age-dependent harvest and mortality rates.

6.6   Change-in-Ratio Methods

The observed changes in the population sex ratio induced by differential harvest of males and females provide an opportunity to extract information on harvest mortality rates. Two- and three-sample approaches have been developed to estimate survival probabilities. The difference between the techniques is whether natural survival will be estimated along with the estimate of harvest survival.

6.6.1   Two-Sample Change-in-Ratio Methods (Paulik and Robson 1969)

Pre- and postharvest sex ratios have long been used in wildlife management for estimating population abundances (Allen 1942, Lauckhart 1950, Dasmann 1952) and harvest mortality (Allen 1942, Lauckhart 1950). Change-in-ratio methods for estimating harvest mortality presented by Paulik and Robson (1969) are essentially the same as those presented by the above authors. Paulik and Robson (1969) viewed harvest as the complement of productivity with regard to the effect of each on changes in the sex ratio of a population. Productivity will change a skewed sex ratio by addition of individuals with an even sex ratio. Alternatively, harvest will alter the ratio of females to males by differential harvest of the sexes. Thus, an estimator of harvest probability can be derived by using the same basic sampling methods used for estimating productivity (Chapter 4) and juvenile survival (Section 5.11).

In the case of a male-only harvest, the harvest probability for males can be estimated by using samples of the sex ratio immediately before and after harvest. The data for the first survey can be represented by using the following notations:

f_1 = number of females observed in first survey;
m_1 = number of males observed in the first survey;
$x_1 = f_1 + m_1$.

We further define

p_1 = probability of detecting an animal during first survey;
N_M = abundance of males during first survey;
N_F = abundance of females during the first survey;
N_1 = abundance of all animals during the first survey $N_M + N_F$.

Assuming a random sample of animals in the first period, the number of females (f_1) and males (m_1) observed can be modeled by the likelihood

$$L(N_M, N_F | p_1, N_1, m_1, f_1) = \binom{N_1}{m_1, f_1} \left(\frac{p_1 N_M}{N_F + N_M} \right)^{m_1} \left(\frac{p_1 N_F}{N_F + N_M} \right)^{f_1} (1 - p_1)^{N_1 - x_1}$$

$$(6.81)$$

where

$$L(p_1, N_1 | x_1) = \binom{N_1}{x_1} p_1^{x_1} (1 - p_1)^{N_1 - x_1}. \tag{6.82}$$

The conditional likelihood of m_1 and f_1, given x_1, can be written as

$$L(N_F, N_M | x_1, f_1, m_1) = \frac{\binom{N}{f_1, m_1} \left(\frac{p_1 N_F}{N_F + N_M}\right)^{f_1} \left(\frac{p_1 N_M}{N_F + N_M}\right)^{m_1} (1 - p_1)^{N_1 - x_1}}{\binom{N_1}{x_1} p_1^{x_1} (1 - p_1)^{N_1 - x_1}}$$

$$= \binom{x_1}{f_1} \left(\frac{N_F}{N_F + N_M}\right)^{f_1} \left(\frac{N_M}{N_F + N_M}\right)^{m_1}. \tag{6.83}$$

In terms of the ratio of females to males, i.e., $R_{F/M} = N_F/N_M$, Eq. (6.83) can be rewritten as

$$L(R_{F/M} | x_1, f_1, m_1) = \binom{x_1}{f_1} \left(\frac{R_{F/M}}{1 + R_{F/M}}\right)^{f_1} \left(\frac{1}{1 + R_{F/M}}\right)^{m_1}. \tag{6.84}$$

After the hunting season, a second survey of the population sex ratio is conducted to assess the consequences of differential harvest probabilities between the sexes. For the second survey, let

p_2 = probability of detecting an animal during the second survey;
S_N = probability of surviving sources of natural mortality;
S_{H-F} = probability of surviving harvest for females;
S_{H-M} = probability of surviving harvest for males;
f_2 = number of females observed during the second survey;
m_2 = number of males observed during the second survey;
$x_2 = f_2 + m_2$.

The likelihood for the second sex ratio survey is written as

$$L(S_{H-F}, S_{H-M}, S_N, N_F, N_M | p_2, N_2, f_2, m_2)$$

$$= \binom{N_2}{f_2, m_2} \left(\frac{p_2 S_{H-F} S_N N_F}{S_{H-F} S_N N_F + S_{H-M} S_N N_M}\right)^{f_2}$$

$$\cdot \left(\frac{p_2 S_{H-M} S_N N_M}{S_{H-F} S_N N_F + S_{H-M} S_N N_M}\right)^{m_2} (1 - p_2)^{N_2 - m_2 - f_2}.$$

The likelihood for the number of observations, x_2, can be written as

$$L(p_2, N_2 | x_2) = \binom{N_2}{x_2} p_2^{x_2} (1 - p_2)^{N_2 - x_2}.$$

The conditional likelihood of f_2 and m_2, given x_2, can then be written as

$$L(S_{H-F}, S_{H-M}, S_N, N_F, N_M | x_2, f_2, m_2) =$$

$$\binom{x_2}{f_2} \left(\frac{S_{H-F} S_N N_F}{S_{H-F} S_N N_F + S_{H-M} S_N N_M}\right)^{f_2} \left(\frac{S_{H-M} S_N N_M}{S_{H-F} S_N N_F + S_{H-M} S_N N_M}\right)^{m_2}. \tag{6.85}$$

Reparameterizing Eq. (6.85) in terms of the preharvest sex ratio, $R_{F/M}$,

$$L = \binom{x_2}{f_2} \left(\frac{S_{H-F} R_{F/M}}{S_{H-F} R_{F/M} + S_{H-M}}\right)^{f_2} \left(\frac{S_{H-M}}{S_{H-F} R_{F/M} + S_{H-M}}\right)^{m_2} \tag{6.86}$$

and letting

$$\gamma = \frac{S_{H-M}}{S_{H-F}},$$

the likelihood can be further simplified to

$$L = \binom{x_2}{f_2}\left(\frac{R_{F/M}}{\gamma + R_{F/M}}\right)^{f_2}\left(\frac{\gamma}{\gamma + R_{F/M}}\right)^{m_2}. \tag{6.87}$$

The joint likelihood for the first (Eq. 6.84) and second (Eq. 6.87) surveys can then be written as

$$L(R_{F/M}, \gamma | x_1, f_1, m_1, x_2, f_2, m_2) = \binom{x_1}{f_1}\left(\frac{R_{F/M}}{1 + R_{F/M}}\right)^{f_1}\left(\frac{1}{1 + R_{F/M}}\right)^{m_1}$$
$$\cdot \binom{x_2}{f_2}\left(\frac{R_{F/M}}{\gamma + R_{F/M}}\right)^{f_2}\left(\frac{\gamma}{\gamma + R_{F/M}}\right)^{m_2}. \tag{6.88}$$

The maximum likelihood estimates for the parameters are

$$\hat{R}_{F/M} = \frac{f_1}{m_1}$$

and

$$\hat{\gamma} = \frac{f_1 m_2}{f_2 m_1}. \tag{6.89}$$

The variance for $\hat{\gamma}$ can be approximated by the expression

$$\mathrm{Var}(\hat{\gamma}) \doteq \left(\frac{\gamma}{R_{F/M}}\right)\left(\frac{\gamma(1 + R_{F/M})^2}{x_1} + \frac{(\gamma + R_{F/M})^2}{x_2}\right) \tag{6.90}$$

with estimated variance

$$\widehat{\mathrm{Var}}(\hat{\gamma}) = \left(\frac{\hat{\gamma}}{\hat{R}_{F/M}}\right)\left(\frac{\hat{\gamma}(1 + \hat{R}_{F/M})^2}{x_1} + \frac{(\hat{\gamma} + \hat{R}_{F/M})^2}{x_2}\right). \tag{6.91}$$

The parameter $\gamma = \dfrac{S_{H-M}}{S_{H-F}}$ expresses the relative survival rate of males to females during harvest. In the special case in which only males are harvested (i.e., $S_{H-F} = 1$), then $\gamma = S_{H-M}$, providing a direct estimate of the harvest survival probability for males.

Assumptions

The assumptions associated with the joint likelihood model (6.88) include the following:

1. All animals, males and females, have independent and equal probabilities of detection in the first survey.
2. All animals, males and females, have independent and equal detection probabilities during the second survey.
3. Detection rates can be different between the first and second periods.

4. The natural survival probability (S_N) is equal for males and females between the two surveys.

5. Immigration, emigration, and recruitment processes are the same for males and females, or the population is otherwise closed.

6. If only males are harvested, then $\gamma = S_{H-M}$.

Assumptions 1 through 3 are associated with binomial sampling. Allen (1942) found that male ring-necked pheasants (*Phasianus colchicus*) flushed more readily than do females and thus were more detectable, violating the assumption of equal detection probabilities between males and females. Some authors (Chapman 1955, Udevitz and Pollock 1992) have suggested using the assumption of a constant detection ratio $\left(\text{i.e., } \eta = \dfrac{p_f}{p_m} \right)$ for males and females. It can be shown that estimator $\hat{\gamma}$ remains valid whether equal or differential detection of the sexes exists when the ratio η is constant between periods. However, differential harvest pressure on males and females may result in different behavioral responses, in which case η would not be constant over time.

Example 6.6: Estimating Harvest Mortality Using CIR for Ring-Necked Pheasants, Rose Lake Wildlife Experiment Station, Michigan

On the opening day of a pheasant hunting season at the Rose Lake Wildlife Experiment Station in Clinton County, Michigan, the sex ratio was recorded from the number of pheasants flushed by hunters (Allen 1942). The number of males observed was 327, and the number of females was 345. Winter field records after harvest showed 132 males and 502 females. Because the hunt was restricted to males only, $\hat{\gamma}$ from Eq. (6.89) estimates S_{H-M} assuming no harvest of females ($S_{H-F} = 1$). The estimate of $\hat{R}_{F/M}$ is calculated as

$$\hat{R}_{F/M} = \frac{f_1}{m_1}$$

$$= \frac{345}{327} = 1.0550.$$

From Eq. (6.89), the estimate of S_{H-M} is computed to be

$$\hat{\gamma} = \hat{S}_{H-M} = \frac{f_1 m_2}{f_2 m_1} = \frac{345 \cdot 132}{502 \cdot 327} = 0.2774,$$

and, from Eq. (6.91),

$$\widehat{\text{Var}}\left(\hat{S}_{H-M}\right) = \left(\frac{\hat{\gamma}}{\hat{R}_{F/M}} \right)\left(\frac{\hat{\gamma}\left(1 + \hat{R}_{F/M}\right)^2}{x_1} + \frac{\left(\hat{\gamma} + \hat{R}_{F/M}\right)^2}{x_2} \right)$$

$$= \left(\frac{0.2774}{1.0550} \right)\left(\frac{0.2774(1 + 1.0550)^2}{672} + \frac{(0.2774 + 1.0550)^2}{634} \right) = 0.00119.$$

The associated standard error of $\hat{S}_{H-M}$ is $\widehat{\text{SE}}\left(\hat{S}_{H-M}\right) = 0.0346$, with an asymptotic 90% confidence interval of CI $(0.2205 \le S_{H-M} \le 0.3347) = 0.90$.

6.6.2 Three-Sample Change-in-Ratio Method
(Selleck and Hart 1957)

Where both sexes are harvested, the joint likelihood model (6.88) estimates the ratio of harvest survival probabilities $\gamma = \dfrac{S_{H-M}}{S_{H-F}}$. Petrides (1954) and Selleck and Hart (1957) presented an extension of the change-in-ratio method by adding a third sex ratio survey of the harvested animals. The third survey of harvested animals provides information on differential harvest that permits separate estimation of S_{H-M} and S_{H-F}.

The pre- and postharvest population surveys are represented by likelihood models (6.84) and (6.87), respectively. The likelihood for the sex ratio survey of harvested carcasses can be written as

$$L(N_F, N_M, S_{H-F}, S_{H-M} | p_3, C, f_3, m_3)$$
$$= \binom{C}{f_3, m_3} \left(\frac{p_3 N_F (1 - S_{H-F})}{N_F (1 - S_{H-F}) + N_M (1 - S_{H-M})} \right)^{f_3}$$
$$\cdot \left(\frac{p_3 N_M (1 - S_{H-M})}{N_F (1 - S_{H-F}) + N_M (1 - S_{H-M})} \right)^{m_3}$$
$$\cdot (1 - p_3)^{C - (f_3 + m_3)}$$

where

f_3 = number of females in the harvest sample;
m_3 = number of males in the harvest sample;
p_3 = probability of surveying a carcass from the harvest;
$x_3 = f_3 + m_3$.

The likelihood for the total number of harvested animals examined, x_3, is written as

$$L(p_3, C | x_3) = \binom{C}{x_3} p_3^{x_3} (1 - p_3)^{C - x_3}.$$

The conditional likelihood of f_3 and m_3, given x_3, can be written as

$$L(N_F, N_M, S_{H-F}, S_{H-M} | x_3, f_3, m_3)$$
$$= \binom{x_3}{f_3} \left(\frac{N_F (1 - S_{H-F})}{N_F (1 - S_{H-F}) + N_M (1 - S_{H-M})} \right)^{f_3}$$
$$\cdot \left(\frac{N_M (1 - S_{H-M})}{N_F (1 - S_{H-F}) + N_M (1 - S_{H-M})} \right)^{m_3},$$

or, in terms of the preharvest ratio, $R_{F/M}$, as

$$L(R_{F/M}, S_{H-F}, S_{H-M} | x_3, f_3, m_3) = \binom{x_3}{f_3} \left(\frac{R_{F/M} (1 - S_{H-F})}{R_{F/M} (1 - S_{H-F}) + (1 - S_{H-M})} \right)^{f_3}$$
$$\cdot \left(\frac{(1 - S_{H-M})}{R_{F/M} (1 - S_{H-F}) + (1 - S_{H-M})} \right)^{m_3}. \tag{6.92}$$

A joint likelihood is used to estimate the harvest probabilities for males and females, based on the observations m_1, f_1, m_2, f_2, m_3, and f_3 and is written as

$$L(R_{F/M}, S_{H-F}, S_{H-M} | x_1, f_1, m_1, x_2, f_2, m_2, x_3, f_3, m_3) =$$
$$\binom{x_1}{f_1}\left(\frac{R_{F/M}}{1+R_{F/M}}\right)^{f_1}\left(\frac{1}{1+R_{F/M}}\right)^{m_1} \cdot \binom{x_2}{f_2}\left(\frac{S_{H-F}R_{F/M}}{S_{H-F}R_{F/M}+S_{H-M}}\right)^{f_2}$$
$$\cdot \left(\frac{S_{H-M}}{S_{H-F}R_{F/M}+S_{H-M}}\right)^{m_2} \cdot \binom{x_3}{f_3}\left(\frac{R_{F/M}(1-S_{H-F})}{R_{F/M}(1-S_{H-F})+(1-S_{H-M})}\right)^{f_3} \quad (6.93)$$
$$\cdot \left(\frac{(1-S_{H-M})}{R_{F/M}(1-S_{H-F})+(1-S_{H-M})}\right)^{m_3}.$$

The MLE for $\hat{R}_{F/M}$ remains unchanged, where

$$\hat{R}_{F/M} = \frac{f_1}{m_1},$$

with

$$\widehat{\text{Var}}(\hat{R}_{F/M}) = \frac{\hat{R}_{F/M}(1+\hat{R}_{F/M})^2}{x_1} \quad (6.94)$$

when sampling with replacement, or

$$\widehat{\text{Var}}(\hat{R}_{F/M}) = \left(1-\frac{x_1}{N}\right)\frac{\hat{R}_{F/M}(1+\hat{R}_{F/M})^2}{(x_1-1)} \quad (6.95)$$

when sampling without replacement. The quantity $\left(1-\dfrac{x_1}{N}\right)$ is the fpc. The male and female harvest survival probabilities are easiest expressed in terms of

$$\hat{R}_1 = \frac{f_1}{m_1},$$
$$\hat{R}_2 = \frac{f_2}{m_2},$$
$$\hat{R}_3 = \frac{f_3}{m_3}.$$

The MLEs for S_{H-M} and S_{H-F} are, respectively,

$$\hat{S}_{H-M} = \frac{(\hat{R}_3 - \hat{R}_1)}{(\hat{R}_3 - \hat{R}_2)} \quad (6.96)$$

and

$$\hat{S}_{H-F} = \frac{\hat{R}_2(\hat{R}_3 - \hat{R}_1)}{\hat{R}_1(\hat{R}_3 - \hat{R}_2)}. \quad (6.97)$$

The variance of $\hat{S}_{H-M}$ is approximated by the delta method as

$$\text{Var}(\hat{S}_{H-M}) \doteq \frac{S_{H-M}(1 - S_{H-M})}{R_{F/M}(S_{H-M} - S_{H-F})^2} \left[\frac{(1 + R_{F/M})^2 S_{H-M}(1 - S_{H-M})}{x_1} \right.$$

$$+ \frac{S_{H-F}(1 - S_{H-M})(S_{H-M} + R_{F/M}S_{H-F})^2}{x_2}$$

$$\left. + \frac{S_{H-M}(1 - S_{H-F})((1 - S_{H-M}) + (1 - S_{H-F})R_{F/M})^2}{x_3} \right] \quad (6.98)$$

with estimated variance

$$\widehat{\text{Var}}(\hat{S}_{H-M}) = \frac{1}{(\hat{R}_3 - \hat{R}_2)^2}$$

$$\cdot \left(\widehat{\text{Var}}(\hat{R}_1) + (\hat{S}_{H-M})^2 \widehat{\text{Var}}(\hat{R}_2) + (1 - \hat{S}_{H-M})^2 \widehat{\text{Var}}(\hat{R}_3) \right), \quad (6.99)$$

where $\widehat{\text{Var}}(\hat{R}_i)$ is estimated by Eqs. (6.94) or (6.95). The variance of $\hat{S}_{H-F}$ is estimated as

$$\text{Var}(\hat{S}_{H-F}) \doteq \frac{1}{R_{F/M}(S_{H-M} - S_{H-F})^2} \left[\frac{(1 + R_{F/M})^2 S_{H-F}^2 (1 - S_{H-F})^2}{x_1} \right.$$

$$+ \frac{S_{H-F}^3 (1 - S_{H-M})^2 (S_{H-M} + R_{F/M}S_{H-F})^2}{x_2 S_{H-M}}$$

$$\left. + \frac{S_{H-M}(1 - S_{H-F})(1 - S_{H-M})((1 - S_{H-M}) + (1 - S_{H-F})R_{F/M})^2}{x_3} \right]$$

$$(6.100)$$

with estimated variance

$$\widehat{\text{Var}}(\hat{S}_{H-F}) = \left(\frac{\hat{R}_2 \hat{R}_3}{\hat{R}_1(\hat{R}_2 - \hat{R}_3)} \right)^2 \left(\widehat{\text{CV}}(\hat{R}_1)^2 + \left(\frac{\hat{R}_3 - \hat{R}_1}{\hat{R}_2 - \hat{R}_3} \right)^2 \widehat{\text{CV}}(\hat{R}_2)^2 \right.$$

$$\left. + \left(\frac{\hat{R}_1 - \hat{R}_2}{\hat{R}_2 - \hat{R}_3} \right)^2 \widehat{\text{CV}}(\hat{R}_3)^2 \right), \quad (6.101)$$

where $\widehat{\text{CV}}(\hat{R}_i)^2 = \dfrac{\widehat{\text{Var}}(\hat{R}_i)}{\hat{R}_i^2}$.

The harvest survival probability for the entire population (S_H) can be found from the expression:

$$S_H = \frac{N_M S_{H-M} + N_F S_{H-F}}{N_M + N_F}$$

$$= \frac{S_{H-M} + R_{F/M} \cdot S_{H-F}}{1 + R_{F/M}},$$

which, by the invariance property of MLE, has the estimator

$$\hat{S}_H = \frac{\hat{S}_{H-M} + \hat{R}_1 \cdot \hat{S}_{H-F}}{1 + \hat{R}_1}. \quad (6.102)$$

Selleck and Hart (1957) provide the algebraically equivalent estimator,

$$\hat{S}_H = \frac{(\hat{R}_2 + 1)(\hat{R}_3 - \hat{R}_1)}{(\hat{R}_1 + 1)(\hat{R}_3 - \hat{R}_2)}. \tag{6.103}$$

Paulik and Robson (1969) give total harvest mortality, or the population exploitation rate, in terms of the percentage of males or females in the harvest and pre- and post-harvest surveys, written as

$$(1 - \hat{S}_H) = \frac{\hat{p}_1 - \hat{p}_2}{\hat{p}_2 - \hat{p}_3}, \tag{6.104}$$

where $\hat{p}_1 = \dfrac{1}{1 + \hat{R}_1}$, $\hat{p}_2 = \dfrac{1}{1 + \hat{R}_2}$, and $\hat{p}_3 = \dfrac{1}{1 + \hat{R}_3}$. Substituting these expressions into Eq. (6.104) and solving for harvest survival, $\hat{S}_H$ leads to Eq. (6.103). The associated variance estimator for $\hat{S}_H$ (Eq. 6.103), based on a delta method approximation, can be written as

$$\widehat{\text{Var}}(\hat{S}_H) = \hat{S}_H^2 \left[\left(\frac{(\hat{R}_3 + 1)}{(\hat{R}_1 + 1)(\hat{R}_3 - \hat{R}_1)} \right)^2 \cdot \widehat{\text{Var}}(\hat{R}_1) \right.$$
$$+ \left(\frac{(\hat{R}_3 + 1)}{(\hat{R}_2 + 1)(\hat{R}_3 - \hat{R}_2)} \right)^2 \cdot \widehat{\text{Var}}(\hat{R}_2) \tag{6.105}$$
$$\left. + \left(\frac{(\hat{R}_3 + 1)}{(\hat{R}_3 + \hat{R}_1)(\hat{R}_3 - \hat{R}_2)} \right)^2 \cdot \widehat{\text{Var}}(\hat{R}_3) \right].$$

Assumptions

The assumptions of the three-sample change-in-ratio method are similar to those of the two-sample change-in-ratio method (Section 6.6.1). The three-sample method relaxes the requirement that only male animals are harvested. However, a random sample of the sex structure of the harvested animals is required.

Example 6.7: Estimating Harvest Mortality Using a CIR for Ring-Necked Pheasants

Selleck and Hart (1957) present hypothetical sample survey data from a pheasant population subject to differential harvest (Table 6.3). The harvest probability for males, $\hat{S}_{H-M}$, Eq. (6.96), is estimated to be

$$\hat{S}_{H-M} = \frac{(\hat{R}_3 - \hat{R}_1)}{(\hat{R}_3 - \hat{R}_2)}$$
$$= \frac{(0.\overline{33} - 1.25)}{(0.\overline{33} - 4)} = 0.25$$

with an estimated variance (Eq. 6.99) calculated as

Table 6.3. Survey data for pre- and postharvest sex ratios and the sex ratio of the harvest from a hypothetical pheasant survey.

Survey	Males	Females	Totals	$\hat{R}_{F/M}$	$\widehat{\text{Var}}(\hat{R})$
Preharvest	$m_1 = 80$	$f_1 = 100$	$x_1 = 180$	$\hat{R}_1 = \dfrac{100}{80}$ $= 1.25$	$\widehat{\text{Var}}(\hat{R}_1) = \dfrac{(1.25)(1+1.25)^2}{180}$ $= 0.03515$
Postharvest	$m_2 = 20$	$f_2 = 80$	$x_2 = 100$	$\hat{R}_2 = \dfrac{80}{20}$ $= 4$	$\widehat{\text{Var}}(\hat{R}_2) = \dfrac{(4)(1+4)^2}{100}$ $= 1$
Harvest	$m_3 = 60$	$f_3 = 20$	$x_3 = 80$	$\hat{R}_3 = \dfrac{20}{60}$ $= 0.33$	$\widehat{\text{Var}}(\hat{R}_3) = \dfrac{(0.\overline{33})(1+0.\overline{33})^2}{80}$ $= 0.0056$

(Data from Selleck and Hart 1957.)

$$\widehat{\text{Var}}(\hat{S}_{H-M}) = \frac{1}{(\hat{R}_3 - \hat{R}_2)^2}\left(\widehat{\text{Var}}(\hat{R}_1) + (\hat{S}_{H-M})^2\,\widehat{\text{Var}}(\hat{R}_2)\right.$$

$$\left. + (1 - \hat{S}_{H-M})^2\,\widehat{\text{Var}}(\hat{R}_3)\right)$$

$$= \frac{1}{(0.3333 - 4)^2}\left(0.03515 + (0.25)^2 \cdot 1 + (1 - 0.25)^2 \cdot 0.0056\right)$$

$$= 0.00750,$$

or $\widehat{\text{SE}}(\hat{S}_{H-M}) = 0.0866.$

For females, the harvest probability (Eq. 6.97) is calculated as

$$\hat{S}_{H-F} = \frac{\hat{R}_2(\hat{R}_3 - \hat{R}_1)}{\hat{R}_1(\hat{R}_3 - \hat{R}_2)}$$

$$= \frac{4(0.3333 - 1.25)}{1.25(0.3333 - 4)} = 0.8,$$

with associated variance estimator (Eq. 6.101)

$$\widehat{\text{Var}}(\hat{S}_{H-F}) = \left(\frac{\hat{R}_2\hat{R}_3}{\hat{R}_1(\hat{R}_2 - \hat{R}_3)}\right)^2\left(\widehat{\text{CV}}(\hat{R}_1)^2 + \left(\frac{\hat{R}_3 - \hat{R}_1}{\hat{R}_2 - \hat{R}_3}\right)^2\widehat{\text{CV}}(\hat{R}_2)^2\right.$$

$$\left. + \left(\frac{\hat{R}_1 - \hat{R}_2}{\hat{R}_2 - \hat{R}_3}\right)^2\widehat{\text{CV}}(\hat{R}_3)^2\right)$$

$$= \left(\frac{4(0.3333)}{1.25(4 - 0.3333)}\right)^2\left(\frac{0.03515}{(1.25)^2} + \left(\frac{0.3333 - 1.25}{4 - 0.3333}\right)^2\left(\frac{1}{4^2}\right)\right.$$

$$\left. + \left(\frac{1.25 - 4}{4 - 0.3333}\right)^2\left(\frac{0.0056}{(0.3333)^2}\right)\right)$$

$$\widehat{\text{Var}}(\hat{S}_{H-F}) = 0.00463,$$

and associated standard error of $\widehat{\text{SE}}(\hat{S}_{H-F}) = 0.0681$. By use of Eq. (6.103), the population-wide harvest survival probability is estimated as

$$\hat{S}_H = \frac{(\hat{R}_2+1)(\hat{R}_3-\hat{R}_1)}{(\hat{R}_1+1)(\hat{R}_3-\hat{R}_2)}$$

$$= \frac{(4+1)(0.3333-1.25)}{(1.25+1)(0.3333-4)} = 0.5556,$$

with associated variance estimate (Eq. 6.105) and standard error calculated, respectively, as

$$\widehat{\text{Var}}(\hat{S}_H) = \hat{S}_H^2 \left[\left(\frac{(\hat{R}_3+1)}{(\hat{R}_1+1)(\hat{R}_3-\hat{R}_1)} \right)^2 \cdot \widehat{\text{Var}}(\hat{R}_1) + \left(\frac{(\hat{R}_3+1)}{(\hat{R}_2+1)(\hat{R}_3-\hat{R}_2)} \right)^2 \right.$$

$$\left. \cdot \widehat{\text{Var}}(\hat{R}_2) + \left(\frac{(\hat{R}_3+1)}{(\hat{R}_3+\hat{R}_1)(\hat{R}_3-\hat{R}_2)} \right)^2 \cdot \widehat{\text{Var}}(\hat{R}_3) \right]$$

$$= 0.5556^2 \left[\left(\frac{(0.3333+1)}{(1.25+1)(0.3333-1.25)} \right)^2 \cdot (0.03515) \right.$$

$$+ \left(\frac{(0.3333+1)}{(4+1)(0.3333-4)} \right)^2$$

$$\left. \cdot (1) + \left(\frac{(0.3333+1)}{(0.3333+1.25)(0.3333-4)} \right)^2 \cdot (0.0056) \right]$$

$$\widehat{\text{Var}}(\hat{S}_H) = 0.00626,$$

or $$\widehat{\text{SE}}(\hat{S}_H) = 0.0791.$$

Application of Program USER to find the maximum likelihood estimates and standard errors for this example can be found in Appendix C.

Discussion of Utility

The ability to estimate the sex ratio of the harvest greatly extends the utility of change-in-ratio methods for estimating harvest mortality. With the addition of carcass surveys, the change-in-ratio method can be used for a much wider variety of scenarios. The method no longer needs to be restricted to a single-sex harvest. With carcass surveys, both sexes can be harvested and estimates of sex-specific harvest survival probabilities can be calculated. Examination of Eqs. (6.96) and (6.97) indicates, however, differential harvest must occur between the sexes. If harvest is nonselective, then $R_1 = R_2 = R_3$, and the estimators are no longer defined.

Selleck and Hart (1957) also used the above formulas to estimate overwinter deeryard mortality rates based on age-ratio data of adults and subadults. A survey of the harvested animals is substituted by a carcass survey in the deer yard prior to spring dispersal. This method was subsequently used by White et al. (1996) to estimate overwinter survival of mule deer (*Odocoileus hemionus*).

6.7   Index-Removal Method: Petrides (1949) and Eberhardt (1982)

Petrides (1949) and Eberhardt (1982) present a method of estimating preharvest abundance if the number of animals harvested (C) is known. The method requires a preharvest survey index (I_1), followed by another index survey (I_2) after harvest. The same information can also be used to estimate harvest mortality (M_H) or its complement, the harvest survival probability (S_H).

The method in its simplest form assumes the index is proportional to population abundance, where

$$I_i = \beta N_i \quad (i = 1, 2)$$

and the proportionality coefficient (β) remains constant across the two survey periods. The index could be based on counts of animal sign (e.g., track counts, pellet counts, call counts) or counts of some subclass of the population (i.e., $E(I_i) = pN_i$, where p = detection probability).

A likelihood model for the index-removal method can be readily constructed for the case in which the proportionality coefficient is a detection rate (p) and the index I_i is an animal count. The joint likelihood for the two index surveys can be written as

$$L(N, p | C, I_1, I_2) = \binom{N}{I_1} p^{I_1} (1-p)^{N-I_1} \cdot \binom{N-C}{I_2} p^{I_2} (1-p)^{N-C-I_2} \quad (6.106)$$

where

N = initial population abundance at time of the first survey;
C = number of animals harvested;
I_i = index count (i = 1, 2);
p = probability of detecting an animal during an index count.

The MLEs for p and N are

$$\hat{p} = \frac{I_1 - I_2}{C}$$

and

$$\hat{N} = \frac{C I_1}{I_1 - I_2}.$$

Of more interest is estimation of the probability of harvest mortality ($M_H = 1 - S_H$). The MLE for M_H, based on the invariance property of MLEs, is

$$\hat{M}_H = \frac{C}{\hat{N}} = \frac{I_1 - I_2}{I_1}.$$

Alternatively, S_H can be estimated by the MLE

$$\hat{S}_H = \frac{I_2}{I_1}. \quad (6.107)$$

Knowledge of C is not necessary to estimate harvest mortality or survival. The variance of $\hat{S}_H$, which is equal to the variance of $\hat{M}_H$, can be derived by using the delta method, where

$$\text{Var}(\hat{S}_H) \doteq S_H^2 \left[\frac{\text{Var}(I_1)}{E(I_1)^2} + \frac{\text{Var}(I_2)}{E(I_2)^2} \right] \tag{6.108}$$

$$\doteq S_H^2 \left[CV(\hat{I}_1)^2 + CV(\hat{I}_2)^2 \right],$$

with estimated variance

$$\widehat{\text{Var}}(\hat{S}_H) = \hat{S}_H^2 \left[\frac{\widehat{\text{Var}}(I_1)}{I_1^2} + \frac{\widehat{\text{Var}}(I_2)}{I_2^2} \right]. \tag{6.109}$$

Variance formula (6.108) is applicable for any valid index survey, where $I_i = \beta N_i$ ($i = 1,2$). In the case of animal sign, $\text{Var}(I_i)$ may be estimated from the sample survey design (e.g., Cochran 1977, Thompson 2002) used to generate I_i.

In the special case in which the proportionality coefficient β is a detection or capture probability, then

$$\text{Var}(I_1) = Np(1 - p),$$
$$E(I_1) = Np,$$
$$\text{Var}(I_2) = (N - C)p(1 - p),$$
$$E(I_2) = (N - C)p,$$

leading to the variance expression

$$\text{Var}(\hat{S}_H) \doteq S_H^2 \left(\frac{1-p}{p} \right) \left(\frac{1}{N} + \frac{1}{(N-C)} \right). \tag{6.110}$$

Only in this special case of binomial sampling is knowledge of C needed in variance calculations.

Assumptions

There are three general assumptions associated with the index-removal estimator (6.107) and variance estimator (6.108):

1. The population is closed with only harvest mortality operating between the two surveys (i.e., natural survival $S_N = 1$).
2. The relationship between the index (I_i) and actual animal abundance (N_i) remains constant between survey periods (i.e., $I_i = \beta N_i$).
3. The number of animals harvested (C) is known without error.

In the special case in which the index is an animal count and variance estimator (6.110) is used, there is an additional assumption:

4. Each animal has an independent and equal detection probability within, as well as between, survey periods.

If hunting a species causes behavioral changes that affect the probability of detection, the assumptions of equal probability of detection within and across surveys will be violated. Violations of the third assumption will result in underestimation of survival if the probability of detection is lower in the second sampling period. Alternatively, when the detection probability is greater during the second survey, survival estimates will be positively biased. Often different age and sex classes have seasonal behaviors that cause violation of the assumption of equal detection probabilities in both surveys (Roseberry and Woolf 1991).

Example 6.8: Estimating Removal Rate of Feral Horses, Cold Springs, Oregon

Eberhardt et al. (1982) used surveys of wild adult horses before and after a known number of removals to estimate abundance and removal rates. The preremoval index survey yielded a count of $I_1 = 301$, and the postremoval index survey yielded a count of $I_2 = 76$ after $C = 357$ adult horses were removed from the Cold Springs population. The harvest or removal rate is estimated to be

$$\hat{M}_H = \frac{I_1 - I_2}{I_1}$$

$$= \frac{301 - 76}{301} = 0.7475,$$

or, alternatively, the harvest survival rate is estimated (Eq. 6.107) to be

$$\hat{S}_H = \frac{I_2}{I_1}$$

$$= \frac{76}{301} = 0.2525.$$

The preremoval abundance for the feral horse population is estimated to be

$$\hat{N}_{\text{Pre}} = \frac{CI_1}{I_1 - I_2}$$

$$= \frac{357(301)}{301 - 76} = 477.587,$$

with a common detection probability estimated to be

$$\hat{p} = \frac{I_1 - I_2}{C}$$

$$= \frac{301 - 76}{357} = 0.6303.$$

Equation (6.110) can be used to estimate the variance associated with $\hat{S}_H$ as

$$\widehat{\text{Var}}(\hat{S}_H) \doteq \hat{S}_H^2 \left(\frac{1 - \hat{p}}{\hat{p}} \right) \left(\frac{1}{\hat{N}} + \frac{1}{(\hat{N} - C)} \right)$$

$$= (0.2525)^2 \left(\frac{1 - 0.6303}{0.6303} \right) \left(\frac{1}{477.587} + \frac{1}{477.587 - 357} \right) = 0.000388,$$

with an estimated standard error of $\widehat{\text{SE}}(\hat{S}_H) = 0.0197$.

Eberhardt (1982) alternatively suggested the count indices may be Poisson-distributed, in which case,

$$\mathrm{Var}(I_1) = E(I_1) = Np \triangleq 301$$
$$\mathrm{Var}(I_2) = E(I_2) = (N - c)p \triangleq 76.$$

The variance of $\hat{S}_H$ can then be estimated (Eq. 6.109), where

$$\widehat{\mathrm{Var}}(\hat{S}_H) = \hat{S}_H^2 \left[\frac{\widehat{\mathrm{Var}}(I_1)}{I_1^2} + \frac{\widehat{\mathrm{Var}}(I_2)}{I_2^2} \right]$$

$$= (0.2525)^2 \left[\frac{301}{(301)^2} + \frac{76}{(76)^2} \right] = 0.00105$$

or an associated standard error of $\widehat{\mathrm{SE}}(\hat{S}_H) = 0.0324$. Use of Poisson sampling error will yield larger variance estimates for $\hat{S}_H$ than will the assumption of binomial sampling error.

Sample Size Calculations

Sampling precision for the Petrides (1949)–Eberhardt (1982) estimator of harvest survival will be defined as

$$P(|\hat{S}_H - S_H| < \varepsilon) = 1 - \alpha$$

where the absolute error in estimation (i.e., $|\hat{S}_H - S_H|$) is less than ε, $(1 - \alpha) \cdot 100\%$ of the time. Assuming $\hat{S}_H$ is normally distributed, the absolute error in estimation for a given α-level is

$$\varepsilon = Z_{1-\frac{\alpha}{2}} \sqrt{\mathrm{Var}(\hat{S}_H)}$$

or, equivalently,

$$\varepsilon = Z_{1-\alpha} \sqrt{S_H^2 \left(\frac{1-p}{p} \right) \left(\frac{1}{N} + \frac{1}{(N-C)} \right)}.$$

The absolute error (ε) is expressed as a function of the preharvest population size (N), harvest survival ($\hat{S}_H$), and the common detection probability (p). Sample size charts (Fig. 6.5) are presented for the index removal method, as a function of the detection probability (p). Lower harvest survival probabilities are estimated more precisely than are higher values. This is expected because as more animals are removed, differences in pre- and postharvest indices will be larger and, hence, easier to detect. In all cases, the absolute error decreases with increased sampling effort (i.e., detection probability).

Discussion of Utility

The relative ease and low cost of collecting index data makes the index-removal technique appealing. However, the validity of the estimates depends heavily on the assumption of constant proportionality between the indices and abundance over time (i.e., $I_i = \beta N_i$). Unfortunately, the structure of index data rarely permits testing the assumption that β is constant. Therefore, investigators need to be especially careful to ensure the assumption of constant proportionality is correct. Yet even with the best intentions and precautions, the assumption of constant proportionality may be violated without the investigator's knowledge.

a. $S_H = 0.2$
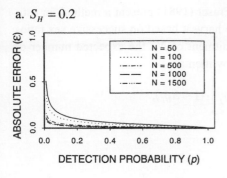

b. $S_H = 0.4$
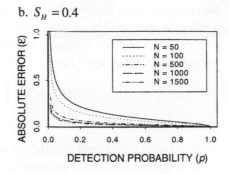

c. $S_H = 0.6$
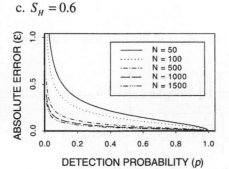

d. $S_H = 0.8$
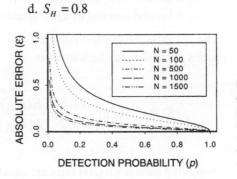

Figure 6.5. Absolute error ε in estimating harvest survival (S_H) by using the index removal method for several values of p, N, and S_H for $1 - \alpha = 0.90$.

6.8 Catch-Effort Methods

First used by Leslie and Davis (1939) and later by DeLury (1947), catch-effort models are based on the assumption the instantaneous harvest mortality rate is a function of harvest effort and the susceptibility of an animal to harvest. Harvest survival can be written as

$$S_H = e^{-cg},$$

where

c = Poisson catchability or vulnerability coefficient;
g = effort.

Harvest mortality is then expressed as

$$M_H = 1 - e^{-cg}. \tag{6.111}$$

There is a large body of literature in fisheries sciences (Leslie and Davis 1939; DeLury 1947, 1951; Ricker 1958; Seber 1982:296–352; Schnute 1985; Quinn and Deriso 1999:15–49) devoted to catch-effort analyses. Much of the literature has been focused on abundance estimation. Much less attention has been devoted to catch-effort techniques in wildlife science.

6.8.1 Successive Sex Ratios: Paloheimo and Fraser (1981)

The observed sex ratio of a cohort can be expected to continually shift over time for a long-lived species subject to differential harvest pressures between the sexes. This observed shift can be used to extract information on harvest mortality rates. Expressing harvest survival by Eq. (6.111), Paloheimo and Fraser (1981) present a method for estimating the vulnerability of males and females to harvest based on successive sex ratios of a cohort over time. Assuming an initial recruitment of N_0, the expected number of animals harvested at age i from a cohort can be written as

$$E(C_i) = N_0 \left[\prod_{j=0}^{i-1} S_{N_j} S_{H_j} \right] \cdot (1 - S_{H_i}),$$

where

N_0 = initial recruitment into the cohort;
C_i = number of animals harvested at age i;
S_{N_i} = natural survival of age i animals;
S_{H_i} = harvest survival of age i animals.

The sex ratio for the harvested animals at age i (i.e., R_{F/M_i}) to the first term of a Taylor series expansion can be written as

$$E(R_{FIM_i}) = E\left(\frac{N_{F_i}}{N_{M_i}} \right) \doteq \frac{E(N_{F_i})}{E(N_{M_i})} = \frac{N_{F_0} \left[\prod_{j=0}^{i-1} S_{N-F_j} S_{H-F_j} \right] (1 - S_{H-F_i})}{N_{M_0} \left[\prod_{j=0}^{i-1} S_{N-M_j} S_{H-M_j} \right] (1 - S_{H-M_i})}.$$

Assuming natural survival rates are equal for both sexes, i.e., $S_{N-F_j} = S_{N-M_j} = S_N$, the equation can be rewritten as

$$E\left(\frac{N_{F_i}}{N_{M_i}} \right) \doteq \frac{N_{F_0} \left[\prod_{j=0}^{i-1} S_{H-F_j} \right] (1 - S_{H-F_i})}{N_{M_0} \left[\prod_{j=0}^{i-1} S_{H-M_j} \right] (1 - S_{H-M_i})}.$$

By expressing the equation in terms of the mortality function (Eq. 6.111), the sex ratio for the harvested animals at age i is expressed as

$$E\left(\frac{N_{F_i}}{N_{M_i}} \right) \doteq R_0 \frac{e^{-c_f \sum_{j=0}^{i-1} g_j} \left(1 - e^{-c_f g_i} \right)}{e^{-c_m \sum_{j=0}^{i-1} g_j} \left(1 - e^{-c_m g_i} \right)}, \qquad (6.112)$$

where

$R_0 = N_{F_0}/N_{M_0}$, the sex ratio at time 0;
c_f = female vulnerability coefficient;
c_m = male vulnerability coefficient;
g_i = effort in the ith hunting season.

The vulnerability parameters, c_f and c_m, and the initial sex ratio, R_0, can be estimated using weighted nonlinear least squares (WNLLS) of the form

$$SS = \frac{\sum_{i=1}^{n} w_i \left(\hat{R}_i - R_0 \cdot e^{(c_m - c_f) \sum_{j=0}^{i-1} g_j} \frac{\left(1 - e^{-c_f g_i} \right)}{\left(1 - e^{-c_m g_i} \right)} \right)^2}{\sum_{i=0}^{n} w_i}, \qquad (6.113)$$

where

$$w_i = \frac{1}{\widehat{\text{Var}}(R_i)}.$$

The least-squares method (Eq. 6.113) assumes additive errors on the original scale. If the errors are multiplicative, the parameter estimates obtained from Eq. (6.113) will be biased. An alternative is the log-transformation (Paloheimo and Fraser 1981) of Eq. (6.112):

$$\ln(R_i) = \ln(R_o) + (c_m - c_f)\sum_{j=0}^{i-1} g_j + \ln\left(\frac{\left(1 - e^{-c_f g_i}\right)}{\left(1 - e^{-c_m g_i}\right)}\right).$$

Weighted nonlinear least squares would be used to obtain estimates of c_f and c_m by minimizing the expression:

$$SS = \frac{\sum_{i=1}^{n} w_i\left[\ln(R_i) - \ln(R_o) - (c_m - c_f)\sum_{j=0}^{i-1} g_j - \ln\left(\frac{\left(1 - e^{-c_f g_i}\right)}{\left(1 - e^{-c_m g_i}\right)}\right)\right]^2}{\sum_{i=1}^{n} w_i}, \quad (6.114)$$

where

$$w_i = \frac{1}{\text{Var}(\ln R_i)} \approx \frac{1}{\text{CV}(R_i)^2}.$$

As with all estimates obtained from nonlinear least squares (or WNLLS), estimates of $\hat{c}_f$ and $\hat{c}_m$ are not guaranteed to be unbiased with minimum variance but, rather, asymptotically unbiased with almost minimum variance (Neter et al. 1996:549). However, Eq. (6.114) will be more robust than will Eq. (6.113) to a multiplicative error structure. Paloheimo and Frasier (1981) used WNLLS based on Eq. (6.114) to obtain estimates of $\hat{c}_f$ and $\hat{c}_m$.

Alternatively, the cohort harvest data can be modeled as a multinomial likelihood, and the parameters c_f and c_m can be estimated by using maximum likelihood methods. The likelihood model for the harvest data is written as

$$L(c_f, c_m, R_0, S_N | x, m_i, f_i, g_i) \propto \left(\frac{R_0\left(1 - e^{-c_f g_0}\right)}{\Psi}\right)^{f_0} \left(\frac{\left(1 - e^{-c_m g_0}\right)}{\Psi}\right)^{m_0}$$

$$\cdot \left(\frac{R_0 S_N e^{-c_f g_0}\left(1 - e^{-c_f g_1}\right)}{\Psi}\right)^{f_1}$$

$$\cdot \left(\frac{S_N e^{-c_m g_0}\left(1 - e^{-c_m g_1}\right)}{\Psi}\right)^{m_1} \cdots$$

$$\cdot \left(\frac{R_0 S_N^w \cdot e^{-c_f \sum_{j=0}^{i-1} g_j}\left(1 - e^{-c_f g_w}\right)}{\Psi}\right)^{f_w}$$

$$\cdot \left(\frac{S_N^w \cdot e^{-c_m \sum_{j=0}^{i-1} g_j}\left(1 - e^{-c_m g_w}\right)}{\Psi}\right)^{m_w} \quad (6.115)$$

where

$$\Psi = \sum_{i=0}^{w} R_0 S_N^i e^{-c_f \sum_{j=0}^{i-1} g_j} \left(1 - e^{-c_f g_i}\right) + S_N^i e^{-c_m \sum_{j=0}^{i-1} g_j} \left(1 - e^{-c_m g_i}\right);$$

$w = $ the oldest age class represented in the sample.

The likelihood (6.115) has two parameters, and $2(w+1) - 1$ minimally sufficient statistics. Hence, the MLEs and associated variances must be obtained by using iterative methods.

The estimator for female harvest survival in the ith year is

$$\hat{S}_{H-F_i} = e^{-\hat{c}_f g_i}, \tag{6.116}$$

with variance approximated by the delta method as

$$\widehat{Var}\left(\hat{S}_{H-F_i}\right) = \widehat{Var}(\hat{c}_f) g_i^2 e^{-2\hat{c}_f g_i}. \tag{6.117}$$

Male harvest survival and associated variance are estimated analogously.

Assumptions

There are several assumptions associated with the estimators of c_f and c_m using the Paloheimo and Fraser (1981) regression model:

1. Samples provide unbiased estimates of the sex ratios.
2. Constant vulnerability, c_f and c_m, extends across all age classes.
3. Harvest mortality is a function of effort (i.e., g_i) only.
4. Annual harvest effort is known and quantifiable.
5. Natural survival is equal for males and females.
6. The population is closed to differential emigration or immigration by the sexes.
7. The estimates $\ln(\hat{R}_i)$ are independently, normally distributed if using Eq. (6.114).

Paloheimo and Fraser (1981) assumed an initial sex ratio of one (i.e., $R_0 = 1$) but examined the bias in c_f and c_m resulting from violation of this assumption. As expected, if the initial sex ratio, R_0, was skewed toward males, then the catch rate of males, c_m, was positively biased and that of females, c_f, was underestimated. Similarly, if R_0 was skewed toward females, c_f was overestimated and c_m, underestimated. Harris and Metzgar (1987) also found the initial sex ratio had to be known or estimable to obtain unbiased estimates of the catch rate. However, assumptions about the value of R_0 are unnecessary because the parameter can be estimated by using least-squares methods. Estimates of harvest survival based on Eq. (6.115) only require unbiased estimates of the sex ratio of the harvest. Whether to assume $R_0 = 1$ or not is a tradeoff between precision and accuracy. If R_0 is indeed one, then using that information in Eqs. (6.114) or (6.115) will improve the precision of the estimates of c. However, if that assumption is wrong, the parameter estimates will be biased. Similarly, for likelihood model (6.115), if auxiliary information can be used to estimate natural survival (S_N), precision of the vulnerability coefficients will be improved.

The assumption of constant vulnerability implies that changes in harvest probabilities are a result of effort alone. The assumption can be violated because of changes in harvest regulations and bag limits or the interaction of hunters with weather conditions during harvest. The vulnerability coefficient can be expected to differ between, say, bow and other "primitive" weapon hunters and firearms users. Technological changes in bow and black-powder hunting over time may also change the vulnerability coefficient. Subtle and not-so-subtle changes in habitat structures can also affect hunter success. Behavioral differences between juveniles and adults may result in higher vulnerability of the younger age classes. Harris and Metzgar (1987) and Fraser et al. (1982) discussed the higher harvest vulnerability in the younger age classes of subadult black bears. If behavioral differences affect harvest probabilities, the assumption that harvest mortality is a function of effort alone is clearly invalid. However, it is unclear whether this violation would induce a negative or positive bias, as the estimates are averages across all years. If natural mortality is unequal between males and females, the method of Paloheimo and Fraser (1981) is inappropriate because differences in the harvest sex ratios are not wholly attributable to differences in vulnerability.

Because different management areas are often subject to different hunting pressures, analyzing sex ratios on a local level may be preferable. However, hunting effort in these separate regions must be known. Investigators should use caution in defining locales because movement by animals between areas can violate the assumption of a closed population.

The method of estimating sex-specific vulnerability can also be used during a single harvest season, where the season is divided into discrete and equal periods, and samples of the harvest are collected to estimate successive sex ratios. The hunting effort for each period would also need to be measured. Shupe et al. (1990) and Roseberry and Klimstra (1992) analyzed differential vulnerability in age classes by regressing the percentage of juveniles in the kill for successive periods in a harvest season against time. However, Roseberry and Klimstra (1992) used cumulative percentages in the regression analysis, an unfortunate practice that violates the assumption of independent observations.

Example 6.9: Estimating Harvest Mortality from Changing Sex Ratios for Moose (*Alces alces*), Thunder Bay District, Ontario, Canada

Male and female vulnerability coefficients, c_m and c_f, were estimated by using both WNLLS of the log-transformed expression (Eq. 6.114) and the multinomial likelihood model of Eq. (6.115) from the moose cohort data of Paloheimo and Fraser (1981) (Table 6.4). The initial cohort sex ratio, R_0, and natural survival, S_N, were estimated from the multinomial likelihood only. For Eq. (6.114), R_0 was set to 1, per Paloheimo and Fraser (1981).

The MLE model estimated $R_0 = 1.3271$ ($\widehat{SE} = 0.6185$) and $S_N = 1$ ($\widehat{SE} = 0.00009$). Figure 6.6 provides a graphical comparison of the predicted harvest sex ratios from Eqs. (6.114) and (6.115) with the observed data. The fitted curves are almost but not quite parallel (Fig. 6.6). The difference in the lines for the two methods is attributable to the estimate of $R_0 \neq 1$ and higher values of the vulnerability coefficients for WNLLS (Table 6.5). Paloheimo and Fraser (1981) mentioned that any bias in R_0 will be almost equal to the bias in the vulnerability coefficients.

The average hunting effort was 6.24 hunters/year ($\times 1000$) from 1964 through 1972. By using the vulnerability coefficients from the multinomial MLE, average harvest mortality for cow moose is estimated to be

$$\hat{M}_{H-F} = 1 - e^{-0.0086(6.24)} = 0.0522$$

Table 6.4. Cohort data for moose born in 1963 from the Thunder Bay District, Ontario Canada.

Year	Age	Age class	Males	Females	Ratio (R_i)	Effort (g_i)	Weights
1964	1.5	0	23	14	0.6087	4.63	23.4884
1965	2.5	1	31	12	0.3871	5.32	57.7345
1966	3.5	2	39	19	0.4872	6.51	53.8285
1967	4.5	3	43	34	0.7907	7.36	30.3694
1968	5.5	4	26	18	0.6923	5.52	22.1919
1969	6.5	5	21	20	0.9524	7.12	11.2939
1970	7.5	6	15	16	1.0667	5.94	6.8044
1971	8.5	7	12	14	1.1667	5.98	4.7473
1972	9.5	8	17	13	0.7647	7.80	12.5974

Effort is measured as thousands of hunters $\left(g_i = \dfrac{\text{Hunters}}{1000} \right)$. (Data from Paloheimo and Fraser 1981.)

Table 6.5. Estimates and associated standard errors of $\hat{R}_0$, $\hat{c}_f$, and $\hat{c}_m$ from the WNLLS method and maximum likelihood model of the Thunder Bay moose data.

Method	$\hat{R}_0$	$\widehat{SE}(\hat{R}_0)$	$\hat{c}_f$	$\widehat{SE}(\hat{c}_f)$	$\hat{c}_m$	$\widehat{SE}(\hat{c}_m)$
WNLLS	—	—	0.0129	0.0034	0.0285	0.0003
MLE	1.3271	0.6185	0.0086	0.0049	0.0228	0.0043

(Data from Paloheimo and Fraser 1981.)

with an estimated variance (Eq. 6.117) of

$$\widehat{\text{Var}}\left(\hat{S}_{H-F} \right) = \widehat{\text{Var}}\left(\hat{M}_{H-F} \right)$$

$$= \widehat{\text{Var}}(\hat{c}_f) g^2 e^{-2\hat{c}_f g}$$

$$\widehat{\text{Var}}\left(\hat{S}_{H-F} \right) = (0.0049)^2 (6.24)^2 e^{-2(0.0086)(6.24)} = 0.0008397$$

or $\widehat{SE}(\hat{M}_{H-F}) = 0.0290$. For bull moose, the average harvest mortality rate is estimated as

$$\hat{M}_{H-M} = 1 - e^{-0.0228(6.24)} = 0.1326$$

with associated standard error of $\widehat{SE}(\hat{M}_{H-M}) = 0.0233$.

6.8.2 Age-at-Harvest Data: Paloheimo and Fraser (1981), Harris and Metzgar (1987)

The method of Paloheimo and Fraser (1981) requires successive harvest sex ratios from a cohort over time to estimate sex-specific vulnerabilities. With the added assumption of a stable and stationary population, Paloheimo and Fraser (1981) and Harris and Metzgar (1987) developed a method to estimate sex-specific vulnerabilities from age- and sex-structures of a single harvest sample. The number of females (f_i) at age i in the harvest was assumed to be Poisson-distributed with the probability density function (pdf) being

$$P(C_{F_i} = f_i) = \frac{(\lambda_{F_i})^{f_i} e^{-\lambda_{F_i}}}{f_i!}$$

and number of males (m_i), similarly with the pdf,

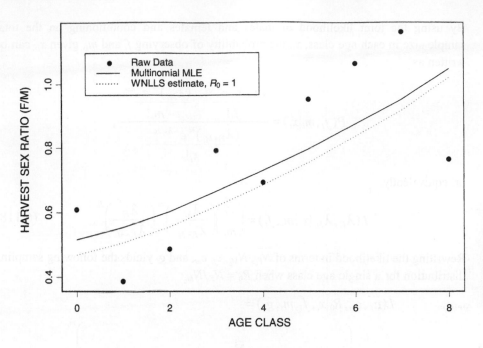

Figure 6.6. Observed moose sex ratios and the predicted curves reported by Paloheimo and Frasier (1981) using weighted nonlinear least squares (WNLLS) (Eq. 6.114), and multinomial likelihood (Eq. 6.115).

$$P(C_{M_i} = m_i) = \frac{(\lambda_{M_i})^{m_i} e^{-\lambda_{M_i}}}{m_i!}$$

where

$$\lambda_{F_i} = N_{F_0} \left(\prod_{j=0}^{i-1} S_{N_i} \right) e^{-c_f \sum_{j=0}^{i-1} g_j} (1 - e^{-c_f g_i});$$

$$\lambda_{M_i} = N_{M_0} \left(\prod_{j=0}^{i-1} S_{N_i} \right) e^{-c_m \sum_{j=0}^{i-1} g_j} (1 - e^{-c_m g_i});$$

N_{F_0} = number of females entering the population at time 0;
N_{M_0} = number of males entering the population at time 0;
S_{N_i} = natural survival for age class i;
C_{Fi} = random variable for the number of females of age class i harvested;
C_{Mi} = random variable for the number of males of age class i harvested;
f_i = number of females of age class i harvested;
m_i = number of males of age class i harvested;
$x_i = m_i + f_i$.

Noting the sum of independent Poisson random variables is also Poisson-distributed, the total number of individuals in each age class has the distribution

$$P(C_{F_i} + C_{M_i} = x_i) = \frac{(\lambda_{F_i + M_i})^{x_i} e^{-\lambda_{F_i + M_i}}}{x_i!},$$

where

$$\lambda_{F_i + M_i} = \left(\prod_{i=0}^{i-1} S_{N_i} \right) \left(N_{F_0} e^{-c_f \sum_{j=0}^{i-1} g_j} (1 - e^{-c_f g_i}) + N_{M_0} e^{-c_m \sum_{j=0}^{i-1} g_j} (1 - e^{-c_m g_i}) \right).$$

By using the joint likelihood of males and females and conditioning on the total sample size in each age class, x_i, the probability of observing f_i and m_i, given x_i, can be written as

$$P(f_i, m_i | x_i) = \frac{\dfrac{(\lambda_{F_i})^{f_i} e^{-\lambda_{F_i}}}{f_i!} \cdot \dfrac{(\lambda_{M_i})^{m_i} e^{-\lambda_{M_i}}}{m_i!}}{\dfrac{(\lambda_{F_i+M_i})^{x_i} e^{-\lambda_{F_i+M_i}}}{x_i!}},$$

or, equivalently,

$$L(\lambda_{F_i}, \lambda_{M_i} | x_i, m_i, f_i) = \binom{x_i}{m_i} \left(\frac{\lambda_{M_i}}{\lambda_{F_i+M_i}}\right)^{m_i} \left(\frac{\lambda_{F_i}}{\lambda_{F_i+M_i}}\right)^{f_i}. \qquad (6.118)$$

Rewriting the likelihood in terms of N_{F_0}, N_{M_0}, c_f, c_m, and g_i yields the following sampling distribution for a single age class when $R_0 = N_{F_0}/N_{M_0}$:

$$L(c_f, c_m, R_0 | x_i, f_i, m_i, g_i) =$$

$$\binom{x_i}{m_i} \left(\frac{e^{-c_m \sum_{j=0}^{i-1} g_j} \left(1 - e^{-c_m g_i}\right)}{\left(R_0 e^{-c_f \sum_{j=0}^{i-1} g_j} \left(1 - e^{-c_f g_i}\right) + e^{-c_m-1 \sum_{j=0}^{i-1} g_j} \left(1 - e^{-c_m g_i}\right) \right)} \right)^{m_i} \qquad (6.119)$$

$$\cdot \left(\frac{R_0 e^{-c_f \sum_{j=0}^{i-1} g_j} \left(1 - e^{-c_f g_i}\right)}{R_0 e^{-c_f \sum_{j=0}^{i-1} g_j} \left(1 - e^{-c_f g_i}\right) + e^{-c_m \sum_{j=0}^{i-1} g_j} \left(1 - e^{-c_m g_i}\right)} \right)^{f_i}.$$

Assuming all animals are harvested independently, the joint likelihood for all age classes harvested can be written as

$$L(c_f, c_m, R_0 | \underset{\sim}{x}, \underset{\sim}{f}, \underset{\sim}{m}, \underset{\sim}{g}) = \prod_{i=1}^{n} L(c_f, c_m, R_0 | x_i, f_i, m_i, g_i). \qquad (6.120)$$

The MLEs and associated variances of c_f, c_m, and R_0 can be calculated by using numerical methods. Paloheimo and Frasier (1981) used the regression Eq. (6.114) for both the cohort data and the analysis of a vertical sample of age-at-harvest data.

Assumptions

The six assumptions for models (6.119) and (6.120) are similar to those used in Section 6.8.1 with the addition of the population being stable and stationary. The model assumptions include the following:

1. The population is stable and stationary.
2. The sample provides an unbiased estimate of the age- and sex-structure of the harvest data.
3. Harvest mortality is a function of effort only.

Table 6.6. Age-structure data and estimated hunting effort for a moose harvest in Thunder Bay, Ontario, 1972.

Age (years)	Age class	Year of entry	Males	Females	Sex ratio $(R_{F/M})$	Hunting effort in year of entry (g_i)	Cumulative effort $\sum_{j=0}^{i-1} g_j$
1.5	0	1972	131	107	0.8168	7.80	0
2.5	1	1971	104	65	0.6250	5.98	7.80
3.5	2	1970	63	40	0.6349	5.94	13.78
4.5	3	1969	25	20	0.8000	7.12	19.72
5.5	4	1968	20	14	0.7000	5.52	26.84
6.5	5	1967	18	20	1.1111	7.36	32.36
7.5	6	1966	15	12	0.8000	6.51	39.72
8.5	7	1965	10	11	1.1000	5.32	46.23
9.5	8	1964	17	13	0.7647	4.63	51.55
10.5	9	1963	9	17	1.8889	4.73	56.18
11.5	10	1962	6	5	0.8333	4.70	60.91
12.5	11	1961	3	6	2.0000	3.76	65.61
13.5	12	1960	4	3	0.7500	3.23	69.37
14.5	13	1959	2	2	1.0000	2.40	72.60

(Data from Paloheimo and Fraser 1981.)

4. Harvest effort is known over time.
5. Natural survival is equal for males and females and across years.
6. The population is closed to immigration and emigration.

By using simulated data, Harris and Metzgar (1987) demonstrated that for a population with 21 age classes, 10 to 15 years of constant age class survivals and recruitment were needed before the age structure sufficiently stabilized so that estimated harvest probabilities were reliable.

Example 6.10: Time-Specific Analysis of Sex Ratios From Moose, Thunder Bay District, Ontario

Paloheimo and Fraser (1981) presented age-structure data on numbers of male and female moose harvested during the 1972 season in the Thunder Bay District, Ontario (Table 6.6). To illustrate the use of age-structure data to estimate vulnerability, maximum likelihood estimates were obtained for c_f, c_m, and R_0 by using Eq. (6.120). The parameter estimates and associated standard errors are:

$$\hat{c}_f = 0.0008 \quad \widehat{SE}(\hat{c}_f) = 0.0075,$$

$$\hat{c}_m = 0.0042 \quad \widehat{SE}(\hat{c}_m) = 0.0078,$$

$$\hat{R}_0 = 3.7294 \quad \widehat{SE}(\hat{R}_0) = 29.8673.$$

These estimates are different from the $\hat{c}_f = 0.0158$ and $\hat{c}_m = 0.0232$ obtained by Paloheimo and Fraser (1981) using Eq. (6.114), under the assumption $R_0 = 1$. Maximum likelihood estimates and standard errors were obtained numerically from Eq. (6.120). Under the assumption that $R_0 = 1$, the MLEs are recalculated to be:

$$\hat{c}_f = 0.0074, \quad \widehat{SE}(\hat{c}_f) = 0.0028,$$

$$\hat{c}_m = 0.0108, \quad \widehat{SE}(\hat{c}_m) = 0.0045.$$

The differential results (Fig. 6.7) between the MLE and WNLLS are due to more than simply whether or not R_0 is estimated. The results of Paloheimo and Fraser (1981) are a consequence of giving more weight to the sex ratios of the older age classes. However, most of the older age classes have small sample sizes with correspondingly larger variance estimates for the sex ratio. Regression, with weights inversely proportional to the variance, should have given these points less influence. However, the variance of $\hat{R}$ is also a function of $\hat{R}$ itself. Weighted regression has difficulties when weights are correlated with the responses being analyzed. Likelihood equations seem to give more weight to the ratios observed in the younger age classes with correspondingly higher sample sizes.

Average hunter effort during 1959 to 1972 was 5.36 hunters/per year ($\times 1000$). By using the MLEs for the vulnerability coefficients, harvest mortality for males under average hunting pressure is estimated to be

$$M_{H-M} = 1 - e^{-0.0042(5.36)} = 0.0223,$$

and, for females,

$$M_{H-F} = 1 - e^{-0.0008(5.36)} = 0.0043.$$

The Paloheimo and Fraser (1981) coefficients translate to harvest mortalities under average hunting pressure of $M_{H-M} = 0.1169$ and $M_{H-F} = 0.0812$. The statistical model used obviously has a tremendous effect on the interpretation of the sex ratios and subsequent harvest mortality rates. The lack of robustness to the numerical approach raises concerns about making any strong inferences regarding harvest mortality in this particular example.

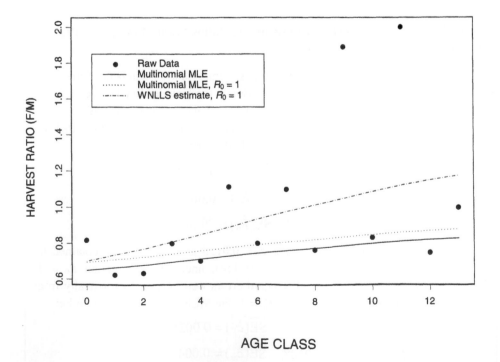

Figure 6.7. Sex ratio and fitted models obtained using weighted nonlinear least squares (WNLLS) (Eq. 6.114) and maximum likelihood estimators (MLEs) (Eq. 6.120), when R_0 is estimated and when setting $R_0 = 1$ to the moose harvest data in Paloheimo and Frasier (1981).

6.9   Proportion of Mortality Owing to Harvest (Gulland 1955)

For some population reconstruction methods based on harvest, an estimate of the proportion of total annual mortality that results from harvest must be available. Expressing survival in terms of instantaneous natural mortality (μ_N) and instantaneous harvest (μ_H) rates,

$$S(t) = e^{-(\mu_N + \mu_H)t},$$

the proportion of the mortality owing to harvest is

$$p_H = \frac{\mu_H}{\mu_N + \mu_H}. \tag{6.121}$$

The expression for p_H can also be rewritten in terms of the ratio of the instantaneous mortality rates as

$$p_H = \frac{\dfrac{\mu_H}{\mu_N}}{1 + \dfrac{\mu_H}{\mu_N}}. \tag{6.122}$$

Gulland (1955), Chapman (1961), and Paulik (1963) present a likelihood model for a single release of marked animals followed by continuous harvest over time, with four assumptions:

1. The population is closed to immigration or emigration.
2. Instantaneous harvest and natural mortality rates, μ_H and μ_N, respectively, are constant over time.
3. Loss of marks is nonexistent, and all marked animals are reported.
4. Sampling is continuous until the marked cohort is extinct.

We define

M_0 = the number of marked individuals released at the beginning of the study;
T_i = the time of death for the ith individual harvested ($i = 1, \ldots, M_0$);
n = the total number of marked animals harvested.

The number of marked animals recovered in the harvest is assumed to have a binomial distribution

$$f(n) = \binom{M_0}{n}\left(\frac{\mu_H}{\mu_N + \mu_H}\right)^n \left(\frac{\mu_N}{\mu_N + \mu_H}\right)^{M_0 - n}.$$

For an animal with an instantaneous mortality rate of $\mu = \mu_N + \mu_H$, the probability it is alive at time T_i is

$$P(T \geq T_i) = e^{-(\mu_N + \mu_H)T_i}.$$

The death times are then exponentially distributed with pdf of

$$f(T_i) = (\mu_N + \mu_H)e^{-(\mu_N + \mu_H)T_i}.$$

The joint likelihood for the number of marked animals observed in the harvest and their recovery times is written as

$$L = f(n) \cdot \prod_{i=1}^{n} f(T_i)$$

or

$$L(\mu_H, \mu | M_0, n, T_i) = \binom{M_0}{n} \left(\frac{\mu_H}{\mu_N + \mu_H} \right)^{M_0 - n} (\mu_H)^n e^{-(\mu_N + \mu_H) \sum_{i=1}^{n} T_i}. \qquad (6.123)$$

Gulland (1955) derived MLEs for μ_N and μ_H, as well as their ratio,

$$\frac{\hat{\mu}_H}{\hat{\mu}_N} = \frac{n}{M_0 - n},$$

and their proportion,

$$\hat{p}_H = \frac{\hat{\mu}_H}{\hat{\mu}_N + \hat{\mu}_H} = \frac{n}{M_0}. \qquad (6.124)$$

Hence, for any well-defined cohort (M_0) of marked or unmarked animals, p_H is estimated by the fraction of animals that is ultimately harvested. Gulland (1955) derived the estimator conceptualizing a cohort of marked individuals. However, estimator (6.124) applies to any well-defined group of individuals. For known M_0, the variance of $\hat{p}_H$ can be estimated by

$$\widehat{\mathrm{Var}}(\hat{p}_H) = \frac{\hat{p}_H (1 - \hat{p}_H)}{M_0}. \qquad (6.125)$$

In this case, where M_0 must be estimated, the estimator of p_H can be expressed as

$$\hat{p}_H = \frac{n}{\hat{M}_0}. \qquad (6.126)$$

and the variance of (6.126) can be found in stages and approximated by the expression

$$\widehat{\mathrm{Var}}(\hat{p}_H) = \frac{\hat{p}_H (1 - \hat{p}_H)}{\hat{M}_0} + \left(\frac{\hat{p}_H (1 - \hat{p}_H)}{\hat{M}_0} - \frac{\hat{p}_H^2}{\hat{M}_0^2} \right) \frac{\widehat{\mathrm{Var}}(\hat{M}_0)}{\hat{M}_0^4}. \qquad (6.127)$$

It should be noted the estimator of $\mu = \mu_N + \mu_H$, based on likelihood (6.123), does not depend on knowing M_0, where

$$\hat{\mu} = \hat{\mu}_N + \hat{\mu}_H = \frac{n-1}{\sum_{i=1}^{n} T_i} \qquad (6.128)$$

is the minimum variance unbiased estimator (Chapman 1961:156). The variance of $\hat{\mu}$ is

$$\mathrm{Var}(\hat{\mu}) = \frac{\mu^2}{n-2}.$$

All that is required for the use of Eq. (6.128) is knowledge the individuals harvested belong to a discrete, well-defined cohort.

Discussion of Utility

For small values of μ, survival can be approximated by the expression

$$S = e^{-\mu} \approx \frac{1}{1+\mu}.$$

Substituting in the Gulland (1955) estimator for μ (Eq. 6.128) into this approximation yields the Chapman-Robson (1960) catch-curve estimator of survival (Eq. 5.112)

$$\hat{S} = \frac{X}{n+X-1},$$

where

$$X = \sum_{i=1}^{n} T_i.$$

The result should not be surprising, given the assumptions of the Gulland (1955) model and the catch-curve analysis are essentially the same.

The Gulland (1955) likelihood is useful for species for which harvest seasons are long and few individuals are harvested, permitting death times to be recorded. Predators such as cougar (*Puma concolor*), bobcat (*Lynx rufus*), fox (*Vulpes* spp.), and coyote (*Canis latrans*) may fall into this category. A low level of constant harvest pressure is often applied to manage predator levels. Age-at-harvest information would have to be collected to place individuals within selected cohorts when estimating μ.

The assumption of constant harvest mortality may be violated if management goals are to maintain constant animal abundance. In which case, hunting regulations and harvest effort may vary directly as both predator and game population increase or decrease owing to fluctuations in recruitment. Natural mortality may vary owing to both extrinsic and intrinsic factors. Therefore, the method of Gulland (1955) should be used cautiously.

The use of Eq. (6.124) or (6.126) to estimate p_H requires either knowledge or an estimate of the initial size of the cohort (M_0) to be monitored over time. Knowing M_0, therefore, is the major limitation in the use of estimators (6.124) and (6.126). Considerable effort would be needed to identify and enumerate the initial abundance of a cohort. For example, a cohort could be defined as 3.5-year-old elk in a population during 2004. Age-at-harvest information would then be used to enumerate harvested numbers from that cohort in subsequent years (n). But the size of the cohort in 2004 must also be derived. Hence, the utility of the method is limited. Perhaps the greatest value of Eq. (6.124) is the realization that when using age-structure data, p_H cannot be gleaned from a single harvest event but only through monitoring the fate of an entire cohort to its ultimate demise.

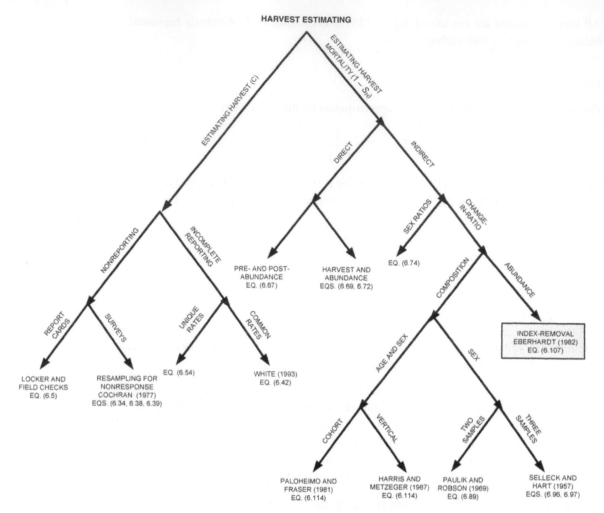

Figure 6.8. Summary of various approaches to harvest and harvest mortality estimation. Methods with sample size calculations highlighted in gray.

6.10 Summary

Two different but related tasks (Fig. 6.8) of estimating harvest numbers and harvest mortality were considered in this chapter. Harvest numbers (C) can be considered absolute take, whereas harvest mortality ($M_H = 1 - S_H$) is the relative take of animals within a population, related through equation

$$\frac{C}{N} = 1 - S_H \quad \text{or} \quad \frac{C}{N} = M_H,$$

where N is population abundance. Knowledge of both the harvest mortality probability and number harvested, in turn, permits abundance estimation, where

$$\hat{N} = \frac{\hat{C}}{1 - \hat{S}_H} = \frac{\hat{C}}{\hat{M}_H}.$$

Figure 6.8 summarizes the various approaches to harvest and harvest mortality estimation presented in this chapter.

Estimates of harvest numbers should consider and adjust for incomplete reporting (Section 6.2.1), survey nonresponse (Section 6.2.2), and incomplete survey information (Section 6.3) from hunters. Approaches to working with incomplete hunter response have been illustrated, but the approaches presented are not exhaustive. Field checks, hunter surveys, and questionnaire designs should be integrated into a coordinated approach to harvest estimation. This coordination should center around the harvest parameters desired and the estimators and variances of those required quantities. The nature of the information (Sections 6.3.1, 6.3.2) acquired from postseason hunter questionnaires has a major impact on the assumptions and realism of subsequent harvest parameters. For realistic estimates of hunter success in separate game-management units, information from both successful and nonsuccessful hunters by area of harvest is required. Far too often, hunter questionnaires and surveys are implemented before estimators of total harvest have been thoroughly derived. The importance of accurate harvest information mandates a change in this strategy.

In the absence of animal marking data, index-removal and change-in-ratio methods are the predominant tools available to field biologists to estimate harvest mortality probabilities. These methods generally require the detection process to be stationary before, during, and after the period of harvest. Unfortunately, the data usually collected by these techniques are insufficient alone to assess the validity of the assumptions of a stationary detection process. Auxiliary information, such as detection distances need to be recorded in order to evaluate the detection function (Buckland et al. 2001).

The paucity of techniques readily available to field biologists to estimate harvest mortality has led many wildlife agencies to conduct auxiliary marking studies focused on survival estimation and to partition sources of mortality. Well-conducted radiotelemetry studies can be used to estimate mortality probabilities to augment other life-history information such as age-at-harvest, sex ratio, and harvest numbers. Together, marking and nonmarking data can be used jointly to improve the assessment of population status and trends.

Estimating the Rate of Population Change

7

Chapter Outline

7.1 Introduction

An unchangeable and basic fact of wildlife demography is that current population size is a function of past abundance and the demographic processes of survival, productivity, immigration, and emigration in the present (Begon et al. 1990). Changes in productivity, immigration and emigration rates, and survival probabilities, or the interaction between these processes, will change population growth. The finite rate of change (λ) or the instantaneous rate of change (r) characterizes the relative change in animal numbers over time. This rate of change integrates the combined effects of additions and deletions in animal numbers into a single resultant effect on abundance over time.

In practice, there are two similar sounding but very different concepts to consider. The intrinsic rate of population change characterizes population growth rate under the most favorable conditions for maximal growth. This is the rate of change of a population during exponential growth in an unlimited environment. Chapman (1928:114) used the term "biotic potential" to designate this maximum reproductive potential and considered it to be an inherent property of an animal to reproduce, survive, and increase in abundance. The biotic potential or intrinsic rate can be expressed as either an annual rate (λ) or an instantaneous rate (r). The other concept is the realized or observed rate of population change, which characterizes the actual rate of change occurring in the population under the prevailing environmental and demographic conditions.

The realized rate of change is not a static value but, rather, depends on the actual trend in abundance from year to year. The realized rate of population change can be expressed as either a finite rate (λ) or an instantaneous rate (r). We use λ_{MAX} and r_{MAX} to distinguish the intrinsic rates of change from those of the realized rates of change (i.e., λ_{REAL} and r_{REAL}). For some applications, it is also convenient to express the rate of change as $R_{REAL} = \lambda_{REAL} - 1$ or $R_{MAX} = \lambda_{MAX} - 1$. The value R expresses the fractional increase or decrease in abundance from one year to the next, where

$$R = \frac{N_{t+1} - N_t}{N_t}$$
$$= \lambda - 1.$$

The distinction between intrinsic and realized rates of change has important management implications. A population with an intrinsic rate of $\lambda_{MAX} > 1$ and a realized rate of $\lambda_{REAL} = 1.0$ suggests a healthy population at equilibrium with its environment. A population with an intrinsic rate of $\lambda_{MAX} > 1$ and a realized rate of $\lambda_{REAL} > 1$ suggests a healthy population growing and not fully bounded by its environment. A population with an intrinsic rate of $\lambda_{MAX} > 1$ but a realized rate of $\lambda_{REAL} \leq 1$ indicates a viable population hindered by environmental (e.g., resource limitations) or anthropogenic (e.g., harvest) effects, with the potential to rebound should conditions once again become favorable. Understanding both the population's biotic potential and current abundance trends is essential to assessing the long-term viability of the population. Figure 7.1 illustrates abundance trends for the same population under exponential and logistic growth. As the population approaches its carrying capacity (Fig. 7.1), the realized rate of change asymptotes to a value of $\lambda_{REAL} = 1$, whereas the intrinsic rate remains at $\lambda_{MAX} = 1.2$.

Different statistical approaches may need to be used to estimate intrinsic and realized rates of population change. Some methods are better suited for estimating one parameter or the other. Similarly, some analytical approaches are better for estimating the annual or finite rate of population change (λ), other methods are better suited for estimating the instantaneous rate of change (r). A method that provides an unbiased estimator of r will *not* provide an unbiased estimator of λ, and vice versa. Therefore, a strategic choice of estimation techniques depends on both available data and the parameter of interest.

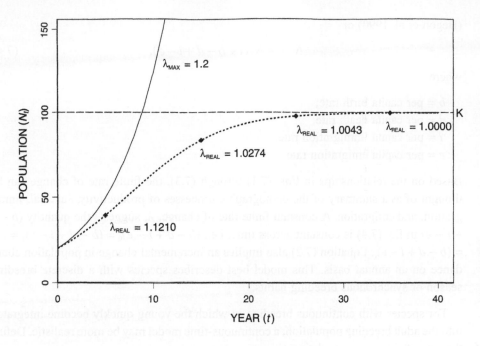

Figure 7.1. Schematic of a population under conditions of exponential (solid line) and logistic growth (dotted line). The plots illustrate an intrinsic annual rate of change of $\lambda_{MAX} = 1.2$, $K = 100$, $N_0 = 25$, and different realized annual rates of change over the course of time. Realized annual rates of change (λ_{REAL}) under logistic growth were calculated at $t = 5$, 15, 25, and 35 years.

7.2   Basic Concepts and Definitions

A population at any time t can be expressed in terms of successive ratios of abundance from time 0 to t as

$$N_t = N_0 \cdot \left(\frac{N_1}{N_0} \cdot \frac{N_2}{N_1} \cdot \frac{N_3}{N_2} \cdot \ldots \cdot \frac{N_t}{N_{t-1}} \right). \tag{7.1}$$

Assuming the finite rate of change, λ, is constant over time implies the successive abundance ratios

$$\frac{N_1}{N_0} = \frac{N_2}{N_1} = \frac{N_3}{N_2} = \ldots = \frac{N_t}{N_{t-1}} = \lambda,$$

or

$$N_t = \lambda N_{t-1}. \tag{7.2}$$

Thus, Eq. (7.1) may be expressed in terms of the population size at time 0 and λ, where

$$N_t = N_0 \lambda^t. \tag{7.3}$$

The finite rate of change takes on positive values only, with increasing populations having a value of $\lambda > 1$; decreasing populations, $\lambda < 1$. The percentage change in a population from time t to $t + 1$ is $(\lambda - 1)100\%$ or $R \cdot 100\%$.

The number of individuals in a population at time $t + 1$, N_{t+1}, is a function of the number of individuals at time t (N_t) and the number of births (B), deaths (D), immigration (I), and emigration (E) between times t and $t + 1$. Mathematically, this can be expressed as

$$N_{t+1} = N_t + B - D + I - E$$

(Begon et al. 1990) or

$$N_{t+1} = N_t(1+b-d+i-e), \tag{7.4}$$

where

b = per capita birth rate;
d = per capita death rate;
i = per capita immigration rate;
e = per capita emigration rate.

Based on the relationships in Eqs. (7.1) through (7.3), the finite rate of change can be thought of as a summary of the demographic processes of productivity, survival, immigration, and emigration. A constant finite rate of change, λ, suggests the quantity $(b - d + i - e)$ in Eq. (7.4) is constant across time, i.e., $(b - d + i - e)_0 = (b - d + i - e)_1 = \ldots = (b - d + i - e)_i$. Equation (7.2) also implies an incremental change in population abundance on an annual basis. This model best describes species with a discrete breeding season or synchronous breeding habits.

For species with continuous breeding in which the young quickly become integrated into the adult breeding population, a continuous-time model may be more realistic. Define the rate of change in population size as

$$r = \frac{N_{t+1} - N_t}{N_t((t+1)-t)},$$

which for one unit of time simplifies to

$$r = \frac{N_{t+1} - N_t}{N_t},$$

implying

$$N_{t+1} = N_t + N_t r$$
$$= N_t(1+r).$$

In the next time interval,

$$N_{t+2} = N_{t+1} + N_{t+1}r$$
$$= N_t(1+r) + N_t(1+r)r$$
$$= N_t(1+r)^2$$

or, in general,

$$N_t = N_0(1+r)^t. \tag{7.5}$$

Now, rather than having population recruitment once a year, allow it to occur twice annually with "compounding interest," then

$$N_t = N_0\left(1 + \frac{r}{2}\right)^{2t}.$$

The compounding interest refers to allowing juveniles of the previous breeding season to contribute to the adult breeding population of the next season. Furthermore, letting this compounding occur k times a year yields the general expression

$$N_{t+1} = N_0 \left(1 + \frac{r}{k}\right)^{kt}.$$

For continuously breeding populations, we let the compounding occur ever so frequently (i.e., $k \to \infty$), in which case,

$$\lim_{k \to \infty} \left(1 + \frac{r}{k}\right)^k = e^r.$$

Thus, for continuously breeding populations,

$$N_t = N_0 (e^r)^t = N_0 e^{rt}. \tag{7.6}$$

Equation (7.6) can also be derived from the differential equation

$$\frac{dN_t}{dt} = rN_t,$$

which assumes the rate of population change depends on the current population size (N_t) and a constant rate of growth (r), which is independent of population size. Equating Eqs. (7.3) and (7.6), one finds that

$$\lambda = e^r \tag{7.7}$$

or, conversely,

$$r = \ln \lambda. \tag{7.8}$$

Both Eqs. (7.3) and (7.6) describe a population with exponential growth where λ is the annual rate of increase and r is the continuous rate of increase. For $|r| < 0.2$ or 0.3, the expression $r \doteq \lambda - 1$ is a reasonable approximation (Eberhardt and Simmons 1992).

The nonlinear relationship between λ and r (Eq. 7.8) has implications for parameter estimation. For nonlinear functions, the expected value of a function of a random variable is not equal to the function of the expected value of the random variable. Thus, if $\hat{\lambda}$ is an unbiased estimator of λ (i.e., $E(\hat{\lambda}) = \lambda$),

$$E(\ln \hat{\lambda}) \neq r,$$

and $\ln \hat{\lambda}$ will be a biased estimator of r. To the first three terms of Taylor series expansion,

$$E(\ln \hat{\lambda}) \doteq \ln \lambda - \frac{\mathrm{Var}(\hat{\lambda})}{2\lambda^2}$$

$$\doteq r - \frac{\mathrm{CV}(\hat{\lambda})^2}{2}. \tag{7.9}$$

Similarly, if $\hat{r}$ is an unbiased estimator of r (i.e., $E(\hat{r}) = r$), then

$$E(e^{\hat{r}}) \neq \lambda,$$

and, to the first three terms of a Taylor series,

$$E(e^{\hat{r}}) \doteq e^r + \frac{e^r \text{Var}(\hat{r})}{2}$$
$$\doteq \lambda \left(1 + \frac{\text{Var}(\hat{r})}{2}\right). \tag{7.10}$$

Because of the potential biases induced by transforming $\hat{\lambda}$ to $\hat{r}$ and vice versa, it is preferable to select an estimation method directly suited for estimating the parameter of choice. In the case of both Eqs. (7.9) and (7.10), the magnitude of the bias is a function of the size of the sampling error of $\hat{\lambda}$ or $\hat{r}$, respectively. In typical field investigations, these sampling errors can be reasonably large with serious potential for bias.

7.3   Two-Sample Methods for Estimating r and λ

The finite rate of change (Eq. 7.2) can be estimated by the ratio

$$\hat{\lambda}_t = \frac{\hat{N}_{t+1}}{\hat{N}_t}, \tag{7.11}$$

where

$\hat{\lambda}_t$ = estimated finite rate of population change from time t to $t + 1$;
$\hat{N}_t$ = estimate of population abundance at time t.

The estimator $\hat{\lambda}_t$ in Eq. (7.11) is unbiased only to the first term of a Taylor series, provided the abundance estimates are unbiased. To the third term of a Taylor series,

$$E(\hat{\lambda}_t) = \lambda_t \left(1 + \text{CV}(\hat{N}_t)^2\right).$$

The variance of $\hat{\lambda}_t$ can be approximated by the delta method, where

$$\text{Var}(\hat{\lambda}_t) \doteq \lambda_t^2 \left[\frac{\text{Var}(\hat{N}_t)}{N_t^2} + \frac{\text{Var}(\hat{N}_{t+1})}{N_{t+1}^2} - \frac{2\text{Cov}(\hat{N}_t, \hat{N}_{t+1})}{N_t N_{t+1}}\right]. \tag{7.12}$$

Should the successive annual estimates of abundance be independent, then

$$\text{Var}(\hat{\lambda}_t) \doteq \lambda_t^2 \left[\text{CV}(\hat{N}_t)^2 + \text{CV}(\hat{N}_{t+1})^2\right]. \tag{7.13}$$

Often, estimates of abundance are difficult to obtain. If indices, I_t, are used to monitor a population such that

$$I_t = \beta N_t \quad \text{for} \quad t = 1, 2$$

then λ can also be estimated by the ratio

$$\hat{\lambda}_t = \frac{I_{t+1}}{I_t} \tag{7.14}$$

with associated variance estimator

$$\text{Var}(\hat{\lambda}_t) \doteq \lambda_t^2 \left[\text{CV}(I_t)^2 + \text{CV}(I_{t+1})^2\right], \tag{7.15}$$

assuming independence between I_t and I_{t+1}. To use Eq. (7.15), it is assumed the field surveys were conducted in such a manner that a measure of sampling error associated with the annual indices was available. Typically, this implies the index surveys were based on probabilistic principles of sample survey design and not on opportunistic data collection. The use of indices in wildlife studies is discussed by Overton (1969), Eberhardt (1978a), and in Chapter 8. Functional relationships between indices and absolute abundances, as well as the effect of variability in wildlife studies, are discussed by Eberhardt (1978a).

Eberhardt and Simmons (1992) mention estimating the finite rate of change from beginning and end counts in a sequence of abundance estimates. McCullough (1982, 1983) used abundance estimates in 1975 and 1981 as beginning and end counts to calculate the instantaneous rate of increase for the George Reserve deer herd. The estimator of r used by McCullough (1982) is based on the exponential-growth model $\hat{N}_t = \hat{N}_0 e^{rt}$, and is written as

$$\hat{r} = \frac{\ln(\hat{N}_t) - \ln(\hat{N}_0)}{t}, \qquad (7.16)$$

where

$\hat{N}_0$ = initial abundance estimate for the period of interest;
$\hat{N}_t$ = final abundance estimate for the period;
t = number of years, or time periods, between 0 and t.

Assuming the abundance estimates are independent, then $\mathrm{Cov}(\hat{N}_t, \hat{N}_0) = 0$ and the variance estimator simplifies to

$$\widehat{\mathrm{Var}}(\hat{r}|t) = \frac{1}{t^2}\left[\widehat{\mathrm{CV}}(\hat{N}_t)^2 + \widehat{\mathrm{CV}}(\hat{N}_0)^2\right]. \qquad (7.17)$$

An alternative to Eq. (7.16) is to estimate the finite rate of change, λ, based on the model $\hat{N}_t = \hat{N}_0 \lambda^t$. The estimator for the finite rate of change is written as

$$\hat{\lambda} = \sqrt[t]{\frac{\hat{N}_t}{\hat{N}_0}}. \qquad (7.18)$$

Assuming independence between $\hat{N}_0$ and $\hat{N}_t$, the variance for the estimator of the finite rate of change (Eq. 7.18) can be approximated by the delta method as

$$\mathrm{Var}(\hat{\lambda}) \doteq \mathrm{Var}(\hat{N}_t)\left(\frac{N_t^{\frac{1}{t}-1}}{tN_0^{\frac{1}{t}}}\right)^2 + \mathrm{Var}(\hat{N}_0)\left(\frac{N_t^{\frac{1}{t}}}{tN_0^{\frac{1}{t}+1}}\right)^2. \qquad (7.19)$$

The variance estimators in Eqs. (7.12), (7.13), (7.17), and (7.19) only include the sampling variance of the abundance estimates. However, the stochastic variation in abundance from year to year can substantially affect the accuracy and precision of the estimates of λ. Year-to-year variability in rates of population change can only be assessed using 3 years or more of data.

Example 7.1: Estimating Rates of Change from Two Abundance Estimates for White-Tailed Deer (*Odocoileus virginianus*), George Reserve, Michigan

After a forced reduction in white-tailed deer abundance to 10 individuals in 1974–1975, the herd was allowed to increase unchecked until 1980–1981, when the herd numbered 212 (McCullough 1982). Using Eq. (7.16), an estimate of the instantaneous intrinsic rate of increase is calculated as

$$\hat{r}_{\text{MAX}} = \frac{\ln(\hat{N}_t) - \ln(\hat{N}_0)}{t}$$

$$= \frac{\ln(212) - \ln(10)}{(1981 - 1975)} = 0.5090.$$

When McCullough (1982) first performed the calculations, he inadvertently used $\log_{10}$ instead of the natural logarithm (i.e., ln or $\log_e$) and reported the wrong value (McCullough 1982, Van Ballenberghe 1983).

By using Eq. (7.18), an estimate of the annual intrinsic rate of increase is calculated to be

$$\hat{\lambda}_{\text{MAX}} = \sqrt[t]{\frac{\hat{N}_t}{\hat{N}_0}}$$

$$= \sqrt[6]{\frac{212}{10}} = 1.6636.$$

In this case, the algebraic relationship $\hat{N}_t = \hat{N}_0 e^{rt} = \hat{N}_0 \lambda^t$ assures that $e^{0.5090} = 1.6636$.

Discussion of Utility

The utility of Eq. (7.11) for estimating λ is limited. In estimating the realized rate of change (λ_{REAL}) from one particular year to the next, it has value. However, Eq. (7.11) provides a poor estimate of the population trend over time. The method essentially attempts to describe a trend by drawing a straight line between two points. Any stochastic error at either point is directly translated into the estimate of the trend (i.e., slope). The individual abundance values are subject to environmental and demographic stochasticity and to sampling error. All of these factors will directly influence the estimates of change when only two points are used. These factors are dampened when a regression line is fit through multiple annual abundance values.

The sole advantage of using estimator (7.16) and (7.18) over more comprehensive methods resides in their computational ease, as the estimators are not consistent nor are the variances unbiased. Any confidence interval estimate or estimate of standard error must be treated with skepticism, because the interannual variance in abundance is ignored. The variance calculations consider only the sampling error in the abundance estimates, and not the stochastic variation in animal numbers over time. Thus, these methods are not recommended for anything other than a cursory inspection of data trends.

7.4   Exponential-Growth Models

The nonlinear Eqs. (7.3) and (7.6) can be used to estimate the intrinsic rate of change should a population be in exponential growth. The equations can be used early in the growth phase to estimate intrinsic rate of change when density-dependent effects have not yet become established. The equations can also be used with a short sequence of annual abundance values when density-dependent effects are operating to estimate the rate of change under prevailing conditions, i.e., the realized rate of change. However, data plots and residual analysis should assess the fit of the models when density-dependent processes may be operating. Because the exponential model can be used to estimate both the intrinsic and realized rates of change, it is important to make the appropriate inferences from the scenario being analyzed. Reporting whether λ_{MAX} or λ_{REAL} is estimated is essential.

7.4.1   Nonlinear Regression for Estimating r or λ

A common approach to estimating the rate of change is to use the relationships

$$N_t = N_0 e^{rt} = N_0 \lambda^t.$$

Assuming the model

$$N_t = N_0 e^{rt} + \varepsilon_t \tag{7.20}$$

where $\varepsilon_t \sim N(0, \sigma^2)$, nonlinear least squares can be used to estimate r by minimizing the sum of squares (SS) as

$$SS = \sum_{i=1}^{n} \left(N_i - N_0 e^{rt_i} \right)^2, \tag{7.21}$$

where n = number of abundance values being analyzed. Alternatively, assuming the model

$$N_t = N_0 \lambda^t + \varepsilon_t, \tag{7.22}$$

where $\varepsilon_t \sim N(0, \sigma^2)$, nonlinear least squares can be used to estimate λ by minimizing the SS as

$$SS = \sum_{i=1}^{n} \left(N_i - N_0 \lambda^{t_i} \right)^2. \tag{7.23}$$

Equations (7.21) and (7.23) can be used to find unbiased estimates of r and λ, respectively. However, transforming $\hat{r}$ to $\hat{\lambda}$, and vice versa, is not recommended (Section 7.2) because of the bias induced by the nonlinear transformation of a random variable.

For most applications, actual abundance values (N_t) will not be known but rather estimated ($\hat{N}_t$). In many animal tagging studies, the variance of an abundance estimate is proportional to N_i. Thus, the assumption of constant variance is violated, suggesting instead the need to use weighted nonlinear least squares (WNLLS). In this case, the finite rate of increase can be estimated by minimizing the SS formula, where

$$SS = \sum_{i=1}^{n} w_i \left(\hat{N}_i - N_0 \lambda^{t_i} \right)^2, \tag{7.24}$$

and where

$$w_i = \frac{1}{\widehat{Var}\left(\hat{N}_i | N_i\right)}.$$

By using matrix notation, the variance-covariance matrix of $\hat{N}_0$ and $\hat{\lambda}$ is estimated as

$$\Sigma = \frac{\left(\underset{\sim}{N}_t - \hat{\underset{\sim}{N}}_t\right)' \mathbf{W}\left(\underset{\sim}{N}_t - \hat{\underset{\sim}{N}}_t\right)}{n-2}(\mathbf{X}'\mathbf{W}\mathbf{X})^{-1}$$

$$= \begin{bmatrix} \widehat{Var}(\hat{N}_0) & \widehat{Cov}(\hat{N}_0, \hat{\lambda}) \\ \widehat{Cov}(\hat{N}_0, \hat{\lambda}) & \widehat{Var}(\hat{\lambda}) \end{bmatrix},$$

(7.25)

where

$\underset{\sim}{N}_t$ = vector of observed abundances;

$\hat{\underset{\sim}{N}}_t$ = vector of fitted values from the estimates of $\hat{N}_0$ and $\hat{\lambda}$;

$\mathbf{W}$ = diagonal matrix with elements w_i;

$\mathbf{X}$ = sensitivity matrix of dimension $(t+1) \times 2$ of the form

$$\mathbf{X} = \begin{bmatrix} \dfrac{\partial \hat{N}_t(0)}{\partial N_0} & \dfrac{\partial \hat{N}_t(0)}{\partial \lambda} \\[2mm] \dfrac{\partial \hat{N}_t(1)}{\partial N_0} & \dfrac{\partial \hat{N}_t(1)}{\partial \lambda} \\[2mm] \vdots & \vdots \\[2mm] \dfrac{\partial \hat{N}_t(t)}{\partial N_0} & \dfrac{\partial \hat{N}_t(t)}{\partial \lambda} \end{bmatrix} = \begin{bmatrix} \hat{\lambda}^0 & 0\hat{N}_0\hat{\lambda}^{-1} \\ \hat{\lambda}^1 & 1\hat{N}_0\hat{\lambda}^0 \\ \vdots & \vdots \\ \hat{\lambda}^t & t\hat{N}_0\hat{\lambda}^{t-1} \end{bmatrix}.$$

The instantaneous rate of change r can also be estimated directly by using WNLLS of the form

$$SS = \sum_{i=1}^{n} w_i\left(\hat{N}_i - N_0 e^{r t_i}\right)^2$$

(7.26)

where

$$w_i = \frac{1}{\widehat{Var}\left(\hat{N}_i | N_i\right)}.$$

The variance of $\hat{r}$ obtained from WNLLS (Eq. 7.26) is also estimated by using Eq. (7.25). The vector of fitted values, $\hat{N}_t$, is obtained by using the expression $\hat{N}_t = \hat{N}_0 e^{\hat{r}t}$, and the sensitivity matrix, $\mathbf{X}$, is calculated as

$$\mathbf{X} = \begin{bmatrix} \dfrac{\partial \hat{N}_t(0)}{\partial N_0} & \dfrac{\partial \hat{N}_t(0)}{\partial r} \\[2mm] \dfrac{\partial \hat{N}_t(1)}{\partial N_0} & \dfrac{\partial \hat{N}_t(1)}{\partial r} \\[2mm] \vdots & \vdots \\[2mm] \dfrac{\partial \hat{N}_t(t)}{\partial N_0} & \dfrac{\partial \hat{N}_t(t)}{\partial r} \end{bmatrix} = \begin{bmatrix} e^{\hat{r}\cdot 0} & 0\hat{N}_0 e^{\hat{r}\cdot 0} \\ e^{\hat{r}\cdot 1} & 1\hat{N}_0 e^{\hat{r}\cdot 1} \\ \vdots & \vdots \\ e^{\hat{r}\cdot t} & t\hat{N}_0 e^{\hat{r}\cdot t} \end{bmatrix}.$$

Assumptions

The estimators based on WNLLS (Eqs. 7.24, 7.26) are not guaranteed to be unbiased with minimum variance but, rather, are asymptotically unbiased with almost minimum variance (Neter et al. 1996:549). The assumptions associated with the estimates of $\hat{\lambda}$ and $\hat{r}$ obtained from WNLLS regression include the following:

1. There is a constant rate of population change.
2. Successive values of $\hat{N}_t$, $t = 0, 1, \ldots$, are estimated independently (uncorrelated errors).
3. Additive errors with mean 0, i.e., $N_t = \hat{N}_0 e^{rt} + \varepsilon_t$ and $E(\varepsilon_t) = 0$.
4. Errors are normally distributed, i.e., $\varepsilon_t \sim N(0, \sigma^2 \propto \mathrm{Var}(\hat{N}_t))$.
5. Independent variables are known without error (i.e., times t_i known).
6. $\hat{N}_t$ is an unbiased estimate of the population abundance at time t.

When the assumption of constant variance is violated in ordinary least squares, the estimators will not have minimum variance (Neter et al. 1996:401). The analysis further assumes the annual estimates of abundance are calculated independently. Thus, the value of $\hat{N}_t$ has no influence on the value of other abundance estimates, and vice versa.

If a yearly sequence of indices is used to estimate $\hat{\lambda}$ or $\hat{r}$, there is the added assumption of a constant proportionality between the abundance index and true abundance, i.e., $I_t = \beta N_t$, for all years. By using the number of family groups as an index for population abundance, Knight et al. (1995) estimated $\hat{\lambda}$ for a population of grizzly bears (*Ursus arctos horribilis*) in Yellowstone National Park, Wyoming. Estimates were obtained from both the raw counts of group numbers and a sightability-adjusted count. The latter produced a slightly lower estimate of the finite rate of population change ($\hat{\lambda}$).

Example 7.2: Estimating r_{MAX} and λ_{MAX} Using Nonlinear Regression for Elk (*Cervus elaphus*), Hanford, Washington

McCorquodale et al. (1988) reported on the growth of an isolated elk population that colonized an arid shrub-steppe environment in eastern Washington State during 1975–1986 (Table 7.1). The herd grew without human intervention through 1986. Aerial counts were used to obtain abundance estimates for the years 1975–1981. Radio telemetry and visual counts were used to enumerate abundance during 1982–1986. The absence of tree cover made accurate counts of all elk possible.

Nonlinear least squares using the equation

$$N_i = N_0 e^{r t_i} + \varepsilon_i,$$

yielded $\hat{N}_0 = 5.9515$ ($\widehat{SE} = 0.8867$), with the instantaneous intrinsic rate of change estimated to be $\hat{r}_{MAX} = 0.2453$ ($\widehat{SE} = 0.0153$).

The alternative nonlinear equation,

$$N_i = N_0 \lambda^{t_i} + \varepsilon_i,$$

yielded $\hat{N}_0 = 5.9515$ ($\widehat{SE} = 0.8867$), with the annual intrinsic rate of change estimated to be $\hat{\lambda}_{MAX} = 1.2780$ ($\widehat{SE} = 0.0195$). The resulting curves (Fig. 7.2) are equivalent. In addition,

Table 7.1. Abundance of elk at Arid Lands Ecology Reserve, Hanford, Washington, 1975–1986.

Year	Abundance	Coded year
1975	8	0
1976	13	1
1977	15	2
1978	N/A	3
1979	16	4
1980	N/A	5
1981	25	6
1982	27	7
1983	40	8
1984	55	9
1985	71	10
1986	89	11

(Data from McCorquodale et al. 1988.)

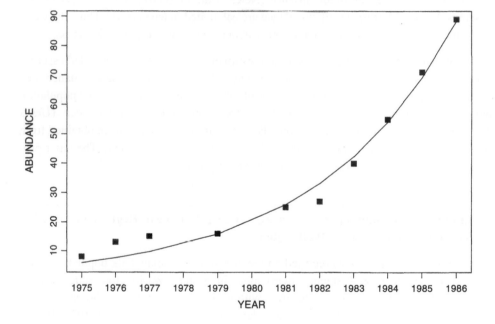

Figure 7.2. Elk numbers at Hanford, Washington, 1975–1986 (McCorquodale et al. 1988), based on aerial and visual counts. The plot of the fitted curve to Eqs. (7.20) and (7.22) is presented.

$$e^{\hat{r}} = \hat{\lambda},$$

or

$$e^{0.2453} = 1.2780,$$

because no data transformation was used. Instead, there was simply a reparameterization of e^r for λ between Eqs. (7.20) and (7.22). Thus, either r or λ can be calculated by using the nonlinear least-squares approach, and one estimate will lead directly to the other estimate (i.e., $e^{\hat{r}} = \hat{\lambda}$). The choice of method should then rely on the desired variance calculation.

The nonlinear least squares will directly compute the standard error of the parameter selected. However, the delta method is needed to convert the variance of one estimate to that of the other. For instance, the variance of $\hat{\lambda}$ obtained from the transformation $\hat{\lambda} = e^{\hat{r}}$ has the approximate following value:

$$\widehat{\text{Var}}\left(e^{\hat{r}}\right) = \widehat{\text{Var}}\left(\hat{r}\right)e^{2\hat{r}}$$

or

$$\widehat{\text{SE}}(e^{\hat{r}}) = \widehat{\text{SE}}(\hat{r})e^{\hat{r}}.$$

In this example,

$$\widehat{\text{Var}}\left(e^{\hat{r}}\right) = (0.0153)(e^{0.2253}) = 0.0192.$$

Alternatively, the variance of $\hat{r}$ obtained from the transformation $\hat{r} = \ln \hat{\lambda}$ has the approximate variance

$$\widehat{\text{Var}}\left(\ln \hat{\lambda}\right) = \frac{\widehat{\text{Var}}\left(\hat{\lambda}\right)}{\hat{\lambda}^2}$$

or

$$\widehat{\text{SE}}\left(\ln \hat{\lambda}\right) = \frac{\widehat{\text{SE}}(\hat{\lambda})}{\hat{\lambda}}$$

$$= \text{CV}(\hat{\lambda}).$$

In this example,

$$\widehat{\text{Var}}\left(\ln \hat{\lambda}\right) = \frac{0.0195}{1.2780} = 0.0153.$$

7.4.2 Log-Linear Regression for Estimating *r*

The exponential-growth models,

$$N_t = N_0 e^{rt} \cdot \varepsilon_t$$

or

$$N_t = N_0 \lambda^t \cdot \varepsilon_t,$$

suggest a log-linear transformation, where

$$\ln N_t = \ln N_0 + rt + \ln \varepsilon_t \tag{7.27}$$

or

$$\ln N_t = \ln N_0 + (\ln \lambda) \cdot t + \ln \varepsilon_t, \tag{7.28}$$

based on multiplicative errors and where $\ln \varepsilon_t \sim N(0,\sigma^2)$. Linear regression of the form

$$y_i = \alpha + \beta x_i + \varepsilon_i$$

can be used without bias to estimate r and $\ln \lambda \ (\equiv r)$ but not λ. Minimizing the SS,

$$\text{SS} = \sum_{i=1}^{n} (\ln N_i - (\alpha + rt_i))^2, \tag{7.29}$$

results in the estimator of r,

$$\hat{r} = \frac{\left[\sum_{i=1}^{n} (\ln N_i) t_i - \frac{\left(\sum_{i=1}^{n} \ln N_i \right)\left(\sum_{i=1}^{n} t_i \right)}{n} \right]}{\left[\sum_{i=1}^{n} (t_i - \bar{t})^2 \right]}, \tag{7.30}$$

and the associated variance estimator,

$$\widehat{\mathrm{Var}}(\hat{r}) = \frac{\left[\frac{\sum_{i=1}^{n} (\ln N_i - (\hat{\alpha} + \hat{r} t_i))^2}{(n-2)} \right]}{\left[\sum_{i=1}^{n} (t_i - \bar{t})^2 \right]}. \tag{7.31}$$

Examples of estimating the instantaneous rate of increase r by using linear least squares and abundance data include a feral horse (*Equus caballus*) population (Eberhardt et al. 1982), merlins (*Falco columbaris*) in Saskatchewan (Oliphant and Huag 1985), and elk (Gogan and Barrett 1987, McCorquodale et al. 1988). In each of these studies, resources did not appear to limit population growth, so that r was interpreted as the maximum rate of population growth r_{MAX} (Caughley and Birch 1971). None of these studies used weighted regression to obtain estimates of r.

Assumptions

There are two assumptions of ordinary least squares:

1. The observations (i.e., N_i's) or equivalently, the ε_i's are independent and normally distributed with constant variance.
2. The data fit the exponential-growth model.

Should population indices rather than abundance be used in the analysis, there is the following additional assumption:

3. $I_i = \beta N_i$ for $i = 0, \ldots, t$.

In the case in which abundance estimates rather than actual abundance values are used in the regression analysis, a weighted regression of the following form should be used:

$$\mathrm{SS} = \sum_{i=1}^{n} w_i \left(\ln \hat{N}_i - (\alpha + r t_i) \right)^2, \tag{7.32}$$

where

$$w_i = \frac{1}{\mathrm{Var}(\ln \hat{N}_i)} \doteq \frac{1}{\mathrm{CV}(\hat{N}_i)^2}.$$

The weighted regression estimator for r is

$$r = \frac{\sum\limits_{i=1}^{n} w_i \left(\ln N_i - \overline{\ln N} \right) (t_i - \bar{t})}{\sum\limits_{i=1}^{n} w_i (t_i - \bar{t})^2}, \qquad (7.33)$$

with associated variance estimator

$$\widehat{\text{Var}}(\hat{r}) = \frac{\sum\limits_{i=1}^{n} w_i \left(\ln \hat{N}_i - (\hat{\alpha} + \hat{r} t_i) \right)^2}{(n-2) \sum\limits_{i=1}^{n} w_i (t_i - \bar{t})^2}. \qquad (7.34)$$

Plotting the $\ln N$ or $\ln \hat{N}$ versus time is a good visual check for the assumption of constant r.

Example 7.3: Estimating r_{MAX} Using Linear Regression for Elk, Hanford, Washington

A reanalysis of the Hanford elk population trend from 1975 through 1986 (McCorquodale et al. 1988) using the log-linear regression model (Eq. 7.27) results in the following equation:

$$\ln N_t = 2.1484 + 0.2001t$$

where the estimate of $\hat{r}_{MAX} = 0.2001$ has a standard error of $\widehat{SE}(\hat{r}_{MAX}) = 0.0145$. The fitted equation had a squared correlation coefficient of 0.9596. A plot of $\ln N_t$ versus time t shows strong linear trend and reasonable fit to the data (Fig. 7.3). The nonlinear least-squares analysis produced a corresponding value of $\hat{r}_{MAX} = 0.2453$ ($\widehat{SE} = 0.0153$).

7.4.3 Ratio Estimators for λ

The relationship

$$N_{i+1} = N_i \lambda + \varepsilon_i \qquad (7.35)$$

suggests using a straight-line regression through the origin of the form $y_i = \beta x_i + \varepsilon_i$ to estimate λ. For a straight-line regression through the origin, the least-squares estimate of λ is

$$\hat{\lambda} = \frac{\sum\limits_{i=1}^{n-1} N_i N_{i+1}}{\sum\limits_{i=1}^{n-1} N_i^2} \qquad (7.36)$$

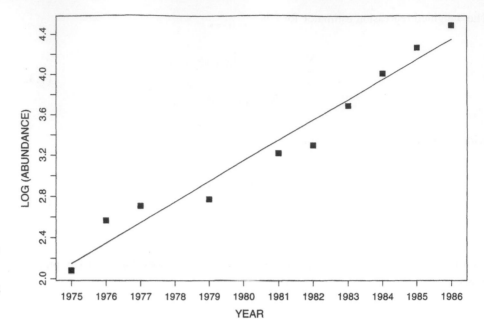

Figure 7.3. Plot of ln-elk numbers at Hanford, Washington, 1975–1986 (McCorquodale et al. 1988), against time along with fitted line to Eq. (7.27).

with variance estimator of

$$\widehat{\text{Var}}(\hat{\lambda}) = \frac{\left[\displaystyle\sum_{i=1}^{n-1} N_{i+1}^2 - \frac{\left(\displaystyle\sum_{i=1}^{n-1} N_i N_{i+1} \right)^2}{\displaystyle\sum_{i=1}^{n-1} N_{i+1}^2} \right]}{\displaystyle\sum_{i=1}^{n-1} N_i^2}. \tag{7.37}$$

Both the independent and dependent variables in Eq. (7.35) will be measured with error when animal abundance is estimated rather than enumerated. Regression analysis, in the situation where both the independent and dependent variables are measured with error, is known as errors-in-variable regression (Casella and Berger 1990:581), measurement error models, or structured equation models (Cheng and Van Ness 1999:4–7). In this case, ordinary least squares will result in a negatively biased estimate of the slope, i.e., $\hat{\lambda}$, where

$$E(\hat{\lambda}) = \frac{\lambda}{1 + \left(\dfrac{\sigma_{SE}^2}{\sigma_N^2 + \mu_N^2} \right)}.$$

Here, σ_{SE}^2 is the sampling error associated with abundance estimation, σ_N^2 is the temporal variance in abundance, and μ_N is the mean abundance. As σ_{SE}^2 increases, the bias increases as well, with $\hat{\lambda}$ more and more underestimating the true value of λ in expectation. At a minimum, weighted least squares of the form

$$\text{SS} = \sum_{i=1}^{n-1} w_{i+1} \left(\hat{N}_{i+1} - \hat{N}_i \lambda \right)^2, \tag{7.38}$$

where

$$w_{i+1} = \frac{1}{\widehat{\text{Var}}(\hat{N}_{i+1}|N_{i+1})},$$

should be used when estimating λ, because the estimates of abundance will most likely have unequal variances. The variance of the weighted least-squares regression estimator is

$$\widehat{\text{Var}}(\hat{\lambda}) = \frac{\sum_{i=1}^{n-1} w_{i+1}(\hat{N}_{i+1} - \hat{\lambda}\hat{N}_i)^2}{\sum_{i=1}^{n}(\hat{N}_i - \bar{\hat{N}}_i)^2}. \tag{7.39}$$

Eberhardt (1987) suggests two other ratio estimators of λ, where

$$\hat{\lambda} = \frac{\sum_{i=1}^{n-1} \hat{N}_{i+1}}{\sum_{i=1}^{n-1} \hat{N}_i} \tag{7.40}$$

and

$$\hat{\lambda} = \frac{1}{n-1}\sum_{i=1}^{n-1} \frac{\hat{N}_{i+1}}{\hat{N}_i}$$

$$= \frac{1}{n-1}\sum_{i=1}^{n-1} \hat{\lambda}_i. \tag{7.41}$$

The variances for the ratio estimator, Eq. (7.40), and the arithmetic mean, Eq. (7.41), are estimated by the variance formula for a ratio (Cochran 1977:31) and the variance for a sample mean (Zar 1984:29), respectively, as

$$\widehat{\text{Var}}(\hat{\lambda}) = \frac{\sum_{i=1}^{n-1}(\hat{N}_{i+1} - \hat{\lambda}\hat{N}_i)^2}{(n-1)\hat{\lambda}^2(n-2)} \tag{7.42}$$

and

$$\widehat{\text{Var}}(\hat{\lambda}) = \frac{\sum_{i=1}^{n-1}(\hat{\lambda}_i - \hat{\lambda})^2}{(n-1)(n-2)}, \tag{7.43}$$

where

$$\hat{\lambda}_i = \frac{\hat{N}_{i+1}}{\hat{N}_i}.$$

However, the variance estimates should be biased because the serial correlation between the abundance estimates has not been considered. Eberhardt (1987) and Eberhardt and

Simmons (1992) suggested using jackknife techniques, successively removing each of the abundance estimates, N_i, to obtain estimates and associated variances from Eqs. (7.36), (7.40), and (7.41).

Example 7.4: Estimating λ_{MAX} Using Straight-Line Regression through the Origin for Elk, Hanford, Washington

The Hanford elk data from 1975–1986 were again analyzed, this time by using regression Eq. (7.35). The two missing values in 1978 and 1980 result in the abundance data from 1978–1981 not being used in the analysis of N_{t+1} versus N_t. The fitted equation is

$$N_{t+1} = 1.2866 N_t,$$

yielding $\hat{\lambda}_{MAX} = 1.2866$ ($\widehat{SE} = 0.0356$). The fitted regression model (Eq. 7.35) had a squared correlation coefficient of 0.9956 (Fig. 7.4).

7.4.4 Time-Series Analysis of λ

If λ is expressed as the ratio

$$\lambda_t = \frac{N_{t+1}}{N_t} \text{ for } t = 0, 1, \ldots, \tag{7.44}$$

the serial correlation between the λ_t's must be considered to properly characterize the variance of the estimate of λ. Eberhardt (1987) viewed the serial correlation as a reason to avoid using a ratio approach to obtain a point estimate and the associated variance estimate for λ. The estimated variances of the regression parameters will be biased if the abundance estimates are correlated.

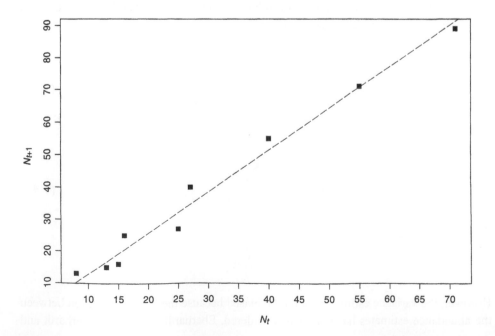

Figure 7.4. Plot of the Hanford, Washington, elk abundance (McCorquodale et al. 1988) in year $t + 1$ against abundance in year t, along with fitted line to Eq. (7.35).

Analyzing sequential values of λ by using a time-series approach offers a way to examine the behavior of λ across years. However, the most valuable use of time-series methods is in estimating the variance of $\hat{\lambda}$ or $\hat{r}$ from a sequence of autocorrelated observations. Starting with the definition in Eq. (7.44), and taking the logarithm of both sides gives

$$\ln(N_{t+1}) - \ln(N_t) = \ln(\lambda_t). \tag{7.45}$$

Eq. (7.45) takes the form of a differenced time series (Brockwell and Davis 1996), which is written as follows:

$$X_t = \ln(\lambda_t) = r_t, \tag{7.46}$$

Differencing time series often produces a stationary series, in which the covariances between observations do not depend on time but only on time lag h (Brockwell and Davis 1996:15). A log-transformation is used to make the abundance estimates behave more normally distributed and multiplicative errors on the original scale additive. Therefore, X_t can be considered to follow a normal distribution, and hence, $\ln(\lambda_t)$ and r_t can also be considered Gaussian for analytical purposes and the resulting patterns described by using time-series methods. This approach is similar to that of the population viability analysis of Dennis et al. (1991).

The mean of X_t (i.e., $\ln(\lambda_t)$) can be estimated by using the arithmetic mean. Letting $\ln(\lambda_t) = X_t$ for simplification, the mean of a stationary series, $\overline{X}_t$ is estimated by

$$\overline{X} = \frac{1}{n}\sum_{t=0}^{n} X_t. \tag{7.47}$$

The variance of $\overline{X}$ is estimated by

$$\mathrm{Var}(\overline{X}_t) = \frac{\sum_{h=-n}^{n}\left(1 - \frac{|h|}{n}\right)\gamma(h)}{n}, \tag{7.48}$$

where

$$h = h\text{th lag;}$$
$$\gamma(h) = \mathrm{Cov}(X_t, X_{t+h}),\ h = 0, \pm1, \pm2, \ldots;$$

which is the autocovariance function (ACVF) for the series, X_t (Brockwell and Davis 1996:56). Time-series methods are used to estimate the autocovariance function, $\gamma(h)$ and, hence, the variance of the mean of r_t (i.e., $\ln(\lambda_t)$).

Stationary time series are usually modeled as an autoregressive, moving average (ARMA) process. However, calculating the ACVF, $\gamma(h)$, for ARMA models is algebraically tedious (Chatfield 1989:40). Modeling the series strictly as a moving average (MA) process makes calculating the ACVF and, hence, the variance of $\hat{r}$ (i.e., $\ln\hat{\lambda}$) much easier. The general equation for the MA model is written as

$$X_t = \mu + Z_t + \sum_{i=1}^{q}\theta_i Z_{t-i}, \tag{7.49}$$

where

μ = mean of the series;

θ_i = coefficient for the MA process for $i > 1$ lag, $(\theta_0 = 1)$;

Z_t = white-noise series with zero mean and variance σ^2, i.e., $Z_t \sim WN(0, \sigma^2)$;

q = number of lags for which $\gamma(h) \neq 0$ (Brockwell and Davis 1996).

The ACVF for the MA process is written as

$$\gamma(h) = \begin{cases} \gamma(h) = \sigma^2 \sum_{i=0}^{q-h} \theta_i \theta_{i+|h|} & \text{if } h = 0, \pm 1, \pm 2, \ldots \\ \gamma(h) = 0 & \text{if } |h| > q \end{cases} \tag{7.50}$$

(Chatfield 1989). The ACVF function is then used in Eq. (7.48) to estimate $(\overline{X})$. Any stationary time series with correlated observations at lag h can be modeled as an MA(q) process, in which $h \leq q$ (usually called a q-correlated time series) (Brockwell and Davis 1996:48). This is a useful result because the variance of $\hat{r}$ (i.e., $\ln(\lambda)$) from any time series of abundance or index data can be estimated using Eqs. (7.49) and (7.50).

Coefficients can be estimated by using standard statistical software, specifying an MA process. Often, more parameters are needed to specify the time series as a pure MA process than an ARMA process (Chatfield 1989:58), and the MA parameter estimates are computationally more complex. However, the variance of the mean is easier to calculate for an MA model than an ARMA model, and the MA process is more easily interpreted. The stationary MA model may be interpreted as a random walk about the mean, $\bar{r}$, the average instantaneous rate of population change.

Example 7.5: Time Series Analysis to Estimate λ_{REAL} for Spectacled Eiders (*Somateria fischeri*), Yukon-Kuskokwim Delta, Alaska

Data on the abundance of spectacled eiders from the Yukon-Kuskokwim Delta, Alaska, were analyzed (Taylor et al. 1996) to illustrate the use of time series in estimating the realized rate of population change. Yearly abundances of eiders were estimated by using aerial transect surveys from 1957–1995 (Fig. 7.5). The total number of eiders was based on twice the number of single male eiders observed plus all other birds observed in pairs and flocks. Single males were assumed to indicate a breeding pair, because males were more easily seen than were female eiders. This example compares estimates of $\hat{r}_{REAL}$ and Var($\hat{r}_{REAL}$) by using regression and time series methods.

Although variance estimates for the abundance estimates (N_i) were available, $\hat{r}_{REAL}$ was estimated by using linear and nonlinear least-squares regression without weighting for sake of comparison with the unweighted time series method. Based on the log-linear regression estimator (Eq. 7.32), the estimated instantaneous rate of change is $\hat{r}_{REAL} = -0.063$ with estimated standard error $\widehat{SE}(\hat{r}_{REAL}) = 0.005$. The nonlinear least-squares estimator (Eq. 7.21) of r_{REAL} yielded $\hat{r}_{REAL} = -0.053$, with $\widehat{SE}(\hat{r}_{REAL}) = 0.007$. Asymptotic 90% confidence intervals for the linear and nonlinear least-squares estimates are, respectively, CI $(-0.071 < r_{REAL} < -0.054) = 0.90$ and CI $(-0.065 < r_{REAL} < -0.041) = 0.90$. The difference between the two estimators is attributable to differences in the error structures assumed by LS and NLLS estimation techniques.

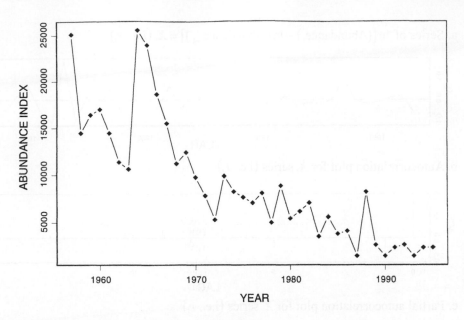

Figure 7.5. The abundance index (number of breeding pairs of birds) versus year for spectacled eiders on the Yukon-Kuskokwim Delta, Alaska, 1957–1995 (Taylor et al. 1996).

The first step in estimating the mean and variance of r_t from the abundance values is to model the time series as a MA process. The abundance and covariances depend on time, t (Fig. 7.5). Taking the natural logarithm of the abundance data and differencing the transformed values as in Eq. (7.45) helps remove any trends and yields a stationary X_t or r_t series (Eq. 7.46). A plot of X_t values (Fig. 7.6a) shows a more even distribution about the mean. Both the ACF (Fig. 7.6b) and partial autocorrelation (PACF) plots (Fig. 7.6c) are more characteristic of a stationary series and suggest the differenced time series, or X_t, is an MA (1) series. The gradual decrease in the PACF plot is indicative of an MA model, and the significant ACF value at lag 1 implies an MA model with $q = 1$. The fitted MA (1) model for the X_t series is

$$X_t = r_t = -0.064 + Z_t - 0.594Z_{t-1},$$

where $Z_t \sim WN(0, 0.179)$ (a white-noise series). Hence, the estimate of r_{REAL}, the sample mean, is -0.064. The ACVF of the MA (1) process is written as

$$\gamma(h) = \begin{cases} \gamma(0) = \sigma^2(1+\theta^2) & \text{if } h = 0 \\ \gamma(1) = \sigma^2\theta & \text{if } h = \pm 1 \\ \gamma(h) = 0 & \text{if } |h| > 1. \end{cases}$$

The variance of $\hat{r}_{\text{REAL}}$ is estimated by Eq. (7.48) as

$$\widehat{\text{Var}}(\overline{X}_t) = \frac{\displaystyle\sum_{h=-n}^{n}\left(1 - \frac{|h|}{n}\right)\gamma(h)}{n}$$

$$= \frac{\left(1 - \dfrac{0}{37}\right)(0.179)\left(1 + (-0.594)^2\right) + 2\left(1 - \dfrac{1}{37}\right)(0.179)(-0.594)}{37}$$

$$= 0.00095,$$

or $\widehat{\text{SE}}(\overline{X}_t) = 0.0309.$

a. Series of $\ln\left[(\text{Abundance}_t) - \ln\left(\text{Abundance}_{t-1}\right)\right] = X_t \ (i.e., r_t)$

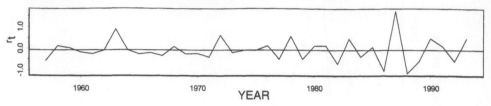

b. Autocorrelation plot for X_t series $(i.e., r_t)$

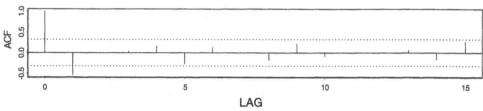

c. Partial autocorrelation plot for X_t series $(i.e., r_t)$

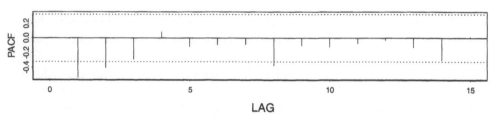

Figure 7.6. The (a) time series for spectacled eiders, (b) autocorrelation function (ACF), and (c) partial autocorrelation (PACF) of X_t or r.

An asymptotic 90% confidence interval for r_{REAL} is $\text{CI}(-0.115 < r_{\text{REAL}} < -0.013) = 0.90$. The variance and standard error are greater than those estimated by using regression techniques, because the correlation between successive data points was taken into account.

Although estimates of $\hat{\lambda}_{\text{REAL}}$ and $\widehat{\text{Var}}(\hat{\lambda}_{\text{REAL}})$ will be slightly biased because of data transformations, the estimates are, respectively,

$$\hat{\lambda} = e^{\hat{r}} = e^{-0.064} = 0.938$$

and

$$\widehat{\text{Var}}(e^{\hat{r}}) \doteq \widehat{\text{Var}}(\hat{r})e^{2\hat{r}},$$

$$\widehat{\text{Var}}(\hat{\lambda}_{\text{REAL}}) = 0.00065e^{2(-0.064)} = 0.000572,$$

or

$$\widehat{\text{SE}}(\hat{\lambda}_{\text{REAL}}) = 0.0239.$$

An asymptotic 90% confidence interval for λ_{REAL} is $0.938 \pm 0.0239(1.645)$ or $\text{CI}(0.899 \le \lambda_{\text{REAL}} \le 0.977) = 0.90$.

The estimates and confidence interval estimates of $\hat{r}_{\text{REAL}}$ agree with the observed trend (Fig. 7.5), which clearly indicates the population of eiders is declining. However, 36 observations are generally considered a short time series, and the variance may be overestimated. The random walk described by the MA (1) process indicates that r_{REAL}

will follow the same trajectory, i.e., $\hat{r}_{REAL} = -0.064$, thus predicting future decline. Eberhardt and Simmons (1992:610) mentioned the stochastic nature of demographic processes would "generate a random walk for an individual population trajectory," which is what the MA process (Eq. 7.49) attempts to describe.

Discussion of Utility

Among the complications in assessing wildlife population trends are the generally short time duration of data collection and the potential for changes in observers, effort in data collection, or data collection protocols. Each of these factors can influence measurement error and/or induce bias in abundance estimates. The effect on a time series is the mean, or the covariance structure may change with time, in which case the series is no longer stationary. In practice, this may be difficult to detect. For example, misidentification of birds to species and differential timing of surveys (with regard to the breeding season) can lead to bias in abundance counts and contribute to sampling error.

Variance estimation with correlated data is not the only way in which time-series techniques can be used in analyzing these data. Dennis and Taper (1994) used an autoregressive (AR) model for estimating the effect of density dependence on population growth rates. The model they used was

$$N_t = N_{t-1} \exp(a + b \log(N_{t-1}) + \sigma Z_t).$$

Taking the log of the equation,

$$\log(N_t) = \log(N_{t-1}) + a + b \log(N_{t-1}) + \sigma Z_t,$$

which is an AR (1) model, with AR coefficient $(1 - b)$. This model assumes the $\log(N_t)$ series is stationary. However, if density dependence effects exist, then the mean and covariances of the abundance series will depend on time, and hence, the time series will be nonstationary.

Several approaches for estimating r and λ based on exponential growth have been presented. Some methods are motivated more by statistical theory; others, by computational ease. As seen in the examples, the various methods do not always yield the same numerical result. What computational method is best? Based on our experiences, simulation studies, and theoretical considerations, we make the following recommendations:

1. Nonlinear regression models

$$N_t = N_0 e^{rt} + \varepsilon_t$$

and

$$N_t = N_0 \lambda^t + \varepsilon_t$$

provide unbiased estimates of r and λ as long as the error structure is additive. However, the nonlinear regression approaches are sensitive to the error structure and will yield biased estimates if errors are multiplicative, i.e.,

$$N_t = N_0 e^{rt} \cdot \varepsilon_t$$

or

$$N_t = N_0 \lambda^t \cdot \varepsilon_t.$$

Unfortunately, a multiplicative error structure is a more likely scenario than additive errors. Survival and production tend to have a multiplicative influence on abundance. Furthermore, many abundance estimates have variances that are proportional to N_t.

2. Based on the population model $N_t = N_0 e^{rt} \varepsilon_t$, the log-linear regression model

$$\ln N_t = \ln N_0 + rt + \ln \varepsilon_t$$

of the form

$$y_i = \alpha + \beta x_i + \varepsilon_i$$

for a straight-line regression, provides an unbiased estimate of r. Additive errors on the log scale are equivalent to multiplicative errors on the nontransformed scale. Hence, this regression model is appropriate for the type of multiplicative errors often seen in nature. The log-linear regression analysis also appears to be rather robust to additive errors on the arithmetic scale. This regression analysis, on the other hand, provides biased estimates of λ.

3. The linear regression through the origin using the relationship

$$N_{t+1} = N_t \lambda + \varepsilon_t$$

is of the form

$$y_i = \beta x_i + \varepsilon_i$$

and will provide unbiased estimates of λ if errors are additive. There are, however, three concerns with this method. First, a more likely error structure is multiplicative errors of the form

$$N_{t+1} = N_t \lambda \cdot \varepsilon_t.$$

Second, if the abundance values are estimated rather than enumerated, the independent variable (i.e., $\hat{N}_t$) is no longer measured without error. The variance in estimation of N_t will tend to induce a negative bias in $\hat{\lambda}$. Third, the regression analysis requires consecutive values of N_t and N_{t+1}. Hence, this regression model is more sensitive to missing values than the log-linear analysis.

4. The time-series analysis produces estimates of λ analogous to the ratio method (Eq. 7.41) of Eberhardt (1987) and Dennis et al. (1991), where

$$\hat{\lambda} = \frac{1}{(n-1)} \sum_{i=1}^{n-1} \frac{N_{i+1}}{N_i}$$

$$= \frac{1}{(n-1)} \sum_{i=1}^{n-1} \hat{\lambda}_i.$$

The focus of time-series analysis is not so much point estimation as it is proper variance estimation. If autocorrelation between successive abundance values exist, ordinary least squares will underestimate or overestimate the true variance and time-series analysis should be used. For long-lived species, whose populations are intrinsically controlled, autocorrelation may be high. On the other hand, for short-lived species whose populations are extrinsically controlled, autocorrelation can be expected to be small. However, in many wildlife applications, abundance estimates rather than actual abundance values are used to estimate λ. Measurement error in these cases may diminish or mask any autocorrelations that may exist. In these circumstances, time-series analysis would be of limited value.

5. If population growth is not exponential, regression models will result in biased estimates of λ. In these circumstances, in which an average rate of growth must be calculated, the estimator (Eq. 7.41)

$$\hat{\lambda} = \frac{1}{n-1} \sum_{i=1}^{n-1} \hat{\lambda}_i$$

may be a more realistic estimate of the true change.

Based on these findings, when a population is experiencing exponential growth, we generally recommend use of the log-linear regression model

$$\ln N_t = \alpha + rt.$$

This approach provides unbiased estimates of r under likely circumstances (i.e., multiplicative errors) and is robust when the actual error structure is additive on N_t. A data plot provides a simple means to assess appropriateness of the fitted models. However, more formal residual analyses are advisable (Belsley et al. 1980). It is important to recall that $e^{\hat{r}}$ will provide a biased estimate of λ. Equation (7.10) suggests the bias-corrected estimator for λ of

$$\hat{\lambda} = e^{\hat{r}} \left(1 + \frac{\widehat{\mathrm{Var}}(\hat{r})}{2} \right)^{-1}. \qquad (7.51)$$

This estimator has not been widely used, and its statistical properties require further examination.

Weighted regression is appropriate if the variance of the dependent variable is not constant. In this case, the general procedure is to weight inversely proportional to the variance. In many wildlife applications, the $\mathrm{Var}(\hat{N}) \propto N$, meaning the variance estimate is correlated with the response variable. As a result, high abundance values tend to have large variances, and low abundance values have smaller variances. Consequently, the regression line is pulled downward, thereby underestimating r or λ. For this reason, we do not generally recommend weighted regression using $\dfrac{1}{\widehat{\mathrm{Var}}(\hat{N}_i)}$. One possibility is to weight inversely proportional to the relative variance, i.e.,

$$w_i = \frac{1}{\left(\dfrac{\widehat{\mathrm{Var}}(\hat{N}_i)}{\hat{N}_i} \right)},$$

thereby eliminating or reducing the correlation between $\widehat{\mathrm{Var}}(\hat{N}_i)$ and $\hat{N}_i$. This adjustment, however, requires knowledge of the structure of $\mathrm{Var}(\hat{N}_i)$ before implementation.

A regression analysis to estimate r_{REAL} or λ_{REAL} is not recommended when exponential growth does not exist. Eberhardt (1982, 1987), among others, often calculated λ based on pair-wise abundance values $\left(\text{i.e., } \hat{\lambda}_i = \dfrac{N_{i+1}}{N_i}\right)$ and averaged them across the replicate estimates (Eq. 7.41). We similarly suggest this approach when estimating λ_{REAL} during the logistic-growth phase of a population.

7.5 Logistic-Growth Models

Eventually, a population will experience density-dependent growth as the population approaches the limit of its environmental resources and its carrying capacity. In these circumstances, the observed rate of population change will be less than the intrinsic rate of increase (Fig. 7.1). Consequently, fitting an exponential-based model, i.e., Eq. (7.3) or (7.6), when the population is no longer in exponential growth, will tend to underestimate the true intrinsic rate of increase (i.e., r_{MAX} or λ_{MAX}).

The standard discrete-time logistic-growth model can be written in terms of the intrinsic rate of increase, λ_{MAX}, where

$$N_{t+1} = N_t + N_t(\lambda_{\text{MAX}} - 1)\left(1 - \frac{N_t}{K}\right). \tag{7.52}$$

Equation (7.52) may be reparameterized in terms of

$$R_{\text{MAX}} = \lambda_{\text{MAX}} - 1$$

for convenience, where

$$N_{t+1} = N_t + N_t R_{\text{MAX}}\left(1 - \frac{N_t}{K}\right). \tag{7.53}$$

Because R_{MAX} is a linear function of λ_{MAX}, an unbiased estimate of one quantity leads to an unbiased estimate of the other, where

$$E(\hat{R}_{\text{MAX}}) = E(\hat{\lambda}_{\text{MAX}} - 1)$$
$$R_{\text{MAX}} = \lambda_{\text{MAX}} - 1 \tag{7.54}$$

and where

$$\text{Var}(\hat{R}_{\text{MAX}}) = \text{Var}(\hat{\lambda}_{\text{MAX}}). \tag{7.55}$$

Sequential values of N_t and N_{t+1} are needed to fit the models (7.52) and (7.53).

Nonlinear least squares can be used, if the abundance values are measured without error, by minimizing the

$$\text{SS} = \sum_{i=1}^{n-1}\left(N_{i+1} - \left(N_i + N_i(\lambda_{\text{MAX}} - 1)\left(1 - \frac{N_i}{K}\right)\right)\right)^2 \tag{7.56}$$

of the general form

$$\sum(\text{Observed}_i - \text{Expected}_i)^2. \tag{7.57}$$

Eberhardt (1987) used a minimum chi-square criterion of

$$\sum \frac{(\text{Observed}_i - \text{Expected}_i)^2}{\text{Expected}_i}.$$

Alternatively, Jeffries et al. (2003) minimized the sum of squared proportional residuals by using the expression

$$\sum \left[\frac{\text{Observed}_i - \text{Expected}_i}{\text{Expected}_i} \right]^2.$$

We recommend using nonlinear least squares with its well-known statistical properties (Seber and Wild 1989).

The continuous-time analogue to the discrete-time model (7.52) can be written as

$$
\begin{aligned}
N_t &= \frac{K}{1 - \left(\dfrac{K - N_0}{N_0} \right) e^{-r_{\text{MAX}} t}} \\[2ex]
&= \frac{K}{1 - \left(\dfrac{K - N_0}{N_0} \right) \lambda_{\text{MAX}}^{-t}}
\end{aligned}
\tag{7.58}
$$

where r_{MAX} is the instantaneous intrinsic rate of increase. The advantage of Eq. (7.58) is that consecutive values of N_t and N_{t+1} are not needed in the regression analysis. More of the data can be used to fit the logistic-growth model with Eq. (7.58) than with Eq. (7.52) or (7.53) if annual abundance values are missing. A disadvantage in using model (7.58) is the need to estimate an additional parameter (N_0).

Numerous models have been proposed to moderate the effect of density dependence (Table 2.1). Pella and Tomlinson (1969) proposed a generalization of the logistic model of the form

$$N_{t+1} = N_t + N_t R_{\text{MAX}} \left(1 - \frac{N_t}{K} \right)^z \tag{7.59}$$

which modifies the speed at which the population approaches its carrying capacity. When the parameter $z > 1$, the population approaches its carrying capacity faster than what a logistic model would predict. When $z < 1$, the rate of growth is lower than a logistic model. Eberhardt (1987) and Jeffries et al. (2003) have subsequently applied the generalized logistic model to an array of wildlife species, often fitting the demographic trends better when $z > 1$.

Example 7.6: Estimating λ_{MAX} under Logistic Growth for Harbor Seals (*Phoca vitulina*), Washington Coast

Jeffries et al. (2003) provided an extensive review of the status and trends of harbor seals in Washington State. Subsequent to the Marine Mammal Protection Act (MMPA) of 1972, harbor seal populations have rebounded in Washington with typical logistic growth patterns. Abundance values for the coastal estuarine population, 1975–1999, follow that logistic growth pattern (Table 7.2).

Table 7.2. Annual abundance estimates of harbor seals in the coastal estuarine environment of Washington State, 1975–1999.

Year	N
1975	1694
1976	1742
1977	2082
1978	2570
1980	2864
1981	4408
1982	5197
1983	4416
1984	4203
1985	6008
1986	4807
1987	7600
1988	6796
1989	6475
1991	8681
1992	7761
1993	8161
1994	5786
1995	6492
1996	7191
1997	7643
1999	7117

(Data from Jeffries et al. 2003.)

The discrete-time models (Eqs. 7.53, 7.59) were fitted to the years of data with consecutive abundance values. The standard logistic model (7.53) produced values of $\hat{R}_{MAX} = 0.5330$ ($\widehat{SE} = 0.186$) and $\hat{K} = 6701$ ($\widehat{SE} = 517$) or $\hat{\lambda}_{MAX} = 1.5530$ ($\widehat{SE} = 0.1836$). The generalized logistic model (7.59) produced values of $\hat{R}_{MAX} = 0.4382$ ($\widehat{SE} = 0.5778$), and $\hat{K} = 6755$ ($\widehat{SE} = 639$) with $z = 1.2720$ ($\widehat{SE} = 1.9945$), yielding $\hat{\lambda}_{MAX} = 1.4382$ ($\widehat{SE} = 0.5778$). The fit was not significantly improved ($P = 0.7666$) using the generalized logistic model, and $\hat{z}$ was not significantly different from 1 ($P = 0.8915$). Plots of the two fitted curves are nearly identical (Fig. 7.7).

Using all years of data, the continuous form of the logistic-growth model (7.58) was also fit to the coastal estuarine harbor seal data. The fitted model produced values of $\hat{r}_{MAX} = 0.2650$ ($\widehat{SE} = 0.0553$), $\hat{K} = 7511$ ($\widehat{SE} = 396$), and $\hat{N}_0 = 1379$ ($\widehat{SE} = 378$). An approximate estimate of λ is then

$$\hat{\lambda}_{MAX} = e^{0.2650}$$
$$= 1.3034,$$

with an estimated standard error based on the delta method of $\widehat{SE}(\hat{\lambda}) = \widehat{SE}(\hat{r}) \cdot \hat{\lambda} = 0.0721$. The residual SS for the continuous model (7.58) with more data of 833.06 (19 degrees of freedom) is smaller than the residual SS for the discrete model (7.53) of 1001.95 (16 degrees of freedom), indicating better fit for the continuous model to the data. The value of $\hat{\lambda}_{MAX} = 1.30$ therefore appears to be a more reasonable estimate for this population.

a. Discrete-time logistic curves

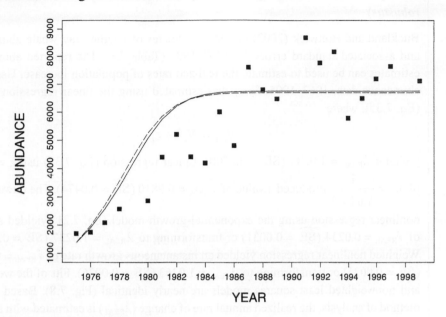

b. Continuous-time logistic curve

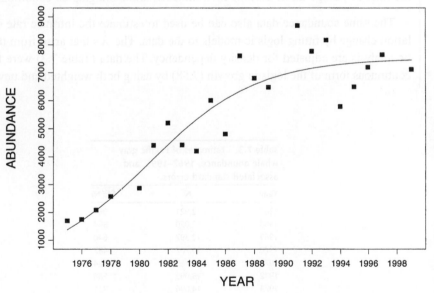

Figure 7.7. Coastal estuarine harbor seal abundance in Washington State, 1975–1999 (Jeffries et al. 2003): (a) fitted to the logistic growth curve (Eq. 7.53; dashed line) and the generalized logistic curves (Eq. 7.59; solid line) and (b) fitted to the continuous logistic growth curve (Eq. 7.58).

The three values of $\hat{\lambda}_{MAX}$ generated from the same data set (1.5530, 1.4382, and 1.3034) illustrate how sensitive estimates of λ_{MAX} are to the estimation approach. Eberhardt (1987) showed similar sensitivity when estimating λ_{MAX} for a variety of species and populations. We recommend a sensitivity analysis as part of any thorough estimation of λ_{MAX}.

Example 7.7: Estimating λ_{REAL} and λ_{MAX} for Pacific Gray Whales (*Eschrichtius robustus*)

Buckland and Breiwick (2002) provided estimates of Pacific gray whale abundance and associated standard errors for 1967–1995 (Table 7.3). The reported abundance estimates can be used to estimate the realized rates of population increase. Using the abundance data, 1967–1995, λ_{REAL} was estimated using the linear regression model (Eq. 7.35), where

$$N_{t+1} = N_t\lambda + \varepsilon_t,$$

yielding $\hat{\lambda}_{REAL} = 1.0284$ ($\widehat{SE} = 0.0500$). Linear regression (Eq. 7.38) using weights of $w_t = \dfrac{1}{\widehat{Var}(\hat{N}_t)}$ produced a value of $\hat{\lambda}_{REAL} = 0.9810$ ($\widehat{SE} = 0.0474$). The unweighted nonlinear regression using the exponential-growth model (Eq. 7.21) yielded a value of $\hat{r}_{REAL} = 0.0234$ ($\widehat{SE} = 0.0031$) or transforming to $\hat{\lambda}_{REAL} = 1.0237$ ($\widehat{SE} = 0.0032$). Weighted nonlinear regression yielded an instantaneous growth rate of $\hat{r}_{REAL} = 0.0260$ ($\widehat{SE} = 0.0035$) or transforming to $\hat{\lambda}_{REAL} = 1.0263$ ($\widehat{SE} = 0.0036$). Fits of the weighted and nonweighted least-squares models are nearly identical (Fig. 7.8). Based on the method of analysis, the realized annual rate of change (λ_{REAL}) is estimated with a range from 0.9810 to 1.0284. The λ_{REAL}'s estimated from the exponential-growth models are the rates of change under current environmental and demographic conditions.

The same abundance data also can be used to estimate the intrinsic rate of population change by fitting logistic models to the data. The $\hat{\lambda}$'s that arise from the logistic analysis are adjusted for density dependency. The data (Table 7.3) were fit to the continuous form of the logistic growth (7.58) by using both weighted and unweighted

Table 7.3. Estimates of Pacific gray whale abundance, 1967–1995, and associated standard errors.

Year	$\hat{N}$	$\widehat{SE}(\hat{N})$
1967	12,921	964
1968	12,070	594
1969	12,597	640
1970	10,707	487
1971	9,760	524
1972	15,099	688
1973	14,696	731
1974	12,955	659
1975	14,520	796
1976	15,304	669
1977	16,879	1,095
1978	13,104	629
1979	16,364	832
1984	21,443	1,182
1985	20,113	927
1987	20,869	913
1992	17,674	1,029
1993	23,109	1,262
1995	22,263	1,078

(Data from Buckland and Breiwick 2002.)

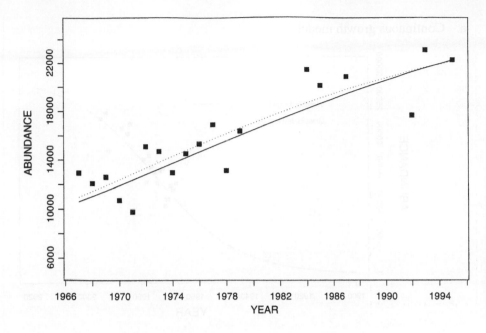

Figure 7.8. Pacific gray whale abundance, 1967–1995 (Buckland and Breiwick 2002), fitted to exponential-growth models using unweighted (Eq. 7.23; dotted line) and weighted (Eq. 7.24; solid line) least squares.

nonlinear least squares. The unweighted analysis yielded values of $\hat{N}_0 = 10967$ ($\widehat{SE} = 993$), $\hat{K} = 26050$ ($\widehat{SE} = 6286$), and $\hat{r}_{MAX} = 0.0737$ ($\widehat{SE} = 0.0375$), translating to a $\hat{\lambda}_{MAX} = 1.0765$ ($\widehat{SE} = 0.0404$). The weighted analysis yielded values of $\hat{N}_0 = 10610$ ($\widehat{SE} = 882$), $\hat{K} = 28388$ ($\widehat{SE} = 11979$), and $\hat{r}_{MAX} = 0.0646$ ($\widehat{SE} = 0.0394$), translating to a $\hat{\lambda}_{MAX} = 1.0667$ ($\widehat{SE} = 0.0420$). Fits of the weighted and unweighted logistic least-squares models are nearly identical (Fig. 7.9) in the observed range of the data. The discrete-time logistic model (7.52) produced values of $\hat{K} = 17842$ ($\widehat{SE} = 5236$), $\hat{R}_{MAX} = 0.2149$ ($\widehat{SE} = 0.2502$), or equivalent $\hat{\lambda}_{MAX} = 1.2149$ ($\widehat{SE} = 0.2502$). The discrete form of the model did not fit the data nearly as well as did the continuous form of the logistic model (7.58). The poor fit was an artifact of having missing consecutive years of abundance data needed for the discrete-time model (Table 7.3).

By using results of the logistic-growth model to estimate $\hat{\lambda}_{MAX}$, an alternative estimate of λ_{REAL} for 1967–1995 can be calculated by the expression

$$\lambda_{REAL} = 1 + (\lambda_{MAX} - 1)\left(1 - \frac{N_t}{K}\right). \tag{7.60}$$

Setting $\hat{\lambda}_{MAX} = 1.0765$ and $\hat{K} = 26050$ and using a midrange value of $N_t = \dfrac{(12921 + 22263)}{2}$, a calculated value of $\hat{\lambda}_{REAL}$ is possible, where

$$\hat{\lambda}_{REAL} = 1 + (1.0765 - 1)\left(1 - \frac{17592}{26050}\right) = 1 + 0.0765(0.3247) = 1.0248.$$

This value is very similar to those yielded by the exponential analyses during the same time period.

a. Continuous growth model

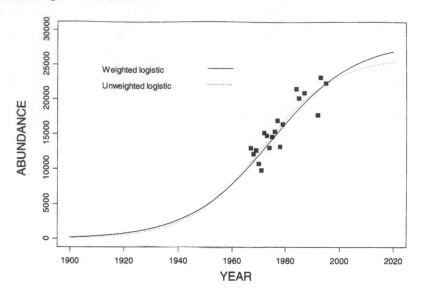

b. Discrete-time growth model

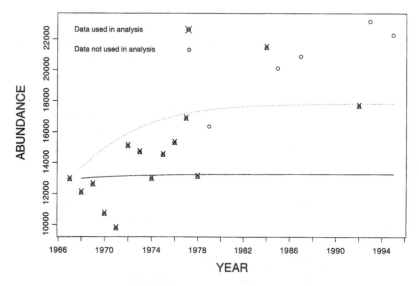

Figure 7.9. Pacific gray whale abundance, 1967–1995 (Buckland and Breiwick 2002): (a) fitted to the continuous logistic-growth model (Eq. 7.58) and (b) fitted to the discrete-time logistic-growth model (Eq. 7.52) using unweighted (dotted line) and weighted (solid line) least squares.

Discussion of Utility

Populations experiencing exponential growth are the exception. Most populations of interest to wildlife managers are beyond the exponential-growth phase and are experiencing some density-dependent growth. For these populations, the logistic-growth equation (7.52) is appropriate. The rate of change estimated from the logistic-growth equation is the intrinsic rate of increase, λ_{MAX}. This is the same rate of population change estimated from the exponential model (Eq. 7.3) when the population is in exponential growth, λ_{MAX}. The choice of model depends on the growth phase of the population. Data plots can help in model selection. When a population is in exponential growth, a plot of $\ln N_t$ versus t will be a straight-line relationship (Fig. 7.3), as will a plot of N_{t+1} versus

N_t (Fig. 7.4). Density-dependent growth will be characterized by concave, downward curves of the same plots.

When a long, uninterrupted time series of abundance values is available, a reliable fit of the logistic model is possible. However, when only a segment of the curve is present for analysis (Fig. 7.9), alternative model values of N_0, K, and λ_{MAX} may fit the data nearly as well. This becomes particularly true for the generalized logistic model (Eq. 7.59) with its four parameters: N_0, K, λ_{MAX}, and z. Typically, the uncertainty in z is expressed by large standard errors. The standard logistic (Eq. 7.52) and generalized logistic (Eq. 7.59) models are nested (i.e., special cases of one another), permitting model selection using both standard statistical methods and Akaike Information Criterion (AIC) (Burnham and Anderson 1998). Residual analyses and plots of the fitted curves to the data should also be an integral part of any analysis.

Ironically, the exponential model (Eq. 7.3) is often used over short spans of time to estimate the realized rate of change (λ_{REAL}) when a population is experiencing density-dependent growth. Regression analysis using the exponential model (Eq. 7.3) is reasonable as long as the data fit the predicted relationship over the interval of interest. As annual abundance approaches the carrying capacity (K) and density dependence more severely dampens $\left(\text{i.e., } \left(1 - \dfrac{N_t}{K}\right) \to 0\right)$ the growth rate, the exponential model will have a difficult time fitting the data. In these circumstances, an estimate of the average, $\lambda_t = \dfrac{N_{t+1}}{N_t}$, over time (Eq. 7.41) is a better representation of the true realized rate of change (λ_{REAL}).

Only under exceptional circumstances should one expect simple logistic models to adequately describe the complex demographic processes of age-specific recruitment, survival, and possible exploitation. Eberhardt (1987:Figs. 3, 4) achieved exceptional fits of model Eqs. (7.52) and (7.59) by reincrementing the model predictions using consecutive annual abundance levels. A regression analysis based on the SS,

$$SS = \sum_{t=1}^{n-1}\left(N_{t+1} - \left(N_t + N_t(\lambda_{\mathrm{MAX}} - 1)\left(1 - \frac{N_t}{K}\right)\right)\right)^2,$$

only concerns the predictions of N_{t+1} from N_t. This fitting underestimates the effects of stochastic variability and environmental changes on population trends. Minimizing this nonlinear SS ignores the actual population trend in the annual sequences of abundance values $N_0, N_1, N_2, \ldots$. Good predictions of N_{t+1} from N_t are possible without adequately fitting the time series of abundance values. On the other hand, the nonlinear SS,

$$SS = \sum_{t=1}^{n}\left(N_t - \left(\frac{K}{1 - \left(\dfrac{K - N_0}{N_0}\right)e^{-r_{\mathrm{MAX}}t}}\right)\right)^2,$$

attempts to directly predict the abundance sequence $N_1, N_2, \ldots$. These differences in the underlying regression models explain the differences in model fits to the Washington State harbor seal data (Fig. 7.7). The continuous-time model (7.58) obviously fits the time series (Fig. 7.7b) better than does the discrete-time logistic curve (Fig. 7.7a). Very different values of λ_{MAX} may arise from the two nonlinear regression analyses. Thus, which approach is better? A value of λ_{MAX} that explains both year-to-year as well as the time series of abundance values seems preferable.

An estimate of carrying capacity (K) is as important a contribution from the logistic model as the estimate of λ_{MAX} itself. An estimate of K is useful in establishing harvest rates and abundance goals for exploited populations. Maximum sustained yield is theoretically achieved by managing animal abundance at half the carrying capacity $\left(\text{i.e., } \dfrac{K}{2}\right)$. Optimal sustainable population goals endeavor to manage population abundance between the levels of $\dfrac{K}{2}$ and K. We next discuss estimates of λ_{MAX} and λ_{REAL} in the presence of animal removals.

7.6   Growth Models with Removals

Sections 7.4 and 7.5 considered population growth in the absence of harvest or other known anthropogenic take. By ignoring the incidence of take, the estimates of the intrinsic rate of increase (λ_{MAX}) will be underestimated. However, under certain circumstances, the models previously discussed may be used to estimate the realized rates of population change. We now introduce modifications to the exponential and logistic-growth models when removals from the population are known.

7.6.1   Exponential-Growth Models

Eberhardt (1987) considered two variations of exponential-growth models, depending on when the removals occurred relative to the time the population was annually surveyed. Starting with the standard exponential-growth model,

$$N_{t+1} = N_t \lambda.$$

Model I assumes the removals, C_{t+1}, occur just before the population survey, in which case

$$N_{t+1} = N_t \lambda - C_{t+1} . \tag{7.61}$$

Solving for λ,

$$\lambda = \frac{N_{t+1} + C_{t+1}}{N_t} . \tag{7.62}$$

In model (7.61), population growth occurs first and the removals thereafter. In Model II (Eberhardt 1987), the removals (C_t) are assumed to have occurred just after the previous population survey, in which case,

$$N_{t+1} = (N_t - C_t)\lambda . \tag{7.63}$$

Solving for λ,

$$\lambda = \frac{N_{t+1}}{(N_t - C_t)} . \tag{7.64}$$

Table 7.4.   Elk abundance data from the Cache Management Unit, Utah, 1952–1966.

t + 1	Current year		t	Previous year	$\hat{\lambda}_t$
	N_{t+1}	C_{t+1}		N_t	
1952	1253	199	1951	1167	1.2442
1958	1377	296	1957	1300	1.2869
1966	970	99	1965	777	1.3758
					$\hat{\bar{\lambda}}_{REAL} = 1.3023$

Estimates of λ_{REAL} were calculated on an annual basis by using Eq. (7.61). (Data from Kimball and Wolfe 1974, as reported by Eberhardt 1987.)

In model (7.63), the removals (C_t) reduce the population abundance (N_t) before population growth is assumed to occur. Least squares can be readily used to estimate the parameter λ in the cases of Eqs. (7.61) and (7.63). Alternatively, Eberhardt (1987) used Eqs. (7.62) and (7.64) on an annual basis, averaging the resultant $\hat{\lambda}$'s across years.

Example 7.8: Estimating λ_{REAL} in the Presence of Removals for Elk, Cache Management Unit, Utah

Eberhardt (1987) provided an example using Eq. (7.61) for estimating λ_{REAL} in the presence of removals for a Utah elk herd. Elk abundance ranged annually from 939 to 1377, and was considered to be beyond the exponential-growth phase. Only data (Kimball and Wolfe 1974) from consecutive years are reproduced here (Table 7.4). The least-squares estimate based on minimizing the SS is

$$SS = \sum_{i=1}^{n-1}(N_{i+1} - (N_i\lambda - C_{i+1}))^2$$
$$= \sum_{i=1}^{n-1}((N_{i+1} + C_{i+1}) - N_i\lambda)^2, \qquad (7.65)$$

with $\hat{\lambda}_{REAL} = 1.2857$ ($\widehat{SE} = 0.0315$). Annual estimates of $\hat{\lambda}_{REAL}$ ranged from 1.2442 to 1.3758, with a mean value of $\hat{\bar{\lambda}}_{REAL} = 1.3023$ ($\widehat{SE} = 0.0388$).

Example 7.9: Estimating λ_{MAX} in the Presence of Removals for Feral Horses, Jackie's Butte, Oregon

Eberhardt (1987) illustrated the use of Eq. (7.63) using feral horse data from Jackie's Butte area of southeastern Oregon, 1969–1980 (Eberhardt et al. 1982). During that period, three removals occurred: 181 horses in 1970, 137 in 1976, and 136 in 1978 (Table 7.5). Horse abundance ranged annually from 94 to 280, and was considered to be in exponential growth between removals. The least-squares estimate of λ, based on minimizing the

$$SS = \sum_{i=1}^{n-1}(N_{i+1} - (N_i - C_i)\lambda)^2, \qquad (7.66)$$

Table 7.5.   Feral horse abundance data, 1969–1980, from Jackie's Butte in southeastern Oregon.

Current year		Previous year			
$t + 1$	N_{t+1}	t	N_t	C_t	$\hat{\lambda}_{MAX}$
1970	263	1969	225	0	1.1689
1972	94	1971	263	181	1.1463
1973	113	1972	94	0	1.2021
1974	140	1973	113	0	1.2389
1976	280	1975	222	0	1.2613
1977	201	1976	280	137	1.4056
1978	235	1977	201	0	1.1692
1979	128	1978	235	136	1.2929
1980	153	1979	128	0	1.1953

$$\hat{\bar{\lambda}}_{MAX} = 1.2312$$

Estimates of λ_{MAX} were calculated on an annual basis by using Eq. (7.64). (Data from Eberhardt et al. 1982, Eberhardt 1987.)

yields $\hat{\lambda}_{MAX} = 1.2250$ ($\widehat{SE} = 0.0256$). Annual estimates ranged from 1.1462 to 1.4056 with a mean value (Eq. 7.41) of $\hat{\bar{\lambda}}_{MAX} = 1.2312$ ($\widehat{SE} = 0.0269$). Note the regression models for both Eqs. (7.65) and (7.66) are straight-line regressions through the origin, where the independent or dependent variables have been adjusted for known removals.

Example 7.10: Estimating the Apparent Rate of Change (λ_{APP}) for Longhorn Cattle, Fort Niobrara National Wildlife Refuge, Nebraska

Fredin (1984a) used data from Gross et al. (1973) and Bartholow (1977) to describe the growth of a longhorn cattle population on the Fort Niobrara National Wildlife Refuge, Nebraska, from 1936–1937 to 1971–1972. The population grew from an initial abundance of 6 animals in 1936–1937 to 39 animals in 1942–1943 before the population was harvested (Table 7.6). From 1942–1943 to 1971–1972, longhorn cattle were harvested. Using data presented by Fredin (1984a), we calculated annual rates of change before and after harvest. Eberhardt et al. (1982) expressed the estimate of λ after accounting for removals as

$$\lambda_{APP} = \frac{N_{t+1} - C_{t+1}}{N_t},$$

and described it as the "apparent" rate of change. This apparent rate of change measures the actual year-to-year change in abundance as the result of natural processes and exploitation. The realized rate of change was calculated before the annual removals, where

$$\lambda_{REAL} = \frac{N_{t+1}}{N_t}.$$

As might be expected, the apparent rate of change is less than the realized rate of change when removals occur. For the period 1950–1954 onward, the average realized rate of change was $\hat{\lambda}_{REAL} = 1.31$ ($\widehat{SE} = 0.014$), whereas the apparent rate of change was reduced to $\hat{\lambda}_{APP} = 1.04$ ($\widehat{SE} = 0.018$). In this example, the population continued to increase an average of 4% per year despite the planned removals.

Table 7.6.   Abundance and removals of longhorn cattle, 1936–1972, at Fort Niobrara National Wildlife Refuge, Nebraska, with realized and apparent rates of annual change.

Year (1 Oct–20 Sep)	No. of cattle at start of year (1 Oct) (N_t)[a]	No. of cattle at end of year (30 Sep) (N_{t+1})	No. of cattle removed (C_{t+1})	$\hat{\lambda}$ after removal $(\hat{\lambda}_{APP})$[b]	$\hat{\lambda}$ before removal $(\hat{\lambda}_{REAL})$
1936–1937	6	10	0	1.67	1.67
1937–1938	10	14	0	1.40	1.40
1938–1939	14	18	0	1.29	1.29
1939–1940	18	25	0	1.39	1.39
1940–1941	25	31	0	1.24	1.24
1941–1942	31	39	0	1.26	1.26
1942–1943	39	50	4	1.18	1.28
1943–1944	46	60	3	1.24	1.30
1944–1945	57	64	8	0.98	1.12
1945–1946	56	79	9	1.25	1.41
1946–1947	70	93	10	1.19	1.33
1947–1948	83	109	17	1.11	1.31
1948–1949	92	121	21	1.09	1.32
1949–1950	100	133	33	1.00	1.33
1950–1951	100	130	30	1.00	1.30
1951–1952	100	129	30	0.99	1.29
1952–1953	99	135	33	1.03	1.36
1953–1954	102	153	46	1.05	1.50
1954–1955	107	157	48	1.02	1.47
1955–1956	109	152	49	0.94	1.39
1956–1957	103	132	33	0.96	1.28
1957–1958	99	124	21	1.04	1.25
1958–1959	103	132	28	1.01	1.28
1959–1960	104	137	33	1.00	1.32
1960–1961	104	139	30	1.05	1.34
1961–1962	109	143	30	1.04	1.31
1962–1963	113	148	18	1.15	1.31
1963–1964	130	171	20	1.16	1.32
1964–1965	151	198	18	1.19	1.31
1965–1966	180	232	6	1.26	1.29
1966–1967	226	287	49	1.05	1.27
1967–1968	238	308	44	1.11	1.29
1968–1969	264	333	71	0.99	1.26
1969–1970	262	332	99	0.89	1.27
1970–1971	233	282	53	0.98	1.21
1971–1972	229	285	59	N/A	1.24

[a]Abundance after removals, which is assumed to have taken place on 1 October.
[b]As reported by Fredin (1984a).
N/A indicates not available. (Data from Fredin 1984a.)

7.6.2   Accounting for Missing Abundance Values

Regression models (7.61) and (7.63) require successive years of abundance surveys to estimate λ. Eberhardt (1987) suggested using polynomial models to estimate λ when gaps in the data exist. For his Model I, a recursive relationship can be used, where

$$N_1 = N_0\lambda - C_1$$
$$N_2 = N_1\lambda - C_2 = N_0\lambda^2 - C_1\lambda - C_2 \qquad (7.67)$$
$$N_3 = N_2\lambda - C_3 = N_0\lambda^3 - C_1\lambda^2 - C_2\lambda - C_3 \qquad (7.68)$$
$$\vdots$$

For example, if one intervening year of data (N_1) is missing, Eq. (7.67) can be used to estimate λ, using abundance data from years N_0 and N_2. Should 2 years of intervening data be missing (e.g., N_1 and N_2), Eq. (7.68) can be used to estimate λ by using the survey results for N_0 and N_3.

Similarly, using the Eberhardt (1987) Model II (Eq. 7.63), the recursive relationship is

$$N_1 = (N_0 - C_0)\lambda$$
$$N_2 = (N_1 - C_1)\lambda = (N_0 - C_0)\lambda^2 - C_1\lambda$$
$$N_3 = (N_2 - C_2)\lambda = (N_0 - C_0)\lambda^3 - C_1\lambda^2 - C_2\lambda$$
$$\vdots$$

and can be used to estimate λ when gaps in survey data exist.

Example 7.11: Estimating λ_{REAL} When Missing Abundance Values Exist for Elk, Cache Management Unit, Utah

We illustrate the estimation of λ_{REAL} when gaps in survey data exist, returning to the Utah elk data set presented by Eberhardt (1987). Eberhardt (1987) reported the following elk data for the years 1955–1957:

Year	Coded year	Abundance (N_t)	Removals (C_t)
1955	0	1360	211
1956	1	–	236
1957	2	1300	310

The estimate of a common λ for 1955–1957 is calculated by using Eq. (7.67), where

$$N_0\lambda^2 - C_1\lambda - C_2 - N_2 = 0$$
$$1360\lambda^2 - 236\lambda - 310 - 1300 = 0$$

or

$$1360\lambda^2 - 236\lambda - 1610 = 0.$$

The solutions to the quadratic equation are 1.1783 and −1.0047, yielding the estimate $\hat{\lambda}_{REAL} = 1.1783$.

7.6.3 Logistic-Growth Models

Extension of Eqs. (7.61) and (7.63) to logistic growth are straightforward. In the case in which removals C_{t+1} occur just before the annual population surveys, then

$$N_{t+1} = N_t + N_t R_{MAX}\left(1 - \frac{N_t}{K}\right) - C_{t+1}, \tag{7.69}$$

which can be generalized to

$$N_{t+1} = N_t + N_t R_{\text{MAX}}\left[1 - \left(\frac{N_t}{K}\right)^z\right] - C_{t+1}. \tag{7.70}$$

When the removals (C_t) are assumed to have occurred just after the previous population survey, then

$$N_{t+1} = (N_t - C_t) + (N_t - C_t)R_{\text{MAX}}\left(1 - \frac{N_t - C_t}{K}\right) \tag{7.71}$$

which can be generalized to

$$N_{t+1} = (N_t - C_t) + (N_t - C_t)R_{\text{MAX}}\left[1 - \left(\frac{N_t - C_t}{K}\right)^z\right]. \tag{7.72}$$

Discussion of Utility

Given constant proportionality through time between a population index and the abundance being monitored (i.e., $I_t = \beta N_t$, for all t), rates of change can be estimated by using either metric. However, this situation drastically changes when animal removals or harvests are considered. Removals (C_t) are expressed in terms of absolute numbers, whereas indices are expressions of relative abundance. Simply substituting I_t for N_t in Eqs. (7.61), (7.63), and (7.69) to (7.72) when calculating rates of change is no longer appropriate. This incompatability of measurement scales illustrates the inherent difficulties of managing exploitation while monitoring populations solely by indices. Eventually, the situation compels many wildlife managers to attempt to calibrate the population index to a measure of absolute abundance. The index removal method (Sections 6.7, 9.4) provides a way of calibrating an index to abundance based on the effect a known number of animal removals has on the change in index.

The longhorn cattle example illustrates another consideration when estimating and reporting rates of change for exploited populations. Estimates of λ_{REAL} from Eqs. (7.61) and (7.63) are the rates of change that would exist under prevailing population and environmental conditions in the absence of exploitation. These are the rates of change that would be experienced at current abundance levels in the absence of exploitation, acknowledging the population is at its current abundance level, in part, because of past exploitations. Thus, this is the rate of change that would have occurred had exploitation ceased that year.

The rate of change calculated for the longhorn cattle by Fredin (1984a), after considering the number of removals, is the "apparent" rate of change under prevailing demographic conditions and exploitation. Here, λ_{APP} measures the actual rate of population change under existing natural and anthropogenic effects. The value of λ_{APP} is a better indicator of existing demographic trends, whereas the estimates from Eqs. (7.61) and (7.63) provide better measures of the population's current growth potential. The difference between the realized and apparent rates of change is a measure of the effect of exploitation on the rate of population growth.

7.7 Productivity-Based Estimator of λ (Kelker 1947)

Kelker (1947) presented a simple method for estimating the finite rate of change for a deer population using estimates of productivity and the percent females in the population. The estimator for λ is based on the following assumptions:

1. There is no mortality of adults.
2. The population experiences one annual breeding season each year.
3. The young mature to reproductive age in 1 year.
4. A constant productivity rate over time.
5. A constant female proportion over time, and hence, a constant sex ratio.

Based on these assumptions, the number of individuals in the population at time $t + 1$ can be written as

$$N_{t+1} = N_t + N_t fP$$
$$= N_t(1 + fP) \tag{7.73}$$

where

N_t = number of adult animals, males and females, in the population at time t, i.e., $N_t = N_{Ft} + N_{Mt}$;

P = average number of young per mature female (i.e., productivity) that survive to the next year;

f = fraction of adults that are female, or $f = \dfrac{R_{F/M}}{1 + R_{F/M}}$, where $R_{F/M} = N_F/N_M$.

Dividing both sides of Eq. (7.73) by N_t, the finite rate of change is estimated by the expression

$$\hat{\lambda} = 1 + \hat{f}\hat{P}. \tag{7.74}$$

The variance estimator of $\hat{\lambda}$ can be derived by the exact variance formula (Goodman 1960) as

$$\widehat{\mathrm{Var}}(\hat{\lambda}) = \widehat{\mathrm{Var}}(\hat{f}\hat{P}) = \hat{P}^2\,\widehat{\mathrm{Var}}(\hat{f}) + \hat{f}^2\,\widehat{\mathrm{Var}}(\hat{P}) - \widehat{\mathrm{Var}}(\hat{f})\widehat{\mathrm{Var}}(\hat{P}). \tag{7.75}$$

Assumption 1 is not realistic, thus limiting the utility of this technique. A more realistic scenario would include adult mortality, where N_{t+1} would be a function of the number of animals surviving from the previous year and juvenile production from the surviving adult females. Thus, N_{t+1} is better expressed as

$$N_{t+1} = SN_t + SN_t fP$$
$$= N_t S(1 + fP).$$

Hence,

$$\hat{\lambda} = S(1 + fP). \tag{7.76}$$

Assuming the population experiences adult mortality and Eq. (7.76) is the true value of λ, the bias of Kelker's (1947) equation is calculated to be

$$\text{Bias}(\hat{\lambda}) = ((1 + fP) - S(1 + fP))$$

$$= \lambda\left(\frac{1}{S} - 1\right).$$

Thus, $\hat{\lambda}$, based on Kelker's (1947) Eq. (7.73), will be overestimated by $\left(\frac{1}{S} - 1\right) \cdot 100\%$, and Kelker's (1947) method should be avoided if at all possible.

Example 7.12: Estimating λ and Productivity for White-Tailed Deer, George Reserve, Michigan

A census conducted in 1928 yielded a count of 2 bucks and 4 pregnant does. A population count in 1933 yielded a total abundance of 160 animals ($N = 160$). Kelker (1947), knowing a 1 : 1 sex ratio was not the case in 1928, made the herd equal to an imaginary 8 animals (4 bucks and 4 does) by adding 2 phantom males to the initial and final herd counts. The estimation of λ was then

$$162 = 8\hat{\lambda}^6$$

$$\hat{\lambda} = 1.651.$$

Next, he estimated P based on the expression

$$\lambda = 1 + fP.$$

Assuming again $f = \frac{1}{2}$, then

$$1.651 = 1 + \frac{1}{2}\hat{P},$$

yielding $\hat{P} = 1.302$ young/adult females.

Alternatively, he could have assumed the final abundance count had a sex ratio of 1:1, in which case

$$\frac{160}{2} = 4\hat{\lambda}^6$$

$$\hat{\lambda} = 1.6480.$$

To estimate productivity in terms of juvenile females/adult female, one can then use the expression

$$1.6480 = 1 + 1\hat{P}$$

for $f = 1$ to estimate $\hat{P} = 0.648$ juvenile females/adult female.

> The formula $\lambda = 1 + fP$ was originally designed to be applied to "pheasants, rabbits, and other small game" and deer that have a maturation and fecundity schedule as described above (Kelker 1947). This method can be used for estimating λ given $\hat{P}$, as well as estimating P given $\hat{\lambda}$.

7.8   Estimating λ Using the Lotka Equation (Cole 1954)

The potential bias in projected population estimates based on Kelker's (1947) estimator of λ illustrates the need to incorporate all the pertinent demographic information to adequately project population trends. Natural survival, harvest mortality, and productivity all influence population growth rates. In turn, overall survival probabilities and productivity will be affected by the demographic structure of a population. Unbiased estimators of population growth rates must consider the interaction between these processes and population structure. Also, age structure and survival probabilities vary throughout the calendar year. Hence, estimators of λ should consider the timing of data collection with regard to parameterization of demographic processes.

7.8.1   General Expression

Henny et al. (1970) derived a computational equation for estimating λ based on the fundamental relationship

$$N_{t+1} = N_{0,t+1} + \underset{\sim}{S}' \cdot \underset{\sim}{N}_t \tag{7.77}$$

where

$N_{0,t+1}$ = abundance of new recruits in year $t + 1$;
$\underset{\sim}{S}'$ = vector of age-specific survival probabilities at time t;

$$= \begin{bmatrix} S_0 \\ S_1 \\ S_2 \\ S_3 \\ \vdots \end{bmatrix};$$

$\underset{\sim}{N}_t$ = vector of age-specific abundance levels at time t;

$$= \begin{bmatrix} N_{0,t} \\ N_{1,t} \\ N_{2,t} \\ \vdots \end{bmatrix};$$

so that $\underset{\sim}{S}'\underset{\sim}{N}_t$ are the numbers of individuals that survived from t to time $t + 1$. The anniversary date of the modeling must be specified. If the anniversary date was the spring breeding population, there would be age classes 1, 2, ..., m and the model would be incremented as

$$N_{t+1} = N_{1,t+1} + [S_1 \ S_2 \ldots] \begin{bmatrix} N_{1,t} \\ N_{2,t} \\ \vdots \end{bmatrix}.$$

Alternatively, if the population being modeled is the fall harvestable population, there would be age classes 0, 1, 2, . . . and the model would be incremented as

$$N_{t+1} = N_{0,t+1} + [S_0 \ S_1 \ldots] \begin{bmatrix} N_{0,t} \\ N_{1,t} \\ \vdots \end{bmatrix}.$$

The number of new recruits in year t (i.e., $N_{0,t}$) is calculated as the product of age-specific fecundity rates, F_x, where

$$N_{0,t} = F_1 N_{1,t} + F_2 N_{2,t} + F_3 N_{3,t} + \cdots \tag{7.78}$$

Next, using the survival process where

$$\begin{aligned} N_{1,t} &= S_0 N_{0,t-1} \\ N_{2,t} &= S_0 S_1 N_{0,t-2} \\ &\vdots \end{aligned} \tag{7.79}$$

and substituting Eq. (7.79) into Eq. (7.78) yields

$$\begin{aligned} N_{0,t} = {} & F_1 S_0 N_{0,t-1} + F_2 S_0 S_1 N_{0,t-2} + F_3 S_0 S_1 S_2 N_{0,t-3} \\ & + \cdots + F_w S_0 S_1 S_2 \ldots S_{w-1} N_{0,t-w} \end{aligned} \tag{7.80}$$

where

w = oldest age;

$w + 1$ = number of age classes if the youngest age class is 0.

Under the conditions of a stable age distribution (SAD),

$$\frac{N_{0,t}}{N_{0,t-1}} = \lambda$$

or, in general,

$$N_{0,t-i} = N_{0,t}\lambda^{-i}. \tag{7.81}$$

Substituting Eq. (7.81) recursively into Eq. (7.80) yields

$$N_{0,t} = F_1 S_0 N_{0,t}\lambda^{-1} + F_2 S_0 S_1 N_{0,t}\lambda^{-2} + \cdots + F_w S_0 S_1 S_2 \ldots S_{w-1} N_{0,t}\lambda^{-w}. \tag{7.82}$$

Dividing both sides of this equation by $N_{0,t}$ produces the fundamental equation

$$1 = F_1 S_0 \lambda^{-1} + F_2 S_0 S_1 \lambda^{-2} + F_3 S_0 S_1 S_2 \lambda^{-3} + \cdots + F_w S_0 S_1 \ldots S_{w-1}\lambda^{-w}, \tag{7.83}$$

which is valid for animals reproducing as 1-year-olds. A more generalized form of the equation is

$$\begin{aligned} 1 = {} & F_a S_{0,a}\lambda^{-a} + F_{a+1} S_{0,a} S_a \lambda^{-(a+1)} + F_{a+2} S_{0,a} S_a S_{a+1}\lambda^{-(a+2)} + \cdots \\ & + F_w S_{0,a} S_a S_{a+1} \ldots S_{w-1}\lambda^{-w}, \end{aligned} \tag{7.84}$$

or

$$1 = F_a S_{0,a} \lambda^{-a} + F_{a+1} S_{0,a+1} \lambda^{-(a+1)} + F_{a+2} S_{0,a+2} \lambda^{-(a+2)} + \cdots + F_w S_{0,w} \lambda^{-w}, \quad (7.85)$$

where

a = age at first breeding;
$S_{0,a}$ = survival from birth to age a.

This is the Lotka equation (Cole 1954), which is usually written in terms of the instantaneous rate of change, $\lambda = e^r$, i.e.,

$$1 = \sum_{x=a}^{w} e^{-rx} F_x S_{0,x}, \quad (7.86)$$

where

F_x = age-specific fecundity, the number of females produced per breeding female of age x;

$S_{0,x}$ = survivorship to age x, calculated as, $S_{0,x} = \prod_{i=0}^{x-1} S_i$, (Chapter 5).

Throughout the rest of the discussion, Eq. (7.86) will be rewritten in terms of λ, where

$$1 = \sum_{x=a}^{w} \lambda^{-x} F_x S_{0,x}. \quad (7.87)$$

The estimate of the finite rate of increase (λ) is a solution to the polynomial (7.87) and is often solved numerically.

Typically, there is no closed form estimator for $\hat{r}$ or $\hat{\lambda}$ from Eq. (7.86) or (7.87). However, these equations provide the basis for examining relationships between birth and death parameters and λ. Uncertainty in the estimation of λ has been largely limited to sensitivity analysis involving changing the estimates of fecundity and survival (McLellan 1989) and calculating a pseudo-distribution for $\hat{\lambda}$ (or $\hat{r}$) by use of Monte Carlo methods (Tait and Bunnell 1980, Van Sickle et al. 1987) or bootstrap methods (Meyer et al. 1986). Henny et al. (1970) used Eq. (7.87) to calculate λ for special cases of $\underset{\sim}{S}$ and $\underset{\sim}{F}$, and to explore demographic relationships between the parameters. The special cases allow simplified expressions of $\hat{\lambda}$ and the calculation of a variance estimator.

Equation (7.84) is often simplified because age-specific fecundity rates for more than a few age classes may not be available, and in some cases, age classes contributing little to population growth (e.g., low birth rates among oldest age classes) are truncated from the equation (Eberhardt 1985). Under the assumptions of constant fecundity and constant adult survival, Eq. (7.84) can be rewritten as

$$1 = \frac{F S_{0,a}}{\lambda^a} \left(1 + \frac{S}{\lambda} + \frac{S^2}{\lambda^2} + \cdots + \frac{S^{w-a}}{\lambda^{w-a}} \right). \quad (7.88)$$

Multiplying both sides of Eq. (7.88) by λ^a yields

$$\lambda^a = F S_{0,a} \left(1 + \frac{S}{\lambda} + \frac{S^2}{\lambda^2} + \cdots + \frac{S^{w-a}}{\lambda^{w-a}} \right). \quad (7.89)$$

Multiplying both sides of Eq. (7.89) by $\left(1 - \dfrac{S}{\lambda}\right)$ and using the algebraic identity

$$(1 - x^{n+1}) = (1 - x)(1 + x + x^2 + \cdots + x^n),$$

where $x = \dfrac{S}{\lambda}$ (Taylor and Mann 1983:566), leads to the expression

$$\lambda^a\left(1 - \frac{S}{\lambda}\right) = FS_{0,a}\left(1 - \left(\frac{S}{\lambda}\right)^{w-a+1}\right),$$

which, when simplified, is Eberhardt's (1982) equation

$$0 = \lambda^a - S\lambda^{a-1} - FS_{0,a}\left(1 - \left(\frac{S}{\lambda}\right)^{w-a+1}\right). \tag{7.90}$$

Equation (7.90) has been used to estimate λ for feral horses (Eberhardt et al. 1982), elk (Eberhardt et al. 1996), grizzly bears (Eberhardt et al. 1994, Hovey and McLellan 1996, Mace and Waller 1998), and sea otters (*Enhydra lutris*) (Eberhardt 1995). Estimation of λ using Eq. (7.90) typically requires numerical methods for the higher-order polynomials.

By using the delta method and implicit differentiation, Eberhardt et al. (1994) derived a variance estimator for λ using Eq. (7.90). The variance estimator (Eberhardt et al. 1994:361) is written as

$$\mathrm{Var}(\lambda) \doteq \mathrm{Var}(\hat{S}_{0,a})\left(\frac{\partial \lambda}{\partial S_{0,a}}\right)^2_{|E(S_{0,a},S,F)} + \mathrm{Var}(\hat{S})\left(\frac{\partial \lambda}{\partial S}\right)^2_{|E(S_{0,a},S,F)}$$
$$+ \mathrm{Var}(\hat{F})\left(\frac{\partial \lambda}{\partial F}\right)^2_{|E(S_{0,a},S,F)}, \tag{7.91}$$

where

$$\frac{\partial \lambda}{\partial S_{0,a}} = \frac{F\lambda^2\left[\left(\dfrac{S}{\lambda}\right)^{w-a+1} - 1\right]}{A},$$

$$\frac{\partial \lambda}{\partial S} = \frac{\lambda\left[(w - a + 1)S_{0,a}F\lambda\left(\dfrac{S}{\lambda}\right)^{w-a+1} - S\lambda^a\right]}{SA},$$

$$\frac{\partial \lambda}{\partial F} = \frac{S_{0,a}\lambda^2\left[\left(\dfrac{S}{\lambda}\right)^{w-a+1} - 1\right]}{A}, \text{ and}$$

$$A = (w - a + 1)S_{0,a}F\lambda\left(\frac{S}{\lambda}\right)^{w-a+1} + \lambda^a(aS - a - a\lambda).$$

In simulation studies conducted by Eberhardt et al. (1994) and Eberhardt (1995), variances estimated by the delta method were found to be similar to those obtained by bootstrapping. The advantage of explicit variance estimators is that they allow researchers to assess the relative contribution of each parameter estimate to the overall size of $\widehat{\text{Var}}(\hat{\lambda})$.

A further simplification of Eq. (7.88) can be made by allowing the number of adult age classes to extend to infinity, i.e., $w \to \infty$. The equation is written as

$$1 = \frac{FS_{0,a}}{\lambda^a}\left(1 + \frac{S}{\lambda} + \frac{S^2}{\lambda^2} + \frac{S^3}{\lambda^3} + \cdots\right),$$

or

$$1 = \frac{FS_{0,a}}{\lambda^a} \cdot \sum_{i=0}^{\infty}\left(\frac{S}{\lambda}\right)^i. \tag{7.92}$$

By using the series expansion

$$\sum_{i=1}^{\infty}\left(\frac{S}{\lambda}\right)^i = \frac{1}{1 - \dfrac{S}{\lambda}}$$

which assumes $0 < \dfrac{S}{\lambda} < 1$ or $S < \lambda$, Eq. (7.92) can be rewritten as

$$\lambda^a - S\lambda^{a-1} = FS_{0,a},$$

or, more simply,

$$\lambda = S + \frac{FS_{0,a}}{\lambda^{a-1}}, \tag{7.93}$$

where again

$$S_{0,a} = \prod_{x=0}^{a-1} S_x.$$

The variance of $\hat{\lambda}$ is approximated by the delta method as

$$\text{Var}(\hat{\lambda}) \doteq \text{Var}(\hat{S})\left(\frac{\partial \lambda}{\partial S}\right)^2 + \text{Var}(\hat{F})\left(\frac{\partial \lambda}{\partial F}\right)^2 + \widehat{\text{Var}}(\hat{S}_{0,a})\left(\frac{\partial \lambda}{\partial S_{0,a}}\right)^2.$$

Rewriting Eq. (7.93) as

$$0 = S\lambda^{a-1} + FS_{0,a} - \lambda^a$$

and using implicit differentiation, the first partial derivatives are

$$\frac{\partial \lambda}{\partial S} = -\frac{\lambda^{(a-1)}}{\lambda^{(a-2)}((a-1)S - a\lambda)},$$

$$\frac{\partial \lambda}{\partial F} = -\frac{S_{0,a}}{\lambda^{(a-2)}((a-1)S - a\lambda)}, \text{ and}$$

$$\frac{\partial \lambda}{\partial S_{0,a}} = -\frac{F}{\lambda^{(a-2)}((a-1)S - a\lambda)}.$$

Substituting the first partial derivatives into the variance expression yields

$$\text{Var}(\hat{\lambda}) \doteq \frac{\left(\text{Var}(\hat{S})\lambda^{2(a-1)} + \text{Var}(\hat{F})S_{0,a}^2 + \text{Var}(\hat{S}_{0,a})F^2\right)}{\left(\lambda^{(a-2)}((a-1)S - a\lambda)\right)^2},$$ (7.94)

assuming F, S, and $S_{a,0}$ are estimated independently. The validity of this assumption will depend on the nature of data collection and analysis used in calculating the demographic parameters.

Equation (7.93) is a general equation for species that first breed at age a, assuming a common fecundity F across all age classes, constant adult survival across all ages, and $S < \lambda$. The last condition may be the hardest to satisfactorily test and will not be met when fecundity is quite low and adult survival is high. If depensation exists (Quinn and Deriso 1999) arising from an Allee effect, fecundity is reduced at low abundances and λ may be less than S (Begon et al. 1990). As will be seen below, however, the assumption of $S < \lambda$ can be relaxed without consequence.

Survival estimates obtained from age-structure data are often calculated under the assumptions of stable and stationary population. Caughley and Birch (1971) caution the resulting estimates of λ based on these survival estimates will be biased if the assumptions are not true. In the situation in which a population is not stable and stationary, survival estimates should be obtained from animal marking studies. However, if a population has reached a SAD but is not stationary (i.e., $\lambda \neq 1$), λ can be estimated by using the method of Udevitz and Ballachey (1998) (Section 5.6.2), given the added condition that λ is constant over time.

7.8.2 Evaluation of Average Fecundity (Henny et al. 1970)

Henny et al. (1970) also used Eq. (7.87) to examine conditions under which population abundance would be constant (i.e., $N_i = N \quad \forall i$, then $\lambda = 1$). Assuming a SAD, and substituting $\lambda = 1$ into Eq. (7.87) yields

$$1 = F_a S_{0,a} + F_{a+1} S_{0,a+1} + \cdots + F_w S_{0,w}.$$ (7.95)

Equation (7.95) represents the condition of stationary abundance, given the population has achieved a SAD. In this case,

$$\begin{bmatrix} S_{0,a} \\ S_{0,a+1} \\ S_{0,a+2} \\ \vdots \end{bmatrix} \propto \begin{bmatrix} N_{1t} \\ N_{2t} \\ N_{3t} \\ \vdots \end{bmatrix}.$$

Thus, over time, the vector of survivorship is proportional to the vector of abundance by age class. Under a SAD, the average fecundity ($\overline{F}$) is by definition

$$\overline{F} = \frac{F_a S_{0,a} + F_{a+1} S_{0,a+1} + F_{a+2} S_{0,a+2} + \cdots}{S_{0,a} + S_{0,a+1} + S_{0,a+2} + \cdots} = \frac{\displaystyle\sum_{i=0}^{w-a} F_{a+i} S_{0,a+i}}{\displaystyle\sum_{i=0}^{w-a} S_{0,a+i}}.$$

Dividing both sides of Eq. (7.95) by $\sum_{i=0}^{w-a} S_{0,a+i}$ yields

$$\frac{1}{\sum_{i=0}^{w-a} S_{0,a+i}} = \frac{F_a S_{0,a} + F_{a+1} S_{0,a+1} + F_{a+2} S_{0,a+2} + \cdots}{\sum_{i=0}^{w-a} S_{0,a+i}},$$

or

$$\frac{1}{\sum_{i=0}^{w-a} S_{0,a+1}} = \overline{F}. \tag{7.96}$$

Henny et al. (1970) defined

$$T = \sum_{i=0}^{w-a} S_{0,a+i},$$

in which case, according to their notation,

$$\frac{1}{T} = \overline{F}, \tag{7.97}$$

where $\overline{F}$ is the average fecundity per female to the anniversary date.

Equation (7.97) suggests a way to assess whether a population is self-sustaining (i.e., $\lambda \geq 1$) or not (i.e., $\lambda < 1$). By using a direct estimate of average fecundity ($\hat{\overline{F}}$), it can be compared to a value of $\frac{1}{\hat{T}}$ derived from a survivorship schedule ($\hat{\underline{S}}$) to assess whether replacements exceed population losses. Two potential situations may exist:

1. If $\overline{F} \geq \frac{1}{T}$, the population is at or above maintenance level (i.e., $\lambda \geq 1$).

2. If $\overline{F} < \frac{1}{T}$, the population is below maintenance level (i.e., $\lambda < 1$).

Because both $\hat{\overline{F}}$ and $\frac{1}{\hat{T}}$ are estimated with error, a simple direct comparison of the point estimates may be misleading and not reflective of the uncertainty in the sampling data. An asymptotic Z-test could be used to compare the independent estimates $\hat{\overline{F}}$ and $f(\hat{\underline{S}}) = \frac{1}{\hat{T}}$ of the form

$$Z = \frac{\hat{\overline{F}} - \frac{1}{\hat{T}}}{\sqrt{\widehat{\operatorname{Var}}\left(\hat{\overline{F}}\right) + \widehat{\operatorname{Var}}\left(\frac{1}{\hat{T}}\right)}} \tag{7.98}$$

to test the hypotheses

$$H_0 : \overline{F} \geq \frac{1}{T} \text{ versus } H_a : \overline{F} < \frac{1}{T}$$

and assess whether the population is self-maintaining (i.e., H_o: $\lambda \geq 1$) versus declining (i.e., H_a: $\lambda < 1$). H_o would be rejected if $Z < Z_\alpha$ where

$$P(Z < Z_\alpha) = \alpha.$$

The $\widehat{\text{Var}}(\hat{\bar{F}})$ depends on how $\hat{\bar{F}}$ is calculated. The most direct approach would involve taking a simple random sample of n adult females, where

$$\hat{\bar{F}} = \frac{\sum\limits_{i=1}^{n} b_i}{n}$$

with

$$\widehat{\text{Var}}(\hat{\bar{F}}) = \frac{\left(1 - \dfrac{n}{N_{AF}}\right) S_{b_i}^2}{n},$$

where

N_{AF} = abundance of adult females;

b_i = net fecundity for the ith female $(i = 1, \ldots, n)$.

Alternatively, $\hat{\bar{F}}$ could be based on age-specific fecundity F_x and age-composition data (i.e., p_i, fraction of breeding females in age class i), where

$$\hat{\bar{F}} = \sum\limits_{i=a}^{w} \hat{\bar{F}}_i \hat{p}_i$$

and where $\hat{\bar{F}}_i$ is the estimate of mean fecundity for the ith age class $(i = 1, \ldots, k)$. Assuming values of $\hat{\bar{F}}_i$ are estimated independently and $\hat{p}_i$'s are multinomial proportions, the variance of $\hat{\bar{F}}$ can be found by using the delta method, where

$$\widehat{\text{Var}}(\hat{\bar{F}}) = \sum\limits_{i=0}^{k} \widehat{\text{Var}}(\hat{\bar{F}}_i) \hat{p}_i^2 + \sum\limits_{i=0}^{k} \widehat{\text{Var}}(\hat{p}_i) \bar{F}_i^2 + 2 \sum\limits_{i=0}^{k} \sum\limits_{\substack{j=0 \\ i>j}}^{k} \hat{\bar{F}}_i \hat{\bar{F}}_j \widehat{\text{Cov}}(\hat{p}_i, \hat{p}_j), \quad (7.99)$$

and where

$$\widehat{\text{Var}}(\hat{\bar{F}}_i) = \frac{s_{b_{ij}}^2}{n_i}$$

for

$$s_{b_{ij}}^2 = \frac{\sum\limits_{j=1}^{n_i} \left(b_{ij} - \hat{\bar{F}}_i\right)^2}{(n_i - 1)},$$

and where

$$\widehat{\text{Var}}(\hat{p}_i) = \frac{\hat{p}_i(1 - \hat{p}_i)}{n}$$

$$\widehat{\text{Cov}}(\hat{p}_i, \hat{p}_j) = \frac{-\hat{p}_i \hat{p}_j}{n},$$

when

n_i = number of animals in age class i;

$$n = \sum_{i=1}^{k} n_i.$$

The variance of $\frac{1}{T}$ would similarly be calculated by using the delta method, considering the proper functional form of T for the population in question and the sources of survival estimates (i.e., $\hat{S}_{0,a+i}$, $i = 0, \ldots, w - a$).

7.8.3 Special Case: Two Age Classes (Henny et al. 1970, Cowardin and Johnson 1979)

Special cases of Eq. (7.93) can be conceived that simplify the relationship between λ and survival and fecundity. The simplest scenario considers only two age classes, juveniles and adults. As with small-game species and white-tailed deer, it will be assumed females breed as 1-year-olds. It will be further assumed that fecundity is age-invariant (i.e., $F_1 = F_2 = \ldots = F$), as is adult survival (i.e., $S_1 = S_2 = \ldots = S$). Survival of juveniles from age 0 to 1 is different from adults with value S_0. Using the approximation of Eq. (7.93) of Henny et al. (1970), the estimator of λ is

$$\lambda = S + \frac{FS_{01}}{\lambda^{1-1}}$$

or

$$\lambda = S + FS_0. \tag{7.100}$$

Alternatively, the same results can be derived by using a simple discrete-time population model,

$$N_{F_{t+1}} = SN_{F_t} + FS_0 N_{F_t}$$
$$= (S + FS_0)N_{F_t}$$

where

N_{F_t} = total female abundance in spring at time t. In this case,

$$\lambda = \frac{N_{F_{t+1}}}{N_{F_t}}$$
$$= S + FS_0. \tag{7.101}$$

This estimator was also presented by Cowardin and Johnson (1979) for a spring-to-spring estimate of $\hat{\lambda}$ for waterfowl. The parameters $\hat{S}$, $\hat{F}$, and $\hat{S}_0$ can be estimated by using the methods discussed in previous chapters. By using the delta method, the variance of $\hat{\lambda}$ (7.101) can be estimated by the quantity

$$\widehat{\text{Var}}(\hat{\lambda}) = \widehat{\text{Var}}(\hat{S}) + \widehat{\text{Var}}(\hat{F}) \cdot \hat{S}_0^2 + \widehat{\text{Var}}(\hat{S}_0) \cdot \hat{F}^2. \tag{7.102}$$

By using the exact variance formula for a product (Goodman 1960), the variance of (7.101) can be more precisely estimated as

$$\widehat{\text{Var}}(\hat{\lambda}) = \widehat{\text{Var}}(\hat{S}) + \widehat{\text{Var}}(\hat{F})\hat{S}_0^2 + \widehat{\text{Var}}(\hat{S}_0)\hat{F}^2 - \widehat{\text{Var}}(\hat{S}_0)\widehat{\text{Var}}(\hat{F}). \tag{7.103}$$

Both Eqs. (7.102) and (7.103) assume the values of $\hat{S}$, $\hat{F}$, and $\hat{S}_0$ are estimated independently.

Assumptions

The various assumptions associated with the λ estimator (7.101) concerning data collection and the life history of the population are as follows:

1. Juvenile survival (S_0) is for the period from one anniversary date to the next and different from adults.
2. Females begin reproducing as 1-year-olds.
3. Fecundity does not depend on the age of the adult.
4. Survival does not depend on the age of the adult.
5. Parameter estimates are unbiased.

The anniversary date refers to the time in the annual cycle or calendar year when survival and productivity are measured. Cowardin and Johnson (1979) used a spring anniversary date. The assumptions about fecundity estimates are important for several reasons. First, juvenile survival (S_0) must be estimated from the time fecundity is measured until the next anniversary date (Fig. 7.10). Second, using Eq. (7.101) and assuming the population is stationary, i.e., $\lambda = 1$, leads to the relationship

$$1 = S + FS_0,$$

$$1 - S = FS_0.$$

"This well known result simply states that the surviving recruits (FS_0) must balance the loss in the adult population ($1 - S$)" for the population to be stationary (Cowardin and Johnson 1979:19). The fecundity to achieve this balance is $F = \dfrac{1-S}{S_0}$.

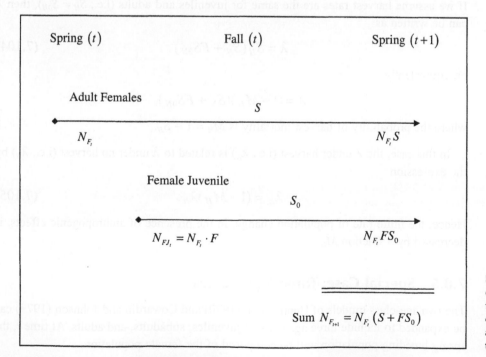

Figure 7.10. The demographic processes for Eq. (7.101) and the change in female abundance from time t to time $t + 1$.

> **Example 7.13: Estimating λ_{SAD} using the Lotka Equation for Mallards (*Anas platyrhynchos*), Chippewa National Forest, Minnesota**
>
> Data from Cowardin and Johnson (1979) on female mallards banded on the Chippewa National Forest, Minnesota, yielded the following parameter estimates:
>
> $$\hat{S} = 0.559$$
> $$\hat{S}_0 = 0.475$$
> $$\hat{F} = 0.590.$$
>
> From these data, λ_{SAD} is estimated (Eq. 7.101) to be
>
> $$\hat{\lambda}_{SAD} = \hat{S} + \hat{F}\hat{S}_0 = 0.559 + 0.590(0.475) = 0.839.$$
>
> A finite rate of change less than 1 indicates the population will tend to decline over time should demographic factors remain unchanged. Because variance estimates for the parameters were not provided, the variance of $\hat{\lambda}_{SAD}$ cannot be calculated using Eq. (7.103) in this example.

7.8.4 Special Case: Two Age Classes with Harvest

The survival parameters in Section 7.8.3 considered total annual survival. Reparameterizing Eq. (7.101) into natural (S_N) and harvest (S_H) survival is useful in examining the effects of harvest mortality on the rate of population change. Starting with Eq. (7.101), the finite rate of change can be reexpressed as

$$\lambda = S_N \cdot S_H + FS_{0N} \cdot S_{0H}.$$

If we assume harvest rates are the same for juveniles and adults (i.e., $S_H = S_{0H}$), then λ can be written as

$$\lambda = S_H(S_N + FS_{0N}) \tag{7.104}$$

or, equivalently,

$$\lambda = (1 - M_H)(S_N + FS_{0N}),$$

where the probability of harvest mortality is $M_H = 1 - S_H$.

In this case, the λ under harvest (i.e., λ_H) is related to λ under no harvest (i.e., λ_N) by the expression

$$\lambda_H = (1 - M_H)\lambda_N. \tag{7.105}$$

Hence, the finite rate of population change, in the presence of anthropogenic effects, is decreased by a fraction M_H.

7.8.5 Special Case: Three Age Classes

The two age-class models of Henny et al. (1970) and Cowardin and Johnson (1979) can be expanded to include three age classes: juveniles, subadults, and adults. At time t, the spring breeding population will be composed of the female population,

$$N_t = N_{A_t} + N_{S_t},$$

where

N_{A_t} = abundance of adult females in population at time t;
N_{S_t} = abundance of subadult females in population at time t.

These individuals are expected to produce a population at time $t + 1$ of

$$N_{t+1} = (N_{A_t}S + N_{S_t}S_1) + N_{A_t}FS_0,$$

assuming only adults breed. At time $t + 2$, the population can be expected to have an abundance of

$$N_{t+2} = N_{A_{t+1}}S + N_{S_{t+1}}S_1 + N_{A_{t+1}}FS_0,$$

or, reexpressed in terms of time t,

$$N_{t+2} = [(N_{A_t}S + N_{S_t}S_1)S + N_{A_t}FS_0S_1] + (N_{A_t}S + N_{S_t}S_1)FS_0.$$

We then define

$$\lambda = \frac{N_{t+2}}{N_{t+1}}$$
$$= \frac{(N_{A_t}S + N_{S_t}S_1 + N_{A_t}FS_0)S + (N_{A_t} + N_{S_t})FS_0S_1}{[N_{A_t}S + N_{S_t}S_1 + N_{A_t}FS_0]}$$
$$= S + \delta FS_0S_1,$$

where

$$\delta = \frac{N_{A_t} + N_{S_t}}{N_{A_t}S + N_{S_t}S_1 + N_{A_t}FS_0}$$
$$= \frac{N_t}{N_{t+1}}$$
$$= \frac{1}{\lambda}.$$

Hence,

$$\lambda = S + \frac{1}{\lambda}FS_0S_1, \tag{7.106}$$

or, in terms of a quadratic equation,

$$\lambda^2 - S\lambda - FS_0S_1 = 0.$$

By using the quadratic formula, solutions to the equation are

$$\lambda = \frac{S \pm \sqrt{S^2 + 4FS_0S_1}}{2}.$$

In particular, we can then estimate λ by the expression

$$\hat{\lambda} = \frac{\hat{S} + \sqrt{\hat{S}^2 + 4\hat{F}\hat{S}_0\hat{S}_1}}{2}, \tag{7.107}$$

where

$\hat{S}$ = estimated annual adult survival probability;

$\hat{S}_1$ = estimated annual subadult survival probability;

$\hat{S}_0$ = estimated juvenile survival from the time fecundity is measured until the anniversary date;

$\hat{F}$ = estimated number of juvenile females produced per adult female.

By using Eq. (7.93) with $a = 2$, the same expression for λ (Eq. 7.106) can be obtained. The variance of $\hat{\lambda}$ approximated by the delta method as

$$\widehat{\mathrm{Var}}\left(\hat{\lambda}\right) \doteq \left(\frac{\hat{F}\hat{S}_0\hat{S}_1^2}{\left(\hat{S}^2 + 4\hat{F}\hat{S}_0\hat{S}_1\right)}\right)\left[\mathrm{CV}(\hat{F})^2 + \mathrm{CV}(\hat{S}_0)^2 + \mathrm{CV}(\hat{S}_1)^2\right]$$
$$+ \widehat{\mathrm{Var}}\left(\hat{S}\right)\left(\frac{(1+\hat{S})^2}{4\left(\hat{S}^2 + 4\hat{F}\hat{S}_0\hat{S}_1\right)}\right).$$

(7.108)

Assumptions

The assumptions leading to Eqs. (7.106) and (7.107) are as follows:

1. All adults (≥ 2 years) have the same fecundity rate (F).
2. All adults (≥ 2 years) have the same survival probability ($S_2 = S_3 = \ldots = S$).
3. Juveniles have a survival probability of S_0, and subadults have a survival probability of S_1.
4. No juveniles or subadults breed.
5. Survival rates are time-invariant and age-specific until adulthood.
6. Fecundity rates are time-invariant.

The fecundity (F) for the adult age classes that is required to maintain $\lambda = 1$ can be calculated based on expression (7.83). Let

$$\underset{\sim}{S} = [S_0, S_1, S, \ldots]$$

and

$$\underset{\sim}{F} = [0, 0, F, \ldots],$$

it then follows that

$$1 = 0S_0 + FS_0S_1 + FS_0S_1S + \cdots$$
$$= FS_0S_1\left(\frac{1}{1-S}\right),$$

leading to

$$F = \frac{1-S}{S_0S_1}.$$

The same result can be obtained directly from Eq. (7.106). The above stability relationship is applicable to any species that produces young at the beginning of its third year of life and has a constant survival probability beginning that same year.

7.8.6   Special Case: Four Age Classes

The population model of Cowardin and Johnson (1979) can be further extended by considering species that have four age classes: adults (N_A), subadults (2-year-olds, N_{S2}; 1-year-olds, N_{S1}), and juveniles. The model assumes the subadult age classes do not breed. Modeling abundance in successive years can begin with the simple process

$$N_t = N_{A_t} + N_{S2_t} + N_{S1_t}$$

for a spring population (i.e., breeding population) just before the breeding season. Abundance 1 year later (i.e., 1 spring later) can be written as

$$N_{t+1} = N_{A_{t+1}} + N_{S2_{t+1}} + N_{S1_{t+1}}$$
$$= \underbrace{(N_{A_t}S + N_{S2_t}S_2)}_{\text{"adults"}} + \underbrace{(N_{S1_t}S_1)}_{\text{"sub-2"}} + \underbrace{(N_{A_t}FS_0)}_{\text{"sub-1"}}.$$

The next year, abundance can be expressed as

$$N_{t+2} = N_{A_{t+2}} + N_{S2_{t+2}} + N_{S1_{t+2}}$$
$$= \underbrace{[(N_{A_t}S + N_{S2_t}S_2)S + N_{S1_t}S_1S_2]}_{\text{"adults"}} + \underbrace{[(N_{A_t}FS_0S_1)]}_{\text{"sub-2"}} + \underbrace{[(N_{A_t}S + N_{S2_t}S_2)FS_0]}_{\text{"sub-1"}}.$$

Similarly, $N_{t+3} = N_{A_{t+3}} + N_{S2_{t+3}} + N_{S1_{t+3}}$, leading to the expression

$$N_{t+3} = \underbrace{\{[(N_{A_t}S + N_{S2_t}S_2)S + N_{S1_t}S_1S_2]S + (N_{A_t})FS_0S_1S_2\}}_{\text{"adults"}}$$
$$+ \underbrace{[(N_{A_t}S + N_{S2_t}S_2)FS_0S_1]}_{\text{"sub-2"}}$$
$$+ \underbrace{\{[(N_{A_t}S + N_{S2_t}S_2)S + N_{S1_t}S_1S_2]FS_0\}}_{\text{"sub-1"}}.$$

An expression for $\lambda = \dfrac{N_{t+3}}{N_{t+2}}$ can be derived as

$$\lambda = S + \frac{[N_{A_t}(FS_0S_1S_2) + N_{S1_t}(FS_0S_1S_2) + N_{S2_t}(FS_0S_1S_2)]}{N_{t+2}}$$

$$= S + \left(\frac{N_t}{N_{t+2}}\right)(FS_0S_1S_2)$$

$$\lambda = S + \frac{1}{\lambda^2}FS_0S_1S_2, \tag{7.109}$$

because

$$\frac{N_t}{N_{t+2}} = \frac{N_t}{N_{t+1}} \cdot \frac{N_{t+1}}{N_{t+2}} = \frac{1}{\lambda} \cdot \frac{1}{\lambda}.$$

Equation (7.109) can also be obtained directly from Eq. (7.93). The value of λ is the largest position real root to the cubic equation

$$\lambda^3 - \lambda^2 S - FS_0S_1S_2 = 0. \tag{7.110}$$

Assumptions of the estimator for λ derived from Eq. (7.110) are analogous to those of Eq. (7.107), taking into account one more subadult age class.

The fecundity for adult age classes required to maintain $\lambda = 1$ can be calculated based on expression (7.110) by setting $\lambda = 1$, in which case

$$F = \frac{1-S}{S_0 S_1 S_2}.$$

7.8.7 General Case

With spring populations composed of only adults, the estimate of λ is derived from first principles as Eq. (7.100), which can be rewritten as

$$\lambda = S + \frac{1}{\lambda^0} FS_0.$$

For spring populations composed of adults and subadults, the finite rate of change can be expressed as (Eq. 7.106)

$$\lambda = S + \frac{1}{\lambda^1} FS_0 S_1.$$

In the case of adults and two subadult age classes, the estimate of λ is derived from Eq. (7.109), where

$$\lambda = S + \frac{1}{\lambda^2} FS_0 S_1 S_2.$$

The general form of the equation used to solve for λ is

$$\lambda = S + F \prod_{x=0}^{a-1} S_x \cdot \frac{1}{\lambda^{a-1}},$$

or, equivalently,

$$\lambda = S + \frac{FS_{0,a}}{\lambda^{a-1}},$$

where

$a =$ age at first reproduction;

$$\prod_{x=0}^{a-1} S_x = S_{0,a}.$$

This is Eq. (7.93), which was previously derived assuming the relationship $S < \lambda$ and an infinite number of adult age classes. By deduction, Eq. (7.93) is now shown to be applicable to a general class of demographic models characterized by unique juvenile and subadult survivals and common adult survival and fecundity rates. The variance of $\hat{\lambda}$ calculated by Eq. (7.93) is given in Eq. (7.94). Thus, Eq. (7.93) implies that in order for a population to be self-sustaining, (i.e., $\lambda \geq 1$),

$$M = 1 - S \leq FS_{0,a}.$$

Hence, adult mortality (M) must be less than the product $FS_{0,a}$.

Discussion of Utility

Equation (7.87) provides a way of deriving a polynomial expression in λ that can be used to solve for the finite rate of population change as a function of age-specific survivals and fecundity. This equation also provides a way to understand the fundamental relationship between survival and fecundity that leads to a growth or decline of a population. In Sections 7.8.4 through 7.8.6, simple demographic models were used to illustrate the derivation of various special cases of Eq. (7.87). By studying those examples, readers should be able to perform similar analyses for other demographic scenarios of interest.

A SAD is the basis of Eq. (7.87) and its derivatives. Therefore, estimates of λ derived from Eq. (7.87) are not the realized rate of change but rather the rate of population change anticipated once a population has achieved a SAD (i.e., λ_{SAD}). The more disparate the actual age distribution of a population is from its SAD, the larger the difference between λ_{REAL} and λ_{SAD}. This distinction is the fundamental choice between using the Lotka equation (7.87) to estimate λ_{SAD} and using a sequence of annual abundance values to estimate λ_{REAL}. There is no practical substitute for a time series of abundance values when λ_{REAL} is sought.

For extrinsically controlled populations, the annual changes in reproductive success and survival are likely to produce a population outside its SAD. In those cases, λ_{REAL} and λ_{SAD} might be expected to be most disparate. On the other hand, large mammal populations governed by intrinsic demographic effects may be close to their SAD. For these populations, the Lotka equation (7.87) may provide a reasonable alternative to regression analyses based on several years of abundance data.

7.9 Estimating λ from a Leslie Matrix (Bernardelli 1941, Leslie 1945, 1948)

Using the Lotka equation to find a close-form expression for λ can be tedious for populations with numerous age classes, each with their own survival and fecundity rates. The Leslie (1945, 1948) projection matrix can be used to find $\hat{\lambda}$ numerically. As in the case of the Lotka equation, $\hat{\lambda}$ calculated from a Leslie matrix is the growth rate attained once the population has established a SAD, i.e., $\hat{\lambda}_{SAD}$. Thus, the value of $\hat{\lambda}$ calculated from a Leslie matrix analysis is a measure of growth potential, not of growth realized. Depending on when the values of survival and fecundity were measured during the population trajectory, the growth potential could describe the exponential (λ_{MAX}), logistic, or asymptotic phase of population growth. The λ calculated is the rate of population change that would be achieved under long-term conditions in which age-specific survivals (S_i) and age-specific fecundities (F_i) are invariant over time.

The standard Leslie matrix model for a four-age-class population can be written as

$$\begin{bmatrix} n_{0,t+1} \\ n_{1,t+1} \\ n_{2,t+1} \\ n_{3,t+1} \end{bmatrix} = \begin{bmatrix} F_0 & F_1 & F_2 & F_3 \\ S_0 & 0 & 0 & 0 \\ 0 & S_1 & 0 & 0 \\ 0 & 0 & S_2 & 0 \end{bmatrix} \begin{bmatrix} n_{0,t} \\ n_{1,t} \\ n_{2,t} \\ n_{3,t} \end{bmatrix}$$

or, more concisely, as

$$n_{t+1} = \mathbf{M} n_t, \tag{7.111}$$

where

n_t = vector for the abundance of females by age classes at time t;
n_{it} = number of females in age class i ($i = 0, 1, 2, \ldots$) at time t.

The $\mathbf{M}$ matrix projects the age distribution that exists at time t to $t + 1$. All matrices of the form $\mathbf{M}$ satisfy the eigenvalue–eigenvector relationship, i.e.,

$$\mathbf{M} n_t = \lambda n_t. \tag{7.112}$$

The eigenvalue–eigenvector relationship illustrates the central concept of a SAD. If each age class in vector n_t is multiplied by the same constant λ, the relative abundance of the different age classes does not change. Equating (7.111) with (7.112) finds that

$$n_{t+1} = \lambda n_t. \tag{7.113}$$

Although the relative abundance by age class does not change, overall abundance changes by the factor λ from time t to time $t + 1$.

The eigenvalues that satisfy the relationship (7.113) are found from the characteristic equation

$$|\mathbf{M} - \lambda \mathbf{I}| = 0, \tag{7.114}$$

where the operator $|\ |$ refers to calculating the determinant of a matrix, in this case, $\mathbf{M} - \lambda \mathbf{I}$ and where $\mathbf{I}$ is the identity matrix. There are as many eigenvalue solutions to Eq. (7.114) as there are dimensions to the Leslie matrix $\mathbf{M}$. For instance, a 4×4 matrix will have four eigenvalues and, correspondingly, four eigenvectors. However, not all eigenvalues need to be unique. By the Frobenius theorem (Gradshteyn and Fyzhik 2000), at least one of the solutions to Eq. (7.114) will be positive. The largest positive eigenvalue solution to Eq. (7.114) is λ, the finite rate of population change. The eigenvector corresponding to this eigenvalue λ provides the means to calculate the SAD of the population (Searle 1966, 1982).

Example 7.14: Eigenvalue Calculations

Consider the 3×3 Leslie projection matrix

$$\mathbf{M} = \begin{bmatrix} 0 & 2 & 3 \\ 0.5 & 0 & 0 \\ 0 & 0.2 & 0 \end{bmatrix} \tag{7.115}$$

with age-specific fecundities of $F_0 = 0$, $F_1 = 2$, $F_2 = 3$ and age-specific survivals of $S_0 = 0.50$ and $S_1 = 0.20$. The characteristic equation based on this projection matrix is

$$\left(\begin{bmatrix} 0 & 2 & 3 \\ 0.5 & 0 & 0 \\ 0 & 0.2 & 0 \end{bmatrix} - \lambda \begin{bmatrix} 1 & 0 & 0 \\ 0 & 1 & 0 \\ 0 & 0 & 1 \end{bmatrix}\right) = 0$$

$$\begin{vmatrix} -\lambda & 2 & 3 \\ 0.5 & -\lambda & 0 \\ 0 & 0.2 & -\lambda \end{vmatrix} = 0.$$

For a 3×3 matrix, a third-order determinant must be found that is a linear function of three second-order determinants. In the general case of a 3×3 matrix of the form

$$\mathbf{A} = \begin{bmatrix} a_{11} & a_{12} & a_{13} \\ a_{21} & a_{22} & a_{23} \\ a_{31} & a_{32} & a_{33} \end{bmatrix},$$

the determinant can be computed as

$$|\mathbf{A}| = a_{11}(-1)^{1+1}\begin{vmatrix} a_{22} & a_{23} \\ a_{32} & a_{33} \end{vmatrix} + a_{12}(-1)^{1+2}\begin{vmatrix} a_{21} & a_{23} \\ a_{31} & a_{33} \end{vmatrix} + a_{13}(-1)^{1+3}\begin{vmatrix} a_{21} & a_{22} \\ a_{31} & a_{32} \end{vmatrix}$$

where a_{11}, a_{12}, and a_{13} are the elements of row 1. Expansion can be performed by using the elements of any row or column of convenience. The exponent of (-1) is the sum of the row and column designations of the expansion element (e.g., for a_{12}, the exponent is $1 + 2$). The determinant for any element is based on the row deletion and column deletion of matrix $\mathbf{A}$ corresponding to the subscripts of the expansion element (e.g., for a_{12}, delete row 1 and column 2 of $\mathbf{A}$).

Readers are encouraged to review Searle (1966, 1982) for detailed descriptions of eigenvalue and eigenvector calculations. Numerous software programs also provide eigenvalue–eigenvector calculations.

By expanding on the elements of the first row of the matrix of interest, then

$$-\lambda(-1)^{1+1}\begin{vmatrix} -\lambda & 0 \\ 0.2 & \lambda \end{vmatrix} + 2(-1)^{1+2}\begin{vmatrix} 0.5 & 0 \\ 0 & -\lambda \end{vmatrix} + 3(-1)^{1+3}\begin{vmatrix} 0.5 & -\lambda \\ 0 & 0.2 \end{vmatrix} = 0.$$

The determinant for a 2×2 matrix is calculated as

$$\begin{vmatrix} a_{11} & a_{12} \\ a_{21} & a_{22} \end{vmatrix} = a_{11} \cdot a_{22} - a_{12} \cdot a_{21}.$$

Using that definition for the determinant of a 2×2 matrix, the example reduces to

$$-\lambda(\lambda^2) - 2(-0.5\lambda) + 3(0.1) = 0$$

or

$$-\lambda^3 + \lambda + 0.3 = 0.$$

Solutions for the cubic equation are $\lambda = -0.7864$, 1.1254, and -0.3389. Hence, the largest positive eigenvalue corresponding to the finite rate of change is $\lambda_{SAD} = 1.1254$.

The next step in the calculations is to find the eigenvector corresponding to eigenvalue $\lambda_{SAD} = 1.1254$. The eigenvector ($\underset{\sim}{w}$) is the solution to the matrix equation

$$[\mathbf{M} - \lambda\mathbf{I}]\underset{\sim}{w} = \underset{\sim}{0}.$$

In the particular case of our projection matrix,

$$\left[\begin{bmatrix} 0 & 2 & 3 \\ 0.5 & 0 & 0 \\ 0 & 2 & 0 \end{bmatrix} - 1.1254 \begin{bmatrix} 1 & 0 & 0 \\ 0 & 1 & 0 \\ 0 & 0 & 1 \end{bmatrix}\right] \begin{bmatrix} a \\ b \\ c \end{bmatrix} = \begin{bmatrix} 0 \\ 0 \\ 0 \end{bmatrix},$$

or

$$\begin{bmatrix} -1.1254 & 2 & 3 \\ 0.5 & -1.1254 & 0 \\ 0 & 0.2 & -1.1254 \end{bmatrix} \begin{bmatrix} a \\ b \\ c \end{bmatrix} = \begin{bmatrix} 0 \\ 0 \\ 0 \end{bmatrix},$$

or, equivalently, the set of equations

$$-1.1254a + 2b + 3c = 0$$
$$0.5a - 1.1254b = 0$$
$$0.2b - 1.1254c = 0.$$

Arbitrarily setting $a = 1$, solutions to the other variables are $b = 0.44429$ and $c = 0.07896$, or in vector notation, the eigenvector is

$$\underset{\sim}{w} = \begin{bmatrix} 1 \\ 0.44429 \\ 0.07896 \end{bmatrix}.$$

Standardizing the vector by the sum

$$1 + 0.44429 + 0.07896 = 1.52325$$

produces an equivalent vector expressed in terms of proportions, where

$$\underset{\sim}{w} = \begin{bmatrix} 0.656 \\ 0.292 \\ 0.052 \end{bmatrix}.$$

The SAD for this population is composed of 65.6% age class 0, 29.2% age class 1, and 5.2% age class 2 individuals. When this SAD is achieved, the population will grow at a finite rate of $\lambda_{SAD} = 1.1254$.

The eigenvalue–eigenvector relationship in Eq. (7.112) can also be exploited to readily find the finite rate of change (λ) for a particular projection matrix. By recursively multiplying the results of the product

$$\underset{\sim}{n}_{t+1} = \mathbf{M}\underset{\sim}{n}_t; \quad t = 0, 1, 2, 3, \ldots,$$

a SAD will ultimately be established and annual abundance will change by a factor of $\lambda_{SAD} = \dfrac{N_{t+1}}{N_t}$. A simple alternative to solving for the eigenvalues and eigenvectors is to repeatedly apply the $\mathbf{M}$ matrix to subsequent vectors of $\underset{\sim}{n}_i$; $t = 0, 1, 2, \ldots$.

Example 7.15: Progressive Matrix Multiplication to Calculate λ

Starting with an arbitrary abundance vector

$$\underset{\sim}{n}_0 = \begin{bmatrix} 30 \\ 20 \\ 10 \end{bmatrix},$$

the projection matrix of Example 7.13 will be used to successively project the population abundance year after year. We start with the relationship

$$\underset{\sim}{n}_1 = \mathbf{M}\underset{\sim}{n}_0,$$

$$\begin{bmatrix} 70 \\ 15 \\ 4 \end{bmatrix} = \begin{bmatrix} 0 & 2 & 3 \\ 0.5 & 0 & 0 \\ 0 & 0.2 & 0 \end{bmatrix} \begin{bmatrix} 30 \\ 20 \\ 10 \end{bmatrix}.$$

Table 7.7 summarizes the demographic trends in successive years when the projection matrix (7.115) is repeatedly applied to the resultant abundance vector. After approximately 20 years, the age composition stabilizes and successive annual abundance values grow at a finite rate of $\lambda_{\text{SAD}} = 1.1254$. The values of λ_{SAD} and the resulting age composition are the same as those calculated by the eigenvalue-eigenvector calculations of Example 7.14.

The involved calculations needed to calculate eigenvalues precludes an easy approach to estimating the variance associated with $\hat{\lambda}$ from a Leslie matrix. Instead, Monte Carlo procedures have been used based on repeated sampling from the distributions for the parameter estimates of $F_x(i = 0, \ldots, w)$ and $S_x(i = 0, \ldots, w-1)$ (Meyer et al. 1986, Alvarez-Buylla and Slatkin 1991, 1994, Taberner et al. 1993, Caswell et al. 1998), or on sensitivity analysis (Mills et al. 1999). Typically, the distribution for the parameter estimates is assumed normal with mean $\hat{\theta}$ and standard deviation $\widehat{\text{SE}}(\hat{\theta}|\theta)$. Alternatively, by using the delta method and with the aid of computer programs, the variance of $\hat{\lambda}$ can

Table 7.7. Successive age-class abundance, composition, total abundance, and annual values of λ for the Leslie matrix model in Example 7.14.

	\multicolumn{11}{c}{Generations (time)}										
	0	1	2	3	...	10	11	...	20	21	...	30	31
Age class abundance													
n_{0t}	30	70	42	79		147.3	170.8		488.4	550.1		1592.7	1792.5
n_{1t}	20	15	35	21		68.3	73.7		217.3	244.2		707.6	796.4
n_{2t}	10	4	3	7		11.4	13.7		38.5	43.5		125.8	141.5
Age class composition													
C_{0t}	0.500	0.787	0.525	0.738		0.649	0.662		0.656	0.657		0.656	0.656
C_{1t}	0.333	0.168	0.438	0.196		0.301	0.285		0.292	0.291		0.292	0.292
C_{2t}	0.167	0.045	0.037	0.065		0.050	0.053		0.052	0.052		0.052	0.052
Total abundance													
N_t	60	89	80	107		227.0	258.1		744.2	837.8		2426.1	2730.4
Relative change													
$\lambda = N_{t+1}/N_t$	—	1.4833	0.8989	1.3375		—	1.1369		—	1.1258		—	1.1254

be approximated from the characteristic equation (Eq. 7.114). By using vector notation, the variance of $\hat{\lambda}$ can be written as

$$\mathrm{Var}\left(\hat{\lambda}\right) = \nabla(g(\lambda))'\,\Sigma\nabla(g(\lambda)), \tag{7.116}$$

where

$\nabla(g(\lambda))$ = gradient of the dominant eigenvalue, i.e., vector of first partial derivatives;
Σ = variance-covariance matrix of the estimates for age-specific survival and fecundity.

Code for a *Mathematica*® (Wolfram 1999) macro to estimate the variance of $\hat{\lambda}$ is provided in Appendix D. The use of the delta method has been suggested by Caswell et al. (1998) and others (Alvarez-Buylla and Slatkin 1991, 1994, Taberner et al. 1993), but none of these authors has provided a convenient means to apply the approach.

Example 7.16: Estimating λ_{SAD} using a Leslie Matrix for Hawaiian Hawks (*Buteo solitarius*)

Klavitter et al. (2003) conducted a study of the Hawaiian hawk to provide demographic information for a population assessment by the U.S. Fish and Wildlife Service. Juvenile and adult survival rates were estimated using radio telemetry. Birds 1 year and older were considered adults. Female Hawaiian hawks first breed as 3-year olds, and fecundity was measured from nest counts. Given the available life-history data, the assumption of an infinite number of adult age classes, estimates of juvenile and adult survival, the following matrix model was constructed:

$$\mathbf{M} = \begin{bmatrix} F_0 & F_1 & F_2 & F_2 \\ S_0 & 0 & 0 & 0 \\ 0 & S & 0 & 0 \\ 0 & 0 & S & S \end{bmatrix} = \begin{bmatrix} 0 & 0 & 0 & \underset{(\widehat{SE}=0.0417)}{0.23} \\ \underset{(\widehat{SE}=0.0981)}{0.50} & 0 & 0 & 0 \\ 0 & \underset{(\widehat{SE}=0.0404)}{0.94} & 0 & 0 \\ 0 & 0 & 0.94 & 0.94 \end{bmatrix}.$$

The finite rate of change was calculated from the above matrix by eigenvalue analysis as 1.0324 ($\hat{\lambda}_{SAD} = 1.0324$). The variance was calculated by using the delta method (Eq. 7.116) to be $\widehat{\mathrm{Var}}(\hat{\lambda}_{SAD}) = 0.0018$, with standard error of $\widehat{SE}(\hat{\lambda}_{SAD}) = 0.0428$.

Alternatively, ($\hat{\lambda}_{SAD}$) can be estimated by using Eq. (7.93), where $a = 3$. The estimate of $\hat{\lambda}_{SAD}$ is calculated as

$$\hat{\lambda} = \hat{S} + \frac{\hat{F}\hat{S}_{0,a}}{\hat{\lambda}^{a-1}}$$

$$\hat{\lambda} = 0.94 + \frac{0.23(0.50 \cdot 0.94 \cdot 0.94)}{\hat{\lambda}^2},$$

leading to the cubic equation

$$\lambda^3 - 0.94\lambda^2 - 0.101614 = 0,$$

with the largest positive root of

$$\hat{\lambda}_{SAD} = 1.0349.$$

The variance is estimated by Eq. (7.94) as

$$\text{Var}(\hat{\lambda}) \doteq \frac{\left(\text{Var}(\hat{S})\lambda^{2(a-1)} + \text{Var}(\hat{F})S_{0,a}^2 + \text{Var}(\hat{S}_{0,a})F^2\right)}{\left(\lambda^{(a-2)}((a-1)S - a\lambda)\right)^2}.$$

The variance of $S_{0,a}$ can be approximated by the exact product formula (Goodman 1960) and the delta method as

$$\widehat{\text{Var}}(S_{0,a}) = \hat{S}^4 \text{Var}(S_0) + \left(4\hat{S}^2\hat{S}_0^2\right)\widehat{\text{Var}}(\hat{S}) - 4\hat{S}^2\widehat{\text{Var}}(\hat{S})\widehat{\text{Var}}(\hat{S}_0)$$

$$= 0.94^4(0.0981)^2 + (4 \cdot 0.94^2 0.5^2)(0.0404)^2$$

$$- 4 \cdot 0.94^2(0.0404 \cdot 0.0981)^2$$

$$\widehat{\text{Var}}(S_{0,a}) = 0.0089,$$

or $\qquad \widehat{\text{SE}}(S_{0,a}) = 0.0943.$

The variance of $\hat{\lambda}$ is calculated as

$$\widehat{\text{Var}}(\hat{\lambda}) \doteq \frac{\left((0.0404)^2(1.0349)^{2(3-1)} + (0.0417)^2(0.4418)^2 + 0.0089(0.23)^2\right)}{\left((1.0349)^{(3-2)}((3-1)0.94 - 3 \cdot 1.039)\right)^2},$$

$$\widehat{\text{Var}}(\hat{\lambda}_{\text{SAD}}) = 0.00170 \text{ or } \widehat{\text{SE}}(\hat{\lambda}_{\text{SAD}}) = 0.0412.$$

The eigenvalue analysis of the Leslie matrix model and general estimator of $\hat{\lambda}$, Eq. (7.93), yield similar point estimates and associated variances.

There are several ways of interpreting the influence of individual survival or fecundity estimates on the calculated value of $\hat{\lambda}_{\text{SAD}}$. Our evaluation begins by considering the influence of measurement error on the estimated value of λ_{SAD}. The estimated standard error of 0.0428 reflects the overall contribution of measurement error in the survival and fecundity parameters on $\hat{\lambda}_{\text{SAD}}$. Monte Carlo simulations were performed to examine the contribution of measurement error from the individual parameters on $\hat{\lambda}_{\text{SAD}}$ (Fig. 7.11). The Monte Carlo simulations estimated the distribution of $\hat{\lambda}_{\text{SAD}}$ when one input parameter was allowed to vary while holding all others constant. In the simulations, the random variable was assumed to be normally distributed with expected value $\hat{\theta}$ and $\sigma = \widehat{\text{SE}}(\hat{\theta})$. Varying adult survival, the resulting standard error in $\hat{\lambda}_{\text{SAD}}$ was 0.0339. The sampling errors associated with varying adult fecundity and juvenile survival produced much smaller standard errors, 0.0134 and 0.0142, respectively. Thus, the measurement error in $\hat{S}$ for adult survival contributes the greatest uncertainty to the estimate of $\hat{\lambda}_{\text{SAD}}$. The greatest improvement in the precision of $\hat{\lambda}_{\text{SAD}}$ would therefore come from a more precise estimate of adult survival.

Klavitter et al. (2003) conducted elasticity analyses (Caswell 1989, 2000) to assess the relative contribution of each demographic parameter to $\hat{\lambda}$. The steps for conducting sensitivity and elasticity analyses were discussed earlier (Section 2.5). For this example, the proportion of individuals by age class in the population when it reaches the SAD (w) is

a. Varying F_3 with resulting standard error of 0.0134.

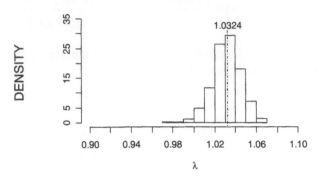

b. Varying S_0 with resulting standard error of 0.0142.

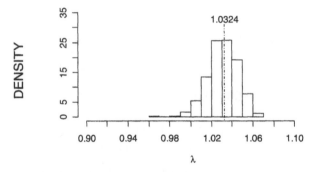

c. Varying S with resulting standard error of 0.0339.

Figure 7.11.
Distribution of $\hat{\lambda}_{SAD}$ for Hawaiian hawks (Klavitter et al. 2003) from Monte Carlo simulations based on varying one parameter (F_3, S_0, or S) of the Leslie matrix at a time.

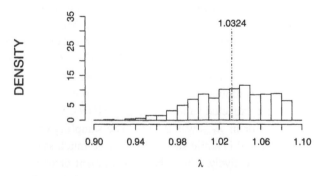

$$w = \begin{bmatrix} 0.1557 \\ 0.0761 \\ 0.0692 \\ 0.6989 \end{bmatrix}.$$

Donovan and Welden (2001) called this vector the right eigenvector of the $\mathbf{M}$ matrix. The left eigenvector of $\mathbf{M}$ is calculated from the transpose of $\mathbf{M}$, where

$$\mathbf{M}' = \begin{bmatrix} 0 & 0.50 & 0 & 0 \\ 0 & 0 & 0.94 & 0 \\ 0 & 0 & 0 & 0.94 \\ 0.23 & 0 & 0 & 0.94 \end{bmatrix}.$$

Using $\mathbf{M}'$, the Leslie model is projected until it reaches a SAD. The right eigenvector of $\mathbf{M}'$ at the SAD provides the reproductive values for each class, called v. In this case,

$$v = \begin{bmatrix} 0.1278 \\ 0.2639 \\ 0.2899 \\ 0.3184 \end{bmatrix}.$$

These reproductive values are often standardized to aid in interpretation. Standardizing these values (dividing each by 0.1278) results in

$$v = \begin{bmatrix} 1.0 \\ 2.1 \\ 2.3 \\ 2.5 \end{bmatrix}.$$

Thus, an individual in the 4+ age class contributes 2.5 times that of a 1-year-old to offspring production.

Caswell (2001) used Eq. (2.42) to assess the sensitivity (s_{ij}) of λ_{SAD} to a change in the value of the demographic parameter a_{ij} in the Leslie matrix, where

$$s_{ij} = \frac{v_i w_j}{v' w},$$

and where v_i and w_j relate to the ith and jth elements in the reproductive vector v and stable age vector w, respectively. The value of $v'w$ is the product of the two vectors w and v, which equals 0.2826. Sensitivity is computed for each entry in the matrix, even those with values of zero. For example, in the case of element a_{11}, we multiply 0.1278 (reproductive value for age class 1) by 0.1557 (the SAD for age class 1) and divide by 0.2826 (i.e., $v'w$), resulting in 0.0705. The entire sensitivity matrix is

$$\begin{bmatrix} 0.0705 & 0.0342 & 0.0311 & 0.3166 \\ 0.1456 & 0.0705 & 0.0642 & 0.6537 \\ 0.1599 & 0.0775 & 0.0705 & 0.7179 \\ 0.1756 & 0.0851 & 0.0775 & 0.7884 \end{bmatrix}.$$

Values in the sensitivity matrix represent changes to λ if we change that parameter and hold all other matrix entries constant. For example, 0.0705 indicates that changing F_1 by one unit changes λ by a factor of 0.0705.

Alternatively, elasticity analysis, which measures the proportional change in the rate of population growth (λ), is often used. Elasticity (e_{ij}) is the proportional change in λ for a proportional change in a matrix element, a_{ij}, calculated as (Eq. 2.43)

$$e_{ij} = \frac{a_{ij}}{\lambda} \frac{\partial \lambda}{\partial a_{ij}} = \frac{a_{ij} s_{ij}}{\lambda}.$$

For example, the elasticity of adult (4+) survival is equal to 0.94 (survival of 4+ in **M**) times 0.7884 (its corresponding sensitivity value in the sensitivity matrix above) divided by 1.0324 ($\hat{\lambda}$), resulting in 0.7178. The entire elasticity matrix is

$$\begin{bmatrix} 0 & 0 & 0 & 0.0705 \\ 0.0705 & 0 & 0 & 0 \\ 0 & 0.0705 & 0 & 0 \\ 0 & 0 & 0.0705 & 0.7178 \end{bmatrix}$$

Klavitter et al. (2003:171) noted that adult (4+) survival was most important to population growth, with an elasticity of 0.72. This value means that a 1% increase in adult survival will cause a 0.72% increase in λ_{SAD} (i.e., $1.0324 \times 1.0072 = 1.0398$). Other parameters have much less influence. A 1% increase in juvenile survival would only result in a 0.07% increase in λ_{SAD}. This example demonstrates how sensitivity and elasticity analyses can be used to help guide management activities to benefit populations. The sensitivity and elasticity analyses suggest λ_{SAD} can be best improved by increasing adult survival. Managers would also be wise to expend resources on improving the estimate of adult (4+) survival should they be interested in improving the precision of $\hat{\lambda}_{SAD}$.

Discussion of Utility

The Leslie matrix can be used to find λ_{SAD} either through eigenvalue calculations or repetitive applications of the projection matrix until a SAD has been achieved. Software programs such as *Mathematica* (Appendix D) can be readily used to find numerical solutions to the eigenvalues. The example of progressive matrix multiplication, on the other hand, emphasizes the resultant eigenvalue solution is only applicable once a population has achieved its SAD. Values of λ_{SAD} may be quite different from the realized rate of change (i.e., λ_{REAL}) and take multiple years to achieve, if ever. Inferring current population growth rates or population vitality from estimates of λ_{SAD} can be risky and unreliable. A value of λ_{SAD} measures future growth potential, whereas an estimated value of λ_{REAL} measures the actual rate under prevailing environmental and demographic conditions.

Both quantities are essential for characterizing the current status and future trends of a population.

The *Mathematica* macro (Appendix D) can be used to calculate the finite rate of change and its variance based on eigenvalue calculations of a Leslie transition matrix. Based on the delta method (Seber 1982:7–9), numerical methods are used to calculate the first derivative of the eigenvalue. The *Mathematica* macro can be used to analyze standard transition matrices of the form

$$\mathbf{M} = \begin{bmatrix} F_0 & F_1 & \cdots & & F_k \\ S_0 & 0 & \cdots & & 0 \\ 0 & S_1 & & & 0 \\ \vdots & & \ddots & & \vdots \\ 0 & \cdots & 0 & S_{k-1} & 0 \end{bmatrix}_{(k+1)\times(k+1),}$$

or for extended age classes $k+$ and beyond, where

$$\mathbf{M} = \begin{bmatrix} F_0 & F_1 & \cdots & & F_k \\ S_0 & 0 & \cdots & & 0 \\ 0 & S_1 & \cdots & & 0 \\ \vdots & & & & \vdots \\ 0 & \cdots & 0 & S_{k-1} & S_{k+} \end{bmatrix}_{(k+1)\times(k+1).}$$

Some of survival and fecundity parameters can be set to zero based on life-history information, as illustrated in the Hawaiian hawk example. In addition, some parameters can be set equal across age classes. By setting parameters equal to one another, it implies biological information is available to support the contention. If the assumption of equality is wrong, the resulting variance estimate for $\hat{\lambda}$ will be seriously underestimated and the estimate of λ biased. The consequences of assuming constant fecundity or survival across older age classes because of lack of age-specific information cannot be captured by the resulting variance calculations. Prudent use of special cases of the Leslie matrix calculations of $\hat{\lambda}$ and $\widehat{\mathrm{Var}}(\hat{\lambda})$ is therefore warranted.

Applying a harvest matrix to the Leslie matrix calculations allows calculations of λ_{SAD} under exploitation. In general, the value of λ_{SAD} expresses the future growth potential of the population once the population has achieved a SAD. The value of λ_{SAD} under exploitation further assumes the level of harvest remains constant over time. If the population has not been previously exploited or if exploitation rates have varied over time, it is reasonable to conclude the population is not at its SAD and, again, λ_{SAD} expresses only an idealized and future growth potential.

7.10   Summary

An array of methods is available to investigators to estimate the finite and instantaneous rates of population change (Fig. 7.12). Some of the methods presented are suited for estimating both the realized and intrinsic rates of change. These methods are generally based on geometric or exponential growth assumptions. The intrinsic rates of change are estimated (i.e., λ_{MAX}, r_{MAX}, or R_{MAX}) by these exponential-based growth models when a population is indeed in the exponential growth phase. Alternatively, when a population

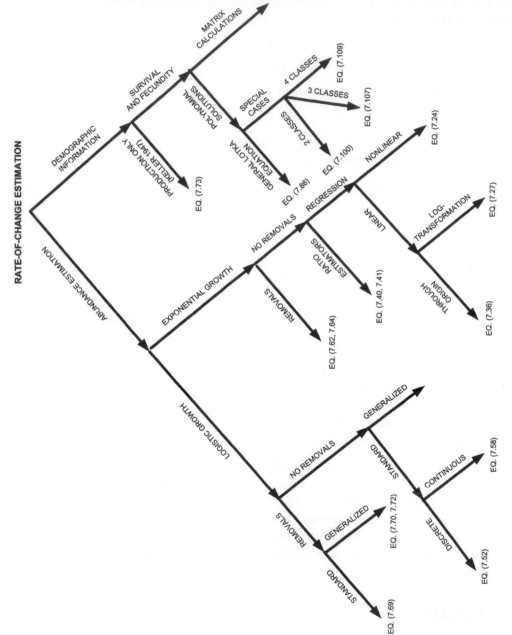

Figure 7.12. Decision tree for the methods of estimating rates of population change presented in Chapter 7.

is experiencing density-dependent growth, the exponential-based methods can only be used to estimate the realized rate of change (i.e., λ_{REAL}, r_{REAL}, or R_{REAL}) over short periods of time. Nevertheless, the assumption of exponential growth will eventually be violated. In situations in which a series of abundance surveys exist during density-dependent growth, the ratio method, i.e.,

$$\hat{\lambda} = \frac{1}{n-1} \sum_{i=1}^{n-1} \left(\frac{N_{i+1}}{N_i} \right),$$

can be expected to provide a better measure of average performance than exponential regression-based methods. Even then, the observed or realized rates of change may systematically vary over the course of time, suggesting the need to periodically report values (Fig. 7.1).

The logistic-based methods, on the other hand, inherently focus on estimating the intrinsic rates of change (i.e., λ_{MAX}, r_{MAX}, or R_{MAX}) in the presence of density-dependent growth. Given that intrinsic rate of change (λ_{MAX}) and carrying capacity (K) have been estimated, it is possible to calculate the observed or realized rate of change at some time t, using the relationship

$$\hat{\lambda}_t = 1 + \left(\hat{\lambda}_{MAX} - 1 \right)\left(1 - \frac{N_t}{\hat{K}} \right)$$

for a standard logistic-growth model or, for the generalized logistic model,

$$\hat{\lambda}_t = 1 + \left(\hat{\lambda}_{MAX} - 1 \right)\left(1 - \left(\frac{N_t}{\hat{K}} \right)^{\hat{z}} \right).$$

There is a negative correlation between $\hat{\lambda}_{MAX}$ and $\hat{z}$ in the generalized logistic model. Consequently, although λ_t is rather insensitive to the fitted value of z, λ_{MAX} is not. For many data sets we have investigated, a wide range of values of z for the generalized logistic model fit the model nearly as well. Standard errors on the z parameters reflect this uncertainty. In a number of instances when the generalized logistic models have been used, the simpler exponential or standard logistic models fit equally or nearly as well. Eberhardt (1987) fit a generalized logistic model with a value of $z = 11$ to longhorn cattle data where we found an exponential growth model to be adequate. Jeffries et al. (2003) used a generalized logistic model with $z = 2.43$ to describe the growth of the harbor seal population in Washington State, whereas we found a standardized logistic model (i.e., $z = 1$) to be adequate. In any thorough analysis of a generalized logistic growth model, a sensitivity analysis should be conducted to assess how values of $\hat{\lambda}_{MAX}$ vary with values of z.

The regression approaches to estimating rates of change are based on observed changes in abundance over time. Inferences to λ are predicated on the appropriate choice of model and data. Using observed abundance values in the calculation of λ results in inferences to the actual population at hand (i.e., λ_{REAL}). This is not true for values of λ_{SAD} calculated from the Lotka or Leslie matrix equations. The values of λ_{SAD} from a Lotka or Leslie matrix analysis are the growth rates potentially attained should the population achieve a SAD. The values of survival and fecundity used in a Lotka or Leslie analysis will also have a strong effect on what λ parameter is being estimated. Using current values of S_x and F_x will yield estimates of λ_{SAD} that could be realized should the population achieve a SAD under prevailing conditions. Using optimal values of survival and fecundity will translate to estimates of λ_{MAX}.

The λ_{REAL} values are important to managers because they convey the current or near-term trends in demography. Estimates of λ_{MAX}, on the other hand, convey essential information on the potential resilience of the population to disturbance. Coupled with carrying capacity (K), λ_{MAX} can be used to examine if a population has reached an optimum sustainable population and the maximum sustainable yield. Current abundance and/or harvest levels can then be compared with these benchmarks. Hence, status and trends monitoring meet in this comparison of demographic benchmarks with real-time abundance values.

Analysis of Population Indices

8

8.1 Introduction

A population index is a measured response presumably related to the actual abundance or density of the population (Dice 1941, Caughley 1977, Caughley and Sinclair 1994). Indices are of two general types, either an incomplete count of animal numbers (e.g.,

Wood et al. 1985, Menkens 1990) or a measure of some trait that can easily be monitored in the environment that is related to animal abundance (e.g., Gates 1966, Sargeant et al. 1998, Hansen and Guthery 2001). Examples of the latter include areal samples of tracks (Van Dyke et al. 1986, Carey and Witt 1991), nests (Kiel 1955), and pellet counts (Fuller 1991, Cavallini 1994). Indices are distinguished from abundance estimators because of their inability to be converted to absolute animal numbers. Typically, the auxiliary information needed to make such a conversion is not collected; otherwise, it would be used and the index converted to absolute abundance.

The absence of auxiliary information, such as detection probabilities, needed to make the absolute conversions has profound consequences (Anderson 2001, 2003). Without this information, it is usually impossible to assess whether a conversion factor remains constant over time or equal across populations. Thus, population indices are, in part, a matter of faith in the stability and robustness of the population processes investigated and the metrics measured. However, science is based not on faith but rather on the interpretation of empirically derived facts. For these reasons, individuals such as Fuller (1991), Anderson (2003), and Ellingson and Lukacs (2003) strongly argue against use of indices in modern wildlife management (although see Engeman 2003, Hutto and Young 2003). Skalski and Robson (1992) recommended using absolute abundance techniques as a fail-safe method in wildlife studies and using indices only if homogeneous detection rates have been found to be true.

Despite the inferential weaknesses of indices, they remain a cornerstone in wildlife science. Their great appeal lies in cost-efficient application over large geographic areas, which may be indexed for the same effort of a much smaller mark-recapture study. In addition, many index methods do not require animal capture and handling, which is also appealing. This noninvasive approach to studying animals is valuable when studying endangered species and species difficult to capture or handle. Furthermore, some ecologists (Caughley and Sinclair 1994, Engeman 2003, Hutto and Young 2003) argue that indices offer a practical alternative for many population monitoring situations. In some situations, a well-constructed and calibrated index might be most appropriate for management (Engeman 2003).

The broad use of population indices is in contrast to the surprisingly little that has been written on the statistical theory underlying these techniques. Much of the existing statistical theory on population indices can be found in Overton (1969:411–432), Seber (1982:52–58), Eberhardt (1976, 1978a), and Williams et al. (2001:257–261). The lack of quantitative guidance has seemingly promoted a cavalier approach to the design of index studies in wildlife science—"They're just indices"—so there is often little concern about important details of the survey design. The sessile nature of many sign-based indices easily allows rigorous sample survey designs (Cochran 1977, Thompson 2000). Visual counts of animals often avail themselves to distance sampling techniques (Buckland et al. 1993). This existing quantitative theory can be the foundation of many index studies, yet it is often ignored.

Perhaps the greatest irony concerning indices is that the studies least robust to assumption violations also receive the least attention to rigorous design and analysis. Absolute abundance techniques can be used to estimate the needed calibration to convert the observed count index to an absolute value of abundance or density. Index studies by their very nature are incapable of estimating the conversion factor and instead rely on the assumption that the factor is invariant over time and/or space. However, this conversion factor is often a complex function of animal behavior, habitat, sampling effort, and human perception. The wildlife literature is replete with examples in which the relationship between an index and animal abundance was not stable but rather affected by weather (McGowan 1953, Gates 1966, Gibbs and Melvin 1993), season (Thompson et al. 1997, Graham 2002), habitat (Reid et al. 1966), year (Reid et al. 1966), or personnel (Bailey 1967, Link and Sauer 2002).

Eberhardt and Simmons (1987) suggest three approaches to using index measures in wildlife investigations, including the following:

1. Direct conversion of the index to an estimate of absolute abundance or density.
2. Calibration of the index through ratio and regression methods, including double sampling.
3. Calculation of an improved index or a prediction equation through use of auxiliary or supplementary information.

To this list, we add the fourth option:

4. The use of indices to compare the relative abundance of populations over time and/or space.

In subsequent sections of this chapter, all four approaches to the use of population indices will be discussed and illustrated.

8.1.1   Relationship between Indices and Abundance

It is often assumed that a linear relationship exists between the index measured and abundance (Gibbs 2000), and this relationship is supported in some fish (e.g., Hall 1986) and wildlife populations (e.g., Rotella and Ratti 1986). A straight-line relationship can be written as

$$x_i = \alpha + \beta N_i$$

where

x_i = index measured at the ith population;
N_i = animal abundance in the ith population.

When we use indices, however, we generally think of relative abundance in whch the ratio of the indices is equivalent to the ratio of the population abundance values, i.e.,

$$E\left(\frac{x_2}{x_1}\right) = \frac{N_2}{N_1}. \tag{8.1}$$

Should the populations each have a unique linear relationship between its index and abundance (Fig. 8.1a), then

$$E\left(\frac{x_2}{x_1}\right) \doteq \frac{E(x_2)}{E(x_1)} = \frac{\alpha_2 + \beta_2 N_2}{\alpha_1 + \beta_1 N_1},$$

and no inference to the rank order of the abundance levels is possible. With common intercepts and/or slopes, the ordination of the populations can be correctly ascertained but not the relative abundance (Fig. 8.1b–e). However, only in the case in which the index

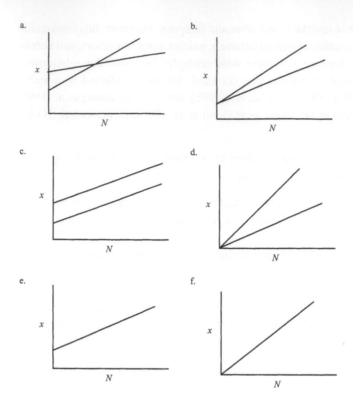

Figure 8.1. Alternative linear relationships between indices and absolute abundance of two populations.

is strictly proportional to abundance and that proportionality (i.e., slope β) is the same across populations (Fig. 8.1f), i.e.,

$$E(x_i) = \beta N_i; \quad i = 1, 2, \tag{8.2}$$

does the ratio of indices correctly represent the ratio of abundance levels, where

$$E\left(\frac{x_2}{x_1}\right) \doteq \frac{E(x_2)}{E(x_1)} = \frac{\beta N_2}{\beta N_1} = \frac{N_2}{N_1}.$$

Therefore, it is very important to understand the theoretical relationship between the index measured and the abundance monitored to properly interpret changes in index values. It is also important to understand that the relationship between an index and abundance might be nonlinear (Gibbs 2000). For example, calling intensity of anurans, studies using presence/absence measures of a species, and bait station surveys all may produce nonlinear relationships (Gibbs 2000). A variety of mathematical tools can be used to interpret the behavior of indices, including differential equations, probability measures, stochastic processes, and process-based models. The goal of model formulation is to examine the environmental, behavior, and anthropogenic influences on the index-abundance relationship to better design and analyze such studies. Because the conversion factor β usually cannot be estimated, understanding how to maintain a stable and proportional relationship between x_i and N_i is crucial to the success of an index study.

8.1.2   Basic Sampling Methods

The study of indices is an exploration into sampling designs. Many indices lend themselves to direct sampling of either spatial, temporal, or time-space sampling units. The resultant metric is then either an estimate of a mean or an index total. In finite sampling theory, the "population" no longer refers to a collection of individual animals. Rather, the target population refers to the collection of possible sampling units in the domain of interest.

There are three related concepts associated with making inferences from a sample survey. These are the target population, the sampling frame, and the sample of units drawn for inspection. As stated above, the target population is the population of inference. It is composed of sampling units, some or all of which are possibly subject to selection. The sampling frame, in turn, is the actual compilation of sampling units from which the sample is drawn. Finally, the sample is the collection of units chosen for characterization and measurement.

The strength and nature of inference depends on the relationship among the target population, sampling frame, and the sample drawn (Fig. 8.2). The strongest statistical inference occurs when the target population and sampling frame are one and the same, and a probabilistic sample is drawn from that sampling frame (Fig. 8.2a). In this case, a statistical inference can be made directly to the target population of interest. A weaker inference exists if the sampling frame is a subset of the larger target population (Fig. 8.2b). Thus, a statistical inference can be made to the sampling frame but not the target population. Investigators must then justify the similarity between the sampling frame and target population in order to make inferences to the target population. This inference is not statistical but rather based on subject matter considerations and open to criticism and possible skepticism. The weakest inference exists when the sampling frame and target population are distinct (Fig. 8.2c). This situation may arise when investigators attempt to control for extraneous sources of variability by choosing a highly controlled environment. Examples include growth chambers, greenhouses, and vole (*Rodentia*) colonies that are not part of the natural environment. In these cases, investigators sacrifice inferential strength for increased precision. No statistical inference can be made from the sample to the target population in this third case. The lessons learned from these studies may be subsequently applied to the design and analysis of more meaningful sampling frames.

Sampling schemes are characterized by the different ways sampling units are drawn from the sampling frame. The more we know about the structure of the sampling frame, the more focused a sampling scheme can be devised to take advantage of that structure and more precisely extract information. Among the considerations include the patterns in response across units, existence of covariate information, and how the sampling units are aggregated in time or space.

A wide range of sampling schemes are available for the design of index surveys (Cochran 1977, Thompson 2000). The three most commonly used schemes are the following:

1. Simple random sampling (SRS).
2. Stratified random sampling (STRS).
3. Systematic sampling (SYS).

The choice of sampling design is predicated on the extent of knowledge about the population under investigation, population structure, and available effort. SRS may be a reasonable choice of design when no prior knowledge exists or the population is randomly

a. Strong statistical inference

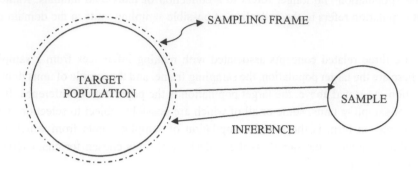

b. Weak statistical inference

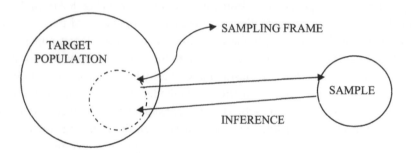

c. No statistical inference

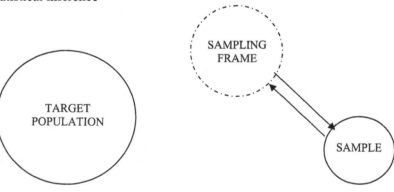

Figure 8.2. Three different conceptual relationships among the target population of inference, the sampling frame, and the sample of units drawn for characterization.

structured. Typically, sampling precision can be improved by using a more sophisticated design. STRS presumes some prior knowledge of the population structure to specify strata. SYS is motivated by logistical ease.

Random Sampling

Random sampling has a very strict meaning in a statistical context. It does not mean haphazard, opportunistic, periodic, convenient, or representative sampling. For a sampling frame with K sampling units of which k are drawn, random sampling implies that all

$$\binom{K}{k} = \frac{K!}{k!(K-k)!}$$

possible unique draws have equal likelihood of selection. Thus, all K sampling units within the sampling frame have equal probability of inclusion in a sample of size k.

SRS without replacement refers to sampling of a target population of finite size without replacement between sample draws. When the first unit is drawn, all K units are equally available for selection. After the first unit is selected, it is set aside and another unit is selected from the $K - 1$ units still available. This process of sampling without replacement continues until the kth unit is drawn. The sampling units are considered to be discrete non-overlapping entities, such that selection of K of K units comprises the entire sampling frame.

The sample mean from an SRS,

$$\frac{1}{k} \sum_{i=1}^{k} x_i,$$

is an unbiased estimator of the true population mean,

$$\overline{X} = \frac{1}{K} \sum_{i=1}^{K} x_i.$$

Typically, the estimator of the population parameter θ is denoted by the symbol $\hat{\theta}$. Hence,

$$\hat{\overline{X}} = \frac{1}{k} \sum_{i=1}^{k} x_i \tag{8.3}$$

is the estimator of the population mean $(\overline{X})$. Should all $\binom{K}{k}$ possible random samples be drawn and $\hat{\overline{X}}$ calculated for each, the variance among these estimates of the population mean would be

$$\mathrm{Var}\left(\hat{\overline{X}}\right) = \frac{1}{k}\left(1 - \frac{k}{K}\right)S^2, \tag{8.4}$$

where $S^2 = \dfrac{\sum_{i=1}^{K}(x_i - \overline{X})}{(K-1)}$ is the between-sampling-unit variability in the population. In practice, however, a single random sample of size k is drawn from the population. From that single sample, the variance of $\hat{\overline{X}}$ can be estimated by the quantity

$$\widehat{\mathrm{Var}}\left(\hat{\overline{X}}\right) = \frac{1}{k}\left(1 - \frac{k}{K}\right)s^2, \tag{8.5}$$

where $s^2 = \dfrac{\sum_{i=1}^{k}\left(x_i - \hat{\overline{X}}\right)}{(k-1)}$ is the between-unit variability within the sample drawn. An estimate of the population total $(X = K\overline{X})$ can be calculated as

$$\hat{X} = K \cdot \hat{\overline{X}}$$

or

$$\hat{X} = \frac{K}{k} \sum_{i=1}^{k} x_i. \tag{8.6}$$

The variance of $\hat{X}$ for SRS is

$$\text{Var}(\hat{X}) = \frac{K^2}{k}\left(1 - \frac{k}{K}\right)S^2 \tag{8.7}$$

and can be estimated by the quantity

$$\widehat{\text{Var}}(\hat{X}) = K^2 \cdot \widehat{\text{Var}}(\hat{\bar{X}})$$
$$= \frac{K^2}{k}\left(1 - \frac{k}{K}\right)s^2. \tag{8.8}$$

The quantity $\left(1 - \frac{k}{K}\right)$ in Eqs. (8.5) and (8.8) is the finite population correction (fpc).

Note that as k approaches K, the fpc goes to zero. This implies that once all K sampling units of a population of size K have been quantified, there is no longer any uncertainty in the values of the population parameters (e.g., $\bar{X}$ or X) and the variance equals zero. This variance relationship is different from that of sampling infinite size populations or sampling finite populations with replacement. In these cases, the variance of the sample mean is

$$\text{Var}(\bar{X}) = \frac{\sigma^2}{n},$$

where in the finite case,

$$\sigma^2 = \frac{\sum_{i=1}^{K}(x_i - \bar{X})}{K},$$

and in the infinite case,

$$\sigma^2 = \int (x_i - \mu)^2 f_X(x)dx,$$

where $f_X(x)$ = probability density function (pdf). The pdf mathematically describes the distribution of possible values (x) for the random variable (X).

SRS can also be used to estimate a population proportion. Let the indicator variable (I_i) be defined as

$$I_i = \begin{cases} 1 \text{ if the } i\text{th sampling unit possesses the trait of interest} \\ 0 \text{ otherwise.} \end{cases}$$

Then

$$M = \sum_{i=1}^{K} I_i$$

is the number of sampling units in the population with the trait, such that the proportion of the population with the trait is then

$$P = \frac{M}{K}. \tag{8.9}$$

For a random sample of size k, the estimator of the population proportion is

$$\hat{P} = \frac{m}{k}, \tag{8.10}$$

where

$$m = \sum_{i=1}^{k} I_i,$$

the number of sampling units drawn with the trait of interest. The variance of $\hat{P}$ can be expressed as

$$\text{Var}(\hat{P}) = \frac{P(1-P)}{k}\left(\frac{K-k}{K-1}\right) \tag{8.11}$$

and estimated by the quantity

$$\widehat{\text{Var}}(\hat{P}) = \frac{\hat{P}(1-\hat{P})}{(k-1)}\left(1-\frac{k}{K}\right). \tag{8.12}$$

Example 8.1: Using Simple Random Sampling to Estimate the Number of Eastern Gray Squirrel (*Sciurus carolinensis*) Nests

A 100-ha woodlot was subdivided into $K = 100$ 1-ha sampling units or plots. A random sample of $k = 20$ plots was selected, and the numbers of eastern gray squirrel nests were enumerated to calculate an index of squirrel abundance. The 20 sites sampled produced the following squirrel nest counts:

$$6, 7, 6, 0, 3, 3, 2, 5, 1, 5$$
$$8, 10, 6, 3, 1, 6, 6, 0, 7, 1.$$

The average number of nests per site (i.e., nest/ha) is estimated (Eq. 8.3) to be

$$\hat{\bar{X}} = \frac{1}{k}\sum_{i=1}^{k} x_i$$

$$= \frac{86}{20} = 4.3,$$

with an estimated variance (Eq. 8.5) of

$$\widehat{\text{Var}}(\hat{\bar{X}}) = \frac{1}{k}\left(1-\frac{k}{K}\right)s^2$$

$$= \frac{1}{20}\left(1-\frac{20}{100}\right)\frac{\sum_{i=1}^{20}(x_i - \hat{\bar{X}})^2}{(20-1)} = \frac{1}{20}\left(1-\frac{20}{100}\right)(8.2210) = 0.3288,$$

or $\widehat{\text{SE}}(\hat{\bar{X}}) = 0.5734$ nests/site. An estimate of total nests across the 100-ha area is calculated as

$$\hat{X} = K \cdot \hat{\bar{X}}$$
$$= 100 \times 4.3 = 430.$$

The variance of $\hat{X}$, in turn, is estimated (Eq. 8.8) to be

$$\widehat{\text{Var}}(\hat{X}) = K^2 \cdot \widehat{\text{Var}}(\hat{\bar{X}})$$
$$= 100^2(0.3288) = 3288,$$

or $\widehat{\text{SE}}(\hat{X}) = 57.34$.

Another metric that could be estimated is the proportion of the 1-ha sampling units with squirrel nests. In which case, by using Eq. (8.10)

$$\hat{P} = \frac{m}{k}$$
$$= \frac{18}{20} = 0.90,$$

with an estimated variance (Eq. 8.12) of

$$\widehat{\text{Var}}(\hat{P}) = \frac{\hat{P}(1-\hat{P})}{k-1}\left(1 - \frac{k}{K}\right)$$
$$= \frac{0.90(1-0.90)}{(20-1)}\left(1 - \frac{20}{100}\right) = 0.003789,$$

or $\widehat{\text{SE}}(\hat{P}) = 0.0616$.

For SRS, confidence intervals can be calculated as

$$\hat{\theta} \pm t_{k-1,1-\frac{\alpha}{2}} \cdot \widehat{\text{SE}}(\hat{\theta}), \tag{8.13}$$

where $t_{k-1,1-\frac{\alpha}{2}}$ is a standard t-statistic with $k - 1$ degrees of freedom such that

$$P(|t_{k-1}| < t) = 1 - \alpha.$$

For estimates of proportion, Cochran (1977:57) recommended the confidence interval estimator

$$\hat{P} \pm t_{k-1,1-\frac{\alpha}{2}} \sqrt{\frac{\hat{P}(1-\hat{P})}{k-1}\left(1 - \frac{k}{K}\right) + \frac{1}{2k}}. \tag{8.14}$$

Stratified Random Sampling

Stratification in sampling is analogous to blocking in experimental design. The purpose of stratification is to subdivide the sampling frame into subpopulations that are internally more homogeneous than is the population as a whole. Through stratification, between-strata variation is eliminated from the variance of the estimate. The stratification need not be perfect to achieve benefits in precision. The elements in the strata simply need to be more similar than are elements in the overall population. After stratification, sampling within each subdivision is handled independently. Selection of the size and shape of sampling units and the sampling scheme within each of the strata can be different. The sampling designs within the strata can be chosen to take advantage of the unique traits of each stratum. The results of the separate strata are subsequently combined to make inference to the overall sampling frame. The simplest scheme, and the one we will consider, is STRS.

We define the following quantities:

x_{ij} = ith observation in the jth strata ($i = 1, \ldots, K_j; j = 1, \ldots, l$);
K_j = number of sampling units in the jth strata ($j = 1, \ldots, l$);
k_j = number of units sampled in the jth strata ($j = 1, \ldots, l$);

$$K = \sum_{j=1}^{l} K_j.$$

Should the sampling units be of equal size, strata weights can be calculated as

$$W_j = \frac{K_j}{K}.$$

The population mean based on STRS is estimated by

$$\hat{\bar{X}} = \sum_{j=1}^{l} W_j \bar{x}_j, \tag{8.15}$$

with variance,

$$\mathrm{Var}(\hat{\bar{X}}) = \sum_{j=1}^{l} W_j^2 \left(1 - \frac{k_j}{K_j}\right) \frac{S_j^2}{k_j}, \tag{8.16}$$

and estimated variance of

$$\widehat{\mathrm{Var}}(\hat{\bar{X}}) = \sum_{j=1}^{l} W_j^2 \left(1 - \frac{k_j}{K_j}\right) \frac{s_j^2}{k_j}. \tag{8.17}$$

The estimate of the population total ($\hat{X}$) follows directly from estimates of the strata means ($\hat{\bar{X}}_j$), where

$$\hat{X} = \sum_{j=1}^{l} K_j \hat{\bar{X}}_j. \tag{8.18}$$

The variance of $\hat{X}$ is

$$\mathrm{Var}(\hat{X}) = \sum_{j=1}^{l} K_j^2 \left(1 - \frac{k_j}{K_j}\right) \frac{S_j^2}{k_j} \tag{8.19}$$

and estimated by the quantity

$$\widehat{\mathrm{Var}}(\hat{X}) = \sum_{j=1}^{l} K_j^2 \left(1 - \frac{k_j}{K_j}\right) \frac{s_j^2}{k_j}, \tag{8.20}$$

where

$$s_j^2 = \frac{\sum_{i=1}^{k_j}(x_{ij} - \bar{x}_j)^2}{(k_j - 1)}$$

and where

$$\bar{x}_j = \frac{1}{k_j} \sum_{i=1}^{k_j} x_{ij}.$$

The proportion of sampling units in the population with a specific trait is estimated as

$$\hat{P} = \frac{1}{K} \sum_{j=1}^{l} K_j \hat{P}_j, \tag{8.21}$$

where $\hat{P}_j = \dfrac{1}{k_j} \sum_{i=0}^{k_j} I_{ij}$ for

$$I_{ij} = \begin{cases} 1 \text{ if the } j\text{th sample in the } i\text{th strata possesses the trait of interest} \\ 0 \text{ otherwise.} \end{cases}$$

The variance of $\hat{P}$ is expressed as

$$\mathrm{Var}(\hat{P}) = \frac{1}{K^2} \sum_{j=1}^{l} \left[K_j^2 \frac{P_j(1-P_j)}{k_j} \left(\frac{K_j - k_j}{K_j - 1} \right) \right] \tag{8.22}$$

with an estimated variance of

$$\widehat{\mathrm{Var}}(\hat{P}) = \frac{1}{K^2} \sum_{j=1}^{l} \left[K_j^2 \frac{\hat{P}_j(1-\hat{P}_j)}{(k_j - 1)} \left(1 - \frac{k_j}{K_j} \right) \right]. \tag{8.23}$$

Often different size sampling units will be used in various strata. In these circumstances, strata weights based on $W_j = \dfrac{K_j}{K}$ are inappropriate. The strata weights in these circumstances need to be computed on an areal basis as

$$W_j = \frac{A_j}{\sum_{j=1}^{l} A_j},$$

where A_j = areal size of the jth spatial stratum ($j = 1, \ldots, l$). When sampling over time, the weights would naturally refer to durations in each stratum such that

$$W_j = \frac{T_j}{\sum_{j=1}^{l} T_j},$$

where T_j = duration of the jth temporal stratum ($j = 1, \ldots, l$).

Example 8.2: Eastern Gray Squirrel Nest Counts—Stratified Sampling

The eastern gray squirrel Example 8.1 is revisited; this time, the 100-ha woodlot is subdivided into two strata corresponding to hardwoods and mixed conifer habitats. The first stratum of hardwoods has $K_1 = 34$ 1-ha plots of which $k_1 = 10$ are selected with nest counts of

$$1, 0, 2, 1, 3, 2, 0, 1, 3, 0.$$

The mixed conifer stratum has $K_2 = 66$ 1-ha plots, of which $k_2 = 14$ are selected with nest counts of

$$6, 2, 7, 5, 8, 3, 4, 8, 7, 4, 5, 6, 3, 9.$$

The estimate of the mean number of nests per 1-ha plot is calculated (Eq. 8.15) as

$$\hat{\bar{X}} = \sum_{j=1}^{2} W_j \bar{x}_j$$

$$= \frac{34}{100}(1.3) + \frac{66}{100}(5.5) = 4.072,$$

with an estimated variance (Eq. 8.17) of

$$\widehat{\mathrm{Var}}\left(\hat{\bar{X}}\right) = \sum_{j=1}^{2} W_j^2 \left(1 - \frac{k_j}{K_j}\right) \frac{s_j^2}{k_j}$$

$$= (0.34)^2 \left(1 - \frac{10}{34}\right) \frac{1.3444}{10} + (0.66)^2 \left(1 - \frac{14}{66}\right) \frac{4.5769}{14}$$

$$\widehat{\mathrm{Var}}\left(\hat{\bar{X}}\right) = 0.01097 + 0.11220 = 0.12317,$$

or $\widehat{\mathrm{SE}}\left(\hat{\bar{X}}\right) = 0.3509.$

As illustrated in Example 8.2, the contribution of each stratum to the overall precision of the study may differ. Optimal allocation considers strata sizes, within-strata variances, and per-unit sampling costs to allocate sampling effort to maximize the precision of a stratified sampling scheme. For total costs (C_{Total}) of performing an STRS scheme expressed by

$$C_{\mathrm{Total}} = c_0 + c_1 k_1 + c_2 k_2 + \cdots + c_l k_l, \tag{8.24}$$

where

c_0 = fixed costs of sampling regardless of sample sizes;
c_j = per-sampling-unit costs in the jth stratum ($i = 1, \ldots, l$);

the optimal allocation of sampling effort across strata is computed as

$$k_j = k. \left[\frac{\dfrac{K_j S_j}{\sqrt{c_j}}}{\displaystyle\sum_{j=1}^{l} \dfrac{K_j S_j}{\sqrt{c_j}}} \right], \tag{8.25}$$

where $k. = \sum_{j=1}^{l} k_j$. For equal costs (i.e., $c_1 = c_2 = \ldots = c_l$), optimal allocation is based on the formula

$$k_j = k. \left[\frac{K_j S_j}{\sum_{j=1}^{l} K_j S_j} \right]. \tag{8.26}$$

Equations (8.25) and (8.26) provide optimal allocations for fixed total sampling effort ($k.$). Equations (8.25) and (8.26) suggest that precision will be optimized, in general, by placing more sampling effort in strata that are (1) larger (i.e., K_j), (2) internally more variable (i.e., S_j^2), and (3) cheaper to sample (i.e., c_j).

To find the optimal values of k_j for fixed total cost C_{Total}, one needs to substitute the values of k_j from either Eq. (8.25) or (8.26) into the cost Eq. (8.24) and solve for $k.$, which yields

$$k. = \frac{(C_{Total} - c_0)}{\sum_{j=1}^{l} \left[c_j \frac{\frac{K_j S_j}{\sqrt{c_j}}}{\sum_{j=1}^{l} \frac{K_j S_j}{\sqrt{c_j}}} \right]}$$

or

$$k. = \frac{(C_{Total} - c_0)}{c \sum_{j=1}^{l} \left[\frac{K_j S_j}{\sum_{j=1}^{l} K_j S_j} \right]},$$

if sampling costs are equal across strata (i.e., $c_1 = c_2 = \ldots = c_l = c$).

Example 8.3: Optimal Allocation—Eastern Gray Squirrel Nest Counts

Optimal allocation between hardwood and mixed conifer strata can be calculated for the squirrel nest Example 8.1. Assuming sampling costs are equal, then for $k. = 24$

$$k_1 = k. \left[\frac{K_j S_j}{\sum_{j=1}^{2} K_j S_j} \right]$$

$$= 24 \left[\frac{34\sqrt{1.3444}}{34\sqrt{1.3444} + 66\sqrt{4.5769}} \right] = 5.24$$

and

$$k_2 = 24 - 5.24 = 18.76$$

or 5 and 19 samples, respectively, for strata 1 and 2.

Systematic Sampling

The logistics of identifying or locating random sampling units across the landscape can be arduous. Furthermore, open-ended sampling over time cannot be randomly performed. Therefore, SYS provides a practical alternative when random sampling is not feasible. The ease of locating periodically spaced sampling units makes SYS appealing. Under certain circumstances, it can also be more precise than is SRS.

Systematic sampling is equivalent to SRS if the response values of the sampling units are independent of order; i.e., the values of the sampling units are random. However, caution is necessary when applying SYS in ecological applications because there is always a risk that sample placement may be synchronized with the periodicity of environmental patterns. In this situation, the sampling variance may seriously underestimate the actual variance of the survey. Unfortunately, from sampling data alone, there is usually no way of ascertaining whether periodicity may exist.

Consider a target population of size K, in which a systematic sample of size k is drawn. In this situation, there are $n = \dfrac{K}{k}$ possible random samples. If $K = nk$, the sample mean is an unbiased estimator of $\bar{X}$. The only randomization involved in SYS is the selection of the starting point. A random selection from the first $1, 2, \ldots, n$ sampling units is the basis for beginning the systematic sample. If k is not a multiple of K, a random number between 1 and K should be drawn as the starting point. Subsequent sampling units are then drawn on either side of the starting position at intervals of size n.

The variance (Cochran 1977:207) for the estimate of the population mean ($\hat{\bar{X}}$) for SYS is

$$\mathrm{Var}\left(\hat{\bar{X}}\right) = \frac{K-1}{K}S^2 - \frac{n(k-1)}{K}S^2_{WSY}, \tag{8.27}$$

where

$$S^2_{WSY} = \frac{1}{n(k-1)}\sum_{i=1}^{n}\sum_{j=1}^{k}(x_{ij} - \bar{x}_i)^2. \tag{8.28}$$

The within-SYS variance (S^2_{WSY}) measures the average variability of sampling units within a systematic sample. The variance S^2 is the between-sample-unit variance for the entire population.

The sample observations can be used to estimate the component S^2_{WSY} from a systematic sample but not S^2. Hence, there is no unbiased estimator for the variance of a systematic sample. Instead, various approximations have been suggested. Wolter (1984) investigated various variance approximations for SYS. If the values of the sampling units are independent of sample location, the variance for SRS can be used where

$$\widehat{\mathrm{Var}}\left(\hat{\bar{X}}\right) = \left(1 - \frac{k}{K}\right)\frac{s^2}{k}. \tag{8.29}$$

Indeed, if $S^2_{WSY} = S^2$, then Eq. (8.27) reduces to Eq. (8.29). Should the population have a stratified structure, Wolter (1984) suggested the estimator

$$\widehat{\mathrm{Var}}\left(\hat{\bar{X}}\right) = \frac{1}{k}\left(1 - \frac{k}{K}\right)\frac{\sum_{i=1}^{k-1}(x_i - x_{i+1})^2}{2(k-1)}, \tag{8.30}$$

and if the population has linear trends along the course of the sampling, then

$$\widehat{\mathrm{Var}}\left(\hat{\bar{X}}\right) = \frac{1}{k}\left(1 - \frac{k}{K}\right)\frac{\sum_{i=1}^{k-2}(x_i - 2x_{i+1} + x_{i+2})^2}{6(k-2)}. \tag{8.31}$$

Typically, the use of the variance formula for SRS (Eq. 8.29) in conjunction with SYS will overestimate the true variance. Variance formulas Eqs. (8.30) and (8.31) will typically reduce the positive bias.

Inspection of Eq. (8.27) indicates the greater the within-systematic-sample variability (S_{WSY}^2), the smaller the overall variance for the estimate of the population mean. Instead, a well-drawn systematic sample will have more internal variability than will the population as a whole (i.e., $S_{WSY}^2 > S^2$). Under these circumstances, SYS will be more precise than SRS. As stated, when $S_{WSY}^2 = S^2$, the variance formula for SYS (Eq. 8.27) reduces to that of SRS (Eq. 8.7). Herein lies the irony of SYS; when done properly, the within-sample variability S_{WSY}^2 of the systematic sample will exceed S^2. However, when estimating the variance, the formula for SRS (i.e., Eq. 8.29) will inflate the variance estimate. Thus, the more precise the SYS, the larger the positive bias of the variance estimate. In other words, the greater the precision of SYS, the less precise the sample results appear. Kish (1965:127–132) describes use of replicated SYS to obtain unbiased variance estimators. However, numerous replicate samples are needed for the Student t-based confidence intervals to be narrow.

8.2   Description of Common Indices

The first rule in design and analysis of an index study is to propose a statistical or mathematical model for the response being measured. Statistical models differ from mathematical models by introducing a stochastic element. This modeling step is essential to understanding the biological, environmental, and anthropogenic factors influencing an index. By understanding influential factors, actions may be taken to control their influence. This information may dictate how the index study should be conducted, what auxiliary information must be collected, and when and where the study should be performed. The model may also suggest uncontrollable factors and the necessary steps to ensure their consistency over time and space. Without model assessment, investigators may inadvertently violate one or more crucial assumptions, thereby relegating the study to more of a learning experience rather than a quantitative assessment of populations. Even more seriously, an influential factor may go unchecked, and spurious conclusions may be drawn from the study.

We now quantitatively describe some common index methods used in wildlife investigations. As important as the methods themselves is the approach used to interpret the behavior of the indices. We hope, through example, that other index methods will receive equal scrutiny when used in the future.

8.2.1   Pellet Counts

Probabilistic samples of pellets, droppings, or fecal material have been performed for numerous species as an index of abundance. The method has been applied to ungulates (Neff 1968, Rowland et al. 1984, Fuller 1991), carnivores (Schauster et al. 2002), lagomorphs (Krebs et al. 2001, Murray et al. 2002), and other species where the droppings are visible and persistent (Vernes 1999). Applications extend to small mammals (Emlen et al. 1957) and gallinaceous birds. However, the most widespread application of the pellet count method has been for deer (Bennett et al. 1940, Neff 1968). The defecation rate for deer has been calculated to convert the mean pellet count to animal density estimates. Rasmussen and Domen (1943) reported 12.7 groups/day in Utah, Eberhardt and Van Etten (1956) reported 12.7 groups/day in Michigan, Rogers et al. (1958) reported a rate of 15.2 groups/day in Colorado, and Smith (1964) reported an average defecation rate of 13.2 groups per day in Utah. Important seasonal changes in defecation rates in deer have been noted (Millspaugh et al. 2002).

The pellet count index (x_i) at a sampling plot can be conceptualized by the simple model

$$E(x_i) = A_i \cdot D_i \cdot F \cdot t_i, \tag{8.32}$$

where

x_i = number of pellet groups at the ith plot;
A_i = area of the ith plot;
D_i = animal density at the ith plot;
F = defecation rate in terms of pellet groups-per-unit-time;
t_i = amount of time pellets were allowed to accumulate at the ith plot.

With knowledge of A_i, t_i, and an estimate of F, animal density can be calculated by the method of moments estimator

$$\hat{D}_i = \frac{x_i}{A_i \hat{F} t_i}. \tag{8.33}$$

The simple nature of the density estimator (Eq. 8.33) makes it the most common candidate for directly converting an index count to an estimate of absolute density.

In mesic environments in which decay rate of pellet groups is high (Lehmkuhl et al. 1994, Nchanji and Plumptre 2001), the response model (Eq. 8.33) can be modified to include the rate of loss. Let

r = instantaneous defecation rate per unit of time;
μ = instantaneous loss rate of fecal pellet groups;
r_t = number of pellets accumulated and still present on the site over t units of time per individual.

A differential equation can used to model the change in pellet groups over time for an individual animal where

$$\frac{dr_t}{dt} = -\mu r_t + r.$$

Solving for r_t and setting $r_0 = 0$ yields the relationship

$$r_t = \frac{r(1 - e^{-\mu t})}{\mu}. \tag{8.34}$$

A revised response model can be constructed, using the value of r_t in place of Ft_i of Eq. (8.32), where

$$E(x_i) = A_i D_i \frac{r(1 - e^{-\mu t_i})}{\mu}. \tag{8.35}$$

The expression $\dfrac{(1 - e^{-\mu t_i})}{\mu}$ (Eq. 8.35) is the effective number of days of actual accumulation at rate r over time t_i. Calibration of the pellet count to animal density now requires information on both the instantaneous defecation (r) and pellet group loss (μ) rates. Use of the index as a measure of relative abundance or density is predicated on these rates as well as the accumulation time t_i being constant across populations being compared.

In practice, the pellet group index is averaged across replicate quadrats probabilistically sampled from the area of inference. A variety of sampling schemes (Section 8.1.2) can be used to validly estimate mean pellet group density. The general form of models (8.32) and (8.35) pertain to either a single plot with areal density D_i or plots within mean areal density $\overline{D}$.

Example 8.4: Estimating Pellet Count Index for Mule Deer (*Odocoileus hemionus*), Washington State

A 9-ha site in south central Washington was subdivided into 144 quadrats of size 25 × 25 m. A simple random sample of 20 quadrats yielded the following pellet group counts for mule deer:

$$5, 7, 0, 11, 3, 1, 6, 2, 9, 4, 8, 0, 2, 1, 5, 8, 6, 0, 1, 5.$$

Mean number of pellet counts per quadrat is estimated by Eq. (8.3) to be

$$\hat{\overline{X}} = \frac{1}{k} \sum_{i=1}^{k} x_i$$

$$= \frac{1}{20}(5 + 7 + 0 + \cdots + 0 + 1 + 5) = 4.2,$$

with an estimated variance (Eq. 8.5) of

$$\widehat{\text{Var}}\left(\hat{\overline{X}}\right) = \frac{\left(1 - \dfrac{k}{K}\right) s^2}{k}$$

$$= \frac{\left(1 - \dfrac{20}{144}\right) 11.0105}{20} = 0.47406$$

or $\widehat{\text{SE}}\left(\hat{\overline{X}}\right) = 0.6885$.

In this study, all pellet groups were counted regardless of their time on the ground. Thus, the index can only be used to express relative abundance. These 20 quadrats would have to have been first cleared of all old droppings, and accumulations allowed to occur for a known duration of time, before the index could be converted to an estimate of abundance.

Sample Size Calculations

The precision of the index survey can be calculated, based on the variance formula (Eq. 8.4) for SRS. We define precision of the mean index value in terms of relative error, where

$$P\left(\left|\frac{\hat{\bar{X}} - \bar{X}}{\bar{X}}\right| < \varepsilon\right) = 1 - \alpha.$$

The desired sampling precision is for the relative error in estimation, $\left|\frac{\hat{\bar{X}} - \bar{X}}{\bar{X}}\right|$, to be less than ε, $(1 - \alpha)100\%$ of the time. The required sample size, with these sampling objectives, is

$$k = \frac{1}{\frac{\varepsilon^2}{Z^2_{1-\frac{\alpha}{2}} CV^2} + \frac{1}{K}}, \tag{8.36}$$

where $CV = \frac{S}{\bar{X}}$. When K is large and $\frac{k}{K}$ small, Eq. (8.36) reduces to the approximation

$$k = \frac{Z^2_{1-\frac{\alpha}{2}} CV^2}{\varepsilon^2} \tag{8.37}$$

Example 8.5: Sample Size Calculations for Pellet Count Survey

The pellet count example (Example 8.4) had a relative precision as measured by the coefficient of variation $(CV(\hat{\bar{X}}))$ of $\frac{0.6885}{4.2} = 0.1639$ or 16.4%. We now consider the sample size required to have a precision of

$$P\left(\left|\frac{\hat{\bar{X}} - \bar{X}}{\bar{X}}\right| < 0.10\right) = 0.95,$$

such that $\varepsilon = 0.10$ and $Z_{1-\frac{\alpha}{2}} = 1.96$. By using the CV for the pellet count data of

$\frac{\sqrt{11.0105}}{4.2} = 0.7900$, the required sampling effort can be computed (Eq. 8.36) as

$$k = \cfrac{1}{\cfrac{\varepsilon^2}{Z^2_{1-\frac{\alpha}{2}}CV^2} + \cfrac{1}{K}}$$

$$= \cfrac{1}{\cfrac{(0.10)^2}{(1.96)^2(0.79)^2} + \cfrac{1}{144}} = 89.97,$$

and then rounding up to $k = 90$ quadrats out of $K = 144$.

Discussion of Utility

The quadrat sampling used to estimate the mean pellet count index could be applied to any animal sign. Other animal sign that lend themselves to spatial sampling includes nests (Kiel 1955), burrows (Severson and Plumb 1998), and tracks (Smallwood and Fitzhugh 1995). For valid inferences, sampling cannot be opportunistic or convenience sampling. Rather, samples need to be drawn probabilistically. Preliminary survey data are needed to obtain initial values of CV for sample size calculations.

The CV values are a function of animal dispersal across the landscape but also the size and configuration of the quadrats. Smith (1964) recommended 100 sq ft (0.0993 ha) quadrats, whereas Eberhardt and Van Etten (1956) recommended 1/50th acre plots (0.008 ha) for pellet count surveys. A component of a well-designed preliminary survey would be a uniformity trial (Federer 1955:61–68) to identify the optimal quadrat size. The Fairfield-Smith empirical variance law (Smith 1938) or other similar variance relationships (Skalski and Robson 1992:41–46) can be used to describe how plot size influences the anticipated size of S^2.

8.2.2   Frequency Index

The frequency index is based on recording the observed proportion ($\hat{P}$) of sampling units within an area that contain one or more animals or their sign (Dice 1952:279, Scattergood 1954, Davis 1963). This method is attractive when enumeration is difficult but ascertaining the presence or absence of individuals is easy. A practical limitation is that quadrats or plots cannot be so large that all 100% possess the species of interest. In that case, useful population comparisons are no longer possible. The survey should be composed of quadrats of different sizes to overcome this problem. Ideally, large quadrats would be subdivided into smaller-sized quadrats, which could be subdivided into yet smaller quadrats for analysis.

Let the areal population be subdivided into K quadrats. A random sample of k quadrats are visually sampled, of which m are found to contain the animal species. The frequency index is the proportion

$$\hat{P} = \frac{m}{k}. \tag{8.38}$$

Suppose, in actuality, M of K quadrats had the species present, in which case, the true value of the index would be

$$P = \frac{M}{K}.$$

Under these conditions, the random variable m is hypergeometrically distributed with likelihood

$$L(m|M,K,k) = \frac{\binom{M}{m}\binom{K-M}{k-m}}{\binom{K}{k}}.$$

The variance of the frequency index is the same as Eq. (8.11).

$$\mathrm{Var}(\hat{P}) = \frac{\left(\frac{M}{K}\right)\left(1-\frac{M}{K}\right)(K-k)}{k(K-1)},$$

with an unbiased estimate (Eq. 8.12) of the variance (Cochran 1977:52)

$$\widehat{\mathrm{Var}}(\hat{P}) = \frac{\hat{P}(1-\hat{P})}{k-1}\left(1-\frac{k}{K}\right), \tag{8.39}$$

where $\left(1-\dfrac{k}{K}\right)$ is the fpc.

Assumptions

The frequency index is based on three assumptions:

1. A random sample of k of K quadrats is selected without replacement for canvassing.
2. The animals or their active sign have a detection probability of 1 if present in the quadrat.
3. Canvassing a quadrat does not affect the presence or distribution of animals on the other quadrats.

The frequency index performs best when the habitat canvassed is relatively homogeneous and the species does not aggregate or live in colonies. The frequency index will poorly track demographic trends if changes in density occur without a corresponding change in the frequency of occurrence. Figure 8.3 illustrates a case when the frequency index goes unchanged despite large changes in density. Here, areas already occupied increase in density without the animals dispersing to other unoccupied sites. It may well be that unoccupied sites are unsuitable for the species, and the increase in abundance is expressed as an increase in density within the suitable habitat only. Dice (1952) suggested as a rule of thumb that the size of the quadrats used in a frequency index be such that rarely will more than one animal be found per quadrat. Greig-Smith (1952) recommended there should not be more quadrats with zero counts than quadrats with just one individual.

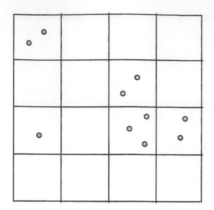

Figure 8.3. Depiction of two populations with the same frequency index of $P = \frac{5}{16}$ but areal density differing by a factor of 3.

Example 8.6: Frequency Index—Eastern Gray Squirrels

The frequency index was applied to monitor eastern gray squirrels on a 100-ha woodlot subdivided into 0.5-ha quadrats. Of the $K = 200$ quadrats, $k = 40$ were randomly selected for canvassing. The response measured was the presence or absence of active squirrel nests. The survey found $m = 12$ quadrats with newly constructed squirrel nests. The estimate of the frequency index is

$$\hat{P} = \frac{m}{k}$$
$$= \frac{12}{40}$$
$$= 0.30.$$

The estimated variance associated with the frequency index is

$$\widehat{\text{Var}}(\hat{P}) = \frac{\hat{P}(1 - \hat{P})}{(k - 1)}\left(1 - \frac{k}{K}\right)$$
$$= \frac{0.30(1 - 0.30)}{(40 - 1)}\left(1 - \frac{40}{200}\right) = 0.00431,$$

with standard error of $\widehat{\text{SE}}(\hat{P}) = 0.0656$.

The survey was repeated the following year with a new selection of $k = 60$ quadrats, of which $m = 14$ had active squirrel nests for a $\hat{P} = 0.2333$ ($\widehat{\text{SE}} = 0.0461$). A 2×2 contingency table can be used to compare the annual frequency indices as follows:

	Year 1	Year 2	
Present	12	14	26
Absent	28	46	74
	40	60	100

with $P(\chi_1^2 > 0.5544) = 0.4565$. The null hypothesis of equal frequency indices is therefore not rejected. It is not necessarily true, however, that squirrel abundance has not changed between years. An observed change in mean squirrel nest count per quadrat would be necessary to begin making direct inferences to a change in abundance. The R × C contingency table (Zar 1996:485–486) is appropriate as long as K is large and $\dfrac{k}{K}$ is small. Otherwise, comparison of frequency indices should be based on interval estimates using variance formula Eq. (8.39).

8.2.3   Auditory Counts

Auditory counts of birds at fixed stations along a drive route or at permanent field locations are commonly used to monitor the abundance of mourning doves (*Zenaida macroura*) (McClure 1939, Baskett et al. 1978, Sauer et al. 1994), ring-necked pheasants (*Phasianus colchicus*) (Gates 1966, Luukkonen et al. 1997, Rice 2003), ruffed grouse (*Bonasa umbellus*) (Dorney et al. 1958, Gullion 1966), and northern bobwhite (*Colinus virginianus*) (Hansen and Guthery 2001) among other species. This technique is potentially useful for any species that indicate their presence through vocalization or other sound-producing activity. Perhaps the best example of use of auditory counts is the Breeding Bird Survey, which began in 1966 as a way to track the status and trends of North American Birds (Bystrak 1981, Geissler and Noon 1981). Each year, volunteers sample over 4100 survey routes across Canada and the United States. Each survey route is 39.4 km long, and observers stop at 0.8-km increments to conduct a point count for 3 minutes. Observers record every bird seen or heard within a 0.4-km radius of each stop. In addition to providing information about status and trends of many bird species (Peterjohn and Sauer 1994), the Breeding Bird Survey helps prioritize research and management activities.

A useful model for auditory counts, which takes into account the hearing acuity of the surveyor, is

$$E(x_{ij}) = \pi\, r_i^2 \cdot D_j \cdot F \cdot t_{ij} \tag{8.40}$$

where

x_{ij} = number of calls recorded by the ith survey or at the jth location;
r_i = radius within which a call may be heard by the ith surveyor;
D_j = bird density at the jth location;
F = frequency of calls per unit time;
t_{ij} = time spent by the ith surveyor at the jth location listening for calls.

Model (8.40) simplifies to

$$E(x_{ij}) = \beta r_i^2 D_j \tag{8.41}$$

if the time used for recording calls is held constant (i.e., $t_{ij} = t$) and the frequency of calls (F) remains the same over time and space.

There are several assumptions associated with the auditory index (Eq. 8.41):

1. Call frequency (F) is independent of animal density and is constant over time and space.

2. The length of time over which calls are recorded is constant, unless time is recorded and the response reported as a rate.
3. Hearing acuity, as measured by r_i, depends on surveyor but not locale.
4. If multiple surveyors are involved, estimates of r_i are available to calibrate the surveyors to one another.

Under these conditions, the ratio of auditory counts at two different locales by the same surveyor has the expected value of

$$E\left(\frac{x_{i2}}{x_{i1}}\right) \doteq \frac{\pi\, r_i^2 D_2 Ft}{\pi\, r_i^2 D_1 Ft}$$

$$= \frac{D_2}{D_1}.$$

The assumption that hearing acuity is a function of the surveyor alone (i.e., r_i) and not a function of both surveyor and locale (i.e., r_{ij}) is crucial to use of this index (Ramsey and Scott 1981). In a hearing test of 274 active birders, Ramsey and Scott (1981) found an order of magnitude difference in hearing ability. This assumption about the nature of r_i may be violated if habitat and environmental factors absorb or mask calls. For example, the hearing radius may be shorter near highways and industrial parks than in unpopulated rural areas. The design of the drive route and selection of call stations should consider these factors. Acoustic instruments may be used to measure ambient noise levels and help in the selection of the call stations.

The call frequency (F) is known to change over the course of the day (McClure 1939, Luukkonen et al. 1997, Hansen and Guthery 2001), vary with weather factors (McClure 1939, Lengagne and Slater 2002), and vary during the season for many species (Nebel and McCaffery 2003, Amrhein et al. 2004). Calling activity of quail is maximized for 10 to 20 minutes after sunrise (Hansen and Guthery 2001). The drumming frequency of ruffed grouse is known to change with proximity to breeding season (Gullion 1966, Archibald 1976). Comparisons of auditory indices between years should consider the timing of the surveys with respect to the peak of the breeding season. Some investigators repeat the survey multiple times during the breeding season, choosing to use the maximal count as the index for that season.

There are multiple stops in a roadside survey, and the average count across sites is used as an index. The index value averaged over n sites has the expected value

$$E(\bar{x}_i) = E\left[\frac{\sum_{j=1}^{n} x_{ij}}{n}\right]$$

$$= \pi\, r_i^2 Ft\, \frac{\sum_{j=1}^{n} D_j}{n}$$

$$= \pi\, r_i^2 Ft\bar{D}$$

$$= \beta \cdot \bar{D},$$

which can be used to compare relative abundance between years or locales. The mean index $\bar{x}_i$ has the variance estimator

$$\widehat{\mathrm{Var}}(\bar{x}_i) = \frac{\sum\limits_{j=1}^{n}(x_{ij}-\bar{x}_i)^2}{n(n-1)} = \frac{s^2}{n}.$$

A statistical t-distribution with $n-1$ degrees of freedom can be used to construct a $(1-\alpha)$ 100% confidence interval of the form

$$\bar{x}_i \pm t_{\frac{\alpha}{2}}\sqrt{\frac{s^2}{n}}. \tag{8.42}$$

Log-transforming model (8.41) produces the log-linear equation

$$\ln x_{ij} = \ln\beta + 2\ln r_i + \ln D_j, \tag{8.43}$$

suggesting the form

$$y_{ij} = \mu + \tau_i + \gamma_j + \varepsilon_{ij} \tag{8.44}$$

for a two-way analysis of variance (ANOVA), where

μ = baseline or mean response level;
$\tau_i = 2\ln r_i$ = effect of the ith surveyor;
$\gamma_j = \ln D_j$ = effect of the jth site;
ε_{ij} = random error term.

Equation (8.44) suggests a survey design, with multiple sites and multiple observers surveying the same sites independently but concurrently, could be used to calibrate the hearing acuity of the surveyors, assess model adequacy, and compare locales. The relative abundance between sites would be estimated by the quotient

$$E\left(\frac{e^{\gamma_2}}{e^{\gamma_1}}\right) \doteq \frac{D_2}{D_1}.$$

Example 8.7: Estimating Surveyor Effect From Auditory Counts of Thrushes, Central Missouri

In central Missouri, auditory counts of thrushes, i.e., wood thrushes (*Hylocichla mustelina*) and American robin (*Turdus migratorius*), were tabulated (Table 8.1) at 14 fixed stations along a driving route in spring 2004. At each stop, two surveyors independently recorded the number of different birds heard. Observers listened for 5 minutes at each site before proceeding to the next station. Counts began at dawn, with stations spaced 0.4 km apart. Both surveyors had experience identifying calls of central Missouri birds.

A generalized linear model (GLM) based on a Poisson error and a log-link was used to analyze the thrush counts based on model Eq. (8.44). An analysis of deviance (ANODEV) table is constructed below, accounting for the effects of stations and surveyors, along with the asymptotic F-tests of significance:

Table 8.1. Auditory counts of thrushes at waystations in central Missouri by two different surveyors.

Station	Surveyor 1	Surveyor 2
1	2	1
2	6	4
3	5	5
4	1	0
5	3	2
6	0	1
7	0	1
8	3	1
9	5	4
10	3	2
11	2	1
12	1	1
13	1	1
14	1	1

Source	DF	DEV	MDEV	F	P
Total$_{Cor}$	27	38.355			
Stations	13	31.550	2.4269	$F_{13,13} = 5.5396$	0.0020
Surveyors	1	1.1070	1.1070	$F_{1,13} = 2.5270$	0.1359
Error	13	5.6948	0.4381		

The station effects ($P = 0.0020$) were statistically significant, but the surveyor differences were not ($P = 0.1359$). Comparison of the observed and fitted values agree well with no residuals significant at $P < 0.05$.

Observed count	Fitted value	Standardized residual
2	1.707	0.244
6	5.690	0.130
5	5.690	−0.289
1	0.569	0.571
3	2.845	0.092
0	0.569	−0.754
0	0.569	−0.754
3	2.276	0.480
5	5.121	−0.053
3	2.845	0.092
2	1.707	0.224
1	1.138	−0.129
1	1.138	−0.129
1	1.138	−0.129
1	1.293	−0.258
4	4.310	−0.149
5	4.310	0.332
0	0.431	−0.657

Observed count	Fitted value	Standardized residual
2	2.155	−0.106
1	0.431	0.867
1	0.431	0.867
1	1.724	−0.551
4	3.879	0.061
2	2.155	−0.016
1	1.293	−0.258
1	0.862	0.149
1	0.862	0.149
1	0.862	0.149

Although the surveyor effect is not significant, it is of interest to see what the count data suggest. The GLM analysis estimates a surveyor 2 parameter value of −0.2776 ($\widehat{SE}$ = 0.2651, at a scale parameter of 1). This effect translated to

$$e^{-0.2776} = 0.7576.$$

Thus, the GLM is estimating surveyor 2 heard only 75.76% as many birds as did surveyor 1. The variance associated with that proportional effect is approximated by the delta method as

$$\widehat{Var}\left(e^{\hat{\theta}}\right) = \widehat{Var}\left(\hat{\theta}\right)e^{2\hat{\beta}} \cdot \text{scale parameter}$$

$$= (0.2651)^2 e^{2(-0.2776)}(0.4381) = 0.01767$$

or, equivalently, a standard error of $\widehat{SE}$ = 0.1329. The relatively large standard error corresponds to the nonsignificance of the surveyor difference. The scale parameter was taken as the mean deviance for error from the ANODEV table (i.e., 0.4381).

If response model (8.43) accurately describes an auditory count survey, then an interaction plot (Fig. 8.4) should indicate parallel lines for the two surveyors. A Tukey test of additivity could be used to formally test the model assumptions (Neter et al. 1990:790–793). Estes and Jameson (1988) used simultaneous counts from two survey teams to census sea otters (*Enhydra lutris*) in California. Based on mapping the observations, a Petersen mark-recapture estimate of abundance was calculated by using the unique observation records of each team and counts of animals observed by both teams (Section 9.2.5).

8.2.4   Visual Counts

Visual count indices take a variety of forms from roadside counts (Hewitt 1967, Overton 1969, Fafarman and Whyte 1979, Bildstein and Grubb 1980, Robinson et al. 2000, Hutto and Young 2002) to circular plot counts (Ramsey and Scott 1978, 1979, Reynolds et al. 1980), and line-transect counts (Buckland et al. 1996, Anderson et al. 2001). For these methods, the area canvassed is specified by the visibility of the observer and animal activity, which are typically unknown. An exception is the strip-transect method in which observations within a prespecified width are recorded, assuming a probability of

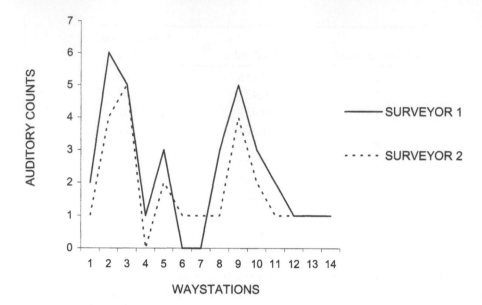

Figure 8.4. An interaction plot for a two-way ANOVA of surveyors-by-site illustrating parallelism.

detection near 1 (Hayne 1949a, Hone 1988). In the general case in which the distance canvassed is unknown, there is no ability to convert the observed counts to absolute abundance or density. The prerequisite information to make this conversion is either the radial sighting distance or the right-angle distance of the animal from the line-transect (Buckland et al. 1996). Without distance information, the assumption of equal detection over time and/or space also cannot be assessed (Burnham and Anderson 1984).

In the most general form, the detection function for the animals observed, $g(x)$, is of the same shape as the histogram of detection distances (Fig. 8.5a). The function $g(x)$ can be fit to the data plot by any of a variety of methods (Buckland et al. 1996). The various line-transect methods differ in how they fit the $g(x)$ curve. Once the $g(x)$ curve has been fit to the histogram data, additional assumptions are imposed. The most typical of these is that $g(0) = 1$, i.e., the probability of detection is 1 on the line (Fig. 8.5b).

Any continuous detection function $g(x)$ can be converted to a pdf, $f_X(x)$. By definition, the area under a pdf is 1, such that

$$\int_0^\infty f_X(x)\,dx = 1.$$

The detection function $g(x)$ can then be converted to a pdf by the relationship

$$f_X(x) = \frac{g(x)}{\int_0^\infty g(x)\,dx} \tag{8.45}$$

since

$$\int_0^\infty \left[\frac{g(x)}{g(x)\,dx}\right]dx = 1.$$

a. Histogram of the detection distances and fitted curve

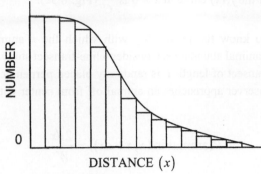

b. Detection function $g(x)$

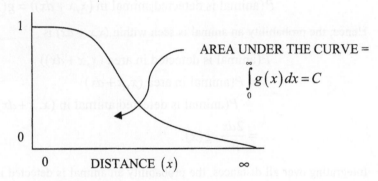

AREA UNDER THE CURVE =

$$\int_0^\infty g(x)\,dx = C$$

c. Probability density function $f_X(x)$

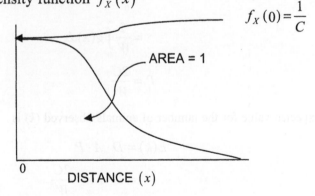

$$f_X(0) = \frac{1}{C}$$

AREA = 1

Figure 8.5. Graphic representation of the analytical steps used in constructing a detection function ($g(x)$) and probability density function ($f_X(x)$) from a histogram of detection distances.

Defining $C = \int_0^\infty g(x)\,dx$, then Eq. (8.45) can be rewritten as

$$f_X(x) = \frac{g(x)}{C}.$$

By letting $x = 0$, it is noted that

$$f_X(0) = \frac{g(0)}{C}$$

$$= \frac{1}{C}.$$

(8.46)

Hence, the height of the $f_X(x)$ curve at $x = 0$ is $\dfrac{1}{C}$ (Fig. 8.5c).

It is necessary to know the probability with which the k animals observed were detected to estimate animal abundance. Consider a line-transect study of an area of length L and width W. A transect of length L is randomly placed perpendicular to the baseline of width W. If an observer approaches an animal off from center with a right-angle distance of x, then

$$P(\text{animal in } (x, x + dx)) = \frac{2 dx}{W}.$$

In turn, the probability of detecting an animal given it was at distance x from the line-transect is

$$P(\text{animal is detected}|\text{animal in } (x, x + dx)) = g(x).$$

Hence, the probability an animal is seen within $(x, x + dx)$ is

$$P(\text{animal is detected in area } (x, x + dx))$$
$$= P(\text{animal in area } (x, x + dx))$$
$$\cdot P(\text{animal is detected}|\text{animal in } (x, x + dx))$$
$$= \frac{2 dx}{W} \cdot g(x). \tag{8.47}$$

Integrating over all distances, the probability an animal is detected is then

$$P = \int_0^\infty \frac{2 g(x)}{W} dx$$
$$= \frac{2}{W} \int_0^\infty g(x) dx$$
$$P = \frac{2C}{W}.$$

The expected value for the number of animals observed (k) is

$$E(k) = D \cdot A \cdot P$$
$$= D \cdot LW \cdot \frac{2C}{W}$$
$$= 2DLC \tag{8.48}$$

or, equivalently,

$$E(k) = \frac{2DL}{f_X(0)}. \tag{8.49}$$

The necessary conditions for the visual counts to be a valid index can be assessed by examining Eq. (8.48), where

$$E\left(\frac{k_2}{k_1}\right) \doteq \frac{2 D_2 L_2 C_2}{2 D_1 L_1 C_1}.$$

For $\dfrac{k_2}{k_1}$ to be a reliable estimate of $\dfrac{D_2}{D_1}$, it then follows the line-transect lengths need to be standardized (i.e., $L_1 = L_2 = L$) and $C_1 = C_2 = C$, implying the detection functions (i.e., $g(x)$) must be equivalent. As illustrated (Fig. 8.5a), the detection functions are modeled by using the histograms of the detection distances. Therefore, tests of homogeneous detection functions can be based on tests of homogeneity of the detection distances. One option is to use $R \times C$ contingency table test of homogeneity based on discrete distance classes.

Alternatively, the Kolmogorov-Smirnov test (Conover 1980:368–384) for equal distributions can be used to test for homogeneity using the continuous distance data. Regardless of approach, detection distances need to be measured and recorded in order to perform tests of assumptions. The distinction between an index count study and line-transect sampling is that detection distances are typically not recorded in the index investigations. Thus, the assumption of homogeneous detection functions cannot be assessed or adjustments made if found false. Although our formulation of the visual indices was based on line-transect studies, the same general remarks also apply to roadside counts (Hewitt 1967, Overton 1969, Fafarman and Whyte 1979, Bildstein and Grubb 1980) and variable-circular plot methods (Ramsey and Scott 1978, 1979) as well.

8.2.5   Catch-per-Unit Effort

Catch-per-unit effort (CPUE) methods (Leslie and Davis 1939, DeLury 1947, 1951, Ricker 1958, Skalski et al. 1983, Laake 1992, Lancia et al. 1996) can be used to estimate absolute abundance of closed populations in the presence of successive removals. This estimation is possible because of the proposed relationship between harvest effort and the probability of capture, as well as the observed decline in catch with successive removal events. A minimum of two samples is necessary for abundance estimation and a minimum of three samples for tests of goodness-of-fit. With only a single-sample catch of size r, the catch represents, under ideal circumstances, an index to abundance where its expected value is

$$E(r_i) = N_i p_i, \tag{8.50}$$

where

r_i = number of animals caught in the ith population;
N_i = animal abundance in the ith population;
p_i = probability of capture exerted on the ith population.

Skalski and Robson (1992:32–38), using a random walk process (Grimmett and Stirzaker 1992:141–150) through a serpentine array of traps, developed an expression for the probability of capture as a function of trapping effort, where

$$p = 1 - e^{-\left[\dfrac{1}{\alpha + \beta\left(\frac{1}{\sqrt{\phi}}\right) + \gamma\left(\frac{1}{\phi}\right)}\right]} \tag{8.51}$$

and where

$\phi = \dfrac{f}{A}$ = trap density;

f = number of traps;
A = area of trapping grid.

Under the special case of $\alpha = \beta = 0$, Eq. (8.51) simplifies to the expression

$$p = 1 - e^{-\frac{1}{\gamma}\phi},$$

or the more familiar equation

$$p = 1 - e^{-c\left(\frac{f}{A}\right)}, \tag{8.52}$$

where c = Poisson catchability or vulnerability coefficient (Seber 1982:296). Note that effort is expressed in terms of trap density not simply the number of trapping devices. Catch-effort Eq. (8.52) describes a declining rate of improvement in the capture probability with increasing trap density as the study area becomes saturated with traps.

Combining Eqs. (8.50) and (8.52), the expected number of animals caught is expressed as

$$E(r_i) = N_i\left(1 - e^{-c\left(\frac{f_i}{A_i}\right)}\right). \tag{8.53}$$

Equation (8.53) illustrates a nonlinear relationship between the catch-index (r_i) and trapping effort (f_i). For values of $\dfrac{cf_i}{A_i} < 0.20$, the capture probability can be approximated by

$$p_i \approx c\left(\frac{f_i}{A_i}\right). \tag{8.54}$$

Thus, Eq. (8.53) can be approximated by

$$E(r_i) \approx cN_i\left(\frac{f_i}{A_i}\right)$$

or, equivalently,

$$E\left(\frac{r_i}{f_i}\right) \approx \frac{cN_i}{A_i} = cD_i \tag{8.55}$$

where D_i is animal density of the ith population.

The conditions under which CPUE data can provide reliable measures of relative abundance can be seen using Eq. (8.55), where

$$E\left[\frac{\left(\frac{r_2}{f_2}\right)}{\left(\frac{r_1}{f_1}\right)}\right] \doteq \frac{\left(\frac{c_2 N_2}{A_2}\right)}{\left(\frac{c_1 N_1}{A_1}\right)}. \tag{8.56}$$

Hence, the CPUE data will provide reliable information on relative abundance when $A_1 = A_2 = A$ and $c_1 = c_2 = c$.

Assumptions of the CPUE index include the following:

1. Units of effort act independently.
2. Capture probabilities are low ($p < 0.20$).
3. The catch coefficient is constant across the populations compared.
4. Trapped areas (A_i) are the same across the populations compared.

The catch coefficient (c) is a function of the interaction of the animals with the trapping gear and the environment. There is no way to assess the assumption of homogeneous catch coefficients with a single trap event. Removal trapping (e.g., Zippin 1956, 1958) or mark-recapture methods across periods are needed to assess the assumption of homogeneous c or, equivalently, homogeneous capture probability p.

The nonlinearity between p and effort $\dfrac{f}{A}$ becomes more evident as capture probabilities increase. Equation (8.51) and, to a lesser extent, Eq. (8.52), indicate diminished return with increasing effort. To visualize this, consider the extreme case in which p has attained a value of 1. Additional effort beyond that point does not change p, only $\dfrac{f}{A}$, with the result of an actual decline in CPUE $\left(\text{i.e., } \dfrac{r}{f}\right)$.

Example 8.8: Comparison of Great Basin Pocket Mice (*Perognathus parvus*) Trap Index Data, Hanford, Washington

Two 1-ha trapping grids were established, each with 100 Sherman live traps. One site was established in native bunchgrass (*Pseudoroegneria spicata*), the other in cheatgrass (*Bromus tectorum*), a nonnative plant in the area. Trapping was conducted at the same time at both sites for two consecutive nights. Great Basin pocket mice caught the first day were marked for possible identification the following day. The results of the two trapping events are summarized below, where n_1 is the number of animals captured and marked on day 1; n_2, the number of animals examined for marks on day 2; and m, the number of mark-recaptures.

Site	n_1	n_2	m	$r = n_1 + n_2 - m$
Cheatgrass	22	25	13	34
Bunchgrass	18	15	3	30

The catch index (r) in this study is the number of distinct individuals caught during the 2 days ($n_1 + n_2 - m = r$). The CPUE data estimates the relative abundance of pocket mice at bunchgrass (N_2) to cheatgrass (N_1) sites of

$$\widehat{\left(\frac{N_2}{N_1}\right)} = \frac{\left(\dfrac{r_2}{f_2}\right)}{\left(\dfrac{r_1}{f_1}\right)}$$

$$= \frac{\left(\dfrac{30}{100}\right)}{\left(\dfrac{34}{100}\right)} = 0.8824.$$

The mean catch per day $\left(\text{i.e., } \dfrac{n_1 + n_2}{2}\right)$ yields similar results, where

$$\widehat{\left(\frac{N_2}{N_1}\right)} = \frac{\left(\dfrac{18+15}{2}\right)}{\left(\dfrac{22+25}{2}\right)} = 0.7021,$$

suggesting, again, less abundance at the bunchgrass site than at the cheatgrass site.

The mark-recapture data can also be used to estimate absolute abundance, based on the Chapman (1951) form of the single mark-recapture method (Seber 1982:60), where

$$\hat{N} = \frac{(n_1+1)(n_2+1)}{(m+1)} - 1$$

with associated variance estimator of

$$\widehat{\text{Var}}(\hat{N}) = \frac{(n_1+1)(n_2+1)(n_1-m)(n_2-m)}{(m+1)^2(m+2)}.$$

Pocket mouse abundance is estimated to be 75.0 $(\widehat{\text{SE}} = 26.153)$ and 41.71 $(\widehat{\text{SE}} = 4.687)$ at the bunchgrass and cheatgrass sites, respectively. Relative abundance of pocket mice at the bunchgrass site to that of the cheatgrass site is calculated to be

$$\left(\frac{\hat{N}_2}{\hat{N}_1}\right) = \frac{75.0}{41.71} = 1.798,$$

now indicating greater abundance at the bunchgrass site.

The nature of the capture data permits a test of homogeneous capture probabilities using an $R \times C$ contingency table (Skalski and Robson 1992:64–65) of the form

	Bunchgrass	Cheatgrass
$n_1 - m$	15	9
$n_2 - m$	12	12
m	3	13
Totals	30	34

which rejects the hypothesis of homogeneity at $P(\chi_2^2 \geq 7.5294) = 0.0232$. Hence, the estimate of relative abundance using the CPUE data is invalid, and the preferred estimate is 1.798 based on the ratio of absolute abundance estimates. Skalski and Robson (1992:97) provide a nearly unbiased estimator for this situation. Overall capture probabilities at the two sites are estimated to be approximately $\dfrac{30}{75} = 0.40$ and $\dfrac{34}{41.71} = 0.8152$ for the bunchgrass and cheatgrass sites, respectively. The difference in capture probabilities confounded the comparison of the CPUE indices in this example, misspecifying even the rank order of the abundance values between sites.

8.2.6   Trap-Line Counts

Trap lines, in contrast to trap girds, are typically used to capture and monitor the abundance of wide-ranging carnivore species (Wood and Odum 1964, Smith et al. 1984). The basic form of the trap line is a linear array of k_i traps at which x_i animals are captured. Seber (1982:57) suggested treating the capture data as a frequency index, where

$$p_i = \frac{x_i}{k_i}.$$

Chi-square tests or GLM tests based on a binomial error and logit-link would then be used to compare indices. However, $\dfrac{x_i}{k_i}$ is simply an expression of CPUE, and as such, many of the earlier remarks (Section 8.2.5) hold here as well.

MacLulich (1951) suggested the area about an assessment line be described by the equation

$$A = R \cdot L + \frac{\pi R^2}{4}, \tag{8.57}$$

where

   L = length of the trap assessment line;

   $\dfrac{R}{2}$ = half-width of the boundary strip about the line.

The boundary strip expresses the effective trapping distance about the assessment line. Hence, the value of R is a function of the interaction of the animals with the trapping gear. The expected number of animals caught along an assessment line can be expressed as

$$E(x_i) = D_i A_i p_i. \tag{8.58}$$

Reparameterizing Eq. (8.58) with Eq. (8.57) for the area trapped and Eq. (8.52) for the probability of capture, the expected catch can be expressed as

$$E(x_i) = D_i \left(RL_i + \frac{\pi R^2}{4} \right) \left(1 - e^{-\left[\frac{cf_i}{RL_i + \frac{\pi R^2}{4}} \right]} \right). \tag{8.59}$$

For low capture probabilities (i.e., $p \le 0.2$),

$$E(x_i) \approx D_i \left(RL_i + \frac{\pi R^2}{4} \right) \frac{cf_1}{\left(RL_i + \frac{\pi R^2}{4} \right)}$$

$$\approx D_i cf_i$$

or, expressed in terms of CPUE,

$$E\left(\frac{x_i}{f_i} \right) \approx D_i c, \tag{8.60}$$

where c is the Poisson catchability or vulnerability coefficient.

Equation (8.60) suggests the CPUE index is an appropriate metric as long as the Poisson catchability coefficient c is constant across the populations being compared and capture probabilities are low ($p \leq 0.20$). Low capture probabilities are common for carnivore populations so the assumption may be reasonable in some cases.

For higher capture probabilities ($p > 0.2$), the catch index (x_i) is an appropriate index as long as effort f_i and trap-line length L_i are held constant (Eq. 8.59). Here, again, the CPUE index assumes that boundary width R and Poisson catchability coefficient c are constant across the populations compared. Unfortunately, the nature of the trap-line data alone is insufficient to assess these assumptions.

8.2.7 Mark-Recapture Estimates as Indices

Catch-per-unit-effort index data may be inappropriate for estimating even the relative abundance of populations (see Example 8.8), in which case, abundance estimates may be required. In many cases, however, absolute abundance estimates from mark-recapture methods should be treated as indices as well. When sampling from a continuous population, the estimate of abundance derived from a mark-recapture trap grid cannot be ascribed to a specific area. Although grid location is known, the area actually trapped extends some unknown distance beyond the boundaries of the trap grid. Special techniques such as the nested-grid method of Otis et al. (1978) or the web-design of Anderson et al. (1983) and Wilson and Anderson (1985) are necessary to estimate the effective area trapped and animal density.

MacLulich (1951) proposed that an unknown boundary strip of width $\frac{1}{2}R$ extends around the trap grid boundary (Fig. 8.6). For a trap grid of size $a_i \times b_i$, the effective area trapped was expressed as

$$(a_i + R)(b_i + R) - \left(1 - \frac{\pi}{4}\right)R^2.$$

The abundance estimated from the mark-recapture trap grid then has the expected value

$$E(\hat{N}_i) = D_i\left[(a_i + R)(b_i + R) - \left(1 - \frac{\pi}{4}\right)R^2\right], \qquad (8.61)$$

where D_i = animal density at the ith site. Relative abundance of two populations, based on the ratio of absolute abundance estimates, has the approximate expected value of

$$E\left(\frac{\hat{N}_2}{\hat{N}_1}\right) \doteq \frac{D_2\left[(a_2 + R)(b_2 + R) - \left(1 - \frac{\pi}{4}\right)R^2\right]}{D_1\left[(a_1 + R)(b_1 + R) - \left(1 - \frac{\pi}{4}\right)R^2\right]}$$

$$= \frac{D_2}{D_1}, \qquad (8.62)$$

if and only if $a_1 = a_2$ and $b_1 = b_2$ for some common R. From Eq. (8.62), it is evident that it is insufficient simply to have study areas of equal size for the two surveys to make inferences to relative density. If inferences to relative density are sought, trap grid con-

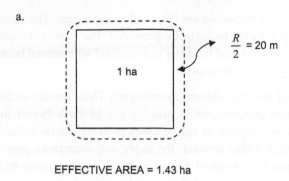

a.

$\frac{R}{2}$ = 20 m

1 ha

EFFECTIVE AREA = 1.43 ha

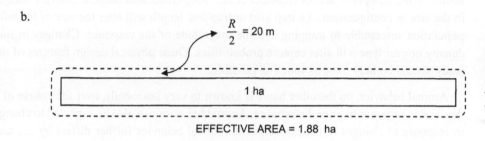

b.

$\frac{R}{2}$ = 20 m

1 ha

EFFECTIVE AREA = 1.88 ha

Figure 8.6. Illustration of two trap grids of different configurations with equal areas and with common boundary strips of width $\frac{R}{2}$.

figurations must also be the same for the populations compared (i.e., common a and b). It must also be assumed the boundary strip widths of $\frac{R}{2}$ are the same across populations.

Beyond the above concerns, there is also the consideration of whether the assumptions of the selected mark-recapture model have been fulfilled. Because every mark-recapture model is a simplification of reality, it is unlikely model assumptions are actually achieved in most cases. Cormack (1980, 1993), for this reason, suggested treating abundance estimates as refined population indices and argued against model-based variance estimates. It is, however, unclear whether population surveys with the same model violations will produce relative abundance estimates that are valid, e.g.,

$$E\left(\frac{\hat{N}_2}{\hat{N}_1}\right) = \frac{N_1}{N_2}.$$

The hope, nevertheless, is that catch-adjusted estimates will be more closely proportional to the true abundance or density than are the CPUE data alone. This may be the case for rather simple model violations but is perhaps not true for more extreme cases of capture heterogeneity.

8.3   Design of Index Studies

Consistency is the hallmark of a well-designed index study. There are two types of consistency that must be considered; one is under direct investigator control, the other is not.

The first is consistency in the way the responses are observed and recorded. The second is consistency in the way the responses are biologically generated. The former is associated with the conduct of the investigation, and the latter is associated with animal behavior and its relationship to the index being measured.

The investigator has control of how the indices are collected. These factors include plot size, equipment, measurement protocols, etc. Changing any of these factors may confound the survey methods with temporal or spatial changes in the actual index. In other words, the proportional relationship between the index and abundance may be affected by a change in the measurement protocol. For example, a change in travel speed during a roadside count survey will alter the probability of detection. A change in personnel may alter detection rates during visual or auditory count surveys based on the acuity of the observers' senses (Kendall et al. 1996, Genet and Sargent 2003). Changes in the size or configuration of a trap grid or trap-line length will alter the size of the wild population susceptible to trapping and the magnitude of the response. Changes in trap density or gear type will alter capture probabilities. These physical design features of the study should be held fixed as much as possible.

Animal behavior, on the other hand, is known to vary seasonally, over the course of a day, and in response to prior disturbance. Animal behavior can also be expected to change in response to changes in the environment. Animal behavior further differs by sex and age. Within these ever varying conditions, investigators must select a subset of conditions to routinely monitor over time and/or space. By standardizing when and where the indices are measured, it is hoped a standard set of animal behaviors are monitored that will, in turn, standardize the index-abundance relationship. However, there is no guarantee. Skalski and Robson (1992) argue that any anthropogenic effect that changes abundance might be expected to also change animal behavior, and for this reason, absolute abundance techniques should always be used. However, necessity and economics may force investigators to use indices when absolute abundance techniques are known to be superior. It is for these situations that this section is written.

Adherence to structured survey designs may help eliminate confounding through the principles of blocking and randomization. Ideally, the same personnel would canvass all populations of interest at the same instance, thereby eliminating personnel and temporal differences in behavioral response. But this is a physical impossibility. Instead, statistical models may be used to perform the feat through model-based adjustments.

8.3.1   Latin-Square Designs

A Latin-square design (Neter et al. 1990:1083–1096) permits two-way blocking. When properly structured, two different design factors can be simultaneously used in blocking, thereby controlling for two factors and, at the same time, reducing the experimental error variance. This experimental design is well-suited when multiple populations are canvassed over several time frames.

Example 8.9: Latin-Square Survey of Dark-Eyed Juncos (*Junco hyemalis*), Colorado

A Latin-square design was used to schedule visual counts of dark-eyed juncos at four different locations in the Piceance Basin of Colorado. The number of birds sighted along a 500-m transect at each of the four sites was the basis of the index count. A single survey crew was available for all of the index counts. Daily differences in weather were expected to alter bird behavior and, consequently, bird counts. Diel patterns in behavior were also expected to effect bird counts.

The expectation of daily and diel differences in bird behavior suggested the need to control for both factors (Table 8.2). The four sites were therefore canvassed on each of 4 days and during one of four time periods each day. Over 4 days, each site was canvassed during each of the four daily time periods. The purpose of the study was to compare the relative bird abundance between sites, blocking on the effects of day and time-of-day. The response model for the analysis can be written as

$$x_{ijk} = \mu \cdot S_i \cdot T_j \cdot D_k \cdot \varepsilon_{ijk}, \qquad (8.63)$$

where

x_{ijk} = bird count at the ith site ($i = 1, \ldots, 4$) during the jth time period ($j = 1, \ldots, 4$) on the kth day ($k = 1, \ldots, 4$);
μ = baseline response;
S_i = effect of the ith site ($i = 1, \ldots, 4$);
T_j = effect of the jth time-of-day period ($j = 1, \ldots, 4$);
D_k = effect of the kth day ($k = 1, \ldots, 4$);
ε_{ijk} = random error term.

A GLM analysis was performed by using a log-link and a Poisson error distribution (for background, see Aitkin et al. 1989, Crawley 1993). The small values of bird counts suggested they might be approximately Poisson-distributed. The multiplicative model suggests the effects will be additive on the log-scale. The log-link permits an analysis based on the log of the expected values without actually log-transforming the data. This feature is useful because one of the observations is the value zero.

An ANODEV table is constructed, accounting for the effects of site, day, and time-of-day along with asymptotic F-tests of significance.

Table 8.2. Counts of dark-eyed juncos, Colorado, by day, time, and site (i.e., S1, S2, S3, and S4) arranged in the Latin-square design.

Date	Survey times			
	8:00–9:15 a.m.	9:45–11:15 a.m.	11:45 a.m.–1:15 p.m.	1:45–3:00 p.m.
October 16	S1–8	S2–1	S3–9	S4–0
October 17	S3–35	S4–5	S1–15	S2–7
October 18	S2–1	S3–24	S4–2	S1–5
October 20	S4–3	S1–11	S2–1	S3–12

Source	DF	DEV	MDEV	F	P
Total$_{Cor}$	15	135.67			
Site	3	92.59	30.68	$F_{3,6} = 22.728$	0.0011
Day	3	29.21	9.74	$F_{3,6} = 7.1734$	0.0207
Time	3	5.722	1.907	$F_{3,6} = 1.4041$	0.3301
Errror	6	8.147	1.358		

The effects of site ($P = 0.0011$) and day ($P = 0.0207$) were statistically significant, whereas time-of-day ($P = 0.3301$) was not. Comparison of observed junco counts with model predictions agrees well with no significant residuals (Table 8.3). The standardized residuals are asymptotically Z-distributed, where large absolute values (e.g., $P(|Z| > 1.96) = 0.05$) may indicate significant lack-of-fit.

The GLM model is an offset model. Instead of modeling the observations relative to the grand mean, it models the observations relative to the first observation. In this case, the junco count at the first site (S1), the first day (October 16), and the first time-of-day period (8:00 to 9:15 a.m.). The choice of baseline is somewhat arbitrary. Neither the significance of effects nor model fit is influenced by this choice. The fitted model estimated the following effects:

Parameter	$\hat{\theta}$	$\widehat{SE}(\hat{\theta})^*$
Baseline	1.792	0.3332
Site 2	−1.346	0.3663
3	0.6410	0.2125
4	−1.531	0.3696
Day 2	1.180	0.2761
3	0.3315	0.3521
4	0.3113	0.3582
Time 2	0.3177	0.2955
3	−0.2552	0.2657
4	−0.2327	0.3190

*Standard errrors based on a scale parameter of 1.

Because the analysis was based on a log-link, the effects are expressed on the natural log-scale and must be back-transformed for interpretation.

The model adjustments can be used to predict the number of juncos that would have been seen, if all sites had been canvassed concurrently. The predictions for site 1, day 1, time 1, is

$$\text{Site 1: } e^{1.792} = 6.0014 \text{ juncos.}$$

In turn, the prediction for site 2, day 1, time 1, is based on using the site 2 adjustment to the previous prediction, where

$$\text{Site 2: } e^{1.792-1.346} = 1.5621.$$

Subsequent sites are modeled similarly, where

Table 8.3. Observed counts of dark-eyed juncos, model-predicted values, and standardized residuals that are approximately Z-distributed.

Observed count	Fitted value	Standardized residual
8	6.001	+0.816
35	37.052	−0.337
1	2.176	−0.797
3	1.772	+0.923
1	2.146	−0.782
5	5.800	−0.332
24	21.800	+0.471
11	11.254	−0.076
9	8.825	+0.059
15	15.122	−0.031
2	1.401	+0.507
1	1.652	−0.507
0	1.028	−1.014
7	4.026	+1.482
5	6.623	−0.631
12	12.322	−0.092

$$\text{Site 3}: \ e^{1.792+0.6410} = 11.3930,$$
$$\text{Site 4}: \ e^{1.792-1.531} = 1.2982.$$

The real interpretive value of the indices, however, is in expressing relative abundance. By using the predictions from day 1 and time 1, the relative abundances at sites 1 to 4 relative to site 1 have the values

$$\frac{\text{Site 1}}{\text{Site 1}} = 1 = e^{0},$$
$$\frac{\text{Site 2}}{\text{Site 1}} = \frac{1.5621}{6.0014} = 0.2603 = e^{-1.346},$$
$$\frac{\text{Site 3}}{\text{Site 1}} = \frac{11.3930}{6.0014} = 1.8984 = e^{0.6410},$$
$$\frac{\text{Site 4}}{\text{Site 1}} = \frac{1.2982}{6.0014} = 0.2163 = e^{-1.531}.$$

These calculations suggest only 0.26 times as many juncos were seen at site 2 compared with site 1. Site 3 had 1.89 times as many junco observations as did site 1, and site 4 had 0.22 times as many junco observations as did site 1. The abundance indices can be calculated relative to any site in a similar manner.

By the multiplicative nature of the response model (8.63), the same relative abundance values will be obtained for any day and time-of-day selected. Bird count predictions will vary by day and time-of-day but not by their relative abundance. To see this relationship, consider the bird counts predicted for day 2, time 1, where

Location	Abundance	Relative abundance
Site 1	$e^{1.792+1.180} = 19.5309$	1
Site 2	$e^{1.792+1.180-1.346} = 5.0835$	0.2603
Site 3	$e^{1.792+1.180+0.6410} = 37.0771$	1.8984
Site 4	$e^{1.792+1.180-1.531} = 4.2249$	0.2163

The log-link GLM analysis provides an ideal framework for estimating relative abundance in this case. Although absolute values of the count may vary over time, the relative relationship of the counts between sites remains unchanged. The goodness-of-fit of the model also suggests this relationship holds in reality.

The ANODEV in this example identified significant differences in mean counts between sites. However, it cannot be directly inferred this indicates significant differences in bird abundance or density. Without analysis of the corresponding right-angle sighting distances, it is unknown whether detection functions were the same or not across sites. It could be that observation distances, on average, were longer at one site than another, and more birds were observed without an actual change in density. The Latin-square design can control for temporal effects on the bird counts but not inherent differences in detection rates at the sites. The only way to ensure the inferences to relative abundance are correct is for auxiliary information on observation distances to be collected.

Discussion of Utility

Personnel effects, in the dark-eye junco example, were controlled by using a single survey crew. What remained were the potential effects of survey timing on the subsequent bird counts. The Latin-square design was therefore used to control for day and time-of-day effects. The number of treatment levels (i.e., sites) in a Latin-square design must equal the number of levels of the other two factors being controlled. In this case, the four sites being compared dictated the need for four levels of day and four levels of time within the day. The size of the Latin-square design increases exponentially with the number of sites, limiting its utility to a few locations.

There are also logistical concerns with the Latin-square design. The structured nature of the design dictates the order in which the sites are canvassed within a day. This meant travel between sites was not always the fastest or shortest route. However, the inferential strength of the Latin-square design often outweighs logistical concerns.

8.3.2   Randomized Block Designs

By using a randomized block design, a single extraneous factor can be controlled, eliminating confounding with treatment effects and a corresponding reduction in error variance. Blocking can be used in a variety of ways to control potential factors influencing index counts. Choice of which design factor to control should be based on logistical concerns and consideration of which extraneous factors may be most influential in affecting the response variable being measured. Should site comparisons be the objective of the study, factors for potential design control include survey crews and temporal effects. The investigator must choose which factor to control and which factors to fold into the error term through the process of randomization. In some instances, not all treatment levels are incorporated into each block, and the result is a randomized incomplete block design (Cochran and Cox 1957:376–395).

Example 8.10: Randomized Block Survey of Indigo Buntings (*Passerina cyanea*), Central Missouri

A rather simple example of a randomized block design consisted of the comparison of relative bird abundance between two sites with the assistance of two survey crews. In central Missouri, counts of male indigo buntings were made along 500-m transect lines at two different sites. The two sites were 10.9 km apart and contained similar habitat types. To standardize all effects, it would be ideal but impossible to canvass both bunting populations simultaneously with the same survey crew. Instead, the two sites were canvassed simultaneously, with one crew at each site. The process was repeated for 4 days, with the crews randomized to the sites for each event. Each survey began at sunrise and lasted approximately 1 hour. A restriction was placed on the randomization, such that each crew surveyed each population twice (Table 8.4).

The response model for the analysis is the multiplicative relationship

$$x_{ij} = \mu \cdot S_i \cdot D_j \cdot \varepsilon_{ij}, \tag{8.64}$$

where

x_{ij} = bird count at the ith site ($i = 1, 2$) during the jth day ($j = 1, \ldots, 4$);
μ = baseline response;
S_i = effect of the ith site ($i = 1, 2$);
D_j = effect of the jth block effect ($j = 1, \ldots, 4$);
ε_{ij} = random error term.

GLM analysis was performed using a log-link and a Poisson error distribution.

An ANODEV table was constructed, accounting for the effects of site and day along with asymptotic F-tests of significance:

Source	DF	DEV	MDEV	F	P
Total$_{Cor}$	7	7.6079			
Site	1	2.5940	2.5940	$F_{3,6} = 20.5873$	0.0200
Day	3	4.636	1.5453	$F_{3,6} = 12.2643$	0.0394
Errror	3	0.3781	0.1260		

The effects of site ($P = 0.0020$) and day ($P = 0.0394$) were statistically significant. Comparison of the observed and fitted values agrees well with no residuals significant at $P < 0.05$.

Table 8.4. Counts of indigo buntings by day, site, and crew (i.e., C1 and C2) in the randomized block design.

Date	Site 1	Site 2
May 1	C1–6	C2–10
May 3	C2–9	C1–14
May 5	C2–5	C2–8
May 6	C1–11	C1–13

Observed count	Fitted value	Standardized residual
6	6.526	−0.206
9	9.382	−0.125
5	5.303	−0.131
11	9.789	+0.387
10	9.474	+0.171
14	13.618	+0.103
8	7.697	+0.109
13	14.211	−0.321

The estimated parameters of the offset model are

Parameter	$\hat{\theta}$	$\widehat{SE}(\hat{\theta})$*
Baseline	1.8760	0.2857
Site 2	0.3727	0.2334
Day 2	0.3629	0.3255
3	−0.2076	0.3734
4	0.4055	0.3227

*Assuming a scale parameter of 1.

The baseline conditions estimate the indigo bunting count at site 1, day 1 to be

$$\text{Site 1: } e^{1.876} = 6.5273.$$

Adjusting the baseline for the site 2 effect estimates that

$$\text{Site 2: } e^{1.876+0.3727} = 9.4754$$

male buntings would have been seen at site 2. The relative male bunting abundance at site 2 compared with that of site 1 is estimated to be

$$\frac{\text{Site 2}}{\text{Site 1}} = \frac{9.4754}{6.5273} = e^{0.3727} = 1.4516.$$

Thus, site 2 is estimated to have 1.45 times as many male indigo bunting observations compared with those of site 1. The variance of the estimate of relative abundance is found by using the delta method, where

$$\widehat{\text{Var}}(e^{\hat{\theta}}) \doteq \widehat{\text{Var}}(\hat{\theta}) \cdot e^{2\hat{\theta}} \cdot (\text{scale parameter})$$

$$= (0.2334)^2 e^{2(0.3727)}(0.1260) = 0.01446,$$

or a standard error of 0.1203. The scale parameter is taken as the mean deviance for error from the ANODEV table (i.e., 0.1260).

As with the dark-eyed junco in Example 8.9, significant differences in mean bird counts were observed between sites. Whether the difference can be inferred to differences in bird densities depends on whether the detection curves are the same for the two populations. Auxiliary right-angle observation distances would be necessary to validate that assumption. If the detection functions are different, line-transect methods of Buckland et al. (1993) would be needed to convert the count indices to estimates of bird density.

8.4 Calibration of Indices

The value of indices lies in their ability to convey information on the relative abundance or density of a species. This outcome is sufficient if a trend in abundance over time or across locales is of interest. This ability depends on the proportionality between the index and absolute abundance remaining constant across the populations compared. When the proportionality does not hold, the indices have little or no value in comparing population levels. Under these circumstances, indices must be converted to estimates of absolute abundance or density, and the comparison performed on the absolute scale. Unfortunately, for most index studies, the auxiliary information needed for this conversion to the absolute scale has not been collected. In marking studies, the capture-recapture histories permit estimation of detection probabilities and subsequent conversion of catch indices to abundance estimates. In animal sighting studies, the information on detection distances permits conversion of visual counts to density estimates.

Conversion of population indices to absolute abundance is also crucial in managing exploited species. Harvest statistics report the number of animals taken in absolute terms. To understand the consequences of the harvest on population dynamics, harvest and abundance values must be on the same absolute scale. Otherwise, managers are left to manage populations on the relative scale for effects measured on the absolute scale.

We now consider the conversion of indices to absolute numbers of animals. We will not focus on mark-recapture (Seber 1982) or distance-sampling (Buckland et al. 1996) methods; they have been addressed adequately elsewhere. Instead, our focus will be on techniques that use auxiliary information readily available to biologists. The pellet-count index is perhaps the best example of this type of calibration. By using information on the rate of sign production (i.e., defecation rate), the index value can be directly converted to an estimate of absolute abundance. Techniques presented will consider change-in-ratio methods, ratio and regression techniques, and double-sampling methods.

8.4.1 Index-Removal Method: Petrides (1949) and Eberhardt (1982)

The index-removal method (Petrides 1949, Eberhardt 1982) was presented earlier (Section 6.7) to estimate the harvest survival probability (S_H). The same information can be used to estimate the pre-harvest abundance of a population, where

$$\hat{N}_{\text{PRE}} = \frac{CI_1}{I_1 - I_2} \tag{8.65}$$

and where

$\hat{N}_{\text{PRE}}$ = estimated population abundance at the time of the first index survey;
I_1 = index count during the first survey prior to harvest;
I_2 = index count during the second survey following harvest;
C = number of animals harvested between the two index surveys.

The postharvest abundance, i.e., $N - C$, can be estimated by the quantity

$$\hat{N}_{\text{POST}} = \frac{CI_2}{I_1 - I_2}. \tag{8.66}$$

By using the delta method, the variance of $\hat{N}_{\mathrm{PRE}}$ can be approximated as

$$\widehat{\mathrm{Var}}\left(\hat{N}_{\mathrm{PRE}}\right) = \hat{N}_{\mathrm{PRE}}^2 \left[\frac{\widehat{\mathrm{Var}}(C)}{C^2} + \frac{I_2^2 \cdot \widehat{\mathrm{Var}}(I_1) + I_1^2 \cdot \widehat{\mathrm{Var}}(I_2)}{I_1^2(I_1 - I_2)^2} \right]. \tag{8.67}$$

Should the harvest or removal number C be known without error, $\widehat{\mathrm{Var}}(C)$ is set to zero in Eq. (8.67). In a similar manner, the variance of the postharvest abundance can be estimated by the expression

$$\widehat{\mathrm{Var}}\left(\hat{N}_{\mathrm{POST}}\right) = \hat{N}_{\mathrm{POST}}^2 \left[\frac{\widehat{\mathrm{Var}}(C)}{C^2} + \frac{I_2^2 \cdot \widehat{\mathrm{Var}}(I_1) + I_1^2 \cdot \widehat{\mathrm{Var}}(I_2)}{I_2^2(I_1 - I_2)^2} \right]. \tag{8.68}$$

The assumptions associated with the index removal method are provided earlier (Section 6.7).

The index counts (Section 6.7) were considered to be binomial random variables (i.e., $E(I_1) = Np$); a function of animal abundance (N) and a sampling probability (p). Eberhardt (1982) considered the index counts to be Poisson random variables with the same expectations. In both cases, the indices were considered to be visual counts of the animal population with probability of detection p. However, the indices need not be animal counts but, instead, can be based on animal sign such as track or pellet counts. Neither the binomial or Poisson error assumptions are correct in the situation of sign counts. For these indices, sample survey variances (see Section 8.1.2 as well as Cochran 1977, Thompson 2000), based on spatial sampling of the landscape should be used in variance calculations of Eqs. (8.67) and (8.68).

Example 8.11: Index-Removal Method for Feral Horses (*Equus caballus*), Cold Springs, Oregon

Surveys of adult feral horses before and after known removals were used by Eberhardt (1982) to estimate abundance. The preremoval visual index was $I_1 = 301$ animals, and the postremoval visual index was $I_2 = 76$ animals following $C = 357$ removals of adult horses in the Cold Springs area of Oregon. The preremoval abundance is estimated by Eq. (8.65) to be

$$\hat{N}_{\mathrm{PRE}} = \frac{CI_1}{I_1 - I_2}$$

$$= \frac{357 \cdot 301}{301 - 76} = 477.587.$$

The postremoval abundance is estimated by Eq. (8.66) as

$$\hat{N}_{\mathrm{POST}} = \hat{N}_{\mathrm{PRE}} - C$$

$$= 477.587 - 357 = 120.587$$

or, equivalently,

$$\hat{N}_{\mathrm{POST}} = \frac{CI_2}{I_1 - I_2}$$

$$= \frac{357 \cdot 76}{301 - 76} = 120.587.$$

The common detection probability (p) for the two surveys is estimated by the expression

$$p = \frac{I_1 - I_2}{C}$$

$$= \frac{301 - 76}{357} = 0.6303.$$

Assuming binomial sampling, then

$$\widehat{\text{Var}}(I_1) = \hat{N}_{\text{PRE}}\,\hat{p}(1 - \hat{p})$$

$$= 477.587(0.6303)(1 - 0.6303)$$

$$= 111.288,$$

$$\widehat{\text{Var}}(I_2) = \hat{N}_{\text{POST}}\,\hat{p}(1 - \hat{p})$$

$$= 120.587(0.6303)(1 - 0.6303)$$

$$= 28.099.$$

The variance of $\hat{N}_{\text{PRE}}$ can be estimated by Eq. (8.67), assuming Var (C) = 0, where

$$\widehat{\text{Var}}(\hat{N}_{\text{PRE}}) = \hat{N}_{\text{PRE}}^2 \left[\frac{\widehat{\text{Var}}(C)}{C^2} + \frac{I_2^2 \cdot \widehat{\text{Var}}(I_1) + I_1^2 \cdot \widehat{\text{Var}}(I_2)}{I_1^2(I_1 - I_2)^2} \right]$$

$$= (477.587)^2 \left[\frac{76^2(111.288) + 301^2(28.099)}{301^2(301 - 76)^2} \right] = 158.5647,$$

or

$$\widehat{\text{SE}}(\hat{N}_{\text{PRE}}) = 12.592.$$

Alternatively, assuming the indices are Poisson-distributed, then

$$\widehat{\text{Var}}(I_1) = E(I_1)$$

$$= N_{\text{PRE}} \cdot p$$

$$= 301,$$

$$\widehat{\text{Var}}(I_2) = E(I_2)$$

$$= N_{\text{POST}} \cdot p$$

$$= 76,$$

and it follows that

$$\widehat{\text{Var}}(\hat{N}_{\text{PRE}}) = (477.587)^2 \left[\frac{76^2 \cdot 301 + 301^2 \cdot 76}{301^2(301 - 76)^2} \right] = 428.8727$$

and

$$\widehat{\text{SE}}(\hat{N}_{\text{PRE}}) = 20.709,$$

which is what Eberhardt (1982) reported.

8.4.2 Intercalibrating Two Indices

When two different types of indices exist or one index will eventually replace another, intercalibration will be necessary. The purpose of the calibration is to convert one index to the value of the other. Consider the first index, which has the form

$$x_{1i} = \beta_1 D_i + \varepsilon_{1i},$$

where

x_{1i} = value of the first index for the ith observation;
D_i = true density of the ith observation;
β_1 = proportionality constant for the first index;
ε_{1i} = random error term.

A second index has a similar form, where

$$x_{2i} = \beta_2 D_i + \varepsilon_{2i}.$$

The purpose of the intercalibration is to translate x_{2i} to the index x_{1i} by using a linear regression equation of the form

$$x_{1i} = a + bx_{2i}. \tag{8.69}$$

The standard method of least squares is used to obtain the coefficients a and b. The regression coefficient b is then estimated as

$$b = \frac{\displaystyle\sum_{i=1}^{n}(x_{1i} - \bar{x}_1)(x_{2i} - \bar{x}_2)}{\displaystyle\sum_{i=1}^{n}(x_{2i} - \bar{x}_2)^2}$$

$$= \frac{\widehat{\text{Cov}}(x_{1i}, x_{2i})}{s_{2i}^2} \tag{8.70}$$

and

$$a = \bar{x}_1 - b\bar{x}_2. \tag{8.71}$$

The difficulty in calibrating indices is that x_{2i}, for example, is typically measured *with* error. However, standard regression analysis assumes x_{2i} is known exactly. The result of the error in the x_{2i}'s can introduce bias into the calibration. Under the least-squares assumption

$$E(b) = \frac{B_1}{B_2}.$$

In the presence of sampling error,

$$E(b) = \frac{B_1}{B_2 + \dfrac{\sigma_{\varepsilon_2}^2}{B_2 \sigma_N^2}}, \tag{8.72}$$

where

$\sigma_{\hat{\varepsilon}_2}^2$ = sampling error associated with index x_{2i};

σ_N^2 = variance in true animal abundance (N_i) between observations used in the calibration exercise.

Inspection of Eq. (8.72) suggests a calibration exercise will provide a good estimate of b only under the one of the following conditions:

1. $\sigma_{\hat{\varepsilon}_2}^2 = 0$.
2. $\sigma_{\hat{\varepsilon}_2}^2$ is small and σ_N^2 is purposefully large by calibrating across a wide range of abundance levels.
3. Both $\sigma_{\hat{\varepsilon}_2}^2$ and σ_N^2 are known.

Discussion of Utility

With any long-term index study, sooner or later methods or personnel will inevitably change. Foreseeing this change, an intercalibration study can be performed to permit interpretation of the new index on the scale of the older method, or vice versa. The inevitable temptation is to perform a "quick and dirty" intercalibration study at the last moment. However, Eq. (8.72) suggests the very opposite. First, a wide range of conditions (i.e., N_i) should be selected for intercalibration. This process may take purposeful preparation and/or patience to avail oneself of a wide range of abundance conditions (i.e., the σ_N^2 is large). Second, instead of a hurried study, considerable care is needed to reduce the measurement error of the second index (i.e., $\sigma_{\hat{\varepsilon}_2}^2$). When $\sigma_{\hat{\varepsilon}_2}^2$ is small and $\sigma_{N_i}^2$ is large, the bias in the intercalibration study can be reduced.

However, intercalibration studies are often done poorly or not at all to save resources for the actual index surveys. The above theory (Eq. 8.72) mandates a well-designed and well-performed intercalibration study. Unless this is done, the ability to carry on the tradition of an earlier index trend study may be lost. The risk is that future changes in index values may be forever confounded with changes in data collection procedures and not be reflective of true population trends.

8.4.3   Ratio Estimators

By definition, indices are generally relatively easy and inexpensive to collect, whereas estimates of absolute abundance or density are more laborious and costly. This is the motivation behind use of most indices in wildlife science. In these situations, in which an inexpensive technique can be applied ubiquitously and a more rigorous abundance procedure applied strategically, ratio estimators may be used to calibrate an index.

Consider a target population composed of K sampling units. The index is measured at all K of K locales. At a random sample of k of these sites, absolute abundance or density is measured.

Define the following terms:

N_i = animal abundance at the ith locale ($i = 1, \ldots, K$);
x_i = observed index at the ith locale ($i = 1, \ldots, K$);
K = total number of sampling units (i.e., locales) in the target population;
k = random sample of locales where both x_i and N_i are measured;

$\bar{x}_k = \dfrac{1}{k}\displaystyle\sum_{i=1}^{k} x_i =$ mean value of the index for the k locales where both metrics are measured;

$\bar{X} = \dfrac{1}{K}\displaystyle\sum_{i=1}^{k} x_i =$ mean value of the index over all K locales;

$\bar{N}_k = \dfrac{1}{k}\displaystyle\sum_{i=1}^{k} N_i =$ mean abundance value for the k locales where both metrics are measured.

The ratio estimator is based on the assumed relationship

$$\frac{\bar{N}}{\bar{X}} = \frac{\bar{N}_k}{\bar{x}_k},$$

leading to the estimator

$$\hat{\bar{N}} = \frac{\bar{N}_k}{\bar{x}_k} \cdot \bar{X}. \tag{8.73}$$

In turn, the estimator of total abundance comes from the relationship

$$\hat{N}_{\text{Total}} = K \cdot \hat{\bar{N}}$$
$$= \frac{\bar{N}_k}{\bar{x}_k}(K\bar{X})$$
$$= \frac{\bar{N}_k}{\bar{x}_k} \cdot X, \tag{8.74}$$

where $X = \displaystyle\sum_{i=1}^{K} x_i =$ the total for the index over all K sampling units. Cochran (1977:153) provides a variance approximation for $\hat{\bar{N}}$ where

$$\mathrm{Var}\left(\hat{\bar{N}}\right) \doteq \left(1 - \frac{k}{K}\right)\frac{\displaystyle\sum_{i=1}^{K}(N_i - Rx_i)^2}{k(K-1)}$$

and where

$$R = \frac{\displaystyle\sum_{i=1}^{K} N_i}{\displaystyle\sum_{i=1}^{K} x_i}.$$

The variance can be estimated by the quantity

$$\widehat{\mathrm{Var}}\left(\hat{\bar{N}}\right) = \left(1 - \frac{k}{K}\right)\frac{\displaystyle\sum_{i=1}^{k}(N_i - \hat{R}x_i)^2}{k(k-1)}, \tag{8.75}$$

where

$$\hat{R} = \frac{\sum_{i=1}^{k} N_i}{\sum_{i=1}^{k} x_i}.$$

For the estimate of total abundance, the variance is based on the relationship

$$\widehat{\mathrm{Var}}(\hat{N}_{\mathrm{Total}}) = \widehat{\mathrm{Var}}(K \cdot \hat{\bar{N}})$$
$$= K^2 \cdot \widehat{\mathrm{Var}}(\hat{\bar{N}}). \qquad (8.76)$$

8.4.4 Regression Estimators

The estimation scheme (Section 8.4.3) can be extended to situations in which the relationship between absolute abundance and the index is a straight-line relationship with a nonzero intercept. The sampling scheme for the regression estimator is the same as that of the preceding ratio estimator (Eq. 8.73). The index is recorded at all K sampling units or sites. Animal abundance is also measured at a random sample of k of K sites.

The regression estimator is based on the linear regression of N_i on x_i for the k locales where both metrics were measured, such that

$$N_i = \hat{\alpha} + \hat{\beta} x_i. \qquad (8.77)$$

The estimate of mean abundance across sites is then estimated by the fitted regression (Eq. 8.77)

$$\hat{\bar{N}} = \hat{\alpha} + \hat{\beta} \bar{X}. \qquad (8.78)$$

Alternatively, because the intercept (i.e., α) is estimated by

$$\hat{\alpha} = \bar{N}_k - \hat{\beta} \bar{x}_k, \qquad (8.79)$$

the regression estimator can be reexpressed as

$$\hat{\bar{N}} = \bar{N}_k + \hat{\beta}(\bar{X} - \bar{x}_k), \qquad (8.80)$$

where

$$\hat{\beta} = \frac{\sum_{i=1}^{k}(N_i - \bar{N}_k)(x_i - \bar{x}_k)}{\sum_{i=1}^{k}(x_i - \bar{x}_k)^2}.$$

The variance of the regression estimator (Cochran 1977:192) is

$$\mathrm{Var}(\hat{\bar{N}}) = \frac{\left(1 - \dfrac{k}{K}\right) S_{N_i}^2 (1 - \rho^2)}{k}, \qquad (8.81)$$

where ρ = the correlation between N_i and x_i across all K locales. The estimated variance of $\hat{\bar{N}}$ can be written as

$$\widehat{\text{Var}}(\hat{\bar{N}}) = \frac{\left(1 - \dfrac{k}{K}\right)}{k(k-2)}\left[\sum_{i=1}^{k}(N_i - \bar{N}_k)^2 - \frac{\left(\sum_{i=1}^{k}(N_i - \bar{N}_k)(x_i - \bar{x}_k)\right)}{\sum_{i=1}^{k}(x_i - \bar{x}_k)^2}\right] \tag{8.82}$$

$$= \frac{\left(1 - \dfrac{k}{K}\right)}{k} \cdot \text{MSE}. \tag{8.83}$$

The MSE is the error mean square from the ANOVA table for the regression analysis of N_i on x_i $(i = 1, \ldots, k)$.

Total abundance across the K locales is estimated as

$$\hat{N}_{\text{Total}} = K \cdot \hat{\bar{N}} \tag{8.84}$$

with associated variance estimator

$$\widehat{\text{Var}}(\hat{N}_{\text{Total}}) = \frac{K^2\left(1 - \dfrac{k}{K}\right)}{k}\,\text{MSE}. \tag{8.85}$$

Assumptions

The assumptions of the regression estimator are a combination of those of SRS and regression analyses:

1. The relationship between N_i and x_i is linear.
2. The abundance values N_i are normally distributed.
3. The index values, x_i, are measured without error.
4. k is a random sample of the K sampling units in the sampling frame.

The regression estimator has bias of order $\dfrac{1}{k}$. If index values are measured with error, the technique induces a negative bias to the slope term β and to the estimator $\hat{\bar{N}}$.

Example 8.12: Ratio Estimator of Abundance for Sea Otters, California Coast

A clever use of ratio estimators was provided by the California Department of Fish and Game when estimating sea otter abundance along the California coast. At strategic locations, ground observers attempted complete enumeration of sea otters along stretches of shoreline (N_i; $i = 1, \ldots$). In conjunction with the ground counts, aerial overflights were conducted to index sea otter abundance at these sites as well as the entire shoreline. Eberhardt (1989) provides the survey data for June 25 to 27, 1974, alluded to in Eberhardt et al. (1979) and Eberhardt and Simmons (1987). The survey occurred over 3 days with paired counts at 31 sites along approximately 240 km of shoreline (Table 8.5). The overflight provided a total index count of $X = 897$ sea otters along the coastline. We plotted the straight-line relationship through the origin between the ground and aerial counts (Fig. 8.7). By using Eq. (8.74) for ratio estimation, total sea otter abundance, based on the 1974 survey data, is estimated to be

$$\hat{N}_{\text{Total}} = \frac{\overline{N}_k}{\overline{x}_k} \cdot X$$

$$= \frac{\left(\dfrac{332}{31}\right)}{\left(\dfrac{202}{31}\right)} \cdot 897 = 1,474.28.$$

Table 8.5. Aerial (x_i) and ground (N_i) counts during the sea otter surveys along the California coast in 1974.

June 25		June 26		June 27	
x_i	N_i	x_i	N_i	x_i	N_i
2	4	1	6	1	1
1	1	9	10	6	14
0	4	0	19	5	7
0	6	13	19	1	6
37	50	12	8	4	5
6	9	5	8	2	8
9	10	6	6	2	1
3	3	5	9	37	47
1	1	0	5		
8	13	1	1		
0	8	11	14		
		14	29		

(Data from Eberhardt 1989.)

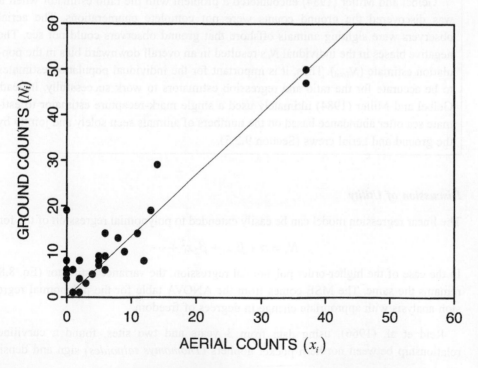

Figure 8.7. Straight-line regression through the origin for sea otter aerial and ground counts (Eberhardt 1989) during the 1974 California coastal survey.

The variance estimate of the abundance total is based on Eqs. (8.75) and (8.76), where

$$\widehat{\text{Var}}\left(\hat{N}_{\text{Total}}\right) = K^2 \left(1 - \frac{k}{K}\right) \frac{\sum_{i=1}^{k}\left(N_i - \hat{R}x_i\right)^2}{k(k-1)}.$$

Unfortunately, K was not specified, but it may be estimated from the relationship

$$X = \overline{X}K,$$

so that

$$\hat{K} = \frac{X}{\overline{x}}$$

$$= \frac{897}{\left(\dfrac{202}{31}\right)} = 137.66 \rightarrow 138.$$

By using the estimated value of K and ignoring the fpc, to be conservative in the absence of true K, the variance is calculated as

$$\widehat{\text{Var}}\left(\hat{N}_{\text{Total}}\right) = 138^2 \frac{\sum_{i=1}^{31}\left(N_i - 1.6436x_i\right)^2}{31(31-1)}$$

$$= \frac{138^2(1159.852)}{31(31-1)} = 23750.776,$$

or an associated standard error of $\widehat{\text{SE}}\left(\hat{N}_{\text{Total}}\right) = 154.11$ for $\hat{R} = 332/202 = 1.6436$.

Geibel and Miller (1984) encountered a problem with the ratio estimator when it was discovered the ground counts were not complete enumerations. The aerial observers were sighting animals offshore that ground observers could not see. The negative biases in the individual N_i's resulted in an overall downward bias in the population estimate (N_{Total}). Thus, it is important for the individual population estimates to be accurate for the ratio and regression estimators to work successfully. Instead, Geibel and Miller (1984) ultimately used a single mark-recapture estimator to estimate sea otter abundance based on the numbers of animals seen solely and jointly by the ground and aerial crews (Section 9.2.5).

Discussion of Utility

The linear regression model can be easily extended to polynomial regression of the form

$$N_i = \alpha + \beta_1 x_i + \beta_2 x_1^2 + \cdots$$

In the case of the higher-order polynomial regression, the variance estimator (Eq. 8.85) remains the same. The MSE comes from the ANOVA table for the polynomial regression analysis with appropriate change in degrees of freedom.

Reid et al. (1966), using data from 3 years and two sites, found a curvilinear relationship between northern pocket gophers (*Thomomys talpoides*) sign and density.

However, further analyses by Eberhardt and Simmons (1987) found the nonlinearity was owing to different sign/gopher ratios between locales. Within locales, the underlying relationship between gopher density and sign density was a straight-line through the origin. The Reid et al. (1966) example illustrates the problem of pooling data without first evaluating the homogeneity of the regression relationship over time and locales. Eberhardt and Simmons (1987) also criticized Floyd et al. (1979) for applying double sampling to one area when using the index-abundance relationship from another locale.

The major limitation of the regression technique is that it requires canvassing all K of K locales to measure the abundance index. When K is large or the index x_i is expensive to collect, the requirement to canvass all locales may be unrealistic. In these circumstances, investigators should consider double-sampling techniques (Sections 8.4.5, 8.4.6). However, if all things are equal, the precision of the ratio and regression estimators will be greater than the corresponding double-sampling methods.

Regression estimators such as $\hat{\bar{N}}$ will have a bias of order $\frac{1}{k}$. However, the variance estimator (Eq. 8.83) is also biased, underestimating the true variance when k is small. Rao (1968) used Monte Carlo simulations to examine this bias for $k \leq 12$, finding underestimation of the variance by 10% to 25%.

8.4.5 Double Sampling for Ratios

The difficulty with the previous ratio and regression estimators (Sections 8.4.3, 8.4.4) is the necessity to canvass all sampling units to know X or $\bar{X}$. Even when indices are relatively fast and easy to collect, the size of the target population may make these methods intractable. Double-sampling methods provide a compromise, balancing precision versus total sampling effort. In general, the previous ratio and regression estimators will be more precise than those of double sampling because the uncertainty associated with the index values is eliminated by a complete census of the population.

The double-sampling scheme consists of two stages of sampling. In the first stage, a large sample of size k' from K is selected where the population index is measured at each sampling unit. In the second stage, a random sample of size k from k' is selected, where animal abundance or density is also measured. Typically, some fraction (v) of k' is chosen in advance, such that

$$k = vk'.$$

In practice, the sampling units where both the index and absolute abundance are to be measured are designated at the time of the first stage and both measurements are taken when the site is initially visited. Thus, although the sampling is viewed as two distinct stages, logistically all measurements can be taken with one pass-through of the sites.

The subsequent use of a ratio or regression estimator depends on whether the relationship between abundance (N_i) and the index (x_i) goes through the origin or not. Data plots, tests of significance, and theoretical considerations may be used to decide whether the intercept is different from zero.

Consider, first, the case in which the relationship between animal abundance and the index is a straight-line relationship through the origin of the form $E(N_i) = \beta x_i$. Define the following quantities:

$\bar{x}_1 = \dfrac{1}{k'} \displaystyle\sum_{i=1}^{k'} x_i =$ mean index value for the sampling units from the first stage of sampling;

$\bar{x}_2 = \dfrac{1}{k} \displaystyle\sum_{i=1}^{k} x_i =$ mean index value from the sampling units in the second stage;

$\bar{N}_2 = \dfrac{1}{k} \displaystyle\sum_{i=1}^{k} N_i =$ mean absolute abundance from the sampling units in the second stage.

The goal of the double-sampling scheme is to estimate $\bar{N}$ or N_{Total} across all K sampling units in the target population.

A ratio estimator can be used to estimate absolute abundance across the entire population in the case where the intercept is zero. The estimator is based on the relationship

$$\frac{\bar{N}}{\bar{x}_1} = \frac{\bar{N}_2}{\bar{x}_2}$$

and solving for mean abundance

$$\hat{\bar{N}} = \frac{\bar{N}_2}{\bar{x}_2} \cdot \bar{x}_1. \tag{8.86}$$

Alternatively, total abundance across all sampling units is estimated by

$$\hat{N}_{\text{Total}} = K \cdot \hat{\bar{N}}$$

$$= K \cdot \frac{\bar{N}_2}{\bar{x}_2} \cdot \bar{x}_1. \tag{8.87}$$

Cochran (1977:344) provides an appropriate expression for the variance of $\hat{\bar{N}}$, where

$$\text{Var}\left(\hat{\bar{N}}\right) = \left(\frac{1}{k'} - \frac{1}{K}\right) S_N^2 + \left(\frac{1}{k} - \frac{1}{k'}\right)(S_N^2 - 2RS_{Nx} + R^2 S_x^2),$$

and where

$$S_N^2 = \frac{\displaystyle\sum_{i=1}^{K}(N_i - \bar{N})^2}{K-1},$$

$$S_x^2 = \frac{\displaystyle\sum_{i=1}^{K}(x_i - \bar{X})^2}{K-1},$$

$$S_{Nx} = \frac{\displaystyle\sum_{i=1}^{K}(N_i - \bar{N})(x_i - \bar{X})}{(K-1)},$$

and

$$R = \frac{\sum_{i=1}^{K} N_i}{\sum_{i=1}^{K} x_i}.$$

The variance of $\hat{\bar{N}}$ can be estimated by the expression

$$\widehat{\text{Var}}\left(\hat{\bar{N}}\right) = \left(\frac{1}{k'} - \frac{1}{K}\right) s_N^2 + \left(\frac{1}{k} - \frac{1}{k'}\right)\left(s_N^2 - 2\hat{R}s_{N_x} + \hat{R}^2 s_x^2\right), \qquad (8.88)$$

where

$$\hat{R} = \frac{\bar{N}_2}{\bar{x}_2},$$

$$s_N^2 = \frac{\sum_{i=1}^{k}(N_i - \bar{N}_2)}{k-1},$$

$$s_x^2 = \frac{\sum_{i=1}^{k}(x_i - \bar{x}_2)}{k-1},$$

$$s_{N_x} = \frac{\sum_{i=1}^{k}(N_i - \bar{N}_2)(x_i - \bar{x}_2)}{(k-1)}.$$

The variance for the estimate of total abundance is estimated by the expression

$$\widehat{\text{Var}}\left(\hat{N}_{\text{Total}}\right) = \widehat{\text{Var}}\left(K\hat{\bar{N}}\right)$$
$$= K^2 \widehat{\text{Var}}\left(\hat{\bar{N}}\right). \qquad (8.89)$$

Example 8.13: Ratio Estimator of Eastern Gray Squirrel Abundance

The nests of gray squirrels provide a conspicuous indicator of squirrel abundance that can be readily canvassed. Determining the number of squirrels associated with the nest count, however, can be time-consuming and costly. Squirrel nest counts, therefore, provide an opportunity to use double sampling (Table 8.6) to collect the relatively inexpensive nest count data and at fewer locales than the more labor-intensive abundance data.

The 100-ha woodlot (Example 8.1) was subdivided into $K = 50$ 2-ha plots. At a random sample of $k' = 20$ plots, the total number of active squirrel nests was recorded. At half of these plots chosen randomly, squirrel abundance was also recorded after detailed surveillance. A plot of squirrel abundance versus nest count (Fig. 8.8) shows a straight-line relationship through the origin with a nonsignificant intercept $(P(|t_8| > 0.4595) = 0.6581)$.

The ratio of squirrels per nest is estimated to be

Table 8.6. Results from a double-sampling design with $k' = 20$ 2-ha plots canvassed for squired nests and $k = 10$ randomly selected where squirrel abundance was also measured.

	Squirrel nests (x_i)	Squirrel abundance (N_i)
	5	
	1	
	4	
	3	
	8	
	0	
	3	
	2	
	6	
	3	
	0	0
	6	10
	2	0
	7	11
	3	6
	3	7
	5	11
	1	5
	7	14
	7	13

$\bar{x}_1 = 3.8$
$\bar{x}_2 = 4.1$, $s_x^2 = 6.9899$ $\bar{N}_2 = 7.7$, $s_N^2 = 24.909$

$k' = 20$ $k = 10$

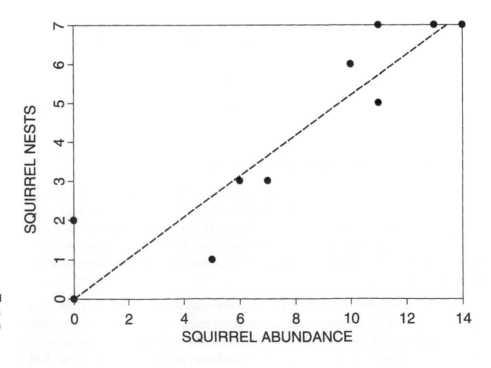

Figure 8.8. The relationship between squirrel abundance and nest count index with a regression line through the origin.

$$\hat{R} = \frac{\sum_{i=1}^{10} N_i}{\sum_{i=1}^{10} x_i}$$

$$= \frac{77}{41} = 1.8780,$$

with mean abundance per plot therefore estimated by Eq. (8.86) to be

$$\hat{\bar{N}} = \frac{\bar{N}_2}{\bar{x}_2} \cdot \bar{x}_1$$

$$= \frac{7.7}{4.1}(3.8) = 7.1366 \text{ squirrels/plot}.$$

Total squirrel abundance is similarly estimated by Eq. (8.87) to be

$$\hat{N}_{\text{Total}} = K\hat{\bar{N}}$$

$$= 50\,(7.1366) = 356.83 \text{ squirrels}$$

on the 100-ha site. The variance of the estimate of total abundance is based on Eqs. (8.88) and (8.89) where

$$\widehat{\text{Var}}\left(\hat{N}_{\text{Total}}\right) = K^2\left[\left(\frac{1}{k'} - \frac{1}{k}\right)s_N^2 + \left(\frac{1}{k} - \frac{1}{k'}\right)\left(s_N^2 - 2\hat{R}s_{N_x} + \hat{R}^2 s_x^2\right)\right].$$

From the sites where both the squirrel abundance and nest index counts were collected, $s_x^2 = 6.9889$, $s_N^2 = 24.90$, and $s_{N_x} = 12.1444$, leading to variance estimate

$$\widehat{\text{Var}}\left(\hat{N}_{\text{Total}}\right) = 50^2\left[\left(\frac{1}{20} - \frac{1}{50}\right)(24.90) + \left(\frac{1}{10} - \frac{1}{20}\right)\right.$$

$$\left. \cdot\left(24.9 - 2(1.8780)(12.1444) + (1.8780)^2(6.9889)\right)\right]$$

$$\widehat{\text{Var}}\left(\hat{N}_{\text{Total}}\right) = 2359.334,$$

or an associated standard error of $\widehat{\text{SE}}\,(\hat{N}_{\text{Total}}) = 48.57$.

8.4.6 Double Sampling for Regression

The previous double-sampling scheme can be readily extended to the situation where there is a linear relationship of the form

$$E(N_i) = \alpha + \beta x_i$$

between abundance and the indices measured. In this case, there is a nonzero intercept (α). The same quantities $\bar{x}_1$, $\bar{x}_2$, and $\bar{N}_2$ are used in the regression estimator as in the double sampling for the ratio method (Section 8.4.5).

A straight-line regression is performed on the k locales where both the abundance and the index are measured, such that

$$N_i = \hat{\alpha} + \hat{\beta} x_i, \tag{8.90}$$

and where

$$\hat{\alpha} = \overline{N}_2 - \hat{\beta} \overline{x}_2,$$

$$\hat{\beta} = \frac{\sum_{i=1}^{k}(N_i - \overline{N}_2)(x_i - \overline{x}_2)}{\sum_{i=1}^{k}(x_i - \overline{x}_2)^2}.$$

The estimate of mean abundance, using Eq. (8.90), is calculated by

$$\hat{\overline{N}} = \hat{\alpha} + \hat{\beta} \overline{x}_1, \tag{8.91}$$

and an estimate of total abundance is computed as

$$\hat{N}_{\text{Total}} = \hat{\alpha} + \hat{\beta}(K\overline{x}_1). \tag{8.92}$$

The variance for the estimate of mean abundance can be expressed (Cochran 1977:339) as a function of the correlation (ρ) between N_i and x_i, where

$$\text{Var}\left(\hat{\overline{N}}\right) = \frac{S_N^2(1-\rho^2)}{k} + \frac{S_N^2 \rho^2}{k'} - \frac{S_N^2}{K} \tag{8.93}$$

for

$$S_N^2 = \frac{\sum_{i=1}^{K}(N_i - \overline{N})^2}{(K-1)}.$$

The variance of $\hat{\overline{N}}$ can be estimated (Cochran 1977:343) by

$$\widehat{\text{Var}}\left(\hat{\overline{N}}\right) = \frac{s_{N\cdot x}^2}{k} + \frac{s_N^2 - s_{N\cdot x}^2}{k'} - \frac{s_N^2}{K}, \tag{8.94}$$

where

$$s_{N\cdot x}^2 = \frac{\sum_{i=1}^{k}(N_i - \overline{N}_2)^2 - \hat{\beta}^2 \sum_{i=1}^{k}(x_i - \overline{x}_2)^2}{(k-2)},$$

which is the MSE from the ANOVA table for the regression of N_i on x_i ($i = 1, \ldots, k$). The variance of $\hat{N}_{\text{Total}}$, from the relationship $\hat{N}_{\text{Total}} = K\hat{\overline{N}}$, can be expressed as

$$\widehat{\text{Var}}\left(\hat{N}_{\text{Total}}\right) = K^2 \widehat{\text{Var}}\left(\hat{\overline{N}}\right).$$

Example 8.14: Regression Estimator of Abundance, White-Tailed Deer (*Odocoileus virginianus*), Wisconsin

McCaffery (1976) calibrated a trail index against density estimates calculated from a sex-age-kill (SAK) model (Section 9.7.1) for white-tailed deer. The index was the average number of deer trails intersecting 0.40-km replicate transects. A regression relationship was established between estimates of deer density for an SAK model and mean number of deer trails/transect, considering numerous deer management units in Wisconsin from 1968–1973 (Table 8.7). The fitted regression model (Fig. 8.9) was

$$D_i = \underset{(\overline{SE}=0.8788)}{-2.8357} + \underset{(\overline{SE}=0.1760)}{2.1644x_i},$$

where

D_i = SAK density estimate (deer/km^2);
x_i = deer trail index count (trails/0.4-km transect).

Unlike typical double-sampling schemes, McCaffery (1976) used the fitted relationship constructed over multiple years and locales to project deer density in new domains. For instance, McCaffery projected deer density (i.e., deer/km^2) for different forest types in Wisconsin using the regression model. In aspen (*Populus tremula*)

Table 8.7. Deer trail index count and estimates of white-tailed deer density (number of deer/km^2) from the sex-age-kill (SAK) model in Wisconsin, 1968–1973.

Management unit	Survey data	Mean number of trails/transect (x_i)	SAK deer density estimate (D_i)
3	May 15, 1968	2.70	2.27
5	May 11, 1970	2.40	4.58
8	May 9, 1973	4.54	6.46
10	Nov 5, 1970	6.02	7.38
12	May 8, 1971	2.26	3.12
19	Oct 23, 1970	3.29	2.73
25	Nov 5, 1968	6.08	6.50
26	Nov 4, 1968	2.98	4.69
31	Nov 11, 1971	3.54	2.54
31	Apr 18, 1973	2.04	2.65
36	Apr 19, 1968	6.50	9.73
36	Nov 10, 1971	3.60	5.73
39	May 13, 1971	2.48	4.54
43	Apr 1, 1968	3.38	4.42
45	Nov 13, 1968	5.46	7.58
49	Apr 10, 1968	6.85	12.15
52	Nov 10, 1970	4.20	5.46
55	Nov 10, 1969	5.52	9.62
58	Nov 10, 1969	4.42	6.62
SWA	Nov 4, 1969	7.12	14.35
SWA	Nov 4, 1970	8.60	18.62
SWA	Nov 7, 1972	8.98	18.50
SWA	Nov 5, 1973	1.54	0.73

(Data from McCaffery 1976.)

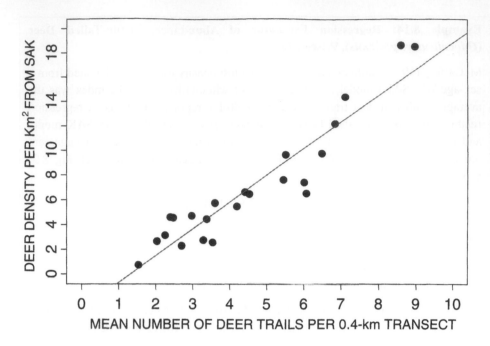

Figure 8.9. The relationship between mean number of deer trails/0.4-km transect and white-tailed deer density/km² (McCaffery 1976).

forests with a mean number of deer trails/0.4-km transect of 6.7, the regression model estimated

$$\hat{D} = -2.8357 + 2.1644(6.7) = 11.67 \text{ deer}/\text{km}^2.$$

In northern hardwood forests with a mean number of deer trails/0.4-km transect of 2.4, deer density was estimated to be

$$\hat{D} = -2.8357 + 2.1644(2.4) = 2.36 \text{ deer}/\text{km}^2.$$

It is valuable to note in this trail count example that the indices do not provide measures of relative abundance because of the linear-regression relationship between deer density and the index count. The relative density of deer in the aspen versus hardwood forests was

$$\frac{\hat{D}_{\text{Aspen}}}{\hat{D}_{\text{Hardwood}}} = \frac{11.67}{2.36} = 4.94.$$

Thus, there was a ratio of approximately 5 : 1 deer in the aspen habitat compared with that of the hardwood habitat. On the other hand, using just the trail count indices, relative abundance would be wrongly estimated as

$$\frac{6.7}{2.4} = 2.79.$$

The predicted deer densities in the aspen and hardwood forests are based on index values, which were not part of the domain used in constructing the regression relationship. Thus, a different approach to variance estimation must be used. Expressing the variance of the prediction in stages,

$$\widehat{\mathrm{Var}}(\hat{D}) = \mathrm{Var}(\hat{\alpha} + \hat{\beta}\hat{x})$$

$$= \mathrm{Var}_{\hat{x}}\left[E(\hat{\alpha} + \hat{\beta}\hat{x}|\hat{x})\right] + E_{\hat{x}}\left[\mathrm{Var}(\hat{\alpha} + \hat{\beta}\hat{x}|\hat{x})\right]$$

$$= \beta^2 \mathrm{Var}(\hat{x}) + \mathrm{MSE}\left[1 + \frac{1}{n} + \frac{(\hat{x} - \bar{x})^2}{\displaystyle\sum_{i=1}^{n} x_i^2}\right], \qquad (8.95)$$

where

Var($\hat{x}$) = variance associated with the index value used in the prediction;

n = number of paired observations used in constructing the regression model;

MSE = error mean square from the regression analysis;

$\bar{x}$ = average index value used in constructing the regression equation.

The second term in Eq. (8.95) comes from standard regression analyses (Sokal and Rohlf 1995:468–469).

McCaffery (1976) did not provide error bounds on the trail count indices used in the projections. However, assuming SE($\hat{x}$) = 0.25 for purposes of illustration, then

$$\widehat{\mathrm{Var}}(\hat{D}) = \hat{\beta}^2 \widehat{\mathrm{Var}}(\hat{x}) + \mathrm{MSE}\left(1 + \frac{1}{n} + \frac{(\hat{x} - \bar{x})^2}{\displaystyle\sum_{i=1}^{n} x_i}\right)$$

$$= (2.1644)^2 (0.25)^2 + 3.0493\left(1 + \frac{1}{23} + \frac{(6.7 - 3.9296)^2}{(405.1322)}\right)$$

$$\widehat{\mathrm{Var}}(\hat{D}) = 0.2728 + 3.2396 = 3.5324,$$

or a standard error of $\widehat{\mathrm{SE}}(\hat{D}) = 1.8795$ for the aspen forest habitat.

Discussion of Utility

Eberhardt et al. (1985) and Eberhardt and Simmons (1987) conducted simulation studies to examine the statistical behavior of double-sampling estimators. They recommended not using the techniques for data sets where $k \leq 5$ when correlation $\rho_{Nx} < 0.85$. They found from simulation studies that double-sampling methods tended to be positively biased (usually less than 5%) even for reasonable data sets.

Cochran (1977:341) offered guidance on optimal allocation of effort between sample sizes k and k', for fixed total costs. Eberhardt and Simmons (1987) illustrated the use of these formulas for optimizing the sampling of mountain pocket gophers.

The variance of $\hat{\bar{N}}$ for double sampling (Eq. 8.93) for regression can be rewritten as

$$\mathrm{Var}(\hat{\bar{N}}) = \frac{\left(1 - \dfrac{k}{K}\right)S_N^2}{k} - \frac{\left(1 - \dfrac{k}{k'}\right)S_N^2 \rho^2}{k}.$$

Inspection indicates the first term is the variance for SRS. When $\rho^2 = 0$, the double-sampling method has the same precision as SRS with size k. The precision of double sampling improves as $|\rho|$ increases. It is therefore important to have a strong correlation between the index x_i and abundance N_i for the sampling method to be an effective alternative to a direct survey of N_i alone.

A potentially valuable alternative to double-sampling techniques is ranked-set sampling, first suggested by McIntyre (1952) and Takahasi and Wakimoto (1968). This method avoids the need to specify a functional relationship between the index (x_i) and absolute abundance (N_i). Instead, the method is based on ranking the initial sampling units and selecting subsequent units for the expensive abundance estimation based on order statistics. The rank-set sampling is more structured than are double-sampling schemes but is robust to nonlinear relationships. In addition, the first stage of rank-set sampling can be based on either rigorous index sampling or simply qualitative judgment. If expert judgment and casual field observations can rank different areas in terms of relative abundance, index sampling can be eliminated in the first stage of sampling, thereby providing cost savings. Kaur et al. (1995) provide a review of rank-set sampling methods. Mode et al. (1999) discussed the relative efficiency of ranked-set sampling and the effect of errors in ranking on precision. Patil et al. (1993) showed regression and double-sampling methods are generally more efficient than ranked-set sampling. However, the potential cost-savings of ranking versus index sampling in the first stage makes ranked-set sampling a viable alternative in some circumstances.

8.5   Analysis of Index Studies

Examples of index studies will be presented to illustrate some of the diverse ways indices are used and analyzed in wildlife investigations. Some analyses have already been presented in previous sections. Examples presented here extend the concepts presented earlier and provide more complex response models for consideration. The focus is on formulation of response models specific to the needs and requirements of the study design. The more tailored a response model is to the needs of an investigation, the more useful the information that can be extracted from the study. Study designs should be structured around an anticipated response model to ensure all necessary design elements, controls, and covariates have been properly considered and measured.

The response models considered in standard applied statistics courses and texts are invariably linear functions of the parameters. A multiplicative response model should be the first consideration in biological studies. Factors that effect recruitment are likely to be multiplicative. For example, change in brood size or litter success will change the relative size of the population by a multiplicative factor, e.g.,

$$N_{\text{Adult+Juveniles}} = N_{\text{Adult}} \cdot (1 + \text{Productivity}).$$

Factors affecting survival also tend to operate multiplicatively, e.g.,

$$N_{t+1} = N_t \cdot S_N \cdot S_H$$

where

S_N = survival due to natural causes;
S_H = survival due to anthropogenic effects.

Sighting or capture probabilities also tend to operate multiplicatively, where the number of animals observed (x) is a function of abundance (N) and detection (p) rates, e.g.,

$$E(x) = Np.$$

Investigators are encouraged to postulate response models for their measured responses that are a function of both the factors they can control and the likely factors they cannot control. This allows investigators to ascertain whether they can estimate the population effects of interest unfettered by extraneous and confounding influences. The response model may also suggest ways to conduct the studies to avoid confounding. It is good practice to propose such models for even the simplest of studies. The practice will eventually lead to the skills needed to conceptualize more formidable studies.

An invaluable tool for wildlife biologists is the ability to use GLMs, a generalization of the linear models presented in standard regression and experimental design courses in college (Aitkin et al. 1989, Crawley 1993). In linear models, the response models are assumed to be a linear function of the parameters and the data normally distributed. The GLM extends this theory to include situations in which some function of the expected value of the dependent variable is linear. The function that makes the expected value a linear function of the parameters is called the link function. For example, the simple exponential model

$$E(N_t) = N_0 e^{rt}$$

has the linear form

$$\ln E(N_t) = \ln N_0 + rt$$

using the log-link. However, GLM analyses do not transform the data. Rather, GLMs transform the expected values, thereby avoiding the unpleasant consequences of undefined response values (e.g., $\ln(0)$). The GLM also extends the analyses to include all members of the exponential family, including normal, binomial, Bernoulli, gamma, and Poisson distributions. This feature increases the opportunities to analyze presence/absence or success/failure data (i.e., Bernoulli), count observations (i.e., Poisson), and proportions (i.e., binomial). Readers are encouraged to learn more about this topic by reviewing Aitkin et al. (1989) and Crawley (1993).

8.5.1   Example: Forest Birds, New South Wales, Australia

Shields (1990) investigated the effects of varying rates of logging (i.e., removal of tree cover) on the abundance of forest bird populations through a replicated and randomized manipulative experiment in New South Wales, Australia. Twelve forested sites designated for logging were randomly assigned to serve as control (0% logging), 50% forest removal, or 90% forest removal sites in a balanced design. In 1987, before logging, bird abundance was surveyed by using circular-transect counts replicated over 4 days during January (Table 8.8). Treatment plots were logged in January 1989, and bird abundance was resurveyed in January 1990 (Table 8.8).

The purpose of the study was to assess whether forest birds were lost proportional to the fraction of trees removed in old-growth forests. The premise was that birds would be retained at a rate greater than a 1 : 1 proportion of birds to trees. The response variable

Table 8.8. Mean count of old-growth species of birds pre- and postlogging, and the fraction of trees logged at old-growth plots, New South Wales, Australia, 1987–1990.

Treatment	N_{POST} (1990)	N_{PRE} (1987)
Control	50	59
	47	49
	69	76
	48	45
50% Logged	52	57
	39	55
	54	60
	57	54
90% Logged	35	67
	29	50
	27	51
	17	26

(Data from Shields 1990.)

was the postlogging bird abundance in 1990. A response model was constructed that assumed the number of birds seen in 1990 at the control plots would be proportional to the baseline counts seen in 1987, where

$$E(N_{Post}) = (\text{Phase effect}) \cdot N_{Pre}.$$

The "phase effect" is the fractional change in bird abundance by the naturally occurring effects of time. At the logged sites, bird abundance was thought to be related not to the fraction of trees removed but rather to the fraction of trees remaining (i.e., more habitat remaining, more birds remaining), such that

$$E(N_{Post}) = (\text{Phase effect}) \cdot N_{PRE} \cdot (1 - \text{Fraction logged})^{\beta}, \qquad (8.96)$$

where β modifies the effect of removal. Specifically, if there is a $1:1$ correspondence between trees removed and birds lost, then $\beta = 1$. If this relationship is not as severe and more birds are retained than the fraction of trees left standing, then $\beta < 1$. The response model suggests the hypotheses,

$$H_o: \beta \geq 1$$
$$H_a: \beta < 1.$$

The null hypothesis states that birds are lost at a rate equal to or greater than the loss of trees; the alternative hypothesis states the rate of bird loss is less.

Log-transforming both sides of Eq. (8.96) produces the log-linear model

$$\ln E(N_{Post}) = \alpha + \ln N_{Pre} + \beta \cdot \ln(1 - \text{Fraction logged}). \qquad (8.97)$$

The intercept will characterize the ln(Phase effect), and the independent variable is $\ln(1 - \text{Fraction logged})$. In addition, model (8.97) has the covariate $\ln N_{Pre}$, but it is not an ordinary covariate. Based on the model, there is no regression coefficient to estimate. Instead, the regression coefficient, by definition, has the value 1. A variable that is entered into the model with a prespecified regression coefficient (or no regression coefficient) is called an "offset" in GLM. No degree of freedom is lost by including these covariates because no regression coefficient is estimated. The GLM analyses were based on a normal error structure, because the bird counts (Table 8.8) were an average of four replicate surveys. The Central Limit theorem suggests the averages may be approximately normally distributed.

The ANOVA table for the New South Wales bird analysis is presented below:

Source	DF	SS	MS	F	P
Total$_{Cor}$	11	0.7461			
Logging	1	0.5838	0.5838	$F_{1,10} = 35.4682$	0.0001
Error	10	0.1623	0.01623		

The mean bird count is significantly related ($P = 0.0001$) to the fraction of trees retained post-logging. The fitted response model is

$$\ln N_{\text{Post}} = \underset{(\widehat{SE}=0.05294)}{-0.02007} + \ln N_{\text{Pre}} + \underset{(\widehat{SE}=0.0381)}{0.2287} \ln(1 - \text{Fraction logged})$$

or back-transforming

$$N_{\text{Post}} = 0.9801 N_{\text{Pre}}(1 - \text{Fraction logged})^{0.2287}.$$

The estimated phase effect of 0.9801 ($\widehat{SE} = 0.05189$) is not significantly different from 1, suggesting bird counts at the control sites (i.e., nonlogged) have not varied between 1987 and 1990. Modeling the response without an intercept results in the simplified model

$$N_{\text{Post}} = N_{\text{Pre}}(1 - \text{Fraction logged})^{0.2391} \qquad (8.98)$$

with $\widehat{SE}(\hat{\beta}) = 0.02544$.

A comparison (below) of the postlogging bird counts with the predicted values from the final regression model (8.98) shows good agreement:

Treatment	N_{Post}	Predicted	Standardized residual
Control	50	59.03	+1.548
	47	49.01	+0.345
	69	76.02	+1.204
	48	45.02	−0.511
50% Logging	52	48.28	−0.638
	39	46.62	+1.307
	54	50.86	−0.539
	57	45.74	−1.931
90% Logging	35	38.63	+0.623
	29	28.85	−0.026
	27	29.40	+0.412
	17	15.00	−0.343

The test of the null hypotheses (i.e., H_o: $\beta \geq 1$) can be performed by using a t-statistic of the form:

$$t_{11} = \frac{0.2391 - 1}{0.02544} = -29.9096,$$

with a significance level of $P(t_{11} \leq -29.9096) \approx 0$. Thus, we conclude that forest birds were retained at a higher fraction than the fraction of old-growth trees left standing.

The fitted response model can also be used to estimate the fraction of birds retained under the different treatment levels, specifically:

$$0 \text{ logging} \qquad (1-0)^{0.2391} = 1,$$
$$50\% \text{ logging} \quad (1-0.50)^{0.2391} = 0.8473,$$
$$90\% \text{ logging} \quad (1-0.90)^{0.2391} = 0.5766.$$

The model suggests that with 50% of the trees removed, 85% of the bird count is retained; for 90% tree removal, 58% of the bird count is retained. Shields (1990) performed a similar analysis examining the count of invading bird species into the old-growth forest.

In this example and others like it, the anthropogenic effect under investigation may affect animal abundance and behavior. There is also the possibility the logging treatment, by virtue of changing the habitat structure, may also have directly affected observation rates. Without auxiliary information on observation distances, there is the possibility that the estimates of relative abundance may be confounding with increases in observational rates owing to opening the forest canopy. Ambiguous and possibly faulty interpretations linger over all index studies regardless of how well they are designed and implemented.

8.5.2 Example: Deer Trail Counts, Wisconsin

McCaffery (1976) reported the use of deer trail counts as an index to white-tailed deer abundance in Wisconsin. In this example, the total deer trail count and number of transect lines of length 0.4 km used in collecting the count data are analyzed from aspen and northern hardwood (i.e., maple [*Acer* spp.], basswood [*Tilia americana*], ash [*Fraxinus* spp.], and elm [*Ulmus* spp.] habitats at eight different deer management units. The goal of the analyses is to estimate relative abundance of deer in northern hardwood forests relative to that in aspen habitat (Table 8.9). The index is the number of deer trails that traverse the line-transects. However, the number of trails counted is a function of not only deer abundance but also the number of transects monitored.

Table 8.9. Number of deer trails and number of 0.4-km transect lines used in aspen and northern hardwood forests in eight different management units in Wisconsin.

Management unit	Aspen		Northern hardwoods	
	Count	Lines	Count	Lines
8	32	7	1	1
25	156	13	59	8
26	40	11	38	19
28	7	5	53	33
36	108	15	9	2
39	8	4	19	16
43	21	7	75	22
45	20	3	48	16

(Data from McCaffery 1976.)

Thus, let x_{ij} = total number of deer trails counted in the ith habitat ($i = 1, 2$) in the jth management unit ($j = 1, \ldots, 8$). The expected number of trail counts will be modeled with the expected value of

$$E(x_{ij}) = (0.4\,\text{km} \times \text{Width} \cdot \text{Mean density}) \cdot \left(\frac{\text{Animal}}{\text{activity level}}\right) \cdot L_{ij} \cdot U_j \cdot H_i \quad (8.99)$$

where

L_{ij} = number of replicate transect lines used in the ith habitat ($i = 1, 2$) in the jth management unit ($j = 1, \ldots, 8$);

U_j = effect of the jth management unit ($j = 1, \ldots, 8$);

H_i = effect of the ith habitat type ($i = 1, 2$).

Log-transforming both sides of Eq. (8.99) produces the log-linear model

$$\ln E(x_{ij}) = \alpha + \ln L_{ij} + \ln U_j + \ln H_i. \quad (8.100)$$

The intercept jointly estimates the effective area of the line-transect, mean deer density, and a proportionality constant associated with animal activity. The term $\ln L_{ij}$ is considered an "offset." Therefore, the value of $\ln L_{ij}$ is treated as a covariate with an implicit regression coefficient of value 1. The deer management unit (U_j) and habitat (H_i) effects are modeled as indicator variables. By using a Poisson-error structure and log-link, the following ANODEV table was constructed:

Source	DF	DEV	MDEV	F	P
Total$_{Cor}$	15	322.540			
Units	7	217.4	31.057	$F_{7,7} = 5.9530$	0.0156
Habitat type	1	68.59	68.59	$F_{1,7} = 13.1467$	0.0084
Error	7	36.521	5.217		

Both deer management unit ($P = 0.0156$) and habitat type ($P = 0.0084$) had a significant effect on the number of deer trails observed after correction for the number of transect lines (Table 8.10). In three of the eight management units, the GLM appreciably over- or underestimated the number of deer trails expected. In the other five units, predictions reasonably matched the observed counts.

The following model parameters were predicted by the GLM:

Parameter	$\hat{\theta}$	$\widehat{SE}$*
Intercept	1.483	0.1742
Unit 25	0.8076	0.1892
26	−0.1341	0.2104
28	−0.4355	0.2249
26	0.5079	0.1971
39	−0.6537	0.2646
43	0.2077	0.2071
45	0.3587	0.2196
Hardwoods	−0.7215	0.0866

*Assuming a scale parameter of 1.

Table 8.10. Observed and predicted deer trail counts from the generalized linear model fit to the McCaffery (1976) index data.

Observed	Predicted	Standardized residual (Z)
32	30.858	+0.206
1	2.142	−0.781
156	128.516	+2.424
59	86.484	−2.953
40	42.403	−0.369
38	35.597	+0.403
7	14.260	−1.922
53	45.740	+1.073
108	109.880	−0.179
9	7.120	+0.704
8	9.171	−0.387
19	17.829	+0.277
21	37.983	−2.756
75	58.018	+2.230
20	18.931	+0.246
40	49.069	−0.153

The intercept is estimating the baseline condition of the aspen habitat at the first management area (i.e., 8). The hardwoods effect is estimating the habitat effect relative to aspen habitat. The relative deer trail abundance in northern hardwoods relative to aspen forests is estimated to be

$$\frac{N_{\text{Hardwood}}}{N_{\text{Aspen}}} = e^{-0.7215} = 0.4860.$$

Thus, there were only 48.6% as many deer trails per unit area in hardwood forests as in aspen forests, and with it, the implication to deer density. The variance of the estimate of relative abundance is calculated, using the delta method, where

$$\text{Var}(e^{\hat{\theta}}) \doteq \text{Var}(\hat{\theta}) \cdot e^{2\hat{\theta}} \cdot \text{scale parameter}$$

$$= (0.08659)^2 \cdot e^{2(-0.7215)} (5.217) = 0.009240,$$

or, equivalently, a standard error of $\widehat{\text{SE}} = 0.0961$, where the scale parameter (i.e., 5.217) is taken as the MSE from the ANODEV table.

8.5.3 Example: Dall Sheep (*Ovis dalli*) Aerial Counts, Arctic National Wildlife Refuge, Alaska

McDonald et al. (1990) presented aerial counts of Dall sheep in 14 different subunits of the Arctic National Wildlife Refuge, Alaska, in 1976, 1979, 1982, and 1986. The purpose of this example is to estimate the relative change in abundance over time for the 12 subunits with consistent histories of surveys over time (Table 8.11). The sheep counts represent a two-way classification by subunit and year. The proposed response model for the analysis is

$$E(x_{ij}) = \mu \cdot L_i \cdot Y_j, \tag{8.101}$$

where

Table 8.11.   Dall sheep reported observed by subunit and year in the Arctic National Wildlife Refuge, Alaska.

Subunit	Year			
	1976	1979	1982	1986
1	35	38	43	86
2	58	49	22	157
3	8	32	57	32
4	33	60	64	104
5	168	121	258	232
6	390	814	583	382
7	339	201	254	382
8	103	112	222	363
10	47	23	71	223
11	192	6	190	372
13	104	72	186	270
14	121	83	125	71

(Date from McDonald et al. 1990.)

x_{ij} = Dall sheep counts in the ith subunit ($i = 1, \ldots, 12$) in the jth year ($j = 1, \ldots, 4$);

μ = baseline response;

L_i = effect of the ith location ($i = 1, \ldots, 12$);

Y_j = effect of the jth year ($j = 1, \ldots, 4$).

The multiplicative equation (8.101) suggests the log-linear response model

$$\ln E(x_{ij}) = \ln \mu + \ln L_i + \ln Y_j. \qquad (8.102)$$

The counts were analyzed by using a log-link assuming a Poisson error structure with the resulting ANODEV table:

Source	DF	DEV	MDEV	F	P
Total$_{Cor}$	47	6157.2			
Year	3	376.0	125.33	$F_{3,33} = 3.1621$	0.0374
Subunit	11	4473.0	406.64	$F_{11,33} = 10.2592$	<0.00001
Error	33	1308.0	39.64		

Both the year ($P = 0.0374$) and the location ($P < 0.00001$) effects were significant with the fitted GLM parameter values:

Parameter	$\hat{\theta}$	$\widehat{SE}(\hat{\theta})*$
Intercept	3.703	0.0738
Year 1979	0.0081	0.0353
Year 1982	0.2612	0.0333
Year 1986	0.5148	0.0316
Subunit 2	0.3477	0.0919
3	−0.4485	0.1127
4	0.2563	0.0937

Parameter	$\hat{\theta}$	$\widehat{SE}(\hat{\theta})$*
5	1.3500	0.0790
6	2.3740	0.0736
7	1.7620	0.0762
8	1.3760	0.0787
10	0.5889	0.0877
11	1.3250	0.0792
13	1.1410	0.0808
14	0.6832	0.0863

*Assuming a scale parameter of 1.

The intercept estimates the expected count in subunit 1 in 1976. All the other parameter estimates are relative adjustments to that initial condition. The estimated year effects must be back-transformed to calculate Dall sheep abundance in subsequent years relative to that in 1976. The relative Dall sheep abundance in 1979 to that of 1976 is estimated to be

$$\frac{\widehat{N_{1979}}}{N_{1976}} = e^{0.008102} = 1.0081\left(\widehat{SE} = 0.2240\right),$$

and the other years as

$$\frac{\widehat{N_{1982}}}{N_{1976}} = e^{0.2612} = 1.2985\left(\widehat{SE} = 0.2721\right),$$

$$\frac{\widehat{N_{1986}}}{N_{1976}} = e^{0.5148} = 1.6733\left(\widehat{SE} = 0.3331\right).$$

The variance of the relative abundance of year 1979:1976 is calculated as

$$\widehat{Var}\left(e^{\hat{\theta}}\right) = Var\left(\hat{\theta}\right)e^{2\hat{\theta}} \cdot (\text{Scale parameter})$$

$$= (0.0353)^2 e^{2(0.008102)}(39.64) = 0.0502,$$

or a standard error of $\widehat{SE} = 0.2240$.

The GLM analysis suggests that Dall sheep abundance from 1976 to 1979 increased by only 0.81%. However, by 1982, the population had increased by 29.85%, and 67.33% by 1986.

8.5.4 Example: White-Tailed Deer Pellet Counts, George Reserve, Michigan

Neff (1968) reports pellet group count data from the George Reserve, Michigan, for white-tailed deer, 1953–1958, as modified from Ryel (1959) (Table 8.12). Numerous 0.02-acre circular plots were sampled each year to monitor deer abundance. The actual number of circular plots used to monitor pellet count groups varied among years. For purposes of illustration, these data will be used to estimate the finite rate of population change (λ) for the George Reserve herd over the 5-year period.

The response model used to analyze the pellet group count data is as follows:

Table 8.12.   Pellet group count data for white-tailed deer, George Reserve, Michigan, 1953–1958.

Year	Time (t)	Number of plots (n)	Total pellet groups counted
1953	0	243	533
1954	1	235	483
1955	2	241	285
1956	3	221	769
1957	4	249	649
1958	5	268	598

(Data from Weff 1968.)

$E(x_i) =$ (Defectation rate $\cdot$ Cumulative time $\cdot$ Plot area $\cdot$ Baseline density)
$\cdot$ (Number of plots) $\cdot \lambda^t$

or, more succinctly,

$$E(x_i) = \alpha \cdot n_i \cdot \lambda^t, \tag{8.103}$$

where

x_i = number of pellet groups counted in year $i (i = 1, \ldots, t)$;
α = initial baseline count per plot;
n = number of plots sampled in year $i (i = 1, \ldots, t)$;
λ = finite rate of population change;
t = time expressed in years.

The log-expected value for the number of pellet groups counted in year i can then be written as follows:

$$\ln E(x_i) = \ln \alpha + \ln(n) + t \ln \lambda$$
$$= \alpha' + \ln(n) + tr, \tag{8.104}$$

where r = the instantaneous rate of population change. In Eq. (8.104), α' is an intercept and $\ln(n)$ is an offset term, where the regression coefficient is implicitly taken to be equal to 1.

By analyzing the pellet count data using a log-link and assuming a Poisson error structure, the following ANODEV table was constructed:

Source	DF	DEV	MDEV	F	P
Total$_{Cor}$	5	292.30			
Year	1	23.66	23.66	$F_{1,4} = 0.3523$	0.5948
Error	4	268.64	67.16		

The ANODEV suggest there was no significant trend in pellet group counts over time ($P = 0.5848$). The fitted model was

$$\ln E(x) = \underset{(\widehat{SE}=0.0314)}{0.6948} + \ln(n) + \underset{(\widehat{SE}=0.01003)}{0.04873t},$$

or back-transforming,

$$E(x) = 2.0033 \cdot n \cdot (1.0499)^t.$$

The regression analysis estimates the instantaneous population rate of change to be $\hat{r} = 0.0487$ or, equivalently, the finite rate of change is estimated to be

$$e^{\hat{r}} = \hat{\lambda} = e^{0.04873} = 1.0499.$$

The variance of $\hat{\lambda}$ is estimated by using the delta method, where

$$\text{Var}(e^{\hat{r}}) \doteq \text{Var}(\hat{r}) \cdot e^{2\hat{r}} \cdot (\text{Scale parameter})$$

$$= (0.01003)^2 \cdot e^{2(0.04873)}(67.16) = 0.00745,$$

or a standard error of $\widehat{\text{SE}}(\hat{\lambda}) = 0.0863$.

Thus, analysis of the pellet group count data for the George Reserve estimates the population increased an average of 4.99% per year during 1953–1958. However, the estimate of λ is not significantly different from the value of $\lambda = 1$, based on the t-test,

$$t_4 = \frac{1.0499 - 1}{0.0863} = 0.5782,$$

$P(|t_4| > 0.5782) = 0.5941$. It should be noted the F-test for $r = 0$ is asymptotically equivalent to the t-test for $\lambda = 1$. Based solely on the pellet group count data, it would be concluded the George Reserve herd was stationary during 1953–1958.

8.6 Summary

There is no single set of methods focused specifically on the analysis of index data. Rather, finite sampling techniques provide a broad statistical basis for collection and characterizations of index data. The observations used in the construction of a population index should be collected in a probabilistic manner that permits inferences to the intended target population. Representative, opportunistic, or convenience sampling are not acceptable substitutions for probabilistic sampling. The sessile or inanimate nature of sign counts avails index data to a myriad of finite sampling methods used in sample survey design (Cochran 1977, Thompson 2000). Strict adherence to finite sampling techniques is a crucial prerequisite to the use of population indices.

A second crucial element of index studies is formulation of a realistic response model that describes the measured indices as a function of biological, environmental, study design, and anthropogenic effects. These response models are essential in identifying driving variables, which are within and outside the control of the investigator. Assumptions are also identified through the modeling exercise that must be true, for population comparisons to be valid. The modeling exercise can also help identify auxiliary data needed to evaluate the assumption of interpopulation homogeneity or to calibrate the indices to absolute measures of animal abundance or density.

What distinguishes population indices from abundance or density estimators is the absence of the prerequisite auxiliary data needed to convert the relative measures to absolute terms. Without the auxiliary data, the assumption of constant proportionality between the index observed and the abundance monitored cannot be tested, nor can the index be converted to absolute abundance. This situation relegates indices, at best, to the role of relative comparisons. This situation also leaves uncertainty whether the indices even provide the ordinal information needed for population comparisons. If the proportionality between the observed indices and the monitored population abundances is not invariant, estimates of relative change in abundance will be biased and inferences to population change invalid.

The nonrobust nature of indices mandates extreme caution in their design, use, and analysis. Even when all precautions have been taken, there will be lingering uncertainty that some crucial design factor was overlooked or not properly considered. Thus, index studies may be adequate for developing working hypotheses or monitoring general trends but insufficient for consummate investigations of critical resource issues.

Estimating Population Abundance

9

Chapter Outline

9.1   Introduction

Population abundance is the culmination of survival, productivity, and migratory processes, both past and present. Animal abundance also reflects the environmental and anthropogenic effects exerted on these processes over time. Thus, population abundance is the ultimate summary of demographic events and the springboard to the future of the population. If past abundance levels characterize the population trend, then current abundance represents its status.

In exploited populations, an estimate of population abundance is usually necessary for interpreting the magnitude of harvest or incidental take. Harvest is usually summarized as the number of animals removed from the population. An abundance estimate can translate harvest numbers into a harvest mortality rate. The joint effects of natural and harvest mortality cannot exceed the reproductive rate without the population declining in a population closed to movement.

In this chapter, we discuss demographic techniques, which largely do not rely on animal tagging. Seber (1982) provides the single best reference on the statistical theory of analyzing mark-recapture and mark-recovery studies. Other references on estimating animal abundance from marking studies include those by Begon (1979), Blower et al. (1981), Davis (1982), and Borchers et al. (2002). There have also been a number of good reviews of abundance techniques for both marking and nonmarking methodology (Cormack 1978, Eberhardt 1978, Pollock 1981, 1991, Seber 1986). There has also been growing interest in recent years in line-transect methods for estimating animal density and abundance, best characterized by Burnham et al. (1980) and Buckland et al. (1993). Estimation methods based on survey counts, change-in-ratios (CIR), catch-effort, and population reconstruction will be the focus of this chapter. In many cases, marking data can be used to augment these methods to provide joint likelihood models, which can yield more precise abundance estimates. Our purpose in this chapter is not to simply provide an alternative to marking methods but instead to provide statistical models for less invasive or opportunistic techniques that may be used in conjunction with modern marking studies.

9.2   Visual Surveys

Visual surveys of animals can provide the least obtrusive way of censusing wild populations. These techniques are also attractive economically. When animals do not need to be captured, handled, or marked, labor and equipment costs are minimized. Visual approaches are also attractive when studying endangered or threatened species, because concerns over incidental or accidental take are minimized. However, the secretive and elusive nature of animals will likely result in undercounts that negatively bias population estimates, severely so in many cases.

To compensate for less than complete enumeration, visual surveys need to adjust observed numbers by an estimate of the probability of detection. By using sufficiently small transect widths, strip transects try to ensure that detection probabilities are close to one. Other methods, such as the bounded-count method (Regier and Robson 1967) and the binomial sampling method of Overton (1969), assume distributional properties for the data that permit inferences to both seen and unseen animals. Yet other methods, such as sightability models (Samuel et al. 1987) and line-transect methods (Buckland et al. 1993), attempt to characterize the detection process to provide correction factors for raw counts. Regardless of the approach, these visual surveys try to estimate absolute animal abundance (N) or density (D) within a prescribed population.

9.2.1   Strip Transects

Strip-transect techniques are a special case of finite sampling methods (Section 8.1.2). The area of inference is subdivided into strip transects of fixed width and length, and a subset of those sample plots is probabilistically selected to make inferences to a population as a whole (Marsh and Sinclair 1989, Miller et al. 1998). The method of selection can be based on a variety of alternative sampling schemes (Section 8.1.2) (Cochran 1977, Thompson 2002).

The key assumption in using strip transects for wildlife surveys is the assertion that all animals within the fixed width of the transect have a detection probability of 1.0. Consequently, within each transect, the animals are enumerated, allowing simple spatial extrapolation to the entire population. The method further requires either of the following assumptions:

1. The animals remain fixed within the area of the strip transects, so animals are not double-counted across replicate transects.
2. The animals redistribute themselves after each strip transect is surveyed; in which case, an average density may be estimated.

The half-width of the strip transect must be selected such that all animals within that distance from the line have absolute assurance of detection. However, sightability generally declines as the distance between the observer and the animal increases. Caughley (1974) described many potential sources of bias for strip-transect methods used in walking and aerial surveys of wildlife, including the following:

As strip width increases:

1. Mean distance between animal and observer increases.
2. Number of obstructions between animal and observer increases.
3. Amount of eye movement to scan strip increases.
4. Time available to locate and recognize an animal decreases.

As travel speed increases:

1. Time available to locate and recognize an animal decreases.
2. Required eye movement increases.

In particular, aerial surveys also have to contend with plane altitude, which further affects sightability. As altitude increases:

1. Mean distance between animal and observer increases.
2. Mean number of obstructions between animal and observer decreases.
3. Required eye movement decreases.

Increases in both strip width and travel speed contribute negative biases to abundance estimates. In aerial surveys, increased altitude contributes to one source of negative bias but reduces negative bias in two other ways. The point Caughley (1974) tried to make was that various design factors may interact to effect bias and that most sources of bias in strip-transect surveys are negative.

In most terrestrial applications of strip transects, it is assumed the animals are physically available for sighting. However, in surveys of marine mammals (McLaren 1961) and burrowing mammals, some fraction of the individuals may be below the surface at time of counting. McLaren (1961) proposed a correction for the abundance estimate from a strip transect based on the probability an animal was available for detection. Define P as the average probability an animal is sighted, given it is within the strip-transect width. Abundance can then be estimated by the formula

$$\hat{N} = \frac{x}{P},$$ (9.1)

where x = number of animals sighted. Abundance estimator (9.1) relies on the ability to estimate the detection probability. McLaren (1961) derived the estimator of P by using information on animal behavior. The probability an animal is observed is modeled as

$$
\begin{aligned}
P &= P(\text{animal observed}) \\
&= P(\text{animal on surface}) + P(\text{animal surfaces}) \\
&= P(\text{animal on surface}) + P(\text{animal surfaces}|\text{submerged}) \\
&\quad \cdot P(\text{animal submerged}).
\end{aligned}
$$ (9.2)

Let

S = the average time an animal spends on the surface;
U = the average time an animal is submerged;

then

$$P(\text{animal on surface}) = \frac{S}{S+U}$$ (9.3)

$$P(\text{animal submerged}) = \frac{U}{S+U}.$$ (9.4)

The conditional probability an animal surfaces given it was submerged is based on a uniform distribution, where

$$P(\text{animal surfaces in } (0,t)|\text{submerged}) = \frac{t}{U}.$$ (9.5)

Evaluation of Eq. (9.2) with Eqs. (9.3) through (9.5) yields

$$
\begin{aligned}
P(\text{animal observed}) &= \frac{S}{S+U} + \frac{t}{U} \cdot \frac{U}{S+U} \\
&= \frac{S+t}{S+U},
\end{aligned}
$$ (9.6)

where t = duration an animal would be visible to an observer.

In practice, the visual detection of marine mammals would be corrected by the average event times for being submerged or on the surface, so that

$$\hat{N} = \frac{x}{\left(\dfrac{\hat{S}+t}{\hat{S}+\hat{U}}\right)} = \frac{x}{\hat{P}}. \tag{9.7}$$

By use of the conditional variance formula,

$$\begin{aligned}
\mathrm{Var}(\hat{N}) &= \mathrm{Var}_P\left[E\left(\frac{x}{\hat{P}}\Big|P\right)\right] + E_P\left[\mathrm{Var}\left(\frac{x}{\hat{P}}\Big|P\right)\right] \\
&\doteq \mathrm{Var}_P\left[\frac{x}{P}\right] + E_P\left[\frac{x^2 \cdot \mathrm{Var}(\hat{P})}{P^4}\right] \\
&\doteq \frac{N(1-P)}{P} + \frac{[N(1-p)+N^2 p]\mathrm{Var}(\hat{P})}{P^3}.
\end{aligned} \tag{9.8}$$

The $\mathrm{Var}(\hat{P})$ can be approximated by the delta method, where

$$\mathrm{Var}(\hat{P}) \doteq \frac{1}{(S+U)^2}\left[(1-P)^2 \cdot \mathrm{Var}(\hat{S}) + \hat{P}^2 \cdot \mathrm{Var}(\hat{U})\right]. \tag{9.9}$$

Eberhardt (1978a) expanded the McLaren (1961) equation for the probability of detecting a marine mammal during the course of a strip transect (Eq. 9.6) by expressing it in terms of vessel velocity. Assuming the course of the ship is randomly positioned relative to the location of the animal, the mean detection probability can be expressed as

$$P = \frac{\pi}{4}\frac{\left(\dfrac{W}{2}\right)}{V(S+U)} + \frac{S}{S+U}, \tag{9.10}$$

where

V = vessel velocity;

$\dfrac{W}{2}$ = half-width of the strip transect.

Discussion of Utility

Strip transects are, in essence, long and narrow sampling units for which the probability of detection is assumed to be one. The strip plots are constructed so that probability of detection is one on either side of the midline as the transect is traversed. Thus, transect width corresponds to the upper shoulder of a detection function. The resulting animal counts are considered to be a complete enumeration of all individuals within the strip plots.

Unfortunately, such enumeration is rarely complete or unbiased. A surveyor is more likely to miss an individual than to imagine or double-count an animal. Thus, the counts are generally negatively biased. The variance formula from finite sampling theory can capture the spatial variability in counts between sampling units but not the systematic error from undercounting. There is insufficient structure to estimate the sampling bias from the survey data alone. For this reason, the results of a strip-transect survey may be better considered as minimal counts or simply indices of abundance. As the detection probability goes to one, the resulting index will converge to the actual animal abundance, leaving only spatial sampling error to be characterized.

9.2.2   Bounded Counts: Robson and Whitlock (1964), Regier and Robson (1967)

Attempts to completely enumerate all individuals within a population often fail despite best efforts. In the situation in which it is conceivable that all animals could be counted during a single canvass of the population, that animals are not counted more than once during a survey event, and that the process can be independently repeated, Regier and Robson (1967) proposed a bounded-count method. The abundance estimator is based on the theory of estimating a truncation point by Robson and Whitlock (1964). Letting N denote the true abundance and m the number of times the population is canvassed, the bounded-count estimator is

$$\hat{N} = x_m + (x_m - x_{m-1})$$
$$= 2x_m - x_{m-1}, \tag{9.11}$$

where

x_m = largest of the m counts obtained;

x_{m-1} = second largest count obtained.

The estimator is based the jackknife method of Quenouille (1956) and has bias of order $\frac{1}{m^2}$. For example, if only two counts are performed, the bias is of order $\frac{1}{4}$, whereas the bias is of order 0.01 when $m = 10$ counts.

An approximate $(1 - \alpha)100\%$ confidence interval for N is bounded below by x_m (i.e., the largest count value) and above by the value

$$\frac{1}{\alpha}(x_m - (1-\alpha)x_{m-1}).$$

The confidence interval method is exact if the counts are from a uniform distribution. Alternatively, if $f(x)$ is an increasing (decreasing) function, then the actual confidence interval coverage is greater (less) than $1 - \alpha$ (Robson and Whitlock 1964).

Assumptions

The assumptions of the bounded-count method include the following:

1. Probability of detection is sufficiently high that it is possible all individuals could be enumerated.
2. The m counts are independent.
3. The probability of detection is constant across all replicate count surveys.
4. Animals are not counted more than once during a canvass.
5. The population is closed during the course of the multiple counts.

This method can be applied in situations in which numerous surveyors have the ability to count the same population concurrently with equal detection. Alternatively, a single surveyor could perform multiple counts of the same population over time if the population is closed during the course of the survey and the detection process is stationary.

Example 9.1: Bounded-Count Method, Brown Bears (*Ursus arctos*), Alaska

Overton (1969) presents an example of the bounded-count method for brown bears along a stream in Alaska (Erickson and Siniff 1963). During a 9-day period, 27 separate counts of bears were performed. The counts were: 94, 67, 118, 81, 16, 34, 62, 40, 91, 81, 43, 95, 65, 44, 113, 86, 48, 70, 54, 29, 76, 54, 30, 72, 61, 18, and 76. The 27 counts yielded values of $x_{27} = 118$ and $x_{26} = 113$, resulting in an abundance estimate (Eq. 9.11) of

$$\hat{N} = 2x_m - x_{m-1}$$
$$= 2(118) - 113 = 123.$$

The 90% confidence interval is calculated with a lower bound of $x_{27} = 118$ and an upper bound of

$$\frac{1}{\alpha}(x_m - (1 - \alpha x_{m-1}))$$

$$\frac{1}{0.10}(118 - (1 - 0.10)113) = 163,$$

such that the CI($118 \leq N \leq 163$) = 0.90.

Discussion of Utility

The key assumption of the bounded-counts method is that detection probabilities are sufficiently high that it is possible to enumerate all individuals during a single canvass of the population. Only in a few circumstances will this assumption be met. Counts of colonial species at rookeries or breeding grounds are one such possible application. Other applications may include carefully conducted territorial mapping of songbirds (Falls 1981, Ferry et al. 1981). When done properly, territorial mapping is expected to have a detection probability of one. Yet other applications may exist in which the animals make their presence conspicuous, for example, counts of prairie-chickens (*Tympanuchus* spp.) on leks during the breeding season. Caution is needed, however, to ensure individual animals are not double-counted because of movement or the inability to recognize individuals. Counting the number of killer whales (*Orcinus orca*) in a specific pod through repeated breaching and diving episodes may be yet another application.

9.2.3   Binomial Method-of-Moments (Overton 1969)

Overton (1969) attributed an abundance estimator based on replicated counts to W. R. Hanson and D. G. Chapman. Interestingly, Doug Chapman denied any involvement in the development of the technique (D. G. Chapman, personal communication). The technique is similar to the previous bounded-count method (Section 9.2.2), in that it assumes independent and identically distributed replicate counts from a population. However, the technique adds the additional assumption that each animal has an independent and equal probability of detection so the various counts are binomially distributed with parameters N and p.

Therefore, let a population survey consist of m replicate counts, where

x_i = number of animals observed during the ith replicate count;
N = actual population abundance;
p = probability of detecting an animal during a single replicate count;
m = number of replicate counts.

The animal counts are then assumed to be independent and identically distributed binomial variables with the joint likelihood

$$L(N, p|\underset{\sim}{x_i}) = \prod_{i=1}^{m} \binom{N}{x_i} p^{x_i} (1-p)^{N-x_i}. \tag{9.12}$$

Based on the binomial distribution,

$$E(x_i) = Np \text{ with } \text{Var}(x_i) = Np(1-p).$$

Furthermore,

$$E(\bar{x}) = Np \tag{9.13}$$

and

$$E(s_{x_i}^2) = Np(1-p), \tag{9.14}$$

where

$$s_{x_i}^2 = \frac{\sum_{i=1}^{m}(x_i - \bar{x})^2}{(m-1)},$$

$$\bar{x} = \frac{1}{m}\sum_{i=1}^{m} x_i.$$

Setting $\bar{x}$ and $s_{x_i}^2$ equal to their expected values and solving for N and p leads to the method-of-moment estimators

$$\hat{N} = \frac{\bar{x}^2}{\bar{x} - s_{x_i}^2} \tag{9.15}$$

and

$$\hat{p} = \frac{\bar{x} - s_{x_i}^2}{\bar{x}}. \tag{9.16}$$

Overton (1969) does not provide a variance estimator, and one will not be presented here. The estimator (9.15) has poor distributional properties, with large sampling variance and biased point estimates.

To the first term of a Taylor series expansion

$$E(\hat{N}) \doteq \frac{E(\bar{x}^2)}{E(\bar{x}) - E(s_{x_i}^2)}$$

$$\doteq \frac{\text{Var}(\bar{x}) + E(\bar{x})^2}{E(\bar{x}) - E(s_{x_i}^2)}$$

$$\doteq \frac{\dfrac{Np(1-p)}{m} + (Np)^2}{Np - Np(1-p)}$$

$$\doteq N + \frac{(1-p)}{pm}.$$

Hence, the Overton (1969) estimator is first-order, positively biased.

The likelihood (9.12) yields the maximum likelihood estimator (MLE) of

$$\hat{p} = \frac{\bar{x}}{\hat{N}}, \tag{9.17}$$

where $\hat{N}$ is the value of N that satisfies the equality

$$\prod_{i=1}^{m}\left(1 - \frac{x_i}{N}\right) = \left(1 - \frac{\bar{x}}{N}\right)^m. \tag{9.18}$$

The solution to Eq. (9.18) is MAX (x_i), the largest observed value of x_i. In this case, the MLE is not a consistent or unbiased estimator of N.

Assumptions

The assumptions of the binomial estimator include the following:

1. Each animal has an equal and independent probability of detection during a survey count.
2. The m counts are independent.
3. The probability of detection is constant across all m replicate survey counts.
4. Animals are not counted more than once during a single survey count.
5. The population is closed during the course of the multiple survey counts.

The assumptions of independent and identical probabilities of detection within and between surveys counts are essential for the survey counts to have the same binomial distribution, leading to Eqs. (9.13) and (9.14). Environmental conditions need to be stable and animal behavior undisturbed over the course of the replicate counts for these assumptions to be true. Further, detection probabilities must be homogeneous among the individual animals in the population. Investigators have the least control of this last factor. High detection rates can minimize to some degree the effect of detection heterogeneity on the abundance estimator.

For the Overton (1969) estimator to be statistically well behaved, a large number of replicate counts (i.e., $m > 20$) with high detection probabilities (i.e., $p > 0.60$) are necessary. Unless these survey conditions are met, the abundance estimator may be severely biased and possess a large sampling variance.

Example 9.2: Binomial Abundance Estimator, Brown Bears, Alaska

By using brown bear data (Example 9.1), the 27 replicate counts have a mean of $\bar{x} =$ 63.63 with associated sampling variance of $s_{x_i}^2 = 728.86$. These survey results yield a negative abundance estimate (Eq. 9.15), because $s_{x_i}^2 > \bar{x}$. The sampling variance in this example is too large for the individual x's to be independently and identically distributed binomial random variables. In the case of independently and identically distributed binomial random variables,

$$\text{Var}(x_i) = Np(1-p) = E(x) \cdot (1-p)$$

or, in other words,

$$\text{Var}(x_i) \le E(x_i).$$

Clearly, this is not true for this brown bear example.

Discussion of Utility

The binomial detection model of Overton (1969) should be used only for the most basic of preliminary surveys. Although the method-of-moment estimator (Eq. 9.15) can provide what may seem to be reasonable point estimates, the MLE is simply the maximum animal count observed. Thus, under the best of circumstances, the survey model is not well defined.

In practice, detection probabilities will likely vary between replicate counts. The result is an inflation in the value of $s_{x_i}^2$, ultimately leading to negatively biased abundance estimates. The bounded-count method (Section 9.2.2) is recommended, in lieu of the binomial sampling model, in all circumstances in which the two approaches may be applicable.

9.2.4   Sightability Models

There is some probability of missing animals in any visual survey, thereby underestimating population size. When animals occur in groups, their visual detection is further complicated because the probability of detection is likely a function of group size (Samuel et al. 1987). Hence, size bias sampling may exist (Thompson 2002:51–53), which further affects abundance estimation.

Population abundance can be estimated if the relationship between the probability of detection and the size of the group can be established. A detection function based on group size and other environmental factors can be calculated, using information derived from radio-tagged animals (Samuel et al. 1987, Unsworth et al. 1990, White and Garrott 1990, Bodie et al. 1995, Anderson et al. 1998, White and Shenk 2001). The probability of detecting an animal group is typically modeled by using a logistic function, where

$$\hat{p}_i = \frac{e^{x_i'\hat{\beta}}}{1 + e^{x_i'\hat{\beta}}} \tag{9.19}$$

and where

$\hat{p}_i$ = estimated probability of detecting the ith animal group;

$\underset{\sim}{x}_i$ = vector of covariates characterizing the conditions of the ith group and its surrounding environment that might affect sightability;

$\hat{\underset{\sim}{\beta}}$ = vector of regression coefficients.

Noting the logistic transformation

$$\ln\left(\frac{\hat{p}_i}{1 - \hat{p}_i}\right) = \underset{\sim}{x}_i'\hat{\underset{\sim}{\beta}}, \tag{9.20}$$

where $\underset{\sim}{x}_i'\hat{\underset{\sim}{\beta}}$ = a linear predictor, generalized linear models (GLM) can be used to estimate the predictive equation (9.19) based on a binomial error structure (more particularly, Bernoulli error structure) and a logit-link (McCullagh and Nelder 1983, Aitkin et al. 1989).

To derive a model for the probability of detection, n radio-tagged animals are released into the population and allowed to mix. Aerial surveys are conducted to subsequently canvass the area. One air crew without the aid of radio receivers canvasses the area to locate animals. For those animal groups detected, group size and associated covariates, such as percentage tree cover, habitat type, animal composition, aspect, and cloud cover, are recorded. A second air crew with the aid of radio receivers also locates animal groups missed by the first survey crew. The first survey crew is assisted in locating these missed groups, and group size and associated covariates are again recorded. With sufficient observations, conditions under which animal groups were and were not observed are characterized. By using data from both successful and unsuccessful trials, GLM is used to construct the sightability model. As few as 30 and as many as several hundred trials are commonly conducted in developing the sightability model. Consequently, a substantial initial investment in radio-tags and aerial surveys is necessary in establishing a reliable sightability model.

With a model for estimating the detection probabilities in hand, animal abundance can be estimated based solely on the groups of animals observed during an aerial survey. The estimate of total abundance (N) is based on the Horwitz-Thompson (Cochran 1977:259–261) estimator, where

$$\hat{N} = \sum_{i=1}^{n} \frac{y_i}{p_i},$$

and

y_i = number of animals sighted in the ith group ($i = 1, \ldots, n$);

n = number of animal groups sighted during the survey;

p_i = probability of detecting the ith group ($i = 1, \ldots, n$).

However, unlike the standard Horwitz-Thompson estimator, the probabilities of inclusion are estimated rather than known. Thus, the abundance estimator is written as

$$\hat{N} = \sum_{i=1}^{n} \frac{y_i}{\hat{p}_i} \tag{9.21}$$

or, substituting in the logistic probability expression,

$$\hat{N} = \sum_{i=1}^{n} \left[y_i \left(1 + e^{-x_i \hat{\beta}} \right) \right]. \tag{9.22}$$

Using the asymptotic normality of the MLE leads to

$$\underset{\sim}{\beta} \sim N(\underset{\sim}{\beta}, \underset{\sim}{\Sigma})$$

where $\underset{\sim}{\Sigma}$ = variance-covariance matrix for the regression coefficients ($\hat{\beta}$). From Evans et al. (1993), $e^{-x_i \hat{\beta}}$ has a log-normal distribution, and

$$E\left(1 + e^{-x_i \hat{\beta}}\right) = 1 + e^{-x_i \beta + \frac{1}{2} x_i \sum x_i}.$$

Steinhorst and Samuel (1989:421) suggest an unbiased estimator for $\dfrac{1}{p_i}$ (i.e., $\hat{\theta}_i$), where

$$\hat{\theta}_i = 1 + e^{-x_i \hat{\beta} - \frac{1}{2} x_i \hat{\Sigma} x_i}, \tag{9.23}$$

resulting in an unbiased abundance estimator,

$$\hat{N} = \sum_{i=1}^{n} y_i \hat{\theta}_i. \tag{9.24}$$

The variance of $\hat{N}$ can be more readily found by redefining estimator (9.24) as

$$\hat{N} = \sum_{i=1}^{G} R_i y_i \hat{\theta}_i,$$

where

$$R_i = \begin{cases} 1 & \text{if the } i\text{th group is detected} \\ 0 & \text{therwise} \end{cases}$$

and where G = total number of animal groups (i.e., herds of size 1 or larger) in the population. The variance of $\hat{N}$ can then be found in stages by using the conditional variance formula

$$\text{Var}(\hat{N}) = \text{Var}_2 \left[E_1 \left(\sum_{i=1}^{G} R_i y_i \hat{\theta}_i \Big| 2 \right) \right] + E_2 \left[\text{Var}_1 \left(\sum_{i=1}^{G} R_i y_i \hat{\theta}_i \Big| 2 \right) \right],$$

where 1 denotes the estimation of p_i and 2 denotes the sampling of the population. The variance is approximated by

$$\text{Var}(\hat{N}) \doteq \sum_{i=1}^{G} \left[\frac{(1 - p_i) y_i^2}{p_i} \right] + \sum_{i=1}^{G} \left[p_i y_i^2 \text{Var}\left(\frac{1}{\hat{p}_i} \right) \right] + 2 \sum_{i=1}^{G} \sum_{\substack{j=1 \\ i<j}}^{G} \left[p_i y_i p_j y_j \text{Cov}\left(\frac{1}{\hat{p}_i}, \frac{1}{\hat{p}_j} \right) \right] \tag{9.25}$$

which is over all G groupings in the population. The variance can be estimated by noting,

for example, $\sum_{i=1}^{G} \frac{(1-p_i)y_i^2}{p_i}$, which can be estimated by $\sum_{i=1}^{n} \left[\frac{\frac{(1-\hat{p}_i)y_i^2}{\hat{p}_i}}{\hat{p}_i} \right]$, or by using the

unbiased estimator $\frac{1}{p_i}$ from Eq. (9.23),

$$\sum_{i=1}^{n} (1-\hat{p}_i)\hat{\theta}_i^2 y_i^2 = \sum_{i=1}^{n} \left(1 - \frac{1}{\hat{\theta}_i}\right)\hat{\theta}_i^2 y_i^2.$$

The other term of Eq. (9.25) can be estimated similarly, so that

$$\widehat{\text{Var}}(\hat{N}) = \sum_{i=1}^{n} \left(1 - \frac{1}{\hat{\theta}_i}\right)\hat{\theta}_i^2 y_i^2 + \sum_{i=1}^{n}\sum_{j=1}^{n} y_i y_j \widehat{\text{Cov}}(\hat{\theta}_i, \hat{\theta}_j), \tag{9.26}$$

which is equivalent to the variance given by Steinhorst and Samuel (1989) for the case of a single population stratum. It should be noted that when $i = j$, $\widehat{\text{Cov}}(\hat{\theta}_i,\hat{\theta}_j)$ is the $\widehat{\text{Var}}(\hat{\theta}_i)$ by definition. By using the properties of a log-normal distribution from Evans et al. (1993),

$$\text{Var}(\hat{\theta}_i) = e^{-2x_i'\beta}\left(e^{x_i'\Sigma x_i} - 1\right)$$

and

$$\text{Cov}(\hat{\theta}_i,\hat{\theta}_j) = e^{-(x_i+x_j)'\beta}\left(e^{x_i'\Sigma x_j} - 1\right).$$

Steinhorst and Samuel (1989:421) give a biased-corrected statistic for the covariance, where

$$\widehat{\text{Cov}}(\hat{\theta}_i,\hat{\theta}_j) = e^{-(x_i+x_j)'\hat{\beta} - \frac{1}{2}(x_i+x_j)'\Sigma(x_i+x_j)}\left(e^{x_i'\Sigma x_j} - 1\right). \tag{9.27}$$

The variance of $\hat{\theta}_i$ is estimated by the quantity

$$\widehat{\text{Var}}(\hat{\theta}_i) = e^{-2x_i'\hat{\beta}}\left(e^{x_i'\Sigma x_i} - 1\right). \tag{9.28}$$

Oh and Scheuren (1983) and Steinhorst and Samuel (1989) extend this method to the situation in which the landscape is probabilistically subsampled. Here, we have assumed the entire location has been canvassed and all animal groups have positive probabilities of detection as described by p_i (Eq. 9.19).

Assumptions

The assumptions of the logistic sightability model include the following:

1. The entire locale is canvassed, and all animal groups have a positive probability of detection.
2. The probability of detection is correctly modeled by the fitted logistic model (Eq. 9.19).
3. All individuals within a detected group are completely enumerated.
4. Separate groups have independent probabilities of detection.

5. The detection model is temporally invariant and correctly describes the detection processes of the current survey.

The logistic detection model is generally constructed by using stepwise regression techniques to identify the best-fit, parsimonious model. Omission of a necessary variable, as with all multiple regression models, will result in estimation bias, whereas inclusion of unnecessary variables will inflate the variance estimates. In application, it is assumed the fitted model remains applicable after development and during actual population surveys (White and Shenk 2001). Thus, sightability models are typically limited to the season and locale where the model was initially developed. Periodic releases of additional radio-tagged animals and additional sightability trials may be used to ensure the logistic model remains relevant over time.

The abundance estimator (Eq. 9.21) will be negatively biased if all animals within a group are not enumerated when detected. There is no mechanism within the technique to account for less than 100% detection of the animals within the groups detected. Similarly, all animal groups within the locale of inference must be potentially capable of detection. This implies the entire area must be canvassed for animal groups. Otherwise, the abundance estimator (Eq. 9.21) is negatively biased. Therefore, sufficient effort must be applied for 100% aerial coverage. If the entire area cannot be canvassed, the area of inference should be probabilistically subsampled and the sightability model applied to each spatial sampling unit. Steinhorst and Samuel (1989) provided generic estimators and a variance formula for spatial subsampling the area of inference followed by application of the sightability model to each unit selected.

Example 9.3: Sightability Model and Hypothetical Elk (*Cervus elaphus*) Data

A simple and hypothetical data set will be used to illustrate the analysis of the sightability model. A training data set of 25 herd observations was used to model the probability of detection (Table 9.1, Fig. 9.1) as a simple function of herd size. The fitted logistic model based on the binary detected or not detected (i.e., 0, 1) observations is

$$\ln\left(\frac{p_i}{1-p_i}\right) = -1.106 + 0.2035x_i, \tag{9.29}$$

where x_i = herd size. The estimated variance-covariance matrix for the regression coefficients $\hat{\beta}_0$ and $\hat{\beta}_1$ is

$$\hat{\Sigma} = \begin{bmatrix} 0.4714 & -0.0536 \\ -0.0536 & 0.0108 \end{bmatrix}.$$

Bias-adjusted estimates of $\hat{\theta}_i$ are then Eq. (9.23)

$$\hat{\theta}_i = 1 + e^{(-1.106+0.2035x_i)-\frac{1}{2}[1,x_i]\begin{bmatrix} 0.4714 & -0.0536 \\ -0.0536 & 0.0108 \end{bmatrix}\begin{bmatrix} 1 \\ x_i \end{bmatrix}}.$$

During the actual survey, $n = 7$ groups of elk (below) were detected with corresponding values of $\hat{\theta}_i$:

Group	Size	$\hat{\theta}_i$
y_1	7	1.6418
y_2	1	3.0435
y_3	3	2.4502
y_4	15	1.0750
y_5	2	2.7308
y_6	5	1.9858
y_7	1	3.0435

Total abundance is estimated to be

$$\hat{N} = \sum_{i=1}^{n} y_i \left(\hat{\theta}_i \right)$$
$$= 7(1.6418) + 1(3.0435) + \cdots + 1(3.0435)$$
$$= 56.45.$$

The variance of $\hat{N}$ is estimated by using Eq. (9.26), along with Eqs. (9.27) and (9.28), where, for example,

$$\widehat{\operatorname{Var}}\left(\hat{\theta}_1 \right) = 0.0907,$$
$$\widehat{\operatorname{Var}}\left(\hat{\theta}_2 \right) = 1.3058,$$
$$\operatorname{Cov}\left(\hat{\theta}_1, \hat{\theta}_2 \right) = 0.1461,$$

leading to $\widehat{\operatorname{Var}}\left(\hat{N} \right) = 274.394$ or a standard error of $\widehat{\operatorname{SE}}\left(\hat{N} \right) = 16.56$.

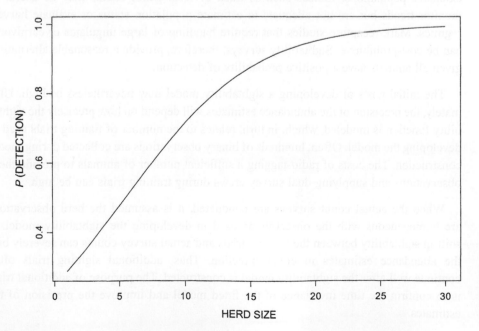

Figure 9.1. Fitted logistic model to the hypothetical elk training data for sightability. The logistic model is solely a function of herd size.

Table 9.1. Training data used to model the probability of
detecting elk groups as a function of herd size, whether the group
was detected or not, and the fitted probability of detection from
the logistic model (Eq. 9.29) for each of the training trials.

Herd size	Observed? (1 = yes, 0 = no)	Fitted probability of observations
1	0	0.2886
1	0	0.2886
1	1	0.2886
1	0	0.2886
1	0	0.2886
1	0	0.2886
2	1	0.3321
2	1	0.3321
2	0	0.3321
2	0	0.3321
3	0	0.3787
3	1	0.3787
3	1	0.3787
4	0	0.4276
8	1	0.6278
8	0	0.6278
8	1	0.6278
9	1	0.6740
9	0	0.6740
10	0	0.7170
12	1	0.7920
12	1	0.7920
17	1	0.9133
19	1	0.9406
28	1	0.9900

Discussion of Utility

There are few options for obtaining reliable abundance estimates for geographically extensive populations. Indices such as track or fecal pellet counts may be useful to monitor trends but are not adequate to estimate population status or evaluate harvest regimes. Mark-recapture studies that require handling of large ungulates or carnivores can be cost-prohibitive. Sightability surveys, therefore, provide a reasonable alternative, given all animals have a positive probability of detection.

The initial costs of developing a sightability model may nevertheless be high. Ultimately, the precision of the abundance estimates will depend on how precisely the sightability function is modeled, which, in turn, relates to the number of training trials used in developing the model. Often, hundreds of binary observations are collected during model construction. The costs of radio-tagging a sufficient number of animals to provide these observations and supplying dual survey crews during training trials can be high.

When the actual count surveys are conducted, it is assumed the herd observations are homogeneous with the observations used in developing the sightability model. A shift in sightability between the training trials and actual survey counts can severely bias the abundance estimates in either direction. Thus, additional sighting trials often continue well after the sightability model is constructed. The purpose of additional trials is to confirm the time invariance of the fitted model and improve the precision of the estimates.

Developed sightability models also tend to be site or location specific. Location-specific differences in habitat, terrain, or animal behavior may invalidate an existing model developed elsewhere (White and Shenk 2001). Training trials, at the very least, within the new locale are necessary to either confirm or refute the spatial invariance of the existing sightability model. In very large areas, a subsample of the region may be probabilistically selected for sightability surveys. Steinhorst and Samuel (1989) extended the development of the sightability model to spatial subsampling. Within those areas selected for canvassing, it is assumed the entire area is traversed and all animal groups have a positive chance of detection. Incomplete canvassing of a selected area will negatively bias abundance estimates. Careful planning and implementation of aerial canvassing is therefore an essential element of a viable sightability survey.

9.2.5   Sight-Resight Method

In situations in which individual animals are readily observed, differentiable, and reasonably static, a variation on the single mark-recapture method (Petersen 1896, Lincoln 1930) may be used to estimate abundance based on two independent visual surveys of a population. During each survey, the number of animals observed is enumerated along with the mapping of the animal locations. After completion of the two independent surveys, observers compare notes to identify those animals detected by either or both of the surveys. Geibel and Miller (1984) and Estes and Jameson (1988) used these techniques to estimate the abundance of sea otters (*Enhydra lutris*) based on canvasses performed by ground and aerial crews. Anthony et al. (1999) used this technique to estimate the abundance of bald eagles (*Haliaeetus leucocephalus*) based on canvasses performed by aerial and combined boat-and-vehicle crews. They called their estimation approach a double-survey method. Other applications have included surveys of African elephants (*Loxodonta africana*) (Caughley 1974), feral horses (*Equus caballus*) and burros (*Equus asinus*) in Australia (Graham and Bell 1989), and osprey (*Pandion haliaetus*) nests (Henny et al. 1977).

The positions of the animals being canvassed should be static, so that both survey crews can map and later relate detections to specific individuals. Consequently, the more the animals move, the greater the need for near coincident surveys. In practice, the two independent surveys need to be conducted almost simultaneously.

The abundance estimator is based on the hypergeometric likelihood for sampling without replacement, where

$$L(N|n_1,n_2,m) = \frac{\binom{n_1}{m}\binom{N-n_1}{n_2-m}}{\binom{N}{n_2}},$$

and

n_1 = number of distinct animals observed by the first survey crew;
n_2 = number of distinct animals observed by the second survey crew;
m = number of animals that were jointly observed by both survey crews.

Although the MLE is

$$\hat{N} = \frac{n_1 n_2}{m},$$

the Chapman (1951) bias-corrected estimator is recommended, where

$$\hat{N} = \frac{(n_1 + 1)(n_2 + 1)}{(m + 1)} - 1, \tag{9.30}$$

which is unbiased when $n_1 + n_2 > N$ (i.e., $m \geq 1$). The variance estimator for $\hat{N}$ is

$$\widehat{\text{Var}}(\hat{N}) = \frac{(n_1 + 1)(n_2 + 1)(n_1 - m)(n_2 - m)}{(m + 1)^2 (m + 2)}. \tag{9.31}$$

The probability of an animal being detected by the first crew (p_1) is estimated by

$$\hat{p}_1 = \frac{m}{n_2}. \tag{9.32}$$

whereas the detection probability for the second survey crew (p_2) is estimated by

$$\hat{p}_2 = \frac{m}{n_1}. \tag{9.33}$$

The exact variance of $\hat{p}_1$ is

$$\text{Var}(\hat{p}_1) = \frac{n_1(N - n_1)(N - n_2)}{n_2 N^2 (N - 1)}, \tag{9.34}$$

with an associated variance estimator of

$$\widehat{\text{Var}}(\hat{p}_1) = \frac{(n_1 - m)(n_2 - m)m}{n_2^2(n_1 n_2 - m)}. \tag{9.35}$$

The variance of $\hat{p}_2$ is

$$\text{Var}(\hat{p}_2) = \frac{n_2(N - n_1)(N - n_2)}{n_1 N^2 (N - 1)}, \tag{9.36}$$

with an associated variance estimator of

$$\widehat{\text{Var}}(\hat{p}_2) = \frac{(n_1 - m)(n_2 - m)m}{n_1^2(n_1 n_2 - m)}. \tag{9.37}$$

Assumptions

Because the sight-resight data will be treated as mark-recapture data, the assumptions of the Lincoln/Petersen single mark-recapture method must hold. These assumptions include the following:

1. The population is closed, with no immigration or emigration during the course of the study.
2. All animals have an equal probability of detection within a survey approach.

3. Detection of an animal by one crew has no effect on the detection of that animal by the other survey crew.
4. All animals are correctly identified as resighted or non-resighted individuals.

The great weakness of this technique is the need to relate detections to specific individuals in order to identify their "recapture" status. Careful mapping of animal positions and near-simultaneous double canvassing of the population are essential for valid estimation. Misidentification of an animal's "recapture" or "resighting" status can bias abundance estimates either high or low. False claims of resighting would bias abundance estimates downward, whereas failure to recognize resighted animals would bias estimates upward. Geibel and Miller (1984) found the aerial and ground crews were not necessarily canvassing the same sea otters. Aerial surveys were locating animals seaward and out of the view of cliff- or ground-based crews. The result was a negative bias in the estimate of the combined near- and offshore population.

Example 9.4: Sight-Resight Survey, Bald Eagles, Lower Columbia River

The November 5, 1984, bald eagle survey along the lower Columbia River shorelines of Oregon and Washington states conducted by Anthony et al. (1999) is used to illustrate the sight-resight method. During that survey, the aerial crew sighted $n_1 = 15$ eagles and the ground/boat crew sighted $n_2 = 20$ eagles. Of the birds seen, $m = 10$ birds were identified as having been sighted by both crews. By using Eq. (9.30), the total eagles abundance is estimated to be

$$\hat{N} = \frac{(n_1 + 1)(n_2 + 1)}{(m + 1)} - 1$$

$$= \frac{(15 + 1)(20 + 1)}{(10 + 1)} - 1 = 29.55 \text{ eagles.}$$

The variance estimate is calculated from Eq. (9.31), where

$$\widehat{\text{Var}}(\hat{N}) = \frac{(n_1 + 1)(n_2 + 1)(n_1 - m)(n_2 - m)}{(m + 1)^2 (m + 2)}$$

$$= \frac{(15 + 1)(20 + 1)(15 - 10)(20 - 10)}{(10 + 1)^2 (10 + 2)} = 11.57,$$

or an associated standard error of $\widehat{\text{SE}}(\hat{N}) = 3.40$.

The probability an eagle was detected by the aerial crew is estimated to be (Eq. 9.32)

$$\hat{p}_1 = \frac{m}{n_2}$$

$$= \frac{10}{20} = 0.50.$$

The variance estimate associated with $\hat{p}_1$ is calculated by using Eq. (9.35), where

$$\widehat{\text{Var}}(\hat{p}_1) = \frac{(n_1 - m)(n_2 - m)m}{n_2^2(n_1 n_2 - m)}$$

$$= \frac{(15 - 10)(20 - 10)10}{(20)^2(15 \cdot 20 - 10)} = 0.00431$$

or an estimated standard error of $\widehat{\text{SE}}(\hat{p}_1) = 0.0657$. For the ground/boat canvass, the probability of detecting an eagle is estimated by Eq. (9.33) to be

$$\hat{p}_2 = \frac{m}{n_1}$$

$$= \frac{10}{15} = 0.6667$$

with an associated standard error of $\widehat{\text{SE}}(\hat{p}_2) = 0.0875$.

Discussion of Utility

The sight-resight method is able to avoid the assumption of constant detection probabilities in the binomial method-of-moment estimator (Section 9.2.3) and the parametric modeling of the detection function in the sightability model (Section 9.2.4). The ability to avoid these model constraints is possible because of the presence of the resighting data. The sight-resight data permit estimation of detection probabilities that permit calibrating the sighting index to absolute abundance. By letting the sighting index be the number of unique animals seen by either crew, i.e., $I = n_1 + n_2 - m$, absolute abundance is then estimated by the expression

$$\hat{N} = \frac{I}{\hat{P}},$$

where $\hat{P} = $ an estimate of the overall probability an animal is detected during the course of the population survey. In the case of a single mark-recapture method, the overall probability of detection is $P = 1 - (1 - p_1)(1 - p_2)$, and can be estimated as a function of Eqs. (9.32) and (9.33). Thus, the abundance estimator is

$$\hat{N} = \frac{n_1 - n_2 - m}{1 - \left(1 - \frac{m}{n_2}\right)\left(1 - \frac{m}{n_1}\right)} = \frac{n_1 n_2}{m},$$

which is the MLE from the single mark-recapture model.

The sight-resight method is ideally suited for surveying sessile objects in the environment. Henny et al. (1977) applied the approach to estimate the number of osprey nests. Liao (1994) used the method to estimate the number of salmon (*Oncorhynchus* spp.) carcasses along the shores of a stream. This approach is advisable for collecting any index for which incomplete counts may occur. The method, however, becomes more difficult to apply and ultimately degrades and becomes inapplicable, as targets of the survey become more mobile relative to the timing of the double survey. Simultaneous counts of animals by the two survey crews are necessary when studying mobile species. Global positioning system (GPS) and careful mapping of prominent land features may be necessary to help identify detections to individual animals. However, the method becomes more difficult to apply as animals become more aggregated. The method seems best suited to solitary species, including some raptors that typically occur alone. Robson and Regier (1964) recommend the expected number of mark-recaptures (*m*) in a Lincoln index be greater than four to minimize the bias of the estimator (9.30) and variance estimator (9.31).

The sight-resight method may be inappropriate for marine mammals or burrowing species. A simultaneous canvassing of such a population by two survey crews can only estimate the abundance of the animals at the surface. To account for submerged or below-ground individuals, the two canvasses of a sight-resight study would need to be staggered in time. The time lag would depend on the duration and fraction of time an animal would be on the surface and available for detection. However, the larger the time lag between the two population canvasses, the more difficult it becomes to relate detections back to specific individuals. Hence, the competing demands of simultaneous as well as staggered canvasses of the population make the sight-resight method unmanageable for marine and burrowing species.

9.3   Line Transects

Line-transect sampling provides a useful and viable alternative to mark-recapture methods of estimating annual abundance. Leopold (1933) attributes the earliest line-transect applications to the unpublished works of R. T. King, who studied ruffed grouse (*Bonasa umbellus*). The method was quickly adapted to surveys of white-tailed deer (*Odocoileus virginianus*) (Fisher 1939) and snowshoe hare (*Lepus americanus*) (Webb 1942). Since then, the technique has been adapted to survey passerine birds (Emlen 1971), raptors (Andersen et al. 1985), marine mammals (Laake 1981), desert tortoises (*Gopherus agassizii*) (Anderson et al. 2001), reef fishes (Bergstedt and Anderson 1990), crustaceans (Patil et al. 1979), and primates (Brockelman 1980).

The appeal of line-transect methods is the ability to canvass large areas with relatively low levels of effort and without needing to handle animals. Thus, the method has a distinct appeal in the study of hard or dangerous-to-handle species and endangered species. Another advantage of line-transect methods is the avoidance of the equal detection assumption common to mark-recapture models. However, to avoid an assumption of a random spatial pattern of individuals, line transects must be randomly located. A systematic pattern of transect placement risks correlation with a systematic pattern in animal abundance and an underestimation of sampling variance.

There are basically two conceptual frameworks for line-transect sampling. The earliest approaches were based on fixed-distance flush surveys. Here, detection of a mobile animal largely depends on the animal's reaction to the approaching surveyor. The animal may jump or "flush," revealing its presence and position. Key measurements are the number of animals flushed and the straight-line distance between the observer and the animals, also known as the radial distance or sighting distance. The second conceptual framework includes the right-angle methods described at length by Burnham et al. (1980) and Buckland et al. (1993, 2001, 2004). Here, detection depends mainly on the observer locating the animal without the help of a flush response. The detection process largely

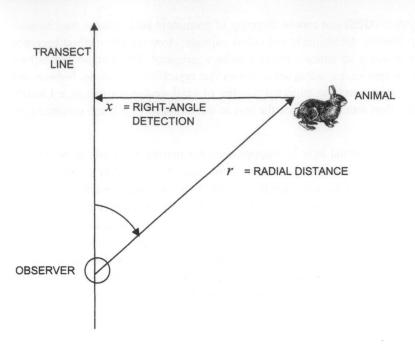

Figure 9.2. Relevant measures taken during line-transect sampling.

depends on visual detection by the searcher. The appropriate measures are the number of animals observed and their right-angle distances to the transect line (Fig. 9.2).

From a logistical perspective, the sighting angle (θ) and the straight-line sighting distance (r) are the easiest to measure during a line-transect survey (Fig. 9.2). The right-angle distance (x) is calculated from the relationship

$$\sin(\theta) = \frac{x}{r}$$

or

$$r\sin(\theta) = x.$$

It is generally assumed in the practice of abundance or density estimation that the distance data collected are measured without error.

9.3.1   Fixed-Distance Methods

The fixed-distance method of Hayne (1949a) is based on geometric probabilities of detection. Each animal is conceptualized as having a unique but fixed flushing distance. Regardless of the angle of approach, the animal is expected to flush when the observer is within distance r_i (Fig. 9.3). Therefore, there is a circular detection zone of diameter $2r_i$ that can vary from one animal to the next.

For simplicity, assume a rectangular area of size L by W is to be surveyed with random placed transects of length L across the baseline W (Fig. 9.3). The probability an animal with flushing radius r_i being detected by a randomly placed transect line is

$$P_i = \frac{2r_i}{W}.$$

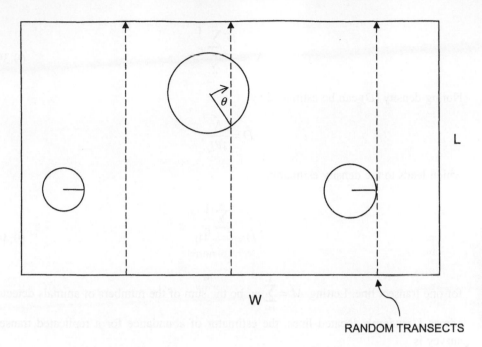

L

W

RANDOM TRANSECTS

Figure 9.3. Schematic of a fixed-distance, line-transect survey with three lines over an area of size $L \times W$. Detection zones of three animals are illustrated with unique radii r_j.

The expected number of animals detected (m) on the transect line is

$$E(m) = \sum_{i=1}^{N} P_i$$

$$= \sum_{i=1}^{N} \frac{2r_i}{W}$$

$$= \frac{2\mu_r N}{W}, \qquad (9.38)$$

where

N = animal abundance;
μ_r = average flushing radius;
m = number of animals detected.

The method-of-moment estimator of animal abundance based on Eq. (9.38) is

$$\hat{N} = \frac{mW}{2\mu_r}.$$

Hayne (1949a) recommended the estimator

$$\frac{1}{\hat{\mu}_r} = \frac{1}{m}\sum_{i=1}^{m}\frac{1}{r_i},$$

leading to the estimator

$$\hat{N} = \frac{W \sum_{i=1}^{m} \frac{1}{r_i}}{2}.$$
(9.39)

Noting density (D) can be estimated by

$$\hat{D} = \frac{\hat{N}}{WL},$$

which leads to the density estimator,

$$\hat{D} = \frac{\sum_{i=1}^{m} \frac{1}{r_i}}{2L},$$
(9.40)

for one transect line. Letting $M = \sum_{i=1}^{k} m_i$ be the sum of the numbers of animals detected across k randomly located lines, the estimator of abundance for a replicated transect survey is

$$\hat{N} = \frac{W \sum_{i=1}^{M} \frac{1}{r_i}}{2k}.$$
(9.41)

Density is estimated across k randomly located transect lines by the expression

$$\hat{D} = \frac{\sum_{i=1}^{M} \frac{1}{r_i}}{2Lk}.$$
(9.42)

The variance of $\hat{N}$ can be found by taking the variance in stages, where

$$\mathrm{Var}(\hat{N}) = \mathrm{Var}_2\left[E_1\left(\frac{M}{\hat{\bar{P}}k}\Big|2\right)\right] + E_2\left[\mathrm{Var}_1\left(\frac{M}{\hat{\bar{P}}k}\Big|2\right)\right]$$

and where "1" denotes the estimation of $\hat{\bar{P}}$ and "2" denotes the sampling of the population. It follows that

$$\mathrm{Var}(\hat{N}) \doteq \mathrm{Var}_2\left[\frac{M}{\bar{P}k}\right] + E_2\left[\frac{M^2}{k^2}\frac{\mathrm{Var}(\hat{\bar{P}})}{\bar{P}^4}\right]$$

$$\doteq \frac{\sigma_{m_i}^2}{\bar{P}^2 k} + \frac{N^2 \mathrm{Var}(\hat{\bar{P}})}{\bar{P}^2}$$

$$\mathrm{Var}(\hat{N}) \doteq \frac{\sigma_{m_i}^2}{\bar{P}^2 k} + \frac{N^2 \sigma_{r_i}^2}{\mu_r^2 M}.$$
(9.43)

In turn, $\sigma_{m_i}^2$ can be estimated by

$$\hat{\sigma}^2_{m_i} = \frac{\sum\limits_{i=1}^{k}(m_i - \bar{m})^2}{k-1}$$

for $\bar{m} = \dfrac{1}{k}\sum\limits_{i=1}^{k}m_i$. Similarly, $\sigma^2_{r_i}$ can be estimated by

$$\hat{\sigma}^2_{r_i} = \frac{\sum\limits_{i=1}^{M}(r_i - \bar{r})^2}{(M-1)}$$

for $\bar{r} = \dfrac{1}{M}\sum\limits_{i=1}^{M}r_i$.

Assumptions

The assumptions of the fixed-distance transect method include the following:

1. Flushing distances are unique but fixed distances for every animal.
2. Flushing distances are measured accurately.
3. Animals act independently.
4. Line transects are randomly distributed.
5. No animal is counted more than once along a line.
6. Flushing distances are stationary during the course of the survey.

These assumptions permit the geometric probabilities of detection to be calculated and abundance extrapolated from the number of animals detected. The assumption of independent detections may be violated if the animals occur in groups (e.g., quail coveys).

In the situation in which animals occur in groups, the line-transect methods need to estimate the number of groups and average group size. Total abundance is then estimated by the product

$$\tilde{N} = \hat{\bar{G}} \cdot \hat{N}, \tag{9.44}$$

where

$\tilde{N}$ = animal abundance;
$\hat{\bar{G}}$ = average group size;
$\hat{N}$ = line-transect estimate of the total number of animal groups.

The variance of $\tilde{N}$ can be estimated by the expression

$$\widehat{\mathrm{Var}}(\tilde{N}) = \widehat{\mathrm{Var}}(\hat{\bar{G}}) \cdot \hat{N}^2 + \widehat{\mathrm{Var}}(\hat{N}) \cdot \hat{\bar{G}}^2 - \widehat{\mathrm{Var}}(\hat{\bar{G}}) \cdot \widehat{\mathrm{Var}}(\hat{N}), \tag{9.45}$$

where

$$\widehat{\mathrm{Var}}(\hat{\bar{G}}) = \frac{\sum\limits_{i=1}^{M}(g_i - \bar{g})^2}{M-1}$$

and where

g_i = size of the ith group detected ($i = 1, \ldots, M$);

$$\bar{g} = \frac{1}{M} \sum_{i=1}^{M} g_i.$$

The process of conducting a line-transect survey may cause animals to move forward along the transect to be counted more than once. Should this occur, the resulting abundance ($\hat{N}$) and density ($\hat{D}$) estimates will be positively biased. Shorter, more numerous transect lines may be a partial solution when such animal behavior occurs. If these conditions persist, there may be little recourse other than to select another estimation technique. Multiple transect lines permit greater precision as well as variance estimation. However, the detection process must be stationary throughout the entire course of the survey. Systematic changes in animal behavior may violate the assumption of stationarity. If diel behavior patterns exist, the estimation process may need to be stratified over time.

Fixed flushing radius and random line transects imply right-angle sighting distances (x_i) will be uniformly distributed between 0 and r_i. Alternatively, the $\sin(\theta) = \dfrac{x_i}{r_i}$ should be uniformly distributed from 0 to 1. The observed empirical distribution of $\sin(\theta) = \dfrac{x_i}{r_i}$ can therefore be compared with a uniform distribution by using a Kolmogorov-Smirnov test (Zar 1984:55–58) to assess goodness-of-fit to the fixed radius model.

Discussion of Utility

Fixed-distance models were the first approaches to line-transect analysis. These methods were motivated by simple geometric detection models that permitted easy estimation of animal abundance and density. However, fixed-distance models make rigid assumptions concerning animal behavior. Regardless of angle of approach or disposition, an animal is assumed to respond when the surveyor is within a fixed but unknown distance to the animal. This method assumes an animal will always respond when approached within that fixed distance. The geometric foundation of the method requires $\sin\theta$ to be uniformly distributed, providing a test of goodness-of-fit to the model. Rejection of the fixed-distance model leaves right-angle distance models as the only viable alternatives.

Although the analyses are more complex, right-angle models are generally recommended for line-transect studies. From a practical perspective, the straight-line distance and flushing angle (θ) should be recorded for each observation. These data permit both calculation of right-angle distances and the ability to assess the suitability of the fixed-distance models. If right-angle methods may ultimately be needed, studies should be designed to obtain sufficient detections in order to adequately model the detection function. A minimum of 30 to 40 observations (Burnham et al. 1980:66) are needed for adequate modeling of the detection function for right-angle methods. In low-density populations, data may need to be pooled across transect lines. Consequently, animal surveys should be designed to help ensure a homogeneous detection process across the replicate transect lines.

9.3.2   Right-Angle Distance Methods

The fixed flushing radius methods are somewhat unrealistic, assuming the flush response is automatic and invariant to direction of approach. If the incidence of flushing is a probabilistic mechanism, a function of the distance between the observer and the animal, then estimation of animal abundance and/or density requires defining a detection function. The variable distance methods make two additional assumptions:

1. All animals have flushing distances drawn from the same pdf.
2. Probability of observing an animal is a simple function ($g(x)$) of the right-angle distance between the animal and the line transect.

These additional assumptions lead to the probability of detection (Section 8.2.4)

$$P = \frac{2}{W}\int_0^\infty g(x)dx,$$

or, equivalently,

$$P = \frac{2}{Wf(0)},$$

where $f(0) = \dfrac{1}{\displaystyle\int_0^\infty g(x)dx}$ is the height of the probability density function (pdf) at distance

$x = 0$. The detection function ($g(x)$) is anticipated to have three basic properties:

1. Monotonic decreasing function of distance (x).
2. Continuous function of distance.
3. Probability of detection of 1 (i.e., $g(0) = 1$) on the transect line.

Early investigators identified specific distributions such as exponential (Gates et al. 1968), logistic (Eberhardt 1968), and power law (Eberhardt 1968). Burnham et al. (1980) generalized the approach, specifying ($g(x)$) by a Fourier series. Buckland et al. (1993) extended the flexibility to include a selection of key functions (i.e., uniform, half-normal, and hazard-rate) and series adjustments (cosine, simple polynomial, and Hermite polynomials). The uniform key function with a cosine adjustment is the Fourier series approach of Burnham et al. (1980). The methods of Buckland et al. (1993, 2001, 2004) provide a suite of semiparametric approaches for modeling the detection function $g(x)$ based on the observed distribution of right-angle detection distances.

In the case of k replicate transect lines, each of length (L), animal abundance is estimated by

$$\hat{N} = \frac{M}{k\left(\dfrac{2}{W\hat{f}(0)}\right)}$$

$$\hat{N} = \frac{MW\hat{f}(0)}{2k}. \qquad (9.46)$$

Density is, in turn, estimated by

$$\hat{D} = \frac{\hat{N}}{WL}$$

$$\hat{D} = \frac{M\hat{f}(0)}{2kL}. \tag{9.47}$$

The central focus of line-transect estimation is identifying a detection function ($g(x)$) that will characterize the empirical distribution of right-angle detection distances and the estimation of $f(0) = \dfrac{1}{\displaystyle\int_0^\theta g(x)dx}$.

Burnham et al. (1980:66) recommended a minimum of 30 to 40 detections in order to estimate $g(x)$. When each replicate transect line has adequate numbers of detections, abundance or density can be estimated independently for each line, in which case,

$$\hat{D} = \frac{1}{k} \sum_{i=1}^{k} \frac{m_i \hat{f}(0)_i}{2L}$$

or, equivalently,

$$\hat{D} = \frac{1}{k} \sum_{i=1}^{k} \hat{D}_i = \bar{\hat{D}}, \tag{9.48}$$

where

$\hat{D}_i$ = density estimated from the ith transect line;
$\hat{f}(0)_i$ = estimate of $f(0)$ based on detections along the ith line alone.

The variance of $\hat{D}$ can be calculated empirically, where

$$\widehat{\text{Var}}(\hat{D}) = \frac{\displaystyle\sum_{i=1}^{k}(\hat{D}_i - \bar{\hat{D}})^2}{k(k-1)}. \tag{9.49}$$

Where sufficient numbers of animals are not detected per line and the detection function is invariant, $f(0)$ may be estimated by pooling all detections. However, the individual per-line estimates of density will no longer be independent. Buckland et al. (1993) recommended using bootstrap (Efron 1979) resampling methods for estimating the variance of $\hat{D}$ or $\hat{N}$.

Precision Calculations

We define precision in terms of the relative error in estimation

$$P\left(\left|\frac{\hat{D} - D}{D}\right| < \varepsilon\right) \geq 1 - \alpha,$$

then

$$\varepsilon \approx Z_{1-\frac{\alpha}{2}} \text{CV}(\hat{D}). \tag{9.50}$$

Assuming transect lines of fixed length (L), the sample size calculations reduce to the question of how many replicate lines (k) are needed to achieve a prescribed level of precision? The variance of $\hat{D}$ can be approximated by the delta method to be

$$\widehat{\mathrm{Var}}(\hat{D}) = \hat{D}^2 \left[\frac{\widehat{\mathrm{CV}}(m_i)^2}{k} + \widehat{\mathrm{CV}}(\hat{f}(0))^2 \right]. \tag{9.51}$$

The squared coefficient of variation $\mathrm{CV}(m_i)^2$ can be empirically estimated from a preliminary survey, where

$$\mathrm{CV}(m_i)^2 = \frac{s_{m_i}^2}{\overline{m}^2}$$

and where

$$s_{m_i}^2 = \frac{\sum_{i=1}^{k}(m_i - \overline{m})^2}{k-1}.$$

The $\mathrm{CV}(\hat{f}(0))$, however, will depend on the total number of animal observations (M) used to estimate $f(0)$ and the particular function ($g(x)$) fit to the data. For a half-normal detection function (Buckland et al. 1993:70–71),

$$\mathrm{Var}(\hat{f}(0)) \doteq \frac{f(0)^2}{2M}. \tag{9.52}$$

In which case,

$$\widehat{\mathrm{Var}}(\hat{D}) \doteq D^2 \left[\frac{\mathrm{CV}(m_i)^2}{k} + \frac{1}{2\overline{m}k} \right]. \tag{9.53}$$

Substituting Eq. (9.53) into Eq. (9.50) leads to

$$k = \frac{Z_{1-\frac{\alpha}{2}}^2 \left[\mathrm{CV}(m_i)^2 + \frac{1}{2\overline{m}} \right]}{\varepsilon^2} \tag{9.54}$$

as the required number of line transects during a population survey.

The variance of $\hat{f}(0)$ is approximately

$$\mathrm{Var}(\hat{f}(0)) \doteq \frac{f(0)^2 \mathrm{CV}(x_i)^2}{M}, \tag{9.55}$$

in the case of an exponential detection function ($g(x)$). Substituting Eq. (9.55) into Eq. (9.51) leads to

$$\mathrm{Var}(\hat{D}) \doteq D^2 \left[\frac{\mathrm{CV}(m_i)^2}{k} + \frac{\mathrm{CV}(x_i)^2}{\overline{m}k} \right]. \tag{9.56}$$

This variance expression for an exponential detection function yields

$$k = \frac{Z_{1-\frac{\alpha}{2}}^2 \left[\mathrm{CV}(m_i)^2 + \frac{\mathrm{CV}(x_i)^2}{\overline{m}} \right]}{\varepsilon^2} \tag{9.57}$$

after substituting Eq. (9.56) in Eq. (9.50).

The half-normal and exponential detection functions do not encompass the breadth of possible models for $g(x)$, but they should provide minimum sample size requirements. The cautious approach would be to use both Eq. (9.54) and Eq. (9.57) and select the largest value of k.

Example 9.5: Line-Transect Sample Size Calculations

Consider the case of a preliminary survey with four replicate lines and the following detections and right-angle detection distances:

Line 1	Line 2	Line 3	Line 4
2 m	9 m	3 m	3 m
5	14	7	14
17		4	53
26		12	
37		19	
47		22	
		28	
		33	
		43	
		67	

The preliminary survey resulted in the mean number of detections per line of $\bar{m} = 5.25$ with $s_m = 3.59$ for a $CV(m_i) = 0.685$. Mean detection distance was $\bar{x} = 22.14$ with $s_x = 18.49$ for a $CV(x_i) = 0.835$.

Thus, for a precision defined by

$$P\left(\left|\frac{\hat{D} - D}{D}\right| < 0.20\right) \geq 0.95,$$

$\varepsilon = 0.20$ and $Z_{1-\frac{\alpha}{2}} = 1.96$. The required number of transect lines is calculated, using Eq. (9.57) for an exponential detection function, as

$$k = \frac{Z_{1-\frac{\alpha}{2}}^2\left(CV(m_i)^2 + \dfrac{CV(x_i)^2}{\bar{m}}\right)}{\varepsilon^2}$$

$$= \frac{(1.96)^2\left[(0.685)^2 + \dfrac{(0.835)^2}{5.25}\right]}{(0.20)^2} = 57.8 \rightarrow 58 \text{ lines}.$$

By using the alternative Eq. (9.54), the required number of transect lines is $k = 55$. The unmanageably large number of required transect lines is dictated principally by the large line-to-line variability in number of observations of animals (i.e., $CV(m_i) = 0.685$) in this example.

Discussion of Utility

Line-transect methods provide a valuable alternative and a means of cross-validating mark-recapture methods. Unlike mark-recapture models, which generally assume homogeneous detection probabilities among individuals, line-transect methods relax the assumptions allowing heterogeneity among individual detection processes. However, to avoid the assumption of randomly distributed animals in space, replicate transect lines should be randomly located. This requirement is in direct control of the investigator and should be exercised. Stratified random sampling should be used if obvious habitat differences exist which might affect animal density.

Statistical software, such as Program DISTANCE (Buckland et al. 1993, 2001), provides a convenient, if not bewildering, set of options for modeling the detection function. Akaike Information Criterion (AIC) can help select an appropriate model that describes the detection process (Burnham and Anderson 1998). Personal experience, nevertheless, is valuable in selecting how to bin, pool, or truncate detection distances in the modeling process.

Detection distances will need to be pooled across replicate transect lines in many applications to obtain adequate sample sizes for modeling the detection process. Pooling is based on the assumption of a homogeneous detection process across the replicate transect lines. Concerns over detection homogeneity become more important as the number of required transect lines increases for a specified level of sampling precision. Differences among survey crews, habitat, and animal behavior can all contribute to detection heterogeneity that may require stratification. Thus, field surveys should be designed from the start to account for potential blocking factors.

Well-designed, right-angle distance methods provide a flexible yet powerful tool for surveying a wide range of wildlife species. Large geographic areas can be potentially surveyed relatively inexpensively and nonobtrusively. The ability to survey wild populations without animal handling lends the technique to both hard-to-handle and dangerous-to-handle species as well as endangered species for which incidental take is a concern.

9.4   Index-Removal Method (Petrides 1949, Eberhardt 1982)

The index-removal method of Petrides (1949) and Eberhardt (1982) was used in Section 8.4.1 to calibrate an index to an estimate of absolute abundance. The change-in-index values before and after a known harvest of individuals (C) was the basis for likelihood Eq. (6.106). However, the index-removal method need not be restricted to situations of harvest or removals. When defining population augmentations as negative removals (i.e., $-C$), the population estimators for N_{Pre} (Eq. 8.65) and N_{Post} (Eq. 8.66) are equally applicable to studies with releases of known numbers of animals.

Example 9.6: Index-Removal Method, Ring-Necked Pheasants (*Phasianus colchicus*)

An initial road survey along a prespecified route resulted in 21 observations of ring-necked pheasants. Seventy game-farm pheasants were then released and, the road survey was repeated, resulting in an observation of 43 individuals. Defining $C = -70$, the population abundance before the game farm release of pheasants is estimated (Eq. 8.65) to be

$$\hat{N}_{Pre} = \frac{(-70)21}{(21-43)}$$

$$= 66.82 \text{ birds,}$$

with postrelease abundance estimated as

$$\hat{N}_{Post} = \hat{N}_{Pre} + 70 = \frac{CI_2}{I_1 - I_2}$$

$$= \frac{-70(43)}{(21-42)}$$

$$= 136.82 \text{ birds.}$$

For variance calculations, see Section 8.4.1.

Discussion of Utility

The index-removal method does not need to be based solely on animal counts before and after harvest. Indices of abundance can also be based on production rates of animal sign. Indices such as auditory calls, pellet group counts, or track counts may also serve to characterize relative abundance. In some instances, animal sign production rates may be robust to behavioral changes associated with harvest. The choice of index to measure should be based on logistics, economics, and behavioral sciences. Whenever possible, sampling variances of the indices should be calculated and incorporated into the abundance assessment.

9.5 Change-in-Ratio Methods

Intentional additions or deletions in animal numbers may be accompanied by changes in the age composition or sex ratios of a population. The less representative these interventions, the more of an effect they will have on the demographic composition of a population. Kelker (1940, 1943) and Petrides (1949) were among the earliest to recognize how changes in population composition before and after interventions could be used to estimate population abundance. Chapman (1954, 1955) was among the first to statistically formulate the sampling models, estimators, and variances. Paulik and Robson (1969) provided an extensive review of CIR methods and were the first to provide sample size guidance.

The first application of the CIR method considered the specific case of male-only harvest on the population sex ratio (Kelker 1940, 1943). Paulik and Robson (1969) considered both additions and deletions to either of the male or female segments of a population. Otis (1980) and Udevitz (1989) extended the method to three demographic categories. CIR methods have also been adapted to open populations (Hanson 1963, Chapman and Murphy 1965) and were used in previous chapters to estimate productivity (Section 4.3), survival (Section 5.11) and harvest mortality (Section 6.6).

9.5.1   Two-Class Model

Consider a closed population composed of X- and Y-type individuals. In an initial random sample survey of size n_1, x_1, and y_1 animals (i.e., $n_1 = x_1 + y_1$) of the X- and Y-type individuals are observed, respectively. Subsequent to the survey, an intervention occurs in which A_X or A_Y animals of the X- and Y-type are added to the population, respectively. Special cases include in which only A_X or A_Y animals are added to the population (i.e., $A_X = 0$ or $A_Y = 0$). Other options include situations in which removals of X- and Y-type animals occur, in which case A values are negative. After the intervention of known size $(A_X + A_Y)$, a second random survey is conducted in which n_2 individuals are observed, with x_2 and y_2 animals (i.e., $n_2 = x_2 + y_2$) of the X- and Y-type individuals, respectively.

Define the following terms as follows:

N_1 = population size at the time of the first survey;
X_1 = number of X-type animals in the population at the time of the first survey;
Y_1 = number of Y-type animals in the population at the time of the first survey (i.e.,
    $Y_1 = N_1 - X_1$);
n_1 = number of animals observed in the first survey;
x_1 = number of X-type observations in the first survey;
y_1 = number of Y-type observations in the first survey;
n_2 = number of animals observed in the second survey;
x_2 = number of X-type observations in the second survey;
y_2 = number of Y-type observations in the second survey;
A_X = number of X-type animals added to the population after the first survey;
A_Y = number of Y-type animals added to the population after the first survey.

The joint likelihood for the two independent surveys can be written as

$$L(X_1,Y_1|A_X,A_Y,n_1,x_1,n_2,x_2) = \binom{n_1}{x_1}\left(\frac{X_1}{X_1+Y_1}\right)^{x_1}\left(\frac{Y_1}{X_1+Y_1}\right)^{n_1-x_1}\binom{n_2}{x_2}$$
$$\cdot\left(\frac{X_1+A_X}{X_1+Y_1+A_X+A_Y}\right)^{x_2}\left(\frac{Y_1+A_Y}{X_1+Y_1+A_X+A_Y}\right)^{n_2-x_2}.$$
$$(9.58)$$

Equation (9.58) is appropriate when sampling with replacement or when the finite population correction (fpc) can be ignored. In the case of sampling without replacement, the joint likelihood can be written as

$$L(X_1,Y_1|A_X,A_Y,n_1,x_1,n_2,x_2) = \frac{\binom{X_1}{x_1}\binom{Y_1}{n_1-x_1}}{\binom{X_1+Y_1}{n_1}}\cdot\frac{\binom{X_1+A_X}{x_2}\binom{Y_1+A_Y}{n_2-x_2}}{\binom{X_1+Y_1+A_X+A_Y}{n_2}}. \quad (9.59)$$

Likelihood models (9.58) and (9.59) can also be reparameterized in terms of N_1 and X_1, or N_1 and Y_1 to directly yield the MLE for abundance. The MLEs for model (9.58) or (9.59) are

$$\hat{X}_1 = \frac{x_1(A_X y_2 - A_Y x_2)}{n_1 x_2 - n_2 x_1},$$ (9.60)

$$\hat{Y}_1 = \frac{(n_1 - x_1)(A_X y_2 - A_Y x_2)}{n_1 x_2 - n_2 x_1},$$ (9.61)

$$\hat{N}_1 = \frac{n_1(A_X y_2 - A_Y x_2)}{n_1 x_2 - n_2 x_1},$$ (9.62)

$$\hat{N}_2 = \hat{N}_1 + (A_X + A_Y).$$ (9.63)

Paulik and Robson (1969) expressed the variance of $\hat{N}_1$ in terms of

$$P_1 = \frac{X_1}{N_1} \text{ and } P_2 = \frac{X_2}{N_2} = \frac{X_1 + A_X}{N_1 + A_X + A_Y},$$

where

$$\text{Var}(\hat{N}_1) = \frac{N_1^2 \text{Var}(\hat{P}_1) + N_2^2 \text{Var}(\hat{P}_2)}{(P_1 - P_2)^2}$$ (9.64)

and where

$$\hat{P}_1 = \frac{x_1}{n_1} \text{ and } \hat{P}_2 = \frac{x_2}{n_2}.$$

In the case of sampling with replacement,

$$\text{Var}(\hat{P}_i) = \frac{P_i(1 - P_i)}{n_i}$$

and is estimated by

$$\widehat{\text{Var}}(\hat{P}_i) = \frac{\hat{P}_i(1 - \hat{P}_i)}{n_i}.$$ (9.65)

In the case of sampling without replacement,

$$\text{Var}(\hat{P}_i) = \frac{P_i(1 - P_i)}{n_i}\left(\frac{N_i - n_i}{N_i - 1}\right)$$

and is estimated by

$$\widehat{\text{Var}}(\hat{P}_i) = \frac{\hat{P}_i(1 - \hat{P}_i)}{(n_i - 1)}\left(1 - \frac{n_i}{N_i}\right).$$ (9.66)

The estimate of $\hat{X}_1$ has an approximate variance (Seber 1982:355) of

$$\text{Var}(\hat{X}_1) \doteq \frac{N_1^2 P_2^2 \text{Var}(\hat{P}_1) + N_2^2 P_1^2 \text{Var}(\hat{P}_2)}{(P_1 - P_2)^2}.$$ (9.67)

The variance of $\hat{N}_2$ is equivalent to the variance of $\hat{N}_1$, whereas the variance of $\hat{X}_2$ is the same as the variance of $\hat{X}_1$. Furthermore, $\widehat{\mathrm{Var}}(\hat{Y}_1)$ can be computed by switching the definitions of the Y- and X-types and using variance formula of Eq. (9.67).

Assumptions

The assumptions of the CIR method include the following:

1. The population is closed except for known additions A_X and A_Y.
2. All animals have the same probability of detection within a sampling period.
3. All animals are correctly classified as either X- or Y-type individuals.
4. The additions A_X and A_Y are known exactly.

The assumptions of closure can be relaxed to allow mortality between periods as long as both X- and Y-type animals have the same survival probability. In this case, $\hat{N}_1$ estimates abundance just before the additions of A_X and A_Y (Seber 1982:357). Not knowing the exact values of additions A_X and A_Y can bias the estimates of abundance in either direction depending on counting errors.

Example 9.7: Two-Category, Change-in-Ratio Estimate of Abundance, Mule Deer (*Odocoileus hemionus*), Colorado

A common scenario for the CIR method is for differential harvest of the sexes between the two survey occasions. Riordan (1948) reported a preliminary survey with $n_1 = 657$ mule deer observed, of which $x_1 = 220$ were males and $y_1 = 457$ were females. A selective hunt removed $A_X = -5500$ males and no females, $A_Y = 0$. The follow-up survey resulted in $n_2 = 1011$ observations, of which $x_2 = 129$ were males and $y_2 = 882$ were females.

Total abundance at time of the first survey is estimated by Eq. (9.62) to be

$$\hat{N}_1 = \frac{n_1(A_X y_2 - A_Y x_2)}{n_1 x_2 - n_2 x_1}$$

$$= \frac{657[(-5500)882 - 0(129)]}{657(129) - 1011(220)} = 23{,}150.8 \text{ deer.}$$

Thus, $\hat{N}_2 = 23{,}150.8 - 5500 = 17{,}650.8$. Assuming sampling with replacement,

$$\widehat{\mathrm{Var}}(\hat{P}_1) = \frac{\left(\dfrac{220}{657}\right)\left(1 - \dfrac{220}{657}\right)}{657} = 0.0003390$$

and

$$\widehat{\mathrm{Var}}(\hat{P}_2) = \frac{\left(\dfrac{129}{1011}\right)\left(1 - \dfrac{129}{1011}\right)}{1011} = 0.0001101.$$

Then by using Eq. (9.64), the variance of $\hat{N}_1$ is estimated to be

$$\widehat{\text{Var}}(\hat{N}_1) = \frac{\hat{N}_1^2 \widehat{\text{Var}}(\hat{P}_1) + \hat{N}_2^2 \widehat{\text{Var}}(\hat{P}_2)}{(\hat{P}_1 - \hat{P}_2)^2}$$

$$= \frac{(23{,}150.8)^2 (0.0003390) + (17{,}650.8)^2 (0.0001101)}{\left(\dfrac{220}{657} - \dfrac{129}{1011}\right)^2} = 5{,}028{,}182.83,$$

with an associated standard error of $\widehat{\text{SE}}(\hat{N}_1) = 2242.4$.

Seber (1982:359) noted that in the situation in which $A_Y = 0$, the estimator of X_1 is robust to unequal detection probabilities of males and females as long as the ratio of their detection probabilities remain constant. However, $\hat{N}_1$ and $\hat{N}_2$ are not robust to unequal detection probabilities. The estimate of X_1 is

$$\hat{X}_1 = \frac{x_1(A_X y_2 - A_Y x_2)}{n_1 x_2 - n_2 x_1} = \hat{P}_1 \hat{N}_1$$

$$= \frac{220[(-5500)882 - 0(129)]}{657(129) - 1011(220)} = 7752.2$$

with an associated standard error of $\widehat{\text{SE}}(\hat{X}_1) = 398.0$.

Precision Calculations

Paulik and Robson (1969) provided sample size charts for CIR methods when the initial ratio of X- to Y-types is 1 : 1. Conner et al. (1986) extended the Paulik and Robson (1969) calculations when P_1 is other than 0.50. Inspections of variance formulas (9.64) and (9.67) indicate the precision of a CIR study increases as the shift in proportion of X-types (i.e., $|P_1 - P_2|$) increases. Thus, the stronger and more obvious the demographic shift in population composition between periods, the more precise the abundance estimates. This implies that population changes as characterized by A_X and A_Y $\left(\text{i.e., } \dfrac{A_X}{A_X + A_Y}\right)$ should be as disparate as possible from P_1 for the CIR method to be effective in estimating population abundance. As sample sizes n_1 and n_2 increase, variance estimates through the effects of $\text{Var}(\hat{P}_1)$ and $\text{Var}(\hat{P}_2)$ also decline.

When interventions are a result of population additions (i.e., A_X and/or A_Y are positive), it is assumed both new and original animals to the population have equal probability of detection. Should the augmented individuals differ in detection probability, the abundance estimates will be biased. Specifically, if the A_X or A_Y have lower detection probability than does the rest of the population, the abundance will be overestimated. Conversely, if the augmented individuals have a higher than normal detection probability, population abundance will be underestimated.

Discussion of Utility

Roseberry and Woolf (1991) had an unfavorable assessment of the CIR method as applied to estimating white-tailed deer abundance. They believed male and female detection probabilities were not equal and that detection probabilities were affected by responses to hunting pressure. The observed sex ratios were not considered representative of the true demographic structure of the population.

Although detection probabilities can purposefully vary between survey periods, the estimates of N_1 and N_2 are not robust to differential detection rates between sex classes. Differential harvest pressure, which facilities the CIR method, may also cause a shift in detection rates between sexes. Aerial surveys may provide more representative estimates for sex ratios than roadside surveys where hunting pressure is often the greatest. However, if the sexes aggregate into herds of different sizes, aerial surveys may be biased. In which case, sightability models may be needed to provide unbiased estimates of the sex ratios (Section 3.2.6) (Samuel et al. 1992).

9.5.2 Three or More Classes

Inspection of likelihoods (9.58) and (9.59) suggest obvious and straightforward extensions of the CIR method to three or more categories of individuals. Define the following:

X_i = number of animals of the i-type in the population at the time of the first survey ($i = 1, \ldots, T$);

A_i = number of i-type animals added to the closed population after the first survey ($i = 1, \ldots, T$);

n_j = number of animals sampled in the jth survey ($j = 1, 2$);

x_{ij} = number of i-type ($i = 1, \ldots, T$) individuals observed in the jth survey ($j = 1, 2$);

T = number of distinct categories of individuals in the population.

In the case of sampling with replacement, the joint likelihood is a product of two multinomial distributions for samples n_1 and n_2 and can be written as

$$L(N_1, X_i | A_i, n_1, n_2, x_{i1}, x_{i2}) = \binom{n_1}{x_{i1}} \prod_{i=1}^{T} P_{i1}^{x_{i1}} \cdot \binom{n_2}{x_{i2}} \prod_{i=1}^{T} P_{i2}^{x_{i2}}, \qquad (9.68)$$

where

$$P_{i1} = \frac{X_i}{N_1} \text{ and } P_{i2} = \frac{X_i + A_i}{N_1 + \sum_{i=1}^{T} A_i} \text{ for } i = 1, \ldots, T-1$$

and where

$$P_{T1} = \frac{N_1 - \sum_{i=1}^{T-1} X_i}{N_1} \text{ and } P_{T2} = \frac{N_1 - \sum_{i=1}^{T-1} X_i + A_T}{N_1 + \sum_{i=1}^{T} A_i}.$$

In the case of sampling without replacement, the joint likelihood is a product of two multi-hypergeometric distributions, where

$$L(N_1, \underset{\sim}{X}_i | \underset{\sim}{A}_i, n_1, n_2, \underset{\sim}{x}_{i1}, \underset{\sim}{x}_{i2}) = \frac{\prod\limits_{i=1}^{T} \binom{X_i}{x_{i1}} \prod\limits_{i=1}^{T} \binom{X_i + A_i}{x_{i2}}}{\binom{N_1}{n_1} \binom{N_1 + \sum\limits_{i=1}^{T} A_i}{n_2}}, \tag{9.69}$$

where, again, the constraints $X_T = N_1 - \sum\limits_{i=1}^{T-1} X_i$ and $X_T + A_T = N_1 - \sum\limits_{i=1}^{T-1} X_i + A_T$ are used to characterize the last cell categories. By using these parameterizations, N_1 is estimated directly along with its variance.

Otis (1980) developed the special case of $T = 3$ categories and provided closed-form estimators for N_1 and N_2. However, variance formulas were not provided, and he recommended using the asymptotic variance estimate obtained from the Newton-Raphson method of numerically solving the likelihood. We similarly recommend using numerical procedures when analyzing CIR studies with three or more categories.

Example 9.8: Three-Category, Change-in-Ratio Estimate of Abundance, Black-Tailed Deer (*Odocoileus hemionus hemionus*), Oregon

A hypothetical example of a three-category CIR method is illustrated based on the harvest regime for black-tailed deer in Oregon. Hunting permits allowed the harvest of animals with forked antlers or larger. A limited number of doe permits included the harvest of spiked bucks. This harvest policy results in three obvious categories of animals; (1) does, (2) spiked bucks, and (3) forked bucks and larger.

A prehunt roadside survey produced $n_1 = 498$ observations composed of $x_{11} = 397$ does, $x_{21} = 22$ spiked bucks, and $x_{31} = 79$ bucks with forked antlers or larger. An intervening harvest resulted in a total of 420 animals with the removal of $A_1 = -205$ does, $A_2 = -17$ spiked bucks, and $A_3 = -198$ forked bucks and larger. A subsequent posthunt survey resulted in $n_2 = 692$ observations of $x_{12} = 573$ does, $x_{22} = 30$ spiked bucks, and $x_{32} = 89$ forked and larger bucks. The survey sampling with replacement suggests the joint likelihood

$$L(N_1, \underset{\sim}{X} | \underset{\sim}{A}, \underset{\sim}{x}_{i1}, \underset{\sim}{x}_{i2}) = \binom{498}{397,22,79} \left(\frac{X_1}{N_1}\right)^{397} \left(\frac{X_2}{N_1}\right)^{22} \left(\frac{N_1 - X_1 - X_2}{N_1}\right)^{79}$$

$$\cdot \binom{697}{573,30,89} \left(\frac{X_1 - 205}{N_1 - 420}\right)^{573} \left(\frac{X_2 - 17}{N_1 - 420}\right)^{30}$$

$$\cdot \left(\frac{N_1 - X_1 - X_2 - 198}{N_1 - 420}\right)^{89}.$$

The MLEs are

$$\hat{N}_1 = 4774.49, \ \widehat{SE}(\hat{N}_1) = 3164.3,$$
$$\hat{X}_1 = 3808.95, \ \widehat{SE}(\hat{X}_1) = 2587.91,$$
$$\hat{X}_2 = 207.71, \ \ \widehat{SE}(\hat{X}_2) = 141.12.$$

The estimate $\hat{X}_3 = 757.83$ is obtained by subtraction with a standard error of $\widehat{SE}(\hat{X}_3)$ = 442.26. Appendix C illustrates the use of Program USER to calculate these MLEs and their associated standard errors.

Postharvest abundance was estimated by

$$\hat{N}_2 = \hat{N}_1 - 420$$
$$= 4354.49,$$

with the same standard error as that of $\hat{N}_1$.

The same data set could also be analyzed by using the dichotomous CIR model (Eq. 9.58) and estimator (Eq. 9.62). For example, by pooling the spike and buck categories, Eq. (9.62) yields a total abundance estimate of $\hat{N}_1 = 4628.64$ with a standard error of $\widehat{SE}(\hat{N}_1) = 3335.34$. Note the trichotomous analysis improves the precision of $\hat{N}_1$ slightly, but neither approach provides a precise estimate of the total abundance (i.e., CVs are 66.3% to 72.1%).

Discussion of Utility

The demographic structure of populations often provides three or more categories of distinguishable individuals. Examples include antlerless, spike, and forked-plus ungulates or hen, drake, and ducklings. Conceivably, age-structure data or species-composition data could also be used in CIR methods. The requirements of equal detection probabilities and distinguishable classes affect the practical limits of CIR methods. Errors in classification can be expected to increase as the number of categories increases. The likelihood models, Eqs. (9.68) and (9.69), do not consider misclassification errors, so any such error will be not be represented in the variance calculations.

It is unlikely numerous categories with relatively smaller changes in composition will increase the precision of CIR methods over dichotomous models. Additional research is needed on the relative efficiency of multicategory versus dichotomous CIR methods. If the data naturally lend themselves to multicategorical analysis, it is relatively easy to perform the analysis. If the sampling assumptions of the multicategorical analysis are true, they will also be true for the traditional dichotomy analysis (Section 9.5.1) of the marginals. Thus, the model that yields the smallest variance estimates should be selected.

9.5.3   Sequential Change-in-Ratio

The two-period CIR methods can be readily extended to three or more sampling periods and two or more population interventions (Pollock et al. 1985, Udevitz 1989). Differential harvest over time could be coupled with repeated population surveys of changing sex ratios over time. Other opportunities may include both additions and removals from the population, coupled with multiple surveys of population composition. Sequential CIR methods may be more precise than are the traditional two-period methods, if the multiple interventions further alter the population composition and the survey sampling is precise.

For illustration, consider the dichotomous case of X- and Y-type animals. Define the following terms:

N_1 = animal abundance at the time of the first survey;

n_i = number of animals sampled during the ith survey;

x_i = number of X-type animals in the ith survey sample;

$y_i = n_i - x_i$ = number of Y-type animals in the ith survey sample;

A_{X_i} = number of X-type animals added to the population during the ith intervention;

A_{Y_i} = number of Y-type animals added to the population during the ith intervention.

In the case of sampling without replacement, the joint likelihood over k surveys and $k - 1$ intervening interventions can be written as

$$L(N_1, X_1 | \underline{A}_X, \underline{A}_Y, \underline{n}, \underline{x}) = \frac{\binom{X_1}{x_1}\binom{N_1 - X_1}{n_1 - x_1}}{\binom{N_1}{n_1}} \cdot \prod_{i=2}^{k} \frac{\binom{X_1 + \sum_{j=1}^{i-1} A_{X_j}}{x_i}\binom{N_1 - X_1 + \sum_{j=1}^{i-1} A_{Y_j}}{n_i - x_i}}{\binom{N + \sum_{j=1}^{i-1} A_{X_j} + \sum_{j=1}^{i-1} A_{Y_j}}{n_i}}.$$

(9.70)

With more than two survey periods, iterative methods are needed to estimate N_1 and X_1. The abundance of Y-type animals at the time of the first survey can be estimated by

$$\hat{Y}_1 = \hat{N}_1 - \hat{X}_1.$$

Animal abundance in subsequent periods ($i = 2, \ldots, k$) is estimated by

$$\hat{N}_i = \hat{N}_1 + \sum_{j=1}^{i-1} A_{X_j} + \sum_{j=1}^{i-1} A_{Y_j}, \qquad (9.71)$$

where

$$\hat{X}_i = \hat{X}_1 + \sum_{j=1}^{i-1} A_{X_j}$$

and

$$\hat{Y}_i = \hat{N}_i - \hat{X}_i.$$

Therefore, $\text{Var}(\hat{N}_i) = \text{Var}(\hat{N}_1)$, $\text{Var}(\hat{X}_i) = \text{Var}(\hat{X}_1)$, and $\text{Var}(\hat{Y}_i) = \text{Var}(\hat{Y}_1)$.

Example 9.9: Sequential Change-in-Ratio Method, Ring-Necked Pheasant

This hypothetical data set illustrates a CIR method based on three sampling periods and two interventions. A preliminary visual survey observed $n_1 = 103$ pheasants, of which $x_1 = 29$ were males and $y_1 = 74$ were hens. A release of $A_{X_1} = 70$ male pheasants followed the initial survey (i.e., $A_{Y_1} = 0$). A second visual survey sighted $n_2 = 122$ birds, of which $x_2 = 57$ were males and $y_2 = 65$ were females. A rooster-only hunting season subsequently followed, in which $A_{X_2} = -51$ birds were harvested. The final element of the population census was a third posthunting visual survey, in which $n_3 = 64$, of which $x_3 = 25$ were males and $y_3 = 39$ were females.

The joint likelihood for the three visual surveys with replacement can therefore be written as

$$L(N, X_1 | \underline{n}, \underline{x}, \underline{A}_X) = \binom{103}{29}\left(\frac{X_1}{N_1}\right)^{29}\left(\frac{N_1 - X_1}{N_1}\right)^{74} \cdot \binom{122}{57}\left(\frac{X_1 + 70}{N_1 + 70}\right)^{57}\left(\frac{N_1 - X_1}{N_1 + 70}\right)^{65}$$

$$\cdot \binom{64}{25}\left(\frac{X_1 + 70 - 51}{N_1 + 70 - 51}\right)^{25}\left(\frac{N_1 - X_1}{N_1 + 70 - 51}\right)^{39}.$$

The resulting MLEs from the joint likelihood are

$$\hat{N}_1 = 201.91, \widehat{SE}(\hat{N}_1) = 85.85$$

and

$$\hat{X}_1 = 59.34, \widehat{SE}(\hat{X}_1) = 31.38.$$

Reparameterization of the likelihood in terms of N_1 and Y_1 yields

$$\hat{Y}_1 = 142.57, \widehat{SE}(\hat{Y}_1) = 55.51.$$

Discussion of Utility

Sequential CIR methods are a natural extension of the two-sample procedures (Sections 9.5.1, 9.5.2). Udevitz (1989) demonstrated the assumption of constant sampling probabilities across categories can be relaxed when the study is composed of three or more periods. He showed that constant probability ratios over time are sufficient for valid estimation. However, interventions that differentially alter animal behavior and/or their detection across subcategories would violate model assumptions.

9.6 Catch-Effort Methods

In exploited or harvested populations, information on animal abundance is contained within the harvested numbers and associated hunting effort. Catch-effort methods model the relationship between the probabilities of harvest and hunting effort to extract that information.

The model most often used to describe the relationship between the probability of harvest and effort is the so-called Poisson catchability model (Seber 1982:296). Harvest probability can be conceptualized beginning with a single unit of effort, which has probability p' of harvesting an animal. If p' is the probability of harvest with one unit of effort, then

$$1 - p'$$

is the probability an animal is not harvested or escapes when exposed to one unit of effort. With g independent units of effort, the probability an animal escapes harvest is

$$(1 - p')^g,$$

and its complement

$$p = 1 - (1 - p')^g \qquad (9.72)$$

is the probability of harvest when g units of effort are exerted. Equation (9.72) is equivalently expressed as

$$p = 1 - e^{g \ln(1-p')}$$

or, for convention,

$$p = 1 - e^{-cg} \tag{9.73}$$

where c is defined as the Poisson catchability coefficient or vulnerability coefficient.

9.6.1 Maximum Likelihood Model

In a catch-effort survey, harvest occurs over time with periodic recording of the numbers of animals harvested and associated harvest effort in each period. Define the following:

n_i = number of animals harvested/removed from the population in the ith period ($i = 1, \ldots, k$);

g_i = units of effort exerted in the ith period ($i = 1, \ldots, k$);

$$x_i = \sum_{j=1}^{i-1} n_j = \text{cumulative number of animals removed up to the time of the } i\text{th period}$$

($i = 1, \ldots, k$).

Here, we will consider only the simple situation, which assumes the population is closed except for the harvest mortality, which is monitored. Seber (1982:328–344) reviewed catch-effort methods for open populations. Within any sampling period, if it is assumed all animals have the same probability of harvest p_i, the numbers of animals subsequently removed n_i ($i = 1, \ldots, k$) have the joint likelihood

$$L(N, p_i | n_i) = \prod_{i=1}^{k} \binom{N - x_i}{n_i} p_i^{n_i} (1 - p_i)^{N - x_i - n_i} \tag{9.74}$$

Reparameterizing p_i in terms of the Poisson catchability model (9.73) and g_i, the likelihood model can be rewritten as follows:

$$L(N, c | n_i, g_i) = \prod_{i=1}^{k} \binom{N - x_i}{n_i} \left(1 - e^{-cg_i}\right)^{n_i} \left(e^{-cg_i}\right)^{N - x_i - n_i}. \tag{9.75}$$

Iterative procedures are necessary to estimate the model parameters N and c.

Assumptions

The assumptions of the generalized catch-effort model include the following:

1. The population is closed except for harvest, which is being monitored.
2. Units of effort are additive.
3. The catch or vulnerability coefficient c is constant over the investigation.
4. Each animal within a period has equal and independent probabilities of harvest.
5. Harvest (n_i) and units of effort (g_i) are correctly recorded.
6. Probability an animal is harvested is conditionally independent across the k sampling periods.

The assumptions of population closure during the hunting season for some big-game species are likely to be approximately true. Recruitment of new individuals through reproduction is usually complete before the scheduled hunting season. Recruitment or loss of individuals through immigration and emigration during the hunting season is, however, possible. Notable examples include elevationally migrating herds of mule deer or flocks of migrating waterfowl.

In applying catch-effort models to wildlife populations, one notable difficulty is in defining a meaningful unit of effort. Effort could be expressed in terms of licensed hunters, hunter days, or hunter hours. However, units of effort may not be strictly additive, as hunter strategies may change with hunter densities. For example, harvest success might be quite different between g hunters acting independently and those same g hunters working cooperatively in game drives. Hunter skills are not equal, thus potentially violating the assumption of additive units of equal effort.

The Poisson catch-coefficient c is also susceptible to changes over time. Animals may become more elusive during the hunting season because of prior experience with hunters. Weather factors over the course of the study may also change both game and hunter behavior, thereby affecting c. A well-documented effect on hunter success of big-game species is the presence or absence of snow cover (Forbes 1945, Robb 1952, Uptegraft 1959, Murphy 1983). Finally, all animals within a population may not have equal probability of harvest. Obvious violations include sex-specific hunting regulations. In these circumstances, the estimation process may need to be stratified, with male and female components of the population estimated separately. Less obvious may be the effects of age and home range locations on the probabilities of an animal being harvested. For these reasons, catch-effort methods have not been applied to wildlife populations as often as fisheries populations, in which gear selectivity and harvest efforts are more systematically controlled and quantified. Nevertheless, the catch-effort method under proper circumstances can be used to quantify wild populations using available harvest information.

Example 9.10: Catch-Effort Abundance Estimation, White-Tailed Deer, Crab Orchard NWR, Illinois

During 10 days between January 1 and 17, 1966, a controlled hunt was conducted on the 18,000-ha closed area at the Crab Orchard National Wildlife Refuge (Roseberry et al. 1969). Hunters were monitored at entry and exit from the refuge and were purposefully distributed at stands throughout the refuge while hunting. Hunters were also instructed not to "trophy hunt" but rather to attempt to take any available deer. Hunter effort (hours hunted) and harvest numbers were recorded daily during the 10 days of harvest (Table 9.2).

The likelihood Eq. (9.75) was used to estimate total abundance (N) and the associated catch coefficient (c). The likelihood maximization produced the estimates

$$\hat{N} = 1197.73, \widehat{\text{SE}}(\hat{N}) = 12.40$$

and

$$\hat{c} = 0.0001036, \widehat{\text{SE}}(\hat{c}) = 0.0000118.$$

Table 9.2. Catch-per-unit-effort data reported for white-tailed deer, Crab Orchard NWR, Illinois, 1966.

Date	Deer harvested (n_i)	Hours hunted (g_i)	CPUE $\left(\dfrac{n_i}{g_i}\right)$	Cumulative removal (x_i)	Cumulative effort (G_i)
1 Jan.	218	1809	0.12051	0	0
2 Jan.	206	2078	0.09913	218	1,809
3 Jan.	139	2114	0.06575	424	3,887
7 Jan.	127	2248	0.05649	563	6,001
8 Jan.	113	2299	0.04915	690	8,249
9 Jan.	76	2488	0.03055	803	10,548
14 Jan.	56	2408	0.02326	879	13,036
15 Jan.	58	2478	0.02341	935	15,444
16 Jan.	47	2134	0.02202	993	17,922
17 Jan.	33	1749	0.01887	1040	20,056
Total	1073				

(Data from Roseberry et al. 1969.)

A chi-square goodness-of-fit statistic of the form

$$\chi^2_{k-2} = \sum_{i=1}^{k} \frac{(n_i - E(n_i))^2}{E(n_i)},$$

where

$$E(n_i) = N\left(1 - e^{-cg_i}\right)e^{-c\sum_{j=1}^{i-1} g_i},$$

was used to assess overall fitness to the model. The resulting test statistic ($P(\chi^2_8 \geq 14.9026) = 0.0611$) found the model fit the data reasonably well ($0.05 < P < 0.10$).

Pollock et al. (1984) found overdispersion when fitting a multinomial capture-recapture model to data from a lobster (*Homarus americanus*) population. To compensate for the lack-of-fit, they increased the MLE estimates of variance by a scale factor similar to that used in GLMs (McCullagh and Nelder 1983). The standard errors were inflated by the quotient

$$\sqrt{\frac{\chi^2_{df}}{df}},$$

based on the chi-square goodness-of-fit statistic. In this case, the scale factor would be 1.3649 and the standard error for $\hat{N}$ adjusted to be $\widehat{SE}(\hat{N}) = 16.92$.

Roseberry et al. (1969) noted an additional 36 deer were crippled or retrieved by officials during the hunt. These incidental losses during the population survey would underestimate actual abundance of the deer population.

Discussion of Utility

The catch-per-unit effort (CPUE) method should be applied so that effort and harvest can be accurately recorded. Thus, controlled or regulated harvests are required to acquire the necessary data. In the Roseberry et al. (1969) study, hunters were purposefully distributed across the study area at designated stands to ensure a consistent and thorough application of effort. However, areas of lower effort existed and deer drives were performed toward the end of the study to harvest deer in these areas. The result was a shift in the CPUE.

MLE provided an estimate of total abundance of $\hat{N} = 1198$ deer in the Crab Orchard white-tailed deer population (Roseberry et al. 1969). A total of 1073 deer were registered during the hunt (Table 9.2). However, the survey likely underestimated abundance. Hunters reported crippling 390 deer besides those actually harvested. A representative survey of 12% of the study area resulted in an estimate of total crippling loss of 350 deer. These numbers represent a crippling loss rate of 24.6% to 26.7% and a corresponding underreporting of effective harvest numbers. The CPUE method, which assumes all harvested animals are reported, will consequently underestimate true abundance in the presence of crippling loss. A more realistic estimate of total abundance would be calculated by adjusting the catch values (i.e., n_i) by the factor $\dfrac{1}{(1-0.246)}$ and reanalyzing the data, in which case, $\hat{N} = 1589.0$. Similar underestimation might be anticipated from any sport hunt in which crippling loss may occur and not be reported. In some cases, crippling loss can be high. Parker (1991) reported unretrieved and crippled black ducks (*Anas rubripes*) represented 25% in one year and 38% in another year of total harvest mortality. Holsworth (1973) reported a crippling rate of 6.8% for white-tailed deer hunts in Ontario.

As in the case of the Roseberry et al. (1969) study, there is a tendency for hunters to change strategies as the hunting season progresses. A change in strategy violates the assumption of a constant vulnerability coefficient (c). Novak et al. (1991) used a competing hazards function to model catch-effort data, permitting both hunting and natural sources of mortality (e.g., crippling loss or other human-induced mortality sources, including natural mortality). It is possible by using a comparable competing risks model to permit two or more types of exploitation with their own vulnerability coefficients.

Variable trapping effort, along with animal removals, can be used to estimate abundance. Lewis and Farrar (1968) used live-trapping and removal to estimate white-tailed deer abundance in Tennessee. Leslie and Davis (1939) used removal trapping to estimate the number of rats (*Rattus rattus*) in a 9.1-ha block of houses in Freetown, Sierra Leone. When amount of trapping can be regulated and held constant across periods, constant effort removal methods (Section 9.6.4) can potentially be used.

9.6.2   Leslie and Davis (1939) Method

Motivated largely by the limitations in numerical capabilities of the 1930s, Leslie and Davis (1939) used a regression approximation to the maximum likelihood method model (9.75). Based on this method, the expected value for the number of animals harvested within a period (n_i), given (x_i) animals have already been removed from the population, can be written as

$$
\begin{aligned}
E(n_i|x_i) &= p_i(N - x_i) \\
&= (1 - e^{-cg_i})(N - x_i).
\end{aligned}
\tag{9.76}
$$

Leslie and Davis (1939) approximated e^{-cg_i} by the first two terms of a Taylor series expansion about 0 (i.e., $e^{-x} \approx 1 - x$), leading to the approximate expression

$$E(n_i|x_i) \approx cg_i(N - x_i).$$

Expressed in terms of CPUE $\left(\text{i.e., } \dfrac{n_i}{g_i}\right)$, the expected value is the linear model

$$E\left(\frac{n_i}{g_i}\Big|x_i\right) = cN - cx_i, \tag{9.77}$$

which is of the form of a straight-line regression

$$y_i = \alpha + \beta x_i.$$

Ordinary least squares assumes the variances of the dependent variables are equal; however, in the case of likelihood (9.75),

$$\text{Var}\left(\frac{n_i}{g_i}\Big|x_i\right) = \frac{1}{g^2}\,\text{Var}(n_i|x_i)$$

$$= \frac{(N - x_i)p_i(1 - p_i)}{g_i^2}.$$

The variance of the dependent variable $\left(\dfrac{n_i}{g_i}\right)$, using the approximation $p_i \approx cg_i$, can be expressed as

$$\text{Var}\left(\frac{n_i}{g_i}\Big|x_i\right) \approx \frac{(N - x_i)c(1 - cg_i)}{g_i}. \tag{9.78}$$

Combining the results of Eqs. (9.77) and (9.78), an iterative reweighted least squares could be performed of the form

$$\text{SS} = \sum_{i=1}^{k} \frac{g_i}{(N - x_i)c(1 - cg_i)}\left(\frac{n_i}{g_i} - c(N - x_i)\right)^2.$$

However, Ricker (1958) recommended using simple linear regression to estimate the parameters, in which case,

$$\hat{c} = \frac{-\sum_{i=1}^{k}(y_i - \bar{y})(x_i - \bar{x})}{\sum_{i=1}^{k}(x_i - \bar{x})^2}. \tag{9.79}$$

where

$$y_i = \frac{n_i}{g_i}$$

and where

$$\hat{N} = \bar{x} + \frac{\bar{y}}{\hat{c}} \tag{9.80}$$

for

$$\bar{x} = \frac{1}{k}\sum_{i=1}^{k} x_i$$

and

$$\bar{y} = \frac{1}{k}\sum_{i=1}^{k} y_i.$$

The variance of $\hat{c}$ can be estimated by

$$\widehat{Var}(\hat{c}) = \frac{MSE}{\sum_{i=1}^{k}(x_i - \bar{x})^2}. \tag{9.81}$$

The variance of $\hat{N}$ can then be estimated following the method of Seber (1982:298) as

$$\widehat{Var}(\hat{N}) = \frac{MSE}{\hat{c}^2}\left[\frac{1}{k} + \frac{(N - \bar{x})^2}{\sum_{i=1}^{k}(x_i - \bar{x})^2}\right], \tag{9.82}$$

where $MSE = \dfrac{\sum_{i=1}^{k}(y_i - \hat{y}_i)^2}{k-2}$ = error mean square from the regression analysis.

Although the linear regression model of Leslie and Davis (1939) is used less frequently, the linear model (9.77) suggests a means to visually check the fit of the catch-effort model. Plotting the CPUE data $\left(\dfrac{n_i}{g_i}\right)$ versus cumulative effort (x_i) should result in a straight-line relationship if the model is correct. Seber (1982:298) noted that a straight-line relationship may be evident even though natural mortality or migration exists.

Assumptions of the Leslie and Davis (1939) model include those of the likelihood model (9.75). However, the regression method also assumes that $1-e^{-cg}$ can be adequately approximated by cg. For capture probabilities $p < 0.20$, the approximation reasonably holds. For wild populations with much higher per-period harvest rates, the MLE approach is recommended. The linear regression estimates of c and N (Eqs. 9.79, 9.80) provide useful initial values for iteratively solving the MLEs of likelihood (9.75).

Example 9.11: Leslie-Davis Abundance Estimator, White-Tailed Deer, Crab Orchard NWR, Illinois

We illustrate the Leslie and Davis (1939) method of estimating population abundance by using the white-tailed deer CPUE data of Roseberry et al. (1969) (Table 9.2). The Leslie-Davis model is the linear regression Eq. (9.77)

$$\left(\frac{n_i}{g_i}\right) = cN - cx_i$$

with the fitted regression coefficients

$$\left(\frac{n_i}{g_i}\right) = 0.1164087 - 0.00010007 x_i$$

and $r^2 = 0.9781$. The regression coefficients translate to a catch coefficient of $c = 0.00010007$ and total abundance of

$$\hat{N} = \bar{x} + \frac{\bar{y}}{\hat{c}}$$

$$= 654.5 + \frac{0.050914}{0.00010007} = 1163.28,$$

where $\bar{y} = \frac{1}{k}\sum_{i=1}^{k}\frac{n_i}{g_i} = 0.50914$ and $\bar{x} = \frac{1}{k}\sum_{i=1}^{x_i} x_i = 654.5$. A plot of the CPUE $\left(\frac{n_i}{g_i}\right)$ versus cumulative removal (x_i) data for the white-tailed deer in this example indicates a strong linear trend (Fig. 9.4). Equation (9.82) can be used to estimate the sampling variance associated with the Leslie-Davis estimator, where

$$\widehat{\text{Var}}(\hat{N}) = \frac{\text{MSE}}{\hat{c}^2}\left[\frac{1}{k} + \frac{(N - \bar{x})^2}{\sum_{i=1}^{k}(x_i - \bar{x})^2}\right]$$

$$= \frac{0.00024614}{(0.00010007)^2}\left[\frac{1}{10} + \frac{(1163.28 - 654.5)^2}{1095990.5}\right] = 8263.298$$

with a standard error of $\widehat{\text{SE}}(\hat{N}) = 90.90$, where MSE $= 0.00024614$ and $\sum_{i=1}^{k}(x_i - \bar{x})^2 = 1,095,990.5$.

9.6.3 DeLury (1947, 1951) Method

The DeLury (1947, 1951) method uses an alternative regression approach to approximate the MLEs of model (9.75). The method is based on the expected catch in the ith period, where

$$E(n_i|N) = N(1 - p_1)(1 - p_2)\ldots(1 - p_{i-1})p_i, \tag{9.83}$$

and substituting the catch-effort model (9.73)

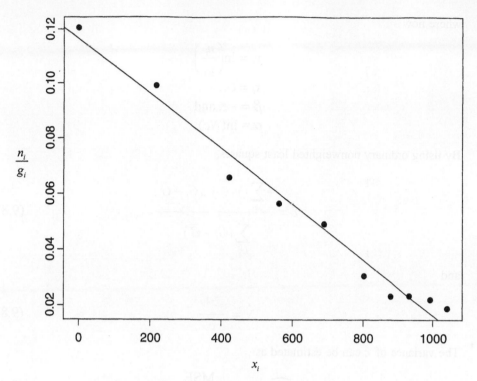

Figure 9.4. Catch-per-unit-effort $\left(\frac{n_i}{g_i}\right)$ versus cumulative animal removal (x_i) for the white-tailed deer data, Crab Orchard NWR, Illinois (Roseberry et al. 1969, Roseberry and Woolf 1991).

$$E(n_i|N) = Ne^{-c\sum_{i=1}^{i-1} g_i} p_i.$$

Defining the cumulative effort to the time of the ith sample where

$$G_i = \sum_{i=1}^{i-1} g_i,$$

the expected value of n_i can be written as

$$E(n_i|N) = Ne^{-cG_i} p_i. \tag{9.84}$$

By reparameterizing p_i by cg_i when p_i is small (i.e., $p_i < 0.20$), the expected value (9.84) can be approximated, in turn, by

$$E(n_i|N) = Ncg_i e^{-cG_i}$$

or, equivalently,

$$E\left(\frac{n_i}{g_i}\Big|N\right) = Nce^{-cG_i}. \tag{9.85}$$

The log of Eq. (9.85) yields the regression model,

$$\ln\left(\frac{n_i}{g_i}\right) = \ln(Nc) - cG_i, \tag{9.86}$$

of the form of a straight-line regression,

$$y_i = \alpha + \beta x_i,$$

where now

$$y_i = \ln\left(\frac{n_i}{g_i}\right),$$
$$x_i = G_i,$$
$$\beta = -c, \text{ and}$$
$$\alpha = \ln(Nc).$$

By using ordinary nonweighted least squares,

$$\hat{c} = \frac{-\sum\limits_{i=1}^{k}(y_i - \bar{y})(G_i - \overline{G})}{\sum\limits_{i=1}^{k}(G_i - \overline{G})^2} \tag{9.87}$$

and

$$\hat{N} = \frac{e^{\bar{y}+\hat{c}\overline{G}}}{\hat{c}}. \tag{9.88}$$

The variance of $\hat{c}$ can be estimated as

$$\widehat{\text{Var}}(\hat{c}) = \frac{\text{MSE}}{\sum\limits_{i=1}^{k}(G_i - \overline{G})^2}. \tag{9.89}$$

Seber (1982:303) provides a variance estimate for $\hat{N}$, where

$$\widehat{\text{Var}}(\hat{N}) = \text{MSE} \cdot \hat{N}^2 \left[\frac{1}{k} + \left(\frac{\hat{c}\overline{G}-1}{\hat{c}}\right)^2 \cdot \frac{1}{\sum\limits_{i=1}^{k}(G_i - \overline{G})^2}\right]. \tag{9.90}$$

Example 9.12: DeLury Abundance Estimator, White-Tailed Deer, Crab Orchard NWR, Illinois

The DeLury (1947, 1951) estimator of population abundance is illustrated by using the white-tailed deer CPUE data from Roseberry et al. (1969) (Table 9.2). The DeLury model is the linear regression equation (9.86), where

$$\ln\left(\frac{n_i}{g_i}\right) = \ln(Nc) - cG_i$$

with fitted regression coefficients,

$$\ln\left(\frac{n_i}{g_i}\right) = -2.26856 - 0.00009428G_i$$

and $r^2 = 0.9434$. The estimated regression coefficients translate to

$$\hat{c} = 0.00009428$$

and

$$\hat{N} = \frac{e^{\bar{y}+\hat{c}\bar{G}}}{\hat{c}}$$

$$= \frac{e^{-3.18258+0.00009428(9695.2)}}{0.00009428} = 1097.43,$$

where

$$\bar{y} = \frac{1}{k}\sum_{i=1}^{k}\ln\left(\frac{n_i}{g_i}\right) = -3.18258,$$

$$\bar{G} = \frac{1}{k}\sum_{i=0}^{k}G_i = 9695.2.$$

A plot of $\left(\dfrac{n_i}{g_i}\right)$ versus cumulative effort G_i indicates a strong linear trend (Fig. 9.5). The variance of $\hat{N}$ is calculated by using Eq. (9.90), where

$$\widehat{\mathrm{Var}}(\hat{N}) = \mathrm{MSE}\cdot\hat{N}^2\left[\frac{1}{k}+\left(\frac{\hat{c}\bar{G}-1}{\hat{c}}\right)^2\cdot\frac{1}{\displaystyle\sum_{i=1}^{k}(G_i-\bar{G})^2}\right]$$

$$= (0.227164)(1097.41)^2$$

$$\cdot\left[\frac{1}{10}+\left(\frac{0.00009428(9695.2)-1}{0.00009428}\right)^2\frac{1}{425626178}\right]$$

$$\widehat{\mathrm{Var}}(\hat{N}) = 27,891.59$$

with a standard error of $\widehat{\mathrm{SE}}(\hat{N}) = 167.01$ when MSE $= 0.227164$ and $\displaystyle\sum_{i=1}^{k}(G_i-\bar{G})^2 =$ 425,626,178.

Discussion of Utility

Basic personal computers have sufficient analytical power to solve for the MLEs from catch-effort studies (Eq. 9.75). The Leslie and Davis (1939) and DeLury (1947) models use Taylor series approximations to the catch-effort model to permit estimation by using linear regression. These approximations may be useful in fisheries studies in which capture probabilities are generally very small. These approximations do not hold when capture rates are above $p \geq 0.20$, situations that can occur in wildlife studies. Thus, few reasons exist to use the linearized models of Leslie and Davis (1939) or DeLury (1947) today.

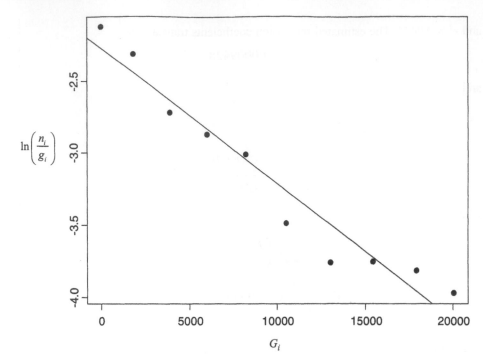

Figure 9.5. Log-catch-per-unit-effort versus cumulative hunting effort (hours) for the white-tailed deer data, Crab Orchard NWR, Illinois (Roseberry et al. 1969).

An argument for the regression models has been the robustness of the variance calculations. In regression analysis, the variance estimates are based on lack-of-fit of the data to the model. The lack-of-fit may be the result of stochastic variability or model misspecification. Because all models are simplifications of reality, some systematic error may exist. Therefore, the regression-based variance estimates may better capture the true uncertainty around the abundance estimates. In the white-tailed deer example from Crab Orchard NW, MLE produced an estimated standard error of 12.40, whereas the regression techniques of Leslie and Davis (1939) and DeLury (1947, 1951) produced standard errors of 90.90 and 167.01, respectively.

A valid variance estimator is as relevant today as in the time of Leslie and Davis (1939). If robust variance estimation is a concern, nonlinear weighted least squares is recommended based on the expected value (Eq. 9.76), where

$$E(n_i|x_i) = (1 - e^{-cg_i})(N - x_i),$$

using weights inversely proportional to the variance. Because the weights will be a function of the estimated parameters N and c, iterative reweighted nonlinear least squares is recommended.

9.6.4   Constant Effort Removal Technique (Zippin 1956, 1958)

When per-period probabilities of harvest or removal are held constant, likelihood Eq. (9.74) reduces to the form

$$L(N, p|\underset{\sim}{n}) = \prod_{i=1}^{k} \binom{N - x_i}{n_i} p^{n_i} (1 - p)^{N - x_i - n_i}. \tag{9.91}$$

This likelihood is known as the constant effort removal technique (Zippin 1956, 1958). A necessary but insufficient condition for this model to be correct is for removal effort to be constant across all k periods. Holding effort constant is necessary but not sufficient if the catch coefficient (c) in Eq. (9.73) varies over time. Hence, both constant effort and constant catchability (c) are necessary for this model.

Likelihood Eq. (9.91) can be rewritten equivalently as

$$L(N, p | r, t_2) = \binom{N}{\underset{\sim}{n}} p^r (1-p)^{K(N-r)+t_2},$$ (9.92)

where

$r = \sum_{i=1}^{k} n_i =$ total number of animals removed;

$t_2 = \sum_{i=1}^{k} (i-1)n_i =$ total number of escapes among the r animals before their removals.

The values r and t_2 are minimum sufficient statistics for the model (Eq. 9.92). Zippin (1956, 1958) provided the MLEs for the constant effort removal model, where

$$\hat{N} = \frac{r}{1-(1-\hat{p})^k}$$

$$= \frac{r}{1-\hat{q}^k}$$ (9.93)

and where $q \, (= 1 - p)$ must be iteratively solved from the equation

$$\frac{q}{1-q} - \frac{kq^k}{1-q^k} = \frac{t_2}{r}.$$ (9.94)

The variance of the abundance estimator is given by Zippin (1956) to be

$$\text{Var}(\hat{N}) = \frac{N(1-q^k)q^k}{(1-q^k)^2 - (pk)^2 q^{k-1}}$$ (9.95)

along with a corresponding variance expression for $\hat{p}$, where

$$\text{Var}(\hat{p}) = \frac{(qp)^2 (1-q^k)}{N\left[q(1-q^k)^2 - (pk)^2 q^k\right]}.$$ (9.96)

Special Case Where k = 2

Seber and LeCren (1967) provide parameter estimates for the removal model when $k = 2$, where

$$\hat{N} = \frac{(r - t_2)^2}{r - 2t_2}$$

or, equivalently,

$$\hat{N} = \frac{n_1^2}{(n_1 - n_2)},$$ (9.97)

with associated variance estimator of

$$\widehat{\text{Var}}(\hat{N}) = \frac{n_1^2 n_2^2 (n_1 + n_2)}{(n_1 - n_2)^4}.$$ (9.98)

The per-period capture probability is estimated from the expression

$$\hat{p} = \frac{r - 2t_2}{r - t_2}$$

or, equivalently,

$$\hat{p} = \frac{n_1 - n_2}{n_1},$$ (9.99)

with associated variance estimator

$$\widehat{\text{Var}}(\hat{p}) = \frac{n_2 (n_1 + n_2)}{n_1^3}.$$ (9.100)

Robson and Regier (1968) recommended using the less biased estimator

$$\hat{N} = \frac{(n_1^2 - n_2)}{(n_1 - n_2)}.$$

If all else is equal, a two-sample, constant-effort removal study will have less precision than will a two-sample, mark-recapture (i.e., Lincoln index) study (Seber 1982:324).

Special Case Where k = 3

Junge and Libosvarsky (1965) provided closed-form estimators for the case of $k = 3$ periods, where, in terms of the minimum sufficient statistics,

$$\hat{N} = \frac{17r^2 - 21rt_2 + 6t_2^2 + r(r^2 + 6rt_2 - 3t_2^2)^{\frac{1}{2}}}{18(r - t_2)}$$ (9.101)

and where

$$\hat{p} = \frac{5r - 3t_2 - (r^2 + 6rt_2 - 3t_2^2)^{\frac{1}{2}}}{4r - 2t_2}.$$ (9.102)

Having obtained the parameter estimates, the sampling variances can be estimated by using Eqs. (9.95) and (9.96), substituting the point estimates for the parameters.

Regression Model

The expected number of animals removed or harvested in the ith period based on the constant effort likelihood Eq. (9.91) is

$$E(n_i) = Np(1-p)^{i-1}.$$ (9.103)

Log-transformation of both sides of Eq. (9.103) yields

$$\ln E(n_i) = \ln Np + (i-1)\ln(1-p)$$ (9.104)

in the form of a linear equation,

$$y_i = \alpha + \beta x_i,$$

where

$$y_i = \ln n_i,$$
$$x_i = (i-1),$$
$$\alpha = \ln Np,$$
$$\beta = \ln(1-p).$$

Hence, a plot of $\ln n_i$ versus period $(i-1)$ provides a visual inspection of goodness-of-fit to model (9.91).

Hayne (1949b) suggested an alternative regression model based on the expected value

$$E(n_i) = p(N - E(x_i))$$ (9.105)

where $x_i = \sum_{j=1}^{i-1} n_j$ is the total number of animals removed from the population to the time of the ith sample. Rearranging Eq. (9.105), then

$$E(x_i) = N - \frac{1}{p} E(n_i),$$ (9.106)

suggesting a linear equation of the form

$$y_i = \alpha + \beta v_i,$$

where this time

$$y_i = x_i,$$
$$\alpha = N,$$
$$\beta = -\frac{1}{p},$$
$$v_i = n_i.$$

The data plot of x_i versus n_i provides yet another way to visually inspect for goodness-of-fit. White et al. (1982), however, strongly recommended against using regression plots or a high r^2-value (i.e., correlation coefficient squared) as sole criteria for assessing goodness-of-fit.

Goodness-of-Fit Test

Skalski and Robson (1979) provide a goodness-of-fit test for the constant effort removal sampling model based on the minimum sufficient statistics. The goodness-of-fit test is based on the hypotheses

$$H_o: p_1 = p_i \text{ for } i = 2, \ldots, k$$

versus

$$H_a: p_1 \neq p_i \text{ for some } i = 2, \ldots, k.$$

The test statistic is asymptotically normally distributed, where

$$\hat{Z} = \frac{\left| n_1 - \hat{E}(n_1 | r, t_2) \right| - \dfrac{1}{2}}{\sqrt{\widehat{\text{Var}}(n_1 | r, t_2)}} \tag{9.107}$$

for

$$\hat{E}(n_1 | r, t_2) = rA - \left[\frac{AC}{(AB - C^2)} \right] (t_2 - rC),$$

$$\widehat{\text{Var}}(n_1 | r, t_2) = r \left[A(1 - A) - \frac{A^2 C^2}{(AB - C^2)} \right],$$

and where

$$A = \frac{\hat{p}}{1 - \hat{q}^k},$$

$$B = \sum_{i=2}^{k} (i - 1)^2 q^{i-1},$$

$$C = \frac{\hat{q}}{\hat{p}} - \frac{k\hat{q}^k}{1 - \hat{q}^k}.$$

The null hypothesis is rejected if

$$P(Z > |\hat{Z}|) \leq \alpha.$$

The testing process begins by using the captures from all periods (i.e., $n_1, n_2, \ldots, n_k$) to test the equality of p_1 to the capture probabilities of the remaining periods. If H_o is rejected, the testing process continues using only counts ($n_2, n_3, \ldots, n_k$) to test the equality of p_2 to the capture probabilities of the remaining periods. When testing the equality of p_2 to the remaining capture probabilities, the counts $n_2, \ldots, n_k$ are relabeled $n_1, \ldots, n_{k-1}$ in recalculating the Z-statistic. This test sequence can be repeated until the null hypothesis is not rejected. The process requires a minimum of three periods to assess goodness-of-fit.

Generalized Removal Model

If heterogeneity in the capture propensity of animals exists, it is likely the more capture-prone animals will be caught in the first one or few periods. Thus, capture probabilities in a removal sample may be inflated during initial period(s), until these animals have been removed from the population. Subsequently, the per-period capture probabilities may be homogeneous. The Z-test of homogeneity (Eq. 9.107) can be used to identify when the capture probabilities become stable and the removal estimator valid. Otis et al. (1978) call this technique the generalized removal model or model M_B in the case of mark-recapture studies in the presence of trap happiness or trap shyness of animals.

The generalized removal estimator can be expressed as

$$\hat{N} = \hat{N}_{v-k} + \sum_{i=1}^{v-1} n_i, \qquad (9.108)$$

where $\hat{N}_{v-k}$ = removal estimator of population abundance using the capture data for periods v, $v + 1, \ldots, k$. The variance of the generalized removal model is simply the variance estimate for the removal technique over periods v to k, where

$$\mathrm{Var}(\hat{N}) = \mathrm{Var}(\hat{N}_{v-k}). \qquad (9.109)$$

Assumptions

The constant-effort removal method has the following assumptions:

1. The population of size N is closed.
2. All animals have equal and independent probabilities of capture within a period.
3. The capture probability is constant across all sampling periods.

The assumption of population closure is more likely to be violated as the duration of the survey increases. Physically removing animals from the study area encourages immigration of animals into the population from the surrounding habitat. Thus, the removal technique is most effective if the survey can be done quickly and with high per-period capture probabilities. In fact, low capture probabilities can jeopardize the success of a population survey. For some capture results, no MLEs exist. The removal method fails if

$$\sum_{i=1}^{k} (k + 1 - 2i) n_i \leq 0.$$

For example, with a $k = 2$ period study, no abundance estimate is possible if $n_1 \leq n_2$. For a $k = 3$ period study, the abundance estimation is not possible if $n_1 \leq n_3$.

Sample Size and Sampling Precision

The removal model is a function of three design variables: N, p, and k. Skalski and Robson (1992) developed sample size curves for the constant-effort removal model. Defining precision as

$$P\left(\left| \frac{\hat{N} - N}{N} \right| < \varepsilon \right) = 1 - \alpha$$

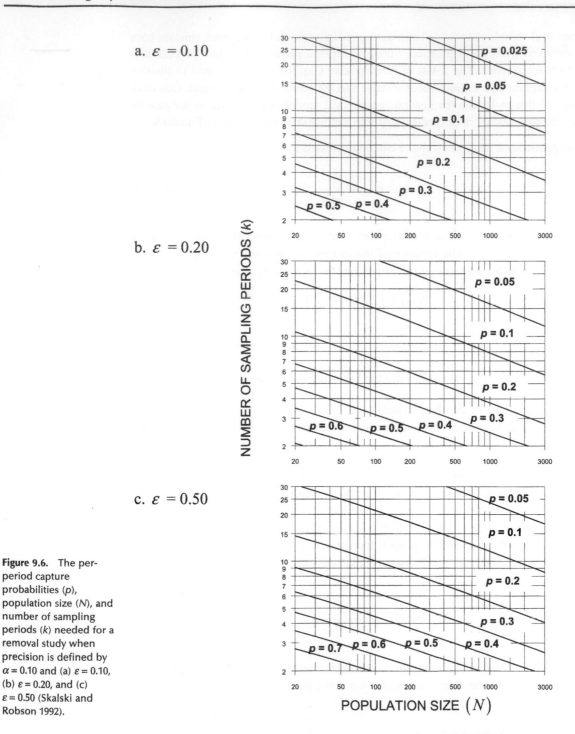

a. $\varepsilon = 0.10$

b. $\varepsilon = 0.20$

c. $\varepsilon = 0.50$

Figure 9.6. The per-period capture probabilities (p), population size (N), and number of sampling periods (k) needed for a removal study when precision is defined by $\alpha = 0.10$ and (a) $\varepsilon = 0.10$, (b) $\varepsilon = 0.20$, and (c) $\varepsilon = 0.50$ (Skalski and Robson 1992).

and assuming $\dfrac{1}{N}$ is approximately normally distributed, Skalski and Robson (1992) constructed precision curves for the cases of $\alpha = 0.10$ and $\varepsilon = 0.10$, 0.20, and 0.50 (Fig. 9.6). As might be expected, sampling precision improves as p and k increase for fixed population size N. The required number of sampling periods k for small values of p increase substantially, risking the possibility of immigration during the study.

Example 9.13: Constant Effort Removal Technique, European Hares (*Lepus europaeus*), Czempiń, Poland

A $k = 4$ period removal study was performed on European hares on a 2,361-ha study area at the Polish Hunting Union Research Station, Czempiń, Poland (Andrzejewski and Jezierski 1966). The four removal samples resulted in the following catch statistics

$$
\begin{aligned}
n_1 &= 722 \\
n_2 &= 191 \\
n_3 &= 69 \\
n_4 &= 36 \\
\hline
r &= 1018
\end{aligned}
$$

with $t_2 = 1(191) + 2(69) + 3(36) = 437$. By using Eq. (9.94),

$$
\frac{q}{1-q} - \frac{kq^k}{1-q^k} = \frac{t_2}{r}
$$

$$
\frac{q}{1-q} - \frac{4q^4}{1-q^4} = \frac{437}{1018},
$$

q was iteratively solved, resulting in a value of

$$q = 0.320659.$$

Population abundance is estimated by using Eq. (9.93), where

$$
\hat{N} = \frac{r}{1-q^k}
$$

$$
= \frac{1018}{1-(0.320659)^4} = 1028.88
$$

with an associated variance estimate (Eq. 9.95) of

$$
\widehat{\mathrm{Var}}(\hat{N}) = \frac{\hat{N}(1-\hat{q}^k)\hat{q}^k}{(1-\hat{q}^k)^2 - (\hat{p}k)^2 \hat{q}^{k-1}}
$$

$$
= \frac{1028.88(1-(0.320659)^4)(0.320659)^4}{(1-(0.320659)^4)^2 - (0.679341(4))^2(0.320659)^{4-1}} = 14.6330
$$

or $\widehat{\mathrm{SE}}(\hat{N}) = 3.8253$.

White et al. (1982), however, found the hare data did not conform to the constant-effort removal model despite a reasonable looking data plot (Fig. 9.7) and high r^2-value. By using the test of homogeneity and goodness-of-fit test (9.107),

$$Z = \frac{|n_1 - \hat{E}(n_1|r, t_2)| - \frac{1}{2}}{\sqrt{\widehat{\text{Var}}(n_1|r, t_2)}}$$

$$= \frac{|722 - 698.9587| - \frac{1}{2}}{\sqrt{49.6453}} = 3.1992$$

where

$$A = \frac{\hat{p}}{1 - q^k} = \frac{0.679341}{1 - (0.320659)^4} = 0.686600,$$

$$B = \sum_{i=2}^{k} (i-1)^2 q^{i-1}$$

$$= (2-1)^2 (0.320659)^1 + (3-1)^2 (0.320659)^2 + (4-1)^2 (0.320659)^3$$

$$= 1.028686,$$

$$C = \frac{\hat{q}}{\hat{p}} - \frac{k\hat{q}^k}{1 - \hat{q}^k} = \frac{0.320659}{0.679341} - \frac{4(0.320659)^4}{1 - (0.320659)^4} = 0.429273,$$

and where

$$\hat{E}(n_1|r, t_2) = rA - \left[\frac{AC}{(AB - C^2)}\right](t_2 - rC)$$

$$= 1018(0.686600) - \left[\frac{(0.686600)(0.429273)}{(0.686600)(1.028686) - (0.429273)^2}\right]$$

$$\cdot (437 - 1018(0.429273))$$

$$\hat{E}(n_1|r, t_2) = 698.9587,$$

$$\widehat{\text{Var}}(n_1|r, t_2) = r\left[A(1-A) - \frac{A^2 C^2}{(AB - C^2)}\right]$$

$$= 1018\left[(0.686600)(1 - 0.686600)\right.$$

$$\left. - \frac{(0.686600)^2 (0.429273)^2}{(0.686600)(1.028686) - (0.429273)^2}\right]$$

$$\widehat{\text{Var}}(n_1|r, t_2) = 49.6453.$$

The null hypothesis of $p_1 = p_i$ ($i = 2, \ldots, 4$) is rejected at $P(|Z| > 3.1992) = 0.0014$.

Recalculating the statistical test of homogeneity using counts $n_1 = 191$, $n_2 = 69$, and $n_3 = 36$ does not reject the null hypothesis at $\alpha = 0.10$. It is concluded that abundance can be estimated by using the generalized removal model, where

$$\hat{N} = n_1 + \hat{N}_{2-4}.$$

To estimate the abundance at the time of the second sample, the special case of $k = 3$ with Eq. (9.101) can be used, where

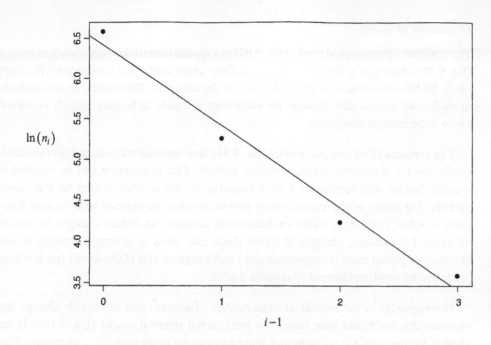

Figure 9.7. Scatterplot (White et al. 1982) of European hare removal data (Andrzejewski and Jezierski 1966) based on the log-linear model (Eq. 9.104).

$$r = 296 \text{ and } t_2 = 141$$

and

$$\hat{N}_{2-4} = \frac{17r^2 - 21rt_2 + 6t_2^2 + r(r^2 + 6rt_2 - 3t_2^2)^{\frac{1}{2}}}{18(r - t_2)}$$

$$= \frac{17(296)^2 - 21(296)(141) + 6(141)^2 + 296(296^2 + 6(296)(141) - 3(141)^2)^{\frac{1}{2}}}{18(296 - 141)}$$

$$= 318.45,$$

resulting in a total abundance estimate of

$$\hat{N} = 722 + 318.45$$
$$= 1040.45.$$

By using Eq. (9.102), the per-period capture probability is estimated to be $\hat{p} = 0.586889$. The variance of $\hat{N}_{2-4}$ and, consequently, $\hat{N}$ can then be estimated by Eq. (9.95), where

$$\text{Var}(\hat{N}) = \frac{N(1 - q^k)q^k}{(1 - q^k)^2 - (pk)^2 q^{k-1}}$$

$$= \frac{318.45(1 - 0.413111^3)(0.413111)^3}{(1 - 0.413111^3)^2 - (0.586889(3))^2(0.413111)^2}$$

$$= 62.3073$$

or a standard error of $\widehat{SE}(\hat{N}) = 7.89$.

Discussion of Utility

The constant-effort removal model (Eq. 9.97) is a special case of the variable-effort model (Eq. 9.75) where $g_i = g$ for all $i = 1, \ldots, k$. Only under controlled conditions will effort likely be held constant. This precludes use of the constant-effort model in the analysis of traditional harvest data. Instead, the catch data will need to be purposefully collected under experimental conditions.

The constant-effort removal model (Eq. 9.91) also assumes the vulnerability or catch coefficient (c) is constant across sampling periods. This assumption can be violated if weather factors, which influence animal behavior, are not constant during the k-removal periods. The timing of the removal study should therefore be planned with weather forecasts in mind. Even then, subtle environmental changes can induce changes in animal behavior. For instance, changes in moon phase can result in dramatic changes in the behavior of pocket mice (*Perognathus* spp.) and kangaroo rats (*Dipodomys* spp.), where predation by owls is enhanced in moonlit nights.

Heterogeneity in the individual vulnerability of animals can effectively change the vulnerability coefficient over time. The generalized removal model (Eq. 9.108) is an attempt to cope with the problem that some animals are more vulnerable to capture than are others. Animals in greater contact with the trapping gear are more likely to be caught in the first one or few periods of a removal study, leaving less vulnerable individuals to be caught in later periods of the study. The key to the generalized removal model is estimating abundance of animals in these later periods when vulnerability is more homogeneous between individuals and over time.

The procedure of studying a population through the process of its demise by removal trapping limits the applicability of the method to select populations. Pest species are the best candidates for removal studies. Animal control can be implemented while estimating both the initial and residual abundance of the population. In the past, removal trapping was commonly used in the study of small mammal populations (Davis 1957, Kikkawa 1964), which can quickly replenish themselves. Removal trapping has also been used in conjunction with pit-fall studies of herptiles (Tilley 1982). Changing public perspectives and animal care policies have redefined when constant-effort removal methods are appropriate. The technique is now best used in analysis of multiple-tag recapture studies under models M_B and M_{BH} of Otis et al. (1978).

9.7 Life-History Models

Life-history models refer to population abundance estimators that are based on simple demographic relationships concerning the structure of populations. All life-history models share in common the process of calibrating some readily measured population response to total abundance. The opportunity for such models is virtually unlimited. The sex-age-kill (SAK) model (Section 9.7.1) begins with harvest numbers of male ungulates to estimate total population abundance. The duck nest survey model (Section 9.7.3) begins with a nest count to estimate total duck abundance. In both cases, readily obtained count data were calibrated to total abundance by using auxiliary information on sex ratios, productivity, etc. A third application of life-history-based abundance estimation of bighorn sheep can be found in Section 10.5.

Life-history models for abundance estimation also provide a framework for deciding which demographic data are valuable to routinely collect. However, when auxiliary parameter estimates are combined to provide abundance estimates, the results can be rather imprecise. In discussing the precision of CIR techniques, Paulik and Robson (1969:13) reported, "when several variable quantities are combined in a mathematical formula as is done to estimate population abundance, the end result may be so uncertain that it is useless for decision-making." Thus, variance expressions and precision calculations are important in assessing how informative these easy-to-construct estimates actually may be.

9.7.1   Sex-Age-Kill Model

The SAK model has been used since the 1950s to estimate population size and monitor white-tailed deer in the midwestern United States (Eberhardt 1960, Creed et al. 1984, Hansen 1998). The model is also known as the Wisconsin method (Roseberry and Woolf 1991). More recently, the approach has also been used to estimate population size of elk (*Cervus elaphus*) (Bender and Spencer 1999), black bear (*Ursus americanus*) (Bender 1998), and black-tailed deer (Shirato 1996, Zahn 1996) populations.

The SAK model is based on a series of independent estimates, including the following:

$\hat{H}$ = male harvest numbers;

$\hat{R}_{F/M}$ = adult sex ratio;

$\hat{R}_{J/F}$ = juvenile-to-adult-female ratio;

$\hat{S}_T$ = total annual survival;

$\hat{P}_H$ = proportion of total mortality associated with harvest.

The fundamental equation for the SAK estimator is

$$\hat{N} = \hat{N}_M + \hat{N}_F + \hat{N}_J,$$

where

$\hat{N}_M$ = estimated abundance of adult males;

$\hat{N}_F$ = estimated abundance of adult females;

$\hat{N}_J$ = estimated abundance of juveniles.

By using the estimated sex ratio (i.e., $\hat{R}_{F/M}$) and juvenile-to-female ratio (i.e., $\hat{R}_{J/F}$), the estimate of total abundance can be reexpressed as

$$\hat{N} = \hat{N}_M \left(1 + \hat{R}_{F/M} + \hat{R}_{F/M} \cdot \hat{R}_{J/F}\right). \tag{9.110}$$

These age and sex ratios may be estimated from spotlight counts (Fafarman and DeYoung 1986, Kie and Boroski 1995), herd composition counts (McCullough 1993), or other means (Jacobson et al. 1997, Koerth et al. 1997). In some cases, the Severinghaus and Maguire (1955) technique is used to estimate adult sex ratios (Creed et al. 1984). The key to the SAK model is the estimate of adult male abundance based on harvest data expressed as

$$\hat{N}_M = \frac{\hat{H}}{1 - \hat{S}_H}, \tag{9.111}$$

where $1-\hat{S}_H$ = estimated probability of harvest mortality. Alternatively, the harvest probability can be estimated by

$$1-\hat{S}_H = (1-S_T)\cdot\hat{P}_H$$
$$= \hat{M}_T \cdot \hat{P}_H,$$

(9.112)

where $\hat{M}_T$ = estimated total annual mortality probability. Combining Eqs. (9.110) through (9.112) yields the SAK model

$$\hat{N} = \frac{\hat{H}}{\hat{M}_T \cdot \hat{P}_H}\left[1+\hat{R}_{F/M}+\hat{R}_{F/M}\cdot\hat{R}_{J/F}\right].$$

(9.113)

The SAK model, as written, has minimal biological assumptions. Instead, the method depends on the ability to obtain unbiased estimates of sex and age ratios, as well as estimates of mortality and the contribution of harvest to total mortality.

The sex ratio, in the case of a stable and stationary population, can be estimated (Section 3.4) by

$$\hat{R}_{F/M} = \frac{1-\hat{S}_M}{1-\hat{S}_F} = \frac{\hat{p}_{YM}}{\hat{p}_{YF}}$$

where

$\hat{p}_{YM}$ = estimated proportion of yearling males in the male segment of the population;

$\hat{p}_{YF}$ = estimated proportion of yearling females in the female segment of the population.

Furthermore, $\hat{M}_T$ can be estimated by $\hat{p}_{YM}$ (Section 5.7.5), in which case, the SAK model can be reparameterized as follows:

$$\hat{N} = \frac{\hat{H}}{\hat{p}_{YM}\cdot\hat{P}_H}\left[1+\frac{\hat{p}_{YM}}{\hat{p}_{FM}}+\left(\frac{\hat{p}_{YM}}{\hat{p}_{FM}}\right)\hat{R}_{J/F}\right].$$

(9.114)

Skalski and Millspaugh (2002) derived generic variance expressions for the SAK abundance estimator. For the estimator of male abundance ($\hat{N}_M$), where

$$\hat{N}_M = \frac{\hat{H}}{\hat{p}_{YM}\cdot\hat{P}_H},$$

(9.115)

the variance can be approximated by the delta method as

$$\text{Var}(\hat{N}_M) \doteq N_M^2\left[\text{CV}(\hat{H})^2+\text{CV}(\hat{p}_{YM})^2+\text{CV}(\hat{P}_H)^2\right].$$

(9.116)

The estimator of adult female abundance ($\hat{N}_F$) can be expressed as

$$\hat{N}_F = \frac{\hat{H}}{\hat{p}_{YM}\cdot\hat{P}_H}\cdot\hat{R}_{F/M}$$

(9.117)

with approximate variance of

$$\text{Var}(\hat{N}_F) \doteq N_F^2 \left[\text{CV}(\hat{H})^2 + \text{CV}(\hat{p}_{YM})^2 + \text{CV}(\hat{P}_H)^2 + \text{CV}(\hat{R}_{F/M})^2 \right]. \quad (9.118)$$

The estimator of juvenile abundance $(\hat{N}_J)$ can be expressed as

$$\hat{N}_J = \frac{\hat{H}}{\hat{p}_{YM} \cdot \hat{P}_H} \cdot \hat{R}_{F/M} \cdot \hat{R}_{J/F}. \quad (9.119)$$

The variance of $\hat{N}_J$ can then be estimated by the expression

$$\text{Var}(\hat{N}_J) \doteq N_J^2 \left[\text{CV}(\hat{H})^2 + \text{CV}(\hat{p}_{YM})^2 + \text{CV}(\hat{P}_H)^2 + \text{CV}(\hat{R}_{F/M})^2 + \text{CV}(\hat{R}_{J/F})^2 \right]. \quad (9.120)$$

The approximate variance expression for the total abundance estimator (Eq. 9.114) can be written as

$$\begin{aligned} \text{Var}(\hat{N}) &\doteq N^2 \left[\text{CV}(\hat{H})^2 + \text{CV}(\hat{p}_{YM})^2 + \text{CV}(\hat{P}_H)^2 \right] \\ &\quad + (N_F + N_J)^2 \cdot \text{CV}(\hat{R}_{F/M})^2 + N_J^2 \cdot \text{CV}(\hat{R}_{J/F})^2 \end{aligned} \quad (9.121)$$

where $\text{CV}(\hat{\theta})$ is expressed as a decimal

$$\text{CV}(\hat{\theta}) = \frac{\sqrt{\text{Var}(\hat{\theta})}}{\theta} \quad (9.122)$$

for any parameter estimator $\hat{\theta}$.

Assumptions

The basic SAK model (Eq. 9.113) has minimal biological assumptions. Instead, the method has four assumptions concerning accurately estimating the input parameters, including the following:

1. Total male harvest (H) is estimated unbiasedly.
2. The sex ratio $(R_{F/M})$ and juvenile-to-female ratio $(R_{J/F})$ are estimated unbiasedly.
3. The annual mortality probability of males (M_T) is estimated unbiasedly.
4. The annual mortality probability can be accurately partitioned into the proportions owing to harvest (P_H) and natural causes $(1 - P_H)$.

The SAK model (Eq. 9.114) imposes upon the previous assumptions two additional requirements:

5. The population is stable and stationary.
6. Natural survival probabilities are constant across all age classes.

These two additional assumptions place restrictions on situations when the SAK model (Eq. 9.114) can be used. Small-game species are unlikely to reach a stable and stationary population. Big-game species, largely governed by intrinsic regulatory factors, may approximate a stable and stationary population. However, harvest mortality must also be constant over time. Successive annual changes in harvest regulations, effort, or performance can lead to a nonstable and nonstationary population. In those circumstances, the more generic model (9.113) should be used over that of Eq. (9.114).

Precision Curves

Inspection of variance Eqs. (9.116), (9.118), and (9.120) indicate that the CVs for $\hat{N}_M$, $\hat{N}_F$, and $\hat{N}_J$ do not depend on the abundance values. Instead, the CVs of the abundance estimates are strictly a function of the CV squares of the input parameters. These variance expressions simply assume the input data sources are independent. It is important to note that when $\hat{R}_{F/M}$ and $\hat{R}_{J/F}$ are obtained from the same sample survey, the covariance between $\hat{R}_{F/M}$ and $\hat{R}_{J/F}$ is not zero. In these circumstances, an additional covariance term must be added to the variance expression for $\hat{N}_J$.

Skalski and Millspaugh (2002) evaluated the relationship between the precision of the input parameters to the SAK model and the resultant precision of the SAK abundance estimates (Fig. 9.8). Defining precision as

$$P\left(\left|\frac{\hat{N} - N}{N}\right| < \varepsilon\right) \geq 1 - \alpha$$

implies that

$$\varepsilon \approx Z_{1-\frac{\alpha}{2}} \cdot \mathrm{CV}(\hat{N}),$$

where $Z_{1-\frac{\alpha}{2}}$ is a standard normal deviate corresponding to $P\left(|Z| > Z_{1-\frac{\alpha}{2}}\right) = \alpha$. For example, a CV of 50% corresponds to a relative error of approximately 100% (i.e., 1.96 (50) = 98). Skalski and Millspaugh (2002) considered the three levels of precision suggested by Robson and Regier (1968) for "rough management" (i.e., $\varepsilon = 0.50$, $\alpha = 0.05$), "accurate management" (i.e., $\varepsilon = 0.25$, $\alpha = 0.05$), and "careful research" (i.e., $\varepsilon = 0.10$,

Figure 9.8. The anticipated coefficient of variation (CV) of population abundance estimates derived from the SAK model under different levels of precision of input variables (Skalski and Millspaugh 2002). Three benchmarks of precision exist at CV = 0.255 ("rough management"), CV = 0.128 ("accurate management"), and CV = 0.051 ("careful research"), as proposed by Robson and Regier (1964).

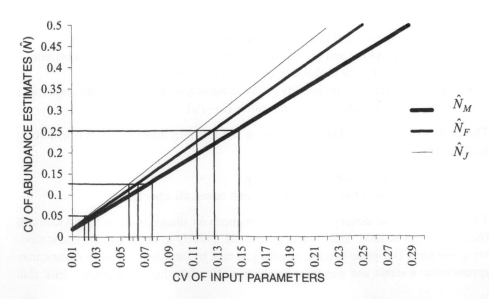

$\alpha = 0.05$). These three levels of precision correspond to CVs of 0.255, 0.128, and 0.051, respectively. Assuming common CV values for all the input parameters, the relationship between input CVs and anticipated CV values for abundance estimates can be plotted (Fig. 9.8). For example, for the estimate of male abundance to have a precision level needed for careful research (i.e., $\varepsilon = 0.10$, $\alpha = 0.05$), each of the three input parameters would need a CV ≤ 0.039. To estimate juvenile abundance with the same level of precision, the five input parameters would need CVs ≤ 0.023. For "rough management" purposes with a precision specified by $\varepsilon = 0.50$, $\alpha = 0.05$, input parameters would need to be estimated with CVs ≤ 0.147 for male abundance estimates and CVs ≤ 0.114 for juvenile abundance estimates.

Example 9.14: Sample Size to Estimate $\hat{R}_{F/M}$ with Required Precision for SAK Model

The information in Fig. 9.8 provides guidance on the level of precision required for the input parameters to the SAK model. In turn, the required level of precision for the input parameters dictate the required sample sizes needed for the estimate of $\hat{R}_{F/M}$, $\hat{R}_{J/F}$, etc. Based on the binomial sampling model

$$L = \binom{n}{f}\left(\frac{N_F}{N_F + N_M}\right)^f \left(\frac{N_M}{N_F + N_M}\right)^{n-f}$$

$$= \binom{n}{f} \cdot \left(\frac{R_{F/M}}{R_{F/M}+1}\right)^f \left(\frac{1}{R_{F/M}+1}\right)^{n-f},$$

the estimate of the sex ratio can be expressed as

$$\hat{R}_{F/M} = \frac{f}{n-f}$$

with associated variances based on the delta method of

$$\mathrm{Var}(\hat{R}_{F/M}) \doteq \frac{R_{F/M}(1+R_{F/M})^2}{n},$$

where n = sample size and f = number of observed females.

The CV for $\hat{R}_{F/M}$ is, in turn, expressed as

$$\mathrm{CV}(\hat{R}_{F/M}) = \frac{\sqrt{\dfrac{R_{F/M}(1+R_{F/M})^2}{n}}}{R_{F/M}}$$

$$= \frac{(1+R_{F/M})}{\sqrt{nR_{F/M}}}.$$

The CV for $\hat{R}_{F/M}$ as a function of n for a variety of values of $R_{F/M}$ can be plotted (Fig. 9.9). For example, in the case of $R_{F/M} = 4$ and the requirement for a CV of 0.029 for a "careful research" estimate of male abundance (N_M), the required sample size to estimate the sex ratio is calculated to be

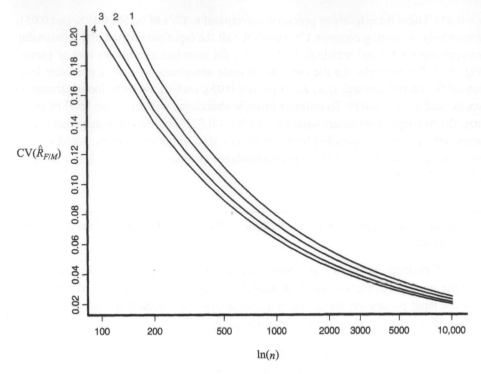

Figure 9.9. Anticipated coefficient of variation (CV) for estimates of the sex ratio $R_{F/M}$ as a function of ln(n) for different values of $R_{F/M}$ = 1, 2, 3, or 4.

$$CV = \frac{(1 + R_{F/M})}{\sqrt{nR_{F/M}}}$$

$$0.029 = \frac{1 + 4}{\sqrt{n4}},$$

$$n = 7431.6 \rightarrow 7432.$$

Thus, 7432 animals should be surveyed to attain a CV of 0.029 when $R_{F/M}$ = 4. These sample size calculations do not consider a fpc to the variance which can be ignored as long as $\frac{n}{N} < 0.10$. In the case in which the fpc cannot be ignored,

$$CV(\hat{R}_{F/M}) = \sqrt{\frac{\left(1 - \dfrac{n}{N}\right)(1 + R_{F/M})^2}{nR_{F/M}}}.$$

9.7.2 SAK-MLE Model

Under the assumptions of a stable and stationary population and of a male-only harvest regime, population abundance can be estimated using readily collected demographic data. The estimate is based on escalating the known number of harvested males (H) by an estimate of their harvest probability, i.e.,

$$\hat{N}_M = \frac{H}{p_H}$$

$$= \frac{H}{1 - S_H},$$

where p_H = probability of harvest = $1 - S_H$. However, rather than estimating the probability of surviving the annual male harvest directly, information on S_H can be obtained from sex ratios and age composition data. Assuming a common natural survival probability for both males and females (S_N) and an annual harvest survival probability of S_H for males only, the sex ratio has the long-term expected value (Section 3.4) of

$$\frac{N_F}{N_M} = R_{F/M}$$

$$= \frac{1 - S_N S_H}{1 - S_N}. \tag{9.123}$$

In the case of a stable and stationary population, the expected proportion of yearling males among the male population (Section 5.7.5) is anticipated to be

$$E(p_Y) = 1 - S_N S_H. \tag{9.124}$$

Hence, a joint likelihood based on harvest numbers (H), information on the population sex ratio ($R_{F/M}$), and proportion of yearling males (p_Y) among the harvest can be used to estimate total male abundance,

$$L = \binom{N_M}{H}(1 - S_H)^H (S_H)^{N_M - H} \cdot \binom{n_1}{m} \left(\frac{N_M}{N_M + N_F} \right)^m \left(\frac{N_F}{N_M + N_F} \right)^{n_1 - m}$$
$$\cdot \binom{n_2}{y}(p_Y)^y (1 - p_Y)^{n_2 - y}, \tag{9.125}$$

where

N_M = abundance of yearling and older males;
N_F = abundance of yearling and older females;
S_H = probability of surviving the male harvest;
S_N = probability of surviving natural sources of mortality;
H = number of males harvested;
p_Y = proportion of yearlings in the male population;
n_1 = number of animals sighted during a visual survey of the sex ratio;
m = number of males sighted among the n_1 animals observed;
n_2 = sample of males used in age classification;
y = number of yearling males among the n_2 examined.

The joint likelihood (Eq. 9.125) is based on harvest numbers of males (H) and a sample of n_1 animals to estimate the population sex ratio by using a visual survey and a subsample of n_2 males harvested to calculate the proportion of yearling males among the harvest. Reparameterizing Eq. (9.125) based on Eqs. (9.123) and (9.124) yields the joint likelihood

$$L = \binom{N_M}{H}(1 - S_H)^H (S_H)^{N_M - H} \binom{n_1}{m} \left(\frac{1 - S_N}{2 - S_N - S_N S_H} \right)^m \left(\frac{1 - S_N S_H}{2 - S_N - S_N S_H} \right)^{n_1 - m}$$
$$\cdot \binom{n_2}{y}(1 - S_N S_H)^y (S_N S_H)^{n_2 - y}. \tag{9.126}$$

The MLEs are

$$\hat{S}_H = \frac{n_2 - y}{\left(n_2 - \dfrac{my}{(n_1 - m)}\right)} = \frac{1 - \hat{p}_Y}{\left(1 - \dfrac{\hat{p}_Y}{\hat{R}_{F/M}}\right)}, \tag{9.127}$$

$$\hat{S}_N = 1 - \frac{my}{n_2(n_1 - m)} = 1 - \frac{\hat{p}_Y}{\hat{R}_{F/M}}, \tag{9.128}$$

$$\hat{N}_M = \frac{H(n_2(n_1 - m) - my)}{(n_1 - 2m)} = \frac{H(\hat{R}_{F/M} - \hat{p}_Y)}{(\hat{R}_{F/M} - 1)\hat{p}_Y}, \tag{9.129}$$

where

$$\hat{R}_{F/M} = \frac{n_1 - m}{m},$$

$$\hat{p}_Y = \frac{y}{n_2}.$$

The variance of $\hat{N}_M$ can be approximated by the delta method, where

$$\widehat{\mathrm{Var}}(\hat{N}_M) = \widehat{\mathrm{Var}}(H)\left[\frac{(\hat{R}_{F/M} - \hat{p}_Y)}{(\hat{R}_{F/M} - 1)\hat{p}_Y}\right]^2 + \widehat{\mathrm{Var}}(\hat{R}_{F/M})\left[\frac{H(1 - \hat{p}_Y)}{(\hat{R}_{F/M} - 1)^2 \hat{p}_Y}\right]^2$$
$$+ \widehat{\mathrm{Var}}(\hat{p}_Y)\left[\frac{H\hat{R}_{F/M}}{(\hat{R}_{F/M} - 1)\hat{p}_Y^2}\right]^2 \tag{9.130}$$

and where

$$\widehat{\mathrm{Var}}(\hat{H}) = \hat{N}_M \hat{S}_H (1 - \hat{S}_H),$$

$$\widehat{\mathrm{Var}}(\hat{R}_{F/M}) = \frac{n_1(n_1 - m)}{m^3},$$

$$\widehat{\mathrm{Var}}(\hat{p}_Y) = \frac{\hat{p}_Y(1 - \hat{p}_Y)}{n_2}.$$

The total adult (i.e., $\geq$ yearling) abundance can be estimated by the quantity

$$\hat{N}_{M+F} = \hat{N}_M(1 + \hat{R}_{F/M})$$
$$= \frac{H(1 + \hat{R}_{F/M})(\hat{R}_{F/M} - \hat{p}_Y)}{\hat{p}_Y(\hat{R}_{F/M} - 1)}$$

or, equivalently,

$$\hat{N}_{M+F} = \frac{H(n_2(n_1 - m) - my)n_1}{(n_1 - 2m)m},$$

(9.131)

with approximate variance

$$\begin{aligned}
\widehat{\text{Var}}(\hat{N}_{M+F}) &\doteq \widehat{\text{Var}}(H)\left[\frac{(\hat{R}_{F/M} + 1)(\hat{R}_{F/M} - \hat{p}_Y)}{(\hat{R}_{F/M} - 1)\hat{p}_Y}\right]^2 \\
&\quad + \widehat{\text{Var}}(\hat{R}_{F/M})\left[\frac{H(\hat{R}_{F/M}^2 - 2\hat{R}_{F/M} + 2\hat{p}_Y - 1)}{(\hat{R}_{F/M} - 1)^2 \hat{p}_Y}\right]^2 \\
&\quad + \widehat{\text{Var}}(\hat{p}_Y)\left[\frac{H(\hat{R}_{F/M} + 1)\hat{R}_{F/M}}{\hat{p}_Y^2(\hat{R}_{F/M} - 1)}\right]^2.
\end{aligned}$$

(9.132)

Assumptions

The abundance estimators of Eqs. (9.129) and (9.131) have the following assumptions:

1. H is known without error.
2. Only males are harvested (with a constant annual harvest probability of $(1 - S_H)$).
3. Males and females have the same natural survival probability (S_N) across all age classes.
4. The population is stable and stationary.
5. A random sample of the population is taken to estimate the sex ratio.
6. A random sample of the harvest (H) is taken to estimate the proportion of yearlings in the male population.

This model strongly depends on the assumptions of a stable and stationary population. Unless these assumptions are true, functional Eqs. (9.123) and (9.124) will not be true. Equations (9.123) and (9.124) are also predicated on the assumption of a constant natural survival probability across all sex and age classes, yearlings and older. The requirement of male-only harvest may be applicable to a number of ungulate species in which antlerless harvest is prohibited. The approach may also be applicable to wild turkeys (*Meleagris gallopavo*) and ring-necked pheasants with male-only harvest. Annual harvest rates must also be constant for relations Eqs. (9.123) and (9.124) to hold.

Example 9.15: Sex-Age-Kill Abundance Estimation, Black-Tailed Deer

In this hypothetical example, black-tailed deer abundance will be estimated by using likelihood (9.126). A harvest of $H = 5800$ bucks 1.5 years and older in a game management area will be assumed. Corresponding to that harvest, $n_2 = 750$ bucks were classified to age, of which $y = 150$ were yearlings. Before the hunting season, a visual survey of $n_1 = 1200$ adult deer were observed, of which $m = 240$ were bucks, the remainder were does.

The male abundance is estimated by Eq. (9.129) as follows:

$$\hat{N}_M = \frac{H(\hat{R}_{F/M} - \hat{p}_Y)}{(\hat{R}_{F/M} - 1)\hat{p}_Y}$$

$$= \frac{5800\left(\dfrac{960}{240} - \dfrac{150}{750}\right)}{\left(\dfrac{960}{240} - 1\right)\dfrac{150}{750}} = 36,733.3$$

for $\hat{R}_{F/M} = 960/240 = 4.0$ and $\hat{p}_Y = 150/750 = 0.20$. By using Eqs. (9.127) and (9.128), the survival estimates are $\hat{S}_H = 0.8421$ and $\hat{S}_N = 0.95$, respectively.

The variance of $\hat{N}_M$ is calculated by using Eq. (9.130), where

$$\widehat{\text{Var}}(\hat{H}) = \hat{N}_M \hat{S}_H (1 - \hat{S}_H)$$

$$= 36,733.3(0.8421)(1 - 0.8421) = 4884.34,$$

$$\widehat{\text{Var}}(\hat{R}_{F/M}) = \frac{n_1(n_1 - m)}{m^3}$$

$$= \frac{1200(1200 - 240)}{240^3} = 0.08333,$$

$$\widehat{\text{Var}}(\hat{p}_Y) = \frac{\hat{p}_Y(1 - \hat{p}_Y)}{n_2}$$

$$= \frac{0.20(1 - 0.20)}{750} = 0.0002133,$$

and where

$$\widehat{\text{Var}}(\hat{N}_M) = \widehat{\text{Var}}(H)\left[\frac{(\hat{R}_{F/M} - \hat{p}_Y)}{(\hat{R}_{F/M} - 1)\hat{p}_Y}\right]^2 + \widehat{\text{Var}}(\hat{R}_{F/M})\left[\frac{H(1 - \hat{p}_Y)}{(\hat{R}_{F/M} - 1)^2 \hat{p}_Y}\right]^2$$

$$+ \widehat{\text{Var}}(\hat{p}_Y)\left[\frac{H\hat{R}_{F/M}}{(\hat{R}_{F/M} - 1)\hat{p}_Y^2}\right]^2$$

$$= 4884.34\left[\frac{(4 - 0.2)}{(4 - 1)0.2}\right]^2 + 0.08333\left[\frac{5800(0.2 - 1)}{(4 - 1)^2 0.2}\right]^2$$

$$+ 0.0002133\left[\frac{5800(4)}{(4 - 1)(0.2)^2}\right]^2$$

$$\widehat{\text{Var}}(\hat{N}_M) = 8,722,319.01$$

or $\widehat{\text{SE}}(\hat{N}_M) = 2953.4$. From the survey data and the results of the analysis, adult female abundance is estimated by

$$\hat{N}_F = \hat{N}_M \cdot \hat{R}_{F/M},$$

which in the case of this example is

$$\hat{N}_F = 36,733.3 \left(\frac{960}{240} \right) = 146,933.2.$$

Finally, total adult abundance would then be estimated by

$$\hat{N}_{M+F} = \hat{N}_M + \hat{N}_F$$
$$= \hat{N}_M \left(1 + \hat{R}_{F/M} \right),$$

which for this example is estimated to be

$$\hat{N}_{M+F} = 183,666.5.$$

By using Eq. (9.132), the estimated standard error for $\hat{N}_{M+F}$ is calculated to be $\widehat{SE}(\hat{N}_{M+F}) = 15,861.7$.

Discussion of Utility

The SAK likelihood model requires a special set of biological circumstances for the model to be applicable. Only in rare cases will the model be completely applicable. However, the relatively simple data requirements of harvest numbers (H), sex ratio ($R_{F/M}$), and an estimate of the proportion of yearlings among harvested males make the method attractive. The method may provide a first approximation to population abundance when little else is available. Cooper (2000) and Cooper et al. (2003) used similar data to estimate elk abundance in Idaho.

The likelihood model (Eq. 9.126) can also be extended to include incomplete reporting by including auxiliary likelihoods of the form

$$\binom{H}{h} p_R^h (1 - p_R)^{H-h} \cdot \binom{n_3}{r} p_R^r (1 - p_R)^{n_3 - r}, \tag{9.133}$$

where

h = number of males reported harvested;

p_R = probability the harvested animal is reported;

n_3 = number of harvested animals observed in field or locker checks;

r = number of n_3 animals subsequently reported harvested by hunters.

Either locker or field checks (Section 6.2.1) can be used to estimate harvest-reporting rates in the auxiliary likelihoods (Eq. 9.133).

Total population abundance (i.e., juveniles plus adults) can also be estimated by having additional knowledge of the fawn/doe or calf/cow ratios (i.e., $R_{J/F}$) such that

$$\hat{N} = \hat{N}_M \left(1 + \hat{R}_{F/M} + \hat{R}_{F/M} \cdot \hat{R}_{J/F} \right).$$

9.7.3   Duck Nest Survey Model

Waterfowl nest count surveys are commonly conducted to monitor productivity and population trends (Cowardin and Johnson 1979). The resulting nest count, however, can also be used as the basis of a population estimate in the presence of auxiliary demographic data. Different approaches exist, but one simple approach begins with the estimated nest count ($\hat{C}_N$) and an estimate of nesting success rate ($\hat{p}_N$) to estimate adult female abundance, where

$$\hat{N}_F = \frac{\hat{C}_N}{\hat{p}_N} \tag{9.134}$$

and where $\hat{p}_N$ = estimated probability a female successfully nests. After the broods have departed the nest but while still remaining together with the hens, an estimate of the juvenile-to-female ratio is obtained from a visual survey. The juvenile abundance is then estimated by the product

$$\hat{N}_J = \hat{N}_F \cdot \hat{R}_{J/F}$$

or, equivalently,

$$\hat{N}_J = \frac{\hat{C}_N \cdot \hat{R}_{J/F}}{\hat{p}_N}. \tag{9.135}$$

Male abundance, in turn, is estimated by using an estimate of the sex ratio, where

$$\hat{N}_M = \hat{N}_F \left(\frac{1}{\hat{R}_{F/M}} \right)$$

or, equivalently,

$$\hat{N}_M = \frac{\hat{C}_N}{\hat{p}_N \hat{R}_{F/M}}. \tag{9.136}$$

Total population abundance is then estimated as the sum of Eqs. (9.134) through (9.136), where

$$\hat{N} = \frac{\hat{C}_N}{\hat{p}_N} \left[1 + \hat{R}_{J/F} + \frac{1}{\hat{R}_{F/M}} \right]. \tag{9.137}$$

By using the delta method and assuming the input parameters are estimated independently, variance expressions can be readily calculated. The variance of hen abundance can be approximated by

$$\text{Var}(\hat{N}_F) = N_F^2 \left[\text{CV}(\hat{C}_N)^2 + \text{CV}(\hat{p}_N)^2 \right]. \tag{9.138}$$

In turn, chick abundance has the approximate variance of

$$\text{Var}(\hat{N}_J) = N_J^2 \left[\text{CV}(\hat{C}_N)^2 + \text{CV}(\hat{p}_N)^2 + \text{CV}(\hat{R}_{J/F})^2 \right]. \tag{9.139}$$

The estimate of drake duck abundance has the approximate variance of

$$\text{Var}(\hat{N}_M) = N_M^2 \left[\text{CV}(\hat{C}_N)^2 + \text{CV}(\hat{p}_N)^2 + \text{CV}(\hat{R}_{F/M})^2 \right]. \tag{9.140}$$

Finally, the estimate of total duck abundance has the approximate variance of

$$Var(\hat{N}) = N^2 \left[CV(\hat{C}_N)^2 + CV(\hat{p}_N)^2 \right] + N_J^2 \cdot CV(\hat{R}_{J/F})^2$$
$$+ N_M^2 \cdot CV(\hat{R}_{F/M})^2. \tag{9.141}$$

Assumptions

The demographic assumptions leading to population estimate (9.137) are minimal. The crucial assumptions are as follows:

1. The juvenile-to-female ratio ($R_{J/F}$) is applicable at the time abundance is inferred.
2. The female-to-male ratio ($R_{F/M}$) is applicable at the time abundance is inferred.

There are three additional assumptions of the nest count abundance estimator (9.137):

3. The estimated sex ($R_{F/M}$) and juvenile-to-adult ($R_{J/F}$) ratios are unbiased.
4. The estimated nest count ($\hat{C}_N$) is unbiased.
5. The estimated proportion of hens nesting is unbiased.

In very large populations, the nest count ($\hat{C}_N$) may be spatially subsampled rather than enumerated. In this case, the variance estimators Eqs. (9.138) through (9.141) account for the sample survey variance of $\hat{C}_N$. In the case of complete enumeration of C_N, the $CV(\hat{C}_N) = 0$ in Eqs. (9.138) through (9.141). The pivotal quantity in estimator (9.137) is the proportion of hens producing nests. A conservative abundance estimator is obtained assuming each hen produces a nest (i.e., $p_N = 1$). To properly complement this assumption of $p_N = 1$, the estimate of average clutch size per hen ($R_{J/F}$) must include both successful and unsuccessful breeding hens.

Discussion of Utility

The nest count abundance estimator is a special case of a much wider variety of life-history-based models. Comparable estimators can be constructed for gray squirrels (*Sciurus carolinensis*) and other species with obvious nesting sites. In species such as tree squirrels, auxiliary information, such as a number of nests per female squirrel, may be necessary. As the number of auxiliary variables required to estimate abundance increases, the anticipated precision of the total abundance estimator decreases for all else being equal (Section 9.7.1, 9.7.3). Practical application of the life-history models is predicated on the strategic choice of easy-to-collect and quantifiable auxiliary variables.

9.8 Age-Structured Population Reconstruction Methods

Population reconstruction methods based on age-at-harvest data have been used for over a half century (Fry 1949, 1957). Their first use was in fisheries management, where catch data are often accessible but other traditional methods of abundance estimation are difficult or impossible to apply. In fisheries science, these methods go by the name of stock assessment techniques; in wildlife management, they are often referred to as population reconstruction methods. In recent years, age-structured assessment techniques have found their way into wildlife science (Lowe 1969, McCullough 1979, Woolf and Harder 1979,

Downing 1980, Fryxell et al. 1988, Ferguson 1993, Gove et al. 2002). The best reviews of the methods nevertheless can still be found in the fisheries literature (Ricker 1975, Gulland 1983, Megrey 1989, Hilborn and Walters 1992, Gallucci et al. 1996, Quinn and Deriso 1999).

Define the following:

h_{ij} = number of animals harvested and reported in the ith year ($i = 1, \ldots, Y$) for the jth age class ($j = 1, \ldots, A$);

N_{ij} = abundance of animals in the ith year ($i = 1, \ldots, Y$) for the jth age class ($j = 1, \ldots, A$).

The goal of these methods is to back-calculate catch data to year-specific, age-specific abundance levels and, from there, to estimate annual abundance by summing over the age classes (Fig. 9.10). Population reconstruction approaches differ in how they obtain and use the necessary information on survival and exploitation rates to perform these calculations.

a.

	Age class					
Year	1	2	3	4	5	6
1	h_{11}	h_{12}	h_{13}	h_{14}	h_{15}	h_{16}
2	h_{21}	h_{22}	h_{23}	h_{24}	h_{25}	h_{26}
3	h_{31}	h_{32}	h_{33}	h_{34}	h_{35}	h_{36}
4	h_{41}	h_{42}	h_{43}	h_{44}	h_{45}	h_{46}
5	h_{51}	h_{52}	h_{53}	h_{54}	h_{55}	h_{56}
6	h_{61}	h_{62}	h_{63}	h_{64}	h_{65}	h_{66}

b.

	Age class						
Year	1	2	3	4	5	6	Totals
1	$\hat{N}_{11}$	$\hat{N}_{12}$	$\hat{N}_{13}$	$\hat{N}_{14}$	$\hat{N}_{15}$	$\hat{N}_{16}$	$\hat{N}_1$
2	$\hat{N}_{21}$	$\hat{N}_{22}$	$\hat{N}_{23}$	$\hat{N}_{24}$	$\hat{N}_{25}$	$\hat{N}_{26}$	$\hat{N}_2$
3	$\hat{N}_{31}$	$\hat{N}_{32}$	$\hat{N}_{33}$	$\hat{N}_{34}$	$\hat{N}_{35}$	$\hat{N}_{36}$	$\hat{N}_3$
4	$\hat{N}_{41}$	$\hat{N}_{42}$	$\hat{N}_{43}$	$\hat{N}_{44}$	$\hat{N}_{45}$	$\hat{N}_{46}$	$\hat{N}_4$
5	$\hat{N}_{51}$	$\hat{N}_{52}$	$\hat{N}_{53}$	$\hat{N}_{54}$	$\hat{N}_{55}$	$\hat{N}_{56}$	$\hat{N}_5$
6	$\hat{N}_{61}$	$\hat{N}_{62}$	$\hat{N}_{63}$	$\hat{N}_{64}$	$\hat{N}_{65}$	$\hat{N}_{66}$	$\hat{N}_6$

Figure 9.10. Matrix of (a) age-at-harvest data and the corresponding matrix of (b) age- and year-specific abundance levels and total population sizes.

If all mortality was because of harvest or exploitation, the age-at-harvest data for a cohort could be summed to estimate the initial abundance of that cohort in any year. Summing across age classes would then provide an estimate of total annual abundance. For cohorts that have worked their way completely through the population, and the last individuals have been harvested, summing would be adequate (Fig. 9.10). However, for cohorts still in the population, the sum of the diagonal terms would not represent the total size of that cohort in any given year.

It is unrealistic to assume all animals are eventually harvested. Instead, the observed harvest numbers represent the animals that survived to a particular age and were eventually harvested, and the take reported. To go from the age-at-harvest data to the age- and year-specific abundance levels (Fig. 9.10), information on age-specific survival rates and harvest probabilities must be known. The age-at-harvest data alone do not have all of the information needed to estimate the prerequisite survival and exploitation rates. Auxiliary information and/or simplified assumptions are necessary to ultimately convert age-at-harvest data to abundance estimates.

9.8.1   Virtual Population Analysis (Fry 1949, 1957)

The Fry (1949, 1957) virtual population analysis (VPA) is a refinement or extension of the earlier method of Derzhavin (1922). In the Fry (1949) method, separate age-composition estimates are used each year, whereas Derzhavin (1922) uses a single composite estimate across all years. This refinement has supplanted the Derzhavin (1922) method. The basic data used by the Fry VPA method are the age-at-harvest values (Fig. 9.10).

The Fry (1949) method estimates minimum population size by summing harvest numbers over the lifetime of the cohort. The estimate of N_{ij}, the abundance of the jth age class in the ith year, is the sum

$$\hat{N}_{ij} = \sum_{k=0}^{\min(A,Y)} h_{i+k,j+k},$$ (9.142)

where

h_{ij} = number of animals harvested in year i of age class j;
A = maximum age class;
Y = maximum number of years of data collection.

For example, the estimate of N_{13} (Fig. 9.10) would be calculated as

$$\hat{N}_{13} = h_{13} + h_{24} + h_{35} + h_{46},$$

whereas N_{24} would be estimated by

$$\hat{N}_{24} = h_{24} + h_{35} + h_{46}.$$

Total abundance in any year would then be estimated by the sum

$$\hat{N}_i = \sum_{j=1}^{A} \hat{N}_{ij}.$$ (9.143)

For example, N_1 (Fig. 9.10) would be estimated by

$$\hat{N}_1 = \hat{N}_{11} + \hat{N}_{12} + \hat{N}_{13} + \hat{N}_{14} + \hat{N}_{15} + \hat{N}_{16}.$$

Summing works well for cohorts that have entirely passed through the age-at-harvest matrix. For the example (Fig. 9.10), only abundance for the first year can be computed by the Fry (1949) VPA method. For long-lived species, considerable lag time may exist between the last viable abundance estimate and the current year. For those years with complete information, the Fry method provides a minimum population size. Only as harvest mortality approaches total annual mortality will the Fry VPA estimate approach actual abundance.

The Fry (1949) method can also be used to estimate a maximum exploitation or harvest mortality rate. By using the quotient

$$\hat{M}_{H_{ij}} = \frac{h_{ij}}{\hat{N}_{ij}}, \tag{9.144}$$

the probability of harvest mortality for the jth age class in year i ($M_{H_{ij}}$) can be estimated. If h_{ij} is the actual age-at-harvest and $\hat{N}_{ij}$ is an underestimate of actual abundance based on VPA, the $\hat{M}_{H_{ij}}$ will be positively biased.

For years and cohorts with incomplete harvest information, several recourses exist. The least desirable is to wait patiently until harvest records are complete. The other alternative is to estimate an exploitation rate that converts the harvest numbers to estimates of prehunt abundance, where

$$\hat{N}_{ij} = \frac{h_{ij}}{\hat{M}_{H_{ij}}}. \tag{9.145}$$

One approach is to estimate $M_{H_{ij}}$ from other age classes and years with complete information (Eq. 9.144). Assuming common harvest mortality probabilities, the abundance for incomplete cohorts in recent years can be estimated. Choice of which historical values to use, or pool across, will depend on data patterns and knowledge of the population. Fryxell et al. (1988) estimated exploitation rates for incomplete cohorts by using a catch-effort relationship (Section 9.8.4).

The minimum population sizes from the Fry (1949) method can be improved by including known deaths from other sources of mortality, including natural mortality and crippling losses. Another approach is to adjust the minimum abundance estimates by an estimate of the lifetime recovery rate or lifetime harvest probability. However, these values are likely to be unavailable or difficult to obtain without auxiliary tagging studies. Roseberry and Woolf (1991:22) state, ". . . normally, the value assigned to this parameter is only an educated guess; therefore, absolute population estimates from reconstruction must be treated accordingly."

If the unknown lifetime recovery rate is constant over time (i.e., harvest mortality constant), the Fry (1949) VPA estimates provide a viable index to track population trends over time. However, changes in hunting regulations or effort can bias the index, leading to false trends.

Assumptions

The assumptions for the Fry (1949) VPA abundance estimates include the following:

1. Age classification is accurate.
2. Harvest numbers are reported accurately.
3. Harvest mortality is the primary source of mortality in the population.
4. Natural mortality is low and constant over time.

If incomplete cohorts are used in the population reconstruction, there is an additional assumption:

5. Harvest mortality is constant, allowing extrapolation from complete to incomplete cohorts.

Roseberry and Woolf (1991) suggested that reconstruction methods tend to underestimate population size owing, in part, to the tendency to underage older adults at check stations. Underreporting of harvested animals and crippling losses will inherently bias the population estimates downward. Harvest numbers should be adjusted for reporting rates and crippling losses before reconstruction analysis. If natural mortality is not constant over time or if harvest mortality is not constant when reconstructing incomplete cohorts, the VPA estimates will be a poor index of abundance and not track true population trends.

Example 9.16: Fry (1949) Virtual Population Analysis Method, Black-Tailed Deer, Champion Tree Farm, Washington

Gilbert (1992) reported age-at-harvest data for female black-tailed deer taken from Champion International's Kapowsin Tree Farm in western Washington State. This restricted access site permitted a direct tally of all animals harvested. Cementum annuli and tooth wear were used to classify the age of harvested animals. Age-at-harvest data (Table 9.3) collected from 1979–1991 were reported by Gilbert (1992). Only the female harvest data are used in this example.

The 13 years of harvest records include age-at-harvest counts for 16 age classes (Table 9.3). Fewer years of monitoring than the number of age classes results in no cohort having complete recovery information for all age classes from 0.5 through 15.5 years. Thus, population reconstruction will need to account for incomplete histories, using estimated harvest mortality rates to escalate the incomplete records. Equation (9.142) can be used in the upper-right diagonal of Table 9.3 to estimate age- and year-class-specific abundance values directly (Table 9.4). For example, $N_{1979,7.5}$ was estimated by the sum

$$\hat{N}_{1979,7.5} = \sum_{k=0}^{\left(\substack{\min 15.5,\\1991}\right)} h_{1979+k,7.5+k}$$

$$= 2+4+1+3+1+1+0+0+0 = 12.$$

Other elements in the upper-right diagonal of Table 9.4 were computed similarly.

The middle-left and lower-left diagonals of Table 9.3 represent cohorts that have not been completely exploited through age class 15.5. For these cohorts, harvest numbers in the last year (i.e., $h_{1991,j}$) need to be converted to estimates of abundance by using Eq. (9.145), where

$$\hat{N}_{1991,j} = \frac{h_{1991,j}}{\hat{M}_{H1991,j}}.$$

For this analysis, age-specific harvest mortalities were calculated by using the harvest numbers in Table 9.3 and abundance estimates in the upper-right diagonal of Table 9.4. For any age class, the harvest mortality rate was estimated by using the harvest numbers and abundance estimates over available years. For instance, age class 3.5 in 1979 had an estimated abundance of 44 (Table 9.4), of which 7 (Table 9.3) were harvested, for an estimated exploitation rate of

$$\hat{M}_{H3.5} = \frac{7}{44} = 0.1591.$$

For age class 4.5 during the years 1979–1980, harvest mortality was estimated to be

$$\hat{M}_{H4.5} = \frac{9+10}{49+37} = 0.2209.$$

In general, for age classes 3.5 through 15.5, harvest mortality was estimated from the following totals of harvest and abundance

	Age Class						
	3.5	4.5	5.5	6.5	7.5	8.5	9.5
Σh	7	19	12	20	22	29	23
ΣN	44	86	95	107	99	88	72
$\hat{M}_{H_j}$	0.1591	0.2209	0.1263	0.1869	0.2222	0.3295	0.3194

	10.5	11.5	12.5	13.5	14.5	15.5
Σh	22	12	10	10	6	3
ΣN	57	38	27	19	9	3
$\hat{M}_{H_j}$	0.3860	0.3158	0.3704	0.5263	0.6667	1

By using observed harvests in 1991 and estimated age-specific harvest mortality probabilities, abundance in 1991 was calculated for age classes 3.5 through 15.5. The middle-left diagonal in Table 9.4 was then calculated by summing across the remaining harvest numbers for years 1979–1990. For example, abundance of age class 5.5 in 1991 was estimated to be

$$\hat{N}_{1991,5.5} = \frac{3}{0.1263} = 23.8,$$

using the estimated age-specific harvest mortality probability of 0.1263 for 5.5-year-olds. The abundance of age class 4.5 in 1990 was then estimated by

$$\hat{N}_{1990,4.5} = \hat{N}_{1991,5.5} + h_{1990,4.5}$$
$$= 23.8 + 10 = 33.8$$

Table 9.3. Age-at-harvest data for female black-tailed deer, Champion Tree Farm, Washington, 1979–1991.

Year	0.5	1.5	2.5	3.5	4.5	5.5	6.5	7.5	8.5	9.5	10.5	11.5	12.5	13.5	14.5	15.5	Totals
1979	9	18	21	7	9	2	7	2	3	2	1	1	0	0	0	0	82
1980	11	17	9	14	10	6	3	4	4	1	4	2	1	0	2	0	88
1981	17	19	14	13	16	4	7	2	3	1	2	1	1	1	1	0	102
1982	28	14	7	18	5	3	3	5	9	5	3	1	2	2	0	0	105
1983	33	46	23	23	20	7	19	9	6	5	5	1	2	3	1	0	203
1984	22	16	25	9	7	6	7	7	4	6	4	0	1	2	1	1	118
1985	11	23	20	17	3	3	3	8	6	3	2	2	0	0	0	0	101
1986	29	18	22	15	11	10	2	6	2	1	1	2	0	0	0	0	119
1987	14	14	11	4	4	3	7	3	1	3	0	2	3	1	0	0	70
1988	21	21	17	8	8	7	4	2	1	2	3	0	0	0	0	0	94
1989	14	22	14	17	10	2	4	7	0	1	1	1	0	1	1	0	95
1990	18	32	20	12	10	8	1	2	1	1	2	0	1	0	0	2	110
1991	22	22	17	10	5	3	0	2	0	0	0	0	1	0	0	0	82

(Data from Gilbert 1992.)

and

$$\hat{N}_{1989,3.5} = \hat{N}_{1991,5.5} + h_{1990,4.5} + h_{1989,3.5}$$
$$= 23.8 + 10 + 17 = 50.8,$$

and so on. The above analysis assumes the age-specific harvest mortalities were the same across all years 1979–1991.

For the lower-left corner of Table 9.4, age-specific harvest mortality rates had to be calculated for age classes 0.5 through 2.5. These mortality probabilities were computed by using the estimated abundance levels from the middle-left diagonal. For example, in age class 2.5 for the years 1979–1990, 203 animals (Table 9.3) were harvested from a total abundance of 725.9 (Table 9.4) for that age class. The harvest probability for age class 2.5 animals is then estimated to be

$$M_{2.5} = \frac{203}{725.9} = 0.2796.$$

The abundance of age class 2.5 in the year 1991 is then estimated to be

$$\hat{N}_{1991,2.5} = \frac{17}{0.2796} = 60.8.$$

The abundance of that cohort the previous year, in turn, is estimated to be

$$\hat{N}_{1990,1.5} = \hat{N}_{1991,2.5} + h_{1990,1.5}$$
$$= 60.8 + 32 = 92.8.$$

Reconstruction of the lower-left corner of Table 9.4 is based on the assumption of constant age-specific harvest mortality.

Finally, annual abundance levels are calculated as the row totals of age-specific abundance values (Table 9.4). For this population, annual abundance ranged from a low 300.1 in 1987 to a high of 493.7 in 1982.

Table 9.4. Estimates of age- and year-class-specific abundance and annual abundance for black-tailed deer, Champion Tree Farm, Washington, 1979–1991, based on the Fry (1949) virtual population analysis.

Year	0.5	1.5	2.5	3.5	4.5	5.5	6.5	7.5	8.5	9.5	10.5	11.5	12.5	13.5	14.5	15.5	Totals
1979	99.7	77	87	44	49	28	24	12	11	13	8	3	1	2	0	0	458.7
1980	79	90.7	59	66	37	40	26	17	10	8	11	7	2	1	2	0	455.7
1981	86	68	73.7	50	52	27	34	23	13	6	7	7	5	1	1	0	453.7
1982	142	69	49	59.7	37	36	23	27	21	10	5	5	6	4	0	0	493.7
1983	101	114	55	42	41.7	32	33	20	22	12	5	2	4	4	2	0	489.7
1984	91	68	68	32	19	21.7	25	14	11	16	7	0	1	2	1	1	377.7
1985	66	69	52	43	23	12	15.7	18	7	7	10	3	0	0	0	0	325.7
1986	110.8	55	46	32	26	20	9	12.7	10	1	4	8	1	0	0	0	335.5
1987	83.6	81.8	37	24	17	15	10	7	6.7	8	0	3	6	1	0	0	300.1
1988	125.9	69.6	67.8	26	20	13	12	3	4	5.7	5	0	1	3	0	0	355.9
1989	106.8	104.9	48.6	50.8	18	12	6	8	1	3	3.7	2	0	1	3	0	368.7
1990	101.7	92.8	82.9	34.6	33.8	8	10	2	1	1	2	2.7	1	0	0	2	375.4
1991	111.1	83.7	60.8	62.9	22.6	23.8	0	9.0	0	0	0	0	2.7	0	0	0	376.5

(Data from Gilbert 1992.)

Discussion of Utility

The Fry (1949) method is of very limited value if only years with complete cohort data are to be analyzed. The longer the lifespan of the animal, the longer would be the time lag in obtaining timely abundance estimates. In lieu of waiting, incomplete cohort data must be converted to abundance estimates by using available information on exploitation rates. Unless auxiliary marking studies have been performed to estimate mortality, this information is unlikely to exist.

The exploitation rates estimated from the Fry VPA analysis will tend to be overestimated because abundance is underestimated. In turn, when these exploitation rates are used to expand the harvest counts of the incomplete cohorts, the age- and year-specific abundance will be underestimated. As noted in the example, if the harvest count is zero in the last year for a cohort, the expansion factor has no effect, leading to further underestimation. This problem can be partially rectified by beginning the abundance estimation by using the last year a cohort had a positive harvest count.

The application of VPA to population reconstruction is usually limited to locations that have controlled access, in which harvest numbers can be accurately tallied. However, the approach can also be used when harvest is subsampled, as long as the sampling fraction is known. The expanded counts from the sample survey could then be used in the population reconstruction. Another limitation is that point estimates of abundance from the Fry method do not have corresponding estimates of variance. Neither sampling error, stochastic variability, nor variance in mortality rates is used to estimate the variance of the projected abundance levels.

9.8.2 Virtual Population Analysis (Gulland 1965)

The VPA approach of Gulland (1965) begins with the same age-specific harvest information (Table 9.3) as that of Fry (1949). The Gulland (1965) method back-calculates cohort abundance by using a nonlinear sequential system of equations. This method expresses the ratio of abundance to catch as a nonlinear function of instantaneous natural (μ_N) and harvest (μ_H) mortality rates.

The VPA of Gulland (1965) is based on two fundamental equations:

$$N_{i+1} = N_i e^{-(\mu_{H_i} + \mu_N)} \tag{9.146}$$

and

$$h_i = N_i \left(\frac{\mu_{H_i}}{\mu_{H_i} + \mu_N} \right) (1 - e^{-(\mu_{H_i} + \mu_N)}), \tag{9.147}$$

where

μ_N = instantaneous natural mortality rate;
μ_{H_i} = instantaneous harvest mortality rate in year i.

The first equation states cohort abundance at age $i + 1$ is a function of cohort abundance at age i and the probability of surviving to the next year. The second equation states the harvest at age i is equal to the cohort abundance at age i times the annual mortality rate and the fraction of mortality due to exploitation. Rearranging Eq. (9.147),

$$N_i = \frac{h_i(\mu_{H_i} + \mu_N)}{\mu_{H_i}(1 - e^{-(\mu_{H_i} + \mu_N)})}. \tag{9.148}$$

Substituting Eq. (9.148) into Eq. (9.146) yields

$$N_{i+1} = \frac{h_i(\mu_{H_i} + \mu_N)e^{-(\mu_{H_i} + \mu_N)}}{\mu_{H_i}(1 - e^{-(\mu_{H_i} + \mu_N)})},$$

or, equivalently,

$$\frac{N_{i+1}}{h_i} = \frac{(\mu_{H_i} + \mu_N)e^{-(\mu_{H_i} + \mu_N)}}{\mu_{H_i}(1 - e^{-(\mu_{H_i} + \mu_N)})}. \tag{9.149}$$

In practice, the estimation procedure begins by providing estimates of the harvest mortality rate for the oldest age class (i.e., $\mu_{H_{AMAX}}$) and the annual instantaneous natural mortality rate (μ_N). By using Eq. (9.147) and assuming mortality is 1, the abundance of a cohort in its last year N_{AMAX} (i.e., $AMAX$ stands for maximum age) is estimated, where

$$N_{AMAX} = \frac{h_{AMAX}}{\left(\dfrac{\mu_{H_{AMAX}}}{\mu_{H_{AMAX}} + \mu_N} \right)}.$$

At this point, Eq. (9.149) is used to iteratively solve for $\mu_{H_{AMAX-1}}$ from the expression

$$\frac{N_{AMAX}}{h_{AMAX-1}} = \frac{(\mu_{H_{AMAX-1}} + \mu_N)e^{-(\mu_{H_{AMAX-1}} + \mu_N)}}{\mu_{H_{AMAX-1}}(1 - e^{-(\mu_{H_{AMAX-1}} + \mu_N)})}.$$

Then $\mu_{H_{AMAX-1}}$ can be used to estimate N_{AMAX-1} by (9.146), where

$$N_{AMAX-1} = N_{AMAX} e^{(\mu_{H_{AMAX-1}} + \mu_N)}$$

or, in general,

$$N_{i-1} = N_i e^{(\mu_{H_{i-1}} + \mu_N)}. \tag{9.150}$$

The process continues with the iterative estimation of $\mu_{H_{AMAX-2}}$, then N_{AMAX-2}, etc. This procedure provides abundance estimates by cohort and year in the upper-right diagonal of the harvest matrix for cohorts that have completely passed through the population.

For incomplete cohorts in the middle-left and lower-left diagonals of the harvest matrix, final cohort abundance is estimated by Eq. (9.147), where

$$N_{MAX} = \frac{h_{MAX}}{\left(\dfrac{\mu_{H_{MAX}}}{\mu_{H_{MAX}} + \mu_N}\right)\left(1 - e^{-(\mu_{H_{MAX}} + \mu_N)}\right)} \tag{9.151}$$

and where $\mu_{H_{MAX}}$ = instantaneous harvest mortality rate for the maximum observed age class (i.e., MAX) of a cohort. Once N_{MAX} has been estimated, $\mu_{H_{MAX-1}}$ is iteratively solved from Eq. (9.149), where

$$\frac{N_{MAX}}{h_{MAX-1}} = \frac{(\mu_{H_{MAX-1}} + \mu_N)e^{-(\mu_{H_{MAX-1}} + \mu_N)}}{\mu_{H_{MAX-1}}\left(1 - e^{-(\mu_{H_{MAX-1}} + \mu_N)}\right)}. \tag{9.152}$$

Then $\mu_{H_{MAX-1}}$ can be used to estimate N_{MAX-1} from Eq. (9.150). This recursive process is continued, estimating the abundance levels of the incomplete cohort back through time.

After the age- and year-specific abundance levels have been estimated, population abundance is estimated by summing the cohort abundance values within a year. The results of the sequential solution of the Gulland model depends on the initial values used for $\mu_{H_{AMAX}}$. Different values of $\mu_{H_{AMAX}}$ can and often do result in very different estimates of cohort abundance N_{ij} and annual abundance N_i.

It is important to note that only initial values of $\mu_{H_{AMAX}}$ and μ_N are needed to analyze the upper-right diagonal of the harvest matrix. In the middle-left and lower-left diagonals of the matrix, age-specific exploitation rates for each age class are needed to complete the analysis. This may be more information than is typically known about a population. One possible source of mortality information is the data from the complete cohorts. For example,

$$E\left(\frac{h}{\hat{N}}\right) \doteq \left(\frac{\mu_H}{\mu_H + \mu_N}\right)\left(1 - e^{-(\mu_H + \mu_N)}\right). \tag{9.153}$$

With the value of μ_N used elsewhere in the analysis, the value of μ_H for that particular cohort and age class can be computed from Eq. (9.153). This approach assumes the subsequent values of μ_H are constant over time.

An iterative or tuned VPA (Pope and Shepherd 1985, Hilborn and Walters 1992) can be used in conjunction with the incomplete cohorts. The process begins with the μ_H values calculated from the complete cohorts, then with a new set of μ_H values based in part on the results from the analysis of the incomplete cohorts.

Assumptions

The assumptions of the Gulland VPA model include the following:

1. Age classification is accurate.
2. Harvest numbers (h_{ij}) are reported accurately.

3. The instantaneous natural mortality rate (μ_N) is constant and known.
4. Terminal ($\mu_{H_{AMAX}}$) is known.

The long history of using VPA has resulted in considerable research on its performance and robustness. The analysis is robust to errors in estimating $\mu_{H_{AMAX}}$ when harvest accounts for 50% or more of total mortality $\left(\text{i.e., } \dfrac{\mu_H}{\mu_H + \mu_N} \geq 0.50\right)$ or when cumulative μ_H over the lifetime of the cohort is greater than $\mu_H + \mu_N$ (Pope 1972, Megrey 1989). Moderate random fluctuations in μ_N are likely to produce relatively small changes in the values of cohort abundance (N_{ij}) and instantaneous harvest mortality (μ_H) (Pope 1972, Ulltang 1977, Megrey 1989). A bias of as much as 25% of μ_H may occur if the mean error in μ_N is of size 0.1 (Agger et al. 1973, Sims 1984). Ulltang (1977) also found the VPA analysis was relatively insensitive to seasonal trends in μ_N and μ_H. Sims (1982) suggested the estimate of N_{ij} is also insensitive to inseason fluctuations in the age composition of the harvest. Megrey (1989) stated that Pope's cohort analysis (Section 9.8.3) shares the same robustness as that of the Gulland VPA analysis.

Example 9.17: Gulland (1965) Virtual Population Analysis Method, Black-Tailed Deer, Champion Tree Farm, Washington

The female black-tailed deer age-at-harvest data of Gilbert (1992) will be subjected to the Gulland (1965) VPA. Based on the previous analysis (Table 9.4), the average exploitation rate across age classes 13.5 to 14.5 was estimated to be 0.5965 (= (0.5263 + 0.6667)/2). This mortality probability value corresponds to

$$M_H = 1 - e^{-\mu_H}$$
$$0.5965 = 1 - e^{-\mu_H}$$

or

$$\mu_H = 0.9076.$$

For this VPA analysis, we will assume the same harvest potential for the last age class as age classes 13.5 to 14.5, setting $\mu_{H_{AMAX}} = 0.9076$ for age class 15.5. It will be further assumed the natural survival probability is 0.85 annually, then

$$0.85 = e^{-\mu_N}$$

or

$$\mu_N = 0.1625.$$

Starting with the harvest of age class 15.5 in 1990 of 2 females, then N_{AMAX} for that cohort is calculated from

$$N_{AMAX} = \frac{h_{AMAX}}{\left(\dfrac{\mu_{H_{AMAX}}}{\mu_{H_{AMAX}} + \mu_N}\right)}$$

$$N_{15.5} = \frac{2}{\left(\dfrac{0.9076}{0.9076 + 0.1625}\right)} = 2.36.$$

Next, $\mu_{H_{14.5}}$ is iteratively solved from the equation

$$\frac{N_{AMAX}}{h_{AMAX-1}} = \frac{\left(\mu_{H_{AMAX-1}} + \mu_N\right)e^{-\left(\mu_{H_{AMAX-1}} + \mu_N\right)}}{\mu_{H_{AMAX-1}}\left(1 - e^{-\left(\mu_{H_{AMAX-1}} + \mu_N\right)}\right)}$$

$$\frac{2.36}{1} = \frac{\left(\mu_{H_{14.5}} + 0.1625\right)e^{-\left(\mu_{H_{14.5}} + 0.1625\right)}}{\mu_{H_{14.5}}\left(1 - e^{-\left(\mu_{H_{14.5}} + 0.1625\right)}\right)}$$

and found to be $\mu_{H_{14.5}} = 0.3282$. Continuing this sequential analysis yields the following parameter estimates for this cohort:

$$
\begin{array}{ll}
N_{15.5} = 2.36 & \mu_{H_{15.5}} = 0.9076 \\
N_{14.5} = 3.85 & \mu_{H_{14.5}} = 0.3282 \\
N_{13.5} = 4.53 & \mu_{H_{13.5}} = 0 \\
N_{12.5} = 8.56 & \mu_{H_{12.5}} = 0.4738 \\
N_{11.5} = 12.23 & \mu_{H_{11.5}} = 0.1944 \\
N_{10.5} = 16.55 & \mu_{H_{10.5}} = 0.1400 \\
N_{9.5} = 25.95 & \mu_{H_{9.5}} = 0.2871 \\
N_{8.5} = 37.01 & \mu_{H_{8.5}} = 0.1926 \\
N_{7.5} = 48.95 & \mu_{H_{7.5}} = 0.1171 \\
N_{6.5} = 65.16 & \mu_{H_{6.5}} = 0.1235 \\
N_{5.5} = 83.15 & \mu_{H_{5.5}} = 0.0813 \\
N_{4.5} = 107.56 & \mu_{H_{4.5}} = 0.0949 \\
\end{array}
$$

By using the same process, the remainder of the upper-right diagonal of Table 9.5 is constructed.

For the middle-left diagonal of Table 9.5, the $\mu_{H_{MAX}}$ values were estimated from the pooled harvest and abundance values calculated from the upper-right diagonal for an age class. Values of $\mu_{H_{MAX}}$ were calculated from Eq. (9.153), assuming $\mu_N = 0.1625$. These $\mu_{H_{MAX}}$ values were then used to estimate the values of N_{ij} for this section of the table. For example, $\mu_{H_{MAX}}$ from age class 12.5 in 1991 was based on the equation

$$\frac{0+1+1+\cdots+3+0}{1.6+2.5+\cdots+1.4} = \left(\frac{\mu_{H_{12.5}}}{\mu_{H_{12.5}} + 0.1625}\right)\left(1 - e^{-\left(\mu_{H_{12.5}} + 0.1625\right)}\right)$$

yielding $\hat{\mu}_{H_{12.5}} = 0.3527$. The abundance for the age 12.5 cohort in 1991 is then estimated by Eq. (9.151), where

$$\hat{N}_{12.5} = \frac{h_{12.5}}{\left(\dfrac{\mu_{H_{12.5}}}{\mu_{H_{12.5}} + \mu_N}\right)\left(1 - e^{-\left(\mu_{H_{12.5}} + \mu_N\right)}\right)}$$

$$= \frac{1}{\left(\dfrac{0.3527}{0.3527 + 0.1625}\right)\left(1 - e^{-\left(0.3527 + 0.1625\right)}\right)} = 3.63.$$

From there, Eq. (9.152) is use to solve for $\mu_{H_{11.5}}$. Then with an estimate of $\mu_{H_{11.5}}$, Eq. (9.150) is used to estimate cohort abundance at age 11.5. This iterative process between Eq. (9.152) and Eq. (9.150) is used to estimate the cohort abundance back through time.

The rest of the middle-left diagonal of Table 9.5 is completed similarly. The lower-left diagonal of Table 9.5 is completed based on the abundance and harvest information in the completed middle-left diagonal in an analogous manner.

Discussion of Utility

The VPA approach of Gulland (1965) is a much more realistic approach to population reconstruction than is the approach of Fry (1949). The Fry (1949) method assumes the vast majority of deaths are owing to harvest, whereas the Gulland (1965) method adjusts cohort abundance for both natural and harvest mortality. Few assumptions (i.e., μ_N and $\mu_{H_{AMAX}}$) are required to reconstruct abundance for cohorts that have completely passed through the population. These assumptions are sufficient if investigators can patiently wait for cohorts to pass through a population before annual abundance is estimated. Longer animal life spans will require more patience if abundance is to be estimated with minimal assumptions.

The practical and compelling alternative to waiting years for cohorts to pass through a population is to reconstruct the incomplete cohort data. This requires age-specific harvest rates. Should the estimates of age-specific harvest rates come from the reconstructed cohort data, there is the additional assumption that these harvest rates are constant over time. Changes in harvest regulations and hunter success can violate this assumption, resulting in greater uncertainty and potential bias in current years. As a result, population estimates in later years may be better treated as population indices than absolute abundance values.

The appeal of population reconstruction is the ready use of commonly collected age-at-harvest data. Expensive animal marking or visual count surveys need not be performed

Table 9.5. Estimates of age- and year-class-specific abundance and annual abundance for black-tailed deer, Champion Tree Farm, Washington, 1979–1991, based on the Gulland (1965) virtual population analysis.

Year	0.5	1.5	2.5	3.5	4.5	5.5	6.5	7.5	8.5	9.5	10.5	11.5	12.5	13.5	14.5	15.5	Totals
1979	183.1	139.6	119.7	85.7	107.6	54.8	38.7	19.1	18.7	21.1	13.3	4.0	1.6	2.8	0.0	0.0	809.6
1980	151.0	140.0	110.4	88.8	66.4	83.2	44.7	26.4	14.4	13.1	16.1	10.4	2.5	1.4	2.4	0.0	771.1
1981	130.6	110.9	106.1	81.9	60.8	47.2	65.2	35.3	18.8	8.6	10.2	10.0	7.0	1.2	1.2	0.0	694.9
1982	182.1	98.1	87.8	73.7	65.0	48.9	36.5	49.0	28.1	13.2	6.4	6.9	7.6	5.0	0.0	0.0	708.3
1983	147.0	112.6	62.3	53.6	44.3	48.8	24.2	28.2	37.0	15.7	6.7	2.7	4.9	4.6	2.5	0.0	595.0
1984	130.0	110.2	72.8	44.7	39.1	32.1	35.1	14.2	15.8	26.0	8.7	0.0	1.4	2.4	1.2	1.2	534.7
1985	357.9	89.3	75.3	46.3	35.2	30.5	24.6	22.5	6.6	9.7	16.6	3.8	0.0	0.0	0.0	0.0	718.2
1986	190.4	287.6	55.8	50.2	29.3	20.8	24.1	15.4	17.3	4.7	5.5	12.2	1.4	0.0	0.0	0.0	714.5
1987	155.0	148.9	234.4	43.7	39.0	22.1	11.3	17.7	12.1	11.9	0.0	3.8	8.6	1.2	0.0	0.0	709.7
1988	263.7	112.4	111.0	191.8	29.8	26.7	15.1	7.7	14.1	8.5	7.4	0.0	1.4	4.5	0.0	0.0	794.2
1989	203.4	203.9	82.7	78.7	153.9	23.5	19.1	6.5	6.6	11.1	6.3	5.3	0.0	1.2	3.8	0.0	805.9
1990	195.8	160.0	154.9	59.3	57.7	123.4	19.1	14.4	4.6	4.7	7.6	5.3	3.6	0.0	0.0	2.4	812.7
1991	213.3	149.9	106.6	122.5	45.8	46.3	0.0	14.4	0.0	0.0	0.0	0.0	3.6	0.0	0.0	0.0	702.4

(Data from Gilbert 1992.)

to obtain population abundance estimates. However, auxiliary data (Section 9.8.5) are needed to relax the assumptions of population reconstruction for incomplete cohorts and provide variance and confidence interval estimates of abundance. The population reconstruction models also provide an excellent framework to decide which demographic parameters need estimation and which auxiliary marking studies should be performed. The Gulland (1965) model assumes concurrent and independent natural and harvest mortality factors operating on a population. If harvest occurs over a relatively short period of time when natural mortality is insignificant, the discrete-time model of Section 9.8.4 may be considered.

9.8.3   Cohort Analysis (Pope 1972)

The cohort analysis of Pope (1972) is based on the same two fundamental equations that formed the basis of the Gulland (1965) VPA analysis (Eqs. 9.146, 9.147). However, Pope (1972) introduced a closed-form approximation to the exponential survival factor. The Pope (1972) method begins with Eq. (9.146)

$$N_{i+1} = N_i e^{-(\mu_{H_i} + \mu_N)}$$

and is rearranged, where

$$N_{i+1} e^{\mu_N} = N_i e^{-\mu_{H_i}}$$

or, equivalently,

$$N_{i+1} e^{\mu_N} = N_i - N_i(1 - e^{-\mu_{H_i}}). \tag{9.154}$$

Furthermore, Eq. (9.147),

$$h_i = N_i \left(\frac{\mu_{H_i}}{\mu_{H_i} + \mu_N} \right)(1 - e^{-(\mu_{H_i} + \mu_N)}),$$

can be rearranged so that

$$N_i = \frac{h_i(\mu_{H_i} + \mu_N)}{\mu_{H_i}(1 - e^{-(\mu_{H_i} + \mu_N)})}. \tag{9.155}$$

Substituting Eq. (9.155) into Eq. (9.154) yields

$$N_{i+1} e^{\mu_N} = N_i - h_i \frac{(\mu_{H_i} + \mu_N)(1 - e^{-\mu_{H_i}})}{\mu_{H_i}(1 - e^{-(\mu_{H_i} + \mu_N)})}. \tag{9.156}$$

Pope (1972) noted that

$$\frac{(\mu_{H_i} + \mu_N)(1 - e^{-\mu_{H_i}})}{\mu_{H_i}(1 - e^{-(\mu_{H_i} + \mu_N)})}$$

can be approximated by $e^{\frac{\mu_N}{2}}$ for values of $\mu_N < 0.3$ and $\mu_{H_i} < 1.2$. Hilborn and Walters (1992:355–356) recommended the approximation when $Z = \mu_{H_i} + \mu_N$ is between 0.1 and 1.7. Use of the approximation in Eq. (9.156) results in

$$N_{i+1} e^{\mu_N} = N_i - h_i e^{\frac{\mu_N}{2}},$$

leading to the estimation equation

$$N_i = N_{i+1}e^{\mu_N} + h_i e^{\frac{\mu_N}{2}}.$$ (9.157)

With values of N_{i+1}, h_i, and μ_N, Eq. (9.157) can be used to estimate N_i. With N_i, h_{i-1}, and μ_N, Eq. (9.157) can be used to estimate N_{i-1}, etc.

MacCall (1986) suggested the alternative approximation

$$\frac{\mu_N}{1 - e^{-\mu_N}}$$

instead of $e^{\frac{\mu_N}{2}}$. In which case,

$$N_{i+1}e^{\mu_N} = N_i - h_i \frac{\mu_N}{1 - e^{-\mu_N}},$$

leading to the estimation equation,

$$N_i = N_{i+1}e^{\mu_N} + h_i \frac{\mu_N}{1 - e^{-\mu_N}}.$$ (9.158)

The MacCall (1986) approximation works better for larger values of μ_N and is more robust to the assumption of when harvest occurs during the year.

The approach to cohort reconstruction depends on whether the cohort has completely passed through the population or not. For a completed cohort, estimation begins with a value for the terminal harvest mortality ($\mu_{H_{AMAX}}$) and the terminal harvest (h_{AMAX}) such that

$$N_{AMAX} = h_{AMAX}\left(\frac{\mu_{H_{AMAX}} + \mu_N}{\mu_{H_{AMAX}}}\right).$$ (9.159)

From there, Eq. (9.157) or Eq. (9.158) is used to estimate cohort abundance in prior years.

For incomplete cohorts, the process begins with a value of ($\mu_{H_{MAX}}$) for the last year of harvest, and abundance is estimated from the equation

$$N_{MAX} = h_{MAX}\left(\frac{\mu_{H_{MAX}} + \mu_N}{\mu_{H_{MAX}}}\right)\frac{1}{1 - e^{-(\mu_{H_{MAX}} + \mu_N)}}.$$ (9.160)

Equation (9.157) or Eq. (9.158) is used to estimate cohort abundance in previous years. Note that Eqs. (9.159) and (9.160) are related. When mortality in the last year of life is assumed to be one, Eq. (9.160) reduces to Eq. (9.159).

A distinction is needed between estimating the upper-right diagonal of the harvest matrix based on complete cohorts from the rest of the matrix. For the completed cohorts, only common values of μ_N and $\mu_{H_{AMAX}}$ are required. For the incomplete cohorts, age-specific exploitation rates are needed for all remaining age classes. Hence, considerably more information is needed if the entire harvest matrix (Fig. 9.10) is to be converted to abundance values.

Assumptions

Assumptions of the Pope (1972) method are basically the same as those of Gulland (1965). The major distinction between the two methods is the use of the approximation $e^{\frac{\mu_N}{2}}$. For a wide range of values of μ_H and μ_N, the approximation is reasonably accurate. Hilborn and Walters (1992) mention that although the error in the approximation may be small, the approximation is used numerous times in population reconstruction, and the errors can accumulate. They recommend avoiding the approximation of Pope (1972) if accuracy is a concern. Instead, the nonlinear function (9.156), which is essentially the Gulland (1965) model, should be used. The value of the Pope method is that the calculations can be readily incorporated into spreadsheet programs, making this method within the technical range of most everyone.

Example 9.18: Pope (1972) Cohort Analysis, Black-Tailed Deer, Champion Tree Farm, Washington

The female black-tailed deer cohort ending with age class 15.5 in 1990 (Table 9.3), previously analyzed with the Gulland (1965) method, is reanalyzed with the methods of Pope (1972) (Table 9.6). In the Gulland VPA Example 9.17, values of $\mu_{H_{AMAX}} = 0.9076$ and $\mu_N = 0.1625$ were used. By using these same parameter values, the Pope method estimates the abundance of $N_{15.5,1990}$ to be

$$N_{15.5} = \frac{h_{15.5}}{\left(\dfrac{\mu_{H_{15.5}}}{\mu_{H_{15.5}} + \mu_N}\right)}$$

$$= \frac{2}{\left(\dfrac{0.9076}{0.9076 + 0.1625}\right)} = 2.36,$$

based on Eq. (9.159). The cohort abundance the prior year is estimated from Eq. (9.157), where

$$N_{14.5} = N_{15.5}e^{\mu_N} + h_{14.5}e^{\frac{\mu_N}{2}}$$

$$= 2.36e^{0.1625} + 1e^{\frac{0.1625}{2}} = 3.86.$$

The abundance in the next preceding year is then

$$N_{13.5} = N_{14.5}e^{\mu_N} + h_{13.5}e^{\frac{\mu_N}{2}}$$

$$= 3.86e^{0.1625} + 0e^{\frac{0.1625}{2}} = 4.54.$$

Continuing the process yields the following abundance levels for the cohort over time:

$$N_{15.5} = 2.36 \quad N_{14.5} = 3.86$$
$$N_{13.5} = 4.54 \quad N_{12.5} = 8.59$$
$$N_{11.5} = 12.28 \quad N_{10.5} = 16.62$$

$$N_{9.5} = 26.06 \quad N_{8.5} = 37.16$$
$$N_{7.5} = 49.14 \quad N_{6.5} = 65.41$$
$$N_{5.5} = 83.45 \quad N_{4.5} = 107.94$$

Comparing these values with the previous example indicates good agreement between the Gulland (1965) and Pope (1972) calculations. In this example, $Z = \mu_H + \mu_N = 1.0701$, which is a Z-value within the range of 0.1 to 1.7 recommended by Hilborn and Walters (1992). The remainder of the upper-right diagonal (Table 9.6) is completed analogously.

The middle-left diagonal of Table 9.6 is completed by using the last years of harvest data and estimates of $\mu_{H_{MAX}}$ for each age class 3.5 to 14.5. The estimates of $\mu_{H_{MAX}}$ are computed identically to the approach in Gulland (1965), in which harvest and abundance values are pooled over time for a specific age class, so that

$$\frac{\Sigma h}{\Sigma \hat{N}} = \left(\frac{\mu_H}{\mu_H + \mu_N} \right)\left(1 - e^{-(\mu_H + \mu_N)}\right)$$

and solved for μ_H. Values of N_{MAX} are estimated analogous to Eq. (9.160) with abundance values in prior years computed by using Eq. (9.157). The lower-left diagonal of Table 9.6 is completed by using the information in the completed middle-left diagonal in an analogous manner.

Discussion of Utility

The Pope (1972) cohort analysis is an appreciable improvement over the Fry (1949) VPA. The Pope (1972) method takes into account both natural survival and terminal harvest rates, resulting in a more realistic and higher estimates of cohort and population abundance. Both the advantages and disadvantages of the Pope (1972) method are in its closed-form approximations to the nonlinear iterative solutions of the Gulland (1965)

Table 9.6. Estimates of age- and year-class-specific abundance and annual abundance for black-tailed deer, Champion Tree Farm, Washington, 1979–1991, based on the Pope (1972) virtual population analysis.

Year	0.5	1.5	2.5	3.5	4.5	5.5	6.5	7.5	8.5	9.5	10.5	11.5	12.5	13.5	14.5	15.5	Totals
1979	183.1	118.3	105.7	85.6	107.9	54.8	36.4	18.6	18.3	21.0	13.4	3.9	1.5	2.6	0.0	0.0	771.2
1980	112.9	139.9	92.3	77.0	66.3	83.5	44.7	24.5	14.0	12.8	16.0	10.5	2.4	1.3	2.2	0.0	700.2
1981	111.0	78.5	106.0	66.5	50.7	47.1	65.4	35.3	17.1	8.2	10.0	9.9	7.1	1.1	1.1	0.0	614.8
1982	166.3	81.4	60.2	73.5	51.9	40.3	36.4	49.1	28.1	11.8	6.0	6.6	7.5	5.1	0.0	0.0	624.4
1983	102.5	98.9	48.0	30.0	44.1	37.7	16.7	28.2	37.2	15.6	5.4	2.4	4.7	4.5	2.5	0.0	478.4
1984	129.6	72.4	61.0	32.5	19.0	31.9	25.6	7.8	15.6	26.1	8.7	0.0	1.1	2.2	1.1	1.2	435.8
1985	79.9	89.0	43.1	36.2	24.9	13.4	24.4	14.3	1.1	9.6	16.6	3.7	0.0	0.0	0.0	0.0	356.1
1986	190.8	51.3	55.4	22.8	20.6	11.9	9.6	15.2	10.3	0.0	5.4	12.3	1.3	0.0	0.0	0.0	406.8
1987	155.3	149.2	33.4	43.4	15.7	14.8	3.7	5.4	12.0	6.0	0.0	3.7	8.6	1.1	0.0	0.0	452.3
1988	263.5	112.7	111.2	21.1	29.5	6.9	8.9	1.3	3.6	8.3	2.4	0.0	1.3	4.5	0.0	0.0	575.1
1989	166.6	203.7	82.9	78.8	8.7	23.2	2.2	1.1	1.1	2.2	6.2	1.1	0.0	1.1	3.9	0.0	582.6
1990	155.2	128.7	154.7	59.4	57.8	0.0	18.8	0.0	0.0	0.0	0.0	5.2	0.0	0.0	0.0	2.4	582.2
1991	168.7	115.3	79.9	122.3	45.9	46.4	0.0	14.2	0.0	0.0	0.0	0.0	3.5	0.0	0.0	0.0	596.1

(Data from Gilbert 1992.)

model. The advantage lies solely in the ease of the resulting spreadsheet computations. The recursive use of the approximation of the model can induce small errors at the cohort level but larger estimation errors at the population level. Unless computational ease is paramount, the Gulland (1965) model should be used instead of the Pope (1972) cohort analysis.

9.8.4   Discrete-Time Virtual Population Analysis (Fryxell et al. 1988)

Fryxell et al. (1988) used discrete-time VPA equations to reconstruct the abundance of moose (*Alces alces*) in Newfoundland, Canada. Gilbert and Raedeke (2004) also used the method to reconstruct black-tailed deer abundance at the Champion Tree Farm in Washington State. The method is based on two fundamental equations:

$$N_{i+1} = (N_i - h_i)S_i \tag{9.161}$$

and

$$h_i = N_i(1 - e^{-cg_i}), \tag{9.162}$$

where

N_i = abundance of the ith age class;
S_i = age-specific natural survival probability for age class i to $i + 1$;
h_i = number of animals of age class i harvested;
c = vulnerability or catch coefficient;
g_i = harvest effort in the ith year.

Equations (9.161) and (9.162) are based on the premise that harvest occurs over a relatively short period of time, so that natural mortality can be assumed to be negligible during the period of exploitation. The cohort analysis of Fryxell et al. (1988) is based on rearranging Eq. (9.161) so that

$$N_i = \frac{N_{i+1}}{S_i} - h_i. \tag{9.163}$$

For cohorts that have passed completely through the population, the cohort analysis begins by assuming that for the last appearance of the cohort,

$$N_{AMAX} = h_{AMAX}.$$

In other words, all animals in the last age class are harvested. For the next-to-last age class of that cohort, abundance is estimated by

$$N_{AMAX-1} = \frac{h_{AMAX}}{S_{AMAX-1}} - h_{AMAX-1}. \tag{9.164}$$

The remaining age classes for that cohort are estimated similarly, by recursively using Eq. (9.164).

For incomplete cohorts, abundance at the time of the last harvest is estimated by rearranging Eq. (9.162), where

$$N_i = \frac{h_i}{\left(1 - e^{-cg_i}\right)}. \tag{9.165}$$

The harvest effort (g_i) is required for the years of concern to use Eq. (9.165). The vulnerability coefficient (c) is estimated from age classes and cohorts that are complete where, from Eq. (9.165),

$$c = \frac{-\ln\left(1 - \dfrac{h_i}{N_i}\right)}{g_i}. \tag{9.166}$$

Fryxell et al. (1988) suggested estimating c by the average vulnerability ($\bar{c}$) calculated across age classes and cohorts. Alternatively, estimation of the vulnerability coefficient can be based on regression analysis and the completed years of cohort analysis. By using the annual abundance estimates ($\hat{N}_i$) and associated harvests (h_i), a nonlinear regression analysis of the form

$$E\left(\frac{h_i}{\hat{N}_i}\bigg| g_i\right) = \left(1 - e^{-cg_i}\right) \tag{9.167}$$

can be used to estimate c. Subsequent use of c to estimate the abundance of incomplete cohorts assumes c is invariant over time and age classes. Gilbert (1992) found different catch-effort relationships from 1979–1991 in a black-tailed deer population in Washington. Changes in hunting strategies, regulations, or hunting conditions can influence the value of the vulnerability coefficient. Therefore, caution must be exerted over the choice of which years to use in the calculation of c.

The other key information needed to use the Fryxell et al. (1988) cohort analysis approach is the age-specific natural survival probabilities (S_i). In many situations, information on age-specific survival probabilities will be absent, and some common value (S) must be used across years and age classes.

Assumptions

The assumptions of the Fryxell et al. (1988) cohort analysis include the following:

1. Age classification is accurate.
2. Harvest numbers are reported accurately.
3. A terminal age exists beyond which few animals survive.
4. Harvest accounts for most if not all of the mortality at the terminal age (i.e., $N_{AMAX} = h_{AMAX}$).
5. Age-specific natural survivorship is available for all age classes or a common natural survival probability exists across all age classes.
6. The vulnerability coefficient is constant across age classes and years.

Unlike the Gulland (1965) and Pope (1972) methods that assume the harvest of animals in the last year of life is only a part of total mortality, the Fryxell et al. (1988) method assumes all deaths in the last year are harvest related. If untrue, this assumption will underestimate subsequent cohort and annual abundance values. The assumption of constant vulnerability should be assessed through data inspections and goodness-of-fit tests. Should Eq. (9.167) adequately fit the model with capture probabilities <0.2, then

$$E\left(\frac{h_i}{\hat{N}_i}\bigg|g_i\right) \doteq cg_i,$$

and a linear relationship between CPUE $\left(\text{i.e.,}\dfrac{h_i}{\hat{N}_i}\right)$ and effort (g_i) should be observed.

As in all reconstruction methods so far reviewed, the subsequent annual abundance estimators are sensitive to the stochastic variability of the final harvest numbers. The consequence of too few or too many kills in the last age class can dramatically affect abundance estimates in population reconstruction.

Discussion of Utility

The Fryxell et al. (1988) approach uses auxiliary CPUE data to provide missing parameter values needed in population reconstruction. Deriso et al. (1985) provided a review of using auxiliary information in conjunction with catch-at-age analyses. The Fryxell et al. (1988) method is a logical and simple extension of these approaches. Fournier and Archibald (1982) provide additional insights into analysis of catch-at-age or age-at-harvest information with auxiliary data.

Although the basic model of Fryxell et al. (1988) specifies the use of age-specific survival probabilities, the general and more tractable approach uses a common estimate of natural survival, not only over age classes but also across years. These simplified assumptions reduce the VPA and cohort population analyses to indices of abundance. LaPointe et al. (1989) discussed the errors in population trends that might occur if there are errors in estimates of harvest and natural mortality rates. Additionally, trends in natural mortality or vulnerability not accounted for in the VPA analyses will also induce spurious population trends in the reconstruction.

9.8.5 Statistical Age-at-Harvest Analysis (Gove et al. 2002)

Traditional VPA or cohort analyses have inherent difficulties in estimating abundance in the presence of incomplete cohorts. Thus, abundance is often underestimated in later years with an unrealistic portrayal of declining abundance trends. Another shortcoming is the absence of variance estimates for either cohort or annual abundance estimates. Still another shortcoming is the necessity to assume mortality and/or vulnerability is constant to estimate abundance from the incomplete cohorts. The result is uncertainty in both the accuracy and precision of VPA or cohort analyses.

Statistical age-at-harvest methods provide a formal framework for population reconstruction of both past and current cohorts. The maximum likelihood framework readily permits the inclusion of auxiliary information needed for abundance estimation and realistic model formulation. More auxiliary information means more flexibility in model construction and higher degree of model specificity. No single statistical model exists but rather there is a suite of options dictated by the demographics and availability of information. Gove et al. (2002) characterized some of these options and illustrated the approach by using age-at-harvest data for elk from northern Idaho. The general form of the joint likelihood model is

$$L_{\text{Joint}} = L_{\text{Age-at-Harvest}} \cdot L_{\text{Auxiliary}} \cdot L_{\text{Reporting}}, \tag{9.168}$$

where

$L_{\text{Age-at-Harvest}}$ = likelihood model describing the cohort data as a function of survival and harvest parameters;

$L_{\text{Auxiliary}}$ = likelihood(s) permitting the estimation of one or more of the demographic parameters in the age-at-harvest likelihood;

$L_{\text{Reporting}}$ = likelihood describing the probability an animal is reported harvested in the age-at-harvest data.

If the reporting rate for harvested animals is 100%, the third likelihood component can be omitted from Eq. (9.168). The locker check or field check models (Section 6.2.1) are an option for the reporting rate likelihood.

Gove (1997) examined four alternative reconstruction models that differed in their assumptions concerning natural survival and harvest parameters. The alternatives included some of the more likely and tractable model formulations and included the following:

Model MpS = Assumes constant natural survival (S) over time and across all age classes, as well as a constant harvest probability (p) over time and all age classes

Model MpS_A = Assumes age-specific natural survival probabilities (S_j) that are constant over time and a harvest probability (p) that is constant over time and across all age classes

Model Mp_YS = Assumes a constant natural survival probability (S) over time and across age classes and harvest probabilities (p_i) that vary between years but constant across age classes

Model (Mp_YS_A) = Assumes age-specific natural survival probabilities (S_j) that are constant over time and harvest probabilities (p_i) that vary between years but are constant across age classes

The number of possible models is almost endless. The most general model, $Mp_{YA}S_{YA}$, has age- and year-specific natural survival and harvest probabilities. However, as these models become more generalized and include more parameters, more of the parameters need to be estimated from auxiliary non-age-at-harvest data. Models MpS and MpS_A each require one parameter be estimated outside of the age-at-harvest likelihood. For models Mp_YS and Mp_YS_A, two parameters must be provided by the auxiliary likelihoods (Gove et al. 2002).

The basic unit of cohort analyses is the cohort itself, each represented by a diagonal in the age-at-harvest matrix (Fig. 9.10). Different cohorts are represented by the elements along the top row and left column of the age-at-harvest matrix (Fig. 9.10b). Each of these cohorts is tracked through time to form the rest of the age-at-harvest matrix. For example, the diagonal elements $h_{11}, h_{22}, \ldots, h_{66}$ of Fig. 9.10a represent a multinomial sample with parameters:

N_{11} = abundance of age class 1 in year 1;

S_{ij} = probability an animal survives natural causes of mortality in the ith age class ($j = 1, \ldots, A$) in the ith year ($i = 1, \ldots, Y$);

p_{ij} = probability an animal is harvested in the jth age class ($j = 1, \ldots, A$) in the ith year ($i = 1, \ldots, Y$);

R_i = probability a harvested animal is reported in the ith year.

Then, for example, the expected number of animals harvested in age class $j = 1$ in year $i = 1$ is

$$E(h_{11}) = N_{11} p_{11} R_1, \tag{9.169}$$

and the expected value of h_{22} is

$$E(h_{22}) = N_{11}(1 - p_{11}) S_{11} p_{22} R_2. \tag{9.170}$$

The multinomial likelihood for cohort N_{11} can be written as

$$
\begin{aligned}
L(N_{11}, \underset{\sim}{p}, \underset{\sim}{S}, \underset{\sim}{R} | \underset{\sim}{h}) = &\binom{N_{11}}{\underset{\sim}{h}} (p_{11} R_1)^{h_{11}} \left((1 - p_{11}) S_{11} p_{22} R_2 \right)^{h_{22}} \\
&\cdot \left((1 - p_{11}) S_{11} (1 - p_{22}) S_{22} p_{33} R_3 \right)^{h_{33}} \\
&\cdot \left((1 - p_{11}) S_{11} (1 - p_{22}) S_{22} (1 - p_{33}) S_{33} p_{44} R_4 \right)^{h_{44}} \\
&\cdot \left((1 - p_{11}) S_{11} (1 - p_{22}) S_{22} (1 - p_{33}) S_{33} (1 - p_{44}) S_{44} p_{55} R_5 \right)^{h_{55}} \\
&\cdot \left((1 - p_{11}) S_{11} (1 - p_{22}) S_{22} (1 - p_{33}) S_{33} (1 - p_{44}) \right. \\
&\quad \left. S_{44} (1 - p_{55}) S_{55} p_{66} R_6 \right)^{h_{66}} \\
&\cdot (1 - \Sigma)^{N_{11} - h_{11} - h_{22} - h_{33} - h_{44} - h_{55} - h_{66}},
\end{aligned}
\tag{9.171}
$$

where Σ is the sum of the other 6 cell probabilities. The age-at-harvest likelihood is then a product of the individual cohort likelihoods, where

$$L_{\text{Age-at-Harvest}} = \prod_{j=1}^{A} L(N_{1j}, \underset{\sim}{p}, \underset{\sim}{S}, \underset{\sim}{R} | \underset{\sim}{h}) \cdot \prod_{i=1}^{Y} L(N_{i1}, \underset{\sim}{p}, \underset{\sim}{S}, \underset{\sim}{R} | \underset{\sim}{h}). \tag{9.172}$$

Alternative model formulations are based on special cases of Eq. (9.171), where natural survival, harvest probabilities, and reporting rates are assumed constant over age classes, time, or both. In constructing the cell probabilities (Eq. 9.171) and expected values of the harvest counts (Eqs. 9.169, 9.170), the harvest season was assumed to be a short, discrete time period during the year when natural mortality causes could be ignored. The formulation of Eq. (9.171) is analogous to the discrete time VPA model of Fryxell et al. (1988). Other formulations similar to Gulland (1965) or Pope (1972) for continuous exploitation and natural mortality could also be used to model the multinomial sampling process.

The MLE is based on Eq. (9.168) and used to directly estimate the cohort abundance in the first row and first column of the population reconstruction matrix (Fig. 9.10b). Abundance levels for the remainder of the reconstruction matrix are estimated from the general expression

$$\hat{N}_{ij} = (\hat{N}_{i-1, j-1} - h_{i-1, j-1}) \cdot \hat{S}_{i-1, j-1}, \tag{9.173}$$

progressively working through time for each cohort. Annual abundance is estimated by summing over the age classes within a year, i.e.,

$$\hat{N}_i = \sum_{j=1}^{A} \hat{N}_{ij}. \tag{9.174}$$

Gove et al. (2002) used profile likelihood methods (Arnold 1990) to estimate confidence intervals for the annual abundance levels.

Assumptions

The basic assumptions of the Gove et al. (2002) method include the following:

1. Age classification is accurate.
2. Harvest numbers are reported accurately or the reporting rate(s) estimated unbiasedly.
3. Survival and harvest processes are modeled correctly.
4. Auxiliary variables are estimated without bias.

Chi-square goodness-of-fit tests, likelihood-ratio tests, and AIC values should be used to select the most appropriate model for the age-at-harvest data. Direct knowledge of the harvest and reporting processes that generated the age-at-harvest data should not be ignored. That direct knowledge should be used to restrict model selection among a set of reasonable choices. An oversimplified model will induce bias, whereas an overly complex model will unnecessarily increase sampling error. The most parsimonious model that adequately describes the harvest data should be selected.

The relative informational content of the three components of the joint likelihood model (Eq. 9.168) can be quite different. Various investigators have suggested tuning (Gallucci et al. 1996:38,225) or weighting (Quinn and Deriso 1999:355,391) the different contributions of the likelihood. A weighted likelihood model could be constructed of the form:

$$\ln L_{\text{Joint}} = f_1 \ln(L_{\text{Age-at-Harvest}}) + f_2 \ln(L_{\text{Auxiliary}}) + f_3 \ln(L_{\text{Reporting}}),$$

where $\sum_{i=1}^{3} f_i = 1$ and $0 < f_i < 1$. Gove (1997) investigated the use of weighted maximum likelihood analysis but found the mean square error (i.e., variance plus squared bias) was always less using nonweighted versus weighted maximum likelihood analysis.

Example 9.19: Statistical Population Reconstruction, Black Bear, Washington State

Age-at-harvest data reported for male black bears for the years 1988–1994 (Bender 1997) and 1995–1996 are analyzed by using the Gove et al. (2002) reconstruction model (Table 9.7). In addition to these data, radiotelemetry results from radio-collared black bears are used as auxiliary data to complete the likelihood model. Because the tagged bears were not classified to age when captured, no age-specific information is available. Thus, the radiotelemetry data will be pooled across years to provide an adequate sample size to estimate constant survival and harvest probabilities (Table 9.8). Consequently, a simple survival model will be assumed for this tagging study, consisting of constant natural survival over time and age classes, as well as a constant annual harvest probability across years and age classes.

The auxiliary likelihood describing the radiotelemetry data is written as

$$L_{\text{Auxiliary}} = \binom{R}{h, n} p_H^h [(1 - p_H)(1 - S_N)]^n [(1 - p_H)S_H]^{R-h-n},$$

where

R = number of bears tagged;
h = number of tagged bears harvested;
n = number of tagged bears that died from nonhunting (i.e., natural) causes;
p = annual probability of harvest;
S_N = probability of surviving natural causes of mortality during the year.

This auxiliary likelihood estimates $\hat{p} = 0.2605$ $(\widehat{SE} = 0.0402)$ and $\hat{S}_N = 0.9773$ $(\widehat{SE} = 0.0159)$.

The limitations of the auxiliary likelihood restrict the form of the age-at-harvest likelihood to include constant natural survival and/or constant annual harvest probabilities (i.e., MpS, MpS_A, or Mp_YS). To better conform to the auxiliary data limitations, only adult male black bears 3.5 years of age or older were used in the analysis (Table 9.7). Older bears were anticipated to more likely have constant harvest and survival probabilities and better conform to the limitations of the auxiliary data.

Annual abundance estimation of adult (3.5 or older) male black bears was computed under the three available models (Table 9.9). Log-likelihood values were computed for each of the alternative models. By using these likelihood values, likelihood-ratio tests were performed and AIC values compared to select the most appropriate model:

Model	Log-likelihood	No. of parameters	AIC
MpS	−778.9353	34	1625.87
Mp_YS	−702.6199	43	1491.24
MpS_A	−750.3889	57	1614.78

The likelihood-ratio test comparing models MpS and Mp_YS was statistically significant $(P\ (\chi_9^2 \geq 152.6308) < 0.0001)$, indicating the more complex model is preferred. A similar comparison between models MpS and MpS_A was also significant $(P\ (\chi_{23}^2 \geq 57.0928) < 0.0001)$, suggesting the selection of the more complex model. Likelihood-ratio tests could not be performed between models Mp_YS and MpS_A. Instead, comparison of the AIC values across the three models recommends model Mp_YS.

By using the selected model, the abundance estimates by age class, year, and annual totals were calculated (Table 9.10). Annual abundance estimates ranged from a low of 1973.5 in 1994 to a high of 2316.1 in 1991. The annual natural survival probability estimated over the joint likelihood model was $\hat{S} = 0.9947$ $(\widehat{SE} = 0.0040)$. The annual probabilities of harvest were estimated as follows:

Year	$\hat{M}_H$	$\widehat{SE}$
1988	0.1606	(0.0082)
1989	0.2548	(0.0101)
1990	0.2480	(0.0108)
1991	0.2391	(0.0105)
1992	0.2760	(0.0124)
1993	0.2969	(0.0146)
1994	0.1935	(0.0132)
1995	0.2502	(0.0175)
1996	0.2688	(0.0181)

The annual harvest probabilities were estimated to range from a low of $\hat{M}_H = 0.1606$ $(\widehat{SE} = 0.0082)$ in 1988 to a high of $\hat{M}_H = 0.2969$ $(\widehat{SE} = 0.0146)$ in 1993. Profile likelihood confidence intervals were calculated for the annual estimates of abundance. Model Mp_YS had an overall chi-square goodness-of-fit statistic of 917.789 with 173 degrees of freedom. The profile likelihood confidence bounds were therefore scaled by a factor of

$$\sqrt{\frac{917.789}{173}} = 2.3033$$

to account for lack-of-fit and overdispersion.

Discussion of Utility

Statistical reconstruction of age-at-harvest data permits variance and confidence interval estimation as well as criteria for model selection. These properties are lacking in the traditional VPA and cohort analyses of Fry (1949), Gulland (1965), and Pope (1972). The analysis also provides MLEs of abundance, survival, and harvest probabilities. The formalized models also provide investigators with guidance on what types and how much auxiliary information must be collected to supplement the age-at-harvest data. Most existing methods of analyzing age-at-harvest data rely on "arbitrary selected lifetime recovery rates" or an "educated guess" to supplement the population reconstruction (Roseberry and Woolf 1991:22–23), thereby precluding rigorous inference to population trends or abundance.

The population reconstruction models also provide a framework upon which to decide which auxiliary field sampling and tagging studies should be performed. The existing statistical theory of tagging studies (e.g., Seber 1982) can be readily incorporated with age-at-harvest likelihoods. The design of tagging studies should therefore be coordinated with long-term population assessment goals. Tagging studies should be designed to provide the age-specific survival or harvest probabilities needed to perform more realistic cohort analyses.

Table 9.7. Male black bear harvest by age class, 1988–1996, in Washington State.

Year	Age																							
	3.5	4.5	5.5	6.5	7.5	8.5	9.5	10.5	11.5	12.5	13.5	14.5	15.5	16.5	17.5	18.5	19.5	20.5	21.5	22.5	23.5	24.5	25.5	>25.5
1988	114	25	63	25	25	32	32	6	6	13	0	0	13	0	0	6	0	0	0	0	0	0	0	0
1989	122	101	53	85	32	26	33	53	16	5	16	11	11	5	5	5	0	0	0	0	0	0	0	0
1990	98	74	67	56	41	26	26	33	15	4	15	9	9	2	4	4	4	2	0	0	0	2	0	2
1991	224	61	50	72	8	19	22	33	11	8	11	17	6	3	6	0	0	3	0	0	0	0	0	0
1992	114	192	46	36	61	18	25	18	29	11	7	0	11	4	4	4	0	4	4	4	0	0	4	4
1993	182	111	111	37	40	26	26	31	17	6	6	6	11	6	11	6	3	11	0	0	3	3	0	3
1994	85	74	33	56	15	15	19	19	7	15	4	7	7	7	0	4	0	11	0	0	0	0	0	4
1995	181	77	120	28	31	12	12	12	15	12	15	15	3	0	6	3	3	3	3	0	0	0	0	0
1996	153	98	77	66	38	49	14	21	17	17	10	0	3	0	0	7	21	3	7	3	0	0	0	0

(Data from Bender 1997.)

Table 9.8.   Radiotelemetry data used in the Washington State black bear population reconstruction analysis.

Data	Count
Total bear-years at risk (R)	119
Harvested (h)	31
Natural mortality (n)	2
Survivals	86

(Data from Bender 1997.)

Table 9.9.   Estimates of annual adult black bear abundance in Washington State under three alternative population reconstruction models.

Year	Models		
	MpS	Mp_YS	MpS_A
1988	2272.3	2241.3	3653.0
1989	2331.4	2271.9	3754.6
1990	2085.7	2011.7	3443.5
1991	2434.6	2316.1	4177.9
1992	2322.1	2173.5	4132.6
1993	2405.1	2208.49	4315.8
1994	2185.7	1973.5	4089.3
1995	2492.5	2201.4	4459.4
1996	2610.4	2245.7	4580.2

Auxiliary studies also need to be designed to have the same relevant time and spatial dimensions as the population reconstruction. The black bear population reconstruction is a good example. Although harvest information was collected statewide, the radiotelemetry survival study was limited to a small part of the state and only over a few years. Nonrepresentative sampling of the survival and/or harvest process can therefore bias overall reconstruction results. Thus, coordination of the auxiliary studies with the population reconstruction effort is important.

In the Washington State black bear example, harvest-reporting rates were assumed to be 100%. Gove et al. (2002) included a likelihood for reporting rates as part of the joint likelihood. Field check or locker check surveys (Section 6.2.1) can be readily incorporated into the joint likelihood model for population reconstruction if reporting is incomplete but estimable. In a similar manner, complete age classification of all harvested animals is not necessary for population reconstruction. If the harvested game is probabilistically sampled and the sampling fraction is known, that element of the survey can also be incorporated into the statistical cohort analysis. For example, if all successful hunters submit a tooth for age classification, those teeth could be randomly subsampled by the agency for age classification and the results incorporated into the joint likelihood model. In this way, the laboratory costs of age classification can be controlled with different sampling fractions each year depending on funding availability.

The statistical cohort analysis can provide a valuable cross-validation with other survey methods. The annual abundance estimates from cohort analyses can also supplement information on population trends in years in which large-scale and costly wildlife

Table 9.10.   Estimated abundance matrix for black bears age 3.5 or older in Washington State using model Mp_tS.

Year	3.5	4.5	5.5	6.5	7.5	8.5	9.5	10.5	11.5	12.5	13.5	14.5	15.5	16.5	17.5	18.5	19.5	20.5	21.5	22.5	23.5	24.5	25.5	>25.5	Total annual abundance	95% CI
1988	538.8	260.9	322.6	187.0	183.7	149.5	140.7	74.5	92.1	75.6	46.9	46.9	35.9	13.8	17.1	16.0	3.1	4.2	8.1	7.7	0.0	3.3	0.0	13.0	2241.3	2218.0–2280.2
1989	413.5	422.5	234.6	258.3	161.1	157.8	116.9	108.1	68.1	85.7	62.3	46.7	46.7	22.8	13.7	17.0	10.0	3.1	4.2	8.1	7.6	0.0	3.3	0.0	2271.9	2256.8–2328.1
1990	327.8	290.0	319.8	180.7	172.3	128.4	131.1	83.4	54.8	51.8	80.2	46.0	35.5	35.5	17.7	8.7	12.0	9.9	3.0	4.1	8.1	7.6	0.0	3.2	2011.7	1974.0–2037.5
1991	806.7	228.6	214.8	251.5	124.0	130.6	101.9	104.6	50.2	39.6	47.6	64.9	36.8	26.3	33.3	13.6	4.7	7.9	7.9	3.0	4.1	8.0	5.5	0.0	2316.1	2293.2–2348.7
1992	420.7	579.6	166.7	164.0	178.5	115.4	111.1	79.5	71.2	39.0	31.4	36.4	47.6	30.7	23.2	27.2	13.5	4.6	4.9	7.8	3.0	4.1	8.0	5.5	2173.5	2167.1–2197.0
1993	644.9	305.1	385.5	120.1	127.3	116.9	96.9	85.6	61.2	42.0	27.8	24.3	36.2	36.4	26.5	19.1	23.0	13.5	0.6	0.9	3.8	3.0	4.1	3.9	2208.5	2183.8–2217.0
1994	430.1	460.4	193.1	273.1	82.7	86.8	90.4	70.5	54.3	43.9	35.8	21.7	18.2	25.1	30.3	15.4	13.0	19.9	2.4	0.6	0.9	0.8	0.0	4.1	1973.5	1966.8–1987.4
1995	618.5	343.3	384.4	159.2	215.9	67.3	71.4	71.1	51.2	47.1	28.8	31.6	14.6	11.1	18.0	30.1	11.4	13.0	8.9	2.4	0.6	0.9	0.8	0.0	2201.4	2176.6–2212.5
1996	604.0	435.1	264.9	263.0	130.5	183.9	55.0	59.1	58.7	36.0	34.9	13.7	16.5	11.6	11.1	11.9	27.0	8.3	9.9	5.8	2.4	0.6	0.9	0.8	2245.7	2240.4–2284.6

surveys are not conducted. Therefore, cohort analyses can be part of a long-term management strategy of both intensive and extensive surveys of wildlife resources. For big-game species with carefully monitored harvests, statistical cohort analysis should be an essential part of the arsenal of survey approaches used to inventory a population.

9.9   Summary

Abundance, being the culmination of birth and death processes, is the ultimate expression of both natural and anthropogenic effects on wild populations. The separate elements of sex ratios, productivity, harvest, and mortality were presented in earlier chapters. These components were integrated in this chapter to estimate animal abundance with such techniques as the SAK model (Section 9.7) and population reconstruction (Section 9.8). An array of other abundance estimation methods was also discussed in this chapter (Fig. 9.11).

Age-at-harvest data are among the most commonly collected and yet ignored data by state wildlife agencies. Nevertheless, the stock assessment methods originally developed for fisheries management (Fry 1949, Pope 1972, Gallucci et al. 1996) are beginning to be used in wildlife science under the general heading of population reconstruction methods (Fryxell et al. 1988). While stochastically based reconstruction models are now the norm in fisheries management (Quinn and Deriso 1999), these principles are only starting to be introduced in wildlife management (Gove et al. 2002). The general area of statistical population reconstruction will become increasingly important in wildlife management. These reconstructive methods hold great promise for monitoring big game populations subject to harvest. Harvest data are relatively easy to collect, and at the same time, agencies can avoid the high costs associated with more direct hands-on survey methods. Age-structure data can also provide crucial information on survival, productivity, and age composition at relatively low cost.

Harvest totals without the corresponding age-composition data still remain a valuable asset. The management of wild populations for sport and subsistence harvest requires knowledge of both animal abundance and harvest success. CPUE data can be used to monitor both hunter satisfaction and population trends. When carefully structured, CPUE data can also be used to estimate absolute abundance (Section 9.6). Harvest counts can also be used in conjunction with CIR methods (Section 9.5) and index-removal methods (Section 9.4) to estimate total abundance. All of these methods use the intervention of harvest removals to induce population responses that can be linked to changes in animal abundance. The CIR and index-removal methods use auxiliary observations to relate harvest numbers to animal abundance. The CPUE methods use changes in success rate with known removals to estimate abundance.

Visual surveys are the most widely used nontagging method to estimate and monitor wildlife abundance. Visual count surveys provide a relatively inexpensive and unintrusive approach to population surveys. Line-transect methods (Section 9.3) have seen a rapid development in statistical theory (Buckland et al. 1993) and breadth of applications. Detection functions estimated from right-angle distance data can be used to both test the assumptions of homogeneous detection probabilities and convert counts to absolute abundance and/or density. The bounded-count method (Section 9.2.2) and the binomial sampling model of Overton (1969) (Section 9.2.3) attempt to estimate

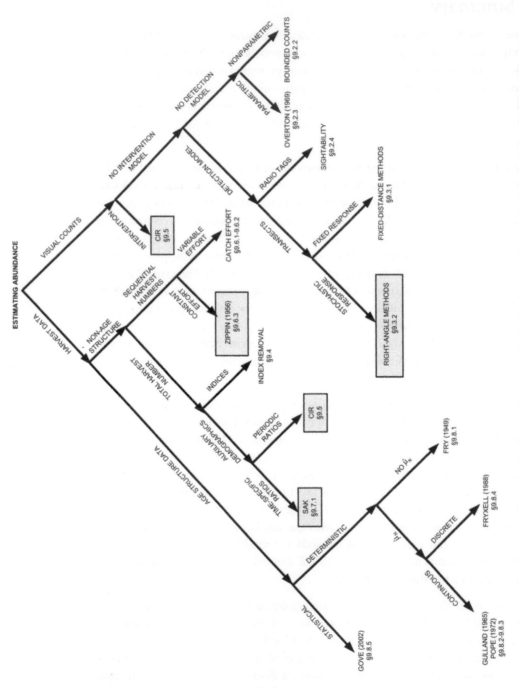

Figure 9.11. Decision tree illustrating the selection of abundance estimation methods. Approaches with sample size calculations highlighted in gray.

abundance without the need for distance information. The Overton (1969) method should generally be avoided, and the bounded-count method should be used only when detection probabilities are known to be extremely high.

The sightability models (Section 9.2.4) of Samuel et al. (1987) provide an interesting transition between the noninvasive methods of visual surveys and animal tagging studies. Radio telemetry is initially used to help construct an empirically based detection function. Once established, the sightability model is used to estimate total abundance in subsequent surveys based on the Horwitz-Thompson estimator (Cochran 1977:259–261) of finite sampling theory. The mark-resighting data are not directly used in formal abundance estimation but are instead used in modeling the detection function.

It is possible to use two or more of the abundance techniques presented in this chapter or use these methods in conjunction with tagging studies (Skalski and Robson 1982, Chen et al. 1998, Alpizar-Jara and Pollock 1999) to refine the accuracy and precision of abundance estimates. A well-designed study should avail itself of all available information and provide a flexible approach to abundance estimation. The likelihood models presented in this chapter can be coupled with the likelihood of traditional tagging models (Seber 1982) to provide a more thorough analysis of demographic data. Case studies in the last chapter of this book will illustrate some of these joint-modeling concepts.

Integration of Analytical Techniques

10

10.1 Introduction and Purpose

Our purpose in this last chapter is to present several examples that highlight integration of field data with the techniques explored in this book for demographic assessment and management. Data analyses and survey designs will be illustrated that combine the use of two or more demographic techniques presented in earlier chapters. This chapter also illustrates the integration of techniques in this book with traditional marking studies (Seber 1982). We also demonstrate the use of these techniques to help monitor and manage populations and the exploitation rates of game species.

The examples we provide are not exhaustive but instead are intended to demonstrate a diversity of possible scenarios and how study objectives might be met by using multiple techniques and approaches to study design. We hope readers will be inspired to use the methods described in this book in more clever and holistic approaches. If there is hope in improving the ability of science to monitor the health and viability of wild populations, it lies in integrating approaches to population assessment. All available demographic tools should be considered, and those selected should provide a robust and comprehensive solution to the problem. Rarely will there only be one simple method or approach available for studying wild animal populations. Also, rarely will the assessment be accomplished by haphazard collection of data. Instead, a successful approach will take into consideration the underlying demographic processes, the demographic parameters that require quantification, and collection of the necessary information. For experimental questions, the study design also must ensure that data are collected to potentially characterize the states of nature under both null and alternative hypotheses.

10.2   Management for Desired Sex Ratios of Elk

For many harvested populations, such as elk (*Cervus elaphus*), hunting regulations and hunter selection for mature males skews sex and age ratios. Observed declines in bull:cow ratios have concerned wildlife managers because yearling bulls might be less efficient breeders. As a result, pregnancy rates might decline and conception dates might be delayed, resulting in lower overwinter calf survival (Prothero 1977, Kimball and Wolfe 1979, Prothero et al. 1979, Squibb et al. 1986, Freddy 1987, Noyes et al. 1996). Although debatable, it is often recognized that fewer than 10 mature bulls per 100 cows might have detrimental impacts to demographic rates (Freddy 1987, Raedeke et al. 2002). Many wildlife agencies have a management goal of a minimum of 25 adult bulls per 100 adult cows in the population.

Given these concerns, management agencies have used several regulations over the past 40 years to manage for "appropriate" sex and age ratios. For example, elk harvest regulations have focused on varying the structure and number of seasons and on placing minimum antler point restrictions on bulls that may be legally harvested (Carpenter 1991). These regulations often greatly influence age and sex ratios of these populations, because harvest comprises the majority of mortality for many large mammal populations. However, appropriate harvest strategies to provide "appropriate" sex and age ratios are hotly debated (Raedeke et al. 2002). For example, many states have instituted antler harvest restrictions, which have been controversial (Carpenter 1991). In this case study, we discuss the analysis of demographic data and management of "appropriate" sex ratios for elk by considering demographic techniques discussed in previous chapters.

The population sex ratio ($R_{F/M}$) is defined as the ratio

$$R_{F/M} = \frac{N_F}{N_M}.$$

Under the assumptions of (1) a constant survival rate across ages for females ($S_F = 1 - M_F$) and for males ($S_M = 1 - M_M$); (2) equal recruitment of both sexes into the population; and (3) stable age and sex structure, the population sex ratio, $R_{F/M}$ can be estimated by the Eq. (3.64):

$$\hat{R}_{F/M} = \frac{1 - \hat{S}_{N-M} \cdot \hat{S}_{H-M}}{1 - \hat{S}_{N-F} \cdot \hat{S}_{H-F}} \tag{10.1}$$

where $\hat{S}_{i-j}$ is the estimated survival of the jth sex ($j = M, F$) for the ith mortality source (i = natural, N; hunting, H). With knowledge of male and female natural mortality rates ($\hat{S}_{N-M}$ and $\hat{S}_{N-F}$), it is possible to estimate what values of $\hat{S}_{H-M}$ and $\hat{S}_{H-F}$ would be necessary to maintain desired sex ratios. That is, we can project the long-term consequences of sex-specific hunting regulations on the sex ratio of a population.

We begin our demographic assessment by keeping the goal of 25 adult bulls per 100 cows in mind and by estimating survival of bull elk using age-at-harvest data from elk in South Dakota. In this case, harvest reporting by hunters was mandatory and elk were classified by age based on cementum annuli (Keiss 1969). In this population, there was no female harvest. Because hunter preference results in differential selection probabilities for all age classes, we will use a survival estimator for left- and right-truncated data

Table 10.1. Age-structure data for bull elk in South Dakota.

Year	Age	Coded age (x)	Harvest (l_x)	$x \cdot l_x$
1989	2.5	0	11	0
1990	3.5	1	10	10
1991	4.5	2	13	26
1992	5.5	3	9	27
1993	6.5	4	7	28
1994	7.5	5	7	35
1995	8.5	6	5	30
Totals			62	156

(Data from J.J. Millspaugh.)

(Chapman and Robson 1960, Robson and Chapman 1961) (Section 5.9.3) to estimate bull elk survival (Table 10.1). Total annual survival S_T is estimated by Eq. (5.121)

$$\frac{T}{l.} = \left(\frac{S_T}{1-S_T}\right) - \frac{(k+1)S_T^{k+1}}{(1-S_T^{k+1})},$$

where $T = \sum_{x=0}^{k} xl_x$, and the term $\dfrac{T}{l.}$ is the average coded age (i.e., 0, 1, 2, . . .) in the sample.

In this example,

$$\frac{156}{62} = \left(\frac{S_T}{1-S_T}\right) - \left(\frac{(6+1)S_T^{6+1}}{1-S_T^{6+1}}\right),$$

where $l. = 11 + 10 + 13 + \cdots + 5 = 62$ and $T = 0(11) + 1(10) + 2(13) + \cdots + 6(5) = 156$. The result is an estimated annual survival of $\hat{S}_T = 0.8847$. The variance of $\hat{S}_T$ is based on Eq. (5.122), where

$$\widehat{\text{Var}}\left(\hat{S}_T\right) = \frac{1}{l.}\left[\frac{1}{\hat{S}_T\left(1-\hat{S}_T\right)^2} - \frac{(k+1)^2\,\hat{S}_T^{k-1}}{\left(1-\hat{S}_T^{k+1}\right)^2}\right]^{-1}$$

$$= \frac{1}{62}\left[\frac{1}{0.8847(1-0.8847)^2} - \frac{(6+1)^2(0.8847)^{6-1}}{\left(1-0.8847^{6+1}\right)^2}\right]^{-1} = 0.003275,$$

for a standard error of $\widehat{\text{SE}}\left(\hat{S}_T\right) = 0.0572$.

In addition, we wish to partition total annual mortality into components associated with natural and hunting mortality for males and females ($\hat{S}_{N-M}$, $\hat{S}_{H-M}$, $\hat{S}_{N-F}$, and $\hat{S}_{H-F}$). Radio telemetry is one procedure that could be used to partition among sources of mortality. The mortalities can be partitioned into hunter and natural causes by using the likelihood

$$L(R, S_H, S_N | h, n, a) = \binom{R}{h, n, a}(1 - S_H)^h (S_H(1 - S_N))^n (S_H S_N)^a, \qquad (10.2)$$

where

S_H = probability of surviving the harvest (i.e., $1 - S_H = M_H$);
S_N = probability of surviving natural causes of mortality (i.e., $1 - S_N = M_N$);

h = number of radio-tagged elk that were harvested;

n = number of radio-tagged elk that died from natural mortality factors (i.e., non-hunter direct mortality);

a = number of radio-tagged elk alive at the end of the year;

$R = h + n + a$.

The above likelihood can be rewritten as

$$L(R,S_H,S_N|h,n,a) = \binom{R}{h,n,a}\left[\left(\frac{1-S_H}{1-S_T}\right)(1-S_T)\right]^h \cdot \left[\left(\frac{S_H - S_H S_N}{1-S_T}\right)(1-S_T)\right]^n (S_T)^a,$$

where S_T = total annual survival (i.e., $S_T = S_N S_H = 1 - M_T$), and still further as

$$L(R,p_H,S_T|h,n,a) = \binom{R}{h,n,a}(p_H(1-S_T))^h \cdot ((1-p_H)(1-S_T))^n (S_T)^a, \quad (10.3)$$

where p_H = proportion of total mortality due to harvest. The maximum likelihood estimates are then

$$\hat{S}_T = 1 - \hat{M}_T = \frac{a}{a+n+h} \quad (10.4)$$

with variance estimator

$$\widehat{\text{Var}}(\hat{S}_T) = \frac{\hat{S}_T(1-\hat{S}_T)}{R}, \quad (10.5)$$

and

$$\hat{p}_H = \frac{h}{h+n} \quad (10.6)$$

with variance estimator

$$\widehat{\text{Var}}(\hat{p}_H) = \frac{\hat{p}_H(1-\hat{p}_H)}{n+h}. \quad (10.7)$$

A total of 18 adult bull elk were radio-collared to document movements and survival. Of those 18 collared bulls, 3 were harvested ($h = 3$) and 1 died of natural causes ($n = 1$). The proportion of total mortality (Eq. 10.6) resulting from harvest (p_H) is estimated as

$$\hat{p}_H = \frac{3}{3+1} = 0.75,$$

with an estimated variance of

$$\widehat{\text{Var}}(\hat{p}_H) = \frac{\hat{p}_H(1-\hat{p}_H)}{n+h}$$
$$= \frac{0.75(1-0.75)}{1+3} = 0.0469,$$

for a standard error of $\widehat{\text{SE}}(\hat{p}_H) = 0.2166$. The radio tags also allow us to estimate total annual survival (Eq. 10.4), $\hat{S}_T = 0.7778$, with a standard error of $\widehat{\text{SE}}(\hat{S}_T) = 0.0980$, based on Eq. (10.5). Note the catch-curve data and analysis provide a more precise estimate of total annual survival (i.e., $\hat{S}_T = 0.8847$, $\widehat{SE}(\hat{S}_T) = 0.0418$).

By use of the estimate of total annual survival ($\hat{S}_T$) from the catch-curve analysis and the estimate of p_H from the radiotelemetry study, harvest mortality is estimated by the equation

$$\hat{M}_{H-M} = (1 - \hat{S}_T)\hat{p}_H$$
$$= (1 - 0.8847)0.75 = 0.0865, \tag{10.8}$$

or $\hat{S}_{H-M} = 0.9135$. The variance of $\hat{M}_{H-M}$ is calculated by using the exact variance estimator for a product (Goodman 1960), where

$$\widehat{Var}(\hat{M}_{H-M}) = \widehat{Var}(\hat{S}_T)\hat{p}_H^2 + \widehat{Var}(\hat{p}_H)(1 - \hat{S}_T) - \widehat{Var}(\hat{S}_T) \cdot \widehat{Var}(\hat{p}_H)$$
$$= (0.001784)(0.75)^2 + 0.0469(1 - 0.8847)^2 - (0.001748)(0.0469)$$
$$= 0.001525,$$

for a standard error of $\widehat{SE}(\hat{M}_{H-M}) = 0.0390$. An estimate of natural mortality is similarly calculated by the equation

$$\hat{M}_{N-M} = (1 - \hat{S}_T)(1 - \hat{p}_H)$$
$$= (1 - 0.8847)(1 - 0.75) = 0.0288, \tag{10.9}$$

or $\hat{S}_{N-M} = 0.9712$. The sampling variance of $\hat{M}_{H-M}$ is again estimated by using the variance for a product, where

$$\widehat{Var}(\hat{M}_{N-M}) = \widehat{Var}(\hat{S}_T)(1 - \hat{p}_H)^2 + \widehat{Var}(\hat{p}_H)(1 - \hat{S}_T)^2 - \widehat{Var}(\hat{S}_T) \cdot \widehat{Var}(\hat{p}_H)$$
$$= (0.001784)(1 - 0.75)^2 + 0.0469(1 - 0.8847)^2 - (0.001748)(0.0469)$$
$$= 0.000651,$$

for a standard error of $\widehat{SE}(\hat{M}_{N-M}) = 0.0255$.

Concurrent with collection of field data from harvested bull elk, a radiotelemetry study was conducted to quantify cow elk survival. Because cow elk were not harvested, we could not use a catch-curve analysis. Instead, between 13 and 21 cow elk were radio-collared each year between 1993 and 1996 (Table 10.2). We used an R × C contingency table analysis (Zar 1996:485–486) to test whether survival probabilities were homogeneous over time. The results indicated the survival probabilities were homogenous ($P(\chi_3^2 \geq 1.7163) = 0.6333$).

Table 10.2. Fate of radio-collared elk in South Dakota where $a =$ number alive and $d =$ number dead.

	1993	1994	1995	1996
a	13	12	16	20
d	1	1	3	1
Totals	14	13	19	21

(Data from J.J. Millspaugh.)

The total annual survival for cow elk from the telemetry study is estimated as

$$\hat{S}_T = \frac{a}{a+d}$$

$$= \frac{61}{61+6} = 0.9104,$$

with variance estimate

$$\widehat{\text{Var}}\left(\hat{S}_T\right) = \frac{\hat{S}_T\left(1-\hat{S}_T\right)}{a+d}$$

$$= \frac{0.9104(1-0.9104)}{61+6} = 0.0012,$$

for $\widehat{\text{SE}}\left(\hat{S}_T\right) = 0.0349$.

By using the parameter estimates $\hat{S}_{N-M} = 0.9712$, $\hat{S}_{N-F} = 0.9104$, $\hat{S}_{H-M} = 0.9135$, and $\hat{S}_{H-F} = 1.00$ (assuming no cows are harvested) and Eq. (10.1), the projected long-term sex ratio of the elk population under prevailing conditions is estimated to be

$$\hat{R}_{F/M} = \frac{1-\hat{S}_{N-M}\cdot\hat{S}_{H-M}}{1-\hat{S}_{N-F}\cdot\hat{S}_{H-F}}$$

$$\hat{R}_{F/M} = \frac{1-(0.9712)(0.9135)}{1-(0.9104)(1)} = 1.259 \text{ females/male},$$

or approximately 5 females for every 4 males.

Alternatively, the harvest morality that can be exerted on the male population to achieve a 4:1 sex ratio can be calculated from the equation

$$R_{F/M} = \frac{1-S_{N-M}\cdot S_{H-M}}{1-S_{N-F}\cdot S_{H-F}}$$

$$4 = \frac{1-0.9712\cdot S_{H-M}}{1-(0.9104)(1)}$$

by solving for S_{H-M}. The solution is a value of $\hat{S}_{H-M} = 0.6606$, or a harvest mortality of 0.3394. This result indicates that harvest mortality of the bull segment of the population ($\hat{S}_{H-M}$) should not exceed 34% to maintain a ratio of 25 bulls per 100 cows. This example (Fig. 10.1) demonstrates how knowledge of survival rates partitioned between natural and hunter-induced mortality can be used to manage for desired sex ratios. Our evaluation suggested male natural survival rates exceeded female natural survival rates, which is uncommon in elk populations (Raedeke et al. 2002). Given these demographics, it becomes apparent why many cow:bull ratios in western states are skewed. Given elk demographics and harvest regulations, including either low or no cow elk permits and open bull seasons, and hunter preferences to harvest adult males, we should expect altered cow:bull ratios. Our results indicate that it takes little harvest mortality in bull elk to skew sex ratios. Managers could use similar approaches for other species and other elk populations provided site-specific information is available.

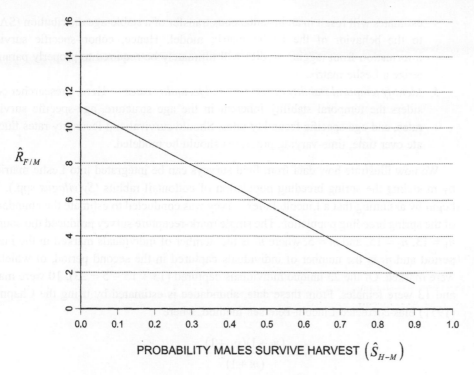

Figure 10.1. The relationship between $\hat{S}_{H-M}$ and $\hat{R}_{F/M}$ assuming that $\hat{S}_{N-M} = 0.9712$, $\hat{S}_{N-F} = 0.9104$, and $\hat{S}_{H-F} = 1.0$ (i.e., no cows were harvested). For example, a ratio of 100 cows : 25 bulls would be $\hat{R}_{F/M} = 4$.

10.3 Combining Field Results with Leslie Matrix Projections

Preceding chapters described techniques for collecting demographic data needed to assess the dynamics of wild animal populations. The necessary steps to construct a Leslie matrix model, regardless of species, include the following:

1. *Identify the population for assessment* — An anniversary date must be selected from the annual cycle for modeling. Choices include modeling the spring breeding population, the fall harvestable population, or the postharvest population. The choice will influence the number of age classes in the model (i.e., with or without young-of-year). The choice also influences the estimates of fecundity rates used in the model. Age-specific fecundities are the number of female young produced that survive to the anniversary date per adult female. The survival values in the model are annual survivals from anniversary date to anniversary date.

2. *Classify age of the female population* — The female abundance must be partitioned into abundance by age classes. Total female abundance and age composition data are needed to derive current abundance conditions.

3. *Calculate age-specific fecundity rates* — The fecundity values used in the Leslie matrix are the net number of female young that are produced and survive to the anniversary date per adult female. Hence, fecundity calculations will require information on numbers of young produced, their sex ratio, and the probability that juveniles survive to the anniversary date of the model. Both the numbers of young produced and juvenile survival may depend on the age of the adult female.

4. *Identify age-specific survival probabilities* — One could obtain survival estimates from a time-specific or vertical life table. However, time-specific life-table analyses assume a stable and stationary population. Subsequent use of these survival

estimates will result in an instantaneous transfer of a stable-age distribution (SAD) to the behavior of the Leslie matrix model. Hence, cohort-specific survival information from tagging studies will ultimately be required to properly parameterize a Leslie matrix.

5. *Identify temporal stability of survival and fecundity rates* — Here, a researcher considers the temporal stability inherent in the age structure, age-specific survival rates, and age-specific fecundity rates. Should survival and fecundity rates fluctuate over time, time-varying processes should be modeled.

We now illustrate how data from field surveys can be integrated into Leslie matrices by modeling the spring breeding population of cottontail rabbits (*Sylvilagus* spp.). We begin by assuming that a Lincoln Index survey was conducted to estimate the abundance of the spring breeding population. The single mark-recapture survey produced the counts: $n_1 = 13$, $n_2 = 15$, and $m = 5$, where n_1 is the number of individuals marked in the initial period and n_2 is the number of individuals captured in the second period, of which m were marked. Of the 23 unique individuals captured $(13 + 15 - 5 = 23)$, 10 were males and 13 were females. From these data, abundance is estimated by using the Chapman (1951) bias-corrected Lincoln-Petersen method, where

$$\hat{N} = \frac{(n_1 + 1)(n_2 + 1)}{(m + 1)} - 1$$

$$\hat{N} = \frac{(13 + 1)(15 + 1)}{(5 + 1)} - 1 = 36.333.$$

The variance (Chapman 1951) is estimated by the formula

$$\widehat{\text{Var}}(\hat{N}) = \frac{(n_1 + 1)(n_2 + 1)(n_1 - m)(n_2 - m)}{(m + 1)^2 (m + 2)}$$

$$= \frac{(13 + 1)(15 + 1)(13 - 5)(15 - 5)}{(5 + 1)^2 (5 + 2)} = 71.111,$$

for a standard error of $\widehat{\text{SE}}(\hat{N}) = 8.433$. Female abundance is estimated as

$$\hat{N}_F = \frac{13}{23}(36.333) = 20.536.$$

Alternatively, we can test whether the proportion of males (p_M) is equal to the proportion of females (p_F), H_0: $p_M = p_F$ and H_A: $p_M \neq p_F$, using a chi-square test,

$$\chi_1^2 = \sum_1^2 \frac{(\text{Observed} - \text{Expected})^2}{\text{Expected}}$$

$$= \frac{(10 - 11.5)^2}{11.5} + \frac{(13 - 11.5)^2}{11.5} = 0.391,$$

which is not significant ($P(\chi_1^2 > 0.391) = 0.532$). It follows that female abundance could be estimated as $\hat{N}_F = 0.5(36.333) = 18.167$. The choice of which technique to use depends on biological information regarding the likelihood of an even sex ratio. In this polygamous population, we use the estimate of $\hat{N}_F = 20.536$.

Next, assume that during the breeding season, 31 adult females were collected from an adjacent area to estimate age composition and determine age-specific fecundity (see below):

Age	Number	Fetuses	Proportion
1	$n = 27$	$\bar{x} = 6.2$	0.871
2	$n = 3$	$\bar{x} = 7.8$	0.097
3	$n = 1$	$\bar{x} = 7.7$	0.032

Projecting the observed age composition to our estimate of female abundance results in

$$\begin{bmatrix} n_{01} \\ n_{02} \\ n_{03} \end{bmatrix} = \begin{bmatrix} 20.536\left(\dfrac{27}{31}\right) \\ 20.536\left(\dfrac{3}{31}\right) \\ 20.536\left(\dfrac{1}{31}\right) \end{bmatrix} = \begin{bmatrix} 17.9 \\ 2.0 \\ 0.6 \end{bmatrix}.$$

To estimate the number of female young produced per adult female by age class, we assume a sex ratio of $1:1$ for offspring. Furthermore, cottontails produce, on average, four litters of young each year. By using the age-specific estimates of production, the numbers of female young produced per adult female annually are summarized below:

Age Class (i)	Offspring
1	$6.2(4)(0.5) = 12.4$
2	$7.8(4)(0.5) = 15.6$
3	$7.7(4)(0.5) = 15.4$

However, these numbers are the number of female offspring born per female, not the number of offspring contributing to next year's age class (i.e., F_i's). Thus, we need the survival rate from birth to April the following breeding season. *Assuming a stable and stationary population*, the total number of female offspring produced in the previous year is estimated as

$$[17.9 \quad 2.0 \quad 0.6] \begin{bmatrix} 12.4 \\ 5.6 \\ 15.4 \end{bmatrix} = 262.4.$$

Of these 262.4 female young produced, only 17.9 survive to the next April, resulting in a survival rate of 0.0682. The net fecundity by age class is then calculated (see below):

Age class (i)	F_i
1	$12.4(0.0682) = 0.846$
2	$15.6(0.0682) = 1.064$
3	$15.4(0.0682) = 1.050$

The survival rates for adult rabbits can be calculated from a time-specific life table, *assuming a stable and stationary population* (see below).

Age class	l_x	d_x	q_x	S_x
0	17.9	15.9	0.888	0.112
1	2.0	1.4	0.700	0.300
2	0.6	0.6	1.000	0
3	0	—	—	—

By using the above results, the following Leslie matrix model is produced:

$$\underset{\sim}{n}_1 = \mathbf{M}\underset{\sim}{n}_0$$

$$\begin{bmatrix} n_{01} \\ n_{11} \\ n_{21} \end{bmatrix} = \begin{bmatrix} 0.846 & 1.064 & 1.050 \\ 0.112 & 0 & 0 \\ 0 & 0.300 & 0 \end{bmatrix} \cdot \begin{bmatrix} 17.9 \\ 2.0 \\ 0.6 \end{bmatrix}.$$

Next, we calculate the eigenvalues for the resulting projection matrix ($\mathbf{M}$) by using the characteristic equation:

$$|\mathbf{M} - \lambda I| = 0$$

$$\begin{vmatrix} 0.846 - \lambda & 1.064 & 1.050 \\ 0.112 & -\lambda & 0 \\ 0 & 0.300 & -\lambda \end{vmatrix} = 0$$

$$(0.846 - \lambda)(-1)^2 \begin{vmatrix} -\lambda & 0 \\ 0.3 & -\lambda \end{vmatrix} + 0.112(-1)^3 \begin{vmatrix} 1.064 & 1.050 \\ 0.300 & -\lambda \end{vmatrix} + 0 = 0$$

$$(0.846 - \lambda)(\lambda^2) - 0.112(-1.064\lambda - 0.315) = 0$$

$$\lambda^3 - 0.846\lambda^2 - 0.119168\lambda - 0.03528 = 0,$$

which has only one real root, $\lambda_{SAD} = 1.000376583$. The corresponding eigenvector is

$$\begin{bmatrix} 0.873 \\ 0.098 \\ 0.029 \end{bmatrix}.$$

Note that $\lambda_{SAD} = 1.00$ within rounding error, and the calculated SAD is the same as the distribution for the adults used in the vertical life-table analysis. Thus, if one assumes a stable and stationary population in deriving the survival and fecundity parameters, it is also assured the population is at the stable age distribution in the Leslie matrix model projection. The preferred approach is to use cohort-specific, tagging studies to estimate the demographic parameters used in a Leslie matrix.

We now repeat the example, this time using radiotelemetry data to estimate adult survival. A total of 35 adult rabbits were radio-marked and monitored for 6 months. In this case, survival is based on animals known to be alive and at risk at age i. The data used in this analysis are summarized below:

Age	1.5	2.5
At risk (l_i)	22	13
Deaths (d_i)	13	9

The estimated survival probability for age 1.5 rabbits would be calculated as

$$\hat{S}_{1.5} = \frac{l_{1.5} - d_{1.5}}{l_{1.5}}$$

$$= \frac{22 - 13}{22} = 0.4091 \text{ for six months.}$$

The estimated variance is based on binomial sampling, where

$$\widehat{Var}(\hat{S}_{1.5}) = \frac{\hat{S}_{1.5}(1 - \hat{S}_{1.5})}{l_{1.5}}$$

$$= \frac{0.4091(1 - 0.4091)}{22} = 0.0110,$$

for a standard error of $\widehat{SE}(\hat{S}_{1.5}) = 0.1048$. Using the same procedure, the survival probability from age 2.5 to 3.5 is estimated to be $\hat{S}_{2.5} = 0.3077$, with a standard error of $\widehat{SE}(\hat{S}_{2.5}) = 0.1280$. Given rabbits were monitored for just 6 months and we need an estimate of total annual survival, these rates must be prorated. If we assume survival is constant throughout the year, the annual survival probability from age 1.5 to 2.5 would be $\hat{S}_{1.5} = 0.4091^2 = 0.1674$; from age 2.5 to 3.5, $\hat{S}_{2.5} = 0.3077^2 = 0.0947$.

By using these survival results, the revised Leslie model can be written as

$$\begin{bmatrix} n_{01} \\ n_{11} \\ n_{21} \end{bmatrix} = \begin{bmatrix} 0.846 & 1.064 & 1.050 \\ 0.1647 & 0 & 0 \\ 0 & 0.0974 & 0 \end{bmatrix} \cdot \begin{bmatrix} 17.9 \\ 2.0 \\ 0.6 \end{bmatrix}.$$

If we project this population through time, the population will grow at a finite rate of $\hat{\lambda}_{SAD} = 1.0338$, when a stable age distribution is established.

These analyses demonstrate basic integration of field data with Leslie matrix models. In addition to outlining the data needed for demographic assessments and how those data might be collected, there are several other important messages illustrated in this example. First, it should be readily apparent that a demographic assessment of a population is a time-consuming process that requires multiple sources of data. Given the diversity of field and analytical techniques available, researchers should carefully consider the inherent assumptions in view of their assessment objectives and available resources. Second, when using Leslie matrix models, it becomes important to consider timing of surveys and what the values in the matrix represent. For example, the fertility values in the top row of the Leslie matrix are not the number of female offspring born per female, a mistake sometimes made (Wethey 1985, Jenkins 1989). Instead, these values represent the net number of offspring produced to the anniversary date of the model. Third, when estimating adult survival rates from a time-specific life table, the table assumes the population is stable and stationary, defeating the purpose of performing the Leslie matrix projection. Instead, alternative methods for survival estimation, such as those presented in Chapter 5, should be considered.

10.4   Comparing and Combining Time- and Cohort-Specific Survival Data

Gove (1997) provides age-at-harvest data and concurrent radiotelemetry survival information for cow elk in northern Idaho during 1988–1993. Harvested cow elk were

Table 10.3. Age-at harvest data and summary statistics $l.$ and T for a catch-curve analysis (Section 5.9.1) for cow elk 4 years and older from northern Idaho, 1988–1990 and 1992–1993.

Age class	Year 1988	1989	1990	1992	1993	Totals
4	8	20	19	18	22	87
5	7	9	15	13	16	60
6	8	11	14	19	15	67
7	6	5	7	5	7	30
8	5	10	7	13	9	44
9	3	3	4	5	10	25
10	3	2	5	9	5	24
11	4	5	0	2	5	16
12	3	0	3	2	5	13
13	2	3	3	1	2	11
14	1	2	3	2	3	11
15	1	0	0	1	3	5
16	1	0	2	1	3	7
17	2	0	0	2	1	5
18	0	0	2	0	1	3
19	0	0	0	0	2	2
20	0	0	1	0	0	1
21	0	0	0	0	0	0
22	0	0	0	0	0	0
23	0	1	0	0	0	1
$l.$	54	71	85	93	109	412
T	223	214	291	305	432	1465

(Data from Gove 1997.)

Table 10.4. Annual results of radiotelemetry survival studies of cow elk in northern Idaho.

Year	Harvest	Natural mortality	Survival	Total tags
1988	1	2	10	13
1989	4	1	20	25
1990	0	1	24	25
1992	4	4	17	25
1993	1	4	18	23
Totals	10	12	89	111

(Data from Gove et al. 2002.)

classified to age by cementum annuli each year to identify age-at-harvest (Table 10.3). Concurrently, radio-collared cow elk provided data on numbers of animals that survived, as well as those that died from natural or harvest sources of mortality each year (Table 10.4). The age-at-harvest data represent a vertical sample of the age distribution of the population. Assuming a stable age distribution, these vertical data provide information on survival. A vertical life-table (Section 5.5.1) analysis can be used if age-specific survivals are different, or a catch-curve analysis (Section 5.9) can be used if the survivals can be assumed to be constant. Only elk 4 years and older were analyzed to permit catch-curve analysis. The 1991 data were excluded from this example because of unexplained heterogeneity in the data.

The likelihood for the catch-curve analysis (Eq. 5.112) is

$$L_{CC} = \binom{l.+T-1}{T}(1-S_T)^{l.} S_T^{T}, \qquad (10.10)$$

where

$l.$ = number of animals in the sample;

$T = \sum_{i=0}^{l.} x_i$ = sum of the coded ages of the animals;

x_i = coded age of the animal $(0, 1, 2, \ldots)$;

S_T = total annual survival.

The likelihood for the radiotelemetry data can be written as

$$L_{RT} = \binom{n}{d}(1 - S_T)^d S_T^a, \tag{10.11}$$

where

$n = a + d$;
a = number of radio-tagged animals that survived the year;
d = number of radio-tagged animals that died regardless of causes during the year.

A test of equal survival across years can be based on the $R \times C$ contingency table test of homogeneity, where

	1988	1989	1990	1992	1993	Totals
T	223	214	291	305	432	1465
$l.$	54	71	85	93	109	412

The contingency table suggests no significant difference in survival between years $(P(\chi_4^2 \geq 4.0211) = 0.4032)$. By use of the pooled catch-curve data across years, total annual survival is estimated by Eq. (5.113), as

$$\hat{S}_T = \frac{T}{l. + T - 1}$$
$$= \frac{1465}{412 + 1465 - 1}$$
$$= 0.7809,$$

with an associated standard error (from Eq. 5.116) of $\widehat{SE}(\hat{S}) = 0.0096$.

Homogeneity of survival across years for the radiotelemetry study can also be tested by using an $R \times C$ contingency table test of the form

	1988	1989	1990	1992	1993	Totals
a	10	20	24	17	18	89
d	3	5	1	8	5	22

The survival probabilities from the radiotelemetry study are also homogeneous $(P(\chi_4^2 \geq 6.4116) = 0.1704)$, permitting pooling across years. The total annual survival from the radiotelemetry study is then estimated by the following equation:

$$\hat{S}_T = \frac{a}{a+d}$$
$$= \frac{89}{89+22}$$
$$= 0.8018,$$

with a binomial variance estimated by

$$\widehat{\mathrm{Var}}(\hat{S}_T) = \frac{\hat{S}_T(1-\hat{S}_T)}{a+d}$$
$$= \frac{0.8018(1-0.8018)}{89+22} = 0.00143,$$

for a standard error of $\widehat{\mathrm{SE}}(\hat{S}_T) = 0.0378$.

The product of Eqs. (10.10) and (10.11) is the joint likelihood for the combined age-at-harvest and radiotelemetry data. A likelihood ratio-test (Zar 1996:469–471, 502) can be used to test for equality of survival between data sets or an asymptotically equivalent test based on the 2×2 contingency table:

Catch-curve	Radiotelemetry
1465	105
412	22

where $(P(\chi_1^2 \geq 1.5010) = 0.2205)$. By using the joint likelihood, the maximum likelihood estimate of the common value of S_T is

$$\hat{S}_T = \frac{T+a}{T+a+l.+d}, \tag{10.12}$$

or, equivalently, the weighted average of the separate catch-curve ($\hat{S}_{T\text{-}CC}$) and radio-telemetry ($\hat{S}_{T\text{-}RT}$) estimates, where

$$\hat{S}_T = \frac{\hat{S}_{T\text{-}CC} \cdot (T+l.) + \hat{S}_{T\text{-}RT}(a+d)}{(T+l.)+(a+d)}. \tag{10.13}$$

In this example,

$$\hat{S}_T = \frac{1465+89}{1465+89+412+22} = 0.7817$$

with an associated binomial variance of

$$\widehat{\mathrm{Var}}(\hat{S}) = \frac{\hat{S}_T(1-\hat{S}_T)}{(T+l.)+(a+d)}$$
$$= \frac{0.7817(1-0.7817)}{(1465+89)+(412+22)} = 0.0000858,$$

for a standard error of $\widehat{\mathrm{SE}}(\hat{S}_T) = 0.0093$. The large sample size of the age-at-harvest data overwhelms the radiotelemetry information, resulting in a pooled estimate very similar to that of the catch-curve analysis alone. However, the similarity of the pooled estimate and that of the catch-curve analysis does not diminish the importance of the radiotelemetry data.

This example illustrates the value of integrating extensive and intensive sampling schemes whenever possible. The cross-validation of the two distinct approaches to survival estimation has significant inferential value. The robustness of the radiotelemetry model, along with the greater geographic scale of the age-at-harvest data, permits inferences that neither technique alone could provide. The radiotelemetry data provide a robust approach to survival estimation for a small segment of the population. The catch-curve analyses provide a model-laden estimate of survival for a much larger population. Used together, the two approaches can provide sound statistical inference to the larger clk population sampled by the age-at-harvest data. Strategic use of both extensive and intensive sampling methods therefore holds the promise of providing reliable and precise demographic assessments of large-scale populations economically.

10.5   A Bighorn Sheep Life-History-Based Abundance Estimator

This example illustrates how seemingly disparate visual count and radiotelemetry data may be combined for assessing population abundance, survival, and productivity of a bighorn sheep (*Ovis canadensis*) herd. The data used in the example are the types commonly collected in present day, large mammal studies. The analysis approach may be best described as a life-history-based model. Simple biological relationships are used in conjunction with field observations to produce an overall assessment, which is greater and more comprehensive than the sum of the individual elements.

In early spring, 19 radio-tagged rams and 27 ewes were present in a population of indeterminate size. An overflight of the population at that time resulted in the following counts:

	Rams	Ewes
Radio-tag sighted	10	7
Non-tagged sighted	12	21
Totals	22	28

The radio-tagged ewes were tracked through the spring, and the 27 radio-tagged adult females produced a total of 23 lambs.

In autumn, another overflight observed a ratio of 7 lambs : 24 ewes. In addition, tracking of the radio-tagged adult sheep over the summer until autumn provided the following survival data:

	Rams	Ewes
Alive	17	23
Dead	2	4
Totals	19	27

By using these demographic observations, the abundance of the bighorn sheep population in the fall of year can be estimated. Nestled within the above data is information on spring adult abundance, productivity, and juvenile and adult survival. The first task is to construct a life-history — based model of fall sheep abundance as a function of the available demographic data.

The fall sheep abundance will be composed of adults that survive to fall and juveniles that are produced and survived to fall, such that

$$N_{Fall} = N_{Adult} + N_{Juveniles}. \tag{10.14}$$

There are sufficient survey data to estimate the spring adult abundance (N_S) and their survival (S_A), so that adult abundance in fall can be estimated as

$$\hat{N}_{Adult} = \hat{N}_S \cdot \hat{S}_A. \tag{10.15}$$

The fall juvenile abundance can be estimated from the adult spring abundance (i.e., N_S), the fraction of adults that are female (p_F), an estimate of productivity (i.e., juveniles per female, P), and an estimate of the juvenile survival probability (S_J), so that

$$\hat{N}_{Juvenile} = \hat{N}_S \cdot \hat{p}_F \cdot \hat{P} \cdot \hat{S}_J. \tag{10.16}$$

However, the change in lamb-to-ewe ratios (Section 5.11.1) between spring and fall surveys can only estimate the ratio of juvenile survival to adult survival ($S_{J/A}$), where

$$S_{J/A} = \frac{S_J}{S_A}$$

in the presence of adult mortality. Hence, to estimate juvenile survival, the product

$$\hat{S}_J = \hat{S}_{J/A} \cdot \hat{S}_A$$

must be used in Eq. (10.16). Therefore, fall juvenile abundance will be estimated by

$$\hat{N}_{Juvenile} = \hat{N}_S \cdot \hat{p}_F \cdot \hat{P} \cdot \hat{S}_{J/A} \cdot \hat{S}_A. \tag{10.17}$$

Combining Eqs. (10.14), (10.15), and (10.17), an estimator of fall population abundance can be written as

$$\hat{N}_{Fall} = \hat{N}_S \hat{S}_A \left(1 + \hat{p}_F \cdot \hat{P} \cdot \hat{S}_{J/A}\right). \tag{10.18}$$

Hence, the life-history-based abundance estimator (Eq. 10.18) is composed of five separate demographic parameters that need to be estimated from the survey data.

Adult bighorn sheep abundance in spring can be estimated from the single mark-recapture model also known as the Lincoln-Petersen method. The mark-resighting data can be summarized in the following R × C contingency table:

		Rams	Ewes	
Capture	10	9	20	29
history	01	12	21	33
	11	10	7	17
		31	48	79

The R × C contingency table summarizes the mark-resighting data based on sex and capture history for the two-period study. A "1" denotes capture in a period, whereas "0" denotes escape or nondetection. For example, of the 31 unique rams detected, 10 were radio-tagged and resighted and another 9 radio-tagged rams were undetected in the overflight. Skalski and Robson (1992) suggested using the above R × C contingency table to test for homogeneity of detection probabilities between subpopulations. The corresponding chi-square test of homogeneity is nonsignificant ($P(\chi_2^2 \geq 3.6680) = 0.1598$), suggesting equal detection probabilities for rams and ewes, as well as the appropriateness of pooling of the data. By use of the pooled counts (i.e., raw totals), the Chapman (1951) bias-corrected estimator of abundance for a two-sample mark-recapture study is

$$\hat{N}_S = \frac{(n_1 + 1)(n_2 + 1)}{(m + 1)} - 1$$

$$= \frac{(46 + 1)(50 + 1)}{(17 + 1)} - 1 = 132.17 \text{ spring adults}$$

for $n_1 = 46$, $n_2 = 50$, and $m = 17$. The variance of $\hat{N}_S$ is calculated as

$$\widehat{\text{Var}}(\hat{N}_S) = \frac{(n_1 + 1)(n_2 + 1)(n_1 - m)(n_2 - m)}{(m + 1)^2 (m + 2)}$$

$$= \frac{(46 + 1)(50 + 1)(46 - 17)(50 - 17)}{(17 + 1)^2 (17 + 2)} = 372.6330,$$

for a standard error of $\widehat{\text{SE}}(\hat{N}_S) = 19.30$.

Based on the $f = 27$ radio-tagged ewes monitored in the population that produced $y = 23$ lambs, productivity (P) is estimated to be

$$\hat{P} = \frac{y}{f}$$

$$= \frac{23}{27} = 0.8519 \text{ lambs/ewe}.$$

Assuming a binomial sampling error, the variance of the productivity estimator can be calculated by using Eq. (3.13), where

$$\widehat{\text{Var}}(\hat{P}) = \frac{\hat{P}(1 + \hat{P})^2}{(y + f)}$$

$$= \frac{\left(\dfrac{23}{27}\right)\left(1 + \dfrac{23}{27}\right)^2}{(23 + 27)} = 0.0584,$$

for an associated standard error of $\widehat{\text{SE}}(\hat{P}) = 0.2417$.

The proportion of females (p_F) in the population at the time of the first overflight, based on the observation of $m = 22$ rams and $f = 28$ ewes, is estimated to be

$$\hat{p}_F = \frac{f}{m+f}$$

$$= \frac{28}{22+28} = 0.56.$$

Again, assuming binomial sampling, the estimated variance of $\hat{p}_F$ is the binomial variance formula

$$\widehat{\mathrm{Var}}(\hat{p}_F) = \frac{\hat{p}_F(1-\hat{p}_F)}{(m+f)}$$

$$= \frac{0.56(1-0.56)}{50} = 0.004928,$$

with an associated standard error of $\widehat{\mathrm{SE}}(\hat{p}_F) = 0.0702$. For both the estimate of productivity $(\hat{P})$ and the proportion of adult females in the population $(\hat{p}_F)$, ignoring the finite sampling correction should tend to overestimate the sampling error.

Adult survival is based on the radio-tag tracking of adult sheep from spring to fall. An R × C contingency table analysis of homogeneous survival

	Rams	Ewes	Totals
Alive	17	23	40
Dead	2	4	6
Totals	19	27	46

finds no significant difference in survival between the sexes $(P(\chi_1^2 \geq 0.1808) = 0.6706)$. Pooling the data, adult survival from spring to fall is estimated from the $a = 40$ alive and $d = 6$ dead radio-tagged adults, where

$$\hat{S}_A = \frac{a}{a+d}$$

$$= \frac{40}{40+6} = 0.8696.$$

Again, using a binomial sampling variance,

$$\widehat{\mathrm{Var}}(\hat{S}_A) = \frac{\hat{S}_A(1-\hat{S}_A)}{a+d}$$

$$= \frac{\left(\frac{40}{46}\right)\left(1-\frac{40}{46}\right)}{40+6} = 0.0025,$$

for a standard error of $\widehat{\mathrm{SE}}(\hat{S}_A) = 0.0497$.

An estimate of the ratio of juvenile to adult survival can be calculated from change-in-ratio (CIR) methods (Section 5.11.1), where Eq. (5.142),

$$\hat{S}_{J/A} = \frac{a_1 y_2}{y_1 a_2}$$

$$= \frac{27 \cdot 7}{23 \cdot 24} = 0.3424,$$

and the sampling variance is estimated by Eq. (5.144), where

$$\widehat{\text{Var}}(\hat{S}_{J/A}) = \frac{\hat{S}_{J/A}}{\hat{R}_{J/A}} \left(\frac{\hat{S}_{J/A}(1+\hat{R}_{J/A})^2}{x_1} + \frac{(1+\hat{S}_{J/A}\hat{R}_{J/A})^2}{x_2} \right)$$

$$= \frac{0.3424}{\left(\frac{23}{27}\right)} \left[\frac{0.3424\left(1+\frac{23}{27}\right)^2}{50} + \frac{\left(1+(0.3424)\left(\frac{23}{27}\right)\right)^2}{31} \right] = 0.0311,$$

and where $\hat{R}_{J/A} = \frac{y_1}{a_1} = \frac{23}{27} \doteq 0.8519, x_1 = a_1 + y_1 = 50, x_2 = 24 + 7 = 31$. The ratio of the juvenile to adult survival has an estimated standard error of $\widehat{\text{SE}}(\hat{S}_{J/A}) = 0.1763$.

By combining the results, the life-history-based model (Eq. 10.18) estimates fall abundance to be

$$\hat{N}_{\text{Fall}} = \hat{N}_S \hat{S}_A (1 + \hat{p}_F \cdot \hat{P} \cdot \hat{S}_{J/A})$$

$$= 132.17(0.8696)(1 + 0.56(0.8519)(0.3424)) = 133.71.$$

Thus, the fall population is composed of an estimated 114.93 adults and 18.77 juveniles. By use of the delta method, the variance of $\hat{N}_{\text{Fall}}$ is approximated by the following formula:

$$\widehat{\text{Var}}(\hat{N}_{\text{Fall}}) = \widehat{\text{Var}}(\hat{N}_S)(\hat{S}_A(1+\hat{p}_F \hat{P}\hat{S}_{J/A}))^2 + \widehat{\text{Var}}(\hat{S}_A)(\hat{N}_S(1+\hat{p}_F \hat{P}\hat{S}_{J/A}))^2$$

$$+ \widehat{\text{Var}}(\hat{p}_F)(\hat{N}_S \hat{S}_A \hat{P}\hat{S}_{J/A})^2 + \widehat{\text{Var}}(\hat{P})(\hat{N}_S \hat{S}_A \hat{p}_F \hat{S}_{J/A})^2$$

$$+ \widehat{\text{Var}}(\hat{S}_J)(\hat{N}_S \hat{S}_A \hat{p}_F \hat{P})^2$$

$$\widehat{\text{Var}}(\hat{N}_{\text{Fall}}) = \hat{N}_{\text{Fall}}^2 \left(\text{CV}(\hat{N}_S)^2 + \text{CV}(\hat{S}_A)^2 \right)$$

$$+ \hat{N}_{\text{Fall-Juv}}^2 \left(\text{CV}(\hat{p}_F)^2 + \text{CV}(\hat{P})^2 + \text{CV}(\hat{S}_{J/A})^2 \right), \tag{10.19}$$

where $N_{\text{Fall-Juv}}$ denotes the abundance of juveniles in the fall season. From the survey data and the previous calculations,

$$\widehat{\text{Var}}(\hat{N}_{\text{Fall}}) = (133.71)^2 \left[\left(\frac{19.30}{132.17} \right)^2 + \left(\frac{0.0497}{0.8696} \right)^2 \right]$$

$$+ (18.77)^2 \left[\left(\frac{0.0702}{0.56} \right)^2 + \left(\frac{0.2417}{0.8519} \right)^2 + \left(\frac{0.1763}{0.3424} \right)^2 \right]$$

$$= 566.9196,$$

for a standard error of $\widehat{\text{SE}}(\hat{N}_{\text{Fall}}) = 23.81$.

For many species, small-scale radiotelemetry studies can be combined with visual counting techniques for demographic assessment. As this example demonstrates, repeated visual surveys (e.g., once in spring, another in fall) and concomitant radiotelemetry studies can provide the information to estimate abundance, productivity, and survival. This approach, however, will require biologists to construct *a priori* life-history models specific to the populations under investigation. Based on the population model, an investigator can then design a study to collect the necessary data to complete the demographic assessment. This approach does not imply biologists should collect a haphazard set of visual count and radiotelemetry data and expect to perform a quality *post hoc* demographic assessment. Just the opposite, study objectives and sampling precision need to be considered early in the planning stages if the demographic assessment is to be successful. With creative life-history models and careful planning, it is possible to estimate demographic parameters both efficiently and economically.

10.6 Partitioning Harvest and Natural Mortality

Partitioning total mortality into natural and anthropogenic components is essential in managing exploited species well. However, the task is usually not possible using harvest information alone (see Chapter 6). Rather, at least two elements from among harvest, natural, and total mortality are needed to partition mortality sources. This example illustrates the use of harvest numbers and their age compositions to estimate total annual mortality and harvest mortality and, in turn, estimate the component of natural mortality. The analysis is based on exploiting the assumption of a stable and stationary population.

Consider a hypothetical example of a white-tailed deer (*Odocoileus virginianus*) harvest. Weekly harvest and effort are recorded for four 5-day hunting periods as follows:

Period	Male harvest (n_i)	Effort (g_i)
1	170	500
2	97	300
3	89	300
4	61	200

In addition to the harvest count, tooth eruption and wear (Taber 1969) were used to differentiate yearling from older adults. Of the 417 animals harvested, 102 were yearlings. These data can be used to estimate total annual survival and the probabilities of natural and harvest mortality under the assumptions of a SAD.

Incremental harvest counts and effort can be used to construct a catch-effort likelihood (Eq. 9.75) of the form

$$L_{CPUE} = \prod_{i=1}^{4} \binom{N - x_i}{n_i} \left(1 - e^{-cg_i}\right)^{n_i} \left(e^{-cg_i}\right)^{N - x_i - n_i}, \qquad (10.20)$$

where

N = initial abundance of deer;
c = vulnerability coefficient;
g_i = hunting effort in the ith period;

n_i = number of animals harvested in the ith period;

$x_i = \sum_{j=1}^{i-1} n_j$ = number of animals removed from the population to the time of the ith

period.

k = number of sampling periods (e.g., 4).

From the proportion of yearlings harvested and the Heincke-Burgoyne method (Eq. 5.90), total annual survival can be estimated from the likelihood

$$L = \binom{x_{k+1}}{l_0}(1 - S_T)^{l_0}(S_T)^{x_{k+1}-l_0},\qquad(10.21)$$

where l_0 = the number of yearlings in the harvest. Total annual survival S_T in Eq. (10.21) can be reparameterized as

$$S_T = S_N \cdot S_H,$$

or, in turn,

$$S = S_N \cdot e^{-cG},$$

based on the catch-effort model, where $G = \sum_{i=1}^{k} g_i$ is the total harvest effort. Hence, the Heincke-Burgoyne likelihood can be rewritten as

$$L = \binom{x_{k+1}}{l_0}(1 - S_N e^{-cG})^{l_0}(S_N e^{-cG})^{x_{k+1}-l_0}.\qquad(10.22)$$

The product of Eqs. (10.20) and (10.22) becomes a joint likelihood, permitting the estimation of S_N, c, and initial male deer abundance (N). Iteratively solving the joint likelihood yields

$$\hat{S}_N = 0.9263, \widehat{SE} = 0.0258,$$
$$\hat{c} = 0.000157, \widehat{SE} = 0.000024,$$
$$\hat{N} = 2259.71, \widehat{SE} = 111.82.$$

By using the invariance property of maximum likelihood estimation, the probability of a deer surviving the hunt is estimated by the expression

$$\hat{S}_H = e^{-\hat{c}G}$$
$$= e^{-(0.000157)1300} = 0.8154.$$

By use of the delta method, the variance of $\hat{S}_H$ can be estimated by

$$\widehat{Var}(\hat{S}_H) = \widehat{Var}(\hat{c})e^{-2\hat{c}G}G^2$$
$$= (0.000024)^2 e^{-2(0.000157)1300}(1300)^2 = 0.0006472,$$

or, equivalently, a standard error of $\widehat{SE}(\hat{S}_H) = 0.0254$. Hence, natural survival is estimated to be $\hat{S}_N = 0.9263$, with a harvest survival probability of $\hat{S}_H = 0.8154$.

For harvested populations, it is important to know the relative contributions of natural and harvest mortality. Even for nongame species, it is important to understand the sources of mortality (e.g., predation and human-induced factors, such as power-line collisions) affecting population demographics. This information forms the basis for effective conservation plans. In this example, harvest and natural mortality were partitioned for white-tailed deer by assuming a stable and stationary population. Unlike the example (Section 10.2) that used radio telemetry to partition mortality sources, harvest counts and hunter effort data were used. This procedure is appealing because the necessary data may be easily obtained during closely monitored hunts, without the need for costly radiotelemetry studies. However, the restrictive stable and stationary assumptions should be closely considered. This approach has other similar applications. For example, an investigator could partition predation mortality from other mortality sources, should information about total kills and predator effort be collected.

10.7 Ring-Necked Pheasant Multisurvey Study

The assumptions of an abundance survey can usually be assessed only after the data have been collected. Thus, abundance surveys should be designed to accommodate multiple techniques and alternative survey models. Tests of assumptions and model selection should be used to identify the most appropriate and parsimonious model to describe the data. A hypothetical ring-necked pheasant (*Phasianus colchicus*) example will be used to illustrate the incorporation of sequential CIR (Section 9.5.3), catch-per-unit effort (CPUE) (Section 9.6.1), and single mark-recapture (Section 9.2.5) methods into a single survey study.

The pheasant population to be surveyed will be subject to a rooster-only hunting season. A controlled-access hunt permits the recording of hunting effort and numbers of birds harvested on a weekly basis. Before the hunting season, marked game-farm — raised rooster pheasants will be released into the population. Exit surveys of the hunter harvest will therefore permit collection of information on numbers of marked and unmarked birds harvested. Periodic visual surveys throughout the course of the investigation will also be performed to collect sex ratio information.

Four sex-ratio surveys were performed. An initial survey to characterize the baseline sex ratio resulted in $x_1 = 75$ observations, with $m_1 = 30$ males and $f_1 = 45$ hens. A release of $A_1 = 125$ marked rooster pheasants was then performed. A subsequent second sex-ratio survey resulted in $x_2 = 110$ observations, with $m_2 = 52$ males and $f_2 = 58$ females. In the first week of sport harvest, $A_2 = -115$ roosters were removed with $g_1 = 200$ units of hunting effort. A third sex-ratio survey after the first week of hunting produced $x_3 = 65$ observations, with $m_3 = 24$ males and $f_3 = 41$ females. The second week of hunting yielded $A_3 = -50$ roosters with $g_2 = 125$ units of hunting effort. The final sex-ratio survey after the second week of hunting resulted in $x_4 = 90$ bird sightings, with $m_4 = 33$ males and $f_4 = 57$ females. Of the total harvest of $n_2 = 165$ rooster pheasants, $m = 54$ marked birds were recovered from the initial release of $n_1 = 125$ marked roosters.

From these survey data, initial pheasant abundance (N_1), initial rooster pheasant abundance (X_1), and the number of hens (i.e., $N_1 - X_1$) at the start of the study can be estimated. With the additional knowledge of the number of birds added and removed from the population, end-of-study abundance can also be estimated, along with harvest mortality. The joint likelihood of the abundance study is the product of the independent likelihoods for the CIR, CPUE, and single mark-recapture surveys, where

$$L = L_{CIR} \cdot L_{CPUE} \cdot L_{MR}. \tag{10.23}$$

The likelihood for the CIR study is based on a four-period sequential CIR model (Section 9.5.3). The likelihood for the CPUE study is based on a two-period, variable-effort removal model (Eq. 9.75). In general, a two-period CPUE study is not recommended, but as an auxiliary likelihood, it provides additional information to improve the precision of the joint likelihood estimates. The mark-recapture model is based on a hypergeometric likelihood (Section 9.2.5). The individual components of the joint likelihood are as follows:

$$
L_{CIR} = \binom{75}{30,45}\left(\frac{X_1}{N_1}\right)^{30}\left(\frac{N_1-X_1}{N_1}\right)^{45}
$$
$$
\cdot\binom{110}{52,58}\left(\frac{X_1+125}{N_1+125}\right)^{52}\left(\frac{N_1-X_1}{N_1+125}\right)^{58}
$$
$$
\cdot\binom{65}{24,41}\left(\frac{X_1+125-115}{N_1+125-115}\right)^{24}\left(\frac{N_1-X_1}{N_1+125-115}\right)^{41} \qquad (10.24)
$$
$$
\cdot\binom{90}{33,57}\left(\frac{X_1+125-115-50}{N_1+125-115-50}\right)^{33}\left(\frac{N_1-X_1}{N_1+125-115-50}\right)^{57},
$$

$$
L_{CPUE} = \binom{X_1+125}{115}\left(1-e^{-200c}\right)^{115}\left(e^{-200c}\right)^{X_1+125-115}
$$
$$
\cdot\binom{X_1+125-115}{50}\left(1-e^{-125c}\right)^{50}\left(e^{-125c}\right)^{X_1+125-115-50}, \qquad (10.25)
$$

$$
L_{MR} = \frac{\binom{X_1}{111}\binom{125}{54}}{\binom{X_1+125}{165}}. \qquad (10.26)
$$

Each of the three abundance methods can be used to separately estimate some or all of the abundance parameters. Analyzing the CIR likelihood alone provides the estimates

$$
\hat{N}_1 = 851.07, \widehat{SE} = 557.58,
$$
$$
\hat{X}_1 = 332.06, \widehat{SE} = 224.98.
$$

The CPUE likelihood provides the estimates

$$
\hat{X}_1 = 181.60, \widehat{SE} = 24.23,
$$
$$
c = 0.0024, \widehat{SE} = 0.0002.
$$

Finally, the mark-recapture likelihood provides the maximum likelihood estimate of $\hat{X}_1$ = 256.94, or the Chapman (1951) biased-corrected estimator (Eq. 9.30)

$$
\hat{X}_1 = 254.29, \widehat{SE} = 31.19.
$$

The intent of the joint likelihood model (10.23) is to provide a more precise estimate of N_1, the initial pheasant abundance, than any one technique might alone. Using the survey data from all three likelihoods, the joint likelihood model (10.23) estimates

$$\hat{N}_1 = 662.70, \widehat{SE} = 81.65,$$
$$\hat{X}_1 = 255.80, \widehat{SE} = 27.46.$$
$$c = 0.0018, \widehat{SE} = 0.0002.$$

In turn, the estimate of total pheasant abundance at the completion of the demographic survey can be estimated as

$$\hat{N}_2 = \hat{N}_1 + 125 - 115 - 50$$
$$= 622.70.$$

with a standard error of $\widehat{SE}(\hat{N}_2) = 81.65$, the same as that of $\widehat{SE}(\hat{N}_1)$, because the additions and deletions to the population are known constants.

Another piece of information that can be extracted from the survey is the exploitation rate of roosters, expressed as the fraction of males harvested. In the case of this study, the harvest mortality rate (M_H) can be estimated as

$$\hat{M}_H = \frac{115 + 50}{\hat{X}_1 + 125}$$
$$= 0.4333.$$

The variance of $\hat{M}_H$ can be approximated by the delta method, where

$$\widehat{Var}(\hat{M}_H) = \widehat{Var}(\hat{X}_1)\left(\frac{(165)^2}{(\hat{X}_1 + 125)^4}\right)$$
$$= 0.000976,$$

or, equivalently, a standard error of $\widehat{SE}(\hat{M}_H) = 0.0312$.

Our last example integrates three different techniques for a single demographic assessment. It demonstrates how periodic visual surveys can be coupled with data from controlled hunts and other interventions (e.g., supplemental releases) to evaluate abundance and harvest mortality. As with other examples, a joint likelihood was developed that combined these disparate sources of data, thus allowing a more precise estimate of abundance to be computed, compared with that of any individual technique. We believe the union of data in this way is key to future advancements in the field of wildlife assessment. In this case, multiple sighting surveys, with carefully monitored harvests, provided a robust estimate of abundance. Intermediate evaluations with visual surveys could also be performed to assess whether harvest or other mortality forms (e.g., predation) influence established and released bird populations. From the data collected, it was also possible to estimate the harvest mortality rate, which can aid future harvest management (e.g., number of hunters allowed and number of game birds to release to meet management objectives).

Appendix A
Statistical Concepts and Theory

Statistical distributions and concepts are used throughout the book to estimate demographic parameters. In this appendix, some of the key quantitative concepts used in parameter estimation and data analysis are described in more detail. Readers are encouraged to read more about maximum likelihood estimation (MLE) in Edwards (1992), regression estimation in Draper and Smith (1998), and parameter estimation in Beck and Arnold (1977).

Maximum Likelihood Estimation

A well-known and widely used method of estimation is MLE, developed by Fisher (1922, 1925). This method provides a statistically efficient and powerful method of deriving parameter estimates and standard errors. The approach also provides the ability to test hypotheses concerning model parameters and the ability to construct confidence intervals. The MLE method has a number of asymptotic (i.e., large sample size) statistical properties, including normality, maximum efficiency, and a lack of bias (Wilks 1962, Rao 1973, Lehmann 1983), that make the method very appealing.

The MLE approach is based on defining a probability distribution(s) that describes the observed random variables (i.e., the data) based on the sampling design and parameters that generated the observations. This function describes the probability of observing the random variables $\underset{\sim}{x}$ for a particular parameterization. The probability function can be denoted by $f(\underset{\sim}{x}|\theta)$, where θ is the vector of parameters and $\underset{\sim}{x}$ is the vector of observed random variables. In turn, the probability function can be conceptualized as a function of θ given $\underset{\sim}{x}$, known as the likelihood function and denoted as $L(\theta|\underset{\sim}{x})$. Here, $L(\theta|\underset{\sim}{x})$, gives the probability or "likelihood" of the unknown parameters θ, given the observed sample data $\underset{\sim}{x}$. The likelihood function is often a product of several probability density functions. The objective of MLE is to find the values of θ that "most likely" generated the sample observations. The most likely parameter values are those that maximize the likelihood function for a given set of data. The values of θ that maximize the likelihood function are known as the "maximum likelihood estimates" and denoted by $\hat{\theta}$.

Under certain regularity conditions, the vector of values $\hat{\theta}$ that maximize the likelihood function can be found by simultaneously solving the system of equations:

$$\frac{\partial L(\underset{\sim}{\theta}|\underset{\sim}{x})}{\partial \theta_1} = 0$$

$$\frac{\partial L(\underset{\sim}{\theta}|\underset{\sim}{x})}{\partial \theta_2} = 0$$

$$\vdots$$

$$\frac{\partial L(\underset{\sim}{\theta}|\underset{\sim}{x})}{\partial \theta_H} = 0,$$

where there are H parameters in the likelihood model $L(\underset{\sim}{\theta}|\underset{\sim}{x})$. It is often easier to work with the log-likelihood, $\ln L(\underset{\sim}{\theta}|\underset{\sim}{x})$, as the values of $\underset{\sim}{\theta}$ that maximize the log-likelihood will also maximize the likelihood.

The system of equations obtained by setting the first derivations of the log-likelihood with respect to θ_i $(i = 1, \ldots, H)$ equal to zero, i.e.,

$$\frac{\partial \ln L(\underset{\sim}{\theta}|\underset{\sim}{x})}{\partial \theta_1} = 0$$

$$\frac{\partial \ln L(\underset{\sim}{\theta}|\underset{\sim}{x})}{\partial \theta_2} = 0$$

$$\vdots$$

$$\frac{\partial \ln L(\underset{\sim}{\theta}|\underset{\sim}{x})}{\partial \theta_H} = 0,$$

is known as the likelihood equations. When the system of likelihood equations can be solved algebraically, the solutions $\hat{\underset{\sim}{\theta}}$ are said to be of "closed form." When the number of minimum sufficient statistics equals the number of parameters, the MLEs will be equivalent to the method-of-moment estimators based on the minimum sufficient statistics. When the number of minimum sufficient statistics is less than the number of model parameters, some or all of the parameters may not be estimable. Finally, when the number of minimum sufficient statistics exceeds the number of model parameters, the MLEs must be solved numerically by using iterative procedures.

A variety of numerical optimization techniques exist to iteratively solve for the MLEs. The well-known Newton-Raphson method (Seber 1982:16–18) is based on use of both first and second partial derivatives of the likelihood model. Quasi-Newton methods avoid the difficulty of deriving and coding the second partial derivates and, in some instances, even the first partial derivatives. The Broyden-Fletcher-Goldfarb-Shanno method is based on only the first derivatives (Press et al. 1986). The FLETCH algorithm uses no explicit expression for either the first or second partial derivatives but, instead, is based solely on coding the log-likelihood function (Fletcher 1970).

Being a function of the random variables $\underset{\sim}{x}$, the MLEs are random variables that might be expected to vary from one sample to another. There is, therefore, a sampling variance associated with each MLE. Any two parameter estimates from the same likelihood function might also be expected to depend on one another. Hence, there is also an associated sample covariance between parameter estimates. These variances and covariances for the MLEs can be represented by a variance-covariance matrix $\underset{\sim}{\Sigma}$ of the form:

$$\Sigma = \begin{bmatrix} \mathrm{Var}(\theta_1) & \mathrm{Cov}(\theta_1,\theta_2) & \cdots & \mathrm{Cov}(\theta_1,\theta_H) \\ \mathrm{Cov}(\theta_2,\theta_1) & \mathrm{Var}(\theta_2) & & \vdots \\ \vdots & & \ddots & \\ \mathrm{Cov}(\theta_H,\theta_1) & \mathrm{Cov}(\theta_H,\theta_2) & \cdots & \mathrm{Var}(\theta_H) \end{bmatrix}_{H \times H}.$$

The $\mathrm{Cov}(\theta_i,\theta_j) = \mathrm{Cov}(\theta_j,\theta_i)$; hence, Σ is a symmetric matrix of dimensions $H \times H$.

The likelihood model provides a convenient means to derive the variance-covariance matrix for the MLEs. The variance-covariance matrix can be derived as the inverse of the information matrix or the Hessian matrix $I(\theta)$. The ijth element of the information matrix is

$$-E\left[\frac{\partial^2 \ln L(\theta|x)}{\partial \theta_i \partial \theta_j}\right].$$

The variance-covariance matrix is then

$$\Sigma = [I(\theta)]^{-1}.$$

When the information matrix is evaluated using the model parameters, $I(\theta)$, the variance-covariance matrix is derived. When estimated parameter values $\hat{\theta}$ are used in evaluating $I(\hat{\theta})$, the result is the estimated variance-covariance matrix. Variances based on the inverse Hessian matrix are asymptotic variances based on the Cramér-Rao minimum variance bound (Rao 1945, Cramér 1946). The variances are the large sample size approximations to the variances of the parameter estimates.

Despite the relative conceptual simplicity of the variance-covariance matrix, the actual derivation can be algebraically complex. The complexity only increases as the number of model parameters increase. Fortunately, most numerical optimization methods used to solve for the MLEs use the inverse of the information matrix in the process. The last iteration of the numerical method therefore provides an estimated variance-covariance matrix.

An alternative approach to variance calculations is to directly calculate the variance of $\hat{\theta}_i$ based on its form and function of the random variables (x). Should $\hat{\theta}_i$ be a linear function of x, the variance calculation is rather straightforward. Alternatively, if $\hat{\theta}$ is a nonlinear function of x, the delta method approach to variance approximation can be used.

Interval Estimation

An asymptotic $(1 - \alpha)100\%$ confidence interval for θ_i can be derived, using the asymptotic normality property of MLEs, from the expression

$$\hat{\theta}_i \pm Z_{1-\frac{\alpha}{2}} \sqrt{\widehat{\mathrm{Var}}(\hat{\theta}_i)},$$

where Z is a standard normal random variable such that $P\left(Z_{\frac{\alpha}{2}} < Z < Z_{1-\frac{\alpha}{2}}\right) = 1 - \alpha$. The asymptotic confidence interval is appropriate when large sample sizes exist.

The likelihood model can also be used to more directly yield a confidence interval for θ_i. The profile likelihood interval method was first suggested by Fisher (1956) and discussed by Box and Cox (1964), and Hudson (1971). The basis of the method is the likelihood-ratio test (LRT), where

$$-2\ln\left(\frac{L_{H_a}}{L_{H_o}}\right) \approx \chi^2_{df}$$

and where

L_{H_a} = value for the likelihood function under the alternative hypothesis;
L_{H_o} = value of the likelihood function under the null hypothesis;
df = degrees of freedom equal to the difference in the number of model parameters
     between the H_o and H_a nested models;

is asymptotically chi-squared distributed. The advantage of the profile likelihood approach is that the LRT attains its asymptotic χ^2 distribution more readily and with smaller sample sizes than the MLEs approach normality. The profile-likelihood confidence interval is then defined by

$$CI(\theta) = \left\{\theta: \quad -2\ln\left(\frac{L(\theta)}{L(\hat{\theta})_{MLE}}\right) \leq \chi^2_{1,1-\alpha}\right\},$$

where

$L(\hat{\theta})$ = value of the likelihood evaluated at the MLE $\hat{\theta}$;
$L(\theta)$ = value of the likelihood evaluated at a value of θ other than the MLE.

Thus, the $(1 - \alpha)100\%$ confidence interval for θ is constructed from all those values of $\hat{\theta}$ that satisfy the inequality

$$-2\ln\left(\frac{L(\theta)}{L(\hat{\theta})}\right) \leq \chi^2_{1,1-\alpha}.$$

In the cases of multiparameter likelihoods, the confidence interval for θ_i is constructed for those values of θ_i that satisfy the inequality

$$-2\ln\left(\frac{L(\theta_i,\hat{\underset{\sim}{\theta}}')}{L(\hat{\underset{\sim}{\theta}})}\right) \leq \chi^2_{1,1-\alpha}$$

where

$L(\hat{\underset{\sim}{\theta}})$ = value for the likelihood evaluated at the MLEs for the vector $\underset{\sim}{\theta}$;
$L(\theta_i,\hat{\underset{\sim}{\theta}}')$ = value of the likelihood evaluated at a value of θ_i other than the MLE and
            at new MLE values for the remaining parameters, $\hat{\underset{\sim}{\theta}}'$, that maximize the
            likelihood for a given value of θ_i.

The profile likelihood confidence intervals are therefore numerically intense and have not been widely used until recently.

Hypothesis Testing

The LRT provides a convenient and asymptotically most powerful means of comparing two competing models. Two models are "nested" or "hierarchical" when one model is a special case of the other. This nesting can occur if one model omits a parameter(s) the other model possesses. Nested models can also occur if one model equates two or more parameters the other model holds unique.

The selection between two alternative and nested models is the choice between the model with more parameters (i.e., the "full model") and the model with fewer parameters (i.e., the "reduced model"). In this case, the LRT is of the form

$$-2\ln\left(\frac{L(\hat{\theta}_R)}{L(\hat{\theta}_F)}\right) \approx \chi^2_{df},$$

where

$L(\hat{\theta}_R)$ = value of the reduced likelihood at the values of its MLEs;

$L(\hat{\theta}_F)$ = value of the full likelihood at the value of its MLEs;

χ^2_{df} = chi-square statistic with df equal to the difference in the number of parameters between the full and reduced models.

The LRT is used to test the set of hypotheses:

H_o: Full model fits the data no better than the reduced model.

H_a: Full model fits the data better than the reduced model.

In many situations, these hypotheses can be translated in terms of model parameters such as

$$H_o: \quad \theta_i = 0$$
$$\text{versus}$$
$$H_a: \quad \theta_i \neq 0$$

or

$$H_o: \quad \theta_i = \theta_j$$
$$\text{versus}$$
$$H_a: \quad \theta_i \neq \theta_j.$$

The null hypothesis is rejected if the probability of observing the chi-square LRT statistic is less than α.

The Wald (1943) test statistic is another option for testing hypotheses concerning model parameters. This test statistic is based on the asymptotic multivariate normality of the MLEs. The test statistic can be used to test the significance of any set of parameters or linear combinations of the parameters. If E is a matrix composed of linear contrasts, and $\hat{\theta}$, the vector of MLEs for the likelihood model, then $E\hat{\theta}$ is a set of linear combinations of the parameters that should also be asymptotically normally distributed. Then the Wald test statistic,

$$W = \left(E\hat{\theta}\right)' \left[E\hat{\Sigma}E'\right]^{-1}\left(E\hat{\theta}\right),$$

is asymptotically chi-squared distributed with degrees of freedom equal to the dimension of E (Johnson and Wichern 1982). The Wald test statistic tests the set of hypotheses

$$H_0: \quad \boldsymbol{E}'\underset{\sim}{\theta} = 0$$

versus

$$H_a: \quad \boldsymbol{E}'\underset{\sim}{\theta} \neq 0.$$

The matrix $\hat{\boldsymbol{\Sigma}}$ is the estimated variance-covariance matrix for the MLEs. The Wald statistic can also be used to test the significance of a single parameter, in which case,

$$W = \frac{\hat{\theta}_i^2}{\widehat{\mathrm{Var}}(\hat{\theta}_i)} = Z^2,$$

is the square of the univariate Z-statistic (i.e., chi-square with one degree of freedom).

An alternative to formal statistical selection between models or parameter values is use of information criterion. One criterion that has recently received considerable attention (Burnham and Anderson 1998) is the Akaike's Information Criterion (AIC) (Akaike 1973). Model selection and parameter inference is based on the quantity

$$\mathrm{AIC} = -2\ln L(\hat{\underset{\sim}{\theta}}) + 2H,$$

where H = number of model parameters. The smaller the AIC value, the better the fit of the model to the data. The model with the smallest AIC is selected among the alternative models investigated. The AIC translates the model selection process into a univariate minimization problem (Burnham and Anderson 1998).

A similar model selection criterion is the Bayesian Information Criterion (BIC) (Schwartz 1993, Shono 2000), expressed as

$$\mathrm{BIC} = -2\ln L(\hat{\theta}) + H\ln(n),$$

where n = the number of datums in the data set used in fitting the model. Again, the model with the smallest BIC value is considered to fit the data best. It is difficult to compare or select among non-nested models. For those who try, the AIC and BIC provide guidance as long as the fundamental structure of the models is not too disparate and the same data are used in the models compared.

Delta Method

The delta method provides a very useful technique for approximating the expected value, variance, and covariance of nonlinear functions of random variables. Of particular interest is the nonlinear function of parameter estimates. The delta method is based on the definition of a Taylor series, where for some function $f(x)$,

$$f(x) = \sum_{n=0}^{\infty} \frac{f(a)^n (x-a)^n}{n!}$$

and where

$$f(a)^n = \left. \frac{\partial^n}{\partial x^n} f(x) \right|_{x=a}.$$

The function $f(x)$ needs to be a continuous function of x and have derivatives to the $(n + 1)$st order. Another way of expressing the Taylor series is

$$f(x) = f(a) + \frac{f'(a)(x-a)^1}{1!} + \frac{f''(a)(x-a)^2}{2!} + \frac{f'''(a)(x-a)^3}{3!} + \cdots$$

Letting x be a random variable with mean μ, the Taylor series expansion of $f(x)$ about the value of μ is then

$$f(x) = f(\mu) + f'(\mu)(x-\mu) + \frac{f''(\mu)(x-\mu)^2}{2} + \cdots$$

Taking the expected value of $f(x)$ to the first three terms of the Taylor series results in

$$E(f(x)) \doteq f(\mu) + \frac{f''(\mu)\mathrm{Var}(x)}{2}.$$

Similarly, taking the variance of $f(x)$ to the first two terms of the Taylor series results in

$$\mathrm{Var}(f(x)) \doteq \mathrm{Var}(x) \cdot f'(\mu)^2.$$

The Taylor series can also be applied to some function of multiple random variables. This leads to the general expressions

$$E(f(\underset{\sim}{x})) \doteq f(\underset{\sim}{\mu}) + \frac{1}{2}\sum_{i=1}^{H}\mathrm{Var}(x_i)\left(\frac{\partial^2 f}{\partial x_i}\right)_{|\mu} + \sum_{i=1}^{H}\sum_{\substack{j=1 \\ i<j}}^{H}\mathrm{Cov}(x_i,x_j)\left(\frac{\partial^2 f}{\partial x_i \partial x_j}\right)_{|\mu}$$

and also

$$\mathrm{Var}(f(\underset{\sim}{x})) \doteq \sum_{i=1}^{H}\mathrm{Var}(x_i)\left(\frac{\partial f}{\partial x_i}\right)_{|\mu}^2 + 2\sum_{i=1}^{H}\sum_{\substack{j=1 \\ i<j}}^{H}\mathrm{Cov}(x_i,x_j)\left(\frac{\partial f}{\partial x_i}\right)_{|\mu}\left(\frac{\partial f}{\partial x_j}\right)_{|\mu},$$

and

$$\mathrm{Cov}(f(\underset{\sim}{x}),g(\underset{\sim}{x})) \doteq \sum_{i=1}^{H}\sum_{j=1}^{H}\mathrm{Cov}(x_i,x_j)\left(\frac{\partial f}{\partial x_i}\right)_{|\mu}\left(\frac{\partial g}{\partial x_j}\right)_{|\mu}$$

from Seber (1982:7–9).

Variance Component Formula

For X and Y jointly distributed random variables, the variance of Y can be decomposed into two components

$$\mathrm{Var}(Y) = \mathrm{Var}_X[E(Y|X)] + E_X[\mathrm{Var}(Y|X)].$$

The first term represents the variance of the conditional mean of Y given X, whereas the second term represents the average conditional variance of Y given X. This formula can be used to derive the variance of Y in stages by conditionalizing on any convenient variable or stage in sampling. The variance component formula is also valuable in assessing the contributions of different sources of error to the overall variance in Y.

Other formulas useful in statistical evaluations are the expected value of some random variable Y,

$$E(Y) = E_X[E(Y|X)],$$

and the covariance between X and Y, where

$$Cov(X,Y) = E_Z[Cov(X,Y|Z)] + Cov_Z[E(X|Z), E(Y|Z)].$$

Linear Regression

For straight-line relationships described by the model

$$y_i = \alpha + \beta x_i + \varepsilon_i$$

where

y_i = dependent variable ($i = 1, \ldots, n$);
x_i = independent variable ($i = 1, \ldots, n$);
α = intercept;
β = slope;
ε_i = random error term ($i = 1, \ldots, n$);

the least squares regression can be used to estimate the parameters by minimizing the sum of squares $\sum_{i=1}^{n}(y_i - (\alpha + \beta x_i))^2$. The slope is estimated by the quantity,

$$\hat{\beta} = \frac{\sum_{i=1}^{n}(x_i - \bar{x})(y_i - \bar{y})}{\sum_{i=1}^{n}(x_i - \bar{x})^2}.$$

The variance of the slope parameter can be estimated by the quantity

$$\widehat{Var}(\hat{\beta}) = \frac{MSE}{\sum_{i=1}^{n}(x_i - \bar{x})^2},$$

where MSE = variance about the regression = $\dfrac{\sum_{i=1}^{n}(y_i - \hat{y}_i)^2}{n-2}$ for $\hat{y}_i = \hat{\alpha} + \hat{\beta}x_i$. In turn, the intercept can be estimated as

$$\hat{\alpha} = \bar{y} - \hat{\beta}\bar{x}$$

with the associated variance estimator

$$\widehat{\mathrm{Var}}(\hat{\alpha}) = \mathrm{MSE} \cdot \left[\frac{1}{n} + \frac{\bar{x}^2}{\sum\limits_{i=1}^{n}(x_i - \bar{x})^2} \right]$$

$$= \frac{\mathrm{MSE} \cdot \sum\limits_{i=1}^{n} x_i^2}{n \sum\limits_{i=1}^{n}(x_i - \bar{x})^2}.$$

The covariance between $\hat{\alpha}$ and $\hat{\beta}$ can be estimated by

$$\widehat{\mathrm{Cov}}(\hat{\alpha}, \hat{\beta}) = \frac{-\bar{x} \cdot \mathrm{MSE}}{\sum\limits_{i=1}^{n}(x_i - \bar{x})^2}.$$

Appendix B
Glossary of Symbols

Each chapter in this book defines algebraic symbols specific to the unique applications of the subject matter. Some of these symbols are consistently used with the same definitions throughout much of the fisheries and wildlife literature. For these symbols, a consistent definition across all book chapters was possible. However, for many terms, algebraic symbols vary from one publication to another. For these terms, it was not possible to be consistent with the primary literature in all cases. In some cases, a symbol had to be chosen that was not redundant with other uses elsewhere in the book and yet still reflected common usage as much as possible. The problem becomes heightened when covering numerous topics. For this reason, chapter-specific definitions of symbols were sometimes necessary.

Readers are encouraged to look within chapters to find the definitions of random variables or parameters. In this glossary, terms commonly used in two or more chapters of the book are defined.

English Alphabet

Symbol	Definition
A_i	Number of i-type animals added to the population after the first survey of a change-in-ratio study.
A_{X_i}	Number of X-type animals added to the population during the ith intervention of a change-in-ratio study.
A_{Y_i}	Number of Y-type animals added to the population during the ith intervention of a change-in-ratio study.
c_h	Number of animals harvested in the hth stratum.
C_h	Total harvest in the hth stratum.
C_{Total}	Total costs associated with performing a sample survey or study.
CV	Coefficient of variation $\left(CV(x) = \dfrac{\sqrt{\operatorname{Var}(x)}}{E(x)} \right)$.
d_t	Number of animals that died in the time interval t to $t + 1$.
f_i	Number of females in the ith sample.
F_i	Fecundity of the ith age class. The net number of female offspring produced per female in the age class i $\left(F = \dfrac{N_{FJ}}{N_F} \right)$.
$f_X(x)$	Probability density function (pdf), $\displaystyle\int_0^x f_X(x)d_x = F_X(x)$.

$F_X(x)$ Cumulative distribution function at the value of x for the random variable X ($F_X(x) = P(X \le x)$).

$\overline{G}$ Average group size observed during a line-transect survey.

g_i Hunting effort in the ith period.

$g(x)$ Detection function, the probability of detecting an object at distance x from a transect line.

H Total number of license holders.

$\mathbf{H}$ Harvest matrix, comprised of the age-specific probabilities of surviving harvest along the diagonal of the matrix.

$h(x)$ $\dfrac{f_X(x)}{S_X(x)}$ = hazard function, the rate of death in the next instance x, given an animal survived to time x.

$H(t)$ Cumulative hazard function at time t.

I Population index or relative abundance (i.e., $I \propto N$).

$\mathbf{I}$ Identity matrix, with values of one along the diagonal and zeros elsewhere.

K Carrying capacity of a population. The maximum number of animals that can be supported on an area without reduction in habitat quality. (In Chapter 8, refers to size of the sampling frame).

K' Moderator of population growth in the Ricker (1954) and the Beverton and Holt (1957) models.

L The length of a line transect or trap assessment line.

l_t Number of animals alive and at risk of death at time t.

M Sum of the number of animals detected across k replicate line transects.

$\mathbf{M}$ Leslie projection (transition) matrix.

M_H Probability of death from harvest sources of mortality ($M_H = 1 - S_H$).

m_i Number of males in the ith sample.

M_N Probability of death from natural sources of mortality ($M_N = 1 - S_N$).

N Absolute population abundance (without a qualifier, the term usually refers to all animals in a population).

N_A $N_M + N_F$ = total abundance of adults.

N_{AMAX} Animal abundance in the oldest age class (A) of a population.

N_F Abundance of all adult females.

N_{FJ} Abundance of juvenile females.

N_{FP} $N_F + N_{JF}$ = abundance of females, both juvenile and adult females, in the population.

N_J Abundance of juveniles.

N_M Abundance of all adult males.

N_{MJ} Abundance of juvenile males.

N_{MP} $N_M + N_{MJ}$ = abundance of males, both juvenile and adult males, in the population.

P Productivity, ratio of total juveniles (males and females) produced per adult female $\left(P = \dfrac{N_J}{N_F} \right)$.

r Instantaneous rate of population change ($r = \ln \lambda$).

r_{MAX} Maximum instantaneous rate of population change under exponential growth ($r_{MAX} = \ln \lambda_{MAX}$).

r_{REAL} Realized instantaneous rate of population change under prevailing environmental and demographic conditions ($r_{REAL} = \ln \lambda_{REAL}$).

r_{SAD} Instantaneous rate of population change when the population has attained a stable age distribution under prevailing demographic conditions of survival and fecundity ($r_{SAD} = \ln \lambda_{SAD}$).

$R_{F/M}$ Sex ratio of females per male in the population.

R_J Sex ratio of juveniles in terms of females per male $\left(R_J = \dfrac{N_{FJ}}{N_{MJ}} \right)$.

S_{F_i} Total annual survival of females from age i to $i + 1$.

S_{H-F} Annual survival probability for females from harvest sources of mortality ($S_{H-F} = 1 - M_{H-F}$).

S_{H-M} Annual survival probability for males from harvest sources of mortality ($S_{N-M} = 1 - M_{H-M}$).

S_i Survival; probability individuals of age class i survive to age class $i + 1$.

S_{M_i} Total annual survival of males from age i to $i + 1$.

S_N Probability of surviving natural sources of mortality.

S_{N-F} Annual survival probability for females from natural causes ($S_{N-F} = 1 - M_{N-F}$).

S_{N-M} Annual survival probability for males from natural causes ($S_{N-M} = 1 - M_{N-M}$).

S_T Probability of surviving all sources of mortality, both natural and anthropogenic (e.g., harvest).

S_x Probability an animal survives from age x to age $x + 1$ ($S_X = P\,(X \geq x + 1 | x)$).

$S_{x,y}$ Probability of survival from age or time x to age or time y.

$S_X(x)$ Survival function expressing the probability an animal is alive at time or age x ($P\,(X > x) = 1 - F_X(x)$).

t_2 Total number of escapes among distinct animals caught during a constant effort removal study $\left(t_2 = \displaystyle\sum_{i=1}^{k} (i-1)n_i \right)$.

U Ratio of total juveniles (males and females) to total adults (males and females) in the population. Considered by Peterson (1949:70) to be the average annual increment to the adult population $\left(U = \dfrac{N_J}{N_A} \right)$.

y_i Number of juveniles (i.e., young) in the ith sample.

W Width of the area surveyed by a line transect.

$\mathbf{W}$ Weighting matrix.

$\mathbf{X}$ Sensitivity matrix.

X_i Number of animals of the i-type in the population at the time of the first survey for a change-in-ratio study.

z Scaling parameter in the generalized logistic-growth model.

Greek Alphabet

Symbol	Definition
λ	Finite rate of population change ($\lambda = e^r$).
λ_{APP}	Apparent finite rate of population change (Eberhardt et al. 1982), which occurs under prevailing conditions and exploitation.
λ_{MAX}	Maximum finite rate of population change under exponential growth ($\lambda_{MAX} = e^{r_{MAX}}$).
λ_{REAL}	Realized finite rate of population change under prevailing environmental and demographic conditions ($\lambda_{REAL} = e^{r_{REAL}}$).

λ_{SAD} Finite rate of population change when the population has attained a stable age distribution under prevailing conditions ($\lambda_{\text{SAD}} = e^{r_{\text{SAD}}}$).

μ Instantaneous total mortality rate ($\mu = \mu_N + \mu_H$).

$\mu_{H_{\text{AMAX}}}$ Instantaneous harvest mortality rate for the oldest age class (A) in a population.

μ_H Instantaneous harvest/hunting mortality rate.

μ_N Instantaneous natural mortality rate.

ρ Correlation coefficient.

$\mathbf{\Sigma}$ Variance-covariance matrix.

ϕ_x Expected value of a Poisson random variable x ($\phi_x = E(x)$).

Appendix C
Program USER

Program USER 3.0 (<u>U</u>ser <u>S</u>pecified <u>E</u>stimation <u>R</u>outine) provides an interactive computing environment to find maximum likelihood estimates for binomial, multinomial, or product multinomial likelihoods. The model parameters must reside in the cell probabilities and not in the binomial or multinomial coefficients. Many of the statistical models in this book are of that form. This free statistical software can be obtained along with an online user's manual at http://www.cbr.washington.edu/paramEst/USER//. The purpose of this appendix is not to provide a comprehensive manual on instruction but rather to illustrate the capabilities of the software. Interested parties should obtain the complete user's manual for instruction prior to use. In this appendix, only annotated data input and parameter output for selected examples will be illustrated.

Illustration 1: Productivity Estimation, Section 4.3.1, Example 4.4

The Gambel quail (*Callipepla gambelii*) data (Section 4.3.1) consisted of an initial survey of $x_1 = 575$ adults of which $m_1 = 325$ males and $f_1 = 250$ females were observed. The second sample after the breeding season resulted in a sample size of $x_2 = 1045$ birds, of which $a_2 = 210$ adults and $y_2 = 835$ subadults were observed. The model input to Program USER is as follows.

Entering Likelihood 1

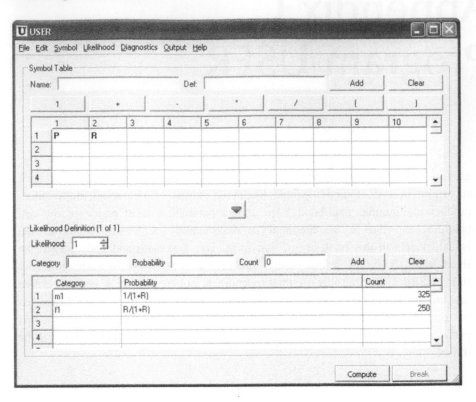

Entering Likelihood 2

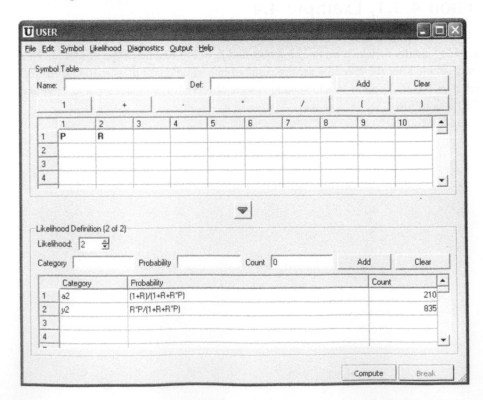

The computer output from Program USER is as follows:

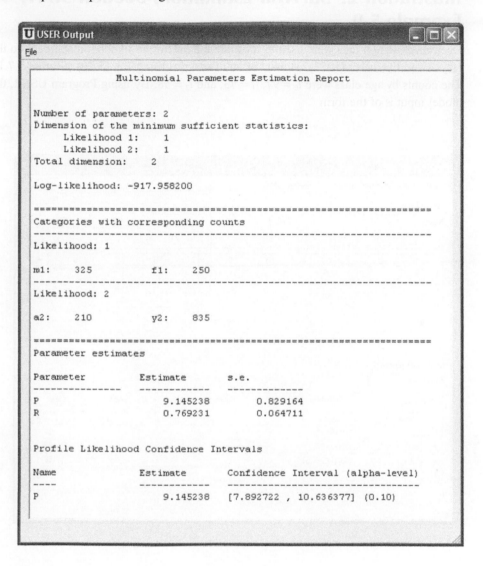

```
┌─────────────────────────────────────────────────────────────────────┐
│ U USER Output                                                _ □ ✕    │
├─────────────────────────────────────────────────────────────────────┤
│ File                                                                  │
│                                                                       │
│               Multinomial Parameters Estimation Report                │
│                                                                       │
│                                                                       │
│  Number of parameters: 2                                              │
│  Dimension of the minimum sufficient statistics:                      │
│       Likelihood 1:    1                                              │
│       Likelihood 2:    1                                              │
│  Total dimension:   2                                                 │
│                                                                       │
│  Log-likelihood: -917.958200                                          │
│                                                                       │
│  =================================================================    │
│  Categories with corresponding counts                                 │
│  -------------------------------------------------------------------  │
│  Likelihood: 1                                                        │
│                                                                       │
│  m1:    325          f1:    250                                       │
│  -------------------------------------------------------------------  │
│  Likelihood: 2                                                        │
│                                                                       │
│  a2:    210          y2:    835                                       │
│                                                                       │
│  =================================================================    │
│  Parameter estimates                                                  │
│                                                                       │
│  Parameter          Estimate        s.e.                              │
│  --------------     ------------    -----------                       │
│  P                      9.145238        0.829164                      │
│  R                      0.769231        0.064711                      │
│                                                                       │
│                                                                       │
│  Profile Likelihood Confidence Intervals                              │
│                                                                       │
│  Name               Estimate        Confidence Interval (alpha-level) │
│  ----               ------------    ---------------------------------- │
│  P                      9.145238   [7.892722 , 10.636377] (0.10)      │
│                                                                       │
└─────────────────────────────────────────────────────────────────────┘
```

The profile likelihood confidence intervals for this example are similar but not identical to the asymptotic confidence intervals of 7.7812 to 10.5092 (Section 4.3.1).

Illustration 2: Survival Estimation, Section 5.7.1, Example 5.9

Deer check station data were used to illustrate the maximum likelihood approach to the Hayne and Eberhardt (1952) estimate of adult survival using Eq. (5.70) (Section 5.7.1). The counts by age class were $l_0 = 97$, $l_1 = 73$, and $l_2 = 58$. By using Program USER, the model input is of the form

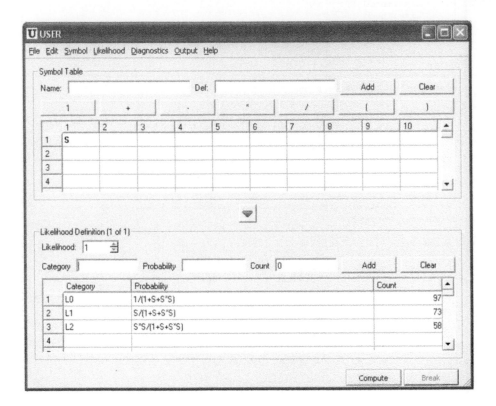

The computer output from Program USER is as follows:

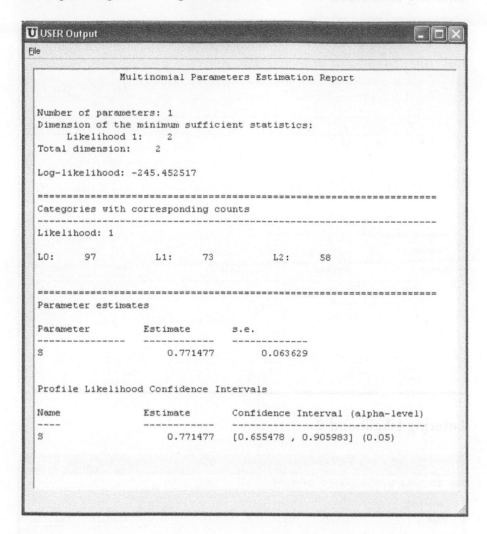

```
U USER Output                                                    _ □ X
File

                  Multinomial Parameters Estimation Report


Number of parameters: 1
Dimension of the minimum sufficient statistics:
        Likelihood 1:    2
Total dimension:    2

Log-likelihood: -245.452517

=================================================================
Categories with corresponding counts
-----------------------------------------------------------------
Likelihood: 1

L0:     97          L1:    73          L2:    58


=================================================================
Parameter estimates

Parameter        Estimate        s.e.
---------------  ------------    ------------
S                    0.771477        0.063629


Profile Likelihood Confidence Intervals

Name             Estimate        Confidence Interval (alpha-level)
----             ------------    ------------------------------------
S                    0.771477    [0.655478 , 0.905983] (0.05)
```

The asymptotic statistical error estimate for this example from the Newton-Raphson method is slightly different from the closed form approximation of 0.0642 (Section 5.7.1).

Illustration 3: Estimating Harvest Survival, Section 6.6.2, Example 6.7

The three-sample method of Selleck and Hart (1957) was illustrated by using hypothetical data from a pheasant survey. The prehunt survey resulted in a sample size of $x_1 = 180$, with $m_1 = 80$ males and $f_1 = 100$ females. The posthunt survey had $x_2 = 100$ observations, with $m_2 = 20$ males and $f_2 = 80$ females. A survey of harvested birds yielded $x_3 = 80$ observations, of which $m_3 = 60$ were males and $f_3 = 20$ were females. Program USER will use likelihood Eq. (6.93) to estimate the male (S_M) and female (S_F) probabilities of surviving the harvest. The model input to Program USER consisted of the following.

Entering Likelihood 1

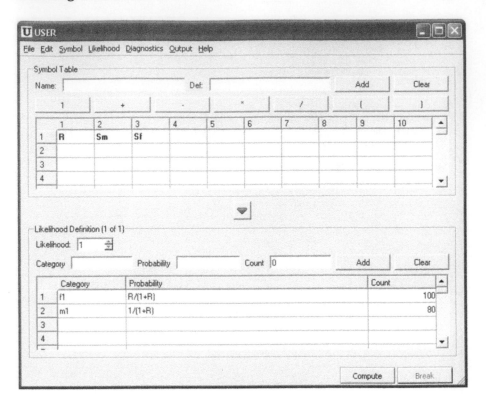

Entering Likelihood 2

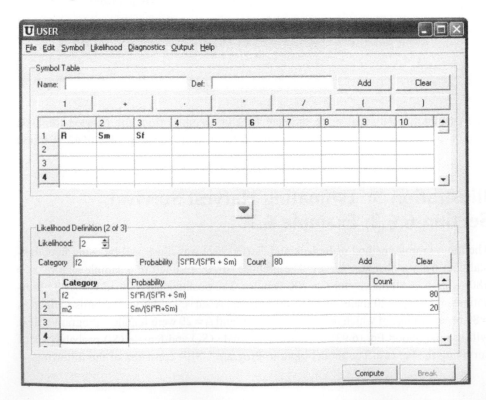

Entering Likelihood 3

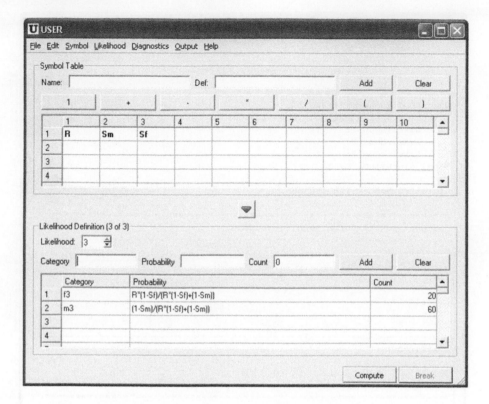

The computer output from Program USER is as follows:

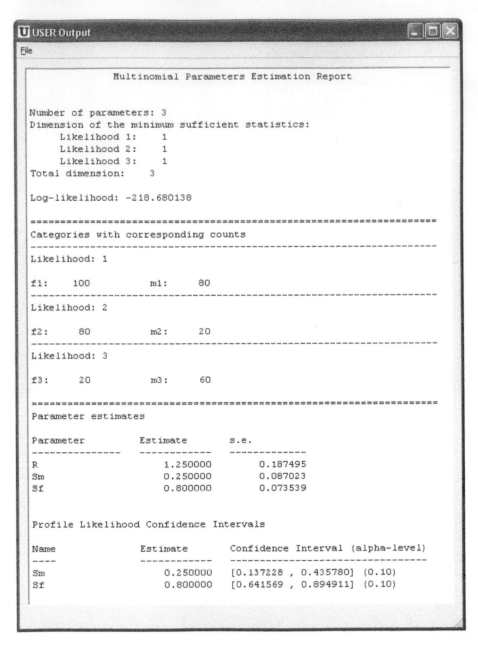

```
U USER Output                                                        _ □ ×
File

                    Multinomial Parameters Estimation Report


Number of parameters: 3
Dimension of the minimum sufficient statistics:
     Likelihood 1:    1
     Likelihood 2:    1
     Likelihood 3:    1
Total dimension:    3

Log-likelihood: -218.680138

==================================================================
Categories with corresponding counts
------------------------------------------------------------------
Likelihood: 1

f1:    100       m1:    80
------------------------------------------------------------------
Likelihood: 2

f2:    80        m2:    20
------------------------------------------------------------------
Likelihood: 3

f3:    20        m3:    60

==================================================================
Parameter estimates

Parameter          Estimate        s.e.
----------------   ------------    ------------
R                  1.250000        0.187495
Sm                 0.250000        0.087023
Sf                 0.800000        0.073539


Profile Likelihood Confidence Intervals

Name               Estimate       Confidence Interval (alpha-level)
----               ------------   ---------------------------------
Sm                 0.250000       [0.137228 , 0.435780]  (0.10)
Sf                 0.800000       [0.641569 , 0.894911]  (0.10)
```

An additional parameter was defined as a construct of the other model parameters in USER, to estimate the overall harvest survival probability (Eq. 6.102), where

$$S_H = \frac{S_M + RS_F}{1+R}$$

with the following commands:

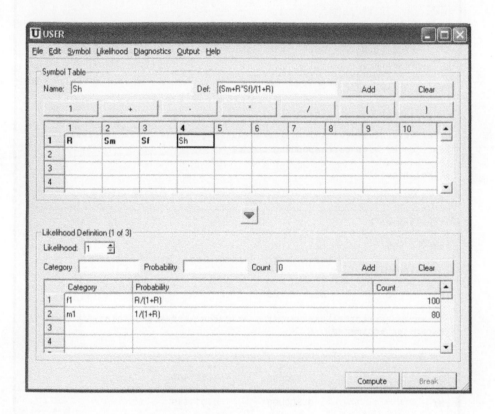

A profile likelihood confidence interval for S_H can then be requested, with the following output:

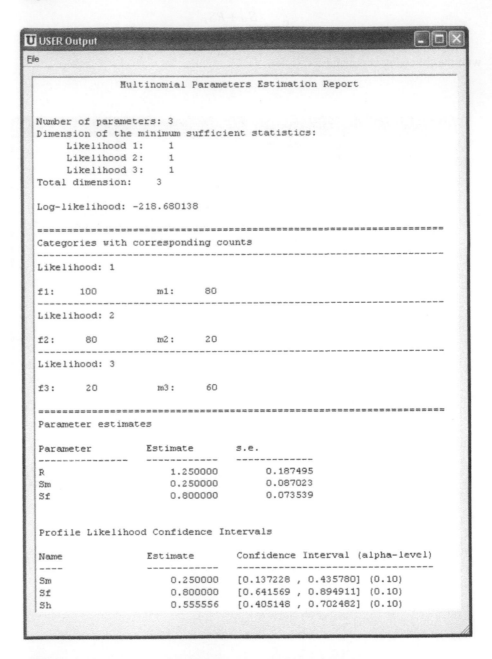

```
Multinomial Parameters Estimation Report


Number of parameters: 3
Dimension of the minimum sufficient statistics:
     Likelihood 1:    1
     Likelihood 2:    1
     Likelihood 3:    1
Total dimension:    3

Log-likelihood: -218.680138

===============================================================
Categories with corresponding counts
---------------------------------------------------------------
Likelihood: 1

f1:    100          m1:     80
---------------------------------------------------------------
Likelihood: 2

f2:     80          m2:     20
---------------------------------------------------------------
Likelihood: 3

f3:     20          m3:     60

===============================================================
Parameter estimates

Parameter          Estimate        s.e.
---------------    ------------    -------------
R                   1.250000        0.187495
Sm                  0.250000        0.087023
Sf                  0.800000        0.073539


Profile Likelihood Confidence Intervals

Name               Estimate        Confidence Interval (alpha-level)
----               ------------    --------------------------------
Sm                  0.250000       [0.137228 , 0.435780] (0.10)
Sf                  0.800000       [0.641569 , 0.894911] (0.10)
Sh                  0.555556       [0.405148 , 0.702482] (0.10)
```

Illustration 4: Abundance Estimation, Section 9.5.2, Example 9.8

The trinomial change-in-ratio method (Section 9.5.2) will be analyzed using Program USER. The joint likelihood for the analysis of the hypothetical black-tailed deer (*Odocoileus hemionus hemionus*) data is entered into the program as follows.

First Likelihood

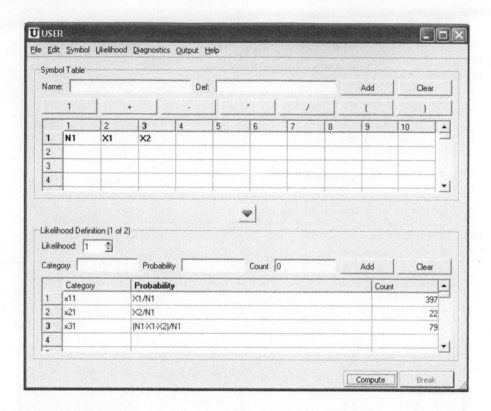

Second Likelihood

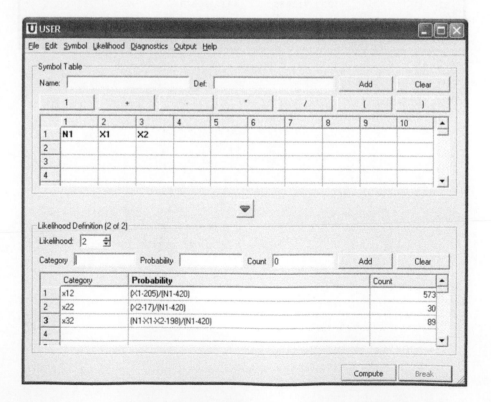

The output from Program USER is as follows:

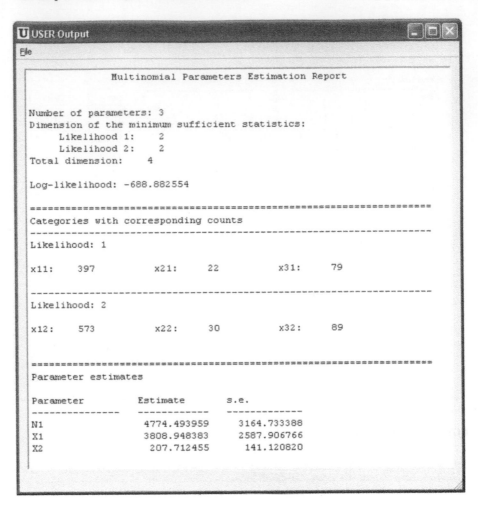

```
Multinomial Parameters Estimation Report


Number of parameters: 3
Dimension of the minimum sufficient statistics:
     Likelihood 1:    2
     Likelihood 2:    2
Total dimension:    4

Log-likelihood: -688.882554

========================================================================
Categories with corresponding counts
------------------------------------------------------------------------
Likelihood: 1

x11:    397         x21:    22         x31:    79

------------------------------------------------------------------------
Likelihood: 2

x12:    573         x22:    30         x32:    89


========================================================================
Parameter estimates

Parameter          Estimate        s.e.
---------------    ------------    ------------
N1                 4774.493959     3164.733388
X1                 3808.948383     2587.906766
X2                  207.712455      141.120820
```

Appendix D
Mathematica Code for Calculating the Variance of the Finite Rate of Population Change, Var($\hat{\lambda}$), from a Matrix Population Model

D

This appendix presents general *Mathematica* code (Wolfram 1999) for calculating the variance of $\hat{\lambda}$ from a $k \times k$ matrix population model using the delta method (Eq. 7.116). We will use a 4×4 matrix from an analysis of Hawaiian hawks (*Buteo solitarius*) as an illustration (see Section 7.9, Example 7.15). Little knowledge of *Mathematica* is assumed, although the reader is encouraged to work through *Mathematica*'s built-in tutorials for beginners.

Mathematica *.nb* files operate by using a method of input and output. After opening *Mathematica*, readers should verify their ability to create output. For example, to calculate $2 + 2 = 4$, type

$$2 + 2$$

and then press the Shift-Enter keys, which instructs *Mathematica* to execute the command. The user will see

$$2 + 2$$
$$4$$

To initialize the matrix library, type (and then press Shift-Enter)

```
<< LinearAlgebra 'MatrixManipulation'
```

Next, enter the size of the Leslie matrix ($k = 4$ for our example), and note the output is "4":

```
k = 4
4
```

To suppress the numeral 4 from being output unnecessarily and to allow multiple programming statements to be entered at once, a semicolon may be used as follows:

```
k = 4;
```

We wish to create the general $k \times k$ Leslie matrix (in this case, $k = 4$):

```
Clear[F];
f = Table[F[i], {i, k}];
Clear[P];
p = Table[P[i], {i, k - 1}] // DiagonalMatrix;
remainders = Table[0, {i, k - 1}];
remainders[[k - 1]] = P[k];
most = AppendColumns[p, {remainders}] // Transpose;
generalLeslie = AppendColumns[{f}, most];
generalLeslie // MatrixForm
```

After pressing Shift-Enter after the last command, the following will be output:

$$\begin{pmatrix} F[1] & F(2) & F(3) & F(4) \\ P[1] & 0 & 0 & 0 \\ 0 & P[2] & 0 & 0 \\ 0 & 0 & P[3] & P[4] \end{pmatrix}$$

In the analysis of Hawaiian hawks, the Leslie matrix was assumed to follow the form

$$\begin{bmatrix} 0 & 0 & 0 & f \\ s_0 & 0 & 0 & 0 \\ 0 & s & 0 & 0 \\ 0 & 0 & s & s \end{bmatrix}, \text{ or, in the notation used above, } \begin{bmatrix} 0 & 0 & 0 & F[4] \\ P[1] & 0 & 0 & 0 \\ 0 & P[2] & 0 & 0 \\ 0 & 0 & P[2] & P[2] \end{bmatrix}$$

We will adjust for the differences between the general Leslie matrix and the reduced form of the example, and enter the point estimates and their associated standard errors. The estimates used in this analysis are $f = 0.23$ (SE = 0.0417), $s_0 = 0.5$ (SE = 0.0981), and $s = 0.94$ (SE = 0.404). These estimates were also assumed to be independent (e.g., Cov $(f, s) = 0$). We first enter the point estimates to provide this information to *Mathematica*. The program also differentiates between a theoretical form of the Leslie matrix (i.e., one that has only variable names) and a form with estimated values entered. Both forms are required for calculating eigenvectors. Thus, this program assigns estimated values for F[i] to the variable fec[i], and assigns estimates of P[i] to surv[i]:

```
fec[4] = 0.23; surv[1] = 0.5; surv[2] = 0.94;
```

Next, enter the theoretically fixed values

$$F[1] = 0; \quad F[2] = 0; \quad F[3] = 0$$

and continue with the parameter relationships

$$P[3] = P[2]; \quad P[4] = P[2];$$

The value we wish to keep in the model (P[2]) is entered second. The equal sign is an assignment operator. Consequently, P[3] = P[2] indicates that P[3] is assigned the value P[2], so the order of entry effects the variable names included in the reduced model. Finally, enter the covariance matrix (the diagonal of which is the square of the standard errors in this example), first entering variance estimates for fecundity parameters and then for survival parameters.

```
myCov = DiagonalMatrix[{.0417^2, .0981^2, 0.0404^2}];
```

To move from the general Leslie matrix to the reduced form that incorporates parameter relationships and fixed (not estimated) values, type

```
Leslie = generalLeslie;
Leslie // MatrixForm // Print;
```

This will produce the correct form of the Leslie matrix for the Hawaiian hawk example:

$$\begin{pmatrix} 0 & 0 & 0 & F[4] \\ P[1] & 0 & 0 & 0 \\ 0 & P[2] & 0 & 0 \\ 0 & 0 & P[2] & P[2] \end{pmatrix}$$

The program calculates the number of fecundity and survival parameters (*numfec* and *numsurv*, respectively) used in the reduced model, and vectors indicating whether each expected parameter from the general Leslie matrix is included in the reduced matrix. For example, the general Leslie matrix expects four survival parameters (P[1] to P[4]), but only two (P[1] and P[2]) are in the reduced matrix.

```
Clear[fecparam];
Clear[survparam];
fecparam = Table[0, {i, k}];
survparam = Table[0, {i, k}];
numfec = 0;
numsurv = 0;
Print["i Expected Actual"];
For[i = 1, i ≤ k, i++,
  expectedValue = "F[" <> ToString[i] <> "]";
  Print[i, " ", expectedValue, "      ", Leslie[[1, i]]];
  If[ToString[Leslie[[1, i]]] != expectedValue,
   Print["Not a parameter!"],
   (* else *)
   Print["A parameter!"];
   numfec++;
   fecparam[[i]] = 1]];
```

```
Print["Number of distinct fecundity parameters: ", numfec];
Print["Vector indicating if F[i] is a distinct parameter: ",
  fecparam];
For[i = 1, i ≤ k - 1, i++,
  expectedValue = "P[" <> ToString[i] <> "]";
  If[ToString[Leslie[[i + 1, i]] ] == expectedValue,
    numsurv++;
    survparam[[i]] = 1]];
(* P[k], on the diagonal, is a special case *)
expectedValue = "P[" <> ToString[k] <> "]";
If[ToString[Leslie[[k, k]] ] == expectedValue,
 numsurv++;
 survparam[[i]] = 1];
Print["Number of distinct survival parameters: ", numsurv];
Print["Vector indicating if P[i] is a distinct parameter: ",
  survparam];
```

which has the output

```
i Expected Actual
1 F[1]      0
Not a parameter!
2 F[2]      0
Not a parameter!
3 F[3]      0
Not a parameter!
4 F[4]      F[4]
A parameter!
Number of distinct fecundity parameters: 1
Vector indicating if F[i] is a distinct parameter:
  {0, 0, 0, 1}
Number of distinct survival parameters: 2
Vector indicating if P[i] is a distinct parameter:
  {1, 1, 0, 0}
```

Next, we assign estimated values to the Leslie matrix and calculate its eigenvalues:

```
For[i = 1, i ≤ k, i++, F[i] = fec[i]] ;
For[i = 1, i ≤ k, i++, P[i] = surv[i]];
"The Leslie matrix: " // Print;
Leslie // MatrixForm // Print;
(* Print the eigenvalues *)
"...and its corresponding eigenvalues." // Print;
Leslie // Eigenvalues // Print;
```

The accuracy of the produced Leslie matrix should be verified:

The Leslie matrix:

$$\begin{pmatrix} 0 & 0 & 0 & 0.23 \\ 0.5 & 0 & 0 & 0 \\ 0 & 0.94 & 0 & 0 \\ 0 & 0 & 0.94 & 0.94 \end{pmatrix}$$

...and its corresponding eigenvalues.
{1.03236, 0.164358 + 0.454692 i, 0.164358 − 0.454692 i, −0.421073}

The program looks for nonpositive or complex eigenvalues equal in magnitude to the dominant real eigenvalue. If they exist, it signals an oscillating population structure; however, this is not particularly common.

```
(* Calculate the eigenvalues of the Leslie matrix *)
actualEigenvalues = Eigenvalues[Leslie];
(* Calculate the dominant of the real part of the eigenvalue. *)
maxmyEigen = actualEigenvalues // Re // Max;
magnitudeLeslieEigenvalues = Eigenvalues[Leslie] // Abs;
magnitudeDominantEigenvalue = Max[magnitudeLeslieEigenvalues];
(* Initialize eigenstatus variable, then update *)
eigenStatus = "No positive, real dominant eigenvalue";
If[maxmyEigen == magnitudeDominantEigenvalue,
   eigenStatus =
   "There exists a positive, real dominant eigenvalue."];
For[i = 1, i ≤ k, i++,
   If[magnitudeLeslieEigenvalues[[i]] ≥ maxmyEigen
         && actualEigenvalues[[i]] ≠ maxmyEigen && maxmyEigen > 0,
      eigenStatus =
      eigenStatus <>
       " There also exists a complex or negative eigenvalue
         of equal magnitude."; i = k] ];
If[
 eigenStatus == "There exists a positive, real dominant eigenvalue.",
 eigenStatus =
 "There exists a unique dominant eigenvalue, and it is
    positive and real."];
Print[ eigenStatus];
```

In this example, as in most cases, the dominant eigenvalue is positive and real.

```
There exists a unique dominant eigenvalue, and it is
    positive and real.
```

Numerical derivatives are calculated by using the approximation

$$f'(x) \approx \frac{f(x + \Delta x) - f(x - \Delta x)}{2\Delta x}.$$

Smaller values of Δx provided more accurate approximations and can be changed by altering

```
delta = 0.0001;
```

Continuing,

```
(* Initialize vector of derivatives *)
Clear[dF];
df = Table[dF[i], {i, numfec}];
Clear[dP];
dp = Table[dP[i], {i, numsurv}];
```

Press Shift-Enter to initialize values, and then type

```
For[i = 1; parnum = 0, i ≤ k, i++,
    If[fecparam[[i]] == 1,
    parnum++;
    F[i] = fec[i] + delta;
    eigenplus = Leslie // Eigenvalues // Re // Max;
    F[i] = fec[i] - delta;
    eigenminus = Leslie // Eigenvalues // Re // Max;
    F[i] = fec[i];
    dF[parnum] = (eigenplus - eigenminus) / (2 * delta) ] ];
```

and

```
For[i = 1; parnum = 0, i ≤ k, i++,
    If[survparam[[i]] == 1,
    parnum++;
    P[i] = surv[i] + delta;
    eigenplus = Leslie // Eigenvalues // Re // Max;
    P[i] = surv[i] - delta;
    eigenminus = Leslie // Eigenvalues // Re // Max;
    P[i] = surv[i];
    dP[parnum] = (eigenplus - eigenminus) / (2 * delta) ] ];
```

to calculate fecundity and survival partial derivatives. These are combined by

```
partials = Flatten[Append[df, dp]];
"The partial derivatives are:" // Print;
partials // Print;
```

with output

```
The partial derivatives are:
{0.316582, 0.145628, 0.943328}
```

Quickly check that the covariance matrix is correct:

```
"The covariance matrix for this Leslie matrix is:" // Print;
myCov // MatrixForm // Print;
```

The covariance matrix for this Lesile matrix is:

$$\begin{pmatrix} 0.00173889 & 0 & 0 \\ 0 & 0.00962361 & 0 \\ 0 & 0 & 0.00163216 \end{pmatrix}$$

Because the covariance matrix is correct, we use the delta method to calculate and print the approximate variance and standard error of $\hat{\lambda}$:

```
lambdavar = partials.myCov.Transpose[{partials}];
Print[eigenStatus]; Print["Lambda = ", maxmyEigen];
Print["The (approx) variance of lambda = ", lambdavar];
Print["The (approx) standard error of lambda is ",
 lambdavar // Sqrt];
Print["The eigenvalues for this data are: ", actualEigenvalues];
There exists a unique dominant eigenvalue, and it is positive and real.
Lambda = 1.03236
The (approx) variance of lambda = {0.00183078}
The (approx) standard error of lambda is {0.0427876}
The eigenvalues for this data are: {1.03236, 0.164358 + 0.454692 i, 0.164358 - 0.454692 i, -0.421073}
```

Thus, $\hat{\lambda} = 1.032$ and $\widehat{\text{Var}}(\hat{\lambda}) = 0.00183$ or $\widehat{\text{SE}} = 0.0428$.

References

Aalen, O. O. 1978. Nonparametric inference for a family of counting processes. Annals of Statistics 6:701–726.

Afton, A. D., and M. G. Anderson. 2001. Declining scaup populations: a retrospective analysis of long-term population and harvest survey data. Journal of Wildlife Management 65:781–796.

Agger, P., I. Boetius, and H. Lassen. 1973. Error in the virtual population: the effect of uncertainties in the natural mortality coefficient. Journal du Conseil, Conseil International pour l'Exploration de la Mer 35:93.

Aitkin, M., D. Anderson, B. Francis, and J. Hinde. 1989. Statistical modelling in GLIM. Oxford University Press, Oxford, United Kingdom.

Akaike, H. 1973. Information theory and an extension of the maximum likelihood principle. Proceedings of Second International Symposium on Information Theory. Akademia Kaido, Budapest, Hungary 267–281.

Allee, W. C., O. Park, A. E. Emerson, T. Park, and K. P. Schmidt. 1949. Principles of animal ecology. W. B. Saunders Co., Philadelphia, Pennsylvania, USA.

Allen, D. L. 1942. A pheasant inventory method based upon kill records and sex ratios. Transactions of the North American Wildlife Conference 7:329–332.

Allen, D. L. 1954. Our wildlife legacy. Funk & Wagnalls, New York, New York, USA.

Alpizar-Jara, R., and K. H. Pollock. 1999. Combining line transect and capture-recapture for mark-resighting studies. In Marine mammal survey and assessment methods: proceedings of the symposium on marine mammal survey and assessment methods, Seattle, WA, February 21–25, 1998 (L. L. McDonald, J. L. Laake, D. G. Robertson, S. C. Amstup, and G. W. Garner, eds.), pages 99–114. Aa Balkema, Rotterdam, The Netherlands.

Alvarez-Buylla, E. R., and M. Slatkin. 1991. Review: finding confidence limits on population growth rates. Trends in Ecology and Evolution 6:221–224.

Alvarez-Buylla, E. R., and M. Slatkin. 1994. Finding confidence limits on population growth rates: three real examples revised. Ecology 75:255–260.

Amarasekare, P. 1998. Allee effects in metapopulation dynamics. American Naturalist 152: 298–302.

Amrhein, V., H. P. Kunc, and M. Naguib. 2004. Seasonal patterns of singing activity vary with time of day in the nightingale (Luscinia megarhynchos). Auk 121:110–117.

Andersen, D. E., O. J. Rongstad, and W. R. Mytton. 1985. Line transect analysis of raptor abundance along roads. Wildlife Society Bulletin 13:533–539.

Anderson Jr., C. R., D. S. Moody, B. L. Smith, F. G. Lindzey, and R. P. Lanka. 1998. Development and evaluation of sightability models for summer elk surveys. Journal of Wildlife Management 62:1055–1066.

Anderson, D. R. 2001. The need to get the basics right in wildlife field studies. Wildlife Society Bulletin 29:1294–1297.

Anderson, D. R. 2003. Response to Engeman: index values rarely constitute reliable information. Wildlife Society Bulletin 31:288–291.

Anderson, D. R., and K. P. Burnham. 1976. Population ecology of the mallard, volume 1: the effect of exploitation on survival. U.S. Fish and Wildlife Service, Washington, D.C., USA. Report 125.

Anderson, D. R., K. P. Burnham, G. C. White, and D. L. Otis. 1983. Density estimation of small mammal populations using a trapping web and distance sampling methods. Ecology 64:674–680.

Anderson, D. R., K. P. Burnham, B. C. Lubow, L. Thomas, P. S. Corn, P. A. Medica, and R. W. Marlow. 2001. Field trials of line transect methods applied to estimation of desert tortoise abundance. Journal of Wildlife Management 65:583–597.

Anderson, S. H. 1985. Managing our wildlife resources. C. E. Merrill Publishing Co., Columbus, Ohio, USA.

Andrewartha, H. G., and L. C. Birch. 1954. The distribution and abundance of animals. University of Chicago Press, Chicago, Illinois, USA.

Andrzejewski, R., and W. Jezierski. 1966. Studies on the European hare, XI: estimation of population density and attempt to plan the yearly take of harvest. Acta Theriologica 11:433–448.

Anthony, R. G., M. G. Garrett, and F. B. Isaacs. 1999. Double-survey estimates of bald eagle populations in Oregon. Journal of Wildlife Management 63:794–802.

Arcese, P., and J. N. M. Smith. 1988. Effects of population density and supplemental food on reproduction in song sparrows. Journal of Animal Ecology 57:119–136.

Archibald, H. L. 1976. Spring drumming patterns of ruffed grouse. Auk 93:808–829.

Arnold, S. F. 1990. Mathematical statistics. Prentice-Hall, Englewood Cliffs, New Jersey, USA.

Bailey, J. A. 1984. Principles of wildlife management. John Wiley & Sons, New York, New York, USA.

Bailey, R. S. 1967. An index of bird population changes on farmland. Bird Study 14:195–209.

Barker, R. J. 1991. Nonresponse bias in New Zealand waterfowl harvest surveys. Journal of Wildlife Management 55:126–131.

Barlow, J. 1991. The utility of demographic models in marine mammal management. Report of the International Whaling Commission 41:573–577.

Bart, J. 1977. Impact of human visitations on avian nesting success. Living Bird 16:187–192.

Bart, J., and D. S. Robson. 1982. Estimating survivorship when the subjects are visited periodically. Ecology 63:1078–1090.

Bartholow, J. M. 1977. Fort Niobrara Refuge: big game management modeling [thesis]. Colorado State University, Fort Collins, Colorado, USA.

Baskett, T. S., M. J. Armbruster, and M. W. Sayre. 1978. Biological perspectives for the mourning dove call-count survey. Transactions of the North American Wildlife and Natural Resources Conference 43:163–180.

Beck, J. V., and K. J. Arnold. 1977. Parameter estimation in engineering and science. John Wiley & Sons, New York, New York, USA.

Beddington, J. R. 1974. Age structure, sex ratio and population density in the harvesting of natural animal populations. Journal of Applied Statistics 11:915–924.

Begon, M. 1979. Investigating animal abundance: capture-recapture for biologists. University Park Press, Baltimore, Maryland, USA.

Begon, M., J. L. Harper, and C. R. Townsend. 1990. Ecology: individuals, populations and communities. Blackwell Scientific Publications, Brookline Village, Massachusetts, USA.

Bellrose Jr., F. C. 1947. Analysis of methods used in determining game kill. Journal of Wildlife Management 11:105–119.

Belsley, D. A., E. Kuch, R. John, and E. Welch. 1980. Regression diagnostics: identifying influential data and sources of collinearity. John Wiley & Sons, New York, New York, USA.

Bender, L. C. 1997. Estimating black bear population size and trend in Washington State. Proceedings of Sixth Western Workshop on Black Bear Management and Research 6:63–68.

Bender, L. C. 1998. Estimating black bear population size and trends in Washington state. Washington Department of Fish and Game, Vancouver, Washington, USA.

Bender, L. C., and P. J. Miller. 1999. Effects of elk harvest strategy on bull demographics and herd composition. Wildlife Society Bulletin 27:1032–1037.

Bender, L. C., and R. D. Spencer. 1999. Estimating elk population size by reconstruction from harvest data and herd ratios. Wildlife Society Bulletin 27:636–645.

Bennett, L. J., P. F. English, and R. McCain. 1940. A study of deer populations by use of pellet group counts. Journal of Wildlife Management 4:398–403.

Bergerud, A. T. 1985. The additive effect of hunting mortality on the natural mortality ranges of grouse. *In* Game harvest management (S. L. Beasom and S. F. Roberson, eds.), pages 345–366. Caesar Kleberg Wildlife Research Institute, Kingsville, Texas, USA.

Bergstedt, R. A., and D. R. Anderson. 1990. Evaluation of the line transect sampling based on remotely sensed data from underwater video. Transactions of the American Fisheries Society 119:86–91.

Bernadelli, H. 1941. Population waves. Journal of the Burma Research Society 31:1–18.

Berryman, A. A. 1999. Principles of population dynamics and their application. Stanley Thornes, Inc., Cheltenham, United Kingdom.

Beverton, R. J. H., and S. J. Holt. 1957. On the dynamics of exploited fish populations. Chapman and Hall, London, United Kingdom.

Bildstein, K. L., and T. C. Grubb Jr. 1980. Roadside raptor count in eastern Texas. Bulletin of the Texas Ornithological Society 13:20–22.

Bleich, V. C., C. S. Y. Chun, R. W. Anthes, T. E. Evans, and J. K. Fischer. 2001. Visibility bias and development of sightability model for Tule elk. Alces 37:315–327.

Blower, J. G., L. M. Cook, and J. A. Bishop. 1981. Estimating the size of animal populations. George Allen and Unwin Ltd., London, United Kingdom.

Bodie, W. L., E. O. Garton, E. R. Taylor, and M. McCoy. 1995. A sightability model for bighorn sheep in canyon habitats. Journal of Wildlife Management 59:832–840.

Bodkin, J. L., D. Mulcahy, and C. J. Lensink. 1993. Age-specific reproductive in female sea otters (*Enhydra lutris*) in south-central Alaska: analysis of reproductive tracts. Canadian Journal of Zoology 71:1811–1815.

Boer, A. H. 1988. Mortality rates of moose in New Brunswick: a life table analysis. Journal of Wildlife Management 52:21–25.

Bolen, W. L., and E. G. Robinson. 1995. Wildlife ecology and management. Prentice Hall, Englewood Cliffs, New Jersey, USA.

Borchers, D. L., S. T. Buckland, and W. Zucchini. 2002. Estimating animal abundance: closed populations. Springer-Verlag, Inc., London, United Kingdom.

Bordage, D., N. Plante, A. Bourget, and S. Paradis. 1998. Use of ratio estimators to estimate the size of common eider populations in winter. Journal of Wildlife Management 62:185–192.

Bowden, D. C., A. E. Anderson, and D. E. Medin. 1984. Sampling plans for mule deer sex and age ratios. Journal of Wildlife Management 48:500–509.

Box, G. E. P., and D. R. Cox. 1964. An analysis of transformations. Journal of the Royal Statistical Society, Series B 26:211–252.

Boyce, M. S., A. R. E. Sinclair, and G. C. White. 1999. Seasonal compensation of predation and harvesting. Oikos 87:419–426.

Brockelman, W. Y. 1980. The use of the line transect sampling method for forest primates. *In* Tropical ecology and development (J. I. Furtado, ed.), pages 367–371. The International Society of Tropical Ecology, Kuala Lumpur, Malaysia.

Brockwell, P. J., and R. A. Davis. 1996. Introduction to time series and forecasting. Springer-Verlag, Inc., New York, New York, USA.

Buckland, S. T., and J. M. Breiwick. 2002. Estimating trends in abundance of eastern Pacific gray whales from shore counts (1967/68 to 1995/96). Journal of Cetacean Research and Management 4:41–48.

Buckland, S. T., D. R. Anderson, K. P. Burnham, and J. L. Laake. 1993. Distance sampling: estimating abundance of biological populations. Chapman & Hall, London, United Kingdom.

Buckland, S. T., D. R. Anderson, K. P. Burnham, J. L. Laake, D. L. Borchers, and L. Thomas. 2001. Introduction to distance sampling: estimating abundance of biological populations. Oxford University Press, Oxford, United Kingdom.

Buckland, S. T., D. R. Anderson, K. P. Burnham, J. L. Laake, D. L. Borchers, and L. Thomas. 2004. Advanced distance sampling. Oxford University Press, Oxford, United Kingdom.

Burgoyne Jr., G. E. 1981. Observations of a heavily exploited deer population. *In* Dynamics of large mammal populations (C. W. Fowler and T. D. Smith, eds.), pages 403–413. John Wiley & Sons, New York, New York, USA.

Burnham, K. P., and D. R. Anderson. 1984*a*. The need for distance data in transect counts. Journal of Wildlife Management 48:1248–1254.

Burnham, K. P., and D. R. Anderson. 1984*b*. Tests of compensatory vs. additive hypotheses of mortality in mallards. Ecology 65:105–112.

Burnham, K. P., and D. R. Anderson. 1998. Model selection and inference: a practical information—theoretic approach. Springer-Verlag, Inc., New York, New York, USA.

Burnham, K. P., D. R. Anderson, and J. L. Laake. 1980. Estimation of density from line transect sampling of biological populations. Wildlife Monographs 72.

Burnham, K. P., D. R. Anderson, G. C. White, C. Brownie, and K. H. Pollock. 1987. Design and analysis methods for fish survival experiments based on release-recapture. American Fisheries Society Monographs 5:1–437.

Byers, T., and D. L. Dickson. 2001. Spring migration and subsistence hunting of king and common eiders at Holman, Northwest Territories, 1996–1998. Arctic 54:122–134.

Bystrak, D. 1981. The North American breeding bird survey. Studies in Avian Biology 6:34–41.

Cada, J. D. 1985. Evaluations of the telephone and mail survey methods of obtaining harvest data from licensed sportsmen in Montana. *In* Game harvest management (S. L. Beasom and S. F. Roberson, eds.), pages 117–128. Caesar Kleberg Research Institute, Kingsville, Texas, USA.

Carey, A. B., and J. W. Witt. 1991. Track counts as indices to abundances of arboreal rodents. Journal of Mammology 72:192–194.

Carpenter, L. H. 1991. Elk hunting regulations, the Colorado experience. *In* Elk vulnerability symposium (A. G. Christensen, L. J. Lyon, and T. N. Lonner, eds.), pages 16–22. Montana State University, Bozeman, Montana, USA.

Casella, G., and R. L. Berger. 1990. Statistical inference. Duxbury Press, Belmont, California, USA.

Caswell, H. 1989. Matrix population models. Sinauer Associates, Sunderland, Massachusetts, USA.

Caswell, H. 2000. Prospective and retrospective perturbation analyses: their roles in conservation biology. Ecology 81:619–627.

Caswell, H. 2001. Matrix population models. Construction, analysis, and interpretation. Sinauer Associates, Sunderland, Massachusetts, USA.

Caswell, H., S. Brault, A. J. Read, and T. D. Smith. 1998. Harbor porpoise and fisheries: an uncertainty analysis of incidental mortality. Ecological Applications 8:1226–1238.

Caughley, G. 1966. Mortality patterns in mammals. Ecology 47:906–918.

Caughley, G. 1974. Bias in aerial surveys. Journal of Wildlife Management 38:921–933.

Caughley, G. 1977. Analysis of vertebrate populations. John Wiley & Sons, London, United Kingdom.

Caughley, G., and L. C. Birch. 1971. Rate of increase. Journal of Wildlife Management 35:658–663.

Caughley, G., and A. R. E. Sinclair. 1994. Wildlife ecology and management. Blackwell Science Publications, Inc., Cambridge, Massachusetts, USA.

Cavallini, P. 1994. Faeces count as an index of fox abundance. Acta Theriologica 39:417–424.

Chapman, D. G. 1951. Some properties of the hypergeometric distribution with applications to zoological censuses. University of California Publications in Statistics 1:131–160.

Chapman, D. G. 1954. The estimation of biological populations. Annals of Mathematical Statistics 25:1–15.

Chapman, D. G. 1955. Population estimation based on change of composition caused by a selective removal. Journal of Wildlife Management 42:279–290.

Chapman, D. G. 1961. Statistical problems in the dynamics of exploited fish populations. Proceedings of the Fourth Berkeley Symposium 4:153–168.

Chapman, D. G., and G. I. Murphy. 1965. Estimates of mortality and population from survey-removal records. Biometrics 21:921–935.

Chapman, D. G., and D. S. Robson. 1960. The analysis of a catch-curve. Biometrics 16:354–368.

Chapman, D. G., W. S. Overton, and A. L. Finkner. 1959. Methods of estimating dove kill. Institute of Statistics, North Carolina State College, Raleigh, North Carolina, USA. Mimeo Series, No. 264.

Chapman, R. N. 1928. The qualitative analysis of environmental factors. Ecology 9:111–122.

Chatfield, C. 1989. The analysis of time series: an introduction. John Wiley & Sons, New York, New York, USA.

Chen, C., K. H. Pollock, and J. M. Hoenig. 1998. Combining change-in-ratio, index-removal, and removal models for estimating population size. Biometrics 54:815–827.

Cheng, C. L., and S. W. Van Ness. 1999. Statistical regression with measurement error. Arnold, London, United Kingdom.

Chiang, C. L. 1960a. A stochastic study of the life table and its applications, I. Probability distributions of the biometric functions. Biometrics 16:618–635.

Chiang, C. L. 1960b. A stochastic study of the life table and its applications, II. Sample variance of the observed expectation of life and other biometric functions. Human Biology 32:221–238.

Chitty, D. 1960. Population processes in the vole and their relevance to general theory. Canadian Journal of Zoology 38:99–113.

Clark, W. R. 1987. Effect of harvest on muskrats. Journal of Wildlife Management 51:265–272.

Cochran, W. G. 1977. Sampling techniques. John Wiley & Sons, New York, New York, USA.

Cochran, W. G., and G. M. Cox. 1957. Experimental designs. John Wiley & Sons, New York, New York, USA.

Coe, R. J., R. L. Downing, and B. S. McGinnes. 1980. Sex and age bias in hunter-killed white-tailed deer. Journal of Wildlife Management 44:245–249.

Cogan, R. D., and D. R. Diefenbach. 1998. Effect of undercounting and model selection on a sightability-adjustment estimator for elk. Journal of Wildlife Management 62:269–279.

Cole, L. C. 1954. The population consequences of life history phenomena. Quarterly Review of Biology 29:103–137.

Conner, M. C., R. A. Lancia, and K. H. Pollock. 1986. Precision of the change-in-ratio technique for deer population management. Journal of Wildlife Management 50:125–129.

Conover, W. J. 1980. Practical nonparametric statistics. John Wiley & Sons, New York, New York, USA.

Cooper, A. B. 2000. The development and application of simulation models to aid in wildlife management decision-making [dissertation]. University of Washington, Seattle, Washington, USA.

Cooper, A. B., R. Hilborn, and J. W. Unsworth. 2003. An approach for population assessment in the absence of abundance indices. Ecological Applications 13:814–828.

Cormack, R. M. 1964. Estimates of survival from the sightings of marked animals. Biometrika 51:429–438.

Cormack, R. M. 1978. Models for capture-recapture. *In* Sampling biological populations, 5 (R. M. Cormack, G. P. Patil, and D. S. Robson, eds.), pages 217–255. International Co-operative Publishing House, Fairland, Maryland, USA.

Cormack, R. M. 1980. Examples of the use of GLIM to analyse capture-recapture studies. *In* Statistics in ornithology (D. Brillinger, S. Fienberg, J. Gani, J. Hartigan, and K. Krickeberg, eds.), pages 243–273. Springer-Verlag, Inc., Berlin, Germany.

Cormack, R. M. 1993. The flexibility of GLIM analyses of multiple recapture or resighting data. *In* Marked individuals in the study of bird populations (J. D. Lebreton and P. M. North, eds.), pages 39–49. Birkhauser-Verlag, Inc., Boston, Massachusetts, USA.

Coughenour, M. B., and F. J. Singer. 1996. Elk population processes in Yellowstone National Park under the policy of natural regulation. Ecological Applications 6:573–593.

Courchamp, F., T. Clutton-Brock, and B. Grenfell. 1999. Inverse density dependence and the Allee effect. Trends in Ecology and Evolution 14:405–410.

Courchamp, F., T. Clutton-Brock, and B. Grenfell. 2000. Multipack dynamics and the Allee effect in the African wild dog, *Lycaon pictus*. Animal Conservation 3:277–285.

Cowardin, L. M., and D. H. Johnson. 1979. Mathematics and mallard management. Journal of Wildlife Management 43:18–35.

Cox, H. 1923. The problem of population. G. P. Putnam's Sons, New York, New York, USA.

Craighead, J. J., J. R. Varney, and F. C. Craighead Jr. 1974. A population analysis of Yellowstone grizzly bears. Montana Forest and Conservation Experiment Station, University of Montana, Missoula, Montana, USA. Bulletin No. 40.

Cramér, H. 1946. Mathematical methods of statistics. Princeton University Press, Princeton, New Jersey, USA.

Crawford, J. A., and E. G. Bolen. 1975. Spring lek activity of the lesser prairie chicken in west Texas. Auk 92:808–810.

Crawley, M. J. 1993. GLIM for ecologists. Blackwell Scientific Publications, Oxford, United Kingdom.

Creed, W. A., F. Haberland, B. E. Kohn, and K. R. McCaffery. 1984. Harvest management: the Wisconsin experience. *In* White-tailed deer: ecology and management (L. K. Halls, ed.), pages 243–260. Stackpole Books, Harrisburg, Pennsylvania, USA.

Crouse, D. T., L. B. Crowder, and H. Caswell. 1987. A stage-based population model for loggerhead sea turtles and implications for conservation. Ecology 68:1412–1423.

Crowe, D. M. 1975. A model for exploited bobcat populations of Wyoming. Journal of Wildlife Management 39:408–415.

Dale, F. H. 1952. Sex ratios in pheasant research and management. Journal of Wildlife Management 16:156–163.

Dasmann, R. 1976. Wildlife ecology and management. John Wiley & Sons, New York, New York, USA.

Dasmann, R. F. 1952. Methods for estimating deer populations from kill data. California Fish and Game 38:225–233.

Davis, D. E. 1957. Observations on the abundance of Korean mice. Journal of Mammology 38:374–377.

Davis, D. E. 1960. A chart for estimation of life expectancy. Journal of Wildlife Management 24:344–348.

Davis, D. E. 1963. Estimating the numbers of game populations. *In* Wildlife investigational techniques, 2nd edition (H. S. Mosby, ed.), pages 89–118. The Wildlife Society, Washington, D.C., USA.

Davis, D. E. 1982. CRC Handbook of census methods for terrestrial vertebrates. CRC Press, Inc., Boca Raton, Florida, USA.

de Kroon, H., J. van Groenendael, and J. Ehrlen. 2000. Elasticities: a review of methods and model limitations. Ecology 81:607–618.

de la Mare, W. K. 1985. On the estimation of mortality rates from whole age data, with particular reference to Minke whales (*Baleanoptera acutorostrata*) in the Southern hemisphere. Report of the International Whaling Commission 35:239–250.

Deevey Jr., E. S. 1947. Life tables for natural populations of animals. Quarterly Review of Biology 22:283–314.

DelGiudice, G., M. R. Riggs, P. Joly, and W. Pan. 2002. Winter severity, survival, and cause-specific mortality of female white-tailed deer in north-central Minnesota. Journal of Wildlife Management 66:698–717.

DeLury, D. B. 1947. On the estimation of biological populations. Biometrics 3:145–167.

DeLury, D. B. 1951. On the planning of experiments for the estimation of fish populations. Journal of the Fisheries Research Board of Canada 8:281–307.

DeMaster, D. P. 1981. Incorporation of density dependence and harvest into a general population model for seals. *In* Dynamics of large mammal populations (C. W. Fowler and T. D. Smith, eds.), pages 389–401. John Wiley & Sons, New York, New York, USA.

Dennis, B., P. L. Munholland, and J. M. Scott. 1991. Estimation of growth and extinction parameters for endangered species. Ecological Monographs 61:115–143.

Dennis, B., and M. L. Taper. 1994. Density dependence in time series observations of natural populations: estimation and testing. Ecological Monographs 64:205–224.

Deriso, R. B., T. J. Quinn II, and P. R. Neal. 1985. Catch-age analysis with auxiliary information. Canadian Journal of Fisheries and Aquatic Sciences 42:815–824.

Derzhavin, A. N. 1922. The stellate sturgeon (*Acipenser stallatus pallus*), a biological sketch. Byulletin Bakinskoi Ikhtiologicheskoi Stantsii 1:1–393.

Dice, L. R. 1941. Methods for estimating populations of mammals. Journal of Wildlife Management 5:398–407.

Dice, L. R. 1952. Natural communities. University of Michigan Press, Ann Arbor, Michigan, USA.

Donovan, T. M., and C. W. Welden. 2001. Spreadsheet exercises in conservation biology and landscape ecology. Sinauer Associates, Sunderland, Massachusetts, USA.

Dorney, R. S., D. R. Thompson, J. B. Hale, and R. F. Wendt. 1958. An evaluation of ruffed grouse drumming counts. Journal of Wildlife Management 22:35–40.

Downing, R. L. 1980. Vital statistics of animal populations. *In* Wildlife management techniques manual, 4th edition (S. D. Schemnitz, ed.), pages 247–267. The Wildlife Society, Washington, D.C., USA.

Draper, N. R., and H. Smith. 1998. Applied regression analysis. John Wiley & Sons, New York, New York, USA.

Eberhardt, L. E., L. L. Eberhardt, B. L. Tiller, and L. L. Cadwell. 1996. Growth of an isolated elk population. Journal of Wildlife Management 60:369–373.

Eberhardt, L. L. 1960. Estimation of vital characteristics of Michigan deer herds. Michigan Department of Conservation, Game Division Report 2282.

Eberhardt, L. L. 1968. A preliminary appraisal of line transects. Journal of Wildlife Management 32:82–88.

Eberhardt, L. L. 1969. Population analysis. *In* Wildlife management techniques, 3rd edition (R. H. Giles Jr., ed.), pages 457–495. The Wildlife Society, Washington, D.C., USA.

Eberhardt, L. L. 1976. Quantitative ecology and impact assessment. Journal of Environmental Management 4:27–70.

Eberhardt, L. L. 1977. "Optimal" management policies for marine mammals. Wildlife Society Bulletin 5:162–169.

Eberhardt, L. L. 1978. Appraising variability in population studies. Journal of Wildlife Management 42:207–238.

Eberhardt, L. L. 1982. Calibrating an index by using removal data. Journal of Wildlife Management 46:734–740.

Eberhardt, L. L. 1985. Assessing the dynamics of wild populations. Journal of Wildlife Management 49:997–1012.

Eberhardt, L. L. 1987. Population projections from simple models. Journal of Applied Ecology 24:103–118.

Eberhardt, L. L. 1995. Using the Lotka-Leslie model for sea otters. Journal of Wildlife Management 59:222–227.

Eberhardt, L. L., and K. B. Schneider. 1994. Estimating sea otter reproductive rates. Marine Mammal Science 10:31–37.

Eberhardt, L. L., and M. A. Simmons. 1987. Calibrating population indices by double sampling. Journal of Wildlife Management 51:665–675.

Eberhardt, L. L., and M. A. Simmons. 1992. Assessing rates of increase from trend data. Journal of Wildlife Management 56:603–610.

Eberhardt, L. L., and R. C. Van Etten. 1956. Evaluation of the pellet group count as a deer census method. Journal of Wildlife Management 20:70–74.

Eberhardt, L. L., D. G. Chapman, and J. R. Gilbert. 1979. A review of marine mammal census methods. Wildlife Monographs 63.

Eberhardt, L. L., A. K. Majorowicz, and J. A. Wilcox. 1982. Apparent rates of increase for two feral horse herds. Journal of Wildlife Management 46:367–374.

Eberhardt, L. L., M. A. Simmons, and S. M. Thomas. 1985. Ratio methods for cost-effective field sampling of commercial radioactive low-level wastes. US Nuclear Regulatory Commission, Washington, D.C., USA. US NUREG/CR-4268.

Eberhardt, L. L., B. M. Blanchard, and R. R. Knight. 1994. Population trend of the Yellowstone grizzly bear as estimated from reproductive and survival rates. Canadian Journal of Zoology 72:360–363.

Eberhardt, L. L., R. A. Garrott, P. J. White, and P. J. Gogan. 1998. Alternative approaches to aerial censusing of elk. Journal of Wildlife Management 62:1046–1055.

Edwards, A. W. F. 1992. Likelihood. Johns Hopkins University Press, Baltimore, Maryland, USA.

Efron, B. 1979. Bootstrap methods: another look at the jackknife. Annals of Statistics 7:1–16.

Elandt-Johnson, R. C., and N. L. Johnson. 1980. Survival models and data analysis. John Wiley & Sons, New York, New York, USA.

Ellingson, A. R., and P. M. Lukacs. 2003. Improving methods for regional landbird monitoring: a reply to Hutto and Young. Wildlife Society Bulletin 31:896–902.

Elmhagen, G., and A. Angerbjorn. 2001. The applicability of metapopulation theory to large mammals. Oikos 94:89–100.

Elowe, K. D., D. P. Fuller, and T. K. Fuller. 1991. Survival and cause-specific mortality rates of black bears in western Massachusetts. Transactions of the Northeast Section of The Wildlife Society 48:76–80.

Emlen, J. M. 1970. Age specificity and ecological theory. Ecology 51:588–601.

Emlen, J. M. 1971. Population densities of birds derived from transect counts. Auk 88:323–342.

Emlen, J. T., R. L. Hine, W. A. Fuller, and P. Alfonso. 1957. Dropping boards for population studies of small mammals. Journal of Wildlife Management 21:300–314.

Engeman, R. M. 2003. More on the need to get the basics right: population indices. Wildlife Society Bulletin 31:286–287.

Erickson, A. W., and D. B. Siniff. 1963. A statistical evaluation of factors influencing aerial survey results on brown bears. Transactions of the North American Wildlife and Natural Resources Conference 28:391–408.

Ericsson, G., K. Wallin, J. P. Ball, and M. Broberg. 2001. Age-related reproductive effort and senescence in free-ranging moose, *Alces alces*. Ecology 82:1613–1620.

Errington, P. L. 1945. Some contributions of a fifteen-year study of the northern bobwhite to a knowledge of population phenomena. Ecological Monographs 15:1–34.

Errington, P. L. 1946. Predation and vertebrate populations. Quarterly Review of Biology 21:144–177, 221–245.

Errington, P. L. 1956. Factors limiting higher vertebrate populations. Science 124:304–307.

Errington, P. L., and F. N. Hamerstrom Jr. 1937. The evaluation of nesting losses and juvenile mortality of the ring-necked pheasant. Journal of Wildlife Management 1:3–20.

Estes, J. A., and R. J. Jameson. 1988. A double-sampling estimate for sighting probability of sea otters in California. Journal of Wildlife Management 52:70–76.

Evans, M., N. Hastings, and B. Peacock. 1993. Statistical distributions. John Wiley & Sons, New York, New York, USA.

Fafarman, K. R., and C. A. DeYoung. 1986. Evaluation of spotlight counts of deer in south Texas. Wildlife Society Bulletin 14:180–185.

Fafarman, K. R., and R. J. Whyte. 1979. Factors influencing nighttime roadside counts of cottontail rabbits. Journal of Wildlife Management 43:765–767.

Falls, J. B. 1981. Mapping territories with playback: an accurate census method for songbirds. Studies in Avian Biology 6:86–91.

Federer, W. T. 1955. Experimental design: theory and applications. Oxford & IBH Publishing, Calcutta, India.

Feller, W. 1968. An introduction to probability theory and its application. John Wiley & Sons, New York, New York, USA.

Ferguson, S. H. 1993. Use of cohort analysis to estimate abundance, recruitment and survivorship for Newfoundland moose. Alces 29:99–113.

Ferguson, S. H. 2002. Using survivorship curves to estimate age of first reproduction in moose (*Alces alces*). Wildlife Biology 8:129–136.

Ferry, C., B. Frochot, and Y. Leruth. 1981. Territory and home range of the blackcap (*Sylvia atricapilla*) and some other passerines, assessed and compared by mapping and capture-recapture. Studies in Avian Biology 6:119–120.

Fillion, F. L. 1975. Estimating bias due to nonresponse mail surveys. Public Opinion Quarterly 39:482–492.

Fillion, F. L. 1980. Human surveys in wildlife management. *In* Wildlife management techniques manual, 4th edition. (S. D. Schemnitz, ed.), pages 441–454. The Wildlife Society, Washington, D.C., USA.

Fisher, D. O., S. D. Hoyle, and S. P. Blomberg. 2000. Population dynamics and survival of an endangered wallaby: a comparison of four methods. Ecological Applications 10:901–910.

Fisher, L. W. 1939. Studies of the eastern ruffed grouse (*Bonasa umbellus umbellus*) in Michigan. Michigan State College, Agricultural College Experiment Station, East Lansing, Michigan, USA.

Fisher, R. A. 1922. On the mathematical foundations of theoretical statistics. Philosophical Transactions of the Royal Society of London, Series A 222:309–368.

Fisher, R. A. 1925. Theory of statistical estimation. Proceedings of the Cambridge Philosophical Society 22:700–725.

Fisher, R. A. 1930. The genetical theory of natural selection. Oxford University Press, London, United Kingdom.

Fisher, R. A. 1956. Statistical methods and scientific inference. Hafner Press, New York, New York, USA.

Fletcher, R. 1970. A new approach to variable metric algorithms. Computer Journal 13:317–322.

Flipse, E., and E. J. M. Veling. 1984. An application of the Leslie matrix model to the population dynamics of the hooded seal, *Cystophora cristata erxleben*. Ecological Modelling 24:43–59.

Floyd, T. J., L. D. Mech, and M. E. Nelson. 1979. An improved method of censusing deer in deciduous-coniferous forests. Journal of Wildlife Management 43:258–261.

Follis, T. B., W. C. Foote, and J. J. Spillett. 1972. Observation of genitalia in elk by laparotomy. Journal of Wildlife Management 4:347–358.

Forbes, C. B. 1945. Weather and the kill of white-tailed deer in Maine. Journal of Wildlife Management 9:76–78.

Fournier, D., and C. P. Archibald. 1982. A general theory for analyzing catch at age data. Canadian Journal of Fisheries and Aquatic Sciences 39:1195–1207.

Fowler, C. W. 1981a. Density dependence as related to life history strategy. Ecology 6:602–610.

Fowler, C. W. 1981b. Comparative population dynamics in large mammals. *In* Dynamics of large mammal populations (C. W. Fowler and T. D. Smith, eds.), pages 437–456. John Wiley & Sons, New York, New York, USA.

Fowler, C. W. 1987. A review of density dependence on populations of large mammals. Current Mammalogy 1:401–441.

Fowler, C. W., and J. D. Baker. 1991. A review of animal population dynamics at extremely reduced population levels. International Whaling Commission Report 41:545–554.

Fowler, C. W., and W. J. Barmore. 1979. A population model of the northern Yellowstone elk herd. *In* Proceedings of the First Conference on Scientific Research in National Parks, volume 1. R. M. Linn, ed.), pages 427–434. National Park Service and Proceedings Series, Washington, D.C., USA.

Fraser, D., J. F. Gardner, G. B. Kolenosky, and S. Strathearn. 1982. Estimation of harvest rate of black bears from age and sex data. Wildlife Society Bulletin 10:53–57.

Freddy, D. J. 1987. Effect of elk harvest systems on elk breeding biology. Colorado Division of Wildlife, Fort Collins, Colorado, USA. Report 01-30-047.

Fredin, R. A. 1984. Levels of maximum net productivity in populations of large terrestrial mammals. *In* Reproduction in whales, dolphins and porpoises. Reports of the International Whaling Commission, special issue 6 (W. F. Perrin, R. L. Brownell Jr., and D. P. DeMaster, eds.), pages 381–387. International Whaling Commission, Cambridge, United Kingdom.

Fritzell, E. K., G. F. Hubert Jr., B. E. Meyen, and G. C. Sanderson. 1985. Age-specific reproduction in Illinois and Missouri raccoons. Journal of Wildlife Management 49:901–905.

Fry, F. E. J. 1949. Statistics of a lake trout fishery. Biometrics 5:26–67.

Fry, F. E. J. 1957. Assessment of mortalities by use of the virtual population. Proceedings of Joint Scientific Meeting of the ICNAF (International Commission for Northwest Atlantic Fisheries), ICES (International Council for the Exploration of the Sea), and FAO (Food and Agriculture Organization of the United Nations) on Fishing Effort, the Effects of Fishing on Resources and the Selectivity of Fishing Gear.

Fryxell, J. M., W. E. Mercer, and R. B. Gellately. 1988. Population dynamics of Newfoundland moose using cohort analysis. Journal of Wildlife Management 52:14–21.

Fuller, T. K. 1991. Do pellet counts index white-tailed deer numbers and population change? Journal of Wildlife Management 55:393–396.

Gabriel, W., and R. Buerger. 1992. Survival of small populations under demographic stochasticity. Theoretical Population Biology 41:44–71.

Gallucci, V. G., S. B. Saila, D. J. Gustafson, and B. J. Rothschild. 1996. Stock assessment: quantitative methods and applications for small-scale fisheries. Lewis Publishers, Boca Raton, Florida, USA.

Gambell, R. 1975. Variations in reproduction parameters associated with whale stock sizes. Report of the International Whaling Commission 25:182–189.

Gates, C. E., W. H. Marshall, and D. P. Olson. 1968. Line transect method of estimating grouse population densities. Biometrics 24:135–145.

Gates, J. M. 1966. Crowing counts as indices to cock pheasant populations in Wisconsin. Journal of Wildlife Management 30:735–744.

Gause, G. F. 1934. The struggle for existence. Williams & Wilkins, Baltimore, Maryland, USA.

Geibel, J. J., and D. L. Miller. 1984. Estimation of sea otter, *Enhydra lutris*, population with confidence bounds, from air and ground counts. California Fish and Game 70:225–233.

Geis, A. D., and E. L. Atwood. 1961. Proportion of recovered waterfowl bands reported. Journal of Wildlife Management 25:154–159.

Geissler, P. H., and B. R. Noon. 1981. Estimates of avian population trends from the North American breeding bird survey. Studies in Avian Biology 6:42–51.

Geist, V. 1982. Adaptive behavioral strategies. *In* Elk of North America (J. W. Thomas and D. E. Toweill, eds.), pages 219–277. Stackpole Books, Harrisburg, Pennsylvania, USA.

Genet, K. S., and L. G. Sargent. 2003. Evaluation of methods and data quality from a volunteer-based amphibian call survey. Wildlife Society Bulletin 31:703–714.

Georgiadis, N., M. Hack, and K. Turpin. 2003. The influence of rainfall on zebra population dynamics: implications for management. Journal of Animal Ecology 40:125–136.

Gerrodette, T., and D. P. DeMaster. 1990. Quantitative determination of optimum sustainable population level. Marine Mammal Science 6:1–16.

Gibbs, J. P. 2000. Monitoring populations: controversies and consequences. *In* Research techniques in animal ecology (L. Boitani and T. K. Fuller, eds.), pages 213–253. Columbia University Press, New York, New York, USA.

Gibbs, J. P., and S. M. Melvin. 1993. Call-response surveys for monitoring breeding waterbirds. Journal of Wildlife Management 61:1262–1267.

Gilbert, B. A. 1992. Long term population dynamics of a black-tailed deer herd on commercial forestland in western Washington [thesis]. University of Washington, Seattle, WA, USA.

Gilbert, B. A., and K. J. Raedeke. 2004. Recruitment dynamics of black-tailed deer in the western Cascades. Journal of Wildlife Management 68:120–128.

Gilpin, M. E., and F. J. Ayala. 1973. Global models of population growth and competition. Proceedings of the National Academy of Sciences of the USA 70:3590–3593.

Gilpin, M. E., T. J. Case, and F. J. Ayala. 1976. θ-selection. Mathematical Biosciences 32:131–139.

Godfrey, M. H., and N. Mrosovsky. 2001. Relative importance of thermal and nonthermal factors on the incubation period of sea turtle eggs. Chelonian Conservation and Biology 4:217–218.

Gogan, P. J. P., and R. H. Barrett. 1987. Comparative dynamics of introduced Tule elk populations. Journal of Wildlife Management 51:20–27.

Gompertz, B. 1825. On the nature of function expressive of the law of human mortality and on a new mode of determining life contingencies. Philosophical Transactions of the Royal Society of London, Series A 115:513–585.

Goodman, L. A. 1960. On the exact variance of products. Journal of the American Statistical Association 55:708–713.

Gotelli, N. J. 2001. A primer of ecology. Sinauer Associates, Sunderland, Massachusetts, USA.

Gove, N. E. 1997. Using age-at-harvest data to estimate demographic parameters for wildlife populations [thesis]. University of Washington, Seattle, Washington, USA.

Gove, N. E., J. R. Skalski, P. Zager, and R. L. Townsend. 2002. Statistical models for population reconstruction using age-at-harvest data. Journal of Wildlife Management 66:310–320.

Gradshteyn, I. S., and I. M. Fyzhik. 2000. Tables of integrals, series, and productions. Academic Press, San Diego, California, USA.

Graham, A., and R. Bell. 1989. Investigating observer bias in aerial survey by simultaneous double-counts. Journal of Wildlife Management 53:1009–1016.

Graham, I. M. 2002. Estimating weasel (*Mustela nivalis*) abundance from tunnel tracking indices at fluctuating field vole (*Microtus agrestis*) density. Wildlife Biology 8:279–287.

Green, R. G., and C. A. Evans. 1940. Studies on a population cycle of snowshoe hares on the Lake Alexander area, III. Effect of reproduction and mortality of young hares on the cycle. Journal of Wildlife Management 4:347–358.

Greenwood, M. 1926. The natural duration of cancer. Reports on Public Health and Medical Subjects, volume 33. Her Majesty's Stationary Office, London, United Kingdom.

Grieg-Smith, P. 1952. The use of random and contiguous quadrats in the study of the structure of plant communities. Annals of Botany, New Series 16:293–316.

Grimmett, G. R., and D. R. Stirzaker. 1992. Probability and random processes. Oxford University Press, Oxford, United Kingdom.

Gross, J. E., J. E. Roelle, and G. L. Williams. 1973. Program ONEPOP and information processor: a systems modeling and communications project. Progress report. Colorado Cooperative Wildlife Research Unit, Fort Collins, Colorado, USA.

Grubb, P. 1974. Population dynamics of the soay sheep. *In* Island survivors: the ecology of the soay sheep of St. Kilda (P. A. Jewell, C. Milner, and J. M. Boyd, eds.), pages 242–272. Athlone Press, Atlantic Highlands, New Jersey, USA.

Gulland, J. A. 1955. On the estimation of population parameters from marked members. Biometrika 42:269–270.

Gulland, J. A. 1965. Estimation of mortality rates. Annex to Arctic Fisheries Working Group Report, document no. 3. International Council for the Exploration of the Sea, Copenhagen, Denmark.

Gulland, J. A. 1983. Fish stock assessment: a manual of basic methods. John Wiley & Sons, New York, New York, USA.

Gullion, G. W. 1966. The use of drumming behavior in ruffed grouse population studies. Journal of Wildlife Management 30:717–729.

Hailey, A. 2000. Implications of high intrinsic growth rate of a tortoise population for conservation. Animal Conservation 3:185–189.

Hall, T. J. 1986. Electrofishing catch per hour as an indicator of largemouth bass density in Ohio impoundments. North American Journal of Fisheries Management 6:397–400.

Hansen, H. M., and F. S. Guthery. 2001. Calling behavior of bobwhite males and the call-count index. Wildlife Society Bulletin 29:145–152.

Hansen, K. M. M. 1998. Integration of archery white-tailed deer (*Odocoileus virginianus*) harvest data into a sex-age-kill population model [thesis]. Michigan State University, East Lansing, Michigan, USA.

Hanski, K., and O. Ovaskainen. 2003. Metapopulation theory for fragmented landscapes. Theoretical Population Biology 64:119–127.

Hanson, W. R. 1963. Calculation of productivity, survival, and abundance of selected vertebrates from sex and age ratios. Wildlife Monographs 9.

Harder, J. D., and R. L. Kirkpatrick. 1994. Physiological methods in wildlife research. *In* Research and management techniques for wildlife and habitats, 5th edition (T. A. Bookhout, ed.), pages 275–306. The Wildlife Society, Bethesda, Maryland, USA.

Hardy, I. C. W. 1997. Possible factors influencing vertebrate sex ratios: an introductory overview. Applied Animal Behavior Science 51:217–241.

Harris, R. B., and L. H. Metzgar. 1987. Estimating harvest rates of bears from sex ratio changes. Journal of Wildlife Management 51:802–811.

Hastings, A. 1997. Population biology: concepts and models. Springer-Verlag, Inc., New York, New York, USA.

Hatter, I. W. 1998. A Bayesian approach to moose population assessment and harvest decisions. Alces 34:47–58.

Hatter, I. W. 1999. An evaluation of moose harvest management in central and northern British Columbia. Alces 35:91–103.

Hayne, D. W. 1949*a*. An examination of the strip census method for estimating animal populations. Journal of Wildlife Management 13:147–157.

Hayne, D. W. 1949*b*. Two methods for estimating populations from trapping records. Journal of Mammology 30:399–411.

Hayne, D. W., and L. L. Eberhardt. 1952. Notes on the estimation of survival rates from age distributions of deer. Fourteenth Midwest Wildlife Conference, Des Moines, Iowa, USA.

Heincke, F. 1913. Investigations on the plaice. General report 1. The plaice fishery and protective measures. Preliminary brief summary of the most important points of the report. Rapports et procés-verbaux des reunions—Conseil permanent international pour l'exploration de la mer 16.

Hellgren, E. C., R. T. Kazmaier, D. C. Ruthven III, and D. R. Synatzske. 2000. Variation in tortoise life history: demography of *Gopherus berlandieri*. Ecology 81:1297–1310.

Henny, C. J. 1967. Estimating band-reporting rates from banding and crippling loss data. Journal of Wildlife Management 31:533–538.

Henny, C. J., W. S. Overton, and H. M. Wight. 1970. Determining parameters for populations by using structural models. Journal of Wildlife Management 34:690–703.

Henny, C. J., M. A. Byrd, J. A. Jacobs, P. D. McLean, M. R. Todd, and B. F. Hall. 1977. Mid-Atlantic coast osprey population: present numbers, productivity, pollutant contamination, and status. Journal of Wildlife Management 41:254–265.

Hensler, G. L., and J. D. Nichols. 1981. The Mayfield method of estimating nest success: a model, estimators, and simulation results. Wilson Bulletin 93:42–53.

Herkert, J. R., C. M. Nixon, and L. P. Hansen. 1992. Dynamics of exploited and unexploited fox squirrel (*Sciurus niger*) populations in the midwestern United States. *In* Wildlife 2001: populations (D. R. McCullough and R. H. Barrett, eds.), pages 864–874. Elsevier Applied Science, London, United Kingdom.

Hesselton, W. T., C. W. Severinghaus, and J. E. Tanck. 1965. Population dynamics of deer at the Seneca Army Depot. New York Fish and Game Journal 12:17–30.

Hewitt, O. H. 1967. A road-count index to breeding populations of red-winged blackbirds. Journal of Wildlife Management 31:39–47.

Hickey, J. J. 1955. Some American population research on gallinaceous birds. *In* Recent studies in avian biology (A. Wolfson, ed.), pages 326–396. University of Illinois Press, Urbana, Illinois, USA.

Hilborn, R., and C. J. Walters. 1992. Quantitative fisheries stock assessment: choice, dynamics, and uncertainty. Chapman and Hall, New York, New York, USA.

Hoffmann, A. 1993. Quantifying selection in wild populations using known-rate and mark-recapture data [dissertation]. University of Washington, Seattle, Washington, USA.

Holsworth, W. N. 1973. Hunting efficiency and white-tailed deer density. Journal of Wildlife Management 37:336–342.

Hone, J. 1988. A test of the accuracy of line and strip transect estimators in aerial surveys. Australian Wildlife Research 15:493–497.

Hopper, R. M., and H. D. Funk. 1970. Reliability of the mallard wing age-determination technique for field use. Journal of Wildlife Management 34:333–339.

Horvitz, D. G., and D. J. Thompson. 1952. A generalization of sampling without replacement from a finite universe. Journal of the American Statistical Association 47:663–685.

Hovey, F. W., and G. N. McLellan. 1996. Estimating population growth of grizzly bears from the Flathead River drainage using computer simulations of reproduction and survival rates. Canadian Journal of Zoology 74:1409–1416.

Hudson, D. J. 1971. Interval estimation from the likelihood function. Journal of the Royal Statistical Society, Series B 33:256–262.

Hutchinson, G. E. 1978. An introduction to population ecology. Yale University Press, New Haven, Connecticut, USA.

Hutto, R. L., and J. S. Young. 2002. Regional landbird monitoring: perspectives from the northern Rocky Mountains. Wildlife Society Bulletin 30:738–750.

Hutto, R. L., and J. S. Young. 2003. On the design of monitoring programs and the use of population indices: a reply to Ellingson and Lukacs. Wildlife Society Bulletin 31:903–910.

Jacobson, H. A., and R. J. Renier. 1989. Estimating age of white-tailed deer: tooth wear versus cementum annulli. Proceedings of the Annual Conference of the Southeastern Association of Fish and Wildlife Agencies 43:286–291.

Jacobson, H. A., J. C. Kroll, R. W. Browning, B. H. Koerth, and M. H. Conway. 1997. Infrared-triggered cameras for censusing white-tailed deer. Wildlife Society Bulletin 25:547–556.

Jeffries, S., H. Huber, J. Calambokidis, and J. Laake. 2003. Trends and status of harbor seals in Washington state: 1978–1999. Journal of Wildlife Management 67:207–218.

Jenkins, S. H. 1989. Comments on an inappropriate population model for feral burros. Journal of Mammalogy 70:667–670.

Jensen, A. L. 1995. Simple density dependent matrix model for population projection. Ecological Modelling 77:43–48.

Johnsgard, P. A. 1973. Grouse and quail of North America. University of Nebraska Press, Lincoln, Nebraska, USA.

Johnson, D. H. 1994. Population analysis. *In* Research and management techniques for wildlife and habitats, 5th edition (T. A. Bookhout, ed.), pages 419–444. The Wildlife Society, Bethesda, Maryland, USA.

Johnson, D. H., and A. B. Sargeant. 1977. Impact of red fox predation on the sex ratio of prairie mallards. U.S. Fish and Wildlife Service. Research report 6.

Johnson, R. A., and D. W. Wichern. 1982. Applied multivariate statistical analysis. Prentice-Hall, Englewood Cliffs, New Jersey, USA.

Jumber, J. F., H. O. Hartley, E. L. Kozicky, and A. M. Johnson. 1957. A technique for sampling mourning dove production. Journal of Wildlife Management 21:226–229.

Junge, C. O., and J. Libosvarsky. 1965. Effects of size selectivity on population estimates based on successive removals with electrical fishing gear. Zoologicke Listy 14:171–178.

Kalbfleisch, J. D., and D. A. Sprott. 1970. Application of likelihood methods to models involving large numbers of parameters. Journal of the Royal Statistical Society, Series B 32:175–208.

Kaplan, E. L., and P. Meier. 1958. Non-parametric estimation from incomplete observations. Journal of the American Statistical Association 53:457–481.

Kaur, A., G. P. Patil, A. K. Sinha, and C. Taillie. 1995. Ranked set sampling: an annotated bibliography. Environmental and Ecological Statistics 2:25–54.

Kear, J. 1965. The internal food reserves of hatchling mallard ducklings. Journal of Wildlife Management 29:523–528.

Keiss, R. E. 1969. Comparison of eruption-wear patterns and cementum annuli as age criteria in elk. Journal of Wildlife Management 33:175–180.

Keith, L. B., and L. A. Windberg. 1978. A demographic analysis of snowshoe hare cycle. Wildlife Monographs 58.

Kelker, G. H. 1940. Estimating deer populations by a differential hunting loss in sexes. Utah Academy of Sciences, Arts and Letters 17:65–69.

Kelker, G. H. 1943. Sex-ratio equations and formulas for determining wildlife populations. Utah Academy of Sciences, Arts and Letters 20:189–198.

Kelker, G. H. 1947. Computing the rate of increase for deer. Journal of Wildlife Management 11:177–183.

Kendall, W. L., B. G. Peterjohn, and J. R. Sauer. 1996. First-time observer effects in the North American breeding bird survey. Auk 113:823–829.

Keyfitz, N., and E. M. Murphy. 1967. Matrix and multiple decrement in population analysis. Biometrics 23:485–503.

Kie, J. G., and B. B. Boroski. 1995. Using spotlight counts to estimate mule deer population size and trends. California Fish and Game 81:55–70.

Kiel Jr., W. H. 1955. Nesting studies of the coot in southwestern Manitoba. Journal of Wildlife Management 19:189–198.

Kikkawa, J. 1964. Movement, activity, and distribution of the small rodents *Clethrionomys glareolus* and *Apodemus sylvaticus* in woodland. Journal of Animal Ecology 33:259–299.

Kimball Jr., J. F., and M. L. Wolfe. 1974. Population analysis of a northern Utah elk herd. Journal of Wildlife Management 38:161–174.

Kimball Jr., J. F., and M. L. Wolfe. 1979. Continuing studies of the demographics of a northern Utah elk population. *In* North American elk: ecology, behavior, and management (M. S. Boyce and L. D. Hayden-Wing, eds.), pages 20–28. University of Wyoming, Laramie, Wyoming, USA.

Kingsland, S. E. 1985. Modeling nature: episodes in the history of population ecology. University of Chicago, Chicago, Illinois, USA.

Kish, L. 1965. Survey sampling. John Wiley & Sons, New York, New York, USA.

Klavitter, J. L., J. M. Marzluff, and M. S. Vekasay. 2003. Abundance and demography of the Hawaiian hawk: is delisting warranted? Journal of Wildlife Management 67:165–176.

Klein, J. P., and M. L. Moeshberger. 1997. Survival analysis: techniques for censored and truncated data. Springer-Verlag, Inc., New York, New York, USA.

Kleinbaum, D. G. 1996. Survival analysis: a self-learning text. Springer-Verlag, Inc., New York, New York, USA.

Klett, A. T., and D. H. Johnson. 1982. Variability in nest survival rates and implications to nesting studies. Auk 99:77–87.

Knight, R. R., B. M. Blanchard, and L. L. Eberhardt. 1995. Appraising status of the Yellowstone grizzly bear population by counting females with cubs-of-the-year. Wildlife Society Bulletin 23:245–248.

Koerth, B. H., C. D. McKown, and J. C. Kroll. 1997. Infrared-triggered camera versus helicopter counts of white-tailed deer. Wildlife Society Bulletin 25:557–562.

Kohlmann, S. G., R. L. Green, and C. E. Trainer. 1999. Effects of collection method on sex and age composition of black bear (*Ursus americanus*) harvest in Oregon. Northwest Science 73:34–38.

Korschgen, C. E., K. P. Kenow, W. L. Green, D. H. Johnson, M. D. Samuel, and L. Sileo. 1996. Survival of radiomarked canvasback ducklings in northwestern Minnesota. Journal of Wildlife Management 60:120–132.

Krausman, P. R. 2001. Introduction to wildlife management. Prentice Hall, Upper Saddle River, New Jersey, USA.

Krebs, C. J. 1978. A review of the Chitty hypothesis of population regulation. Canadian Journal of Zoology 56:2463–2480.

Krebs, C. J. 1994. Ecology: the experimental analysis of distribution and abundance. HarperCollins College Publishers, New York, New York, USA.

Krebs, C. J., R. Boonstra, V. Nams, M. O'Donoghue, K. E. Hodges, and S. Boutin. 2001. Estimating snowshoe hare population density from pellet plots: a further evaluation. Canadian Journal of Zoology 79:1–4.

Kruuk, E. B. L., T. H. Clutton-Brock, S. D. Albon, J. M. Pemberton, and F. E. Guinness. 1999. Population density affects sex ratio variation in red deer. Nature 399:459–461.

Laake, J. L. 1981. Abundance estimation of dolphins in the eastern Pacific with line transect sampling—a comparison of the techniques and suggestions for future research. *In* Report of the Workshop on Tuna-Dolphin Interactions, special report 4 (P. S. Hammond, ed.), pages 56–95. Inter-American Tropical Tuna Commission, La Jolla, California, USA.

Laake, J. L. 1992. Catch-per-unit-effort models: an application to an elk population in Colorado. *In* Wildlife 2001: populations (D. R. McCullough and R. H. Barrett, eds.), pages 44–55. Elsevier Applied Science, London, United Kingdom.

Lancia, R. A., K. H. Pollock, J. W. Bishir, and M. C. Conner. 1988. A white-tailed deer harvesting strategy. Journal of Wildlife Management 52:589–595.

Lancia, R. A., J. W. Bishir, M. C. Conner, and C. S. Roseberry. 1996. Use of catch-effort to estimate population size. Wildlife Society Bulletin 24:731–737.

Lande, R. 1993. Risks of population extinction from demographic and environmental stochasticity and random catastrophes. American Naturalist 142:911–927.

Lande, R., S. Engen, and B. E. Saether. 2003. Stochastic population dynamics in ecology and conservation. Oxford University Press, Oxford, United Kingdom.

Lang, L. M., and G. W. Wood. 1976. Manipulation of the Pennsylvania deer herd. Wildlife Society Bulletin 4:159–165.

LaPointe, M. F., R. M. Peterman, and A. D. MacCall. 1989. Trends in fish mortality rate along with errors in natural mortality rate can cause spurious time trends in fish stock abundance estimated by virtual population analysis (VPA). Canadian Journal of Fisheries and Aquatic Sciences 46:2129–2139.

Lauckhart, J. B. 1950. Determining the big-game population from the kill. Transactions of the North American Wildlife Conference 15:644–649.

Lebreton, J. D., K. P. Burnham, J. Clobert, and D. R. Anderson. 1992. Modeling survival and testing biological hypotheses using marked animals: a unified approach with case studies. Ecological Monographs 62:67–118.

Lee, E. T. 1992. Statistical methods for survival data analysis. John Wiley & Sons, New York, New York, USA.

Lefkovitch, L. P. 1965. The study of population growth in organisms grouped by stage. Biometrics 21:1–18.

Lehmann, E. L. 1983. Theory of point estimation. John Wiley & Sons, New York, New York, USA.

Lehmkuhl, J. F., C. A. Hansen, and K. Sloan. 1994. Elk pellet-group decomposition and detectability in coastal forests of Washington. Journal of Wildlife Management 58:664–669.

Lengagne, T., and P. J. B. Slater. 2002. The effects of rain on acoustic communication: tawny owls have good reason for calling less in wet weather. Proceedings of the Royal Society of London, Biological Sciences 269:2121–2125.

Leopold, A. 1933. Game management. Charles Scribner's Sons, New York, New York, USA.

Leslie, P. H. 1945. On the use of matrices in certain population mathematics. Biometrika 33:182–212.

Leslie, P. H. 1948. Some further notes on the use of matrices in population mathematics. Biometrika 35:213–245.

Leslie, P. H., and D. H. S. Davis. 1939. An attempt to determine the absolute number of rats on a given area. Journal of Animal Ecology 8:94–113.

Levins, R. 1969. Some demographic and genetic consequences of environmental heterogeneity for biological control. Bulletin of the Entomological Society of America 15:237–240.

Lewis, J. C., and J. W. Farrar. 1968. An attempt to use the Leslie census method on deer. Journal of Wildlife Management 32:760–764.

Liao, S. 1994. Statistical models for estimating salmon escapement and stream residence time based on stream survey data [dissertation]. University of Washington, Seattle, Washington, USA.

Liermann, M., and R. Hilborn. 1997. Depensation in fish stocks: a hierarchic Bayesian meta-analysis. Canadian Journal of Fisheries and Aquatic Sciences 54:1976–1984.

Lincoln, F. C. 1930. Calculating waterfowl abundances on the basis of banding returns. Circular 118. U.S. Department of Agriculture, Washington, D.C., USA.

Link, W. A., and J. R. Sauer. 2002. A hierarchical analysis of population change with application to cerulean warblers. Ecology 83:2832–2840.

Lotka, A. J. 1925. Elements of physical biology. Williams and Wilkins, Baltimore, Maryland, USA.

Lovejoy, T. E. 1996. Beyond the concept of sustained yield. Ecological Applications 6:363.

Lowe, V. P. W. 1969. Population dynamics of the red deer (*Cervus elaphus* L.) on Rhum. Journal of Animal Ecology 38:425–457.

Lubow, B. C., G. C. White, and D. R. Anderson. 1996. Evaluation of a linked sex harvest strategy for cervid populations. Journal of Wildlife Management 60:787–796.

Luukkonen, D. R., H. H. Prince, and I. L. Mao. 1997. Evaluation of pheasant crowing rates as a population index. Journal of Wildlife Management 61:1338–1344.

Lyrholm, T., S. Leatherwood, and J. Sigurjonsson. 1987. Photo-identification of killer whales (*Orcinus orca*) off Iceland, October 1985. Cetology 52.

MacCall, A. D. 1986. Virtual population analysis (VPA) equations for nonhomogeneous populations, and a family of approximations including improvements on Pope's cohort analysis. Canadian Journal of Fisheries and Aquatic Sciences 43:2406–2409.

MacDonald, D., and E. G. Dillman. 1968. Techniques for estimating non-statistical bias in big game harvest surveys. Journal of Wildlife Management 32:119–129.

Mace, R. D., and J. S. Waller. 1998. Demography and population trend of grizzly bears in the Swan Mountains, Montana. Conservation Biology 12:1005–1016.

MacLulich, D. A. 1951. A new technique of animal census, with example. Journal of Mammology 32:318–328.

Malthus, T. R. 1798. An essay on the principle of population, as it affects the future improvement of society. Microfilm. Connecticut Research Publications, Inc., Woodbridge, Connecticut, USA.

Manly, B. F., and J. A. Schmutz. 2001. Estimation of brood and nest survival: comparative methods in the presence of heterogeneity. Journal of Wildlife Management 65:258–270.

Manly, B. F. J., and G. A. F. Seber. 1973. Animal life tables from capture-recapture data. Biometrics 29:487–500.

Marsh, H., and D. F. Sinclair. 1989. Correcting for visibility bias in strip transect aerial surveys of aquatic fauna. Wildlife Society Bulletin 53:1017–1024.

Martinson, R. K., and J. A. McCann. 1966. Proportion of recovered goose and brant bands that are reported. Journal of Wildlife Management 30:856–858.

May, R. M. 1973. Stability and complexity in model ecosystems. Princeton University Press, Princeton, New Jersey, USA.

May, R. M. 1974. Biological populations with nonoverlapping generations: stable points, stable cycles, and chaos. Science 18:645–647.

May, R. M. 1976. Models for single populations. *In* Theoretical ecology: principles and applications (R. M. May, ed.), pages 4–25. W. B. Saunders, Philadelphia, Pennsylvania, USA.

May, R. M., and G. F. Oster. 1976. Bifurcations and dynamic complexity in single ecological models. American Naturalist 110:573–599.

Mayfield, H. F. 1961. Nesting success calculated from exposure. Wilson Bulletin 73:255–261.

Mayfield, H. F. 1975. Suggestions for calculating nest success. Wilson Bulletin 87:456–466.

McCaffery, K. R. 1976. Deer trail counts as an index to populations and habitat use. Journal of Wildlife Management 40:308–316.

McClure, H. E. 1939. Cooing activity and censusing of the mourning dove. Journal of Wildlife Management 3:323–328.

McCorquodale, S. M., L. L. Eberhardt, and L. E. Eberhardt. 1988. Dynamics of a colonizing elk population. Journal of Wildlife Management 52:309–313.

McCullagh, P., and J. A. Nelder. 1983. Generalized linear models. Chapman and Hall, London, United Kingdom.

McCullough, D. R. 1979. The George Reserve deer herd—population ecology of a K-selected species. University of Michigan Press, Ann Arbor, Michigan, USA.

McCullough, D. R. 1982. Population growth rate of the George Reserve deer herd. Journal of Wildlife Management 46:1079–1083.

McCullough, D. R. 1983. Rate of increase of white-tailed deer on the George Reserve: a response. Journal of Wildlife Management 47:1248–1250.

McCullough, D. R. 1984. Lessons from the George Reserve, Michigan. *In* White-tailed deer: ecology and management (L. K. Halls, ed.), pages 211–242. Stackpole Books, Harrisburg, Pennsylvania, USA.

McCullough, D. R. 1993. Variation in black-tailed deer herd composition counts. Journal of Wildlife Management 57:890–897.

McCullough, D. R. 1996. Spatially structured populations and harvest theory. Journal of Wildlife Management 60:1–9.

McCullough, D. R., D. S. Pine, D. L. Whitmore, T. M. Mansfield, and R. H. Decker. 1990. Linked sex harvest strategy for big game management with a test case on black-tailed deer. Wildlife Monographs 112.

McDonald, L. L., H. B. Harvey, F. J. Mauer, and A. W. Brackney. 1990. Design of aerial surveys for Dall sheep in the Arctic National Wildlife Refuge, Alaska. Proceedings of Biennial Symposium of the Northern Wild Sheep and Goat Council 7:176–193.

McGowan, T. A. 1953. The call-count as a census method for breeding mourning doves in Georgia. Journal of Wildlife Management 17:437–445.

McIntyre, G. A. 1952. A method for unbiased selective sampling, using ranked sets. Australian Journal of Agricultural Research 3:385–390.

McLaren, I. A. 1961. Methods of determining the numbers and availability of ring seals in the eastern Canadian Arctic. Arctic 14:162–175.

McLellan, B. N. 1989. Dynamics of a grizzly bear population during a period of industrial resource extraction, III. Natality and rate of increase. Canadian Journal of Zoology 67:1865–1868.

Megrey, B. A. 1989. Review and comparison of age-structured stock assessment models from theoretical and applied points of view. American Fisheries Society Symposium 6:8–48.

Melchiors, M. A., R. E. Thackston, and D. G. Stobaugh. 1985. Results of spotlight and helicopter deer surveys. Proceedings of the Annual Conference of the Southeastern Association of the Fish and Game Wildlife Agencies 39:506–511.

Menkens Jr., G. E., D. E. Biggins, and S. H. Anderson. 1990. Visual counts as an index of white-tailed prairie dog density. Wildlife Society Bulletin 18:290–296.

Merrill, E. H., and M. S. Boyce. 1991. Summer range and elk population dynamics in Yellowstone National Park. *In* The Greater Yellowstone ecosystem: redefining America's wilderness heritage (R. B. Keiter and M. S. Boyce, eds.), pages 263–274. Yale University Press, New Haven, Connecticut, USA.

Mertz, D. B. 1970. Notes on methods used in life-history studies. *In* Readings in ecology and ecological genetics (J. H. Connell, D. B. Mertz, and W. W. Murdoch, eds.), pages 4–17. Harper and Row, New York, New York, USA.

Merwin, D. S. 2000. Comparing levels and factors of lamb mortality between two herds of Rocky Mountain bighorn sheep in the Black Hills, South Dakota, USA [thesis]. University of Washington, Seattle, Washington, USA.

Meyer, J. S., C. G. Ingersoll, L. L. McDonald, and M. S. Boyce. 1986. Estimating uncertainty in population growth rates: jackknife vs. bootstrap techniques. Ecology 67:1156–1166.

Miller, C. A., and W. L. Anderson. 2002. Digit preference in reported harvest among Illinois waterfowl hunters. Human Dimensions of Wildlife 7:55–65.

Miller, D. H., A. L. Jensen, and J. H. Hammill. 1995. Density dependent matrix model for gray wolf population projection. Ecological Modelling 151:271–278.

Miller, K. E., B. B. Ackerman, L. W. Lefebvre, and K. B. Clifton. 1998. An evaluation of strip-transect aerial survey methods for monitoring manatee population in Florida. Wildlife Society Bulletin 26:561–570.

Mills, L. S., D. F. Doak, and M. J. Wisdom. 1999. Reliability of conservation actions based on elasticity analysis of matrix models. Conservation Biology 13:815–829.

Millspaugh, J. J., and J. M. Marzluff. 2001. Radio tracking and animal populations. Academic Press, San Diego, California, USA.

Millspaugh, J. J., B. E. Washburn, M. A. Milanick, J. Beringer, L. Hansen, and T. M. Meyer. 2002. Non-invasive techniques for stress assessment in white-tailed deer. Wildlife Society Bulletin 30:899–907.

Mode, N. A., L. L. Conquest, and D. A. Marker. 1999. Ranked set sampling for ecological research: accounting for the total cost of sampling. Environmetrics 10:179–194.

Moller, A. P., and S. Legendre. 2001. Allee effect, sexual selection, and demographic stochasticity. Oikos 92:27–34.

Morris, D. W. 2002. Measuring the Allee effect: positive density dependence in small mammals. Ecology 83:14–20.

Mosby, H. S. 1969. The influence of hunting on the population dynamics of a woodlot gray squirrel population. Journal of Wildlife Management 33:59–73.

Mosby, H. S., and W. S. Overton. 1950. Fluctuations in the quail population on the Virginia Polytechnic Institute farms. Transactions of the North American Wildlife Conference 15:347–353.

Murphy, K. M. 1983. Relationships between a mountain lion population and hunting pressure in western Montana [thesis]. University of Montana, Missoula, Montana, USA.

Murray, D. L., J. D. Roth, E. Ellsworth, A. J. Wirsing, and T. D. Steury. 2002. Estimating low-density snowshoe hare populations using fecal counts. Canadian Journal of Zoology 80:771–781.

Mysterud, A., N. G. Yoccoz, N. C. Stenseth, and R. Langvatn. 2000. Relationships between sex ratio, climate and density in red deer: the importance of spatial scale. Journal of Animal Ecology 69:959–974.

Nchanji, A. C., and A. J. Plumptre. 2001. Seasonality in elephant dung decay and implications for censusing and population monitoring in south-western Cameroon. African Journal of Ecology 39:24–32.

Nebel, S., and B. J. McCaffery. 2003. Vocalization activity of breeding shorebirds: documentation of its seasonal decline and applications for breeding bird surveys. Canadian Journal of Zoology 81:1702–1708.

Neff, D. J. 1968. The pellet-group count technique for big game trend, census and distribution: a review. Journal of Wildlife Management 32:597–614.

Nelson, W. 1972. Theory and applications of hazard plotting for censored failure data. Technometrics 14:946–965.

Neter, J., W. Wasserman, and M. H. Kutner. 1990. Applied linear statistical models, 3rd edition. Irwin, Homewood, Illinois, USA.

Neter, J., M. H. Kutner, C. J. Nachtsheim, and W. Wasserman. 1996. Applied linear statistical models, 4th edition. Irwin, Chicago, Illinois, USA.

Newton, I., P. Rothery, and L. C. Dale. 1998. Density-dependence in the bird populations of an oak wood over 22 years. Ibis 140:131–136.

Nixon, C. M., R. W. Donohoe, and T. Nash. 1974. Overharvest of fox squirrels from woodlots in western Ohio. Journal of Wildlife Management 39:1–25.

Novak, J. M., K. T. Scribner, W. D. Dupont, and M. H. Smith. 1991. Catch-effort estimation of white-tailed deer population size. Journal of Wildlife Management 55:31–38.

Noyes, J. H., B. K. Johnson, L. D. Bryant, S. L. Findholt, and J. W. Thomas. 1996. Effects of bull age on conception dates and pregnancy rates of cow elk. Journal of Wildlife Management 60:508–517.

Oddie, K. R. 2000. Size matters: competition between male and female great tit offspring. Journal of Animal Ecology 69:903–912.

Oh, H. L., and F. J. Scheuren. 1983. Weighting adjustment for unit nonresponse. In Incomplete data in sample surveys (W. G. Modov, I. Olkin, and D. B. Rubin, eds.), pages 143–184. Academic Press, New York, New York, USA.

Oliphant, L. W., and E. Huag. 1985. Productivity, population density and rate of increase of an expanding merlin population. Raptor Research 19:56–59.

Otis, D. L. 1980. An extension of the change-in-ratio method. Biometrics 36:141–147.

Otis, D. L. 2001. Quantitative training of wildlife graduate students. Wildlife Society Bulletin 29:1043–1048.

Otis, D. L., K. P. Burnham, G. C. White, and D. R. Anderson. 1978. Statistical inference for capture data from closed populations. Wildlife Monographs 62.

Overton, W. S. 1969. Estimating the numbers of animals in wildlife populations. In Wildlife management techniques, 3rd edition (R. H. Giles, ed.), pages 403–455. The Wildlife Society, Washington, D.C., USA.

Paisley, R. N., R. G. Wright, J. F. Kubisiak, and R. E. Rolley. 1998. Reproductive ecology of eastern wild turkeys in southwestern Wisconsin. Journal of Wildlife Management 62:911–916.

Paloheimo, J. E., and D. Fraser. 1981. Estimation of harvest rate and vulnerability from age and sex data. Journal of Wildlife Management 45:948–958.

Pan, W., and R. A. Chappell. 1998. A nonparametric estimator of survival function for arbitrarily truncated and censored data. Lifetime Data Analysis 4:187–202.

Parker, G. R. 1991. Survival of juvenile American black ducks on a managed wetland in New Brunswick. Journal of Wildlife Management 55:466–470.

Parker, J., F. Rosell, T. A. Hermansen, G. Sorlokk, and M. Staerk. 2002. Sex and age composition of spring-hunted Eurasian beaver in Norway. Journal of Wildlife Management 66:1164–1170.

Parmar, M. K., and D. Marchin. 1995. Survival analysis, a practical approach. John Wiley & Sons, New York, New York, USA.

Patil, G. P., C. Taillie, and R. L. Wigley. 1979. Transect sampling methods and their application to deep-sea red crab. *In* Environmental biomonitoring, assessment, prediction, and management— certain case studies and related quantitative issues (J. Cairns Jr., G. P. Patil, and W. E. Waters, eds.), pages 51–75. International Co-operative Publishing House, Fairland, Maryland, USA.

Patil, G. P., A. K. Sinha, and C. Taillie. 1993. Relative precision of ranked set sampling: a comparison with the regression estimator. Environmetrics 4:399–412.

Paulik, G. J. 1963. Estimates of mortality rates from tag recoveries. Biometrics 19:28–57.

Paulik, G. J., and D. S. Robson. 1969. Statistical calculations for change-in-ratio estimators of population parameters. Journal of Wildlife Management 33:1–27.

Pearl, R. 1925. The biology of population growth. A. A. Knopf, New York, New York, USA.

Pearl, R. 1927. The growth of populations. Quarterly Review of Biology 2:532–548.

Pella, J. J., and P. K. Tomlinson. 1969. A general stock production model. Bulletin of the Inter-American Tropical Tuna Commission 13:419–496.

Pendleton, G. W. 1992. Nonresponse patterns in the federal waterfowl hunter questionnaire survey. Journal of Wildlife Management 56:344–348.

Peterjohn, B. G., and J. R. Sauer. 1994. Population trends of woodland birds from the North American Breeding Bird Survey. Wildlife Society Bulletin 22:155–164.

Peterle, T. J., and W. R. Fouch. 1969. Exploitation of a fox squirrel population on a public shooting area. Michigan Department of Conservation, Report 2251.

Petersen, C. G. J. 1896. The yearly immigration of young plaice into the Limfjord from the German Sea. Report of the Danish Biological Station 6:1–77.

Peterson, R. L. 1949. Management of moose. Proceedings of the Convention of the International Association of Game, Fish and Conservation Commissioners 39:71–75.

Peterson, R. L. 1955. North American moose. University of Toronto Press, Toronto, Ontario, Canada.

Petrides, G. A. 1949. Viewpoints on the analysis of open season sex and age ratios. Transactions of the North American Wildlife Conference 14:391–410.

Petrides, G. A. 1954. Estimating the percentage kill in ringnecked pheasants and other game species. Journal of Wildlife Management 18:294–297.

Pielou, E. C. 1969. An introduction to mathematical ecology. John Wiley & Sons, New York, New York, USA.

Pietz, H. H. 1972. Locker and field checks of antelope and analysis of hunter report card data. South Dakota Department of Game, Fish and Parks, Pierre, South Dakota, USA. Project Number SD W-095-R-5/Job 2-F.

Pimlott, D. H. 1959. Reproduction and productivity of Newfoundland moose. Journal of Wildlife Management 23:381–401.

Pinder III, J. E., J. G. Wiener, and M. H. Smith. 1978. The Weibull distribution: a new method of summarizing survivorship data. Ecology 59:175–179.

Polacheck, T. 1985. The sampling distribution of age-specific survival estimates from an age distribution. Journal of Wildlife Management 49:180–184.

Pollock, K. H. 1981. Capture-recapture models: a review of current methods, assumptions, and experimental design. Studies in Avian Biology 6:426–435.

Pollock, K. H. 1991. Modeling capture, recapture, and removal statistics for estimation of demographic parameters for fish and wildlife populations: past, present, and future. Journal of the American Statistical Association 86:225–238.

Pollock, K. H., J. E. Hines, and J. D. Nichols. 1984. The use of auxillary variables in capture-recapture and removal experiments. Biometrics 40:329–340.

Pollock, K. H., R. A. Lancia, M. C. Conner, and B. L. Wood. 1985. A new change-in-ratio procedure robust to unequal catchability of types of animal. Biometrics 41:653–662.

Pollock, K. H., S. R. Winterstein, C. M. Bunck, and P. D. Curtis. 1989. Survival analysis in telemetry studies: the staggered entry design. Journal of Wildlife Management 53:7–15.

Pope, J. G. 1972. An investigation of the accuracy of virtual population analyses using cohort analysis. International Commission for the Northwest Atlantic Fisheries Research Bulletin 9:65–74.

Pope, J. G., and J. G. Shepherd. 1985. A comparison of the performance of various methods for tuning VPAs using effort data. Journal du Conseil, Conseil International pour l'Exploration de la Mer 42:129–151.

Press, W. H., B. P. Flannery, S. A. Teukolsky, and W. T. Vetterling. 1986. Numerical recipes. Cambridge University Press, Cambridge, United Kingdom.

Prothero, W. L. 1977. Rutting behavior in elk [thesis]. Utah State University, Logan, Utah, USA.

Prothero, W. L., J. J. Spillett, and D. R. Balph. 1979. Rutting behavior of yearling and mature bull elk: some implications for open bull hunting. *In* North American elk: ecology, behavior, and management (M. S. Boyce and L. D. Hayden-Wing, eds.), pages 160–165. University of Wyoming, Laramie, Wyoming, USA.

Punt, A. E., and A. D. M. Smith. 2001. The gospel of maximum sustainable yield in fisheries management: birth, crucifixion, and reincarnation. *In* Conservation of exploited species (J. D. Reynolds, G. M. Mace, K. H. Redford, and J. G. Robinson, eds.), pages 41–66. Cambridge University Press, Cambridge, United Kingdom.

Quenouille, M. H. 1956. Notes on bias reduction. Biometrika 43:353–360.

Quick, H. F. 1963. Animal population analysis. Pages 190–228 *in* H. S. Mosby, editor. Wildlife investigational techniques, 2nd edition. Edwards Brothers, Inc., Ann Arbor, Michigan, USA.

Quinn II, T. J., and R. B. Deriso. 1999. Quantitative fish dynamics. Oxford University Press, Oxford, United Kingdom.

Raedeke, K. J., J. J. Millspaugh, and P. E. Clark. 2002. Population characteristics. *In* North American elk: ecology and management (D. E. Toweill and J. W. Thomas, eds.), pages 449–491. Smithsonian Institution Press, Washington, D.C., USA.

Ramsey, F. L., and J. M. Scott. 1978. Use of circular plot surveys in estimating the density of a population with Poisson scattering. Oregon State University, Department of Statistics, Corvallis, Oregon, USA.

Ramsey, F. L., and J. M. Scott. 1979. Estimating population densities from variable circular plot surveys. *In* Sampling biological populations (R. M. Cormack, G. P. Patil, and D. S. Robson, eds.), pages 155–181. International Co-operative Publishing House, Fairland, Maryland, USA.

Ramsey, F. L., and J. M. Scott. 1981. Tests of hearing ability. Studies in Avian Biology 6:341–345.

Rao, C. R. 1945. Information and accuracy attainable in the estimation of statistical parameters. Bulletin of the Calcutta Mathematical Society 37:81–91.

Rao, C. R. 1973. Linear statistical inference and its applications. John Wiley & Sons, New York, New York, USA.

Rao, J. N. K. 1965. On two simple schemes of unequal probability sampling without replacement. Journal of Indian Statistical Association 3:173–180.

Rao, J. N. K. 1968. Some small sample results in ratio and regression estimation. Journal of the Indian Statistical Association 6:160–168.

Rasmussen, D. I., and E. R. Doman. 1943. Census methods and their application in the management of mule deer. Transactions of the North American Wildlife Conference 8:369–380.

Regier, H. A., and D. S. Robson. 1967. Estimating population number and mortality rates. *In* The biological basis of freshwater fish production (S. D. Gerking, ed.), pages 31–66. Blackwell Scientific Publications, Oxford, United Kingdom.

Reid, V. H., R. M. Hansen, and A. L. Ward. 1966. Counting mounds and earth plugs to census mountain pocket gophers. Journal of Wildlife Management 30:327–334.

Renner, M., and L. S. Davis. 2001. Survival analysis of little penguin (*Eudyptula minor*) chicks on Motuara Island, New Zealand. Ibis 143:369–379.

Reynolds, R. T., J. M. Scott, and R. A. Nussbaum. 1980. A variable circular-plot method for estimating bird numbers. Condor 82:309–313.

Rice, C. G. 2003. Utility of pheasant call counts and brooding counts for monitoring population density and predicting harvest. Western North American Naturalist 63:178–188.

Ricker, W. E. 1954. Stock and recruitment. Journal of the Fisheries Research Board of Canada 11:559–623.

Ricker, W. E. 1958. Handbook of computations of biological statistics of fish populations. Fisheries Research Board of Canada, Ottawa, Ontario, Canada.

Ricker, W. E. 1975. Computation and interpretation of biological statistics of fish populations. Fisheries Research Board of Canada, Ottawa, Ontario, Canada. Bulletin 191.

Riordan, L. E. 1948. The sexing of deer and elk by airplane in Colorado. Transactions of the North American Wildlife Conference 13:409–428.

Robb, D. 1952. Any deer and lots of snow. Missouri Conservationist 13:2–3, 13.

Robinette, W. L., and O. A. Olsen. 1944. Studies of the productivity of mule deer in central Utah. Transactions of the North American Wildlife Conference 9:156–161.

Robinson Jr., D. A., W. E. Jensen, and R. D. Applegate. 2000. Observer effect on a rural mail carrier survey population index. Wildlife Society Bulletin 28:330–332.

Robson, D. S., and D. G. Chapman. 1961. Catch curves and mortality rates. Transactions of the American Fisheries Society 90:181–189.

Robson, D. S., and H. A. Regier. 1964. Sample size in Petersen mark-recapture experiments. Transactions of the American Fisheries Society 93:215–226.

Robson, D. S., and H. A. Regier. 1968. Estimation of population number and mortality rates. *In* Methods for assessment of fish production in freshwater, IBP Handbook 3 (W. E. Ricker, ed.), pages 124–158. Blackwell Scientific Publications, Oxford, United Kingdom.

Robson, D. S., and J. H. Whitlock. 1964. Estimation of a truncation point. Biometrika 51:33–39.

Roelle, J. E., and J. M. Bartholow. 1977. User documentation: PROGRAM ONEPOP. Colorado Division of Wildlife, Fort Collins, Colorado, USA.

Rogers, G., O. Julander, and W. L. Robinette. 1958. Pellet-group counts for deer censuses and range-use index. Journal of Wildlife Management 22:193–199.

Roseberry, J. L. 1979. Bobwhite population responses to exploitation: real and simulated. Journal of Wildlife Management 43:285–305.

Roseberry, J. L., and W. D. Klimstra. 1992. Further evidence of differential harvest rates among bobwhite sex-age groups. Wildlife Society Bulletin 20:90–94.

Roseberry, J. L., and A. Woolf. 1991. A comparative evaluation of techniques for analyzing white-tailed deer harvest data. Wildlife Monographs 117.

Roseberry, J. L., and A. Woolf. 1998. Habitat-population density relationships for white-tailed deer in Illinois. Wildlife Society Bulletin 26:252–258.

Roseberry, J. L., D. C. Autry, W. D. Klimstra, and J. Mehrhoff. 1969. A controlled deer hunt on Crab Orchard National Wildlife Refuge. Journal of Wildlife Management 33:791–796.

Rotella, J. J., and J. T. Ratti. 1986. Test of a critical density index assumption: a case study with gray partridge. Journal of Wildlife Management 50:532–539.

Roughgarden, J. 1979. Theory of population genetics and evolutionary ecology: an introduction. Macmillan Co., New York, New York, USA.

Rowland, M. M., G. C. White, and E. M. Karlen. 1984. Use of pellet-group plots to measure trends in deer and elk populations. Wildlife Society Bulletin 12:147–155.

Rupp, S. P., W. B. Ballard, and M. C. Wallace. 2000. A nationwide evaluation of deer hunter harvest survey techniques. Wildlife Society Bulletin 28:570–578.

Ryding, K. E. 2002. Estimation of demographic parameters used in assessing wildlife population trends [dissertation]. University of Washington, Seattle, Washington, USA.

Ryel, L. A. 1959. Deer pellet group surveys on an area of known herd size. Michigan Department of Conservation, Lansing, Michigan, USA. Game Division report 2252.

Samuel, M. D., E. O. Garton, M. W. Schlegel, and R. G. Carson. 1987. Visibility bias during aerial surveys of elk in north central Idaho. Journal of Wildlife Management 51:622–630.

Samuel, M. D., R. K. Steinhorst, E. O. Garton, and J. W. Unsworth. 1992. Estimation of wildlife population ratios incorporating survey design and visibility bias. Journal of Wildlife Management 56:718–725.

Sargeant, G. A., D. H. Johnson, and W. E. Berg. 1998. Interpreting carnivore scent-station surveys. Journal of Wildlife Management 62:1235–1245.

Sarno, R. J., W. R. Clark, M. S. Bank, W. S. Prexl, M. J. Behl, W. E. Johnson, and W. L. Franklin. 1999. Juvenile guanaco survival: management and conservation implications. Journal of Applied Ecology 36:937–945.

Sauer, J. R., D. D. Dolton, and S. Droege. 1994. Mourning dove population trend estimates from call-count and North American Breeding Bird Surveys. Journal of Wildlife Management 58:506–515.

Scattergood, L. W. 1954. Estimating fish and wildlife populations: a survey of methods. *In* Statistics and mathematics in biology (O. Kempthorne, ed.), pages 273–285. Iowa State College Press, Ames, Iowa, USA.

Schauster, E. R., E. M. Gese, and A. M. Kitchen. 2002. An evaluation of survey methods for monitoring swift fox abundance. Wildlife Society Bulletin 30:464–477.

Scheffer, V. B. 1951. The rise and fall of a reindeer herd. Scientific Monthly 73:356–362.

Schierbaum, D. L., and D. D. Foley. 1957. Differential age and sex vulnerability of the black duck to gunning. New York Fish and Game Journal 4:88–91.

Schnute, J. 1985. A general theory for analysis of catch and effort data. Canadian Journal of Fisheries and Aquatic Sciences 42:414–429.

Schwartz, C. C. 1993. Constructing simple population models for moose management. Alces 29:235–242.

Searle, S. R. 1966. Matrix algebra for the biological sciences. John Wiley & Sons, New York, New York, USA.

Searle, S. R. 1982. Matrix algebra useful for statistics. John Wiley & Sons, New York, New York, USA.

Seber, G. A. F. 1973. The estimation of animal abundance and related parameters. Griffith, London, United Kingdom.

Seber, G. A. F. 1982. The estimation of animal abundance and related parameters, 2nd edition. Macmillan Co., New York, New York, USA.

Seber, G. A. F. 1986. A review of estimating animal abundance. Biometrics 42:267–292.

Seber, G. A. F., and E. D. LeCren. 1967. Estimating population parameters from catches large relative to the population. Journal of Animal Ecology 36:631–643.

Seber, G. A. F. 1973. The estimation of animal abundance and related parameters. Griffith, London, United Kingdom.

Seber, G. A. F., and C. J. Wild. 1989. Nonlinear regression. John Wiley & Sons, New York, New York, USA.

Selleck, D. M., and C. M. Hart. 1957. Calculating the percentage of kill from sex and age ratios. California Fish and Game 43:309–316.

Sen, A. R. 1971. Some recent developments in waterfowl sample survey techniques. Journal of the Royal Statistical Society, Series C 20:139–147.

Severinghaus, C. W. 1969. Minimum deer populations on the Moose River recreation area. New York Fish and Game Journal 16:19–26.

Severinghaus, C. W., and H. F. Maguire. 1955. Use of age composition data for determining sex ratios among adult deer. New York Fish and Game Journal 2:242–246.

Severson, K. E., and G. E. Plumb. 1998. Comparison of methods to estimate population densities of black-tailed prairie dogs. Wildlife Society Bulletin 26:859–866.

Sherman, P. W., and M. C. Runge. 2003. Demography of a population collapse: the northern Idaho ground squirrel (*Spermophilus brunneus brunneus*). Ecology 83:2816–2831.

Shields, J. M. 1990. The effects of logging on bird populations in southeastern New South Wales [dissertation]. University of Washington, Seattle, Washington.

Shirato, G. 1996. Deer, Region 6. *In* Game Status and Trend Report, pages 50–54. Washington Department of Fish and Game, Wildlife Management Program, Olympia, Washington, USA.

Shono, H. 2000. Efficiency of the finite correction of Akaike's information criteria. Fisheries Science 66:608–610.

Shupe, T. E., F. S. Guthery, and R. L. Bingham. 1990. Vulnerability of bobwhite sex and age classes to harvest. Wildlife Society Bulletin 18:24–26.

Siler, W. 1979. A competing-risk model for animal mortality. Ecology 60:750–757.

Sims, S. E. 1982. The effect of unevenly distributed catches on stock-size estimates using virtual population analysis (cohort analysis). Journal du Conseil, Conseil International pour l'Exploration de la Mer 40:47–52.

Sims, S. E. 1984. On the analysis of the effects of errors in the natural mortality rate on stock size estimates using virtual population analysis (cohort analysis). Journal du Conseil, Conseil International pour l'Exploration de la Mer 41:149–153.

Skalski, J. R., and J. J. Millspaugh. 2002. Generic variance expressions, precision, and sampling optimization for the sex-age-kill model of population reconstruction. Journal of Wildlife Management 66:1308–1316.

Skalski, J. R., and D. S. Robson. 1979. Tests of homogeneity and goodness-of-fit to a truncated geometric model for removal sampling. *In* Sampling biological populations (R. M. Cormack, G. P. Patil, and D. S. Robson, eds.), pages 293–313. International Co-operative Publishing House, Fairland, Maryland, USA.

Skalski, J. R., and D. S. Robson. 1982. A mark and removal field procedure for estimating population abundance. Journal of Wildlife Management 56:742–752.

Skalski, J. R., and D. S. Robson. 1992. Techniques for wildlife investigations: design and analysis of capture data. Academic Press, San Diego, California, USA.

Skalski, J. R., D. S. Robson, and C. L. Matsuzaki. 1983. Competing probabilistic models for catch-effort relationships in wildlife censuses. Ecological Modelling 19:299–307.

Small, R. J., J. C. Hozwart, and D. H. Rusch. 1991. Predation and hunting mortality of ruffed grouse in central Wisconsin. Journal of Wildlife Management 55:512–520.

Smallwood, K. S., and E. L. Fitzhugh. 1995. A track count for estimating mountain lion (*Felis concolor californica*) population trend. Biological Conservation 71:251–259.

Smith, A. D. 1964. Defecation rates of mule deer. Journal of Wildlife Management 28:435–444.

Smith, B. L., and S. H. Anderson. 1998. Juvenile survival and population regulation of the Jackson elk herd. Journal of Wildlife Management 62:1036–1045.

Smith, H. F. 1938. An empirical law describing heterogeneity in the yields of agricultural crops. Journal of Agricultural Science 28:1–23.

Smith, L. M., I. L. Brisbin Jr., and G. C. White. 1984. An evaluation of total trapline captures as estimates of furbearer abundance. Journal of Wildlife Management 48:1452–1455.

Smith, S. G. 1991. Assessing hazards in wild populations using auxiliary variables in tag-release models [dissertation]. University of Washington, Seattle, Washington, USA.

Sokal, R. B., and F. J. Rohlf. 1995. Biometry, the principles and practice of statistics in biological research. W. H. Freeman, New York, New York, USA.

Spinage, C. A. 1972. African ungulate life tables. Ecology 53:645–652.

Squibb, R. C., J. F. Kimball, and D. R. Anderson. 1986. Bimodal distribution of estimated conception dates in Rocky Mountain elk. Journal of Wildlife Management 50:118–122.

Squibb, R. C., R. E. Danvir, J. F. Kimball, S. T. David, and T. D. Bunch. 1991. Ecology of conception in a northern Utah elk herd. Proceedings of a Symposium on Elk Vulnerability 110–118.

Steinert, S. F., H. D. Riffel, and G. C. White. 1994. Comparisons of big game harvest estimates from check station and telephone surveys. Journal of Wildlife Management 58:335–340.

Steinhorst, R. K., and M. D. Samuel. 1989. Sightability adjustment methods for aerial surveys of wildlife populations. Biometrics 45:415–425.

Stephens, P. A., and W. J. Sunderland. 1999. Consequences of the Allee effect for behavior, ecology and conservation. Trends in Ecology and Evolution 14:401–405.

Stokes, A. W. 1954. Population studies of the ring-necked pheasants on Pelee Island, Ontario. Ontario Department of Lands and Forests. Wildlife Series Number 4.

Stolen, M. K., and J. Barlow. 2003. A model life table for bottlenose dolphins (*Tursiops truncatus*) from the Indian River Lagoon system, Florida, U.S.A. Marine Mammal Science 19:630–649.

Strickland, M. D., H. J. Harju, K. R. McCaffrey, H. W. Miller, L. M. Smith, and R. J. Stoll. 1996. Harvest management. *In* Research and management techniques for wildlife and habitats, 5th edition (T. A. Bookhout, ed.), pages 445–473. The Wildlife Society, Bethesda, Maryland, USA.

Strong, D. R. 1986. Density vagueness: abiding the variance in the demography of real populations. *In* Community ecology (J. Diamond and T. J. Case, eds.), pages 257–268. Harper & Row, New York, New York, USA.

Sun, L., R. Shine, Z. Debi, and T. Zhengren. 2001. Biotic and abiotic influences on activity patterns of insular pit-vipers (*Gloudius shedaoensis*) from north-eastern China. Biological Conservation 97:387–398.

Swenson, J. E. 1986. Differential survival by sex in juvenile sage grouse and gray partridge. Ornis Scandanavica 17:14–17.

Taber, R. D. 1969. Criteria of sex and age. *In* Wildlife management techniques, 3rd edition (R. H. Giles, ed.), pages 325–401. The Wildlife Society, Washington, D.C., USA.

Taberner, A., P. Castañera, E. Silvestre, and J. Dopazo. 1993. Estimation of the intrinsic rate of natural increase and its error by both algebraic and resampling approaches. Cabios 9:535–540.

Tait, D. E. N., and F. L. Bunnell. 1980. Estimating rate of increase from age at death. Journal of Wildlife Management 44:296–299.

Takahasi, K., and K. Wakimoto. 1968. On unbiased estimates of the population mean based on the sample stratified by means of ordering. Annals of the Institute of Statistical Mathematics 20:1–31.

Tang, S., and L. Chen. 2002. Density-dependent birth rate, birth pulses and their population dynamic consequences. Journal of Mathematical Biology 44:185–199.

Taper, M. L., and P. J. Gogan. 2002. The northern Yellowstone elk: density dependence and climatic conditions. Journal of Wildlife Management 66:106–122.

Taylor, A. E., and W. R. Mann. 1983. Advanced calculus. John Wiley & Sons, New York, New York, USA.

Taylor, B. L., and D. P. DeMaster. 1993. Implications of non-linear density dependence. Marine Mammal Science 9:360–371.

Taylor, B. L., P. R. Wade, R. A. Stehn, and J. F. Cochrane. 1996. A Bayesian approach to classification criteria for spectacled eiders. Ecological Applications 6:1077–1089.

Taylor, C. E., D. L. Otis, H. S. Hill Jr., and C. R. Ruth. 2000. Design and evaluation of mail surveys to estimate deer harvest parameters. Wildlife Society Bulletin 28:2000.

Teer, J. G., J. W. Thomas, and E. A. Walker. 1965. Ecology and management of white-tailed deer in the Llano Basin of Texas. Wildlife Monographs 15.

Thogmartin, W. E., and J. E. Johnson. 1999. Reproduction in a declining population of wild turkeys in Arkansas. Journal of Wildlife Management 63:1369–1375.

Thompson, P. M., D. J. Tollit, D. Wood, H. M. Corpe, P. S. Hammond, and A. Mackay. 1997. Estimating harbour seal abundance and status in an estuarine habitat in north-east Scotland. Journal of Applied Ecology 34:43–52.

Thompson, S. K. 1992. Sampling. John Wiley & Sons, New York, New York, USA.

Thompson, S. K. 2002. Sampling, 2nd edition. John Wiley & Sons, New York, New York, USA.

Thompson, W. L., G. C. White, and C. Gowan. 1998. Monitoring vertebrate populations. Academic Press, San Diego, California, USA.

Tilley, S. G. 1982. Dusky salamanders. *In* CRC Handbook of census methods for terrestrial vertebrates (D. E. Davis, ed.), pages 13–14. CRC Press, Inc., Boca Raton, Florida, USA.

Trautman, C. G. 1982. History, ecology and management of the ring-necked pheasant in South Dakota. South Dakota Department of Game, Fish and Parks. Wildlife Reseach Bulletin 7.

Trent, T. T., and O. J. Rongstad. 1974. Home range and survival of cottontail rabbits in southwestern Wisconsin. Journal of Wildlife Management 38:459–472.

Tsai, W., N. Jewell, and M. Wang. 1987. A note on the product-limit estimator under right censoring and left truncation. Biometrika 74:883–886.

Udevitz, M. S. 1989. Change-in-ratio methods for estimating the size of closed populations [dissertation]. North Carolina State University, Raleigh, North Carolina, USA.

Udevitz, M. S., and B. E. Ballachey. 1998. Estimating survival rates with age-structure data. Journal of Wildlife Management 62:779–792.

Udevitz, M. S., and K. H. Pollock. 1992. Change-in-ratio methods for estimating population size. *In* Wildlife 2001: populations (D. R. McCullough and R. H. Barrett, eds.), pages 90–101. Elsevier Science Publishers Ltd., London, United Kingdom.

Ulltang, O. 1977. Sources of error in and limitation of virtual population (cohort analysis). Journal du Conseil, Conseil International pour l'Exploration de la Mer 37:249–260.

Unsworth, J. W., L. Kuck, and E. O. Garton. 1990. Elk sightability model validation at the National Bison Range, Montana. Wildlife Society Bulletin 18:113–115.

Unsworth, J. W., D. F. Pac, G. W. White, and R. M. Bartmann. 1999. Mule deer survival in Colorado, Idaho, and Montana. Journal of Wildlife Management 63:315–326.

Uptegraft, D. D. 1959. The ecology and deer-carrying capacity of the Howland Island game management area [thesis]. Syracuse University, Syracuse, New York, USA.

Uzunogullari, U., and J. Wang. 1992. A comparison of hazard rate estimators for left truncated and right censored data. Biometrika 79:297–310.

Van Ballenberghe, V. 1983. Rate of increase of white-tailed deer on the George Reserve: a re-evaluation. Journal of Wildlife Management 47:1245–1247.

Vance, D. R., and J. A. Ellis. 1972. Bobwhite populations and hunting on Illinois public hunting areas. Proceedings of the National Bobwhite Quail Symposium 1:165–174.

Vandermeer, J. H., and D. E. Goldberg. 2003. Population ecology: first principles. Princeton University Press, Princeton, New Jersey, USA.

Van Dyke, F. G., R. H. Brocke, and H. G. Shaw. 1986. Use of road track counts as indices of mountain lion presence. Journal of Wildlife Management 50:102–109.

Van Sickle, J., C. A. M. Attwell, and G. C. Craig. 1987. Estimating population growth rate from an age distribution of natural deaths. Journal of Wildlife Management 51:941–948.

Verhulst, P. F. 1838. Notice sur la loi que la population suit dans son accroissement. Correspondance mathématique et physique de l'Observatoire de Bruxelles Adolphe Quetelet, 1832–1839:113–121.

Vernes, K. 1999. Pellet counts to estimate density of a rainforest kangaroo. Wildlife Society Bulletin 27:991–996.

Wade, P. R. 1998. Calculating limits to the allowable human-caused mortality of cetaceans and pinnipeds. Marine Mammal Science 14:1–37.

Wald, A. 1943. Tests of statistical hypotheses concerning several parameters when the number of observations is large. Transactions of the American Mathematical Society 54:462–482.

Warner, R. E., P. Hubert, P. C. Mankin, and C. A. Gates. 2000. Disturbance and the survival of female ring-necked pheasants in Illinois. Journal of Wildlife Management 64:663–672.

Webb, W. L. 1942. Notes on a method of censusing snowshoe hare populations. Journal of Wildlife Management 6:67–69.

Wethey, D. S. 1985. Catastrophe, extinction, and species diversity: a rocky intertidal example. Ecology 66:445–456.

Whipple, G. C. 1919. Vital statistics: an introduction to the science of demography. John Wiley & Sons, Inc., New York, New York, USA.

White, G. C. 1993. Precision of harvest estimates obtained from incomplete responses. Journal of Wildlife Management 57:129–134.

White, G. C. 1999. Population viability analysis: data requirements and essential analyses. In Research techniques in animal ecology (L. Boitani and T. K. Fuller, eds.), pages 288–331. Columbia University Press, New York, New York, USA.

White, G. C., and R. A. Garrott. 1990. Analysis of wildlife radio-tracking data. Academic Press, San Diego, California, USA.

White, G. C., and B. C. Lubow. 2002. Fitting population models to multiple sources of observed data. Journal of Wildlife Management 66:300–309.

White, G. C., and T. M. Shenk. 2001. Population estimation with radio-marked animals. In Radio tracking and animal populations (J. J. Millspaugh and J. M. Marzluff, eds.), pages 329–350. Academic Press, San Diego, California, USA.

White, G. C., D. R. Anderson, K. P. Burnham, and D. L. Otis. 1982. Capture-recapture and removal methods for sampling closed populations. Los Alamos National Laboratory, Los Alamos, New Mexico, USA.

White, G. C., A. F. Reever, F. G. Lindzey, and K. P. Burnham. 1996. Estimation of mule deer winter mortality from age ratios. Journal of Wildlife Management 60:37–44.

Wiegand, K., S. D. Sarre, K. Henle, T. Stephan, C. Wissel, and R. Brandl. 2001. Demographic stochasticity does not predict persistence of gecko populations. Ecological Applications 11:1738–1749.

Wielgus, R. B. 2002. Minimum viable population and reserve sizes for naturally regulated grizzly bears in British Columbia. Biological Conservation 106:381–388.

Wight, H. M., R. G. Heath, and A. D. Geis. 1965. A method for estimating fall adult sex ratios from production and survival data. Journal of Wildlife Management 29:185–192.

Wilks, S. S. 1962. Mathematical statistics. John Wiley & Sons, New York, New York, USA.

Williams, B. K., J. D. Nichols, and M. J. Conroy. 2001. Analysis and management of animal populations. Academic Press, San Diego, California, USA.

Wilson, K. R., and D. R. Anderson. 1985. Evaluation of a density estimator based on a trapping web and distance sampling theory. Ecology 66:1185–1194.

Wisdom, M. J., and L. S. Mills. 1997. Sensitivity analysis to guide population recovery: prairie-chickens as an example. Journal of Wildlife Management 61:302–312.

Wisdom, M. J., L. S. Mills, and D. F. Doak. 2000. Life stage simulation analysis: estimating vital-rate effects on population growth for conservation. Ecology 81:628–641.

Wolfram, S. 1999. The Mathematica book, 4th edition. Cambridge University Press, New York, New York, USA.

Wolter, K. M. 1984. An investigation of some estimators of variance for systematic sampling. Journal of the American Statistical Association 79:781–790.

Wood, J. E., and E. P. Odum. 1964. A nine-year history of furbearer populations on the AEC Savannah River Plant Area. Journal of Mammalogy 45:540–581.

Wood, J. M., A. R. Woodward, S. R. Humphrey, and T. C. Hines. 1985. Night counts as an index of American alligator population trends. Wildlife Society Bulletin 13:262–273.

Woolf, A., and J. D. Harder. 1979. Population dynamics of a captive white-tailed deer herd with emphasis on reproduction and mortality. Wildlife Monographs 67.

Wright, V. L. 1978. Causes and effects of biases on waterfowl harvest estimates. Journal of Wildlife Management 42:251–262.

Zahn, H. M. 1996. Deer, Region 6. *In* Game status and trend report, pages 81–84. Washington Department of Fish and Game, Wildlife Management Program, Olympia, Washington, USA.

Zar, J. H. 1984. Biostatistical analysis, 2nd edition. Prentice Hall, Englewood Cliffs, New Jersey, USA.

Zar, J. H. 1996. Biostatistical analysis, 3rd edition. Prentice Hall, Upper Saddle River, New Jersey, USA.

Zippin, C. 1956. An evaluation of the removal method of estimating animal populations. Biometrics 12:163–169.

Zippin, C. 1958. The removal method of population estimation. Journal of Wildlife Management 22:82–90.

Index